WILLIAM H. BROWN

Student Study Guide and Problems Book for

ORGANIC CHEMISTRY

Volume 1

Brent L. Iverson and Sheila A. Iverson
University of Texas at Austin
with contributions by William H. Brown

Saunders Golden Sunburst Series

SAUNDERS COLLEGE PUBLISHING
Harcourt Brace College Publishers

Fort Worth • Philadelphia • San Diego • New York • Orlando • Austin
San Antonio • Toronto • Montreal • London • Sydney • Tokyo

Printed in the United States of America.

Iverson/Iverson: Student Study Guide and Problems Book, Volume I, Chapters 1-16 for Organic Chemistry by William H. Brown

ISBN 0-03-097260-4

56 018 98765432

To Carina, Alexandra, and Alanna
with love

This Study Guide and Problems Book is a companion to *Organic Chemistry* by William Brown. Because of the comprehensive nature of this guide, we have printed it in two volumes. Volume I contains Chapters 1-16, while Volume II contains Chapters 17-25. Each volume provides a detailed section-by-section *overview* of the major points covered in the text. Reading these overviews before and after reading a text chapter should help identify and summarize important elements of the material. The overviews are written in a very compact *outline format. Key terms* are printed in boldface the first time they appear. Bracketed sentences in italic print provide *hints for studying and pitfalls to avoid.* Especially important ideas are identified with the symbol "✱".

A *summary of reactions* is presented in tabular form. The starting materials are listed on the vertical axis, while the products of the reaction are listed across the top. Where the two intersect in the table is the section number where a particular reaction can be found in the text. Below the table are generalized descriptions and explanations of the different reactions in the table.

All of the *problems* from the text have been reprinted in this guide, so there is no need to flip back and forth between the text and the guide. Detailed, stepwise *solutions* to all of the problems are provided. This guide was reviewed for accuracy by Brent and Sheila Iverson, Kevin Shreder, and R. Scott Lokey at the University of Texas, and William Brown at Beloit College. If you have any comments or questions, please direct them to Professor Brent Iverson, Department of Chemistry and Biochemistry, the University of Texas at Austin, Austin, Texas 78712. E-mail: biverson@utxvms.cc.utexas.edu.

Brent and Sheila Iverson
The University of Texas at Austin
November 1994

CONTENTS: Volume I

Chapters 17-25 can be found in Volume II of the study guide

CHAPTER 1: COVALENT BONDING

1.0 OVERVIEW

- Organic Chemistry is the study of compounds that contain carbon atoms in combination with other types of atoms such as hydrogen, nitrogen, oxygen, and chlorine.
- The Lewis model of bonding qualitatively describes coordination numbers of atoms and molecular geometries.
- Valence bond theory and molecular orbital theory comprise a more accurate theoretical framework with which to understand relationships between molecular structure and reactivity.
- Organic chemists are concerned primarily with where electrons are located in an atom, molecule, or ion, because then they can understand or predict structure, bonding, and reactivity. ✳

1.1 ELECTRONIC STRUCTURE OF ATOMS

- Electrons are found around atoms in defined regions of space called **atomic orbitals.**
  - An atomic orbital can hold up to two electrons (the **Pauli Exclusion Principle**).
  - Atomic orbitals are classified as s, p, or d.
  - For most of organic chemistry, s and p atomic orbitals are the only types of orbitals that we need to consider.
  - The d atomic orbitals are important for third row elements such as sulfur (S) and phosphorus (P).
- Orbitals with the same **principal quantum number** (**1**, **2**, **3**, etc.) form what is called a **shell of electrons**. For example, the **2**s and **2**p orbitals are in the same shell (the **2**nd shell) while **1**s and **2**s orbitals are in different shells.
  - There is only one s orbital for a given shell
  - There are three p orbitals for shells with principal quantum number **2** and higher. The three **2**p orbitals are orthogonal to each other and are designated $2p_x$, $2p_y$, and $2p_z$.
  - There are five d orbitals for shells with a principal quantum number of **3** or higher.
  - All the orbitals of the same type in a given shell have the same energy; that is, they are said to be **degenerate**. For example, all three **2**p orbitals are degenerate.
- Different elements have different numbers of electrons, and these are placed in orbitals beginning with the lowest energy orbitals (the **Aufbau Principle**).
  - Orbitals with the smallest principal quantum number are lowest in energy and are filled first.
  - For orbitals with the same principal quantum number, s orbitals are filled before p orbitals which are filled before d orbitals.
  - According to **Hund's rule**, one electron is added to each degenerate orbital before two electrons are added to any one of them.
- Chemists are primarily interested in the **valence electrons** of an atom. **Valence electrons** are the electrons in the outermost shell of an atom. ✳
  - For H, the valence electrons are in the **1**st shell, namely the **1**s orbital.
  - For C, N, O, and F the valence electrons are those in the **2**nd shell, namely the **2**s and the three **2**p orbitals.
  - For Si, P, S, and Cl the valence electrons are those in the **3**rd shell, namely the **3**s, and the three **3**p orbitals. For P and S, the **3**d orbitals are important.

1.2 THE LEWIS MODEL OF BONDING

- **Chemical bonds** in molecules are made from electrons in the valence shells of atoms. ✳
  - Bonds do not involve electrons from shells below the valence shell.
  - Atoms overwhelmingly prefer to be surrounded by a filled valence shell of electrons (**noble gas configuration**). ✳

- A filled valence shell for H is 2 electrons, and a filled valence shell for C, N, O, and F is 8 electrons; the "**octet rule.**"
- P and S can have 8 electrons in their valence shell, but the valence shell for S may contain as many as 10 electrons, and the valence shell of P may contain as many as 12 electrons. This is due to the presence of **3**d orbitals on these third row elements.

• Atoms take part in chemical bonds to fill their valence shells. ✱

• In an **ionic bond** an atom transfers one or more electrons to a different atom creating negatively and positively charged ions that then attract each other.
- Ionic bonding is only observed when the electron transfer creates filled valence shells for both of the ions.

• In a **covalent bond** atoms share pairs of electrons, thus increasing the number of electrons around each atom. In this way the valence shell of each atoms is filled.
- Sharing of a pair of electrons holds the two atoms together.
- Organic chemistry is concerned primarily with covalent bonds.
- Noble gases do not normally take part in bonding because their valence shells are already filled.

• Electron pairs taking part in covalent bonds are not necessarily shared evenly between the atoms. The more **electronegative element** taking part in a bond attracts the majority of the electron density of the shared electrons. ✱
- The unequal sharing of electrons (**percent of ionic character**) of a covalent bond can be determined quantitatively using a table of electronegativities of the elements.
- On the periodic table of the elements, **electronegativity** increases from the bottom left hand corner to the upper right hand corner (See Table 1.3 in book).

• Knowing how to identify the electron-rich and electron-poor regions of molecules is the key to understanding and learning organic chemistry. ✱
- The unequal sharing of electrons in covalent bonds forms the basis for reactivity in a molecule, so understanding how electrons are distributed in a given molecule allows chemical reactions to be predicted accurately. *[By the time you have finished studying organic chemistry you should be able to look at the structure of a molecule, and then, based on your understanding of the electron distribution in the molecule, predict chemical reactions. This prediction approach is much more successful than trying to memorize reactions without understanding the reasons they take place.]*

• **Lewis structures** are used to represent molecules. *[Lewis structures are not as hard as they look, but you need a lot of practice to get the hang of them.]*
- In a Lewis structure, a line between two atoms represents a pair of electrons taking part in a bond and a pair of dots represents an unshared or lone pair of electrons.

• **To draw a Lewis structure:**
- **First**, draw single bonds (a line) between all of the atoms known to be connected to each other. Usually the connectivity information must be provided for you in the form of a condensed structural formula (Section 1.4 A). For example, for the condensed structural formula CH_3CHO write down:

```
    H   O
    |   |
H—C—C—H
    |
    H
```

- **Second**, count all the valence electrons of each atom in the structure.

  Remember that in the neutral states, H has 1 valence electron, C has 4 valence electrons, N has 5 valence electrons, O has 6 valence electrons, F has 7 valence electrons.

  Add a valence electron for each unit of negative charge on an atom or ion, and subtract a valence electron for each unit of positive charge.

- **Third**, draw any remaining bonds (double or triple bonds) and add any lone pairs of electrons that may be necessary so that each atom in the molecule is surrounded by a **filled valence shell** of electrons. ✱

  For H, a filled valence shell is 2 electrons (1 bond), and for C, N, O, and F; a filled valence shell is 8 electrons distributed as described in the following table:

Atom	# Of Bonds (Single bonds count as 1 bond double bonds as 2, and triple bonds as 3)	# Of Lone Pairs Of Electrons
C	4	0
N	3	1
O	2	2
F	1	3

  Finishing the example, two lone pairs of electrons are added to the oxygen atom and a double bond is added between the carbon and oxygen atoms to fill all valences and complete the structure:

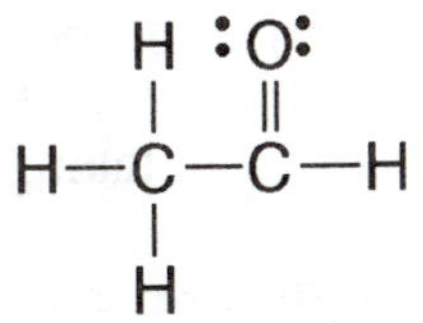

```
     H  :O:
     |   ||
  H—C—C—H
     |
     H
```

  In most organic molecules, the halogens Cl, Br, and I are treated the same as F. On the other hand, the neutral atoms B, P, and S have some unusual bonding patterns as described in the following table.

Atom	# Of Bonds (Single bonds count as 1 bond double bonds as 2, and triple bonds as 3)	# Of Lone Pairs Of Electrons
B	3	0
P	3	1
P *	5	0
S	2	2
S	4	1
S *	6	0

  (* 3d orbitals are involved, so more than 8 valence electrons can be accommodated in the valence shell.)

  If there is a negative formal charge (see below) on an atom add a lone pair of electrons and use one less bond.

  If there is a positive formal charge on an atom other than carbon, add a bond and use one less lone pair of electrons.

  For carbon with a positive formal charge, you cannot add a bond because carbon already has four. In this case, you should use one less bond and no lone pairs (the carbon is surrounded by only six valence electrons).

• **Some Helpful Hints** for drawing Lewis structures:

  - After drawing the single bonds between all connected atoms, draw the lone pairs of electrons. For neutral atoms, the number of lone pairs on a given element does not change from molecule to molecule.

In the neutral, uncharged state, C has zero lone pairs, N has one lone pair, O has two lone pairs and F has three lone pairs. For example, in the molecule CO_2, draw the single bonds and lone pairs of electrons around oxygen as follows:

$$\ddot{\underset{\cdot\cdot}{O}}-C-\ddot{\underset{\cdot\cdot}{O}}$$

- After drawing the lone pairs, fill in multiple bonds as necessary. If there are not enough single bonds to other atoms surrounding an atom to give a filled valence shell, then use multiple bonds.

  In these cases you should draw the multiple bond(s) to the adjacent atom(s) that also need multiple bond(s) to fill their valence shell. The CO_2 example is completed by adding a double bond to each oxygen atom and thereby filling the valence shell of carbon as well:

$$\ddot{\underset{\cdot\cdot}{O}}=C=\ddot{\underset{\cdot\cdot}{O}}$$

• Computation of **formal charges** is a useful bookkeeping method for keeping track of charges on a molecule. ✱
  - For the computation of formal charge, the electrons in bonds are counted as being distributed evenly between the bonded atoms. One electron is counted for each atom taking part in a single bond, two electrons for each atom taking part in a double bond, and three electrons for each atom taking part in a triple bond.
  - Lone pairs of electrons and electrons in shells lower than the valence shell (1s electrons for C, N, O, F etc.) are counted as belonging entirely to the atoms to which they are attached.
  - Total formal charge derives from comparing the number of electrons counted as above to the number of protons in the nucleus of the atom (an extra electron results in a formal charge of -1, an extra proton results in a formal charge of +1, etc.).
  - The following tables show the formal charges associated with certain atoms that might be found in organic molecules and reaction intermediates.

Table Of Atoms With +1 Formal Charge

Atom	# Of Bonds (Single bonds count as 1 bond double bonds as 2, and triple bonds as 3)	# Of Lone Pairs Of Electrons
H	0	0
C	3	0
N	4	0
O	3	1
S	3	1
P	4	0

Table Of Atoms With -1 Formal Charge

Atom	# Of Bonds (Single bonds count as 1 bond double bonds as 2, and triple bonds as 3)	# Of Lone Pairs Of Electrons
C	3	1
N	2	2
O	1	3
S	1	3
F	0	4

1.3 BOND ANGLES AND SHAPES OF MOLECULES

- The **Valence-Shell Electron-Pair Repulsion (VSEPR)** model of molecular structure assumes that *areas* of valence electron density around an atom are distributed to be as far apart as possible in three-dimensional space. ✱
  - When using the VSEPR model, lone pairs of electrons, the two electrons in a single bond, the four electrons in a double bond, and the six electrons in a triple bond are each counted as only a single area of electron density.
  - Four areas of electron density around an atom adopt a tetrahedral shape with bond angles near 109.5° such as in methane, CH_4.
  - Three areas of electron density around an atom adopt a trigonal planar shape with bond angles near 120° such as in formaldehyde, $H_2C{=}O$.
  - Two areas of electron density around an atom adopt a linear shape with bond angles near 180°, such as in acetylene, $HC{\equiv}CH$.
  - The VSEPR model predicts shape but does not explain why the electrons are located where they are. Thus, it is only a useful model and not a theory. Two theories of electronic structure are presented in Sections 1.8 and 1.9.

1.4 FUNCTIONAL GROUPS

- **Condensed structural formulas** are highly abbreviated versions of Lewis structures that are used to describe molecules.
  - The number of hydrogen atoms attached to a given atom are denoted with a subscript. For example $-CH_3$ means there are three hydrogen atoms attached to the carbon atom.
  - Lines are used to denote bonds. A single line (-) between two atoms represents a single bond, a double line (=) denotes a double bond, etc.

    Lone pairs of electrons are not drawn.
  - Parentheses are used to denote branching in a molecule. In other words, whole groups of atoms attached to a given atom are placed in parentheses. For example, $(CH_3)_3CH$ indicates that there are three $-CH_3$ groups attached to the remaining carbon atom.
- Carbon combines with other atoms to form characteristic structural units called **functional groups** such as hydroxyl groups (-OH), carboxyl groups ($-CO_2H$), and carbonyl groups (C=O) that are important for three reasons: ✱
  - **First**, they are sites of chemical reactions, and a particular functional group, in whatever compound it is found, undergoes the same types of chemical reactions.
  - **Second,** they are used to divide organic molecules into classes in terms of their physical properties.
  - **Third**, they are the basis of naming compounds.

1.5 CONSTITUTIONAL ISOMERISM

- **Constitutional isomers** are compounds that have the same molecular formulas, but the atoms are connected to each other differently.
  - When asked to write all possible structures with a given molecular formula, use a combination of a systematic method and creativity. A reasonable approach is to first write all possible carbon skeletons, then place the functional group(s) in all possible locations for each carbon skeleton. *[This is harder than it looks. The only way to get good at writing constitutional isomers is to practice.]*

1.6 THE THEORY OF RESONANCE

- **Resonance theory** is used to depict and understand those special chemical species for which no single Lewis structure provides an adequate description. ✱ *[This also requires a lot of practice.]*
  - Resonance theory is particularly good at helping to understand cases of partial bonding (for example 1.5 bonds between two atoms, etc.) or when a formal charge is distributed between

more than one atom. In these situations, the true structure (referred to as a **resonance hybrid**) is thought of as a composite of two or more **contributing structures.**

In drawing resonance structures, a straight, double headed arrow (↔) is placed between contributing structures. Curved arrows (⌒) are used to indicate how electrons can be redistributed to make one contributing structure from another. Always draw the curved arrows to indicate where a pair of electrons started (tail of arrow) to where the electrons end up (head of arrow).

- No atoms are moved between contributing structures, and only *certain* kinds of electrons are moved. ✱

  Lone pairs of electrons or a pair of electrons taking part in a multiple bond are moved.
  A lone pair electrons can only move a very short distance. A lone pair of electrons from an atom can only move to an adjacent bond to make a multiple bond. A pair of electrons in a multiple bond can only move to an adjacent atom to create a new lone pair of electrons or to an adjacent bond to make a multiple bond.

- Contributing structures cannot have atoms with more than a filled valence shell of electrons.✱

  There can be no more than 2 electrons around H or more than 8 electrons around C, N, O, or F.
  You cannot have more than a filled valence shell because there are no more orbitals in which to place the extra electrons. On the other hand, atoms can have less than the filled valence shell, for example a carbon atom with +1 formal charge only has 6 valence electrons and one empty **2p** orbital.

• Keep in mind that even though we use multiple contributing structures to describe a molecule or ion, the molecule or ion in fact only has one true structure. It does not alternate between the contributing structures. ✱

• The following qualitative rules are used to estimate the **relative importance** of **contributing structures.**

- Equivalent structures (those that have the same patterns of covalent bonding) contribute equally. For example, carbonate ion (CO_3^{2-}) has three equivalent contributing structures.
- Structures in which all atoms have filled valence shells (complete octets) contribute more than those in which one or more valence shells are not filled.
- Structures involving separation of unlike charges contribute less than those that do not involve charge separation.
- Structures that carry a negative charge on the more electronegative element contribute more than those with the negative charge on the less electronegative element.

1.7 QUANTUM OR WAVE MECHANICS

• Electrons have certain properties of **particles** and certain properties of **waves.** ✱

- Electrons have mass and charge like particles.
- Because they are so small and are moving so fast, electrons have no defined position. Their location is best described by wave mechanics and a wave equation called the **Schrödinger equation.**

  There are many solutions to the Schrödinger equation for a given atom.
  Solutions of the Schrödinger equation are called **wave functions** and are represented by the Greek letter ψ (psi).
  Each wave function (ψ) describes a different orbital.

- The **sign of the wave function** ψ can shift from positive (+) to negative (-) in different parts of the same orbital. This is analogous to the way that waves can have positive or negative amplitudes. ✱

  The sign of the wave function does not indicate anything about charge. *[This can be confusing. Make sure that you understand it before you go on.]*

- The value of ψ^2 is proportional to the probability of finding electron density at a given point in an orbital. Note that the sign of ψ^2 is always positive, because the square of even a negative value is still positive.
  In a **2**p orbital, it is just as probable to find electron density in the negative lobe as it is to find electron density in the positive lobe. *[Make sure you understand this statement.]*
- A **node** is any place in an orbital at which the value of ψ and thus ψ^2 is zero.
  A nodal surface or nodal plane are surfaces or planes where ψ and ψ^2 is zero. There is absolutely no electron density at a node, a nodal surface, or a nodal plane.
- The Schrödinger equation can in principle describe covalent bonding, but even with powerful computers the equation is too complicated to be solved exactly for large molecules.
  Because of this complexity, two sets of approximations for the Schrödinger equation are used to describe covalent bonding in molecules; the valence bond approach and the molecular orbital approach.
  These two approaches are complementary, they are each useful for understanding different things.

1.8 VALENCE BOND APPROACH TO COVALENT BONDS

- According to **valence bond theory**, a covalent bond is formed when two atomic orbitals overlap, that is when they occupy the same region in space. The greater the overlap, the stronger the bond. ✻
  - Each electron in a covalent bond is delocalized over the two atomic orbitals that overlap to form the bond. The majority of electron density is thus concentrated between atoms taking part in bonds, consistent with Lewis structures.
- For elements more complicated than hydrogen, it is helpful to combine (**hybridize**) the *valence* atomic orbitals on a given atom before looking for overlap with orbitals from other atoms. ✻
  - For C, N, and O hybridization means the **2**s atomic orbital is combined with one, two, or all three **2**p atomic orbitals.
- The results of the orbital combinations are called **hybrid orbitals**, the number of hybrid orbitals are equal to the number of atomic orbitals combined.
  - An **sp^3 hybrid orbital** is the combination of one **2**s orbital with three **2**p orbitals.
    Four sp^3 orbitals of equivalent energy are created.
    Each sp^3 orbital has one large lobe and a smaller one of opposite sign pointing in the opposite direction (with a node at the nucleus). The large lobes point to a different corners of a tetrahedron. This explains the tetrahedral structure of molecules like methane, CH_4.
  - An **sp^2 hybrid orbital** is the result of combining the **2**s orbital with two **2**p orbitals.
    Three sp^2 orbitals of equivalent energy are created.
    Each sp^2 orbital has one large lobe and a smaller one of opposite sign pointing in the opposite direction (with a node at the nucleus). The large lobes point to a different corner of a triangle. This explains the trigonal planar structure of molecules like formaldehyde, $CH_2{=}O$.
    The left over **2**p orbital lies perpendicular of the plane formed by the three sp^2 orbitals.
  - An **sp hybrid orbital** is the combination of one **2**s orbital with one **2**p orbital.
    Two sp orbitals of equivalent energy are created.
    Each sp orbital has two lobes of opposite sign pointing in opposite directions (with a node at the nucleus).
    The lobes with like sign point in exactly opposite directions (180° with respect to each other). This explains the linear structure of molecules like acetylene, $HC{\equiv}CH$.
    The two left over **2**p orbitals are orthogonal to each other, and orthogonal to the two sp hybrid orbitals as well.

- Carbon atoms in molecules are either sp^3, sp^2, or sp hybridized. **1**s Orbitals are not considered for hybridization with C, N, or O because the **1**s orbitals do not participate in covalent bonding.

• Bonding in complex molecules can be *qualitatively* understood as overlap of hybrid orbitals. ✱

• Organic chemistry is primarily concerned with two types of covalent bonds, namely sigma bonds and pi bonds.

- A **σ (sigma) bond** occurs when the majority of the electron density is found on the bond axis.

A σ bond results from the overlap between an s orbital and any other atomic orbital.

A σ bond also results from the overlap of an sp^3, sp^2, or sp hybrid orbital and any s, sp^3, sp^2, or sp hybrid orbital.

Because rotating a σ bond does not decrease the overlap of the orbitals involved (σ bonds have cylindrical symmetry), a σ bond can rotate freely about the bond axis.

- A **π (pi) bond** occurs when the majority of the electron density is found above and below the bond axis.

A π bond results from the overlap of two **2**p orbitals that are parallel to each other, and orthogonal to the σ bond that exists between the two atoms.

Because rotating a π bond by 90° destroys the orbital overlap, π bonds cannot rotate around the bond axis. *[Understand this before going on.]*

- The hybridization of a given atom (sp^3, sp^2, or sp) determines the geometry and type of bonds made by that atom. The important parameters associated with each hybridization state are listed in the following table.

Carbon Atom Hybridization State Parameters

Hybridiztion State	# Of Hybrid Orbitals	# Of **2**p Orbitals Left Over	# Of Groups Bonded To Carbon	# Of σ Bonds	# Of π Bonds	Geometry Around Carbon
sp^3	**4**	**0**	**4**	**4**	**0**	**Tetrahedral**
sp^2	**3**	**1**	**3**	**3**	**1**	**Trigonal Planar**
sp	**2**	**2**	**2**	**2**	**2**	**Linear**

- To summarize, valence bond theory is based on the approximation that bonding is the result of overlap of atomic orbitals, and that electron density is concentrated between atoms. This approach is intuitively consistent with Lewis structures, and is extremely useful for qualitatively understanding molecular structure and reactivity in organic chemistry. *However*, simple valence bond theory is not usually used for detailed quantitative calculations. Modern calculations are carried out with a more detailed theory called molecular orbital theory.

1.9 MOLECULAR ORBITAL APPROACH TO COVALENT BONDING

• **Molecular orbital theory** assumes individual electron pairs are not localized in bonds between atoms, but are found in molecular orbitals that are distributed over the *entire* molecule.✱

• Molecular orbitals are analogous to atomic orbitals and are described by the following four rules:

- **First**, combination of *n* atomic orbitals in a molecule forms *n* molecular orbitals, each of which extends over the entire molecule.
  The number of molecular orbitals is equal to the number of atomic orbitals combined, because wave functions can be combined by both addition and subtraction.
- **Second**, molecular orbitals just like atomic orbitals, are arranged in order of increasing energy.
- **Third**, filling of molecular orbitals is governed by the same principles as the filling of atomic orbitals. (See Section 1.1)
  Electrons are placed in molecular orbitals starting with the lowest energy orbitals first.
  A molecular orbital can hold no more than two electrons.
  Two electrons in the same molecular orbital have opposite spins.
  When two or more **degenerate** (same energy) molecular orbitals are available, one electron is placed in each before any one of them gets two electrons.
- **Fourth**, bond order is one-half the difference of the number of electrons in bonding molecular orbitals, minus the number of electrons in antibonding molecular orbitals.

• There are some major differences between molecular orbital theory and valence bond theory.
- In molecular orbital theory, each molecular orbital is distributed around the entire molecule, not localized in bonds between atoms.
- For constructing molecular orbitals, even the non-valence atomic orbitals are used such as the 1s orbital of carbon.
- Hybridized atomic orbitals are not used in molecular orbital theory.
- In the case of degenerate bonding molecular orbitals like those found in O_2, you can have one electron in each. Valence bond theory predicts only paired electrons in these situations.

• When two atomic orbitals combine to form a molecular orbital, the wave functions are both added and subtracted to create one **bonding molecular orbital** and one **antibonding molecular orbital**.
- A bonding molecular orbital occurs when the electron density of the orbital is concentrated between the atomic nuclei.
  Electrons in bonding molecular orbitals stabilize covalent bonds because they serve to offset the repulsive forces of the positively charged atomic nuclei.
  The energy of a bonding molecular orbital is lower than the energy of the two uncombined atomic orbitals.
- An antibonding molecular orbital (designated with an *) occurs when the electron density of the orbital is concentrated in regions of space outside the area between the atomic nuclei.
  Electrons in antibonding molecular orbitals do not stabilize covalent bonds because the electrons are *not* positioned to offset the repulsive forces of the positively charged atomic nuclei.
  The energy of an antibonding molecular orbital is higher than the energy of the two atomic orbitals.
- When two s orbitals such as two **1s** orbitals are combined, a σ bonding molecular orbital and a σ^* (antibonding) molecular orbital are produced. *[Make sure you are familiar with the general shapes of all the molecular orbitals for simple diatomic molecules like H_2 and O_2 (Section 1.9A in the book).]*
- When two **$2p_x$** atomic orbitals are combined along the x axis, a σ bonding molecular orbital and a σ^* (antibonding) molecular orbital are produced.
- When the remaining four **$2p_y$** and **$2p_z$** orbitals are combined, two degenerate π bonding molecular orbitals and two degenerate π^* (antibonding) molecular orbitals are produced.
- An **electronic ground state** occurs when all of the electrons are in the molecular orbitals of lowest possible energy.
- An **electronic excited state** occurs when an electron in a lower lying orbital is promoted to an orbital that is higher in energy. This can occur when light is absorbed by a molecule, for example.

- To a first approximation, only atomic orbitals of similar energy are combined to produce molecular orbitals.
  - For example, in homonuclear diatomic molecules like O_2, the **1**s orbitals are combined only with each other, the **2**s orbitals are combined only with each other, etc.
- A major triumph of molecular orbital theory is that it correctly predicts the presence of two unpaired electrons in ground state oxygen, O_2. This has been found to be true experimentally and has some interesting consequences in both chemistry and biochemistry. Deriving orbital diagrams for molecules more complicated than diatomics usually requires a computer, but the results can accurately predict the electronic structure of molecules. To summarize, molecular orbital theory is based on the combination of atomic orbitals to form molecular orbitals that extend over the entire molecule. Molecular orbital calculations give very accurate results concerning electronic energies and electronic distribution in complex molecules.

A FINAL COMMENT.

- You may become confused when valence bond theory and molecular orbital theory are presented together. It is important to keep the following points in mind when learning about these two theories.
- These two theories are complimentary, and are used for different purposes.
  - Valence bond theory and molecular orbital theory are best thought of as different approximations for the electronic structure of molecules.
  - Despite the apparent differences, the approaches almost always yield the same predictions of molecular structure and reactivity, the molecular orbital calculations are simply more detailed.
  - In other words, if you add up the electron densities in all of the calculated molecular orbitals, you find that the majority of electron density is localized between bonded atoms as predicted by valence bond theory.
- The bottom line:
  - Valence bond theory provides a good *qualitative* description of bonding in molecules for use in routine situations. ✱ *[You should use this approach (with electrons centered between bonded atoms and overlap of hybridized atomic orbitals) as a way to think about the molecules and reactions described in the rest of this book.]*
  - Molecular orbital theory is usually reserved for detailed calculations when more *quantitative* results are required.

CHAPTER 1
Solutions to the Problems

Problem 1.1 Show that the following obey the octet rule.
(a) Sulfur (atomic number 16) forms sulfide ion, S^{2-}.

S (16 electrons): $1s^2 2s^2 2p^6 3s^2 3p^4$

S^{2-} (18 electrons): $1s^2 2s^2 2p^6 3s^2 3p^6$

(b) Magnesium (atomic number 12) forms Mg^{2+}.

Mg (12 electrons): $1s^2 2s^2 2p^6 3s^2$

Mg^{2+} (10 electrons): $1s^2 2s^2 2p^6$

Problem 1.2 Calculate the percent ionic character of the following covalent bonds. For each, show which atom bears the partial negative charge and which atom bears the partial positive charge.
(a) N-H (b) C-Mg

$$\frac{(3.0 - 2.1)}{3.0} \times 100 = 30\% \qquad \overset{\delta- \ \delta+}{\text{N-H}}$$

$$\frac{(2.5 - 1.2)}{2.5} \times 100 = 52\% \qquad \overset{\delta- \ \delta+}{\text{C-Mg}}$$

(c) B-H

$$\frac{(2.1 - 2.0)}{2.1} \times 100 = 4.8\% \qquad \overset{\delta+ \ \delta-}{\text{B-H}}$$

Problem 1.3 Draw Lewis structures, showing all valence electrons, for the following covalent molecules.
(a) C_2H_6 (b) CS_2 (c) HCN

```
   H  H
   |  |
H--C--C--H          :S=C=S:          H-C≡N:
   |  |
   H  H
```

Problem 1.4 Draw Lewis structures for the following ions, and state which atom in each bears the formal charge.

(a) CH_3^+ (b) CH_3^- (c) OH^-

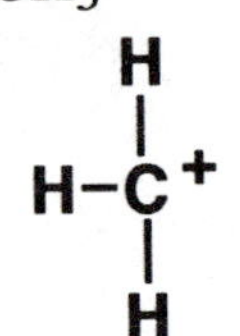

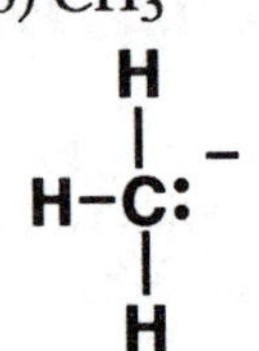

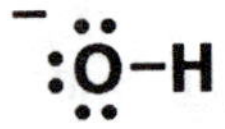

$^-$:Ö-H

Problem 1.5 Assign formal charges in the alternative Lewis structure for phosphoric acid that shows only single bonds to phosphorus and only eight electrons in its valence shell.

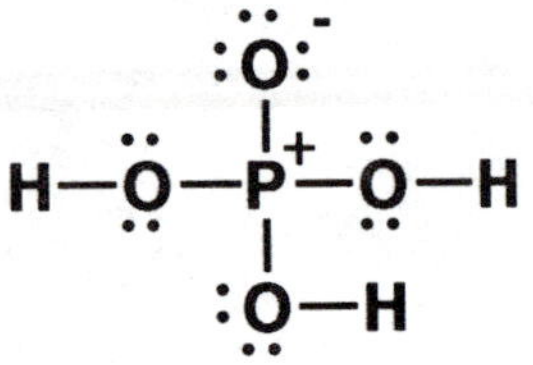

There is a negative formal charge on one oxygen and a positive formal charge on phosphorus.

Problem 1.6 Predict all bond angles for the following molecules and ions.

(a) CH_3OH (b) PF_3 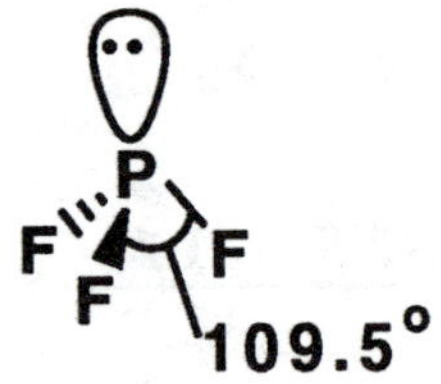 (c) CH_2Cl_2 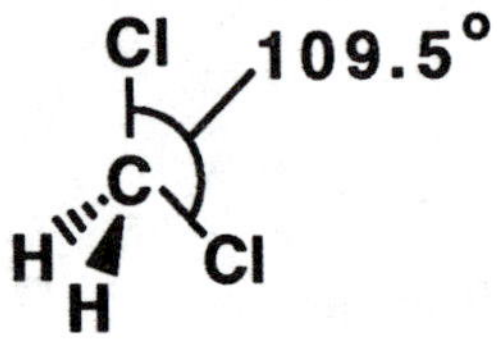 (d) CH_3^+

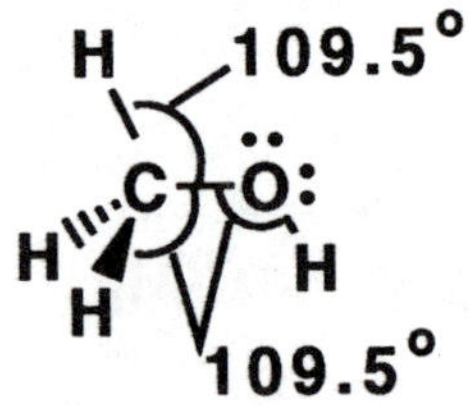

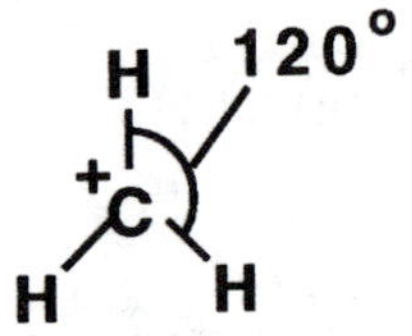

109.5° 109.5° 109.5° 109.5° 120°

Problem 1.7 There is one ether of molecular formula C_3H_8O. Draw a condensed structural formula for this compound.

$CH_3-CH_2-O-CH_3$

Problem 1.8 Draw condensed structural formulas for the three ketones of molecular formula $C_5H_{10}O$.

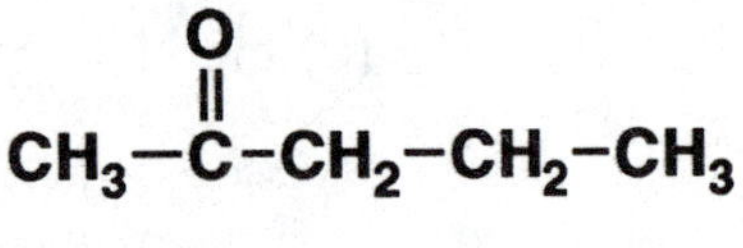

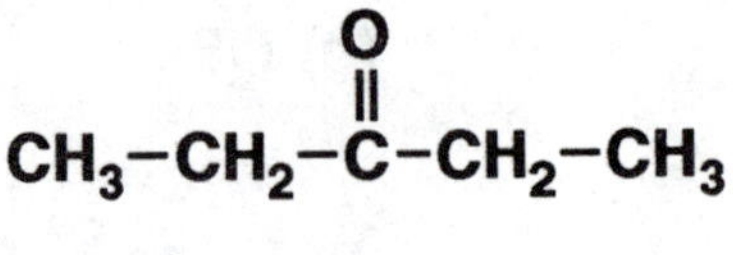

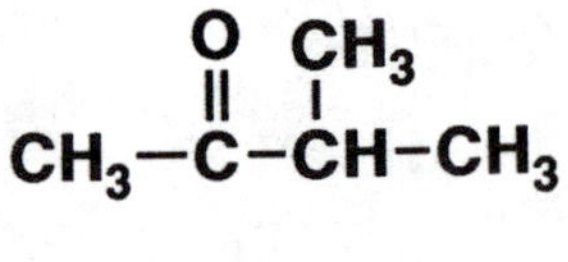

$CH_3-C(=O)-CH_2-CH_2-CH_3$ $CH_3-CH_2-C(=O)-CH_2-CH_3$ $CH_3-C(=O)-CH(CH_3)-CH_3$

Problem 1.9 Draw condensed structural formulas for the two carboxylic acids of molecular formula C_4H_8O.

$CH_3-CH_2-CH_2-CO_2H$ $CH_3-CH(CH_3)-CO_2H$

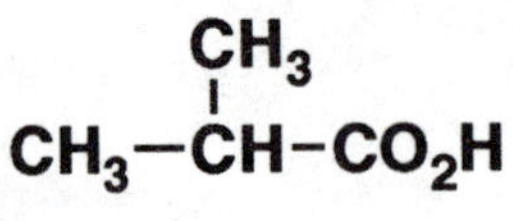

Problem 1.10 Draw condensed structural formulas for:
(a) the aldehyde and the ketone of molecular formula C_3H_6O.

$CH_3-CH_2-C(=O)-H$ $CH_3-C(=O)-CH_3$

(b) the two carboxylic acids of molecular formula $C_4H_8O_2$.

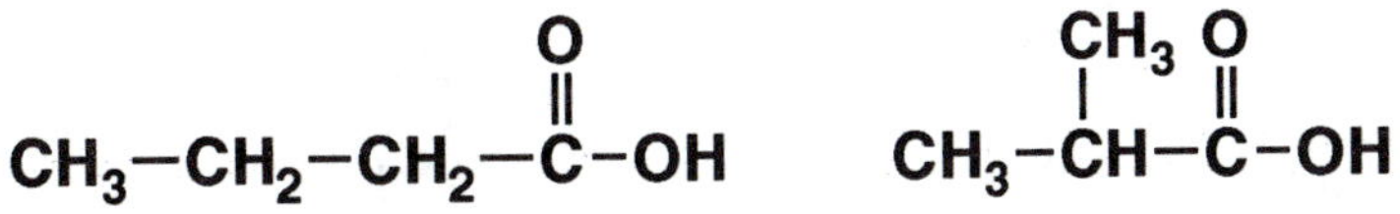

$CH_3-CH_2-CH_2-C(=O)-OH$ $CH_3-CH(CH_3)-C(=O)-OH$

(c) the two esters of molecular formula $C_3H_6O_2$.

$CH_3-C(=O)-O-CH_3$ $H-C(=O)-O-CH_2-CH_3$

Problem 1.11 Divide the following molecules into groups of constitutional isomers.

(a) $CH_2=CH-O-CH=CH_2$

(b) 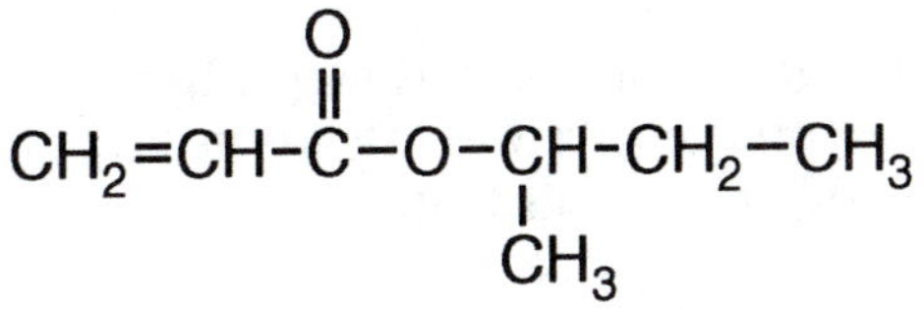 $CH_2=CH-C(=O)-O-CH(CH_3)-CH_2-CH_3$

(c) $CH_3-CH_2-O-C\equiv CH$

(d) 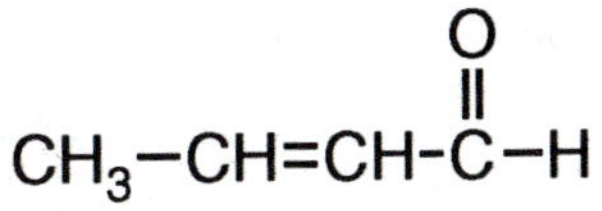 $CH_3-CH=CH-C(=O)-H$

(e) 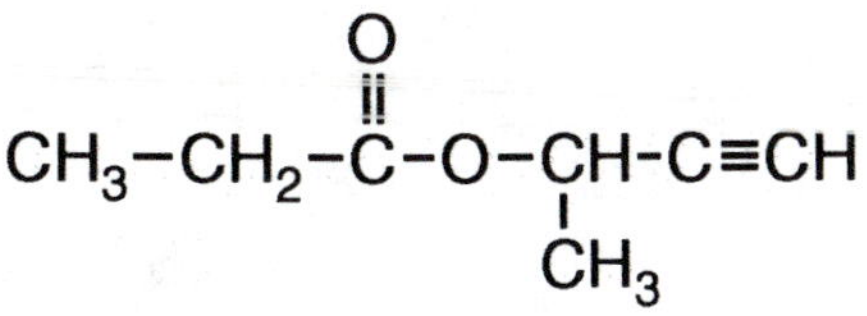 $CH_3-CH_2-C(=O)-O-CH(CH_3)-C\equiv CH$

(f) 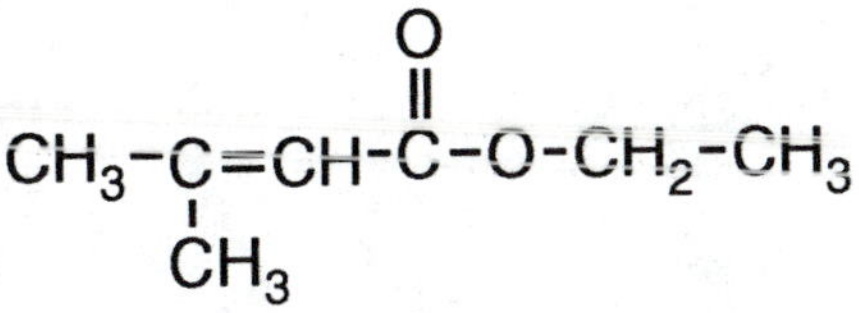 $CH_3-C(CH_3)=CH-C(=O)-O-CH_2-CH_3$

Following are molecular formulas of each compound:
(a)C_4H_6O (b)$C_7H_{12}O_2$ (c)C_4H_6O (d)C_4H_6O (e)$C_7H_{10}O_2$ (f)$C_7H_{12}O_2$
From this information, see that compounds (a), (c) and (d) are constitutional isomers. Compounds (b) and (f) are constitutional isomers as well. Compound (e) is not a constitutional isomer of any other structure given in this problem.

Problem 1.12 Draw the contributing structure indicated by curved arrows. Be certain to show all formal charges.

(a) 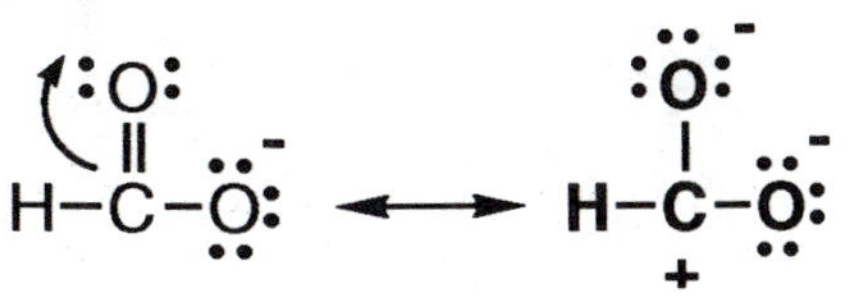 $H-C(=\ddot{O}:)-\ddot{O}:^-$ ⟷ $H-C^+(-\ddot{O}:^-)-\ddot{O}:^-$

(b) 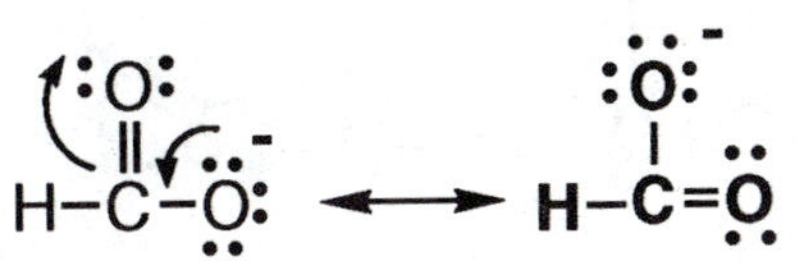 $H-C(=\ddot{O}:)-\ddot{O}:^-$ ⟷ $H-C(-\ddot{O}:^-)=\ddot{O}$

(c) $CH_3-C(=O)-\ddot{O}-CH_3 \longleftrightarrow CH_3-C(-O^-)=\overset{+}{O}-CH_3$

Problem 1.13 Following are pairs of contributing structures. Estimate the relative contribution of each structure to its hybrid.

(a) $H_2C=CH_2 \longleftrightarrow H_2\overset{+}{C}-\overset{-}{C}H_2$ (b) $H-C(=O)-\ddot{O}^- \longleftrightarrow H-\overset{+}{C}(-O^-)-\ddot{O}^-$

(c) $H-C(=O)-\ddot{O}-H \longleftrightarrow H-C(-O^-)=\overset{+}{O}-H$

The first structure makes the greater contribution in (a), (b), and (c). In each case, the second contributing structure involves the disfavored creation and separation of unlike charges.

Problem 1.14 Describe the bonding in the following molecules in terms atomic orbitals involved, and predict all bond angles.

(a) $CH_3-\ddot{O}-CH_3$ (b) $CH_3-CH=CH_2$ (c) $CH_3-\ddot{N}H_2$

(a) H–C–O–C–H: sp3 (C), sp3 (O), sp3 (C); $\sigma_{sp^3\text{-}1s}$; $\sigma_{sp^3\text{-}sp^3}$; 109.5°

(b) H–C–C=C–H: sp3, sp2; $\sigma_{sp^3\text{-}1s}$; $\sigma_{sp^2\text{-}sp^2}$; $\sigma_{sp^3\text{-}sp^2}$; $\pi_{2p\text{-}2p}$; $\sigma_{sp^2\text{-}1s}$; 109.5°; 120°

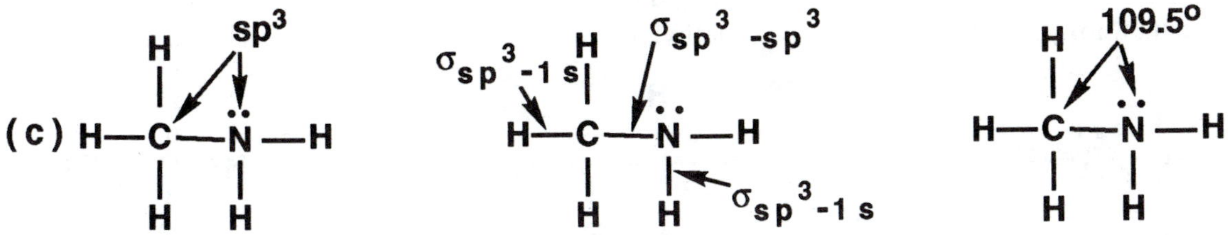

Problem 1.15 Describe the ground-state electron configuration for the helium molecule, He_2, and show how molecular orbital theory accounts for the fact that He_2 is not stable-that helium, instead, exists as a monatomic gas, He.

Combination of a single 1s orbital from each helium atom gives two molecular orbitals; a sigma bonding MO and a sigma antibonding MO. In the ground state electron configuration of He_2, two electrons occupy the lower-lying bonding MO and two electrons occupy the equally higher-lying antibonding MO. Thus, there is no net stabilization of He_2 compared to two isolated He atoms. He_2 is not stable.

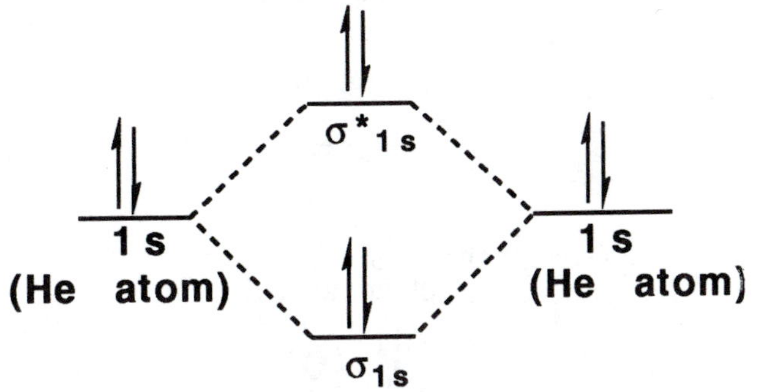

He_2 molecule

Lewis Structures

Problem 1.16 Write Lewis structures for the following molecules. Be certain to show all valence electrons. Note that for the oxygen acids - that is (l), (n), (r), (s), (t), (v) and (w) - each oxygen is attached directly to the central atom (C, N, P or S) and each ionizable hydrogen is attached to an oxygen atom. Also note that none of these compounds contains a ring of atoms. Under each compound is its name so that you can find the compound within the text.

(a) H_2O_2
hydrogen peroxide

(b) N_2H_4
hydrazine

(c) CH_3OH
methanol

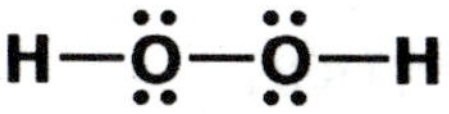

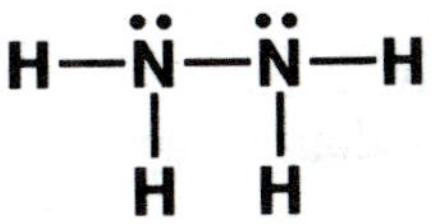

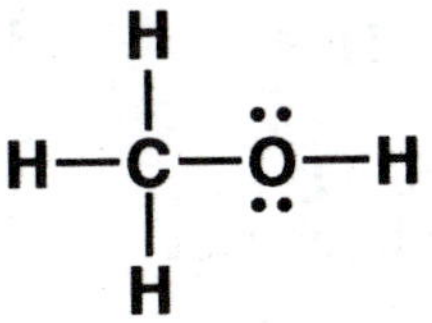

(d) CH_3SH
methanethiol

(e) CH_3NH_2
methylamine

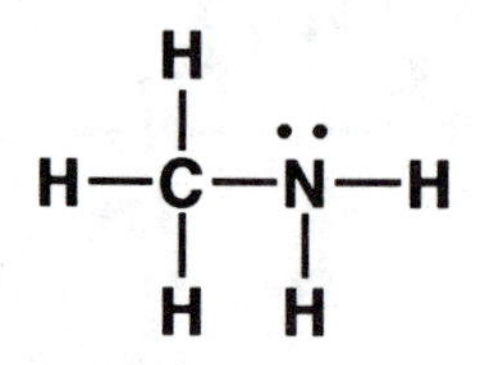

(f) CH_3Cl
chloromethane

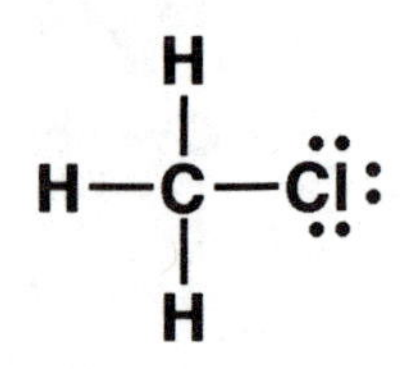

(g) CH_3OCH_3
dimethyl ether

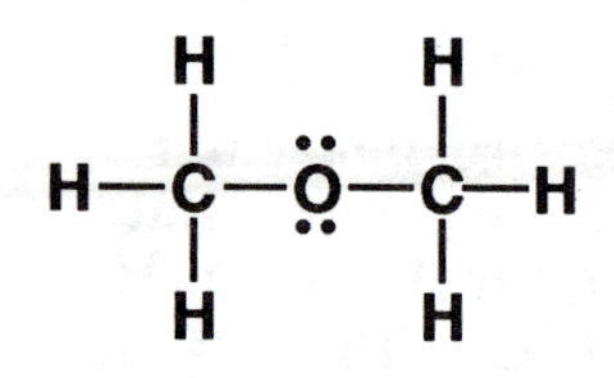

(h) C_2H_6
ethane

(i) C_2H_4
ethene

(j) C_2H_2
ethyne

(k) CO_2
carbon dioxide

(l) H_2CO_3
carbonic acid

(m) CH_2O
methanal

(n) CH_3CO_2H
ethanoic acid

(o) CH_3COCH_3
propanone

(p) CH_3NNCH_3
dimethyldiimine

(q) HCN
hydrogen cyanide

(r) HNO_3
nitric acid

(s) HNO_2
nitrous acid

(t) HCO_2H
methanoic acid

(u) NH_2OH
hydroxylamine

(v) H_2SO_4
sulfuric acid

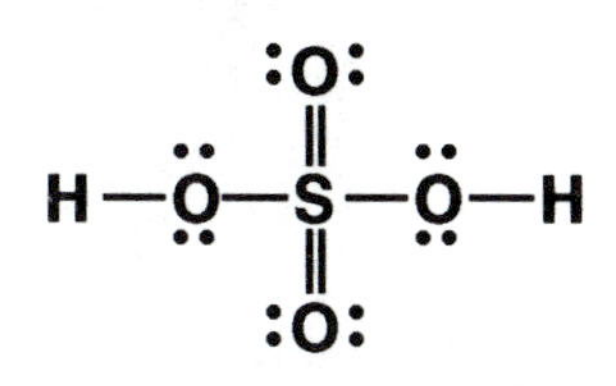

(w) H_3PO_4
phosphoric acid

Problem 1.17 Write Lewis structures for the following ions. Be certain to show all valence electrons and all formal charges. Under each formula is given the name of the ion.

(a) OH^-
hydroxide ion

(b) H_3O^+
hydronium ion

(c) NH_4^+
ammonium ion

(d) NH_2^-
amide ion

(e) HCO_3^-
bicarbonate ion

(f) CO_3^{2-}
carbonate ion

(g) Cl^-
chloride ion

(h) Cl^+
chloronium ion

(i) NO_2^-
nitrite ion

(j) NO_3^-
nitrate ion

(k) CH_3^-
methyl anion

(l) CH_3^+
methyl cation

(m) $CH_3CO_2^-$
ethanoate ion

(n) HCO_2^-
methanoate ion

(o) SO_4^{2-}
sulfate ion

(p) HPO_4^{2-}
hydrogen phosphate ion

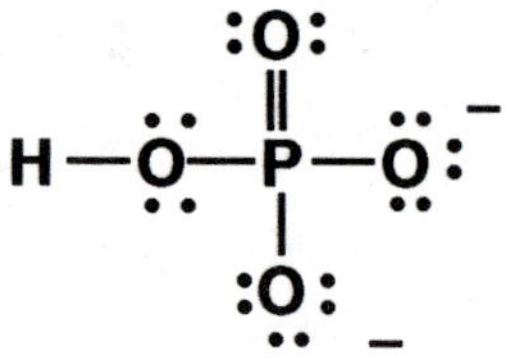

H—O—P(=O)(—O: $^-$)—O: $^-$

Problem 1.18 Following the rule that each atom of carbon, oxygen, nitrogen, and the halogens reacts to achieve a complete outer shell of eight electrons, add unshared pairs of electrons as necessary to complete the valence shells of the following molecules and ions. Assign formal charges as appropriate.

The following structural formulas show all valence electrons and all formal charges.

(a) 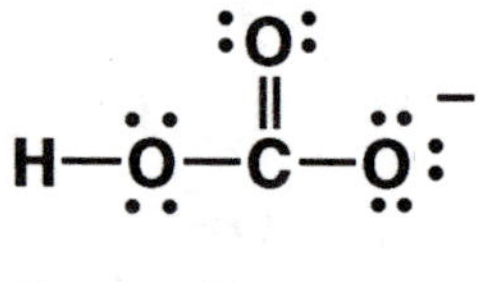 H—O—C(=O)—O: $^-$

(b) H—C(H)(H)—C(H)(H)—O: $^-$

(c) CH_3—N^+(CH_3)(CH_3)—CH_3

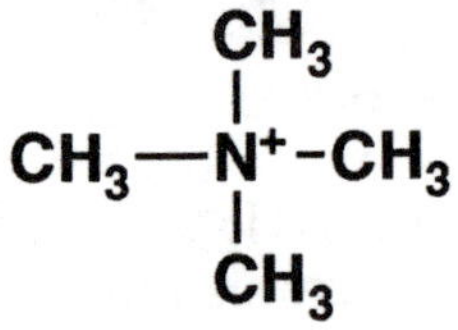

(d) 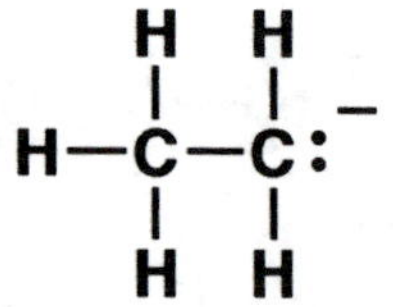 H—C(H)(H)—C(H)(H): $^-$

(e) 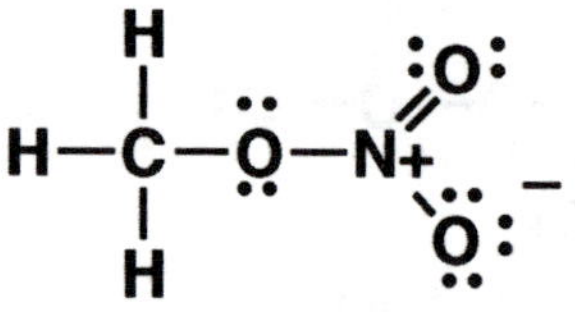 H—C(H)(H)—O—N^+(=O)—O: $^-$

(f) 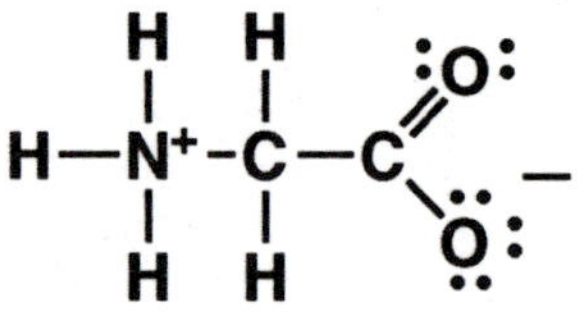 H—N^+(H)(H)—C(H)(H)—C(=O)—O: $^-$

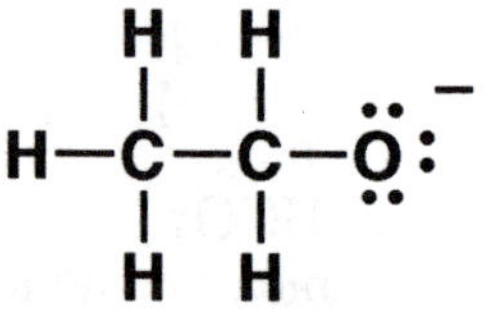

(g) 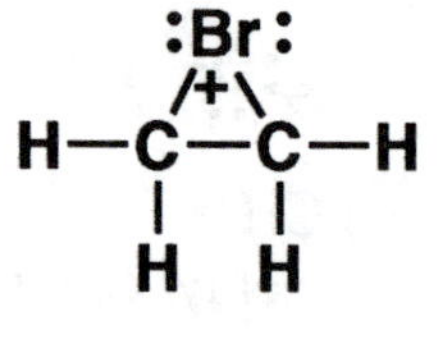 H—C(H)—C(H)—H bridged by :Br: (+)

Problem 1.19 Following are several Lewis structures showing all valence electrons. Assign formal charges to each structure as appropriate.

There is a formal positive charge in parts (a), (e), and (f). There is a formal negative charge in parts (b), (c), and (d).

(a) 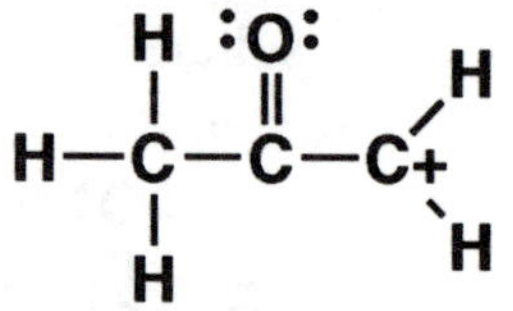 H—C(H)(H)—C(=O)—C^+(H)H

(b) H—N(H)—C(O$^-$)=C(H)—H

(c) H—C^-(H)—C(=O)—H

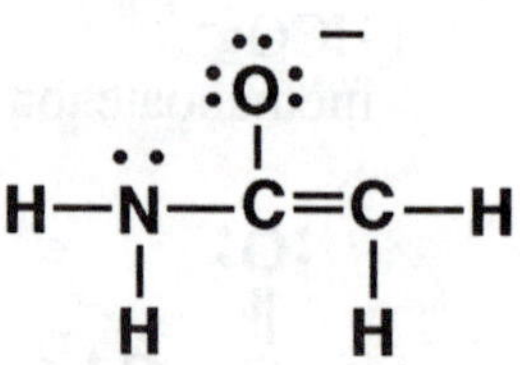

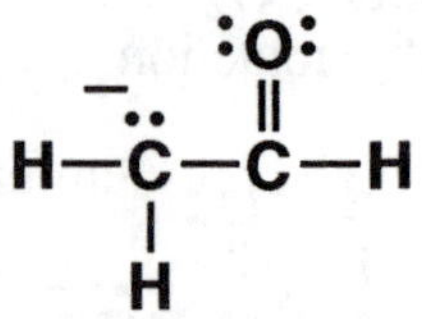

(d) 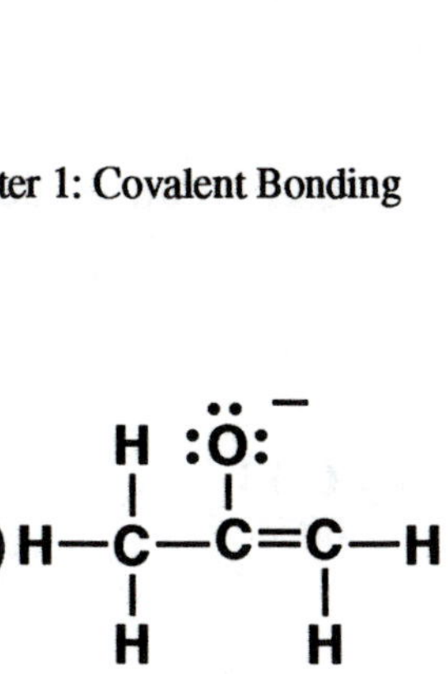 (e) 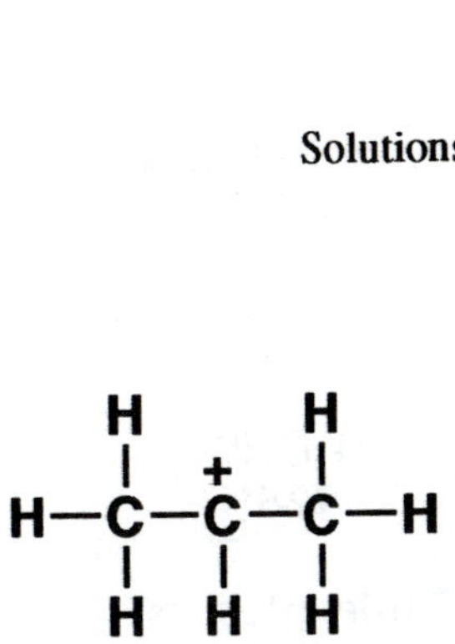 (f) 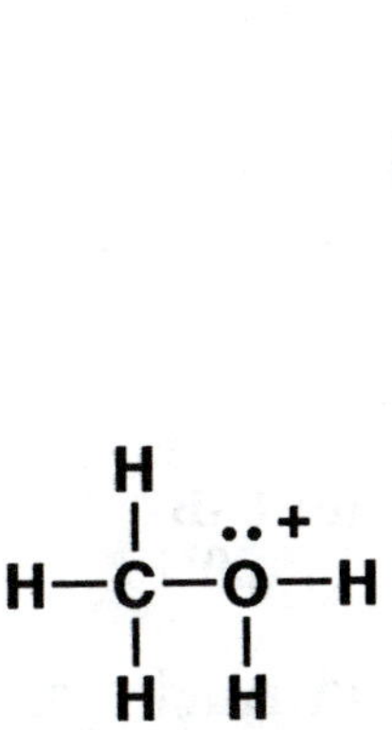

Problem 1.20 Following are compounds containing ionic and covalent bonds. Draw the Lewis structure for each and show by dashes which are covalent bonds and, by indication of charges, which are ionic bonds. Under each formula is given the name of the compound.

(a) CH_3ONa
sodium methoxide

(b) NH_4Cl
ammonium chloride

(c) $NaHCO_3$
sodium bicarbonate

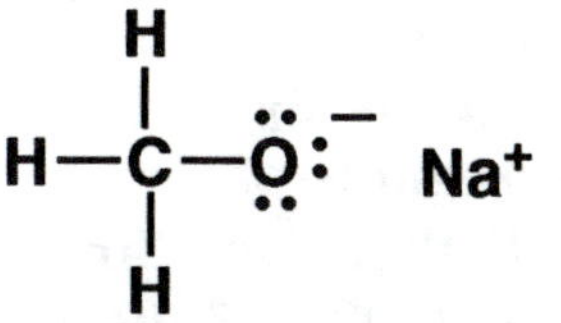

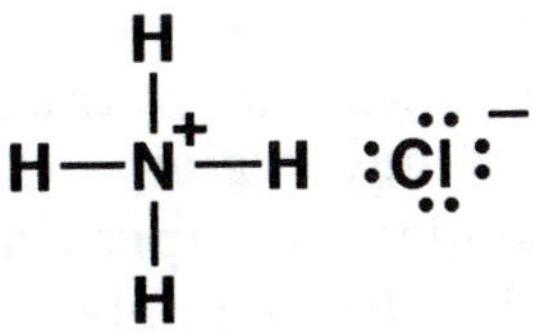

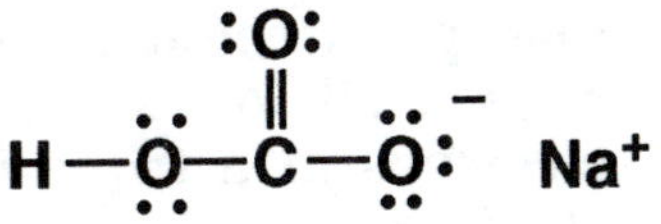

In naming these compounds, the cation is named first followed by the name of the anion.

Partial Ionic Character of Covalent Bonds

Problem 1.21 Arrange the single covalent bonds within each set in order of increasing partial ionic character.

(a) C-H, O-H, N-H
C-H < N-H < O-H
16% 30% 40%

(b) C-H, B-H, O-H
B-H < C-H < O-H
4.8% 16% 40%

(c) C-H, C-Cl, C-I
C-I < C-H < C-Cl
0.0% 16% 17%

(d) C-S, C-O, C-N
C-S < C-N < C-O
0.0% 17% 29%

(e) C-Li, C-B, C-Mg
C-B < C-Mg < C-Li
20% 52% 60%

Problem 1.22 Following are several organometallic compounds. Calculate the percent ionic character of each carbon-metal bond.

(a) 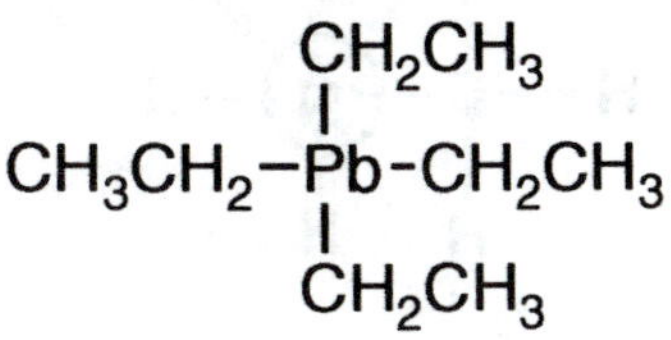
tetraethyllead

(b) 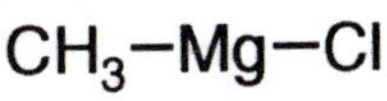
methylmagnesium chloride

(c) 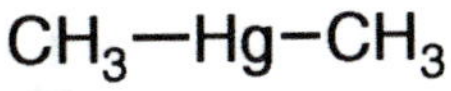
dimethylmercury

(d) 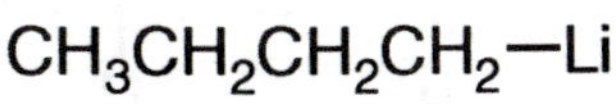
butyllithium

(e) 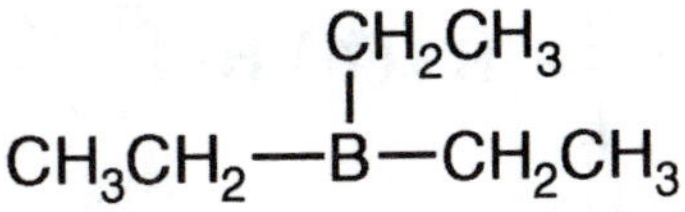
triethylborane

The organometallic bonds in this answer are arranged in order of increasing partial ionic character of the carbon-metal bond. The percent ionic character of each bond is calculated by the formula given in Section 1.2B.

(e) C-B	(c) C-Hg	(a) C-Pb	(b) C-Mg	(d) C-Li
20%	24%	28%	52%	60%

Bond Angles and Shapes of Molecules

Problem 1.23 Explain how the valence-shell electron-pair repulsion (VSEPR) model is used to predict bond angles.

To use the valence-shell electron-pair repulsion (VSEPR) model to predict bond angles, first draw a Lewis structure for the molecule or ion showing all valence electrons. Then count the region (bundles) of electron density in the valence shell of each atom about which you wish to predict bond angles. Remember that lone pairs count as one region, both electrons in a single bond count as one region, all four electrons in a double bond count as one region and all six electrons in a triple bond count as one region. If there are four regions of electron density, predict bond angles of approximately 109.5°. If there are three regions of electron density, predict bond angles of approximately 120°. If there are two regions of electron density, predict bond angles of 180°.

Problem 1.24 Following are Lewis structures for several molecules and ions. Use the VSEPR model to predict bond angles about each highlighted atom.

Approximate bond angles as predicted by the valence-shell electron-pair repulsion model are as shown.

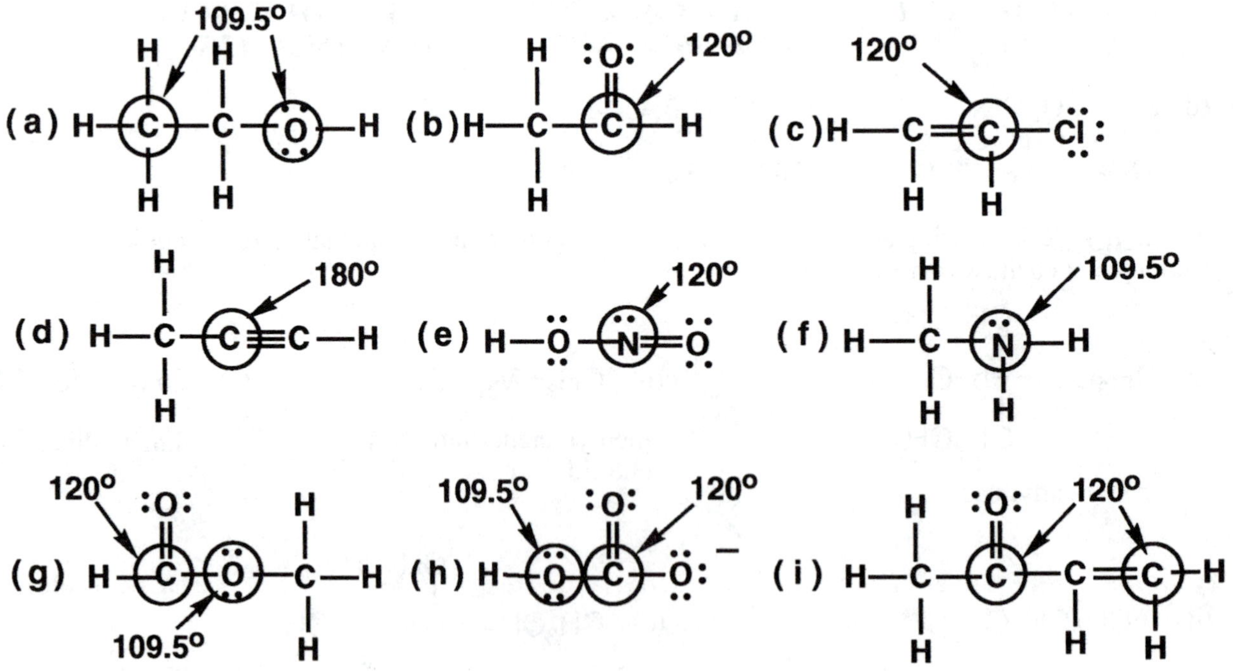

Problem 1.25 Use the VSEPR model to predict bond angles about each atom of carbon, nitrogen, and oxygen in the following molecules. For each molecule that contains atoms of oxygen or nitrogen, be certain that you first show all valence electrons on these atoms.

(a) $CH_3-CH_2-\ddot{O}-H$ 109.5°

(b) $CH_3-CH_2-C(=O)-H$ 109.5° 120°

(c) $CH_3-CH=CH_2$ 109.5° 120°

(d) $CH_3-C\equiv C-H$ 109.5° 180°

(e) $CH_3-\ddot{N}H_2$ 109.5°

(f) $CH_3-\ddot{N}(CH_3)-CH_3$ 109.5° (all 3 C atoms are the same)

(g) $CH_3-CH_2-C(=O)-\ddot{O}-H$ 109.5° 109.5° 120°

(h) $CH_3-C(=O)-\ddot{O}-CH_3$ 109.5° 109.5° 120°

(i) $CH_3-C(=O)-CH_3$ 109.5° 109.5° 120°

(j) $CH_2=C=CH_2$ 120° 180°

(k) $CH_3-CH=\ddot{N}-\ddot{O}-H$ 109.5° 109.5° 120°

Problem 1.26 Use the VSEPR model to predict the geometry of the following molecules:

(a) H_2Se

Selenium is in Group 6 of the periodic table and has six valence electrons. In the compound H_2Se, selenium is surrounded by four regions (bundles) of electron density. Therefore, predict 109.5° for the H-Se-H bond angle. The VSEPR model also predicts 109.5° for the H-S-H bond angle in H_2S. The observed H-S-H bond angle, however, is 93.3°, which indicates a failure of the VSEPR model. The observed H-Se-H bond angle is also nearer to 90°.

(b) CS_2

Review your answer to Problem 1.3(b) for the Lewis structure of carbon disulfide. Since there are only two bundles of electron density surrounding the central carbon atom, predict 180° for the S-C-S bond angle.

(c) SiH_4

Silicon is in Group 4 of the periodic table, and like carbon, has four valence electrons. In silane, SiH_4, silicon is surrounded by four regions of electron density. Therefore, predict all H-Si-H bond angles to be 109.5° and the molecule is thus tetrahedral around Si.

(d) PCl_3

Phosphorus in PCl_3 is surrounded by four regions of electron density, three single bonds and a lone pair of electrons. Therefore, predict all Cl-P-Cl bond angles to be 109.5° and the molecule is thus pyramidal.

Problem 1.27 Use VSEPR model to predict the geometry of the following ions:

(a) NH_2^- (b) NO_2^- (c) NO_2^+

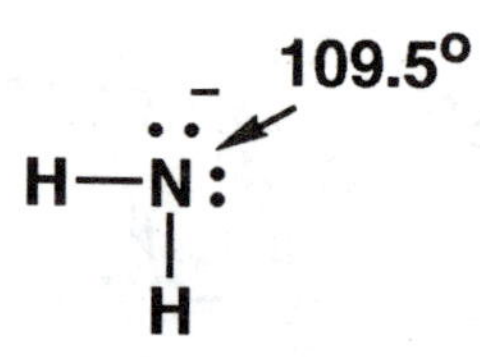

120°

$^-$:O—N=O

180°

O=N=O
+

(d) NO_3^- (e) $CH_3CO_2^-$ (f) CH_3^-

120°

:O:
‖
$^-$:O—N—O:$^-$
+

109.5° 120°

H :O:
H—C—C—O:$^-$
H

109.5°

H
H-C:$^-$
H

(g) $AlCl_4^-$

109.5° :Cl:

:Cl—Al$^-$—Cl:

:Cl:

Functional Groups

Problem 1.28 Draw Lewis structures for the following functional groups. Be certain to show all valence electrons on each.

(a) :O: ‖ —C—

carbonyl group

(b) :O: ‖ —C—O—H

carboxyl group

(c) —O—H

hydroxyl group

Problem 1.29 Draw condensed structural formulas for all compounds of molecular formula C_4H_8O that contain the following functional groups:

(a) A ketone (there is only one)

$$CH_3-\overset{\overset{\large O}{||}}{C}-CH_2-CH_3 \quad \text{also written as} \quad CH_3COCH_2CH_3$$

(b) An aldehyde (there are two)

$$CH_3-CH_2-CH_2-\overset{\overset{\large O}{||}}{C}-H \quad \text{also written as} \quad CH_3CH_2CH_2CHO$$

$$CH_3-\underset{\underset{\large CH_3}{|}}{CH}-\overset{\overset{\large O}{||}}{C}-H \quad \text{also written as} \quad (CH_3)_2CHCHO$$

(c) a carbon-carbon double bond and an ether (there are four)

$$CH_3-O-CH_2-CH=CH_2 \quad CH_3-O-CH=CH-CH_3 \quad CH_3-CH_2-O-CH=CH_2$$

$$CH_3-O-\overset{\overset{\large CH_3}{|}}{C}=CH_2$$

(d) a carbon-carbon double bond and an alcohol (there are eight)

There are three separate but related things to build into this answer; the carbon skeleton (the order of attachment of carbon atoms), the location of the double bond and the location of the -OH group. Here, as in other problems of this type, it is important to have a system and to follow it. As one way to proceed, first decide the number of different carbon skeletons that are possible. A little doodling with paper and pencil should convince you that there are only two.

$$C\text{-}C\text{-}C\text{-}C \quad \text{and} \quad C-\overset{\overset{\large C}{|}}{C}-C$$

Next locate the double bond on these carbon skeletons. There are three possible locations for it.

$$C=C-C-C \quad \text{and} \quad C-C=C-C \quad \text{and} \quad C=\overset{\overset{\large C}{|}}{C}-C$$

Finally, locate the -OH group and then add the remaining seven hydrogens to complete each structural formula. For the first carbon skeleton, there are four possible locations of the -OH group; for the second carbon skeleton there are two possible locations; and for the third, there are also two possible locations. As we shall discuss in Chapter 17, four of these compounds (marked by an asterisk) are not stable and are in equilibrium with a more stable aldehyde or ketone. You need not be concerned, however, with this now. Just concentrate on drawing the required eight structural formulas.

```
 *                           *OH                                OH
HO-CH=CH-CH2-CH3              |                                  |
                       CH2=C-CH2-CH3                   CH2=CH-CH-CH3

                                                             *OH
                                                              |
CH2=CH-CH2-CH2-OH     HO-CH2-CH=CH-CH3                 CH3-C=CH-CH3

            CH3                CH3
 *           |                  |
HO-CH=C-CH3              CH2=C-CH2-OH
```

Problem 1.30 Draw structural formulas for
(a) The eight alcohols of molecular formula $C_5H_{12}O$.

To make it easier for you to see the patterns of carbon skeletons and functional groups, only carbon atoms and hydroxyl groups are shown in the following solutions. To complete these structural formulas, you need only supply enough hydrogen atoms to complete the tetravalence of each carbon.

There are three different carbon skeletons on which the -OH group can be placed:

```
                                       C
                                       |
C-C-C-C-C          C-C-C-C           C-C-C
                     |                 |
                     C                 C
```

Three isomeric alcohols are possible from the first carbon skeleton, four from the second carbon skeleton, and one from the third carbon skeleton.

```
                          OH               OH
                          |                |
HO-C-C-C-C-C          C-C-C-C-C        C-C-C-C-C
    (1)                  (2)              (3)

                 OH          OH                              C
                 |           |                               |
HO-C-C-C-C     C-C-C-C     C-C-C-C     C-C-C-C-OH          C-C-C-OH
     |           |           |           |                   |
     C           C           C           C                   C
   (4)          (5)         (6)         (7)                 (8)
```

(b) The six ethers of molecular formula $C_5H_{12}O$.

Following are structural formulas for the six isomeric ethers of molecular formula $C_5H_{12}O$. They are drawn first with all possible combinations of $(C)_1$-O-$(C)_4$ and then all possible combinations of $(C)_2$-O-$(C)_3$.

```
                          C                     C
                          |                     |
C-O-C-C-C-C           C-O-C-C-C           C-O-C-C-C
    (1)                   (2)                   (3)

      C                                         C
      |                                         |
  C-O-C-C           C-C-O-C-C-C           C-C-O-C-C
      |
      C
     (4)                  (5)                   (6)
```

(c) The eight aldehydes of molecular formula $C_6H_{12}O$.

Following are structural formulas for the eight aldehydes of molecular formula $C_6H_{12}O$. They are drawn starting with the aldehyde group and then attaching the remaining five carbons in a chain (structure 1), then four carbons in a chain and one carbon as a branch on the chain (structures 2, 3, and 4) and finally three carbons in a chain and two carbons as branches (structures 5, 6, 7, and 8).

```
               O                     O                     O
               ||                    ||                    ||
   C-C-C-C-C-C-H         C-C-C-C-C-H           C-C-C-C-C-H
                           |                       |
                           C                       C
      (1)                     (2)                    (3)

           O                C   O                 C   O
           ||               |   ||                |   ||
   C-C-C-C-C-H            C-C-C-C-H             C-C-C-C-H
         |                  |                       |
         C                  C                       C
      (4)                     (5)                    (6)

              C O                        O
              | ||                       ||
          C-C-C-C-H              C-C-C-C-H
              |                      |
              C                      C-C
             (7)                    (8)
```

(d) The six ketones of molecular formula $C_6H_{12}O$.

Following are structural formulas for the six ketones of molecular formula $C_6H_{12}O$. They are drawn first with all combinations of one carbon to the left of the carbonyl group and four carbons to the right (structures 1, 2, 3, and 4) and

then with two carbons to the left and three carbons to the right (structures 5 and 6).

```
    O                 O                 O
    ||                ||                ||
C—C—C—C—C—C       C—C—C—C—C         C—C—C—C—C
                        |                   |
                        C                   C

    (1)               (2)               (3)

    O C                   O                 O C
    || |                  ||                || |
  C—C—C—C           C—C—C—C—C—C         C—C—C—C—C
      |
      C

    (4)                   (5)               (6)
```

(e) The eight carboxylic acids of molecular formula $C_6H_{12}O_2$.

There are eight carboxylic acids of molecular formula $C_6H_{12}O_2$. They have the same carbon skeletons as the eight isomeric aldehydes of molecular formula $C_6H_{12}O$ shown in part (c) of this problem. In place of the aldehyde group, substitute a carboxyl group.

```
             O                  O                    O
             ||                 ||                   ||
C—C—C—C—C—C—OH       C—C—C—C—C—OH         C—C—C—C—C—OH
                       |                      |
                       C                      C

     (1)                  (2)                  (3)

          O             C   O              C   O
          ||            |   ||             |   ||
C—C—C—C—C—OH        C—C—C—C—OH         C—C—C—C—OH
      |                 |                  |
      C                 C                  C

     (4)                  (5)                  (6)

         C O                    O
         | ||                   ||
     C—C—C—C—OH           C—C—C—C—OH
         |                    |
         C                    C—C

        (7)                   (8)
```

(f) The nine esters of molecular formula $C_5H_{10}O_2$.

Following are the nine esters of molecular formula $C_5H_{10}O_2$. First are drawn all esters with hydrogen attached to the carbonyl carbon and four carbons to oxygen

(structures 1,2,3 and 4), then one carbon on the carbonyl carbon and three carbons on the oxygen (structures 5 and 6), then two carbons on the carbonyl carbon and two on the oxygen (structure 7) and finally three carbons on the carbonyl carbon and one carbon on oxygen (structures 8 and 9).

```
  O                    O                    O
  ||                   ||                   ||
H-C-O-C-C-C-C        H-C-O-C-C-C          H-C-O-C-C-C
                           |                      |
                           C                      C

    (1)                  (2)                  (3)

  O   C                O                    O
  ||  |                ||                   ||
H-C-O-C-C            C-C-O-C-C-C          C-C-O-C-C
      |                                         |
      C                                         C

    (4)                  (5)                  (6)

                                                O
    O                      O                    ||
    ||                     ||                 C-C-C-O-C
C-C-C-O-C-C          C-C-C-C-O-C                |
                                                C

    (7)                  (8)                  (9)
```

Constitutional Isomers

Problem 1.31 Which of the following are true about constitutional isomers?
(a) They have the same molecular formula.
(b) They have the same molecular weight.
(c) They have the same order of attachment of atoms.
(d) They have the same physical properties.

Parts (a) and (b) are true. Parts (c) and (d) are false. Constitutional isomers have a different order of attachment of atoms and, therefore, are different compounds. Because constitutional isomers are different compounds, they have different physical and chemical properties.

Problem 1.32 Do the following structural formulas in each set represent constitutional isomers, or are they the same molecule? Name the functional group(s) in each molecule.

```
       CH3      CH3                  CH3
       |        |                    |
(a) CH3-CH-O-CH-CH3    and   CH3-CH-O-CH-CH3
                                      |
                                      CH3
```

These are the same molecule. It has an ether group.

(b) $CH_2{=}CH{-}CH_2{-}CH_3$ and $CH_3{-}CH{=}CH{-}CH_3$

These are constitutional isomers. They both have a carbon-carbon double bond.

(c) $CH_3{-}O{-}C({=}O){-}H$ and $CH_3{-}C(OH){=}O$

These are constitutional isomers. The molecule on the left contains an ester group. The molecule on the right contains a carboxyl group.

(d) $HO{-}CH_2{-}CH(CH_3){-}OH$ and $CH_3{-}CH(CH_2{-}OH){-}OH$

These are the same molecule. It has two hydroxyl groups.

(e) $H{-}C({=}O){-}CH_2{-}CH(CH_3){-}CH_3$ and $(CH_3)_2CH{-}CH_2{-}C({=}O){-}H$

These are the same molecule. It is an aldehyde with a carbonyl group.

(f) $CH_3{-}CH(CH_2{-}CH(CH_3){-}CH_3){-}CH_2{-}CH_3$ and $CH_3{-}CH(CH_3){-}CH_2{-}CH_2{-}CH(CH_3){-}CH_3$

These molecules are constitutional isomers, and they have no functional groups.

<u>Resonance and Contributing Structures</u>

<u>Problem 1.33</u> Draw the contributing structure indicated by the curved arrow(s). Assign formal charges as appropriate.

(a) $H{-}\ddot{O}{-}C({=}\ddot{O}){-}\ddot{O}:^-$ ⟷ $H{-}\ddot{O}{-}C(\ddot{O}:^-){=}\ddot{O}$

(b) $H{-}\ddot{O}{-}C({=}\ddot{O}){-}\ddot{O}:^-$ ⟷ $H{-}\overset{+}{O}{=}C(\ddot{O}:^-){-}\ddot{O}:^-$

(c)  H–O–C(=O)–O$^-$ ⟷ H–O–C$^+$(–O$^-$)–O$^-$

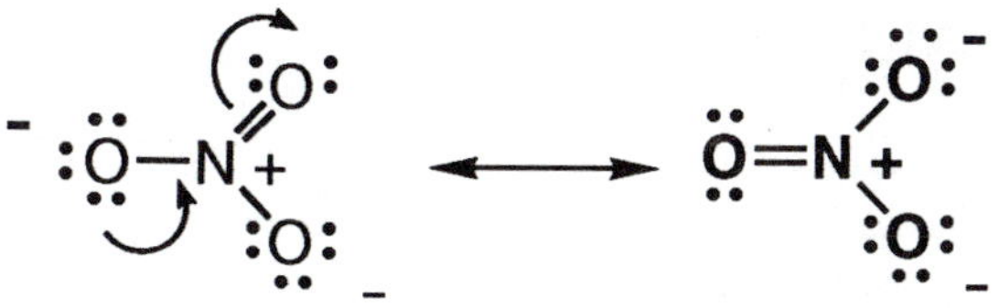

(d) O=C=O ⟷ O=C$^+$–O$^-$

(e) H–O–N=O ⟷ H–O–N$^+$–O$^-$

(f) H–O–N=O ⟷ H–O$^+$=N–O$^-$

(g) $^-$O–N$^+$(=O)–O$^-$ ⟷ O=N$^+$(–O$^-$)–O$^-$

(h) H_2C=O ⟷ H_2C^+–O$^-$

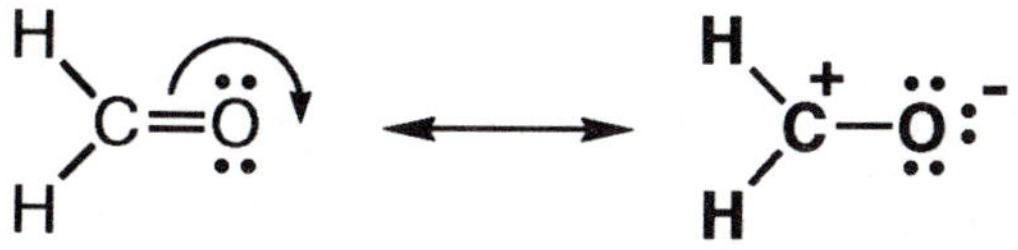

(i)  H–C=C–C(=O)–H (with H, H on the C=C carbons) ⟷ H–C=C–C$^+$(–O$^-$)–H

(j) 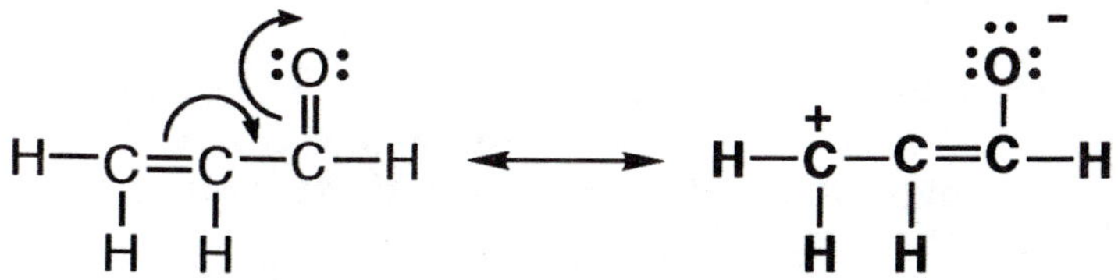 H–C=C–C(=O)–H (with H, H on the C=C carbons) ⟷ H–C$^+$–C=C(–O$^-$)–H

Problem 1.34 In the previous problem you were given one contributing structure and asked to draw another. Label pairs of contributing structures that are equivalent. For pairs of contributing structures that are not equivalent, label the more important contributing structure and explain your reasoning.

(a) The two structures are equivalent.
(b, c, d, e, f) The first structure is more important, because the second involves creation and separation of unlike charges.
(g) The two structures are equivalent.

(h, i, j) The first structure is more important, because the second involves creation and separation of unlike charges.

Problem 1.35 Are the structures in each set valid contributing structures? Explain.

(a)  $H_2C{=}\ddot{\underset{..}{O}} \longleftrightarrow H_2\bar{C}{\equiv}O^+$

The structure on the right is not a valid contributing structure because there are 10 valence electrons around the carbon atom. Besides this, the two structures can not be valid contributing structures because they have a different number of valence electrons. The molecule on the left has 12 valence electrons, and the molecule on the right has 14 valence electrons.

(b)  $H{-}\ddot{N}{=}\overset{+}{N}{=}\overset{-}{\ddot{\underset{..}{N}}} \longleftrightarrow H{-}\overset{-}{\ddot{\underset{..}{N}}}{-}\overset{+}{N}{\equiv}N{:}$

Both of these are valid contributing structures.

(c) 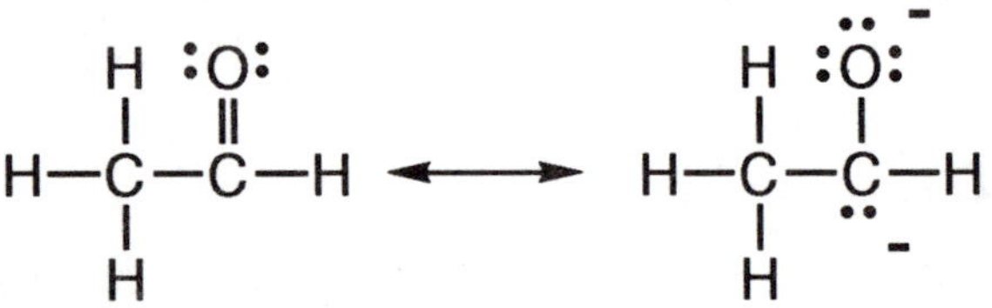 $H_3C{-}C(=O){-}H \longleftrightarrow H_3C{-}\overset{-}{C}(-O^-){-}H$

The structure on the right is not a valid contributing structure because there are two extra electrons and thus it is a completely different species.

(d) 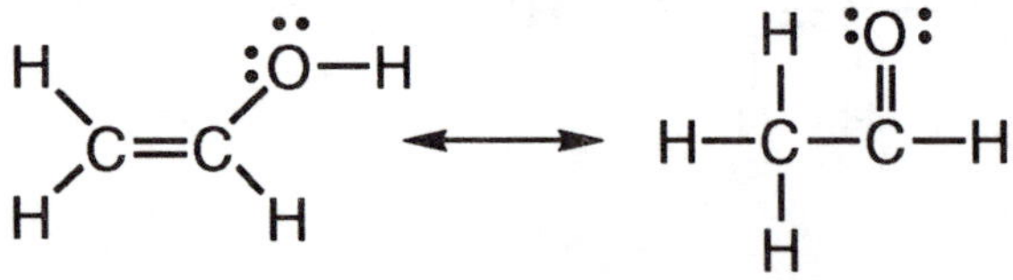 $H_2C{=}CH{-}\ddot{O}{-}H \longleftrightarrow H_3C{-}C(=O){-}H$

Although each is a valid Lewis structure, they are not valid contributing structures for the same resonance hybrid. An atomic nucleus, namely a hydrogen, has changed position. Later you will learn that these two molecules are related to each other, and are called tautomers.

Problem 1.36 Following is the structural formula of an ester. It is one contributing structure. Draw two more contributing structures for this hybrid and show by the use of curved arrows how the first is converted to the second and the second is converted to the third. Be certain to assign all formal charges as appropriate.

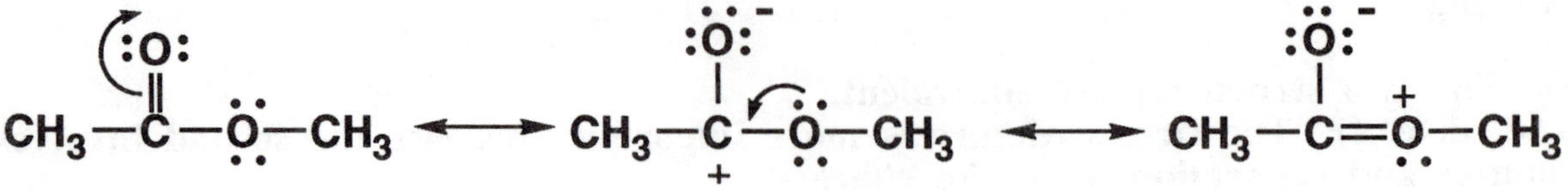

$CH_3{-}C(=O){-}\ddot{O}{-}CH_3 \longleftrightarrow CH_3{-}\overset{+}{C}(-O^-){-}\ddot{O}{-}CH_3 \longleftrightarrow CH_3{-}C(-O^-){=}\overset{+}{O}{-}CH_3$

Valence Bond Theory

Problem 1.37 Following are Lewis structures for several molecules. State the orbital hybridization of each circled atom.

Each circled atom is either sp, sp^2, or sp^3 hybridized.

(a) H—C—C—H with H's above and below; circled C: sp^3

(b) $H_2C=CH_2$; circled C: sp^2

(c) H—C≡C—H; circled C: sp

(d) $H_2C=\ddot{O}$; circled C: sp^2

(e) H—C(=O)—O—H; circled C: sp^2

(f) H_3C—O—H; circled O: sp^3

(g) H_3C—N(H)—H; circled N: sp^3

(h) H_3C—O—C(=O)—H; circled O: sp^3

(i) H—O—N=O; circled N: sp^2

(j) $CH_2=C=CH_2$; circled C: sp

(k) H_3C—CH=N—O—H; circled N: sp^2

Problem 1.38 Following are Lewis structures for several molecules. Describe each circled bond in terms of the overlap of atomic orbitals.

Shown is whether the bond is sigma or pi, as well as the orbitals used to form it.

(a) H—C—C—H with H's above and below; circled C—C bond: σ sp^3-sp^3

(b) $H_2C=CH_2$; circled C=C bond: σ sp^2-sp^2; π 2p-2p

(c) H—C≡C—H; circled C—H bond: σ sp-1s

(d) 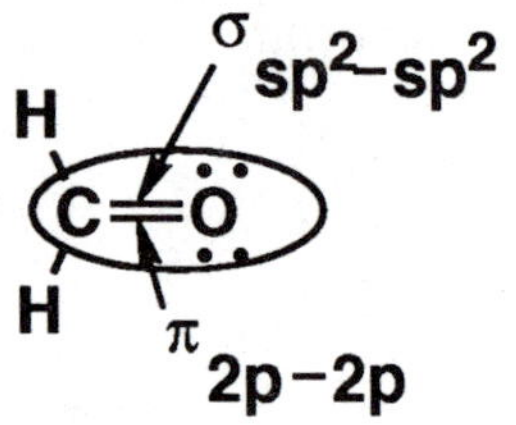 $H_2C=O$: σ sp²–sp²; π 2p–2p

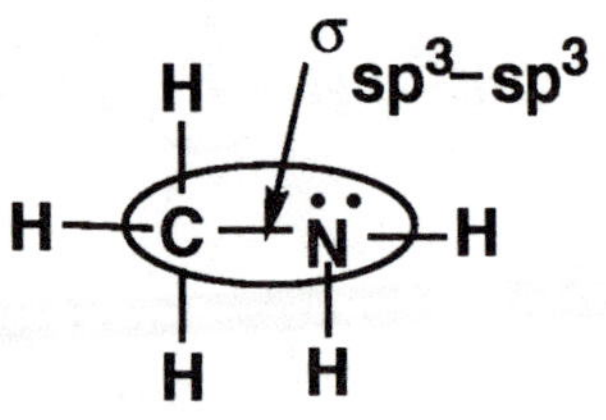

(e) 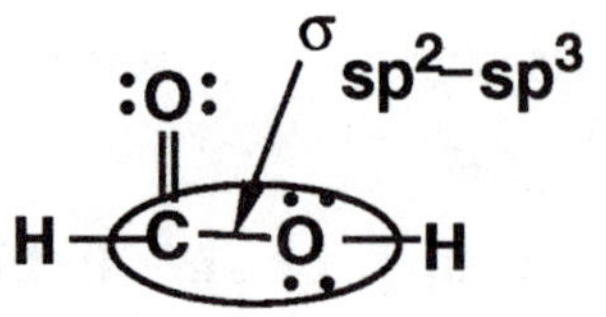 H–C(=O)–O–H: σ sp²–sp³

(f) 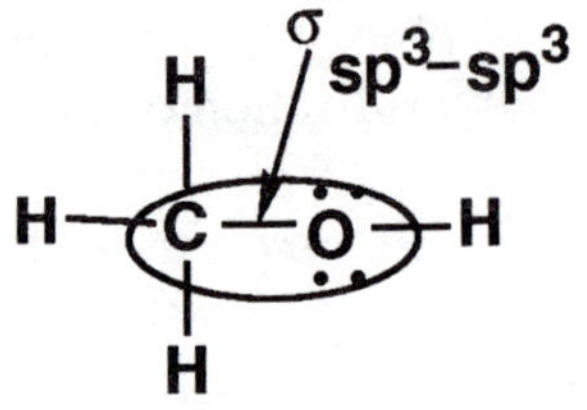 H_3C–O–H: σ sp³–sp³

(g) H_3C–NH_2: σ sp³–sp³

(h) H_3C–O–C(=O)–H: σ sp²–1s

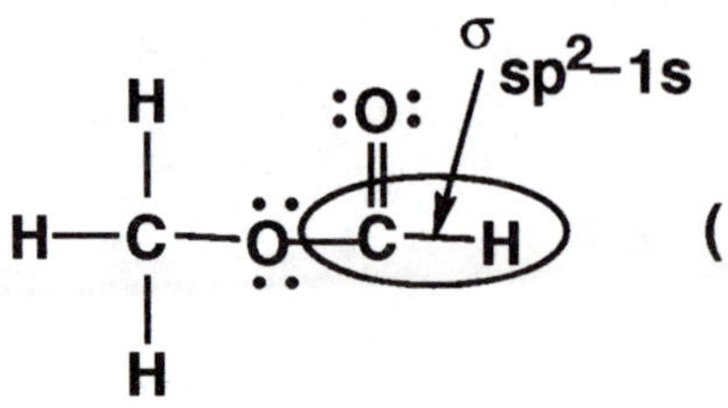

(i) 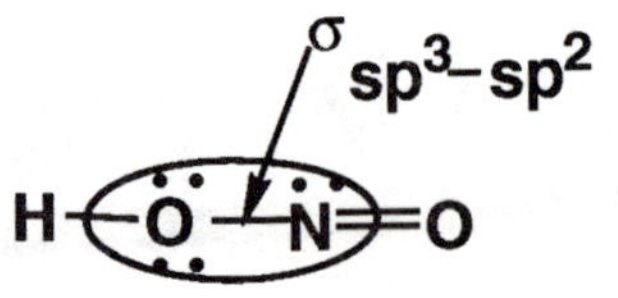 H–O–N=O: σ sp³–sp²

(j) 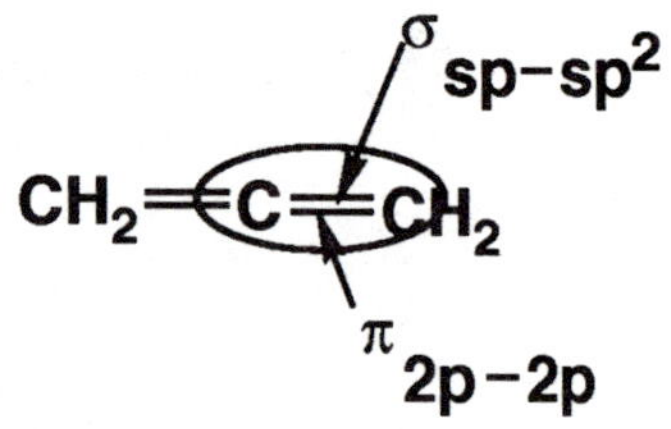 $CH_2=C=CH_2$: σ sp–sp²; π 2p–2p

(k) 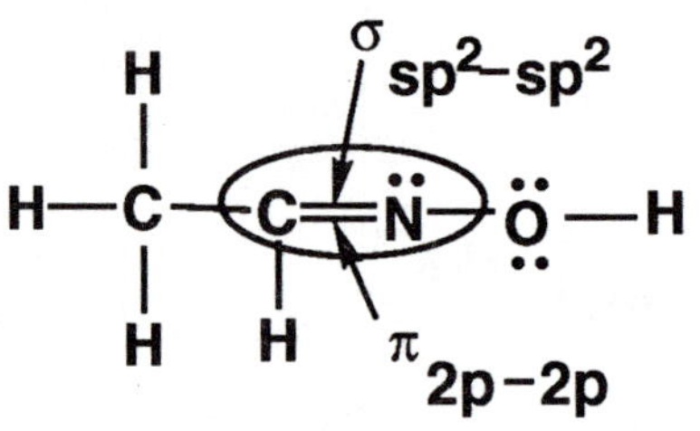 H_3C–CH=N–O–H: σ sp²–sp²; π 2p–2p

Problem 1.39 Following is the structural formula of famotidine, which is manufactured by Merck Sharpe & Dohme under the name Pepcid. Pepcid is a competitive inhibitor of histamine H_2 receptors. The primary clinically important pharmacological activity of famotidine is inhibition of gastric secretion. Both the acid concentration and volume of gastric secretions are suppressed by Pepcid.

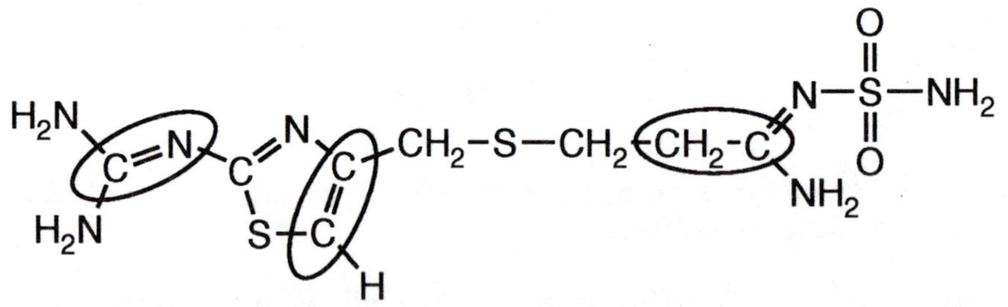

$(H_2N)_2C=N$–(thiazole ring: C=N, C=C–H, S)–CH_2–S–CH_2–CH_2–C(NH_2)=N–$S(=O)_2$–NH_2

(a) Complete the Lewis structure of famotidine showing all valence electrons and any formal positive or negative charges.

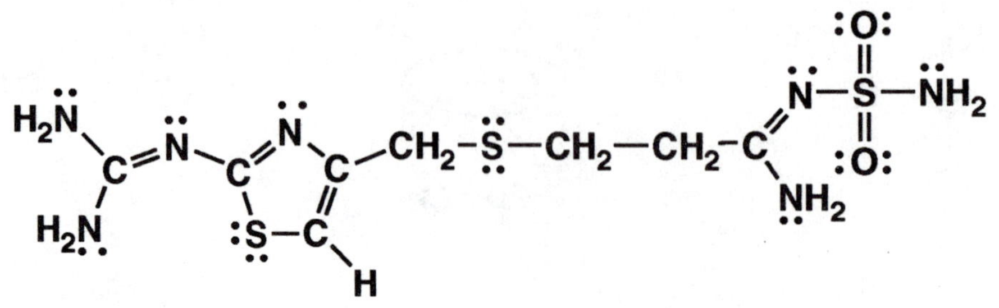

$(H_2\ddot{N})_2C=\ddot{N}$–(thiazole ring)–$CH_2$–$\ddot{S}$–$CH_2$–$CH_2$–C($\ddot{N}H_2$)=$\ddot{N}$–S(=:O:)(=:O:)–$\ddot{N}H_2$

(b) Describe each circled bond in terms of the overlap of atomic orbitals.

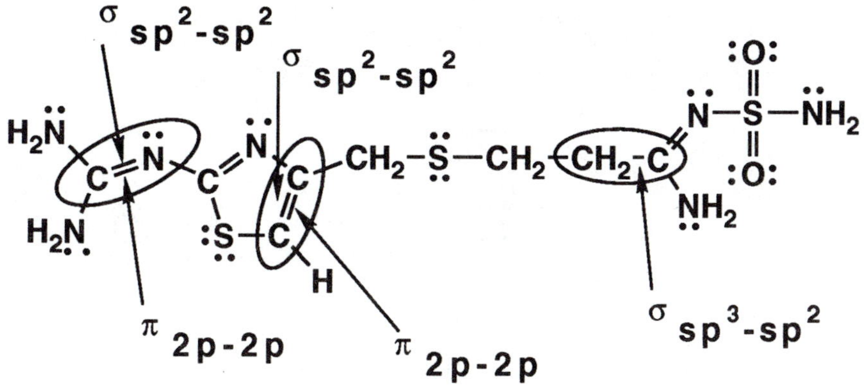

Molecular Orbital Theory

Problem 1.40 Figure 1.29 in Section 1.9 shows a molecular orbital energy-level diagram for the ground state of the oxygen molecule, O_2.

(a) Complete a similar molecular orbital energy diagram for the ground state of the nitrogen molecule, N_2.

In the ground state, electrons in a nitrogen atom occupy 1s, 2s, and 2p atomic orbitals for a total of five atomic orbitals. Combination of five atomic orbitals from each nitrogen in N_2 gives 10 molecular orbitals. In the ground state, the 1s and 2s bonding and antibonding MOs are filled and have no significant effect on bonding. All six electrons from 2p atomic orbitals lie in bonding molecular orbitals.

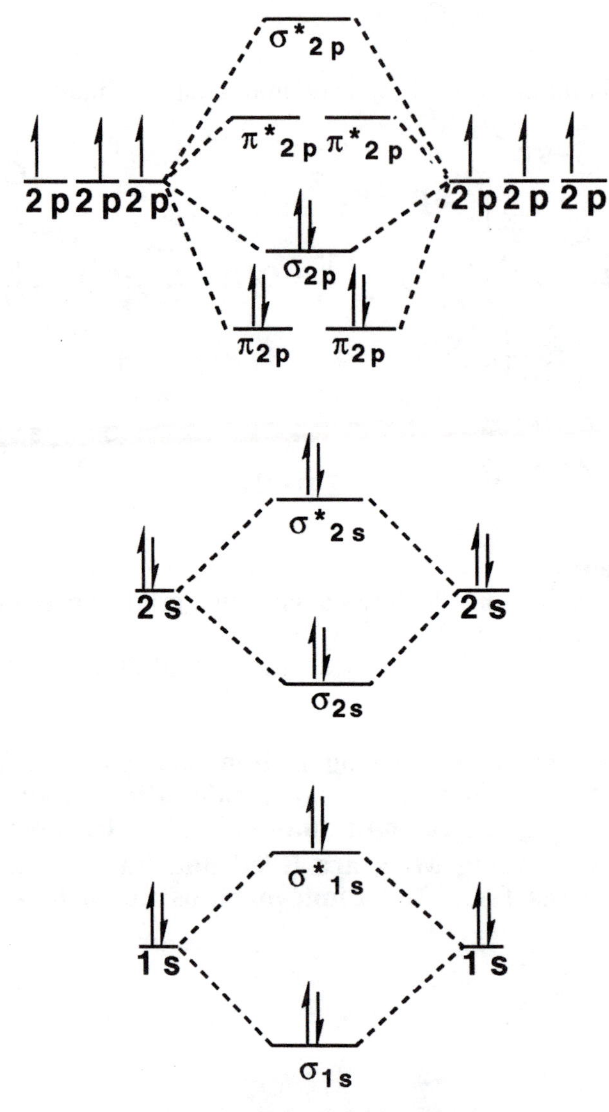

ground state electron configuration for N_2

(b) Do you predict N_2 to be diamagnetic or paramagnetic?

Diamagnetic. There are no unpaired electrons in the ground state.

(c) Determine the bond order in N_2 and compare this with the number of bonds according to valence bond theory.

The bond order is 3, as predicted by the valence bond model.

Problem 1.41 The molecule C_2 exists in the vapor state at high temperature and has been shown to have a bond dissociation energy of 144 kcal/mol (602 kJ/mol). For comparison, the bond dissociation energy of O_2 is 118 kcal/mol (494 kJ/mol) and that of N_2 is 225 kcal/mol (942 kJ/mol).

(a) Describe the ground-state electron configuration of C_2.

Following is a molecular orbital energy diagram for the ground state electron configuration of C_2.

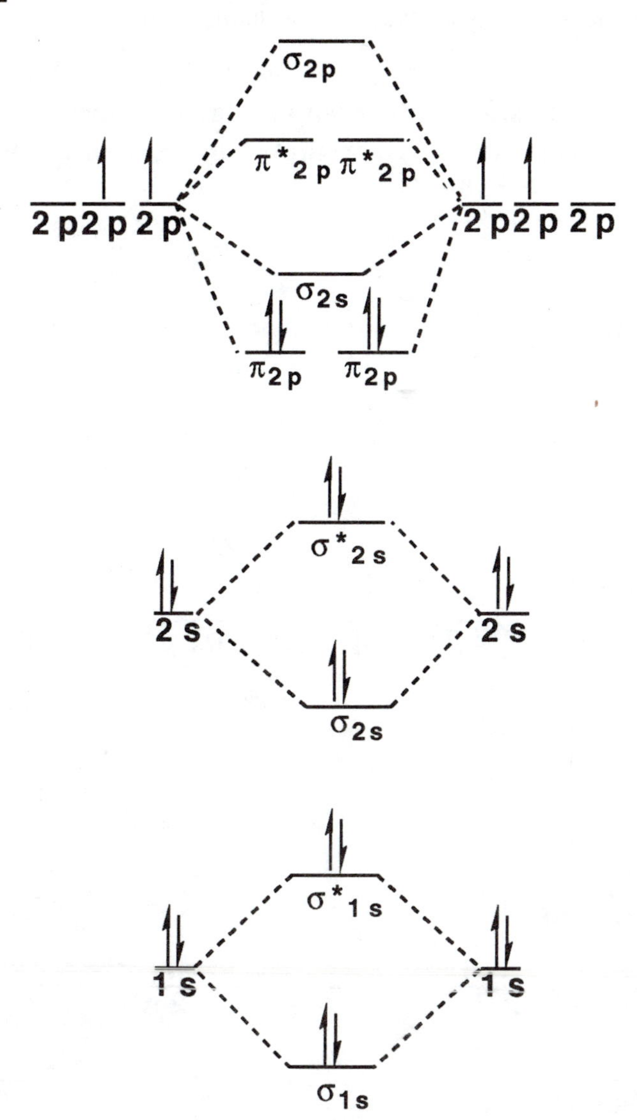

ground state electron configuration for C_2

(b) Would you predict C_2 to be diamagnetic or paramagnetic?

Diamagnetic. There are no unpaired electrons.

(c) Determine the bond order of C_2.

The bond order is 2, since there are four more electrons in bonding orbitals compared with antibonding orbitals.

Problem 1.42 Using molecular orbital theory, describe the ground-state electron configuration for the following molecules and ions. Calculate the bond order for each, decide whether each should be stable and predict whether each is paramagnetic or diamagnetic.
(a) B_2

The boron molecule, B_2, has a bond order of 1 and is paramagnetic (it has two unpaired electrons). This molecule is predicted to be stable and has a bond dissociation energy of 69.3 kcal/mol.

(b) B_2^+

The boron molecule ion, B_2^+, has a bond order of 0.5 which means that a partial bond is formed. It is predicted to be stable only under certain experimental conditions and to be paramagnetic.

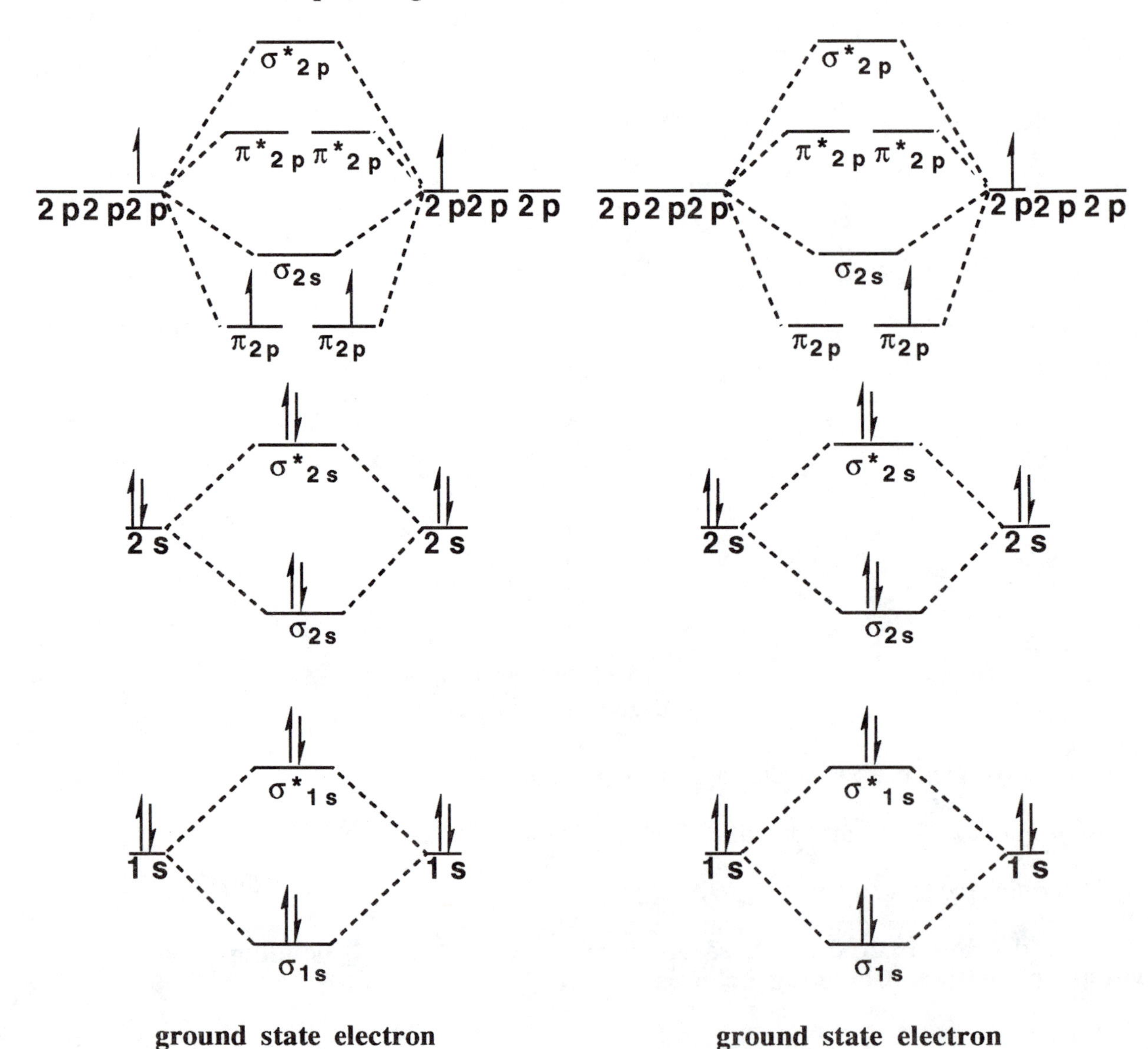

ground state electron configuration for B_2

ground state electron configuration for B_2^+

(c) Be_2

The beryllium molecule, Be_2, has a bond order of zero. Its eight electrons (four from each atom of beryllium) are distributed equally in bonding and antibonding molecular orbitals. No bond is formed and, therefore, Be_2 is not stable.

(d) He_2^+

The helium molecule ion, He_2^+, has three electrons; two in a bonding MO and one in an antibonding MO. The bond order is 0.5. A partial bond is formed, and He_2^+ is predicted to be stable under certain experimental conditions.

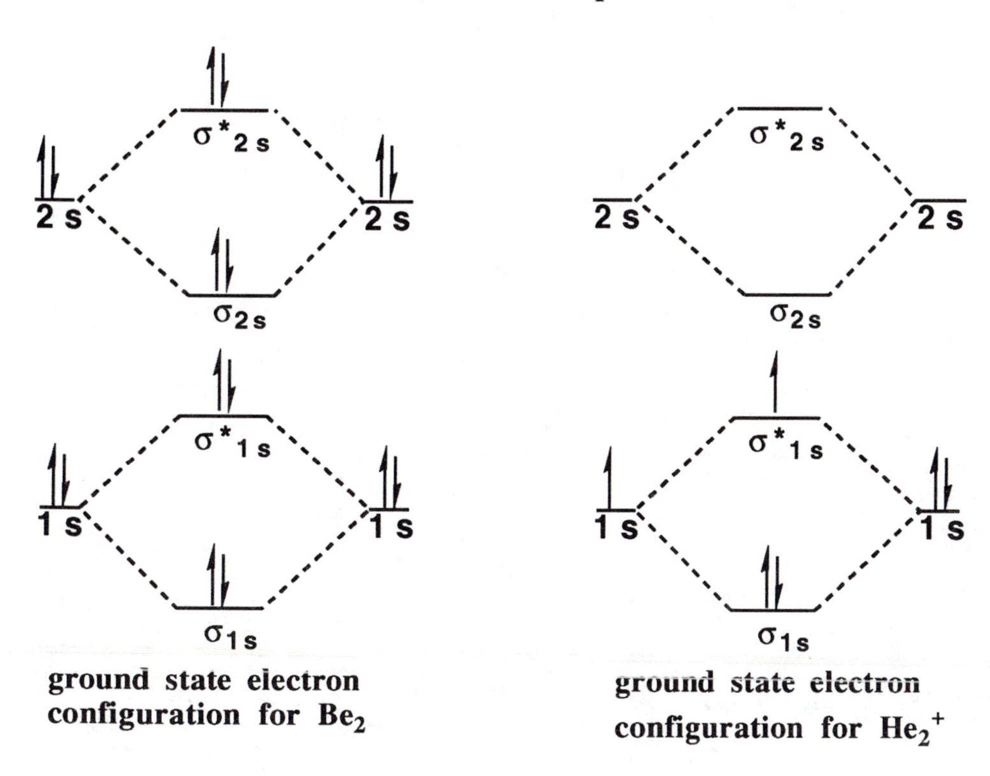

ground state electron configuration for Be_2

ground state electron configuration for He_2^+

(e) O_2^{2-}

The O_2^{2-} molecule ion has a bond order of 1. It is predicted to be stable and diamagnetic.

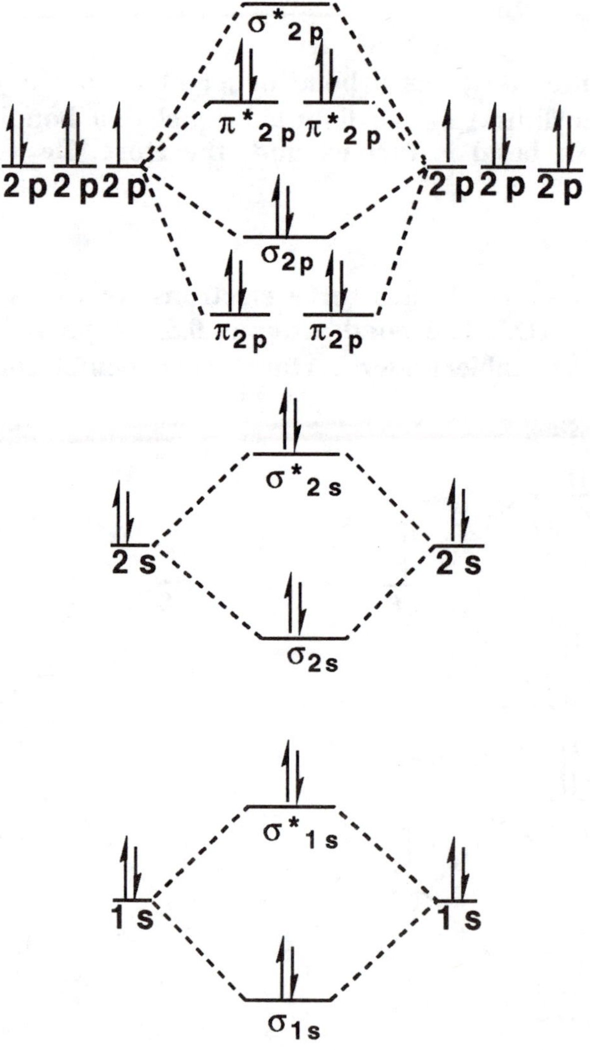

ground state electron configuration for O_2^{2-}

CHAPTER 2: ALKANES AND CYCLOALKANES

SUMMARY OF REACTIONS

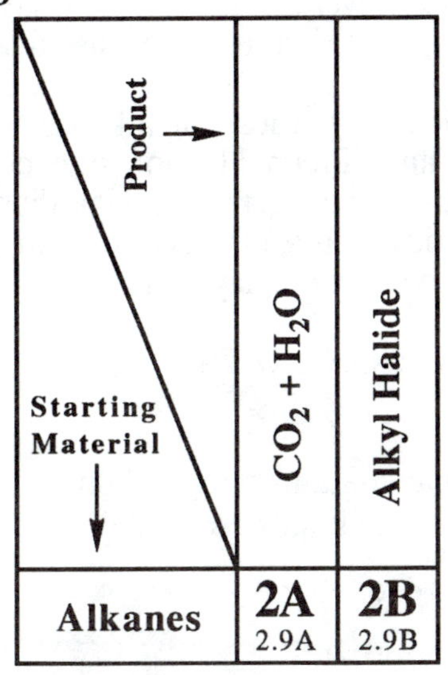

REACTION 2A: OXIDATION (Section 2.9A)

- Alkanes react with O_2 to give CO_2, H_2O, and heat.

$$CH_4 + 2O_2 \longrightarrow CO_2 + 2H_2O$$

- This reaction is the basis for using alkanes as sources for heat and energy.

REACTION 2B: HALOGENATION (Section 2.9B)

- Alkanes react with Cl_2 and Br_2 in the presence of light or heat by a substitution reaction.

$$CH_4 + Cl_2 \longrightarrow CH_3Cl + HCl$$

- In a substitution reaction, one atom or group of atoms is replaced by another atom or group of atoms.

- The reaction can proceed further if more halogen is used.

$$CH_3Cl + Cl_2 \longrightarrow CH_2Cl_2 + HCl$$

-The product **haloalkane** is named by using the same rules as before, but the prefixes **fluoro**, **chloro**, **bromo**, and **iodo** are used for the halogen that is treated like any other substituent.

• When complex alkanes react with halogens several different products are formed, since the halogen can substitute for different hydrogen atoms. *[This may seem simple enough, but it can get very complicated when complex alkanes are considered]*

- The substitutions observed in the products are **regioselective**, that is substitution is not statistically random.

In general, tertiary hydrogen atoms are replaced in preference to secondary hydrogen atoms, and secondary hydrogen atoms are replaced in preference to primary hydrogen atoms. *[Be sure to review these definitions in Section 2.3 if necessary.]*
Bromination is more regioselective than chlorination. *[At this point it is not prudent to explain these selectivities, but you will be able to understand them when the mechanism of the reaction is presented in Chapter 5.]*
- The reaction of alkanes with F_2 is seldom used because it is highly exothermic and difficult to control.
- Because of their unique physical properties, haloalkanes are widely used as solvents (methylene chloride), refrigerants (Freon-11), non-stick coatings (Teflon), artificial blood replacements (perfluorodecalin), and anesthetics (Halothane).
- Recent concern over possible destructive effects of certain haloalkanes like chlorofluorocarbons (CFCs) in the ozone layer has lead to attempts to limit their use.

SUMMARY OF IMPORTANT CONCEPTS

2.0 OVERVIEW

• A **hydrocarbon** is a molecule that contains only carbon and hydrogen, and an **alkane** is a hydrocarbon that contains only single bonds. ✳

2.1 STRUCTURE OF ALKANES

• Alkanes have the general formula C_nH_{2n+2}.
• The carbon atoms of alkanes are sp^3 hybridized and thus tetrahedral, with bond angles of approximately 109.5°.

2.2 CONSTITUTIONAL ISOMERISM IN ALKANES

• **Constitutional isomers** are two or more molecules that have the same molecular formula but the atoms are attached to each other in different ways. (This was discussed previously in Section 1.5).
  - Constitutional isomers have different chemical properties.
  - For methane (CH_4), ethane (C_2H_6), and propane (C_3H_8) there is only one way to attach the carbon atoms to each other, hence there are no constitutional isomers of these alkanes. For alkanes with four or more carbon atoms, the number of constitutional isomers is given by a complex mathematical formula called a generating function.
  - There is no foolproof way to find all constitutional isomers for a given molecular formula, you must use a combination of a systematic method and creativity. *[This is harder than it looks and requires a great deal of practice.]*
  - A reasonable system to answer the constitutional isomer questions is to first write all possible carbon skeletons by starting with the straight chain alkane, then systematically adding appropriate branches.

2.3 NOMENCLATURE OF ALKANES

• To name organic compounds, chemists use systematic nomenclature rules established by the **International Union of Pure and Applied Chemistry (IUPAC)**. ✳
• For simple, unbranched alkanes the name consists of a **prefix** and a **suffix**. ✳
  - The prefix indicates the number of carbon atoms. For example, "**prop**" means three carbon atoms. For a more complete list of prefixes see Table 2.2 in the text. *[It is important to learn these now because the rest of the book assumes you are familiar with these names.]*
  - Following the prefix, the suffix **ane** is used to designate that a compound is an alkane. For example, **propane** is an alkane with three carbon atoms ($CH_3CH_2CH_3$). Later in Section 2.5 you will see that the **ane** suffix is actually composed of the so-called infix "**an**" and the true suffix "**e.**"

- For substituted or **branched alkanes**, the nomenclature is based on viewing the molecule as a chain with substituents. ✳
  - To write a **substituent**, the **ane** suffix of the parent hydrocarbon is dropped and is replaced by the suffix **yl**. For example, **propyl** is used to describe a substituent with three carbon atoms ($CH_3CH_2CH_2$-).
- Branched alkanes are named using the following set of rules.
  - The alkane derived from the longest continuous chain of carbon atoms is taken as the **parent chain**. The **root** or **stem name** of the branched alkane is that of the parent chain. *[This can be tricky, especially when the parent chain is drawn in a crooked fashion.]*
  - Each substituent attached to the parent chain is given a name and a number. Certain common names (see below) can be used for naming substituents such as "isopropyl."
  - The **substituent number** shows the carbon of the parent chain to which the substituent is attached. The numbers are designated on the parent chain according to the following rules. *[This numbering scheme is as complicated as it seems, and requires a lot of practice to master thoroughly.]*

    If there is one substituent, number the parent chain from the end that gives the substituent the lower number. For example, a correct name is 2-methylhexane, *not* 5-methylhexane.

    If there are two or more *identical* substituents, number the parent chain from the end that gives the lower number to the substituent encountered first, and the number for each substituent is given in the final name. Indicate the number of times the same substituent occurs by a special set of prefixes. The prefixes **di**, **tri**, **tetra**, **penta**, or **hexa.** For example, 2,3-dimethylhexane has methyl groups at positions 2 and 3 on the parent hexane chain.

    If there are two or more *different* substituents, list them in **alphabetical order**, and number the parent chain from the end that gives the lower number to the substituent encountered first. For example, 4-ethyl-3-methyloctane is an acceptable name because ethyl comes before methyl in terms of alphabetization.

    If there are *different* substituents in equivalent positions on the parent chain, give the lower number to the substituent of lower alphabetical order.

    Hyphenated prefixes, for example, ***sec-*** and ***tert-*** are not considered when alphabetizing.

    The prefix **iso** is not a hyphenated prefix, and therefore is included when alphabetizing.
- In spite of the precision of the IUPAC system, an unsystematic set of **common names** is still used for certain compounds. *[These names are deeply rooted in organic chemistry and are still widely used. Remember that it is always correct to use an IUPAC name. However, it is also important to learn how to use the common names, because you will run across them often.]*
  - In the **common nomenclature**, the total number of carbon atoms in an alkane, regardless of their arrangement, determines the name. The following terms are used in common nomenclature to indicate a few selected branching patterns.

    Normal or **n-** indicates that all carbons are joined in a continuous chain . For example *n*-butane ($CH_3CH_2CH_2CH_3$)

    Iso is used to indicate that one end of an otherwise continuous chain terminates in a $(CH_3)_2CH$- group. For example isobutane ($(CH_3)_2CHCH_3$).

    Neo is used to indicate that one end of an otherwise continuous chain terminates in a $(CH_3)_3C$- group. For example, neopentane ($(CH_3)_4C$).

    More complicated patterns of branching cannot be accommodated by common nomenclature, so the IUPAC system must be used.

    Use some common names such as *tert*-butyl and isopropyl for substituents, even though the rest of the molecule is named according to IUPAC rules.
- **Classify atoms** according to their **environment**. ✳
  - Classify a carbon atom in an alkane according to the number of alkyl groups bonded to it. *[This is very important when it comes to understanding relative reactivity.]* A carbon atom bonded to a single alkyl group is a **primary carbon** atom, a carbon atom bonded to two

alkyl groups is a **secondary carbon** atom, a carbon atom bonded to three alkyl groups is a **tertiary carbon** atom, and a carbon atom bonded to four alkyl groups is a **quaternary carbon** atom.

- **Hydrogen atoms** are also classified as primary, secondary, or tertiary when they are bonded to a primary, secondary, or tertiary carbon atom, respectively.
- **Equivalent** hydrogen atoms have the same chemical environment. ✱ *[This concept is very important when it comes to spectroscopy, especially nuclear magnetic resonance (NMR) spectroscopy].*

  To determine which hydrogens in a molecule are equivalent, use the following procedure: In your mind, replace each hydrogen with a "**test atom**." If replacement of two different hydrogens by the "test atom" gives the same compound, then the hydrogens are equivalent. If replacement of two different hydrogens by the "test atom" gives different compounds, then the hydrogens are not equivalent.

2.4 CYCLOALKANES

- Organic chemists use **line-angle drawings** as a simple way to represent complex molecules. In line-angle drawings each line represents a C-C bond, each double line represents a C=C bond, and each triple line represents a C≡C bond. The vertex of each angle represents a carbon atom. In this way, only the **carbon framework** of the molecule is shown. It is understood that hydrogen atoms complete the tetravalence of the carbon atoms. For example,  represents pentane, ($CH_3CH_2CH_2CH_2CH_3$).
- A **cycloalkane** is an alkane in which there is a ring of carbon atoms. ✱
  - **IUPAC cycloalkane nomenclature** rules are as follows:

    Use the prefix **cyclo** in front of the name of the alkane with the same number of carbon atoms as the number of carbons in the ring. For example, cyclohexane is a six-membered ring, 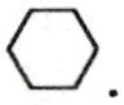 .

    List substituents on the ring by name and number as you would on an open-chain hydrocarbon.

    If there is only a single substituent on the ring, there is no need to give a number. If there are two or more substituents, give each substituent a number to indicate its location on the ring. Number the atoms of the ring beginning with the substituent of lowest alphabetical order.
- A **bicycloalkane** is an alkane with two rings that share one or more atoms in common.
  - **IUPAC bicycloalkane nomenclature** rules are as follows:

    The parent name of a bicycloalkane is that of the alkane with the same number of carbon atoms as are in the bicyclic ring system.

    Numbering begins at one bridgehead carbon and proceeds along the largest ring to the second bridgehead carbon, and then along the next largest ring back to the original bridgehead carbon, and so forth until all atoms of the bicyclic ring are numbered.

    Ring sizes are shown by counting the number of carbon atom linked to the bridgeheads and placing them in decreasing order in brackets between the prefix **bicyclo** and the parent name. For example, bicyclo[2.2.1]heptane 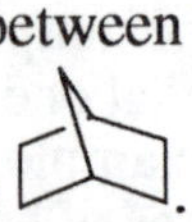 .

    Name and locate of substituents by the rules already described in Section 2.3A.
- A **spiroalkane** is a cycloalkane in which the two rings share only one atom. Name spiroalkanes by the rules described above, except the prefix **spiro** is used. For example, spiro[4.4]nonane  . Numbering begins at the carbon atom on the shorter bridge nearest the spirocarbon atom, around the shorter bridge, through the spirocarbon atom, and around the longer bridge.

2.5 THE IUPAC SYSTEMS-A GENERAL SYSTEM OF NOMENCLATURE

- The name assigned to any compound consists of at least three parts; **the prefix, the infix** and **the suffix.**
  - The prefix tells the number of carbon atoms in the parent chain. See Table 2.2 in the text for examples.
  - The infix (part of the name directly in front of the suffix) tells the nature of the carbon-carbon bonds in the parent chain.
    an means the compound has all single bonds, **en** means one or more double bond, and **yn** means there is one or more triple bond.
  - The suffix tells the class of the compound to which the substance belongs.
    The class of a compound is determined by the functional groups present.
    Important suffixes include **e** for hydrocarbons, **ol** for alcohols, **al** for aldehydes, **one** for ketones, and **oic acid** for carboxylic acids.

2.6 CONFORMATIONS OF ALKANES AND CYCLOALKANES

- The **conformation** of an alkane refers to the *three-dimensional* arrangement of atoms that results from rotation about carbon-carbon bonds. ✱
  - It is convenient to analyze alkane conformations using a **Newman projection.** Although there may be a number of C-C bonds in a molecule to analyze, you can look at only one C-C bond with each Newman projection.
    In a Newman projection, view the molecule along the axis of one C-C bond. ✱
    [Understanding this statement is the key to using Newman projections.]
    Thus a Newman projection examines how the different groups are distributed around only the two adjacent carbon atoms involved with the selected C-C bond.
    When drawing a Newman projection, use a large circle to indicate location. Show the three groups bound to the carbon atom that are *nearer* your eye on lines extending from the *center* of a circle at angles of 120°. Show the three groups bound to the carbon atom that is *farther* from your eye on lines extending from the *circumference* of a circle at angles of 120°. *[Make sure you know how to go between a three dimensional molecular model and a Newman projection. Figures 2.5 and 2.6 are especially helpful.]*

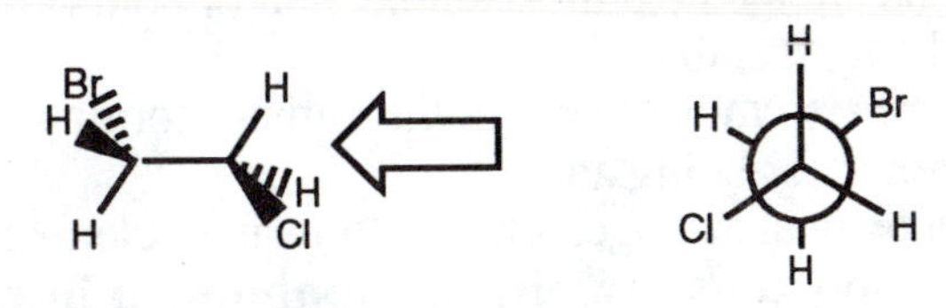

- At room temperature, the C-C bonds can **rotate rapidly**. Thus, an infinite number of conformations are possible around a C-C bond as it rotates. The two extreme conformations are named **eclipsed** and **staggered**.
  - In the **eclipsed conformation**, the groups on the near carbon atom are directly in front of the groups on the far carbon atom.

  - In the **staggered conformation**, the groups on the near carbon atom are as far apart as possible from the groups on the far carbon.

- **A dihedral angle** (θ), is the angle between a given substituent on the near carbon atom and a given substituent on the far carbon of a Newman projection.

  For eclipsed conformations, the dihedral angles are thus 0°, 120°, 240°, for nearest groups, and for staggered conformations, the dihedral angles are thus 60°, 180°, 300° for nearest groups.

• The staggered conformations have lower **potential energy** than eclipsed conformations, probably due to the **repulsion** of **electron pairs** in the bonds resulting in their preferring to be as far apart as possible. Hydrogen atoms are probably not large enough to "crash" into each other even in the eclipsed conformation. ✱

  - This lower potential energy means that alkanes spend the majority of their time in a staggered conformation.

• Taking the bond between the carbons 2 and 3 as reference, there are two types of **staggered conformations for butane**, namely **gauche** and **anti**. ✱

  - The two **gauche** conformations have the two methyl groups adjacent, that is with dihedral angles of 60°.

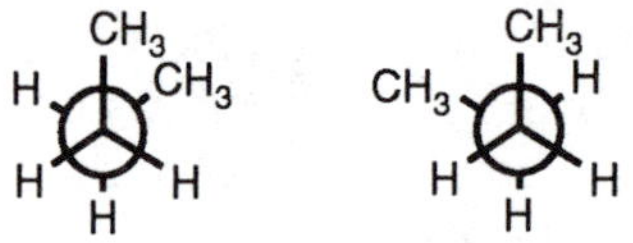

CH_3 H CH_3 H H H CH_3 CH_3 H H H H

  - The **anti** conformation has the two methyl groups as far apart as possible, that is with a dihedral angle of 180°. *[If you do not fully understand this, review the preceding sections before going any further.]*

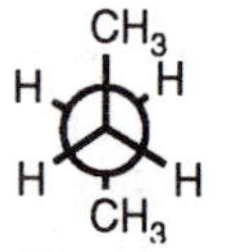

CH_3 H H H H CH_3

  The methyl groups take up a large amount of space and thus the anti conformation is the most stable (lowest potential energy), because the methyl groups are farthest apart. The gauche conformations are the next most stable, and of course all of the eclipsed conformations are the least stable.

  Other large groups are also more stable in the anti conformation. The larger the groups, the larger the preference for being anti. ✱

• The most stable three-dimensional arrangement of atoms in cycloalkanes minimizes the two types of **steric strain**, namely **angle strain** and **nonbonded interaction strain**. ✱ *[Notice the discussion has turned back to cycloalkanes.]*

  - **Angle strain** arises because the geometry of certain cycloalkanes creates bond angles other than the ideal 109.5°.
  - **Nonbonded interaction strain** arises because the geometry of cycloalkanes forces nonbonded atoms or groups into close proximity. As expected, this type of strain is proportionately more important for larger atoms or groups.

• The three carbon atoms of **cyclopropanes** must necessarily lie in a plane.

  - There is a large amount of angle strain in cyclopropane because the bond angles are 60°, a long way from the preferred 109.5°. There is a large amount of nonbonded interaction strain because all of the groups bonded to the central carbon atoms are eclipsed. *[Use a model to prove this to yourself if necessary.]*

• To minimize steric strain, the **larger cycloalkanes** exist in a variety of **puckered** nonplanar conformations. ✱

• The most stable conformation of **cyclobutane** is slightly puckered  .

- The puckered conformation of cyclobutane relieves nonbonded interaction strain , because the hydrogens are no longer fully eclipsed. Note that the puckering does cause a slight increase in angle strain because the angles are decreased to about 88°.

• **An envelope conformation** is the most stable conformation of **cyclopentane.** . There are five possible envelope conformations, each with a different carbon atom that is out of the plane formed by the other four.
  -This puckering relieves nonbonded steric strain by reducing the number of eclipsed atoms in the molecule. In this case the puckering only causes a slight increase in angle strain, since the bond angles are 105°.

• There are a number of different puckered conformations of **cyclohexane**, by far the most important of which is a remarkably stable **chair conformation** . ✳ *[Understanding the following details of cyclohexane chair conformations is very important, because you will need to use these ideas in the future when issues like the relative stabilities of reaction intermediates or carbohydrates are discussed.]*
  - The chair conformation is dramatically more stable than the planar form, because *all* the groups are perfectly staggered in the chair conformation. Furthermore, the chair conformation has *all* bond angles near the ideal 109.5°. Cyclohexane molecules therefore spend the great majority of their time in the chair conformation.

• In the chair conformation of cyclohexane, the 12 different hydrogen atoms attached to the six carbon atoms of the ring can be classified as one of two types, **axial** or **equatorial.** ✳
  - The **six axial positions** are **perpendicular** to the mean plane of the cyclohexane ring; three axial hydrogens point straight up, and three point straight down.
  - The **six equatorial positions** point roughly outward from the cyclohexane ring. *[Models will help you understand the difference between axial and equatorial positions.]*
    There are two different chair conformations of cyclohexane that are in equilibrium with each other. *[Using models, you should verify that interconverting the two possible chair cyclohexane conformations changes all of the axial hydrogens to equatorial hydrogens, and vice versa.]*

• There are several less stable puckered conformations of cyclohexane such as the **boat** and **twist boat**. These conformations are less stable than the chair conformation, because they have either some eclipsed hydrogen atoms or bond angles other than the optimum 109.5°. These less stable conformations of cyclohexane are intermediates in the interconversion of the two chair forms of cyclohexane.

• If one or more of the hydrogens of a cyclohexane are replaced by any larger atom or group, the more stable chair conformation is the one that places the larger atom or group in an equatorial position. ✳ *[This is perhaps the most important concept involved with cyclohexane chair conformations, and you will use it over and over again.]*
  - The larger the atom or group, the greater the preference for it to be in an equatorial position. This preference for large atoms or groups to be equatorial derives from a special kind of nonbonded interaction strain called **1,3-diaxial interactions.**
    Atoms or groups that are axial are relatively close to the two other atoms or groups that are also axial, thus groups will crash into these other axial substituents. This contact occurs between axial atoms or groups in the 1 and 3 positions of the cyclohexane ring, hence the name **1,3-diaxial interactions**. *[Confirm this nonbonded interaction strain for yourself by making a model of methylcyclohexane, and make the chair conformation that places the methyl group in an axial position.]*

• Atoms or groups in equatorial positions are out away from other groups, thus minimizing nonbonded steric strain. ✳ *[Confirm this by converting your model to the other chair conformation, and notice the methyl group, which is now equatorial, is relatively free from nonbonded interaction strain.]*

- Chemists have synthesized a variety of highly strained small-ring compounds including **propellanes**, **cubanes**, and **prismanes**.

2.7 *CIS-TRANS* ISOMERISM IN CYCLOALKANES AND BICYCLOALKANES

- Cycloalkanes with substituents on two or more carbons of the ring show a type of isomerism called ***cis-trans* isomerism.** ✳
  - ***Cis-trans* isomerism** is a type of **geometric isomerism**, that is isomerism that depends on the placement of substituent groups on the atoms of a ring or on a double bond.
    Cis-trans isomerism can be understood by thinking of the cycloalkane as a planar structure. *[This is just a helpful trick, of course the true cycloalkane structures for everything larger than cyclopropane are puckered.]*
    For a cycloalkane with two substituents, the ***cis* isomer** is the one in which the two substituents are on the same side of the ring plane.
    The ***trans* isomer** is the one in which the two substituents are on opposite sides of the ring plane. In other words, for a given constitutional isomer such as 1,2 dimethylcyclopentane, the two methyl groups can be either *cis* or *trans* with respect to each other. *[Note that in order for the cis and trans comparison to be valid, the same constitutional isomer must be considered in both cases.]*
    No matter how much the cycloalkane ring puckers or interchanges between conformations, the two methyl groups of the *cis* isomer will always be on the same side of the ring plane, and the two methyl groups of the *trans* isomer will always remain on opposite sides of the ring plane. Put another way, no amount of conformational change can convert the *cis* isomer into the *trans* isomer, and vice versa. *[Again, making models will save you a lot of time, and it will also make things more clear .]*
  - The analysis of disubstituted cyclohexanes becomes more complicated in the context of chair conformations. For example, *trans*-1,4-dimethylcyclohexane can exist in two chair conformations. *[You should make a model to prove this to yourself.]*
    In one chair conformation, both methyl groups are axial. In the other chair conformation, both methyl groups are equatorial.
    The chair conformation with both methyl groups equatorial is more stable. *[It has fewer 1,3-diaxial interactions]*
  - For *cis*-1,4-dimethylcyclohexane, each chair conformation has one methyl group axial and one methyl group equatorial, so both of these conformations are equally stable. *[Again, a model will be very helpful here.]*
- **Bicycloalkanes** also exhibit *cis-trans* isomerism. *[You should practice analyzing compounds like cis- and trans-decalin with models using the same ideas just discussed for the cyclopentane and cyclohexane derivatives.]*

2.8 PHYSICAL PROPERTIES OF ALKANES AND CYCLOALKANES

- At room temperature, the simple alkanes the size of butane or smaller are **gases**, while pentane through decane are **liquids**.
- At lower temperatures, the alkanes can be frozen into **solids**.
- The fact that these compounds can exist as liquids and solids depends on the existence of **intermolecular forces** that can hold the molecules together.
- All intermolecular forces that hold ions and molecules together, are **electrostatic** in nature, that is, they are based on attraction between oppositely charged groups. Following are the most important types of attractive forces:
  - **Ion-ion interactions** are the result of electrostatic attraction between oppositely charged ions
    This is by far the strongest attractive force, since it involves attraction between species with full positive and negative charges.
    Ion-ion interactions are not common in organic chemistry.

- **Ion-dipole interactions** are the result of electrostatic attraction between an ion and an oppositely charged dipole of a polar molecule.
  Ion-dipole interactions are also not that common in organic chemistry, but they are especially important in solutions of ionic solids in polar solvents like water.
- **Dipole-dipole interactions** are the result of electrostatic attraction between two molecules that have permanent dipole moments.
  The regions of the molecules with opposite partial charge attract each other.
  Dipole-dipole interactions are common in organic chemistry, being present whenever molecules contain polar functional groups.
  The strength of dipole-dipole interactions is inversely proportional to the third power of the distance between the two dipoles, so dipole-dipole interactions are only important when the interacting particles are very close together.
  Hydrogen bonding is an especially important type of dipole-dipole interaction that occurs between molecules with H-O and H-N bonds. Biochemistry has many exquisite examples of hydrogen bonding. ✳
- **Dispersion forces**, the weakest of all the intermolecular forces, are the result of electrostatic attraction between *temporary* dipole moments.
  Even molecules such as alkanes without a large permanent dipole moment have **temporary dipole moments** caused by instantaneous fluctuations of electron density. Averaged over time, the electron distribution is symmetrical, but at any instant there are small dipole moments caused by small non-symmetrical shifts in electron density. Such an instantaneous dipole moment induces an equally instantaneous but opposite dipole moment in adjacent molecules or atoms. These weak, but opposite dipole moments form the basis of a weak electrostatic attraction referred to as dispersion forces.
  The strength of dispersion forces depends on how easily an electron cloud is **polarized**. Small atoms with tightly held electrons show weaker dispersion forces than larger atoms or molecules with less tightly held electrons.

• The properties of alkanes can be understood by considering the attractive forces mentioned above. ✳
• Alkanes are **hydrophobic**, that is they do not dissolve in polar solvents like water.
• **Solubility** is determined by both entropy and attractive forces between molecules. ✳
  -If the forces of attraction between each of two substances in the pure state are replaced by comparable or stronger forces upon mixing, then one will dissolve in the other.
  -Alkanes do not have permanently large dipole moments, so they cannot interact with the water molecules by either dipole-dipole or ion-dipole interactions.
• Alkanes also have very low **boiling points** because they are held together primarily by nothing but weak dispersion forces.
  -In general, larger alkanes (higher molecular weight) have higher boiling points than smaller alkanes (lower molecular weight), and unbranched alkanes have higher boiling points than branched constitutional isomers. The strength of dispersion forces is proportional to the surface area of contact between molecules. Branched molecules are more compact resulting in smaller surface areas than unbranched constitutional isomers. *[The above ideas explain a large amount of experimental data.]*

2.9 REACTIONS OF ALKANES

• Alkanes are relatively unreactive, having small permanent dipole moments and only the relatively strong sigma type of bonds.
• On the other hand, alkanes can react with oxygen and halogens under certain conditions. ✳

2.10 SOURCES OF ALKANES

• The three natural sources of alkanes are **natural gas**, **petroleum** and **coal**.

- **Natural gas** is mostly methane, with some ethane and a small amount of other small alkanes that are gases at room temperature.
- **Petroleum** is an incredibly complex mixture of compounds, and the commercially important fractions are purified by large scale distillation towers. Presently, petroleum is by far the most important source of organic raw materials for products like fuels, lubricants, nylon, dacron, textile fibers, asphalt, and synthetic rubber.
- **Coal** has an extremely complex structure, and a great deal of chemistry is required to produce useful alkanes such as fuels.

CHAPTER 2
Solutions to the Problems

Problem 2.1 Identify the members of each pair as formulas of identical compounds or as formulas of constitutional isomers.

```
        CH2-CH3                                 CH3       CH3
        |                                       |         |
(a) CH3-CH-CH-CH3          and     CH3-CH2-CH-CH2-CH-CH3
           |
           CH2-CH3
```

These molecules are constitutional isomers. Each has six carbons in the longest chain. The first has one-carbon branches on carbons 3 and 4 of the chain; the second has one-carbon branches on carbons 2 and 4 of the chain.

```
        CH3                              CH3
        |                                |
(b) CH3-CH-CH-CH3          and   CH3-CH-CH-CH2-CH3
           |                         |
           CH2-CH3                   CH3
```

These molecules are identical. Each has five carbons in the longest chain, and one-carbon branches on carbons 2 and 3 of the chain.

Problem 2.2 Draw structural formulas for the three constitutional isomers of molecular formula C_5H_{12}.

```
                                    CH3                  CH3
                                    |                    |
CH3-CH2-CH2-CH2-CH3            CH3-CH-CH2-CH3       CH3-C-CH3
                                                         |
                                                         CH3
```

Problem 2.3 Write IUPAC names for the following alkanes.

```
        CH3              CH3                                      CH2-CH2-CH3
        |                |                                        |
(a) CH3-CH-CH2-CH2-CH-CH-CH3       (b) CH3-CH2-CH2-C-CH2-CH2-CH3
                   |                                    |
                   CH2-CH2-CH3                          CH-CH3
                                                        |
                                                        CH3
```

5-isopropyl-2-methyloctane

4-isopropyl-4-propylheptane

(c) 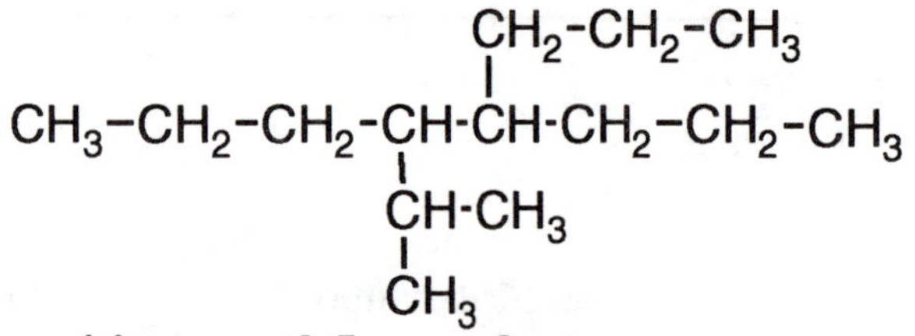 $CH_3-CH_2-CH_2-CH(CH(CH_3)CH_3)-CH(CH_2-CH_2-CH_3)-CH_2-CH_2-CH_3$

4-isopropyl-5-propyloctane

(d) 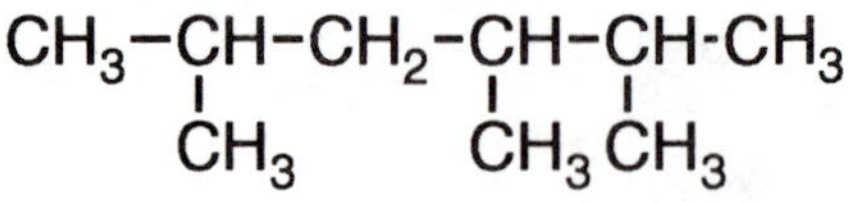 $CH_3-CH(CH_3)-CH_2-CH(CH_3)-CH(CH_3)-CH_3$

2,3,5-trimethylhexane

Problem 2.4 State the number of sets of equivalent hydrogens in each compound and the number of hydrogens in each set.

(a) 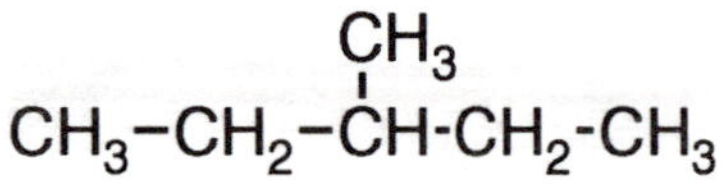 $CH_3-CH_2-CH(CH_3)-CH_2-CH_3$

(b) 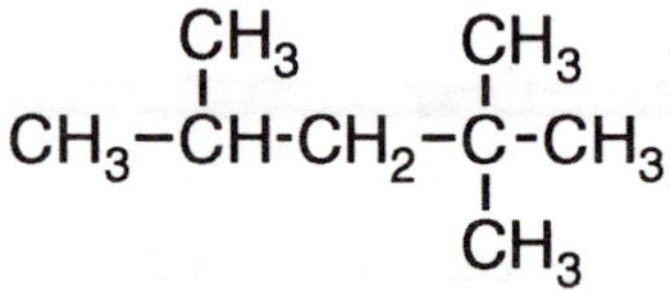 $CH_3-CH(CH_3)-CH_2-C(CH_3)_2-CH_3$

(a) $CH_3-CH_2-CH(CH_3)-CH_2-CH_3$

set of 6 equivalent primary hydrogens

set of 3 equivalent primary hydrogens

set of one tertiary hydrogen

set of 4 equivalent secondary hydrogens

(b) $CH_3-CH(CH_3)-CH_2-C(CH_3)_2-CH_3$

set of 6 equivalent primary hydrogens

set of 9 equivalent primary hydrogens

set of one tertiary hydrogen

set of 2 equivalent secondary hydrogens

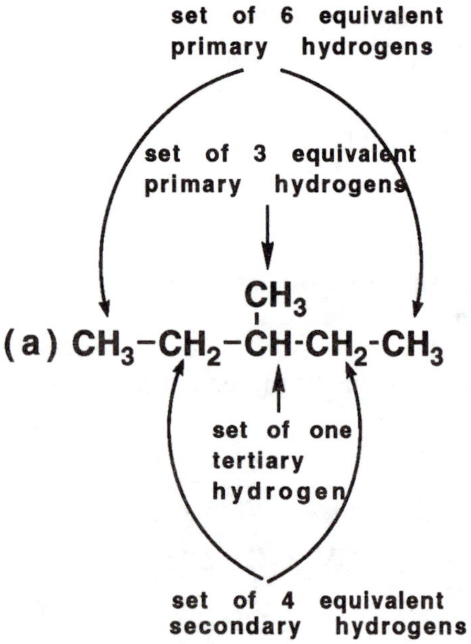

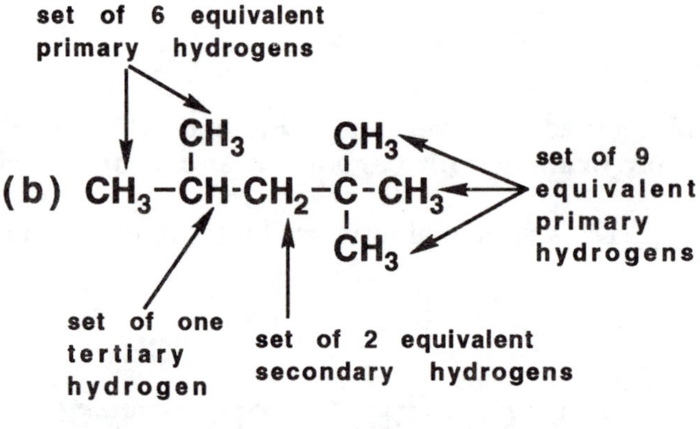

Problem 2.5 Following are line-angle drawings for three cycloalkanes. Write a structural formula and molecular formula for each.

(a)

$(C_5H_9)CH_2CH(CH_3)_2$

isobutylcyclopentane

(b)

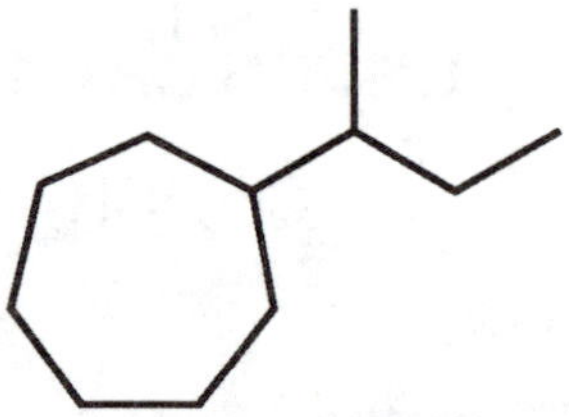

$(C_7H_{13})CH(CH_3)CH_2CH_3$

sec-butylcycloheptane

(c)

H_2C $C(CH_3)CH_2CH_3$ H_2C

1-ethyl-1-methylcyclopropane

Problem 2.6 Draw structural formulas for the following bicycloalkanes and spiroalkanes.

(a) bicyclo[3.1.0]hexane

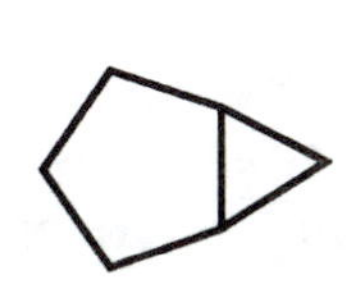

(b) bicyclo[2.2.2]octane

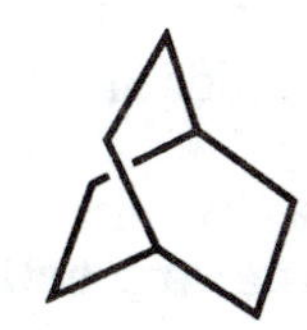

(c) bicyclo[4.2.0]octane

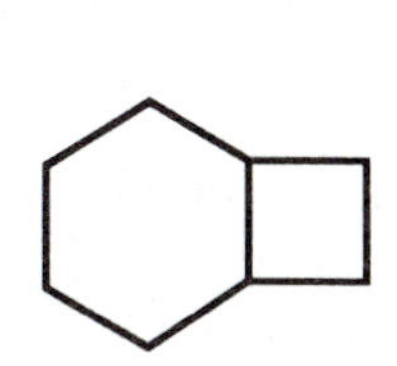

(d) 2,6,6-trimethylbicyclo[3.1.1]heptane

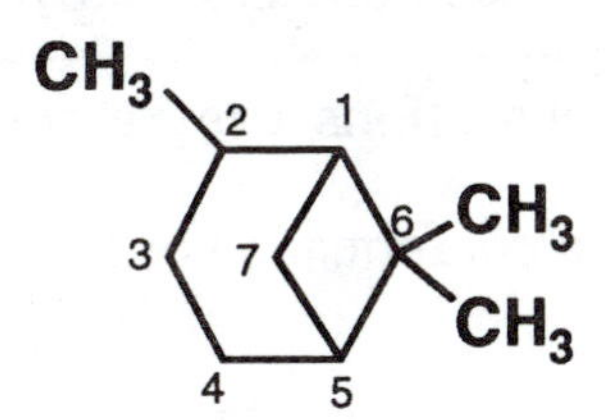

(e) spiro[4.2]heptane

(f) spiro[5.2]octane

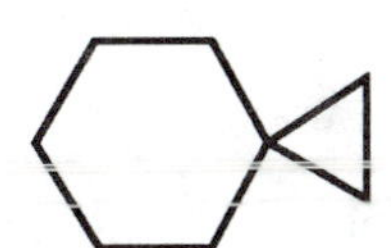

Problem 2.7 Combine the proper prefix, infix and suffix and write the IUPAC name for the following.

(a) $CH_2=CH_2$

ethene

(b) $CH_3\text{-}C\equiv CH$

propyne

(c) $CH_3\text{-}CH_2\text{-}CH_2\text{-}\overset{O}{\overset{\|}{C}}\text{-}CH_3$

2-pentanone

(d) $CH_3\text{-}CH_2\text{-}CH_2\text{-}CH_2\text{-}\overset{O}{\overset{\|}{C}}\text{-}H$

pentanal

(e) O

cyclopentanone

(f)

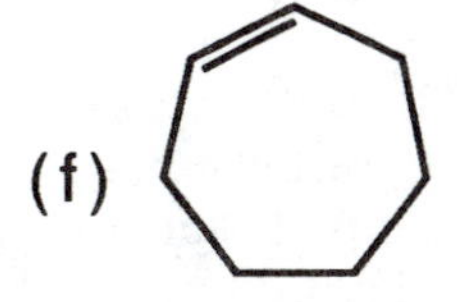

cycloheptene

Problem 2.8
(a) Draw Newman projections for all eclipsed conformations of 1,2-dichloroethane formed by rotation from 0° to 360°.

higher in energy

lower in energy
(related by reflection)

(b) Arrange the eclipsed conformations in order of increasing energy.

Of the three eclipsed conformations, the one drawn on the left is higher in energy because the two larger chlorine atoms are eclipsed. The second and third are equivalent and lower in energy.

(c) Which, if any, of these eclipsed conformations are related by reflection?

The second and third conformations are related by reflection.

Problem 2.9 Following is a chair conformation of cyclohexane with carbon atoms numbered 1 through 6.

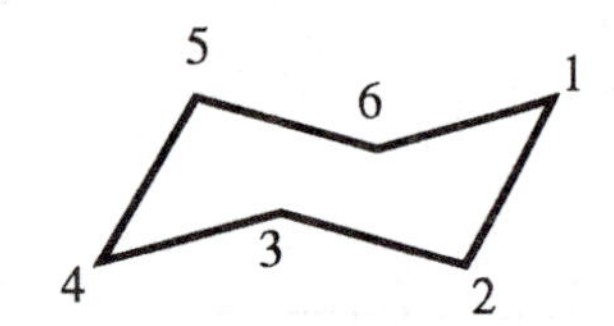

(a) Draw hydrogen atoms that are up (above the plane of the ring) on carbons 1 and 2, and down (below the plane of the ring) on carbon 4.
(b) Which of these hydrogens are equatorial; which are axial?
(c) Draw the other chair conformation. Now, which hydrogens are equatorial; which are axial?

Problem 2.10 The conformational equilibria for methyl, ethyl, and isopropylcyclohexane are all about 95% in favor of the equatorial conformation, but that for *tert*-butylcyclohexane is virtually completely on the equatorial side. Explain by using molecular models and making drawings why the conformational equilibria for the first three compounds are comparable, but why the conformational equilibrium for *tert*-butylcyclohexane lies considerably farther toward the equatorial conformation.

Rotation is possible about the single bond connecting the axial substituent to the ring. Axial methyl, ethyl and isopropyl groups can assume a conformation where a hydrogen creates the 1,3-diaxial interactions. With a *tert*-butyl substituent, however, a bulkier -CH_3 group creates the 1,3-diaxial interaction. Because of the increased steric strain (nonbonded interactions) created by the axial *tert*-butyl, the potential energy of the axial conformation is considerably greater than that for the equatorial conformation.

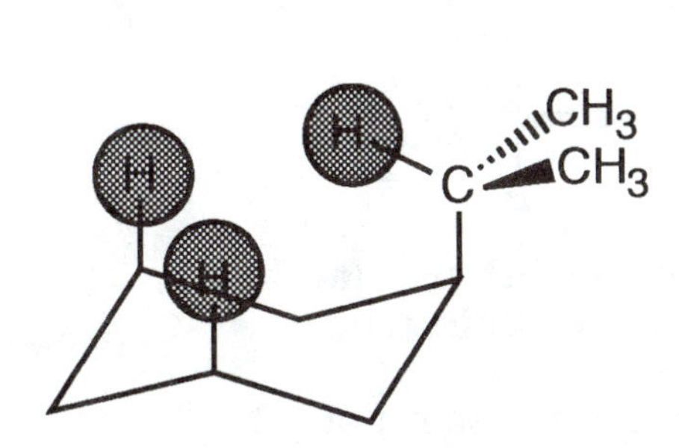

axial isopropyl group moderately severe 1,3 diaxial interaction

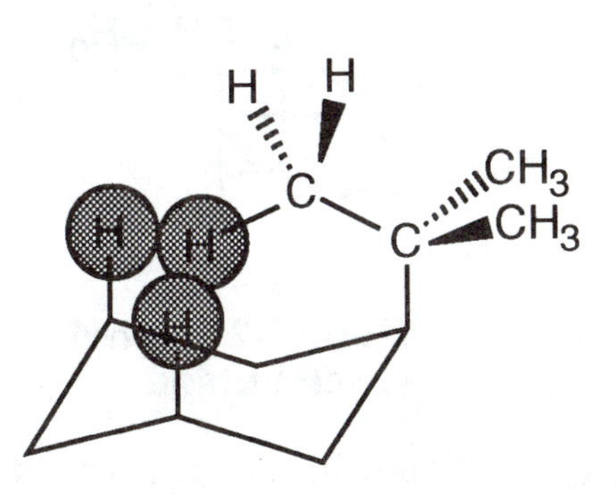

axial *tert*-butyl group very severe 1,3 diaxial interaction

Problem 2.11 Following are several cycloalkanes of molecular formula C_7H_{14}. State which show *cis-trans* isomerism and for each that does, draw the *cis* and *trans* isomers.

(a) CH_3 CH_3

1,3-Dimethylcyclopentane shows *cis-trans* isomerism. In the following drawings, the ring is drawn as a planar pentagon with substituents above and below the plane of the pentagon.

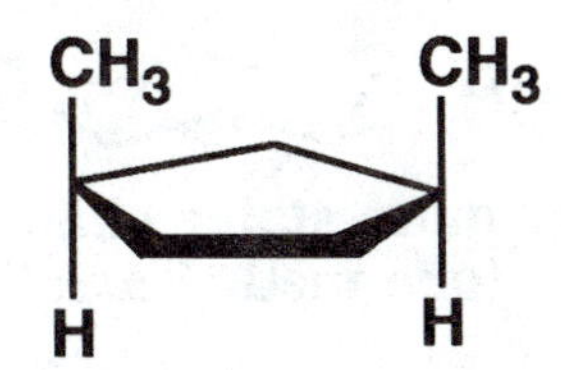

***cis*-1,3-dimethyl-cyclopentane**

CH_3 H

H CH_3

***trans*-1,3-dimethyl-cyclopentane**

(b) CH_2CH_3

Ethylcyclopentane does not show *cis-trans* isomerism.

(c) 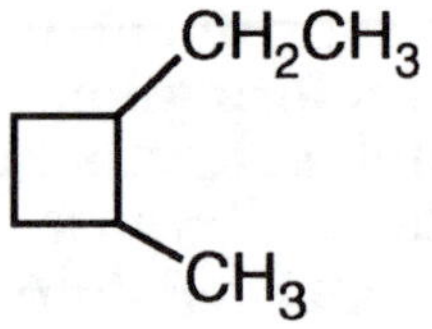 CH_2CH_3 / CH_3

1-Ethyl-2-methylcyclobutane shows *cis-trans* isomerism.

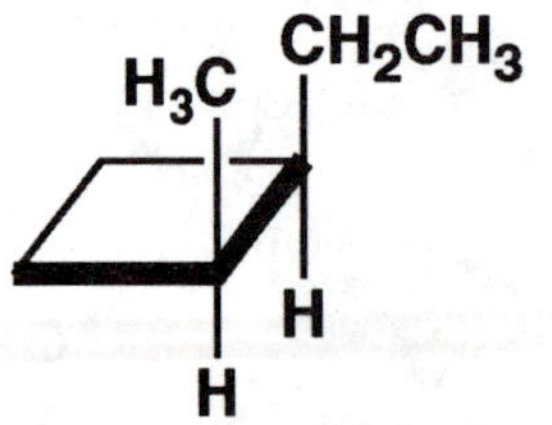

H_3C CH_2CH_3 H H — **cis-1-ethyl-2-methyl-cyclobutane**

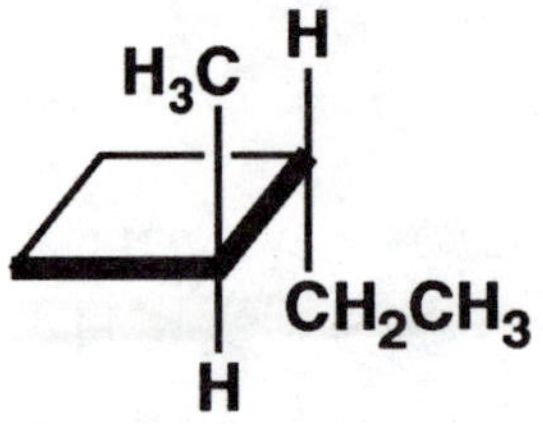

H_3C H CH_2CH_3 H — **$trans$-1-ethyl-2-methyl-cyclobutane**

Problem 2.12 Following is a planar hexagon representation for one isomer of 1,2,4-trimethylcyclohexane. Draw alternative chair conformations of this compound and state which is the more stable.

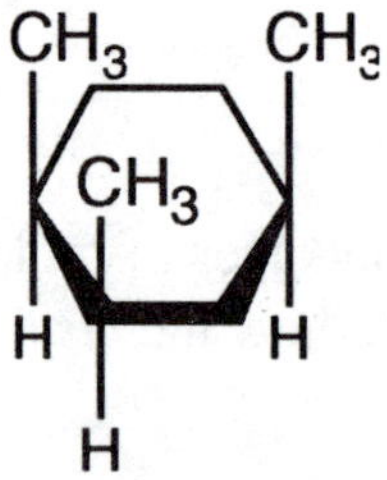

CH_3 CH_3 CH_3 H H H

Following are alternative chair conformations for the all *cis* isomers of 1,2,4-trimethylcyclohexane. The alternative chair conformation on the right is the more stable because it has only one axial methyl group.

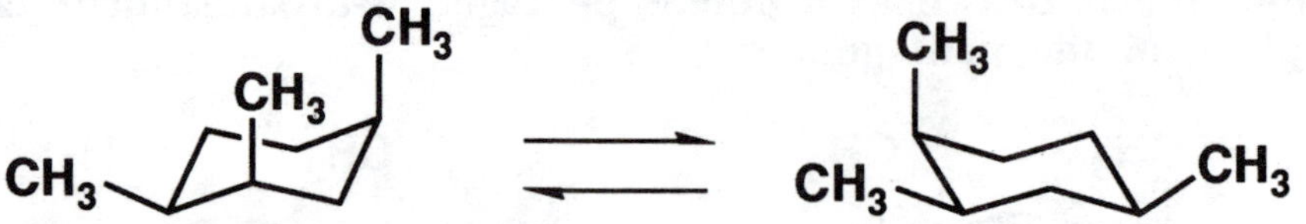

CH_3 CH_3 CH_3 ⇌ CH_3 CH_3 CH_3

less stable chair (two methyls axial)

more stable chair (one methyl axial)

Problem 2.13 Which of the following *cis-trans* isomers is the more stable?

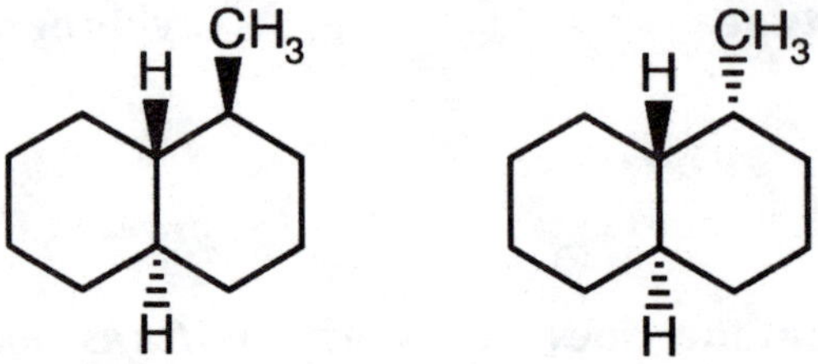

H CH_3 H   H CH_3 H

In the first isomer of *trans*-decalin, the methyl substituent is equatorial and in the second isomer, it is axial. The equatorial methyl isomer is more stable.

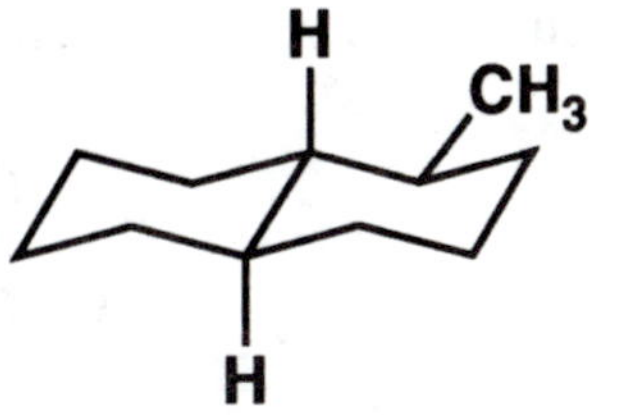

H
CH_3
H

more stable isomer
(methyl substituent is equatorial)

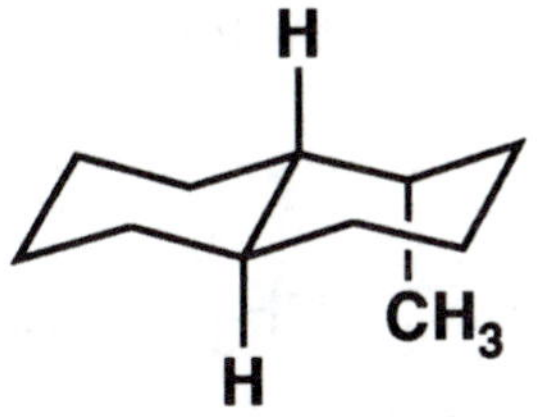

H
CH_3
H

less stable isomer
(methyl substituent is axial)

Problem 2.14 Arrange the following in order of increasing boiling point.
(a) 2-methylbutane, 2,2-dimethylpropane, pentane

2,2-dimethylpropane,	2-methylbutane,	pentane
(bp 9.5°C)	(bp 29°C)	(bp 36°C)

(b) 3,3-dimethylheptane, 2,2,4,4-tetramethylpentane, nonane

2,2,4,4-tetramethylpentane,	3,3-dimethylheptane,	nonane
(bp 99°C)	(bp 137°C)	(bp 151°C)

Problem 2.15 Name and draw structural formulas for all monochlorination products formed by the treatment of butane with Cl_2. Calculate the expected statistical distribution of products and, given the regioselectivity of Cl_2, predict how the actual product distribution might differ from the calculated statistical distribution.

$$CH_3-CH_2-CH_2-CH_3 \; + \; Cl_2 \xrightarrow[\text{heat}]{\text{light or}} \text{monochloroalkanes} \; + \; HCl$$

There are two possible monochlorination products. In butane, there is one set of six equivalent primary hydrogens and one set of four equivalent secondary hydrogens. Therefore, based on statistical distribution, predict 60% 1-chlorobutane and 40% 2-chlorobutane. Based on the known regioselectivity of chlorination (secondary hydrogen > primary hydrogen) predict more than 40% 2-chlorobutane and less than 60% 1-chlorobutane.

	$CH_3\text{-}CH_2\text{-}CH_2\text{-}CH_2\text{-}Cl$	$CH_3-CH_2-CH(Cl)-CH_3$
	1-chlorobutane **(butyl chloride)**	**2-chlorobutane** **(sec-butyl chloride)**
Predict:	**60%**	**40%**
Observed:	**less than 60%**	**greater than 40%**

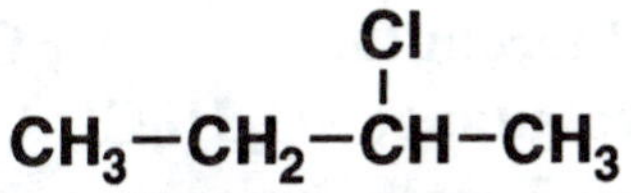

Constitutional Isomerism in Alkanes

Problem 2.16 Which of the following are identical compounds and which are constitutional isomers?

(a) $CH_3-CH_2-CH(Cl)-CH_3$

(b) $CH_3-C(CH_2Cl)(Cl)-CH_3$

(c) $CH_3-CH(Cl)-CH(Cl)-CH_3$

(d) $CH_2(Cl)-CH_2-CH(Cl)-CH_3$

(e) cyclobutane$-Cl$

(f) $Cl-CH_2-$cyclopropane

(g) $CH_3-CH(CH_2-Cl)-CH_3$

(h) $CH_3-CH_2-CH_2-CH_2-Cl$

(i) $Cl-CH_2-CH(CH_3)-CH_3$

(j) $CH_3-CH(Cl)-CH_2-CH_2-Cl$

(k) $CH_3-CH(CH_2-CH_3)-Cl$

(l) $CH_3-C(CH_3)(Cl)-CH_3$

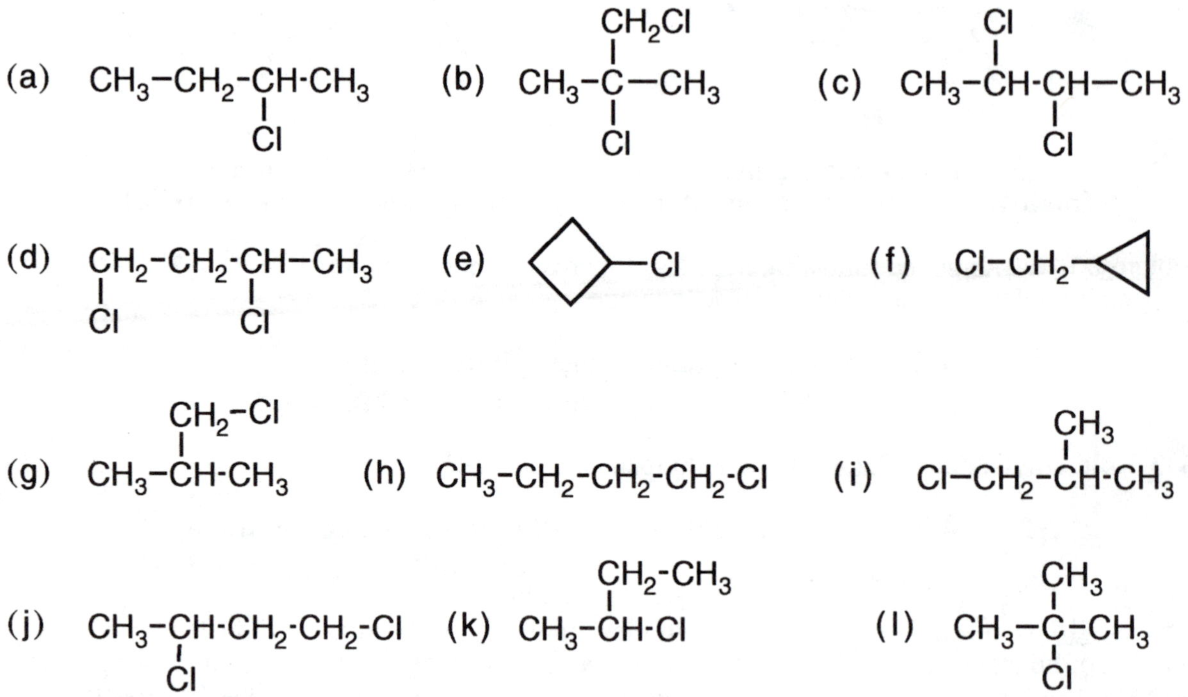

Following are names and molecular formulas of each

(a) 2-chlorobutane; C_4H_9Cl
(b) 1,2-dichloro-2-methylpropane; $C_4H_8Cl_2$
(c) 2,3-dichlorobutane; $C_4H_8Cl_2$
(d) 1,3-dichlorobutane; $C_4H_8Cl_2$
(e) chlorocyclobutane; C_4H_7Cl
(f) chloromethylcyclopropane; C_4H_7Cl
(g) 1-chloro-2-methylpropane; C_4H_9Cl
(h) 1-chlorobutane; C_4H_9Cl
(i) 1-chloro-2-methylpropane; C_4H_9Cl
(j) 1,3-dichlorobutane; $C_4H_8Cl_2$
(k) 2-chlorobutane; C_4H_9Cl
(l) 2-chloro-2-methylpropane; C_4H_9Cl

The following are identical: (a),(k) (d),(j) (g),(i)

The following four compounds are constitutional isomers: 1-chlorobutane (h), 2-chlorobutane (a)(k), 1-chloro-2-methylpropane (g)(i), and 2-chloro-2-methylpropane (l). Chlorocyclobutane (e) and chloromethylcyclopropane (f) are also constitutional isomers.

Problem 2.17 Name and draw structural formulas for all constitutional isomers of molecular formula C_7H_{16}.

There are nine alkanes of molecular formula C_7H_{16}.

$CH_3CH_2CH_2CH_2CH_2CH_2CH_3$
heptane (bp 94.8)

$CH_3CH(CH_3)CH_2CH_2CH_2CH_3$
2-methylhexane (bp 90.0)

$CH_3CH_2CH(CH_3)CH_2CH_2CH_3$
3-methylhexane (bp 92.0)

$CH_3C(CH_3)_2CH_2CH_2CH_3$
2,2-dimethylpentane (bp 79.2)

$CH_3CH(CH_3)CH(CH_3)CH_2CH_3$
2,3-dimethylpentane (bp 89.8)

$CH_3CH(CH_3)CH_2CH(CH_3)CH_3$
2,4-dimethylpentane (bp 80.5)

$CH_3CH_2C(CH_3)_2CH_2CH_3$
3,3-dimethylpentane (bp 86.1)

$CH_3CH_2CH(CH_2CH_3)CH_2CH_3$
3-ethylpentane (bp 93.5)

$CH_3CH(CH_3)C(CH_3)_2CH_3$
2,2,3-trimethylbutane (bp 80.9)

Nomenclature of Alkanes and Cycloalkanes

Problem 2.18 Write IUPAC names for the following alkanes and cycloalkanes.

(a) $CH_3CH(CH_3)CH_2CH_2CH_3$
2-methylpentane (isohexane)

(b) $CH_3CH(CH_3)CH_2CH_2CH(CH_3)CH_3$
2,5-dimethylhexane

(c) $CH_3CH_2CH(CH_3)CH_2CH(CH_2CH_3)CH_3$
3,5-dimethylheptane

(d) $CH_3(CH_2)_4CH(CH_2CH_3)CH_2CH_3$
3-ethyloctane

(e) $CH_3CH_2CH(CH(CH_3)CH_3)CH_2CH_2CH_2CH_3$
3-ethyl-2-methylheptane

(f) $CH_3CH_2CH_2CH(CH_2CH(CH_3)_2)CH_2CH_3$
4-ethyl-2-methylheptane

The longest chain in (e) is seven carbons and, therefore, it is a heptane. Depending on the numbering pattern you pick, it is named either 3-ethyl-2-methylheptane, or 3-isopropylheptane.

3 4 5 6 7
$CH_3-CH_2-CH-CH_2-CH_2-CH_2-CH_3$
1 |2
$CH_3-CH-CH_3$

3-ethyl-2-methylheptane

1 2 3 4 5 6 7
$CH_3-CH_2-CH-CH_2-CH_2-CH_2-CH_3$
|
$CH_3-CH-CH_3$

3-isopropylheptane

While each numbering pattern is unambiguous, the first is the correct name. An additional IUPAC rule states that in cases where two or more carbon chains of equal length might serve as the parent chain, you must choose the one that gives the larger number of substituents. The numbering pattern chosen for the first name leads to a heptane with two substituents, while the numbering pattern for the second name leads to a heptane with only one substituent.

(g) $CH_3(CH_2)_8CH_3$

decane

(h) $(CH_3)_2CHCH_2CH_2C(CH_3)_3$

2,2,5-trimethylhexane

(i) cyclopropane with two CH_3 on one carbon

1,1-dimethylcyclopropane

(j) cyclopentane–$CH_2CH(CH_3)_2$

isobutylcyclopentane

(k) cyclohexane with CH_2CH_3, CH_3, CH_3

2,4-dimethyl-1-ethylcyclohexane

(l) cyclobutane with Br (hashed) and Br (wedged)

***trans*-1,2-dibromo-cyclobutane**

Problem 2.19 Write structural formulas for the following compounds.

(a) 2,2,4-trimethylhexane

CH_3 CH_3
| |
$CH_3CCH_2CHCH_2CH_3$
|
CH_3

(b) 1,1,2-trichlorobutane

Cl
|
$Cl\,CHCHCH_2CH_3$
|
Cl

(c) 2,2-dimethylpropane

CH_3
|
CH_3CCH_3
|
CH_3

(d) 3-ethyl-2,4,5-trimethyloctane

CH_3CH_2 CH_3
| |
$CH_3CHCHCHCHCH_2CH_2CH_3$
| |
CH_3 CH_3

(e) 2-bromo-2,4,6-trimethyloctane

$$\begin{array}{c} \quad Br \quad CH_3 \quad CH_3 \\ CH_3\overset{|}{\underset{|}{C}}CH_2\overset{|}{C}H\,CH_2\overset{|}{C}HCH_2CH_3 \\ CH_3 \end{array}$$

(f) 5-butyl-2,2-dimethylnonane

$$\begin{array}{c} CH_3 \\ CH_3\overset{|}{\underset{|}{C}}CH_2CH_2\underset{|}{C}HCH_2CH_2CH_2CH_3 \\ CH_3 \qquad CH_2CH_2CH_2CH_3 \end{array}$$

(g) 4-isopropyloctane

$$\begin{array}{c} CH_3CH_2CH_2\underset{|}{C}HCH_2CH_2CH_2CH_3 \\ CH_3CHCH_3 \end{array}$$

(h) 3,3-dimethylpentane

$$\begin{array}{c} CH_3 \\ CH_3CH_2\overset{|}{\underset{|}{C}}CH_2CH_3 \\ CH_3 \end{array}$$

(i) 1,1,1-trichloroethane

CH_3CCl_3

(j) *trans*-1,3-dimethylcyclopentane

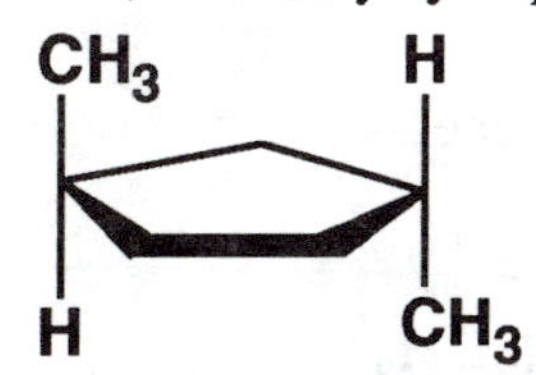

(k) *cis*-1,2-diethylcyclobutane

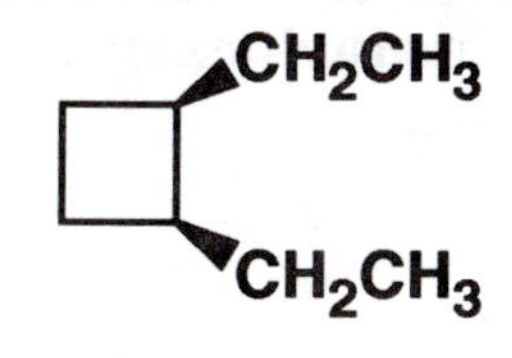

(l) 1,1-dichlorocycloheptane

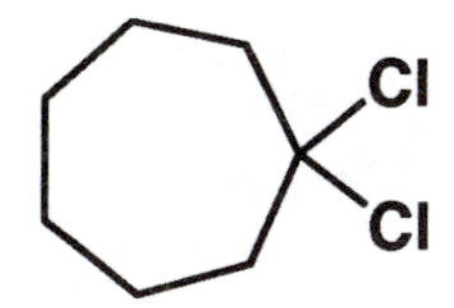

Problem 2.20 Explain why each of the following is an incorrect IUPAC name. Write a correct IUPAC name for the intended compound.

(a) 1,3-dimethylbutane

$$\begin{array}{c} CH_3 \\ CH_3\overset{|}{C}HCH_2CH_2CH_3 \end{array}$$

The longest chain is pentane. Its IUPAC name is 2-methylpentane.

(b) 4-methylpentane

$$\begin{array}{c} CH_3 \\ CH_3\overset{|}{C}HCH_2CH_2CH_3 \end{array}$$

The pentane is numbered incorrectly. Its IUPAC name is 2-methylpentane.

(c) 2,2-diethylbutane

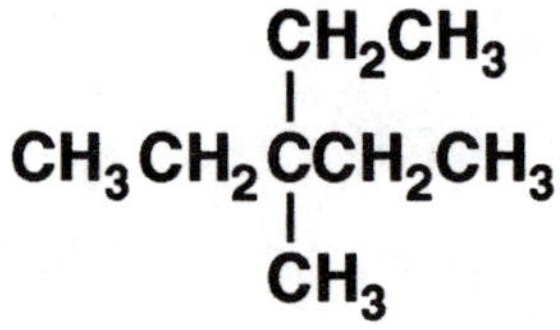

$$\begin{array}{c} CH_2CH_3 \\ | \\ CH_3CH_2CCH_2CH_3 \\ | \\ CH_3 \end{array}$$

The longest chain is pentane. Its IUPAC name is 3-ethyl-3-methylpentane.

(d) 2-ethyl-3-methylpentane

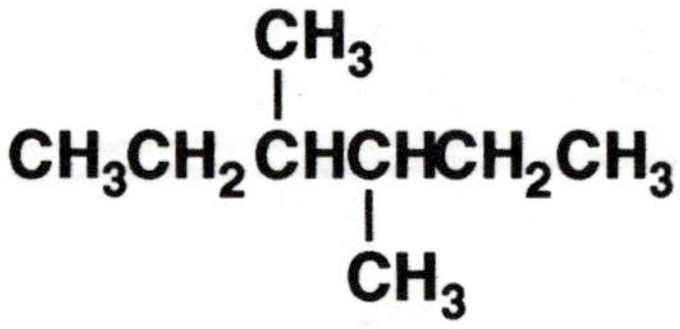

$$\begin{array}{c} CH_3 \\ | \\ CH_3CH_2CHCHCH_2CH_3 \\ | \\ CH_3 \end{array}$$

The longest chain is hexane. Its IUPAC name is 3,4-dimethylhexane.

(e) 4,4-dimethylhexane

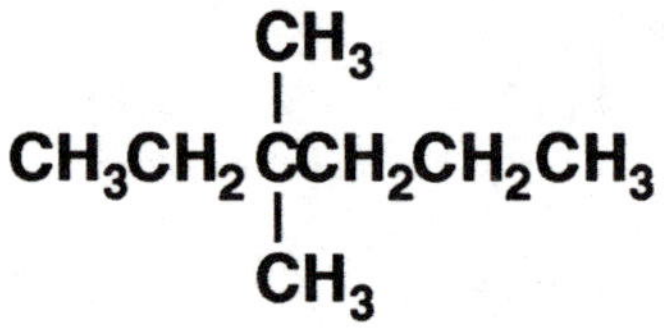

$$\begin{array}{c} CH_3 \\ | \\ CH_3CH_2CCH_2CH_2CH_3 \\ | \\ CH_3 \end{array}$$

The hexane chain is numbered incorrectly. Its IUPAC name is 3,3-dimethylhexane.

(f) 2-propylpentane

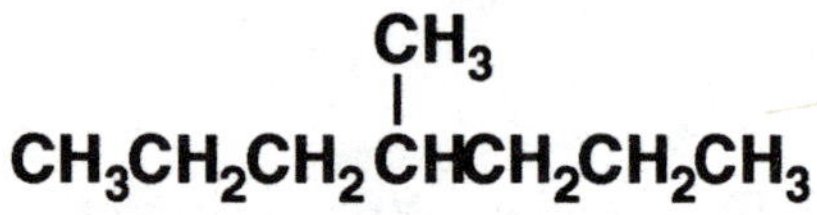

$$\begin{array}{c} CH_3 \\ | \\ CH_3CH_2CH_2CHCH_2CH_2CH_3 \end{array}$$

The longest chain is heptane. Its IUPAC name is 4-methylheptane.

(g) 2,2-diethylheptane

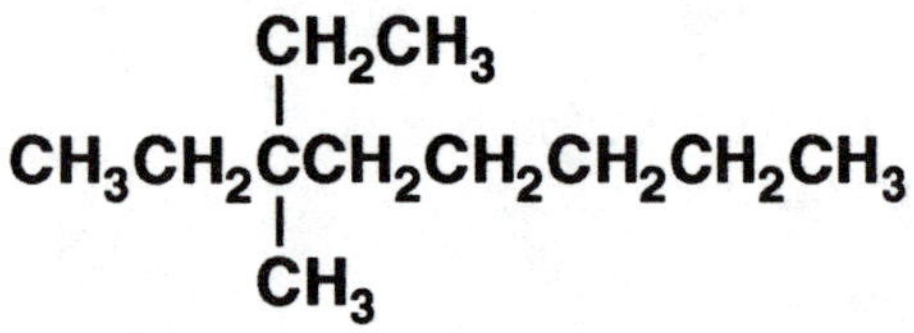

$$\begin{array}{c} CH_2CH_3 \\ | \\ CH_3CH_2CCH_2CH_2CH_2CH_2CH_3 \\ | \\ CH_3 \end{array}$$

The longest chain is octane. Its IUPAC name is 3-ethyl-3-methyloctane.

(h) 5-butyloctane

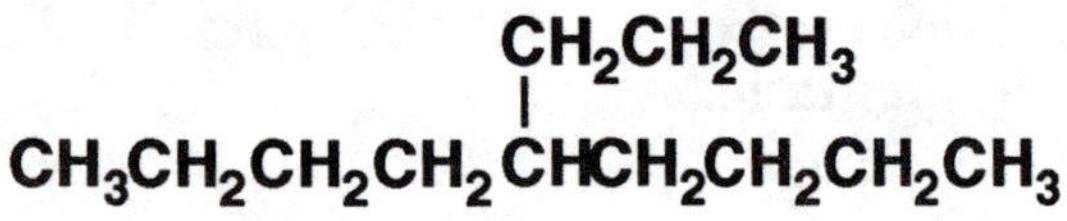

$$\begin{array}{c} CH_2CH_2CH_3 \\ | \\ CH_3CH_2CH_2CH_2CHCH_2CH_2CH_2CH_3 \end{array}$$

The longest chain is nonane. Its IUPAC name is 5-propylnonane.

(i) 2,2-dimethylcyclopropane

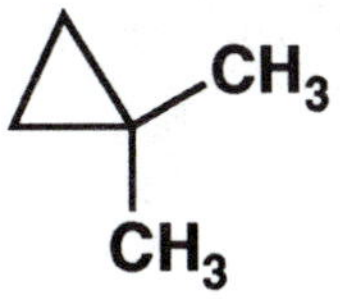

The ring is numbered incorrectly. Its IUPAC name is 1,1-dimethylcyclopropane.

(j) 2-*sec*-butyloctane

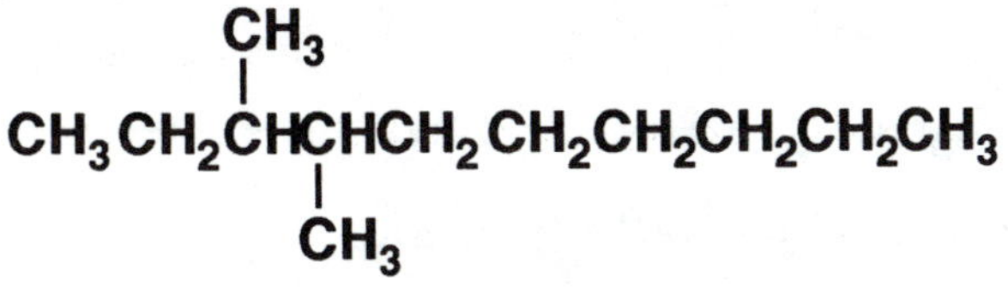

$$CH_3CH_2\underset{CH_3}{\overset{CH_3}{CH\,CH}}CH_2CH_2CH_2CH_2CH_2CH_3$$

The longest chain is decane. Its IUPAC name is 3,4-dimethyldecane.

(k) 4-isopentylheptane

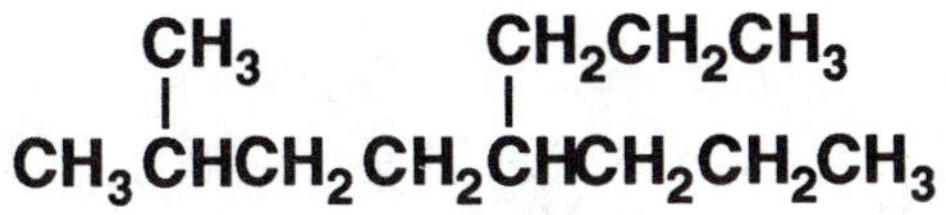

$$CH_3\overset{CH_3}{CH}CH_2CH_2\overset{CH_2CH_2CH_3}{CH}CH_2CH_2CH_3$$

The longest chain is octane. Its IUPAC name is 2-methyl-5-propyloctane.

(l) 1,3-dimethyl-6-ethylcyclohexane

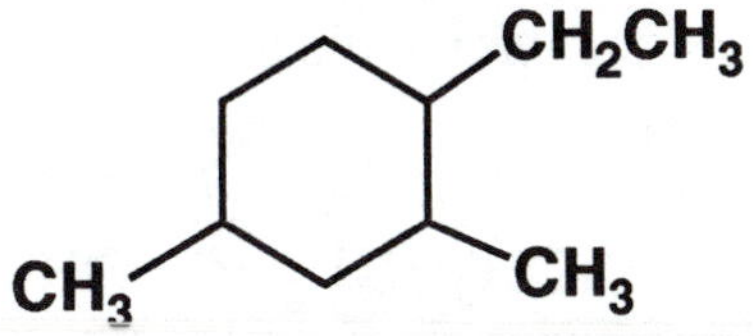

The ring is numbered incorrectly. Its IUPAC name is 2,4-dimethyl-1-ethylcyclohexane.

Problem 2.21 There are 35 constitutional isomers of molecular formula C_9H_{20}. Name and draw structural formulas for the eight that have five carbon atoms in the longest chain.

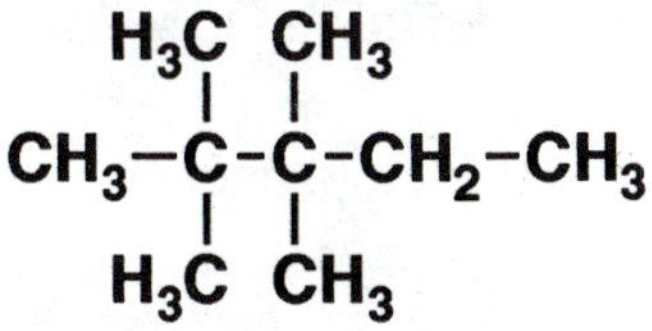

$$CH_3-\underset{H_3C}{\overset{H_3C}{C}}-\underset{CH_3}{\overset{CH_3}{C}}-CH_2-CH_3$$

2,2,3,3-tetramethylpentane

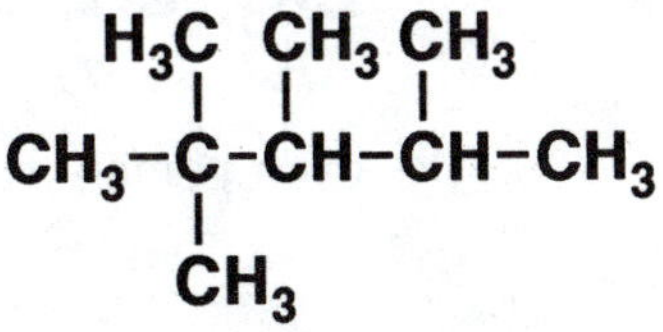

$$CH_3-\underset{CH_3}{\overset{H_3C}{C}}-\overset{CH_3}{CH}-\overset{CH_3}{CH}-CH_3$$

2,2,3,4-tetramethylpentane

$$CH_3-\underset{CH_3}{\overset{CH_3}{C}}-CH_2-\underset{CH_3}{\overset{CH_3}{C}}-CH_3$$

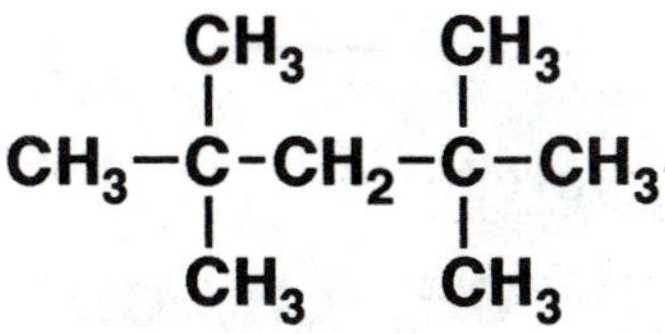

2,2,4,4-tetramethylpentane

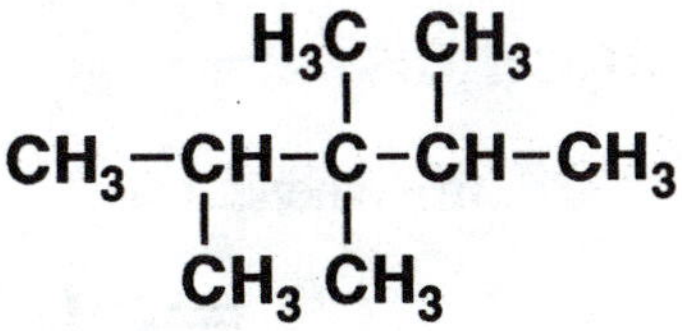

$$CH_3-\underset{CH_3}{CH}-\underset{CH_3}{\overset{H_3C\ CH_3}{C}}-CH-CH_3$$

2,3,3,4-tetramethylpentane

$$\begin{array}{c} CH_3 \quad\;\; CH_3 \\ CH_3-CH-CH-CH-CH_3 \\ CH_2CH_3 \end{array}$$

2,4-dimethyl-3-ethylpentane

$$\begin{array}{c} CH_2CH_3 \\ CH_3-CH_2-C-CH_2-CH_3 \\ CH_2CH_3 \end{array}$$

3,3-diethylpentane

$$\begin{array}{c} CH_3 \\ CH_3-CH_2-C-CH_2-CH_3 \\ CH_3-CH-CH_3 \end{array}$$

2,3-dimethyl-3-ethylpentane

$$\begin{array}{c} CH_3 \\ CH_3-C-CH_3 \\ CH_3-CH_2-CH-CH_2-CH_3 \end{array}$$

2,2-dimethyl-3-ethylpentane

The IUPAC System of Nomenclature

Problem 2.22 For each of the following IUPAC names, draw the corresponding structural formulas.

(a) 3-pentanone

$$\begin{array}{c} O \\ \| \\ CH_3-CH_2-C-CH_2-CH_3 \end{array}$$

(b) 2,2-dimethyl-3-pentanone

$$\begin{array}{c} O \;\; CH_3 \\ \| \;\;\; | \\ CH_3-CH_2-C-C-CH_3 \\ CH_3 \end{array}$$

(c) 2-butanone

$$\begin{array}{c} O \\ \| \\ CH_3-C-CH_2-CH_3 \end{array}$$

(d) ethanoic acid

$$\begin{array}{c} O \\ \| \\ CH_3-C-OH \end{array}$$

(e) hexanoic acid

$$\begin{array}{c} O \\ \| \\ CH_3(CH_2)_4-C-OH \end{array}$$

(f) propanoic acid

$$\begin{array}{c} O \\ \| \\ CH_3-CH_2-C-OH \end{array}$$

(g) propanal

$$\begin{array}{c} O \\ \| \\ CH_3-CH_2-C-H \end{array}$$

(h) 1-propanol

$$CH_3-CH_2-CH_2-OH$$

(i) 2-propanol

$$\begin{array}{c} OH \\ | \\ CH_3-CH-CH_3 \end{array}$$

(j) cyclopentene

(k) cyclopentanol

OH

(l) cyclopentanone

O

(m) cyclopropanol

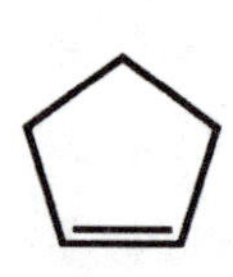

(n) ethene

$$CH_2{=}CH_2$$

(o) ethanol

$$CH_3-CH_2-OH$$

(p) ethanal

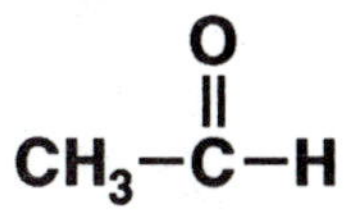

(q) decanoic acid

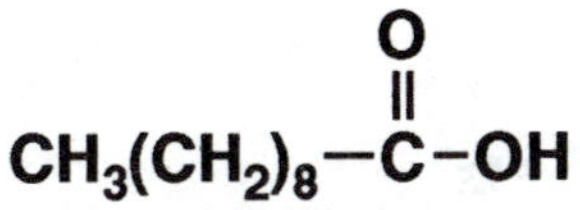

(r) propanone

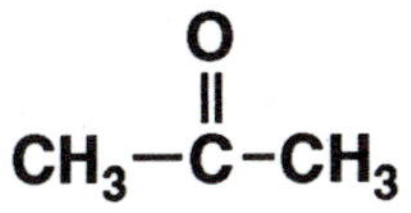

Conformations of Alkanes and Cycloalkanes

Problem 2.23 Consider 1-bromopropane (propyl bromide).

(a) Draw a Newman projection in which $-CH_3$ and -Br are anti (staggered, dihedral angle 180°).

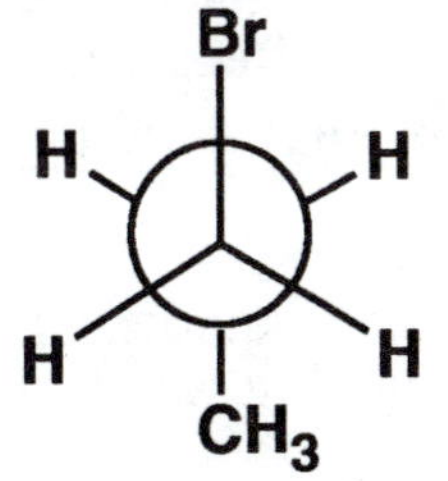

lowest in energy

(b) Draw Newman projections in which $-CH_3$ and -Br are gauche (staggered, dihedral angles 60° and 300°).

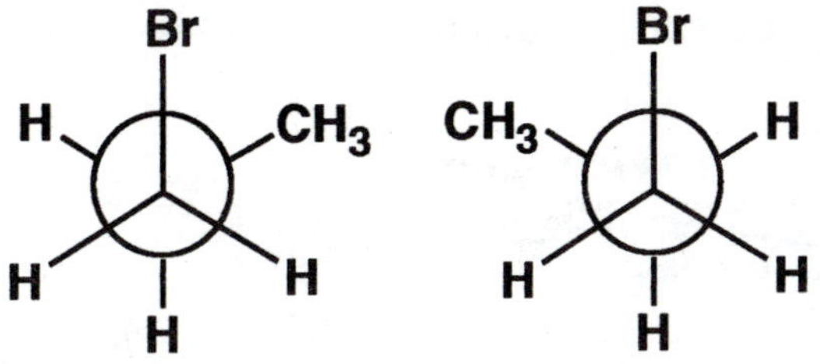

related by reflection

(c) Which of these is the lowest energy conformation.

The anti (staggered, dihedral angle 180°) is the lowest energy conformation.

(d) Which of these conformations, if any, are related by reflection?

The two gauche conformations are of equal energy, and are related by reflection.

Problem 2.24 Consider 1-bromo-2-methylpropane (isobutyl bromide) and draw:

(a) the staggered conformation(s) of lowest energy.

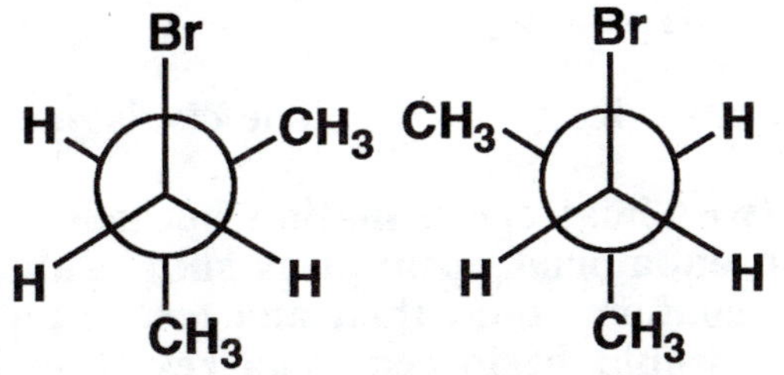

lowest in energy
(related by reflection)

(b) the staggered conformation(s) of highest energy.

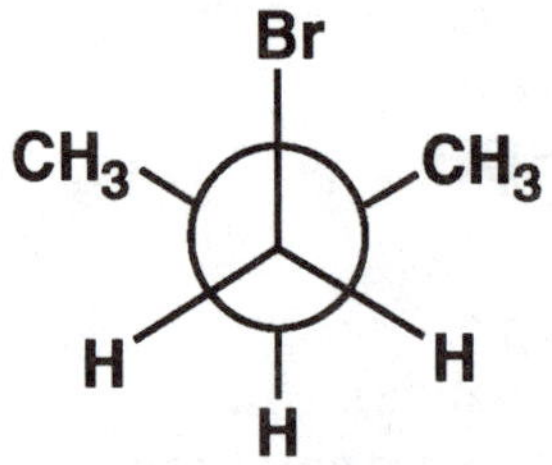

Br
CH_3 CH_3
H H
H

highest in energy

The lower energy staggered conformations have one methyl group anti (dihedral angle 180°) to the bromine and are related by reflection. The staggered conformation with methyl groups at dihedral angles of 60° and 300° are higher in energy.

Problem 2.25 In cyclohexane, an equatorial substituent is equidistant from the axial group and the equatorial group on an adjacent carbon. What is the simplest way to demonstrate this fact?

The best way to see this fact is to draw a Newman projection of one of the carbon-carbon bonds. As can be seen, the axial hydrogen from the carbon atom in the front is in between and thus equidistant to the axial and equatorial hydrogen atoms on the rear carbon atom.

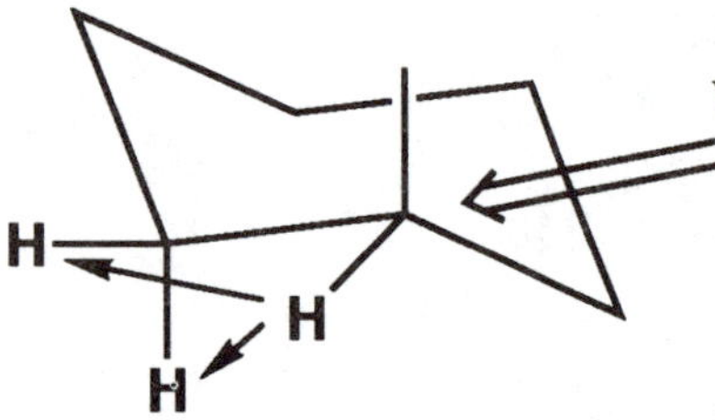

H H H

View along this axis

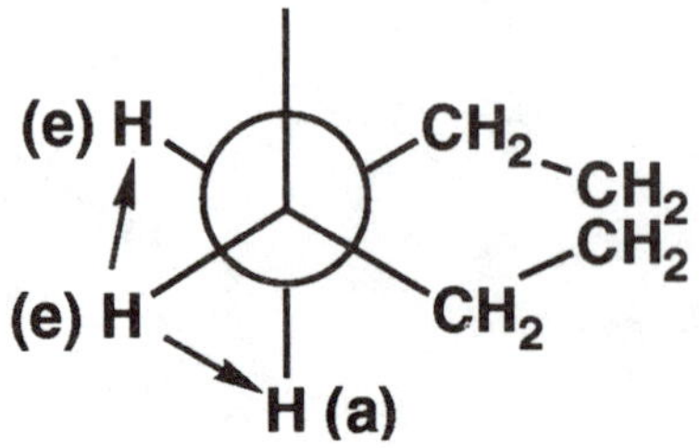

(e) H CH_2 CH_2 CH_2
(e) H CH_2
H (a)

Problem 2.26 *trans*-1,4-Di-*tert*-butylcyclohexane exists in a normal chair conformation. *cis*-1,4-Di-*tert*-butylcyclohexane, however, adopts a twist boat conformation. Draw both isomers and explain why the *cis* isomer is more stable in the twist boat-conformation.

H_3C H
H_3C—C
H_3C
CH_3
C—CH_3
H CH_3

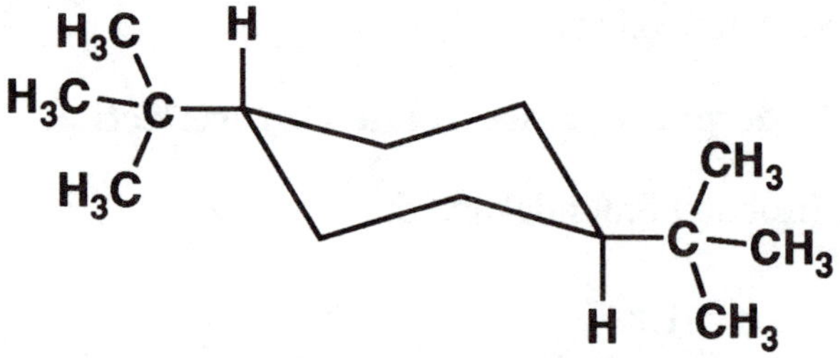

The *trans* isomer in the chair form

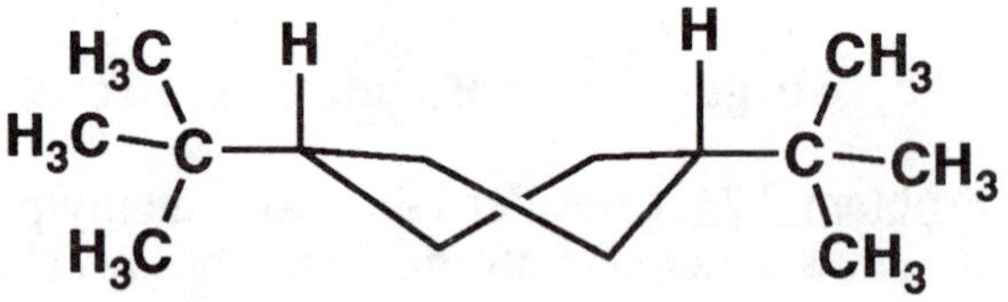

H_3C H H CH_3
H_3C—C C—CH_3
H_3C CH_3

The *cis* isomer in the twist-boat form

The *cis* isomer adopts a twist boat conformation because both of the bulky *tert*-butyl groups can be in an equatorial position as shown above. If the *cis* isomer existed in a normal chair conformation, then one *tert*-butyl group would be equatorial, while the other would be forced axial resulting in a large amount of steric strain.

***Cis-Trans* Isomerism in Cycloalkanes and Bicycloalkanes**

Problem 2.27 Name and draw structural formulas for the *cis* and *trans* isomers of 1,2-dimethylcyclopropane.

***cis*-1,2-dimethyl-cyclopropane** ***trans*-1,2-dimethyl-cyclopropane**

Problem 2.28 Name and draw structural formulas for all cycloalkanes of molecular formula C_5H_{10}. Be certain to include *cis-trans* isomers as well as constitutional isomers.

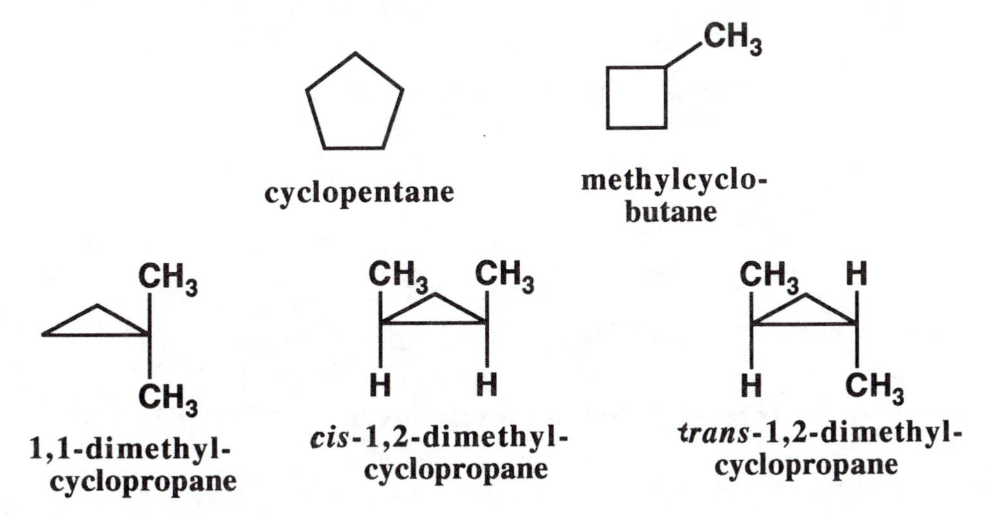

cyclopentane **methylcyclo-butane**

1,1-dimethyl-cyclopropane ***cis*-1,2-dimethyl-cyclopropane** ***trans*-1,2-dimethyl-cyclopropane**

Problem 2.29 Using a planar pentagon representation for the cyclopentane ring, draw structural formulas for the *cis* and *trans* isomers of:

(a) 1,2-dimethylcyclopentane

***cis*-1,2-dimethyl-cyclopentane** ***trans*-1,2-dimethyl-cyclopentane**

(b) 1,3-dimethylcyclopentane.

***cis*-1,3-dimethyl-cyclopentane** ***trans*-1,3-dimethyl-cyclopentane**

2.30 Draw the alternative chair conformations for the *cis* and *trans* isomers of 1,2-dimethylcyclohexane; 1,3-dimethylcyclohexane; and 1,4-dimethylcyclohexane.
(a) Indicate by a label whether each methyl group is axial or equatorial.
(b) For which isomer(s) are the alternative chair conformations of equal stability?
(c) For which isomer(s) is one chair conformation more stable than the other?

***Cis* and *trans* isomers are drawn as pairs. The more stable chair is labeled in cases where there is a difference.**

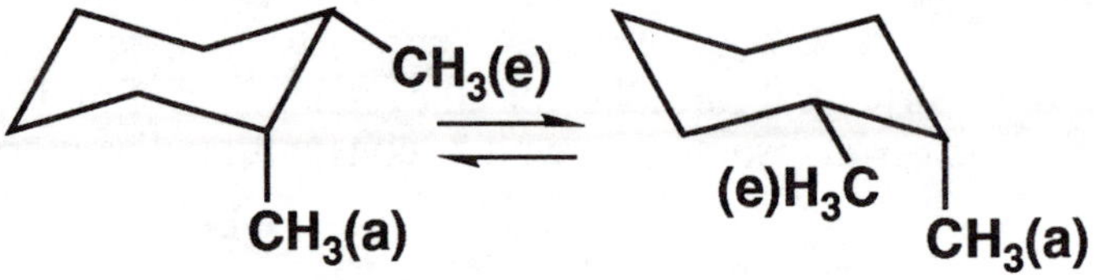

CH_3(e) CH_3(a) ⇌ (e)H_3C CH_3(a)

***cis*-1,2-dimethylcyclohexane**
(chairs of equal stability)

more stable chair

CH_3(e) CH_3(e) ⇌ CH_3(a) CH_3(a)

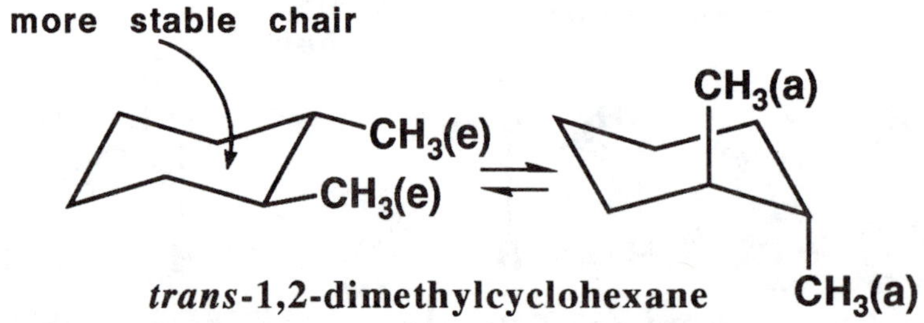

***trans*-1,2-dimethylcyclohexane**

more stable chair

(e)H_3C CH_3(e) ⇌ (a)H_3C CH_3(a)

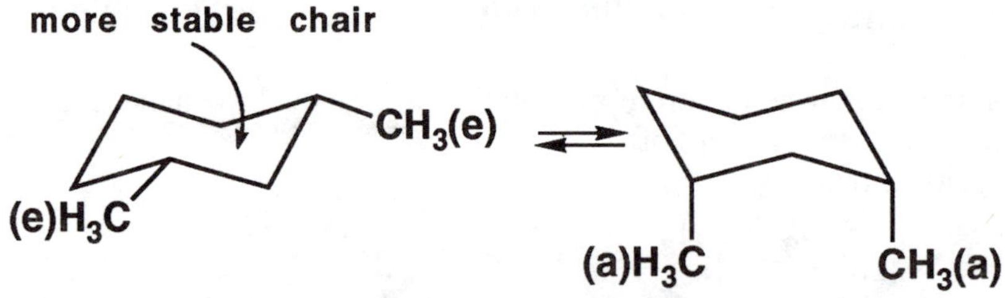

***cis*-1,3-dimethylcyclohexane**

CH_3(a) CH_3(e) ⇌ (e)H_3C CH_3(a)

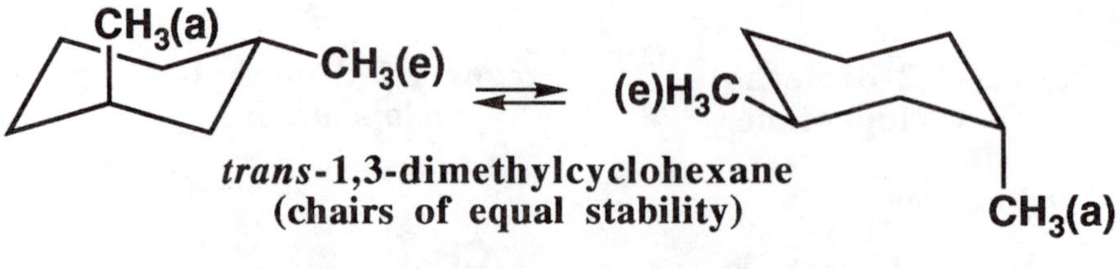

***trans*-1,3-dimethylcyclohexane**
(chairs of equal stability)

(e)H_3C CH_3(a) ⇌ (a)H_3C CH_3(e)

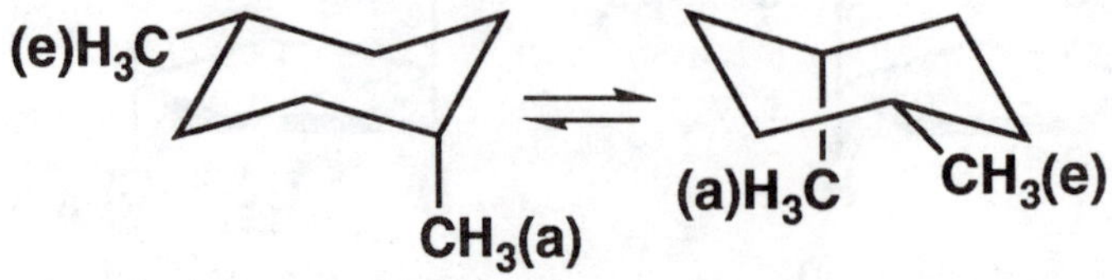

***cis*-1,4-dimethylcyclohexane**
(chairs of equal stability)

more stable chair

(e)H_3C — CH_3(e) ⇌ CH_3(a) / (a)H_3C

***trans*-1,4-dimethylcyclohexane**

Problem 2.31 Complete the following table to show correlations between *cis*, *trans* and axial, equatorial for the disubstituted derivatives of cyclohexane.

These relationships are summarized in the following table.

Position of substitution	Cis	Trans
1,2-	a,e or e,a	e,e or a,a
1,3-	e,e or a,a	a,e or e,a
1,4-	a,e or e,a	e,e or a,a

Problem 2.32 There are four *cis-trans* isomers of 2-isopropyl-5-methylcyclohexanol.

$CH(CH_3)_2$ OH CH_3

2-isopropyl-5-methylcyclohexanol

(a) Using a planar hexagon representation for the cyclohexane ring, draw structural formulas for the four *cis-trans* isomers.
(b) Draw the more stable chair conformation for each of your answers in part (a).
(c) Of the four *cis-trans* isomers, which would you predict to be the most stable? Explain your reasoning. (If you have answered this part correctly, you have picked the isomer found in nature and given the name menthol)

Following are planar hexagon representations for the four *cis-trans* isomers. In each, the isopropyl group is shown by the symbol R-. One way to arrive at these structural formulas is to take one group as a reference and then arrange the other two groups in relation to it. In these drawings, -OH is taken as the reference and placed above the plane of the ring. Once -OH is fixed, there are only two possible arrangements for the isopropyl group on carbon 2; either *cis* or *trans* to -OH. Similarly, there are only two possible arrangements for the methyl group on carbon-5; either *cis* or *trans* to -OH. Note that even if you take another

substituent as a reference, and even if you put the reference below the plane of the ring, there are still only four *cis-trans* isomers for this compound.

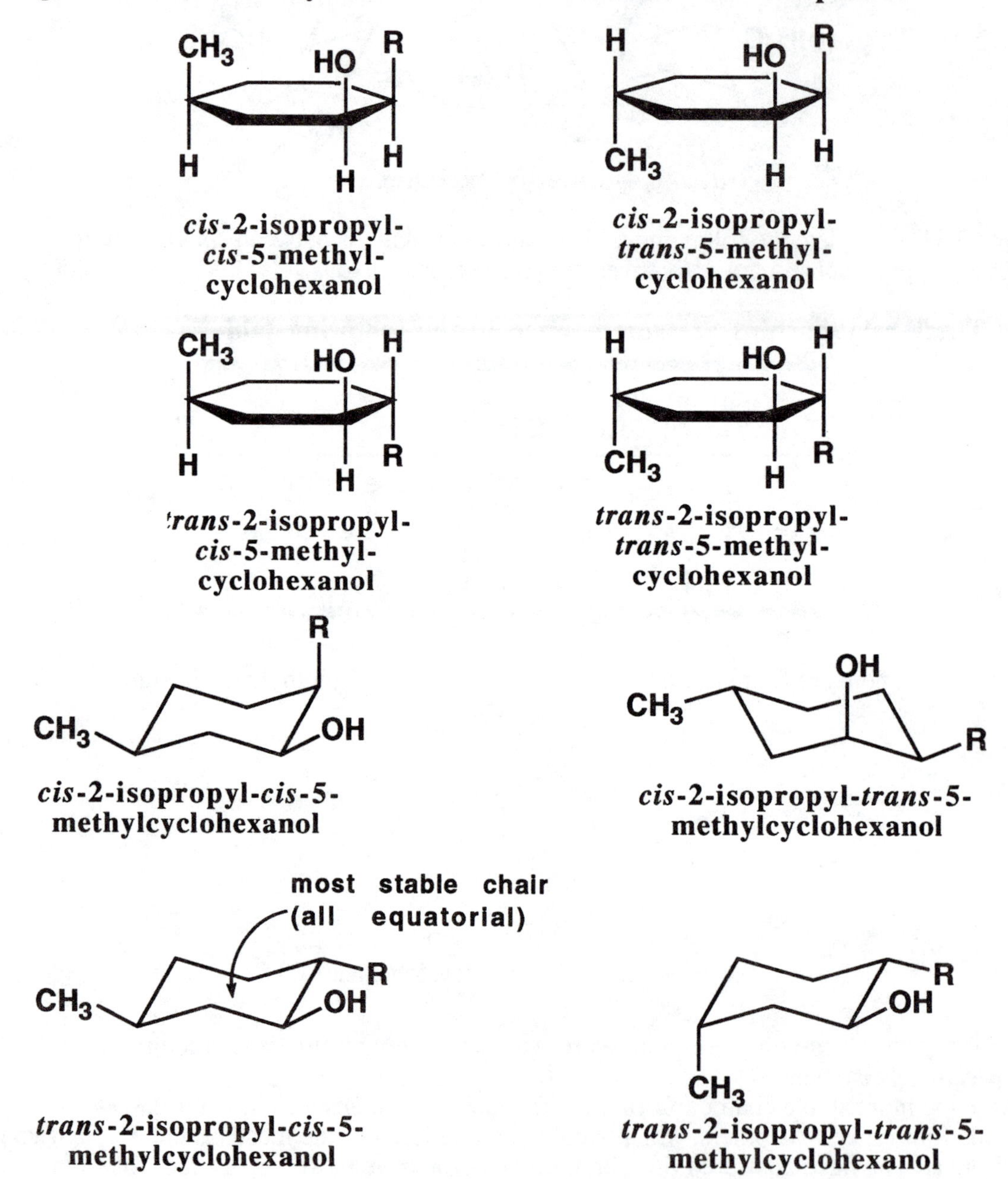

Problem 2.33 Following are planar hexagon representations for several substituted cyclohexanes. For each, draw alternative chair conformations and state which chair is more stable.

(a) HO OH

OH HO

OH (chairs of equal stability) OH

(b) OH OH — OH OH ⇌ HO OH more stable chair

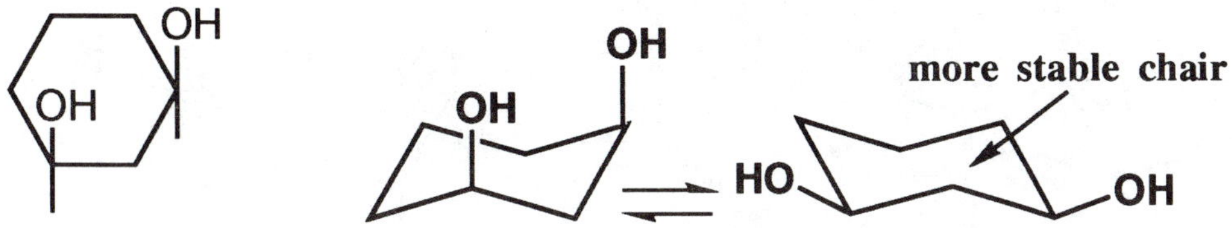

(c) OH CH3 HO — OH CH3 OH ⇌ HO HO CH3 more stable chair

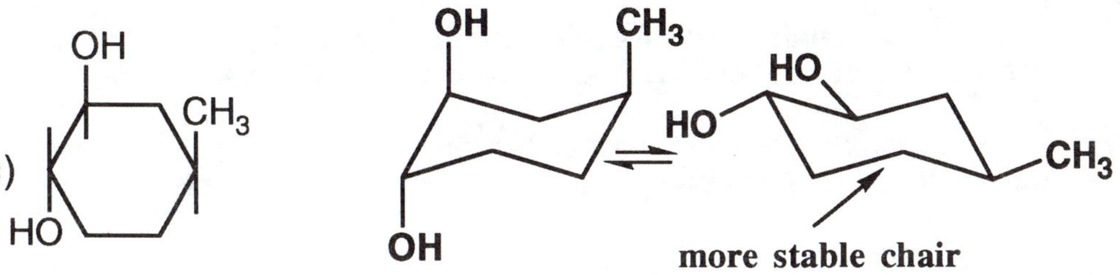

Problem 2.34 1,2,3,4,5,6-Hexachlorocyclohexane shows *cis-trans* isomerism. At one time a crude mixture of these isomers was sold as the insecticide benzene hexachloride (BHC) under the trade names Kwell and Gammexane. The insecticidal properties of the mixture arise from one isomer known as the γ-isomer (gamma-isomer) which is *cis*-1,2,4,5-*trans*-3,6-hexachlorocyclohexane.

(a) Draw a structural formula for 1,2,3,4,5,6-hexachlorocyclohexane disregarding for the moment the existence of *cis-trans* isomerism. What is the molecular formula of this compound?

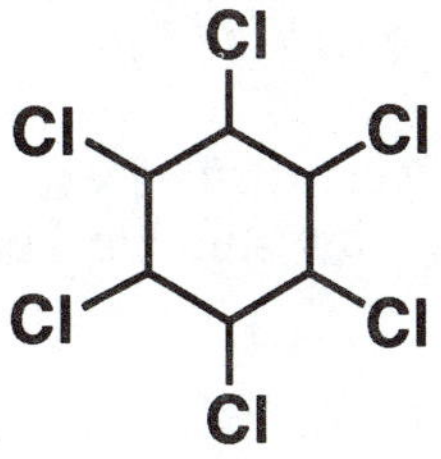

Cl Cl Cl Cl Cl Cl

$C_6H_6Cl_6$

(b) Using a planar hexagon representation for the cyclohexane ring, draw a structural formula for the γ-isomer.

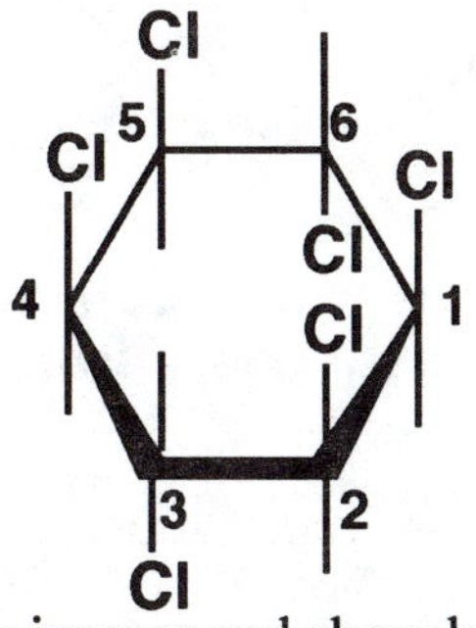

Cl 5 6 Cl Cl Cl Cl 4 1 3 2 Cl

(c) Draw a chair conformation for the γ-isomer and show by labels which chlorine atoms are axial and which are equatorial.

(d) Draw the alternative chair conformation of the γ-isomer and again label which chlorine atoms are axial and which are equatorial.

(e) Which of the alternative chair conformations of the γ-isomer is more stable? Explain.

the two chairs are of equal stability; in each three -Cl atoms are axial and three are equatorial

Problem 2.35 What kinds of conformations do the 6 member rings exhibit in adamantane and twistane? You will find it helpful to build molecular models, particularly of twistane.

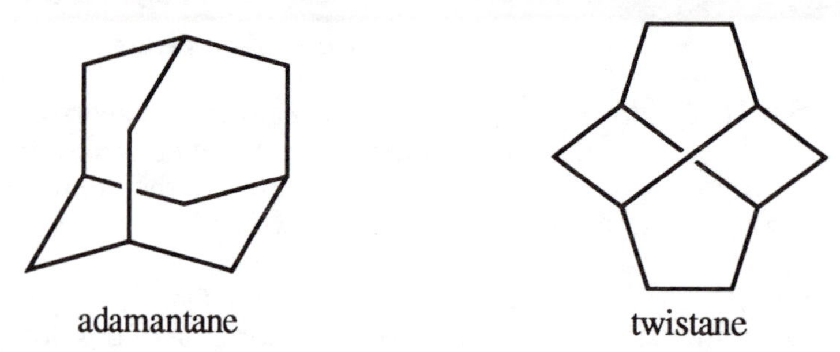

adamantane twistane

In adamantane, the cyclohexane rings all have chair conformations, and in twistane the cyclohexane rings all have twist-boat conformations.

Problem 2.36 Which of the following bicycloalkanes would you expect to show *cis-trans* isomerism? Explain. For each that does, draw suitable stereorepresentations of both *cis* and *trans* isomers.

(a) bicyclo[2.2.2]octane

No cis-trans isomers. It is only possible to fuse a two-carbon bridge to carbons 1 and 4 of a cyclohexane ring if the two-carbon bridge is fused in a *cis* fashion. This molecule is drawn below in two different stereorepresentations.

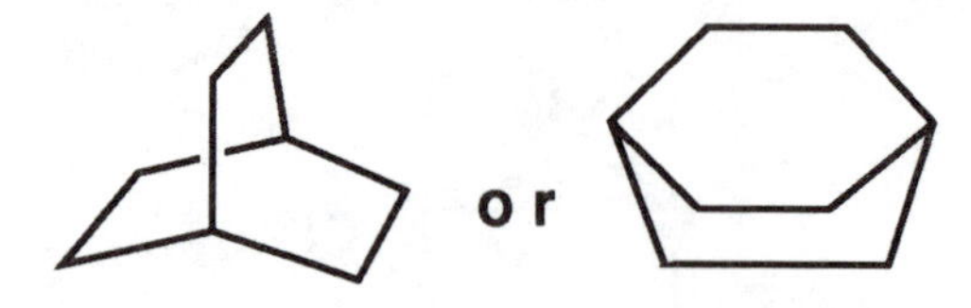

bicyclo[2.2.2]octane (no cis, trans isomers)

(b) bicyclo[4.3.0]nonane

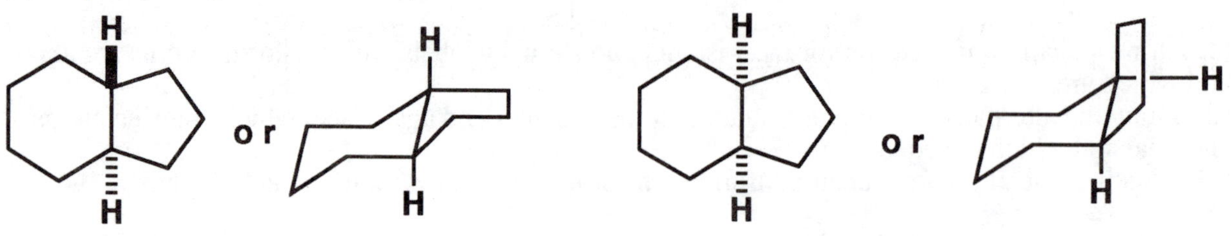

***trans*-bicyclo[4.3.0]nonane** ***cis*-bicyclo[4.3.0]nonane**

(c) 2-methylbicyclo[2.2.1]heptane

2-methylbicyclo[2.2.1]heptane
(a cis and a trans isomer)

(d) 1-chlorobicyclo[2.2.1]heptane

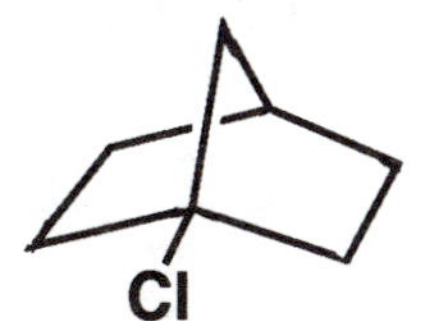

1-chlorobicyclo[2.2.1]heptane
(no cis-trans isomers)

(e) 7-chlorobicyclo[2.2.1]heptane

7-chlorobicyclo[2.2.1]heptane
(no *cis* or *trans* isomers)

Physical Properties

Problem 2.37 In Problem 2.17, you drew structural formulas for all isomeric alkanes of molecular formula C_7H_{16}. Predict which isomer has the lowest boiling point; which has the highest boiling point.

Names and boiling points of these isomers are given in the solution to Problem 2.17 The isomer with the lowest boiling point is 2,2-dimethylpentane, bp 79.2°C. The isomer with the highest boiling point is heptane, bp 94.8°C.

Problem 2.38 What generalizations can you make about the densities of alkanes relative to that of water?

All alkanes that are liquid at room temperature are less dense than water. This is why alkanes such as those in gasoline or petroleum float on water.

Problem 2.39 What unbranched alkane has about the same boiling point as water? (Refer to Table 2.5 on the physical properties of alkanes.) Calculate the molecular weight of this alkane and compare it with that of water.

Heptane, C_7H_{16}, has a boiling point of 98.4°C and a molecular weight of 100. Its molecular weight is approximately 5.5 times that of water. Although considerably smaller, the water molecules are held together by hydrogen bonding while the much larger heptane molecules are held together only by the relatively weak dispersion forces.

Reactions of Alkanes

Problem 2.40 Name and draw structural formulas for all possible monohalogenation products that might be formed in the following reactions.

(a) 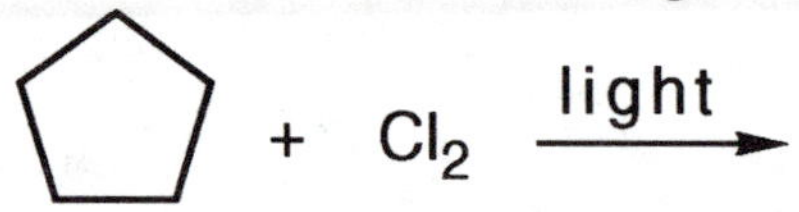 cyclopentane + Cl_2 $\xrightarrow{\text{light}}$

Monochlorination of cyclopentane gives only chlorocyclopentane (cyclopentyl chloride).

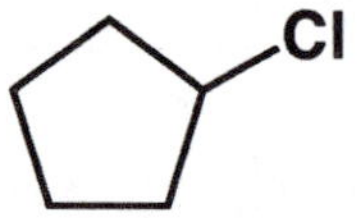

chlorocyclopentane

(b) 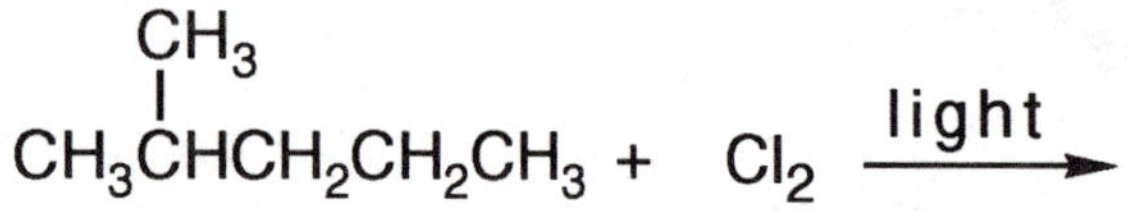 $CH_3CH(CH_3)CH_2CH_2CH_3$ + Cl_2 $\xrightarrow{\text{light}}$

Monochlorination of 2-methylpentane can give a mixture of five constitutional isomers.

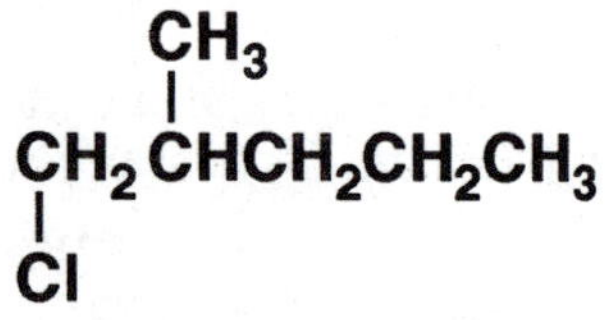

$CH_2(Cl)CH(CH_3)CH_2CH_2CH_3$ — **1-chloro-2-methylpentane**

$CH_3C(CH_3)(Cl)CH_2CH_2CH_3$ — **2-chloro-2-methylpentane**

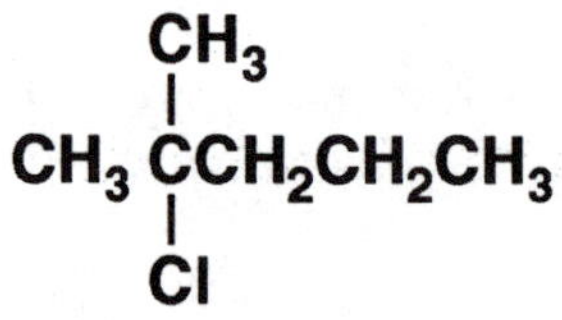

$CH_3CH(CH_3)CH(Cl)CH_2CH_3$ — **3-chloro-2-methylpentane**

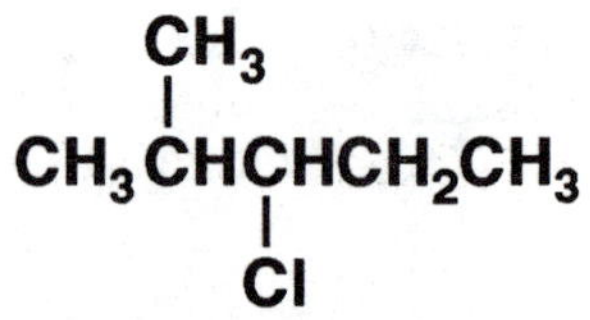

$CH_3CH(CH_3)CH_2CH(Cl)CH_3$ — **2-chloro-4-methylpentane**

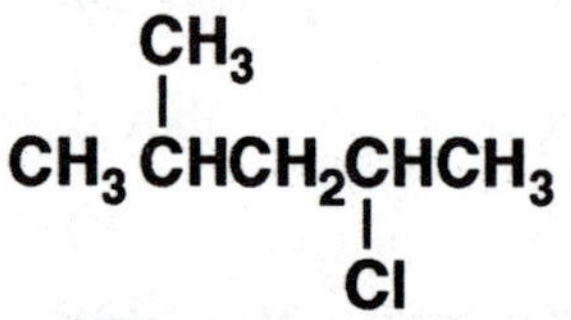

$CH_3CH(CH_3)CH_2CH_2CH_2Cl$ — **1-chloro-4-methylpentane**

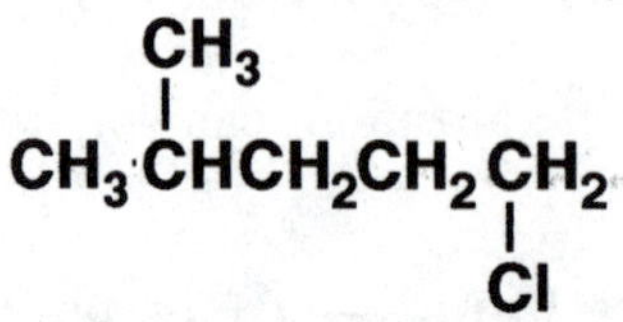

(c) $\underset{\displaystyle CH_3}{\overset{\displaystyle CH_3}{CH_3CHCHCH_3}}$ + Br_2 $\xrightarrow{\text{light}}$

Two monobromination products are possible from 2,3-dimethylbutane.

$H_3C\ CH_3$ / $CH_2CHCHCH_3$ / Br

1-bromo-2,3-dimethyl-butane

H_3CCH_3 / CH_3CCHCH_3 / Br

2-bromo-2,3-dimethyl-butane

(d) + Br_2 $\xrightarrow{\text{light}}$

Only one monobromination product is possible from cyclopropane.

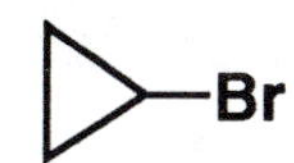

bromocyclopropane

Problem 2.41 Which compounds could be prepared in high yields by regioselective halogenation of an alkane?

(a) 2-chloropentane
(b) chlorocyclobutane
(c) 3-bromo-3,4-dimethylheptane
(d) 1-bromo-1-methylcyclohexane
(e) 2-bromo-2,4,4-trimethylpentane
(f) iodoethane

The compounds that can be prepared in high yields are (b), (d), and (e). In the case of (b), there is only one product possible for monocholorination so the yield will be high. For (d), the desired product is the result of reaction at the only tertiary carbon atom in the methylcyclohexane molecule. Therefore, the large preference of bromination for substitution at tertiary sites will guarantee that the desired product will predominate. For (e), there is also one tertiary carbon atom so yields will be high at that site.
For (a) and (c) there are other products that will be formed in equivalent amounts thus lowering the yield. For (f) iodination simply is not an efficient process.

Problem 2.42 There are three constitutional isomers of molecular formula C_5H_{12}. When treated with chlorine gas at 300°C, isomer A gives a mixture of four monochlorination products. Under the same conditions, isomer B gives a mixture of three monochlorination products and isomer C gives only one monochlorination product. From this information, assign structural formulas to isomers A, B, and C.

Structural formulas for the three alkanes of are:

$$\underset{\text{A}}{CH_3-\underset{|}{\overset{CH_3}{CH}}-CH_2-CH_3} \qquad \underset{\text{B}}{CH_3\text{-}CH_2\text{-}CH_2\text{-}CH_2\text{-}CH_3} \qquad \underset{\text{C}}{CH_3-\underset{CH_3}{\overset{CH_3}{C}}-CH_3}$$

A — **2-methylbutane (isopentane)**

B — **pentane**

C — **2,2-dimethylpropane (neopentane)**

To arrive at the correct assignments of structural formulas, first write formulas for all monochloroalkanes possible from each structural formula. Then compare these numbers with those observed for A, B, and C. Because isomer B gives three monochlorination product, it must be pentane. By the same reasoning, A must be 2-methylbutane, and C must be 2,2-dimethylpropane.

Problem 2.43 Complete and balance the following combustion reactions. Assume that each hydrocarbon is converted completely to carbon dioxide and water.

(a) propane + O_2 $\longrightarrow$

$$CH_3CH_2CH_3 + 5\,O_2 \longrightarrow 3\,CO_2 + 4\,H_2O$$

(b) octane + O_2 $\longrightarrow$

$$2\,CH_3(CH_2)_6CH_3 + 25\,O_2 \longrightarrow 16\,CO_2 + 18\,H_20$$

(c) cyclohexane + O_2 $\longrightarrow$

$$C_6H_{12}\text{ (cyclohexane)} + 9O_2 \longrightarrow 6CO_2 + 6H_2O$$

(c) 2-methylpentane + O_2 $\longrightarrow$

$$2\,CH_3\overset{CH_3}{\overset{|}{C}}HCH_2CH_2CH_3 + 19O_2 \longrightarrow 12CO_2 + 14H_2O$$

Problem 2.44 Following are heats of combustion per mole for methane, propane, and 2,2,4-trimethylpentane. Each is a major source of energy. On a gram-for-gram basis, which of these hydrocarbons is the best source of heat energy.

Hydrocarbon	A major component of	ΔH (kcal/mol)
CH_4	natural gas	-212
$CH_3CH_2CH_3$	LPG	-531
$CH_3C(CH_3)_2CH_2CH(CH_3)CH_3$	gasoline	-1304

On a gram-per-gram basis, methane is the best source of heat energy.

Hydrocarbon	Molecular Weight	Heat of Combustion (kcal/mol)	Heat of Combustion (kcal/gram)
methane	16.04	212	13.3
propane	44.09	531	12.0
2,2,4-trimethylpentane	114.2	1304	11.4

Problem 2.45 Following are heats of combustion for ethane, propane, and pentane at 25°C.

Hydrocarbon	ΔH(kcal/mol)
ethane(g)	-372.8
propane(g)	-530.6
pentane(g)	-845.2

(a) From these data, calculate the average heat of combustion of a methylene ($-CH_2-$) group in a gaseous hydrocarbon.

The average heat of combustion per $-CH_2-$ group is -157.5 kcal/mol. From more extensive data (not given in this problem), the average heat of combustion per methylene group in long chain, unstrained alkanes is -157.4 kcal/mol.

(b) Using the value of the heat of combustion of a methylene group calculated in part (a), estimate the heat of combustion of gaseous cyclopropane.

Using the value from part (a), calculate the heat of combustion of gaseous cyclopropane to be 3 x (-157.4 kcal/mol) = -472.2 kcal/mol.

(c) Compare your estimated value and the experimentally determined value of 499.9 kcal/mol. How might you account for the difference between the two values?

The heat of combustion of cyclopropane is -499.9 - (-472.2) = -27.7 kcal/mol greater. The increased heat of combustion is a measure of steric strain (angle strain and eclipsed hydrogen interactions) present in the cyclopropane molecule.

Problem 2.46 Using the value of -157.4 kcal/mol as the average heat of combustion of a methylene group:
(a) calculate the heat of combustion for the following cycloalkanes.
(b) calculate the total "strain energy" for each cycloalkanes.
(c) calculate the strain energy per methylene group.

Cycloalkane	Calculated heat of combustion (kcal/mol)	Observed heat of combustion (kcal/mol)	Calculated strain energy (kcal/mol)
cyclopropane	________	-499.9	________
cyclobutane	________	-655.9	________
cyclopentane	________	-793.5	________
cyclohexane	________	-944.5	________
cycloheptane	________	-1,108.2	________
cyclooctane	________	-1,269.0	________
cyclononane	________	-1,429.5	________
cyclodecane	________	-1,586.0	________
cycloundecane	________	-1,742.4	________
cyclododecane	________	-1,891.2	________
cyclotetradecane	________	-2,203.6	________

Following is the completed table

Cycloalkane	Calculated Heat of Combustion (kcal/mol)	Observed Heat of Combustion (kcal/mol)	Total Strain Energy (kcal/mol)	Strain Energy Per $-CH_2-$ Group (kcal/mol)
cyclopropane	-472.2	-499.9	27.7	**9.23**
cyclobutane	-629.6	-655.9	26.3	**6.57**
cyclopentane	-787.0	-793.5	6.5	**1.30**
cyclohexane	-944.4	-944.5	0.1	**0.02**
cycloheptane	-1101.8	-1,108.2	6.4	**0.91**
cyclooctane	-1259.2	-1,269.0	9.8	**1.22**
cyclononane	-1416.6	-1,429.5	12.9	**1.43**
cyclodecane	-1574.0	-1,586.0	12.0	**1.20**
cycloundecane	-1731.4	-1,742.4	11.0	**1.00**
cyclododecane	-1888.0	-1,891.2	2.4	**0.20**
cyclotetradecane	-2203.6	-2,203.6	0.0	**0.00**

(d) rank these cycloalkanes in order of most stable to least stable, based on strain energy per methylene group.

The rank order of alkane stabilities listed from most to least stable is as follows: Cyclotetradecane, cyclohexane, cyclododecane, cycloheptane, cycloundecane, cyclodecane, cyclooctane, cyclopentane, cyclononane, cyclobutane, cyclopropane.

Problem 2.47 Hydrocarbons react with strong oxidizing agents. They are quite unreactive, however, toward strong reducing agents such as sodium metal. What do these facts suggest about the energetics of the bonding and antibonding orbitals of hydrocarbons?

As you will learn later, reduction with sodium metal involves transfer of an electron from the sodium to another molecule such as the hydrocarbon. This transferred electron must go into an antibonding orbital, so the higher the energy of this antibonding orbital, the harder to transfer an electron. The low reactivity of hydrocarbons indicates that the antibonding orbitals are relatively high in energy.

$$Na^{\circ} \longrightarrow Na^{+} + e^{-}$$

CHAPTER 3: ACIDS AND BASES

3.0 OVERVIEW

- According to **Svante Arrhenius**, an **acid** is a substance that dissolves in water to increase the concentration of hydrogen ion, H^+. A **base** is a substance that dissolves in water to increase the concentration of hydroxide ion, HO^-.
  - The hydrogen ion, also called a **proton**, is a hydrogen atom stripped of its single electron.

3.1 BRØNSTED-LOWRY ACIDS AND BASES

- According to **Johannes Brønsted** and **Thomas Lowry**, an **acid** is a **proton donor**, and a **base** is a **proton acceptor**. ✱
  - The reaction of an acid with a base thus involves a **proton transfer** from the acid to the base. When an acid transfers a proton to a base, the acid is converted to its **conjugate base**. When a base accepts a proton, the base is converted to its **conjugate acid**.

$$\underset{\text{Acid}}{\text{H-A}} + \underset{\text{Base}}{\text{B}^-} \rightleftharpoons \underset{\text{Conjugate Base}}{\text{A}^-} + \underset{\text{Conjugate Acid}}{\text{H-B}}$$

3.2 QUANTITATIVE MEASURE OF ACID AND BASE STRENGTH

- The strength of an acid or base is a **thermodynamic property**; it is expressed as the position of equilibrium between an acid and a base to form their conjugate acid-conjugate base pairs. ✱
  - A **strong acid** is one that is completely ionized in aqueous solution, in other words there is complete transfer of the proton from the acid to water to form H_3O^+. A **weak acid** is one that is incompletely ionized in aqueous solution. Most organic acids, such as carboxylic acids, are weak acids.
  - A **strong base** is one that is completely ionized in aqueous solution, in other words there is complete proton transfer from H_2O to the base to form HO^-. A **weak base** is one that is incompletely ionized in aqueous solution. Most organic bases are weak bases.
- The strength of a weak acid is described by the acid ionization constant $\mathbf{K_a}$ that is defined by the following equation. ✱

$$K_a = \frac{[H^+][A^-]}{[HA]}$$

Where A^- is the conjugate base of the acid HA.

  - The ionization constants for weak acids have negative exponents, so it is convenient to refer to the $\mathbf{pK_a}$ where $pK_a = -\log_{10}K_a$. In other words, a weak acid with a pK_a of 5.0 has a K_a of 1.0×10^{-5}.
  - Because pK_a is defined as the *negative* log of the K_a, the larger the pK_a, weaker the acid and the smaller the pK_a, the stronger the acid. Note, the strongest acids actually have negative pK_a values.
  - Note that when you are asked to find the pH of an aqueous solution with a known concentration of weak acid, you must set up and solve a quadratic equation. *[This would be a good time to review how to solve quadratic equations.]*

    Let $x = [H^+]$. By definition $[H^+] = [A^-]$, so we can say that $x = [H^+] = [A^-]$. Because the equilibrium concentration of [HA] is equal to [Acid total] - $[A^-]$, then the K_a equation can be written as:

$$K_a = \frac{(x)(x)}{[\text{Acid total} - x]}$$ which can be rearranged to give

$$x^2 + (K_a)(x) - (K_a)([\text{Acid total}]) = 0.$$

This equation can be solved for x using standard algebra, because you will be given the total acid concentration and the pK_a is listed in a table. Be sure to convert the listed pK_a into K_a before trying to solve for x.

Notice that when an acid is very weak, there is not very much ionization so that [HA] = [Acid total] - $[A^-]$ ≈ [Acid total]. In this case the equation simplifies to:

$$x^2 = (K_a)([\text{Acid total}])$$

Thus, the math is significantly easier.

- Because by definition, $pH = -\log_{10}[H^+]$, this can be substituted into the K_a equation and rearranged to give the very useful **Henderson-Hasselbalch equation**:

$$pH = pK_a + \log \frac{[A^-]}{[HA]}$$

- The Henderson-Hasselbalch equation thus relates pK_a to easily measured parameters, and thus provides a useful way to determine pK_a in the laboratory. ✳
- These equations are only valid for weak acids, that is acids that are weaker than H_3O^+ (pK_a of -1.74). Acids stronger than H_3O^+ ionize to form their conjugate base and H_3O^+. Therefore, strong acids appear to have the same acid strength in water, namely that of H_3O^+.
- For the same reasons, any base stronger than HO^- will ionize in water to form its conjugate acid and HO^-. Therefore strong bases will have the same base strength in water, namely that of HO^-.
- In order to measure acid strength of the strong acids, solvents that are themselves stronger acids than water are used. Since these values are measured a different way, they should be used only for qualitative comparisons with the weak acids measured in water. Similarly, bases too strong to be measured in water are measured in solvents more basic than water.
- **Superacids** are extremely reactive mixtures of a protic acid and Lewis acid that have remarkable proton donating power. These are used to protonate species such as alkanes that cannot otherwise be protonated.

3.3 THE POSITION OF EQUILIBRIUM IN ACID-BASE REACTIONS

- The conjugate base of a strong acid is a weak base, and the conjugate base of a weak acid is a strong base. ✳ *[This relationship will make the following rule easier to use]*
- For any proton transfer reaction, the **position of equilibrium** favors the side of the reaction equation that has the weaker acid and weaker base. ✳ *[With this extremely helpful rule, you can predict the outcome of virtually any proton transfer reaction as long as you know the relevant pK_a's.]*

H-A	+	B^-	⇌	A^-	+	H-B
Stronger Acid		Stronger Base		Weaker Base		Weaker Acid

Equilibrium Favors This Side

- The **equilibrium constant (K_{eq})** is the ratio of the K_a for the acid species (H-A) on the left side of the equation as written divided by the K_a for the conjugate acid (H-B) species on the right side of the equation:

$$K_{eq} = \frac{K_a \text{ of acid on left side of equation}}{K_a \text{ of conjugate acid on right side of equation}}$$

- This equation is just a quantitative way to state that the position of equilibrium favors the weaker acid and weaker base. A K_{eq} greater than one favors the species on the right as written, and a K_{eq} less than one favors the species on the left as written.

3.4 LEWIS ACIDS AND BASES

• According to the definition first proposed by G.N. Lewis, a **Lewis Acid** is a species that forms a new covalent bond by accepting a pair of electrons. A **Lewis base** is a species that forms a new covalent bond by donating a pair of electrons. ✳ *[This concept is important, because it describes much more than just proton transfer reactions.]*

- The reaction of a Lewis acid with a Lewis base can be described by the following equation:

$$A + :B \longrightarrow A\text{-}B$$

- Here the "A" represents a Lewis acid and ":B" represents a Lewis base.

• The Lewis definitions of acids and bases are more general than the Brønsted-Lowry definitions: all Brønsted-Lowry acids (proton donors) are also Lewis acids and all Brønsted-Lowry bases (proton acceptors) are also Lewis bases. The Lewis definitions cover reactions other than proton transfer reactions such as diethyl ether reacting with boron trifluoride.

CHAPTER 3
Solutions to the Problems

Problem 3.1 Write the following as proton transfer reactions. Label which reactant is the acid and which the base; which product is the conjugate base and which the conjugate acid. Use curved arrows to show the flow of electrons in each reaction.

(a) $CH_3\text{-}O\text{-}H + NH_4^+ \longrightarrow CH_3\text{-}\overset{+}{O}(H)\text{-}H + NH_3$

$CH_3\text{-}\ddot{O}\text{-}H + H\text{-}\overset{+}{N}H_3 \longrightarrow CH_3\text{-}\overset{+}{O}H_2 + :NH_3$

base **acid** **conjugate acid** **conjugate base**

(b) $CH_3\text{-}O\text{-}H + NH_2^- \longrightarrow CH_3\text{-}O^- + NH_3$

$CH_3\text{-}\ddot{O}\text{-}H + :\ddot{N}H_2^- \longrightarrow CH_3\text{-}\ddot{O}:^- + H\text{-}\ddot{N}(H)\text{-}H$

acid **base** **conjugate base** **conjugate acid**

(c) $CH_3\text{-}S\text{-}H + HCl \longrightarrow CH_3\text{-}\overset{+}{S}H_2 + Cl^-$

$CH_3\text{-}\ddot{S}\text{-}H + H\text{-}\ddot{Cl}: \longrightarrow CH_3\text{-}\overset{+}{S}(H)\text{-}H + :\ddot{Cl}:^-$

base **acid** **conjugate acid** **conjugate base**

Problem 3.2 For each value of [H+], calculate the corresponding value of pH.
(a) blood plasma of $[H^+] = 3.72 \times 10^{-8}$

By definition, pH is the log of the [H+] times -1. Therefore, the pH of this solution is $-1 \times \log(3.72 \times 10^{-8}) = 7.43$.

(b) an acid rain of $[H^+] = 5.89 \times 10^{-5}$

The pH of acid rain is $-1 \times \log(5.89 \times 10^{-5}) = 4.23$.

Problem 3.3 Predict the position of equilibrium and calculate the equilibrium constant for the following reactions.

(a) $CH_3NH_2 + CH_3CO_2H \rightleftharpoons CH_3NH_3^+ + CH_3CO_2^-$

methylamine, acetic acid, methylammonium ion, acetate ion

Acetic acid is the stronger acid; equilibrium lies to the right

$$CH_3NH_2 + CH_3CO_2H \rightleftharpoons CH_3NH_3^+ + CH_3CO_2^-$$

CH_3CO_2H: pK_a 4.76; $CH_3NH_3^+$: pK_a 9.64; $K_{eq} = 7.60 \times 10^4$

(stronger base) (stronger acid) (weaker acid) (weaker base)

(b) $CH_3CH_2O^- + NH_3 \rightleftharpoons CH_3CH_2OH + NH_2^-$

ethoxide ion, ammonia, ethanol, amide ion

Ethanol is the stronger acid; equilibrium lies to the left.

$$CH_3CH_2O^- + NH_3 \rightleftharpoons CH_3CH_2OH + NH_2^-$$

NH_3: pK_a 33; CH_3CH_2OH: pK_a 15.9; $K_{eq} = 7.9 \times 10^{-18}$

(weaker base) (weaker acid) (stronger acid) (stronger base)

Problem 3.4 Write the reaction between each Lewis acid/base pair, showing electron flow by means of curved arrows.

(a) $B(CH_3CH_2)_3 + OH^- \longrightarrow$

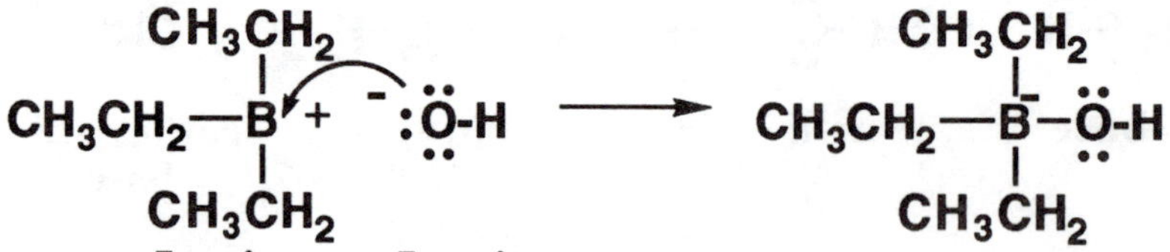

$$(CH_3CH_2)_3B + {}^-{:}\ddot{O}\text{-}H \longrightarrow (CH_3CH_2)_3\overset{-}{B}\text{-}\ddot{O}\text{-}H$$

Lewis acid ($B(CH_3CH_2)_3$), **Lewis base** (OH^-)

(b) $Br^- + AlBr_3 \longrightarrow$

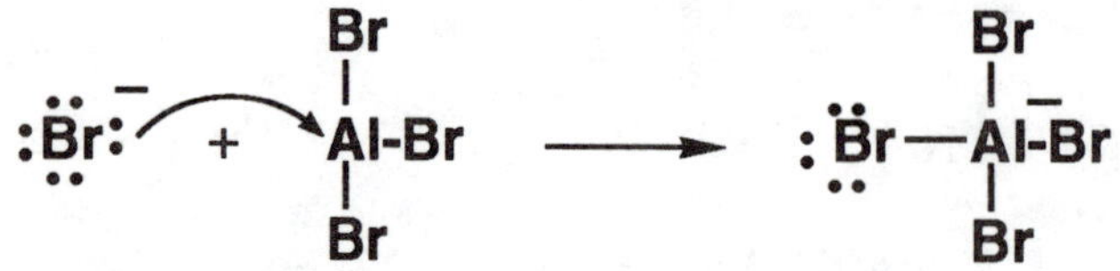

$${:}\ddot{Br}{:}^- + AlBr_3 \longrightarrow {:}\ddot{Br}\text{—}\overset{-}{Al}Br_3$$

Lewis base (Br^-), **Lewis acid** ($AlBr_3$)

Brønsted-Lowry Acids and Bases

Problem 3.5 Complete the following proton-transfer reactions using curved arrows to show the flow of electron pairs in each reaction. In addition, write Lewis structures for all starting materials and products. If you are uncertain about which substance in each equation is the proton donor, refer to Table 3.2 for the relative strengths of proton acids.

(a) $CH_3COOH + OH^- \longrightarrow$

$CH_3-C(=O)-\ddot{O}-H + {}^-:\ddot{O}-H \longrightarrow CH_3-C(=\ddot{O})-\ddot{O}:^- + H-\ddot{O}-H$

acid, base, conjugate base, conjugate acid

(b) $CH_3COOH + CH_3O^- \longrightarrow$

$CH_3-C(=\ddot{O})-\ddot{O}-H + {}^-:\ddot{O}-CH_3 \longrightarrow CH_3-C(=\ddot{O})-\ddot{O}:^- + H-\ddot{O}-CH_3$

acid, base, conjugate base, conjugate acid

(c) $NH_4^+ + OH^- \longrightarrow$

$H_3N^+-H + {}^-:\ddot{O}-H \longrightarrow H_3N: + H-\ddot{O}-H$

acid, base, conjugate base, conjugate acid

(d) $NH_3 + HCl \longrightarrow$

$H_3N: + H-\ddot{Cl}: \longrightarrow H-N^+H_3 + :\ddot{Cl}:^-$

base, acid, conjugate acid, conjugate base

(e) $CH_3\overset{O}{\overset{\|}{C}}O^- + CH_3NH_3^+ \longrightarrow$

$CH_3-C(=O)-O^-$ (base) + $H-N^+H_2-CH_3$ (acid) $\longrightarrow$ $CH_3-C(=O)-O-H$ (conjugate acid) + $:NH_2-CH_3$ (conjugate base)

(f) $CH_3CH_2OH + HCl \longrightarrow$

$CH_3-CH_2-\ddot{O}-H$ (base) + $H-Cl$ (acid) $\longrightarrow$ $CH_3-CH_2-O^+H_2$ (conjugate acid) + $:Cl:^-$ (conjugate base)

(g) $CH_3CH_2OCH_2CH_3 + HCl \longrightarrow$

$CH_3-CH_2-\ddot{O}-CH_2-CH_3$ (base) + $H-Cl$ (acid) $\longrightarrow$ $CH_3-CH_2-O^+(H)-CH_2-CH_3$ (conjugate acid) + $:Cl:^-$ (conjugate base)

(h) $CH_3CH_2^- + H_2O \longrightarrow$

$CH_3-CH_2:^-$ (base) + $H-\ddot{O}-H$ (acid) $\longrightarrow$ CH_3-CH_2-H (conjugate acid) + $^-:\ddot{O}-H$ (conjugate base)

Problem 3.6 For each of the proton-transfer reactions in Problem 3.5, label which starting material is the acid and which product is its conjugate base; which starting material is the base and which product is its conjugate acid.

See the answers to Problem 3.5 for the labels.

Problem 3.7 Each of the following molecules and ions can function as a base. Write the structural formula of the conjugate acid formed by reaction of each with H^+.

(a) CH_3CH_2OH

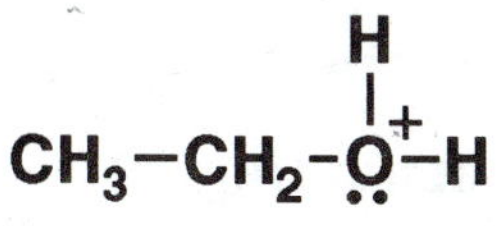

$CH_3-CH_2-\ddot{O}^+(H)-H$

(b)  HCH (with C=O)

$H-C(=\ddot{O}^+-H)-H$

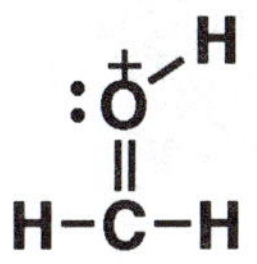

(c) $(CH_3)_2NH$

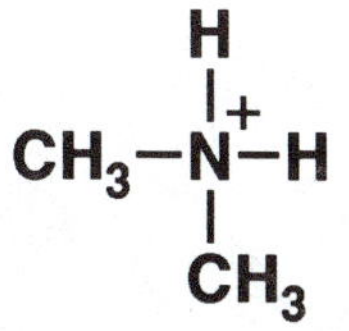

$CH_3-N^+(H)(CH_3)-H$

(d) HCO_3^-

$H-\ddot{O}-C(=O)-\ddot{O}-H$

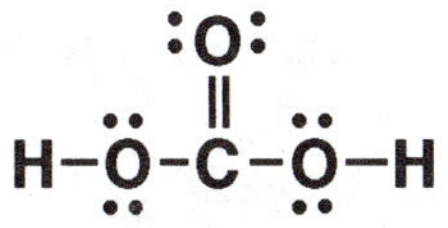

Problem 3.8 In acetic acid, the O-H proton is more acidic than the H_3C protons. Show how the concept of electronegativity can be used to account for this difference in acidity.

In general terms, a proton is more acidic when it is attached to a more electronegative atom. Oxygen is more electronegative than carbon. Furthermore, when the proton is removed from the oxygen atom, a negatively charged oxygen results. This negative charge can be shared with the adjacent oxygen atom of the carboxylic acid via the contributing structures shown below:

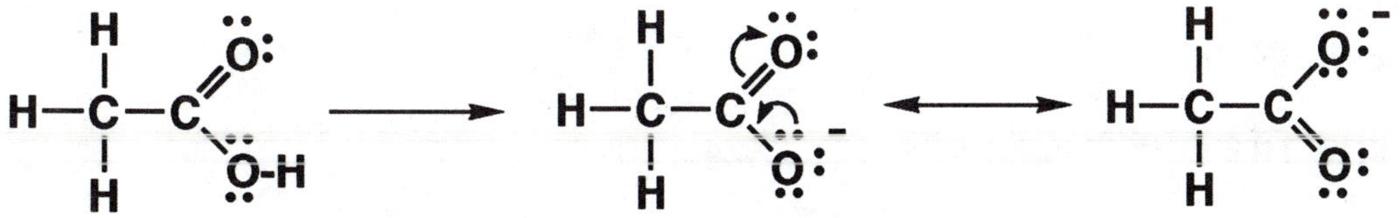

$H_3C-C(=O)-\ddot{O}-H \longrightarrow H_3C-C(=O)-O^- \longleftrightarrow H_3C-C(-O^-)=O$

resonance forms

In both contributing structures, the negative charge is on the more electronegative atom. This resonance stabilization is responsible for the acidity of acetic acid. No such resonance stabilization is possible with the C-H bonds in this molecule.

Problem 3.9 Using the concepts of resonance and electronegativity as appropriate, offer an explanation for the following observations.

(a) H_3O^+ is a stronger acid than NH_4^+.

Oxygen is more electronegative than nitrogen, so the proton is more acidic on H_3O^+.

(b) acetic acid is a stronger acid than ethanol.

The deprotonated acetic acid is stabilized by resonance as described in problem 3.8. There is no such resonance stabilization possible with ethanol.

(c) ethanol and water have approximately the same acidity.

Both have the proton attached to oxygen atoms, and neither deprotonated species can be resonance stabilized.

(d) hydrocyanic acid is a stronger acid than acetylene.

The cyanide anion can be stabilized somewhat be the following contributing structures:

$$H{-}C{\equiv}N: \longrightarrow {}^{-}:C{\equiv}N: \longleftrightarrow :C{=}\ddot{N}^{-}$$

resonance forms

There is no such resonance stabilization possible with acetylene anion.

Quantitative Measure of Acid and Base Strength

Problem 3.10 In each pair, select the stronger acid:

(a) pyruvic acid (pK_a 2.49) or lactic acid (pK_a 3.85)

The stronger acid is the one with the smaller pK_a; the larger value of K_a. Therefore, pyruvic acid is the stronger acid.

(b) citric acid (pK_{a1} 3.08) or phosphoric acid (pK_{a1} 2.10)

phosphoric acid

(c) nicotinic acid (niacin, K_a 1.4×10^{-5}) or acetylsalicylic acid (aspirin, K_a 3.3×10^{-4})

acetylsalicylic acid

(d) phenol (K_a 1.12×10^{-10}) or boric acid (K_a 7.24×10^{-10})

boric acid

Quantitative Position of Equilibrium in Acid/Base Reactions

Problem 3.11 Estimate the value of K_{eq} for the following equilibria and state which lie considerably toward the left, which lie considerably toward the right. The value of pK_a for HCN is 9.31. The values of pK_a for other acids in this problem can be found in table 3.2.

(a) $CH_3CO_2H + HCO_3^- \rightleftharpoons CH_3CO_2^- + H_2CO_3$

acetic acid, bicarbonate ion, acetate ion, carbonic acid

$$CH_3CO_2H + HCO_3^- \rightleftharpoons CH_3CO_2^- + H_2CO_3 \qquad K_{eq} = 4.0 \times 10^1$$

pK_a 4.76 (stronger acid) ... **pK_a 6.36 (weaker acid)**

(b) CH_3CO_2H + NH_3 ⇌ $CH_3CO_2^-$ + NH_4^+

acetic acid, ammonia, acetate ion, ammonium ion

CH_3CO_2H + NH_3 ⇌ $CH_3CO_2^-$ + NH_4^+ $K_{eq} = 3.0 \times 10^4$

CH_3CO_2H: pK_a 4.76 (stronger acid)

NH_4^+: pK_a 9.24 (weaker acid)

(c) C_6H_5OH + CO_3^{2-} ⇌ $C_6H_5O^-$ + HCO_3^-

phenol, carbonate ion, phenoxide ion, bicarbonate ion

C_6H_5OH + CO_3^{2-} ⇌ $C_6H_5O^-$ + HCO_3^- $K_{eq} = 2.4$

C_6H_5OH: pK_a 9.95 (stronger acid)

HCO_3^-: pK_a 10.33 (weaker acid)

(d) CN^- + HCO_3^- ⇌ HCN + CO_3^{2-}

cyanide ion, bicarbonate ion, hydrocyanic acid, carbonate ion

CN^- + HCO_3^- ⇌ HCN + CO_3^{2-} $K_{eq} = 9.6 \times 10^{-2}$

HCO_3^-: pK_a 10.33 (weaker acid)

HCN: pK_a 9.31 (stronger acid)

(e) $C_6H_5CO_2^-$ + H_2O ⇌ $C_6H_5CO_2H$ + OH^-

benzoate ion, water, benzoic acid, hydroxide ion

$C_6H_5CO_2^-$ + H_2O ⇌ $C_6H_5CO_2H$ + OH^- $K_{eq} = 3.10 \times 10^{-12}$

H_2O: pK_a 15.7 (weaker acid)

$C_6H_5CO_2H$: pK_a 4.19 (stronger acid)

Problem 3.12 For an acid-base reaction, one way to determine the predominate species at equilibrium is to say that the reaction arrow points to the acid with the higher value of pK_a. For example:

$$HCN + H_2O \longleftarrow CN^- + H_3O^+$$
pK$_a$ 9.31 pK$_a$ -1.74

$$HCN + OH^- \longrightarrow CN^- + H_2O$$
pK$_a$ 9.31 pK$_a$ 15.7

Explain why this rule works.

The position of equilibrium favors the side that has the weaker acid and weaker base. The weaker acid has a higher pK$_a$.

Problem 3.13 Using pK$_a$ values given in Table 3.2, predict the position of equilibrium in the following acid-base reactions and calculate the K_{eq} for each reaction.

(a) $PO_4^{3-} + CH_3CO_2H \rightleftharpoons HPO_4^{2-} + CH_3CO_2^-$

To determine the equilibrium constant K_{eq}, you use the K_a for the strongest acid on each side of the equation. Divide the K_a for the acid on the left by the K_a for conjugate acid on the right.

$$PO_4^{3-} + CH_3CO_2H \rightleftharpoons HPO_4^{2-} + CH_3CO_2^-$$

pK$_a$ = 4.76 **pK$_a$ = 12.32**

$K_a = 1.74 \times 10^{-5}$ **$K_a = 4.79 \times 10^{-13}$**

$$K_{eq} = \frac{1.74 \times 10^{-5}}{4.79 \times 10^{-13}} = 3.6 \times 10^{7}$$

(b) $NH_4^+ + HCO_3^- \rightleftharpoons NH_3 + H_2CO_3$

$$NH_4^+ + HCO_3^- \rightleftharpoons NH_3 + H_2CO_3 \qquad K_{eq} = 1.3 \times 10^{-3}$$

pK$_a$ = 9.24 **pK$_a$ = 6.36**

$K_a = 5.75 \times 10^{-10}$ **$K_a = 4.37 \times 10^{-7}$**

(c) $NH_4^+ + OH^- \rightleftharpoons NH_3 + H_2O$

$$NH_4^+ + OH^- \rightleftharpoons NH_3 + H_2O \qquad K_{eq} = 2.9 \times 10^{6}$$

pK$_a$ = 9.24 **pK$_a$ = 15.7**

$K_a = 5.75 \times 10^{-10}$ **$K_a = 2.0 \times 10^{-16}$**

(d) $H_3PO_4 + CH_3CH_2OH \rightleftharpoons H_2PO_4^- + CH_3CH_2OH_2^+$

$H_3PO_4 + CH_3CH_2OH \rightleftharpoons H_2PO_4^- + CH_3CH_2OH_2^+$ $K_{eq} = 2.0 \times 10^{-6}$

$pK_a = 2.1$ $pK_a = -3.6$

$K_a = 7.9 \times 10^{-3}$ $K_a = 4.0 \times 10^{3}$

Problem 3.14 Calculate the ratio of acetic acid to sodium acetate concentrations needed to prepare a solution of pH 4.50.

At pH 4.50 the Henderson-Hasselbalch equation becomes 4.50 = 4.76 + log (x). Solving for x gives:

$$x = \frac{[A^-]}{[HA]} = 0.55$$

But the question asked for the ratio of acetic acid to sodium acetate so the final answer is:

$$\frac{[HA]}{[A^-]} = \frac{1}{0.55} = 1.8$$

Lewis Acids and Bases

Problem 3.15 Complete the following reactions between Lewis acid/Lewis base pairs. Label which starting material is the Lewis acid, which the Lewis base, and use curved arrows to show the flow of electrons in each reaction. Note that in doing these problems, it is essential that you show all valence electrons for at least the atoms participating directly in each reaction.

(a) $(CH_3)_2C(CH_3)-\ddot{Cl}:$ + $AlCl_3$ ⟶ $(CH_3)_3C-\ddot{Cl}^{+}-\overset{-}{Al}Cl_3$

Lewis Base (alkyl chloride); Lewis Acid ($AlCl_3$)

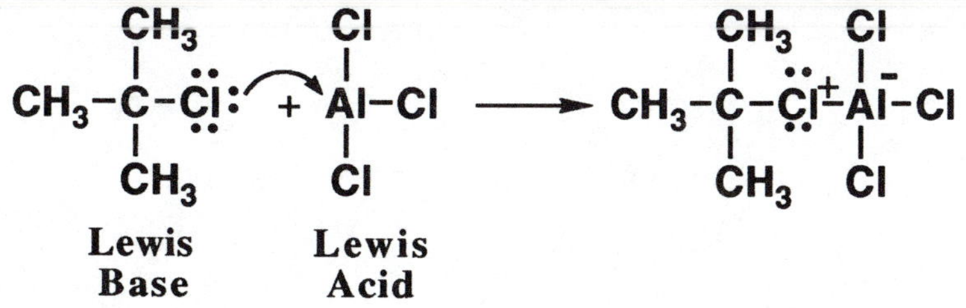

(b) $CH_3-\overset{+}{C}(CH_3)-CH_3$ + $H-\ddot{O}-H$ ⟶ $(CH_3)_3C-\overset{+}{O}H_2$

Lewis Acid (carbocation); Lewis Base (water)

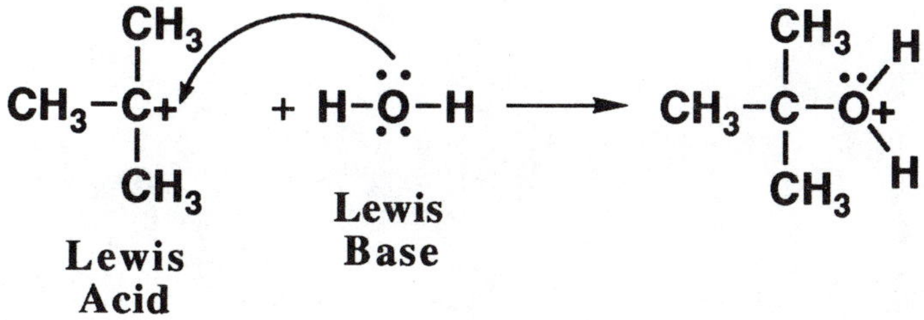

(c) 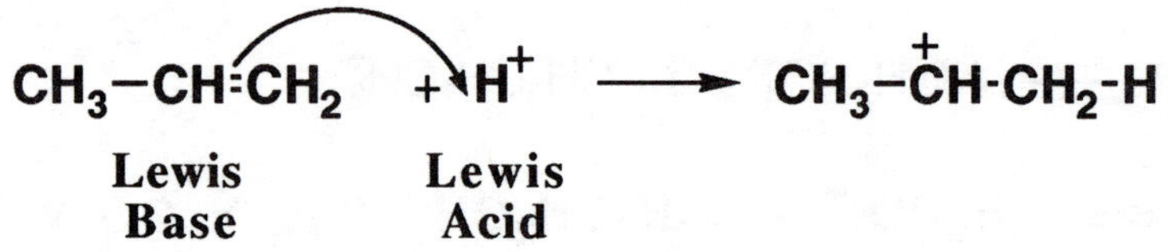 $CH_3-CH=CH_2 + H^+ \longrightarrow CH_3-\overset{+}{C}H-CH_2-H$

Lewis Base, Lewis Acid

(d) 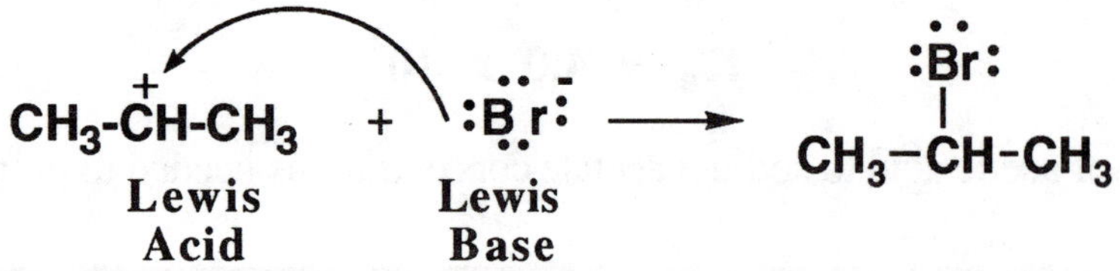 $CH_3-\overset{+}{C}H-CH_3 + :\ddot{B}r:^- \longrightarrow CH_3-CH(Br)-CH_3$

Lewis Acid, Lewis Base

(e) 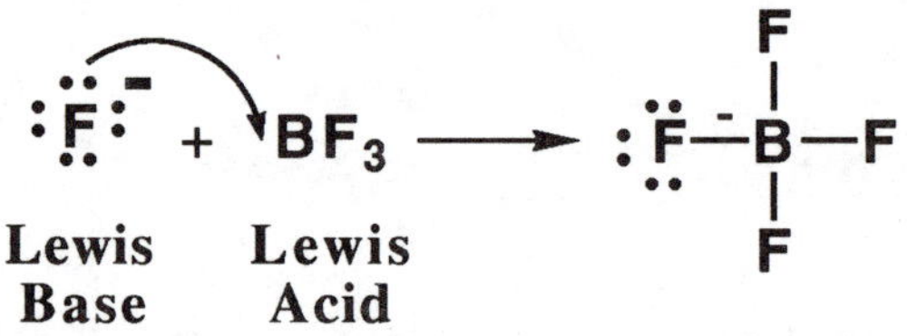 $:\ddot{F}:^- + BF_3 \longrightarrow :\ddot{F}-\overset{-}{B}F_3$

Lewis Base, Lewis Acid

(f) 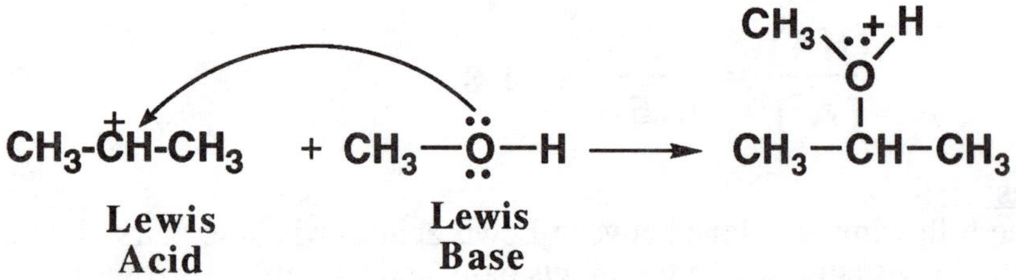 $CH_3-\overset{+}{C}H-CH_3 + CH_3-\ddot{O}-H \longrightarrow CH_3-CH(\overset{+}{O}(CH_3)H)-CH_3$

Lewis Acid, Lewis Base

CHAPTER 4: ALKENES I

SUMMARY OF REACTIONS

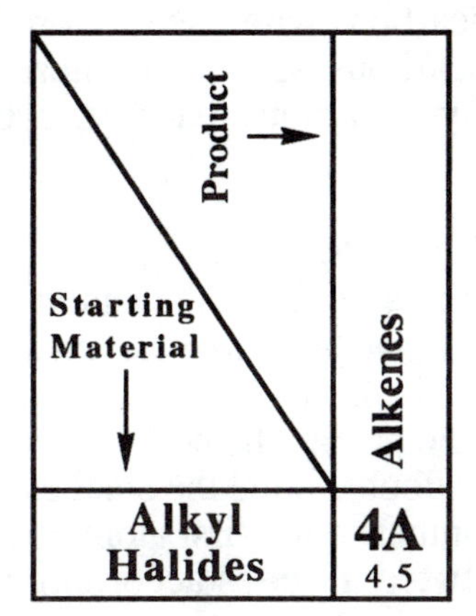

REACTION 4A: DEHYDROHALOGENATION OF HALOALKANES (Section 4.5)

$$\text{X–C–C–H} \xrightarrow{HO^-} \text{C=C}$$

X = halogen

- Alkenes are often prepared from haloalkanes with a β-**elimination** reaction using base. This process is referred to as **dehydrohalogenation** because **H-halogen** is removed from a molecule to leave an alkene.
- Dehydrohalogenation involves the removal of a halogen atom and a hydrogen atom from the carbon atom adjacent to the one attached to the halogen atom. The hydrogen is considered to be on the β carbon atom, hence the name, β-elimination.
- When there is the possibility of forming more than one alkene (hydrogen atoms attached to two or more different β carbon atoms to choose from), **Zaitsev's rule** states that the predominant product is most substituted alkene. The explanation for this will be presented later.

$$(CH_3)_2CH\text{–}CHBr\text{–}CH_3 \xrightarrow[\text{heat}]{KOH} (H_3C)_2C{=}CH(CH_3) + (H_3C)_2CH\text{–}CH{=}CH_2 + KBr + H_2O$$

Major Product (3 alkyl substituents) Minor Product (1 alkyl substituent)

SUMMARY OF IMPORTANT CONCEPTS

4.0 OVERVIEW

- **Unsaturated hydrocarbons** are hydrocarbons that contain one or more carbon-carbon double or triple bond.
  - **Alkenes** are unsaturated hydrocarbons with one or more carbon-carbon double bond.
  - **Alkynes** are unsaturated hydrocarbons with one or more carbon-carbon triple bond.
  - **Aromatic hydrocarbons** are hydrocarbons that have cyclic structures with special patterns of alternating double bonds.

4.1 NOMENCLATURE OF ALKENES

- Form IUPAC names by changing the **an** infix of the parent alkane to the infix **en**. For example ethene and propene are alkenes with two and three carbon atoms respectively. ✱
  - Form names for more complicated alkenes by choosing the longest carbon chain with the carbon-carbon double bond as the parent alkane.

    Number the chain to give the double bond the smallest number, indicate the position of the double bond by using the number of the first atom of the double bond.

    Name branched or **substituted alkenes** according to the same rules discussed for alkanes.

    For **cycloalkenes**, number the carbon atoms of the double bond 1 and 2 in the direction that gives the substituent encountered first the smallest number.
  - Alkenes that contain more than one carbon-carbon double bond are called **dienes** or **trienes**, and so forth.
  - **Unconjugated double bonds** are separated by at least one sp^3 hybridized atom. (, for example).
  - **Conjugated double bonds** are on adjacent pairs of atoms, that is at least two adjacent double bonds with no sp^3 hybridized atoms between them. (, for example).
  - A molecule has **cumulated double bonds** if two double bonds are on the same carbon atom ($CH_2=C=CH_2$, for example).
  - Several alkenes are known by their common names including ethylene, propylene, isobutylene, and butadiene. Substituents are also given common names such as methylene, vinyl, and allyl. *[These must be learned because their use is so widespread.]*

4.2 STRUCTURE OF ALKENES

- According to the valence bond model, a carbon-carbon double bond consists of **one sigma bond** formed by the overlap of sp^2 hybridized orbitals of adjacent carbon atoms and **one pi bond** formed by the overlap of unhybridized 2p orbitals. ✱ *[This picture of the orbitals involved with the carbon-carbon double bond is crucial to your understanding of the reactions and properties of alkenes, so make sure you understand Figure 4.1 in the book.]*
  - The most important implication of this model of carbon-carbon double bonds is that they **can not rotate** because rotation would decrease the extent of 2p-2p overlap. ✱
  - Each carbon atom in a carbon-carbon double bond is sp^2 hybridized so its geometry is trigonal planar with bond angles near 120°. Notice that this means the carbon atoms and the atoms attached directly to them are in the same plane.
  - Carbon-carbon triple bonds are shorter than carbon-carbon double bonds, which are shorter than carbon-carbon single bonds. This is because electrons in the sp orbitals that overlap to form the sigma portion of triple bonds are held closest to the nuclei, because they have a higher percentage of s character. By the same logic, electrons in the sp^2 orbitals that overlap to form the sigma portion of double bonds are held closer to the nuclei, because they have a higher percentage s character than sp^3 orbitals.

- Not surprisingly, carbon-carbon triple bonds are stronger than carbon-carbon double bonds which are stronger than carbon-carbon single bonds.

• According to the molecular orbital model, the ethene molecule has two parallel 2p atomic orbitals on adjacent carbon atoms that overlap to form a filled bonding molecular orbital and an unfilled antibonding molecular orbital.

4.3 *CIS-TRANS* ISOMERISM IN ALKENES

• **Stereoisomers** are molecules that have the same molecular formula, the same connectivity of atoms, but a different arrangement of atoms in space. ✱ *[Stereoisomerism is one of the most important three-dimensional concepts in chemistry, and various types of stereoisomers will be presented throughout the book. The use of molecular models is always helpful in discussions of stereoisomerism.]*

• In cycloalkanes, ***cis-trans*** isomerism depends on the arrangement of substituents with respect to the mean plane of the ring (Section 2.7).

• Because carbon-carbon double bonds can not rotate, alkenes can also display ***cis-trans* isomerization.**

• For alkenes that have one substituent and one hydrogen atom on each double-bonded carbon:

- A ***cis*** alkene is one in which the substituents are on the **same side** of the double bond. For example, *cis*-2-butene. ✱

$$\begin{matrix} H_3C & & CH_3 \\ & C{=}C & \\ H & & H \end{matrix}$$

cis-2-butene

- A ***trans*** alkene is one in which the substituents are on **opposite sides** of the double bond. For example, *trans*-2-butene. ✱

$$\begin{matrix} H_3C & & H \\ & C{=}C & \\ H & & CH_3 \end{matrix}$$

trans-2-butene

• The **E,Z system** developed by Cahn, Ingold, and Prelog is a more comprehensive system of nomenclature.

• The **E,Z system** uses priority rules to rank the two substituents on each carbon atom of the double bond:

- Each atom is assigned a priority based on atomic number; the higher the atomic number the higher the priority.

- If you cannot assign priority differences to the two substituents by comparing the first atoms, then continue down the chains until the first point of difference is reached. *[This can be confusing. Notice the total size of the substituents attached to the double bond is not important, it is the <u>first point of difference in priority</u> that matters. For example, a -CH_2Cl group has a higher priority than $CH_2CH_2CH_2CH_2CH_3$ because the first point of difference, Cl, has a higher priority than any atom on the first carbon of the larger alkyl group.]*

- In the case of double and triple bonds, count the atoms participating in the double or triple bond as if they are bonded to an equivalent number of similar atoms. For example, a -$CH{=}CH_2$ group is counted as $\begin{matrix} C & C \\ -C & -C-H \\ H & H \end{matrix}$ and a -$C{\equiv}CH$ group is counted as $\begin{matrix} C & C \\ -C & -C-H \\ C & C \end{matrix}$. *[The only way to get good at this is to practice.]*

- To assign an alkene as **E** or **Z**, use the following rules:

If the atoms or groups of atoms of higher priority are on the **same side** of the double bond, it is a **Z alkene**. It is easy to remember this as Z for "<u>z</u>ame <u>z</u>ide."

If the higher priority substituents are on **opposite** sides, it is an **E alkene**. *[Caution! Because the letter Z has a zig-zag shape, some students find it tempting to assume that the*

Z stands for higher priority groups on opposite sides of the alkene making a zig-zag shape around the double bond. This is not the way to assign structures, because E is used for higher priority groups on opposite sides of alkenes.]

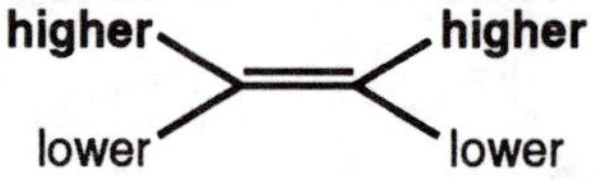

higher / higher / lower / lower — **Z** ("zame zide")

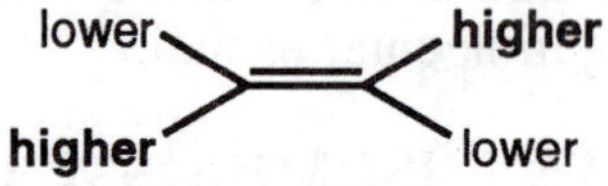

lower / higher / higher / lower — **E**

For alkenes with more than one double bond, each double bond is named as E or Z as applicable.

- Only cycloalkenes with 8 or more carbon atoms in the ring can have a trans geometry, otherwise the angle strain will only allow for *cis* double bonds.
- For alkenes with **n** double bonds, there are up to 2^n possible *cis-trans* isomers. There will be fewer if the molecule contains any symmetry.

4.4 PHYSICAL PROPERTIES OF ALKENES:

- Because alkenes are nonpolar, the only interactions between alkene molecules are dispersion forces, thus their properties are very similar to those of alkanes.

4.6 NATURALLY OCCURRING ALKENES-TERPENE HYDROCARBONS

- Alkenes are common in nature, and comprise a very important set of biological molecules. **Terpenes** are one group of biological alkene molecules that have some interesting features. **Terpenes** are based on carbon skeletons that can be divided into two or more units that have the same carbon skeleton of **isoprene**.

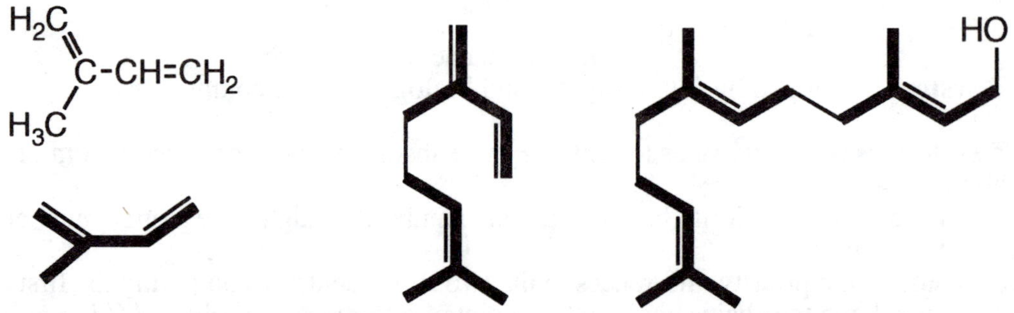

$H_2C=C(CH_3)-CH=CH_2$

HO

Isoprene Myrcene Farnesol

(The isoprene units of these terpenes are shown in bold)

- In nature, terpenes are not synthesized from isoprene, but from the pyrophosphate ester of 3-methyl-3-butene-1-ol.

CHAPTER 4
Solutions to the Problems

Problem 4.1 Draw structural formulas for all alkenes of molecular formula C_6H_{12} that have the following carbon skeletons. Give each alkene an IUPAC name.

(a) 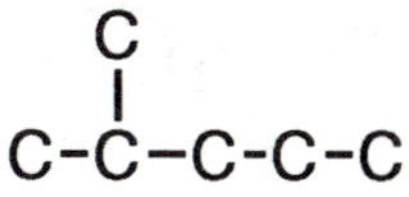 C-C(C)-C-C-C

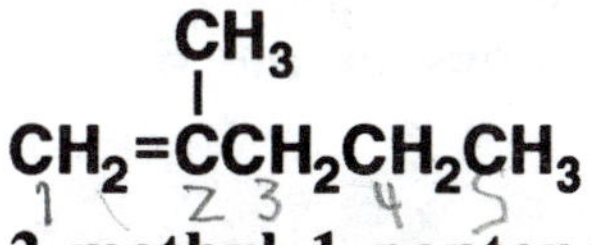

$CH_2{=}C(CH_3)CH_2CH_2CH_3$ — **2-methyl-1-pentene**

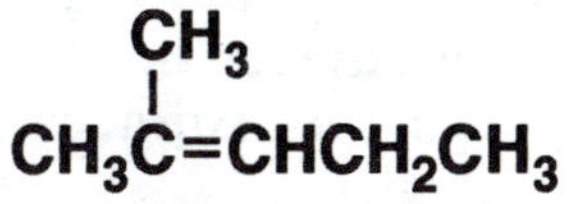

$CH_3C(CH_3){=}CHCH_2CH_3$ — **2-methyl-2-pentene**

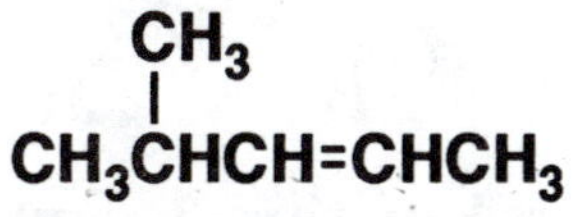

$CH_3CH(CH_3)CH{=}CHCH_3$ — **4-methyl-2-pentene**

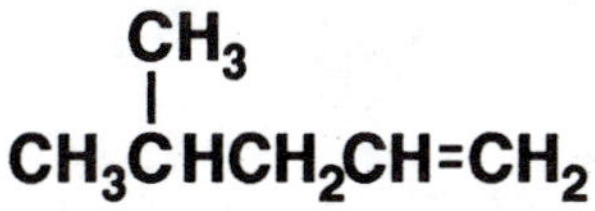

$CH_3CH(CH_3)CH_2CH{=}CH_2$ — **4-methyl-1-pentene**

(b) 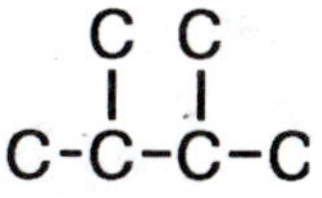 C-C(C)-C(C)-C

$CH_2{=}C(CH_3)CH(CH_3)CH_3$ — **2,3-dimethyl-1-butene**

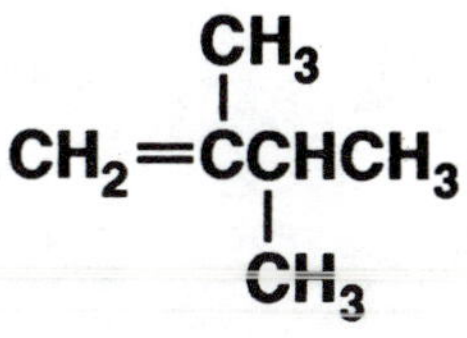

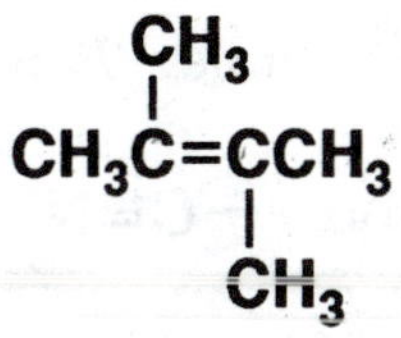

$CH_3C(CH_3){=}C(CH_3)CH_3$ — **2,3-dimethyl-2-butene**

(c) 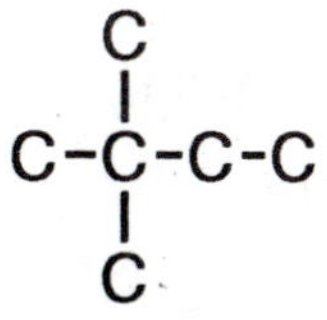 C-C(C)(C)-C-C

$CH_3C(CH_3)_2CH{=}CH_2$ — **3,3-dimethyl-1-butene**

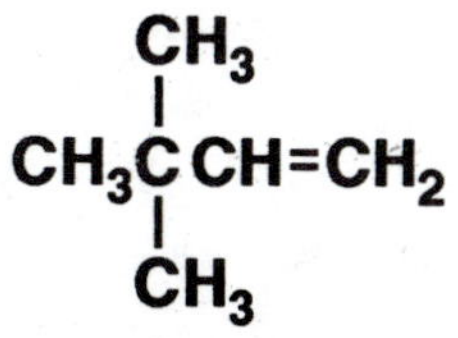

Problem 4.2 Which of the following alkenes show *cis-trans* isomerism? For each that does, draw structural formulas for the isomers.

(a) 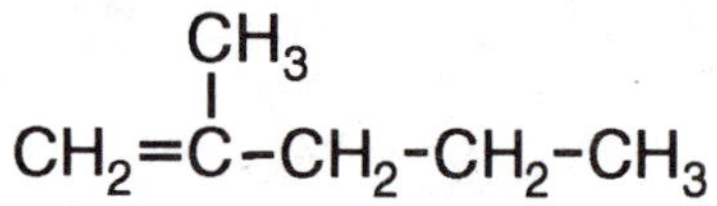 $CH_2{=}C(CH_3){-}CH_2{-}CH_2{-}CH_3$ (b) $CH_3{-}C(CH_3){=}CH{-}CH_2{-}CH_3$

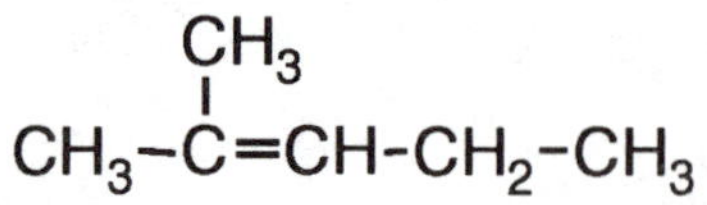

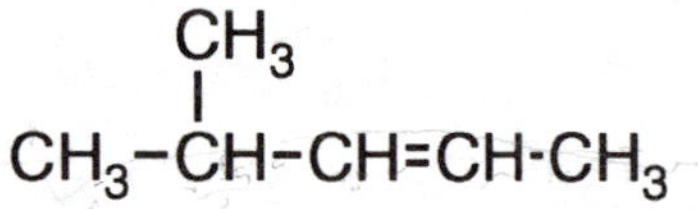

(c) $CH_3{-}CH(CH_3){-}CH{=}CH{-}CH_3$ (d) 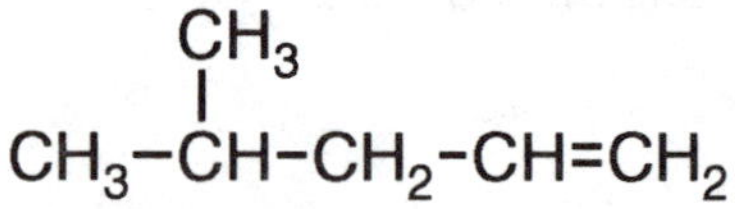 $CH_3{-}CH(CH_3){-}CH_2{-}CH{=}CH_2$

***Cis-trans* isomerism is possible only for the alkene in part (c). Each other alkene has two identical substituents on one of the carbons of the double bond. Following are the *cis-trans* isomers for part (c).**

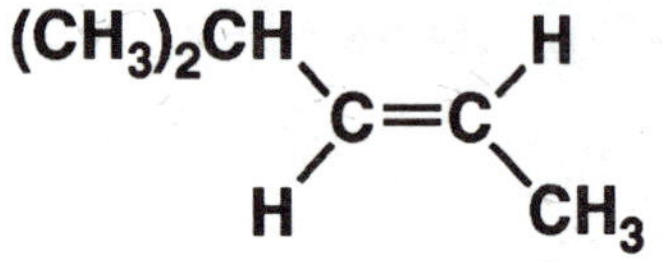

$(CH_3)_2CH(H)C{=}C(H)CH_3$ (trans) $(CH_3)_2CH(H)C{=}C(H)CH_3$ (cis)

***trans*-4-methyl-2-pentene** 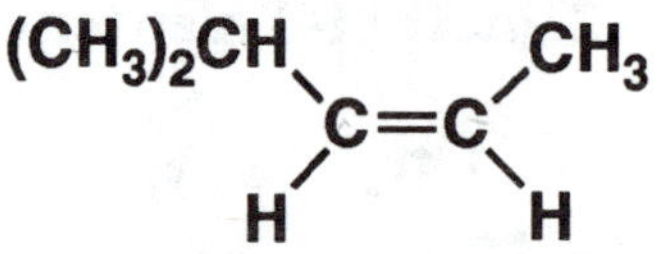 ***cis*-4-methyl-2-pentene**

Problem 4.3 Assign priorities to the following groups and explain the basis for the assignment.

(a) $-CH_2NO_2$ and $-NO_2$

$-NO_2$ > $-CH_2NO_2$

because N (atomic number 7) > C (atomic number 6)

(b) $-C{\equiv}CH$ and $-C{\equiv}N$

$-C{\equiv}N$ > $-C{\equiv}CH$

because C to N (atomic number 7) > C to C (atomic number 6)

Problem 4.4 Name the following alkenes and specify configuration by the E-Z system.

(a) 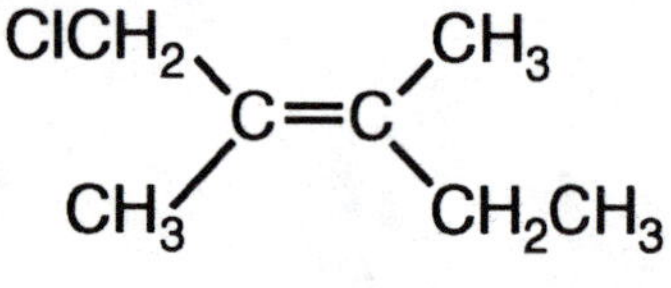 $ClCH_2(CH_3)C{=}C(CH_3)CH_2CH_3$

(E)-1-chloro-2,3-dimethyl-2-pentene

(b) 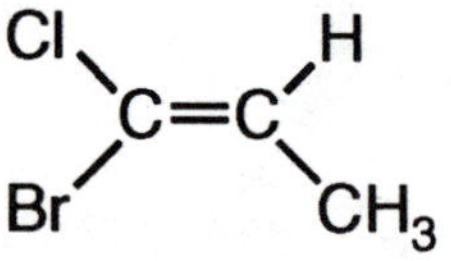 $Cl(Br)C{=}C(H)CH_3$

(Z)-1-bromo-1-chloro-1-propene

(c) 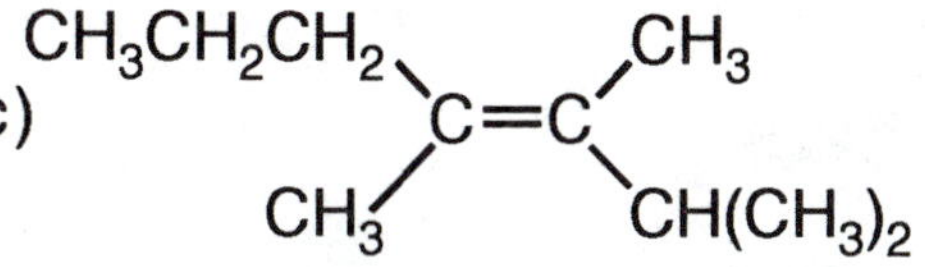 $CH_3CH_2CH_2(CH_3)C=C(CH_3)CH(CH_3)_2$

(E)-2,3,4-trimethyl-3-heptene

Problem 4.5 Draw structural formulas for the alkenes formed on dehydrohalogenation of the following haloalkanes. Where two alkenes are possible, predict which is the major product.

(a) 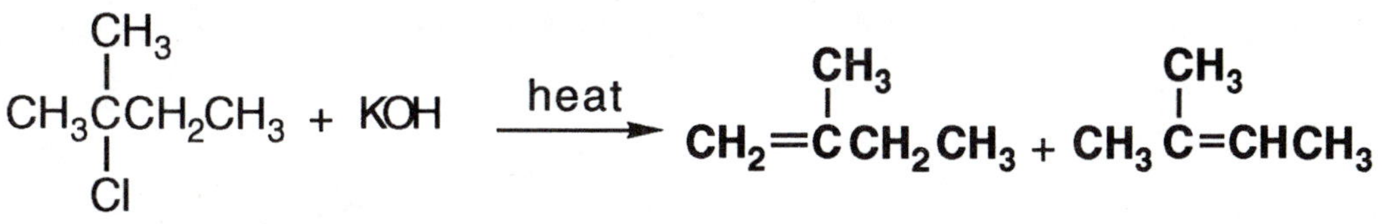 $CH_3C(CH_3)(Cl)CH_2CH_3$ + KOH $\xrightarrow{\text{heat}}$ $CH_2{=}C(CH_3)CH_2CH_3$ + $CH_3C(CH_3){=}CHCH_3$

2-chloro-2-methylbutane — **major product** ($CH_3C(CH_3){=}CHCH_3$)

(b) 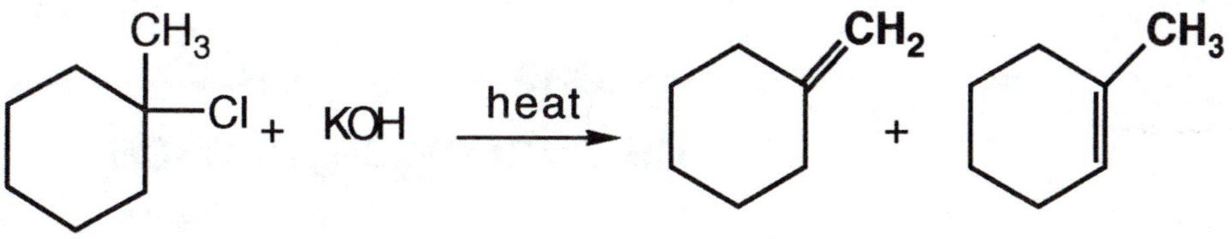 1-chloro-1-methylcyclohexane + KOH $\xrightarrow{\text{heat}}$ methylenecyclohexane ($=CH_2$) + 1-methylcyclohexene (CH_3)

1-chloro-1-methyl-
cyclohexane

major product (1-methylcyclohexene)

Nomenclature

Problem 4.6 Draw structural formulas for the following compounds.

(a) 2-methyl-3-hexene

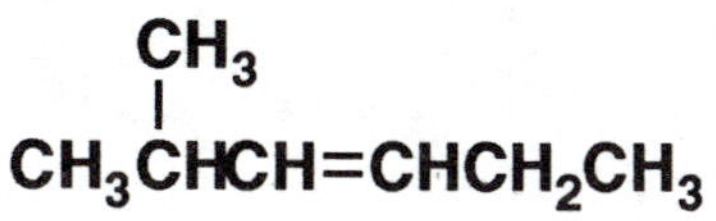

$CH_3CH(CH_3)CH{=}CHCH_2CH_3$

(b) 2-methyl-2-hexene

$CH_3C(CH_3){=}CHCH_2CH_2CH_3$

(c) 2-methyl-1-butene

$CH_2{=}C(CH_3)CH_2CH_3$

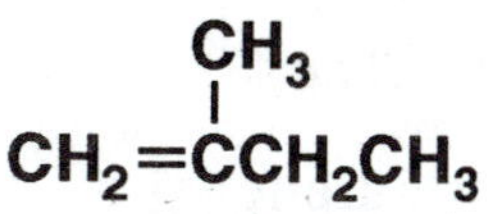

(d) 3-ethyl-3-methyl-1-pentene

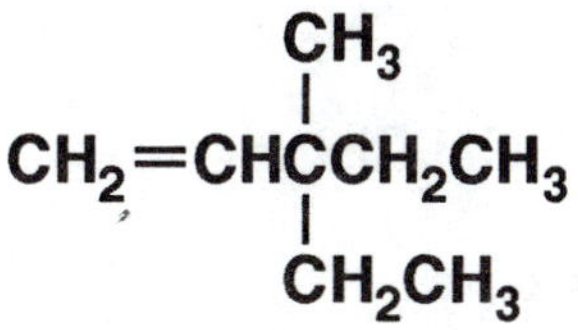

$CH_2{=}CHC(CH_3)(CH_2CH_3)CH_2CH_3$

(e) 2,3-dimethyl-2-butene

$CH_3C(CH_3){=}C(CH_3)CH_3$

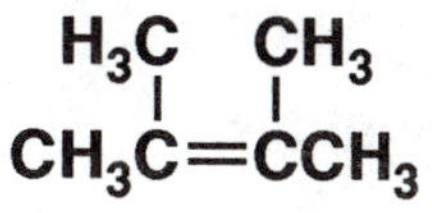

(f) 1-pentene

$CH_2{=}CHCH_2CH_2CH_3$

(g) 2-pentene

$CH_3CH{=}CHCH_2CH_3$

(h) 1-chloropropene

$ClCH{=}CHCH_3$

(i) 2-chloropropene

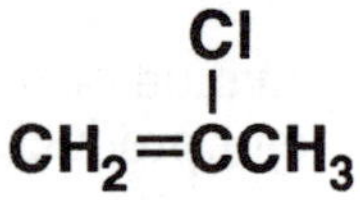

$CH_2{=}C(Cl)CH_3$

(j) 3-methylcyclohexene

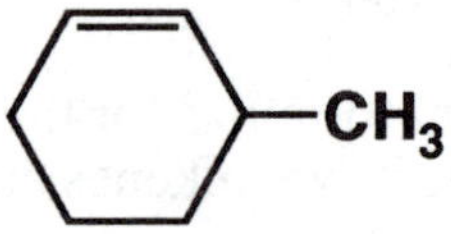

(k) 1-chlorocyclohexene

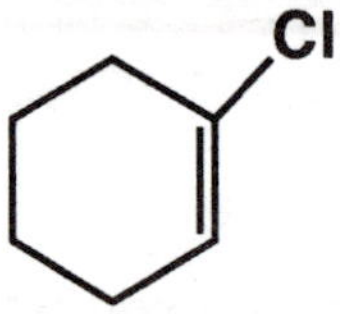

(l) 1-isopropyl-4-methylcyclohexene

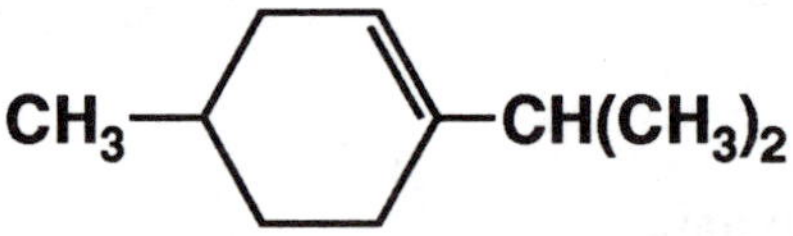

(m) tetrachloroethylene

$CCl_2{=}CCl_2$

(n) 2,6-dimethyl-2,6-octadiene

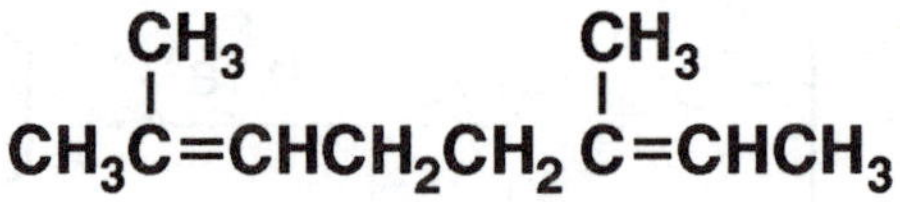

$CH_3C(CH_3){=}CHCH_2CH_2C(CH_3){=}CHCH_3$

(o) allylcyclopropane

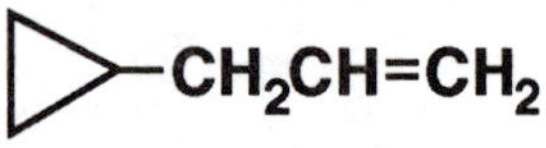

(p) vinylcyclopropane

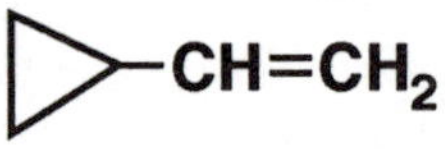

(q) bicyclo[2.2.1]-2-heptene

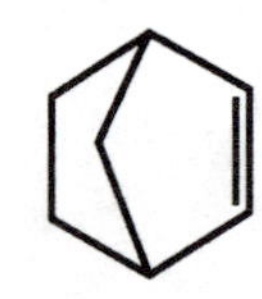

or

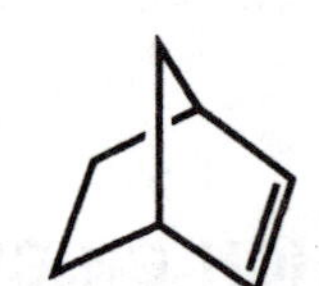

(r) bicyclo[4.4.0]-1-decene

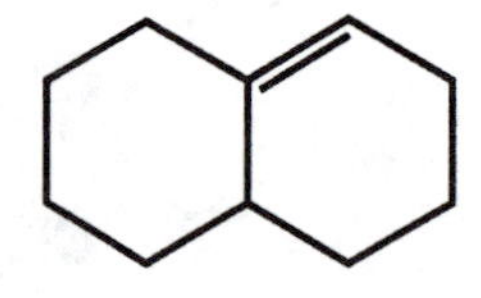

Problem 4.7 Name the following compounds.

(a) 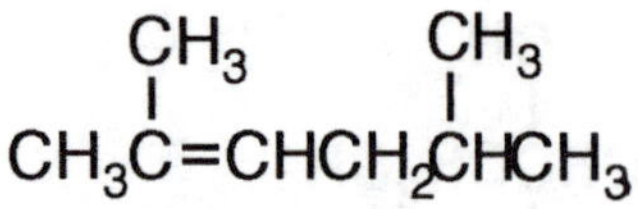 $CH_3C(CH_3){=}CHCH_2CH(CH_3)CH_3$

2,5-dimethyl-2-hexene

(b) 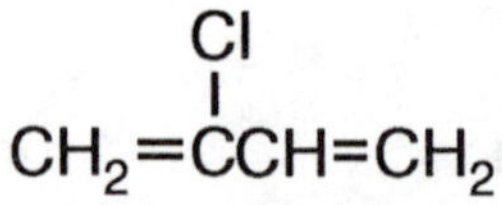 $CH_2{=}C(Cl)CH{=}CH_2$

2-chloro-1,3-butadiene

(c) 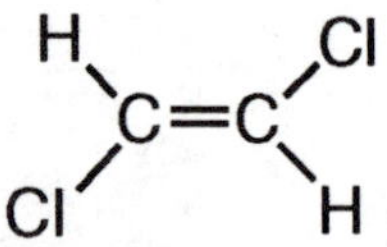 H(Cl)C=C(Cl)H

***trans*-1,2-dichloroethene**
(E)-1,2-dichloroethene

(d) 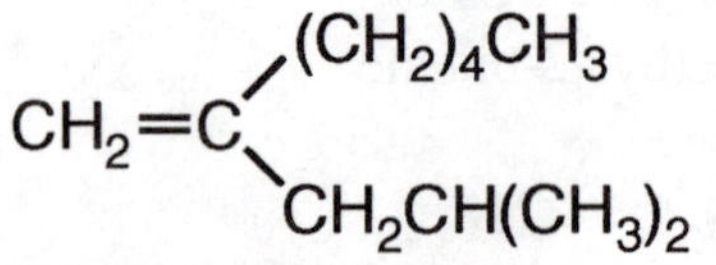 $CH_2{=}C[(CH_2)_4CH_3]CH_2CH(CH_3)_2$

2-isobutyl-1-heptene

(e) 4-chloro-1,4-dimethylcyclopentene

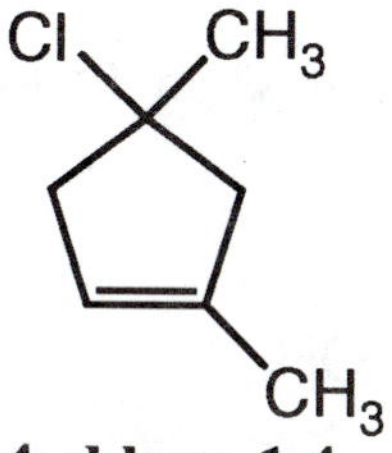

(f) *cis*-1,2-divinylcyclohexane

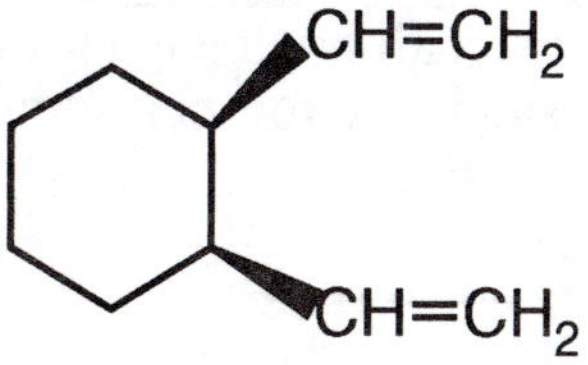

(g) $(CH_3)_2CHCH{=}C(CH_3)_2$

2,4-dimethyl-2-pentene

(h) $ClCH_2(H)C{=}C(H)CH_2Cl$

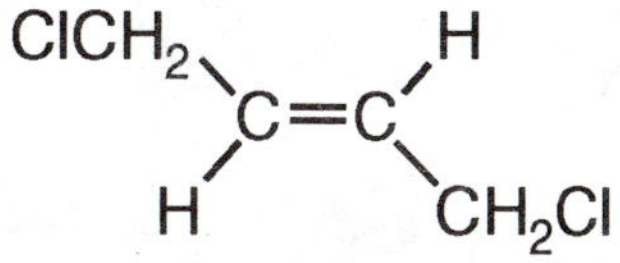

trans-1,4-dichloro-2-butene
(E)-1,4-dichloro-2-butene

(i) $F_2C{=}CF_2$

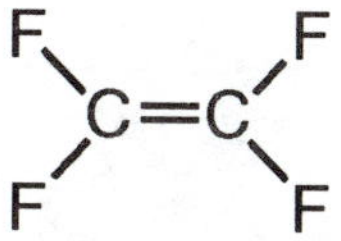

tetrafluoroethene
tetrafluoroethylene

(j) 5-chloro-5-ethyl-1,3-cyclopentadiene

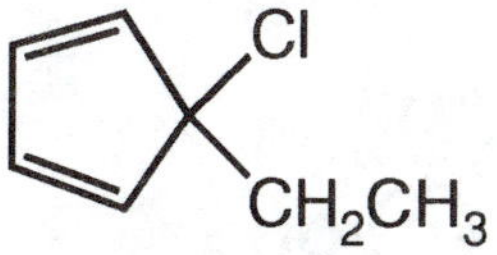

(k) 1,4-cyclohexadiene

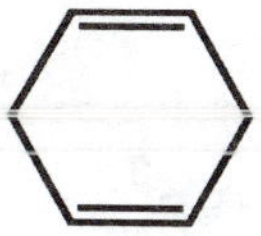

(l) 1,7,7-trimethylbicyclo[2.2.1]-2-heptene

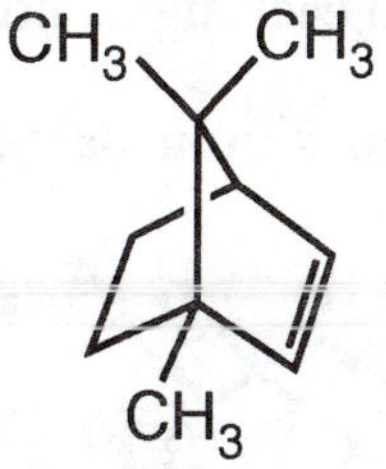

Problem 4.8 Following are structural formulas for four molecules that contain both a carbon-carbon double bond and another functional group. Give each an IUPAC name.

(a) $CH_2{=}CH{-}C({=}O){-}OH$
2-propenoic acid
(acrylic acid)

(b) $CH_2{=}CH{-}C({=}O){-}H$
2-propenal
(acrolein)

(c) $CH_3{-}CH{=}CH{-}C({=}O){-}OH$
2-butenoic acid
(crotonoic acid)

(d) $CH_3C({=}O)CH{=}CH_2$
3-buten-2-one
(methyl vinyl ketone)

Structure: Valence Bond Theory

Problem 4.9 Predict all bond angles about each circled carbon atom. To make these predictions, use the valence shell electron-pair repulsion model (Section 1.3).

(a)

109.5°

120°

(b) CH_2OH

120°

CH_2OH

(c) OH

120°

OH

(d) $HC\equiv C-CH=CH_2$

180°

$HC\equiv C-CH=CH_2$

Problem 4.10 For each circled carbon atom in the previous problem, identify which atomic orbitals are used to form each sigma bond and which are used to form each pi bond.

Each bond is labelled sigma or pi and the orbitals overlapping to form each bond are shown.

(a) $\sigma_{sp^3\text{-}1s}$, $\sigma_{sp^3\text{-}sp^3}$, $\pi_{2p\text{-}2p}$, $\sigma_{sp^2\text{-}sp^3}$, $\sigma_{sp^2\text{-}sp^2}$, $\sigma_{1s\text{-}sp^2}$

(b) $\sigma_{sp^2\text{-}sp^3}$, CH_2OH, $\pi_{2p\text{-}2p}$, $\sigma_{sp^2\text{-}sp^2}$

(c) $\pi_{2p\text{-}2p}$, $\sigma_{sp^2\text{-}sp^2}$, C-OH, $\sigma_{sp^3\text{-}sp^2}$, $\sigma_{sp^2\text{-}sp^3}$

(d) $H-C\equiv C-CH=CH_2$: $\pi_{2p\text{-}2p}$, $\sigma_{sp\text{-}sp}$, $\sigma_{sp\text{-}sp^2}$, $\pi_{2p\text{-}2p}$

Problem 4.11 Following is the structural formula of 1,2-propadiene (allene)

$$CH_2{=}C{=}CH_2$$

(a) State the orbital hybridization of each carbon atom.

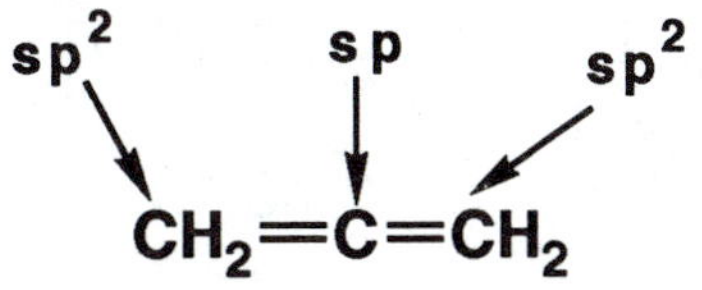

sp^2 sp sp^2

$CH_2{=}C{=}CH_2$

(b) Describe each carbon-carbon double bond in terms of the overlap of atomic orbitals.

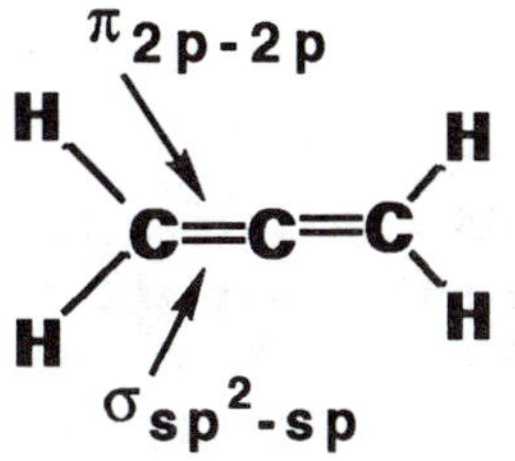

π_{2p-2p}

$H_2C{=}C{=}CH_2$

σ_{sp^2-sp}

(c) Predict all bond angles in allene.

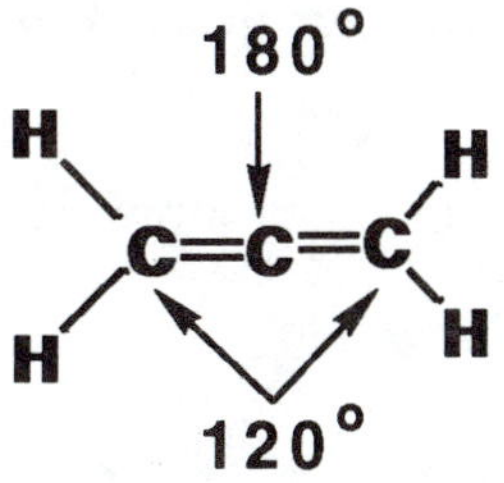

180°

$H_2C{=}C{=}CH_2$

120°

(d) Draw a stereorepresentation showing the shape of this molecule.

The central carbon atom of allene is sp hybridized with bond angles of 180° about it. The terminal carbons are sp^2 hybridized with bond angles of 120° about each. The planes created by H-C-H bonds at the ends of the molecule are perpendicular to each other.

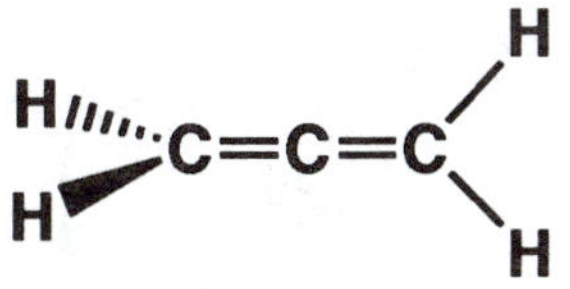

$H_2C{=}C{=}CH_2$

Problem 4.12 Following are lengths for a series of C-C single bonds. Propose an explanation for the differences in bond lengths.

Structure	Length of C-C single bond (nm)
CH_3-CH_3	0.1537
$CH_2=CH-CH_3$	0.1510
$CH_2=CH-CH=CH_2$	0.1465
$HC{\equiv}C-CH_3$	0.1459

The s electrons are on average held closer to atomic nuclei than p electrons. Thus hybrid orbitals with higher s character have the electrons held closer to the nucleus and thus make bonds that are shorter. As shown in the table, a σ bond formed from overlap of an sp^3 orbital with an sp^2 orbital is shorter than a σ bond formed from overlap of an sp^3 orbital with another sp^3 orbital. Similarly, sp^3-sp overlap produces a bond that is shorter than that produced by sp^3-sp^2 overlap.

<u>*Cis-Trans* Isomerism in Alkenes</u>

Problem 4.13 Which of the following molecules show *cis-trans* isomerism? For each that does, draw the *cis* isomer.

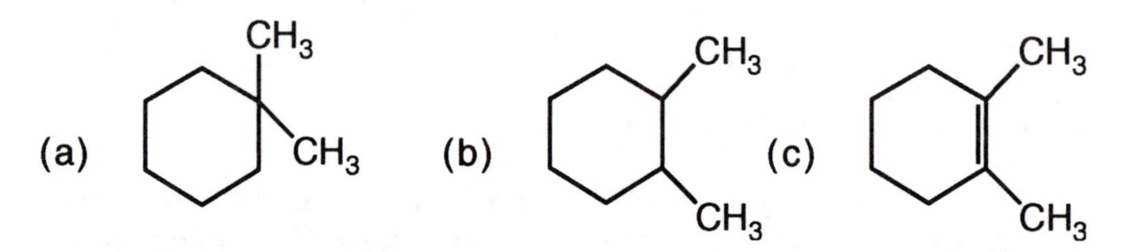

Only part (b) shows *cis-trans* isomerism.

***cis*-1,2-dimethyl-cyclohexane** ***trans*-1,2-dimethyl-cyclohexane**

Problem 4.14 Which of the molecules in Problem 4.6 show *cis-trans* isomerism? For each that does, draw the *trans* isomer.

Only parts (a), (g), (h) and (n) show *cis-trans* isomerism. For part (n) only the double bond between carbons 6-7 shows *cis-trans* isomerism.

(a) 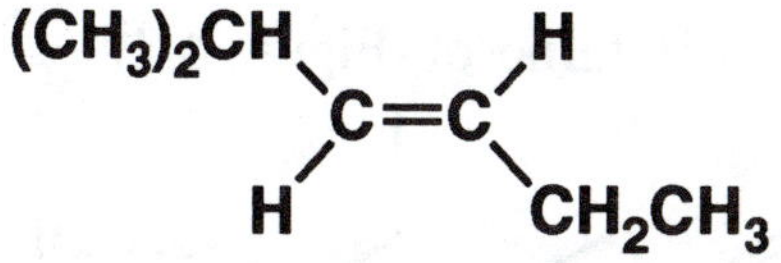 $(CH_3)_2CH(H)C{=}C(H)CH_2CH_3$ (g) 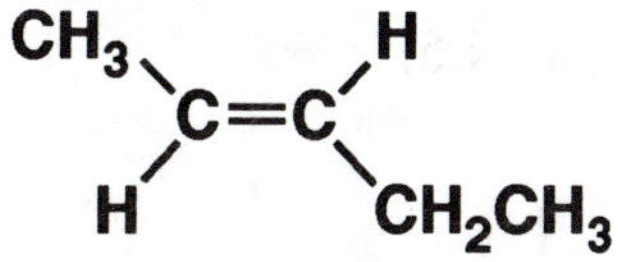 $CH_3(H)C{=}C(H)CH_2CH_3$

(h) 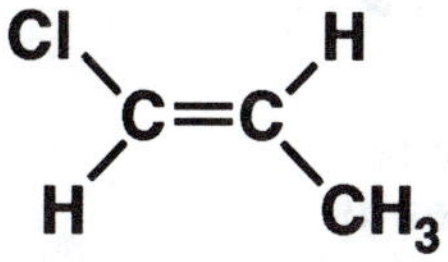 $Cl(H)C{=}C(H)CH_3$ (n) 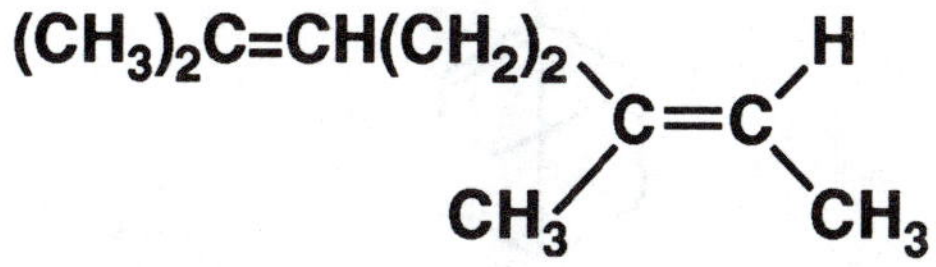 $(CH_3)_2C{=}CH(CH_2)_2(CH_3)C{=}C(H)CH_3$

Problem 4.15 Which of the molecules in Problem 4.7 show *cis-trans* isomerism? For each that does, draw the *trans* isomer.

Only parts (c), (f) and (h) show *cis-trans* isomerism.

(c) 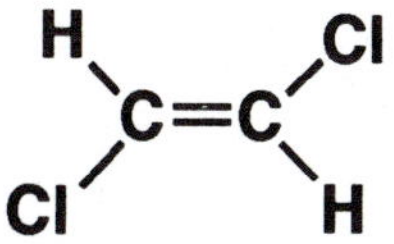 $H(Cl)C{=}C(Cl)H$

***trans*-1,2-dichloroethene**

(f) 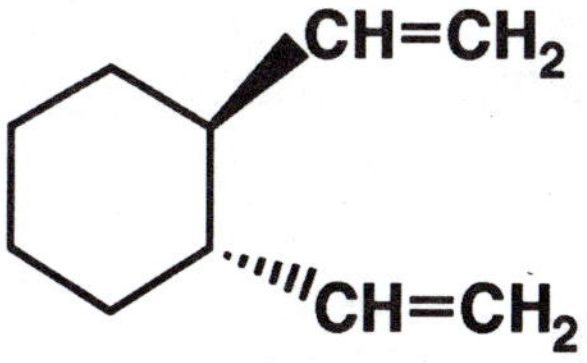 cyclohexane with $CH{=}CH_2$ (wedge) and $CH{=}CH_2$ (dash)

***trans*-1,2-divinylcyclohexane**

(h) 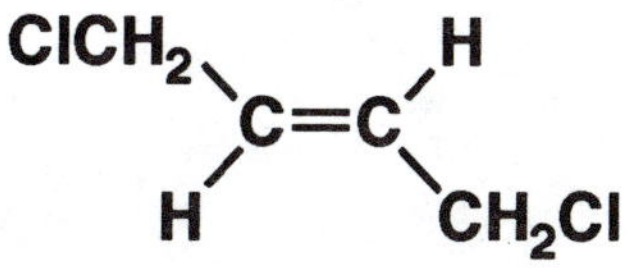 $ClCH_2(H)C{=}C(H)CH_2Cl$

***trans*-1,4-dichloro-2-butene**

Problem 4.16 How many *cis-trans* isomers are possible for the following natural products?
(a) geraniol (Figure 4.6) (b) limonene (Figure 4.6)

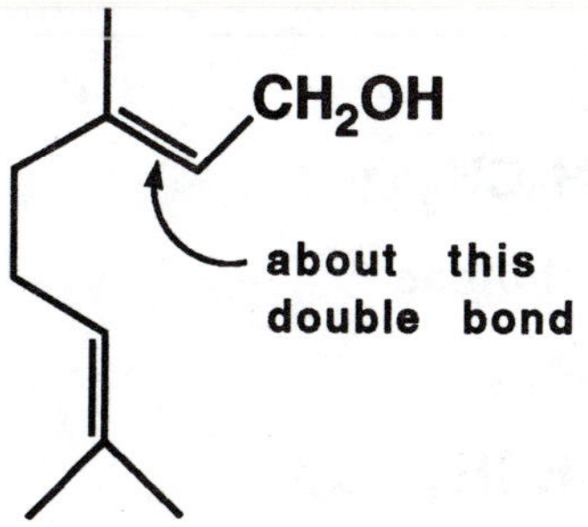

CH_2OH

about this double bond

geraniol
(one pair of *cis-trans* isomers; the trans isomer is shown)

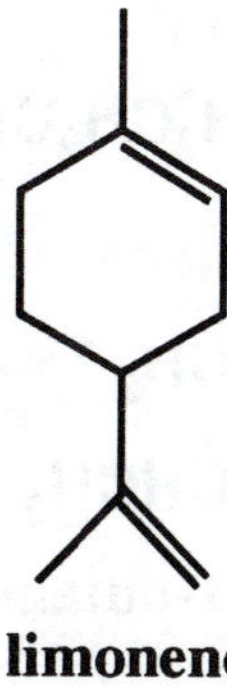

limonene
(no *cis-trans* isomers)

(c) α-pinene (Figure 4.6)

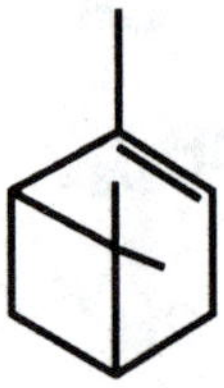

α-Pinene
(no *cis-trans* isomers)

(d) farnesol (Figure 4.7)

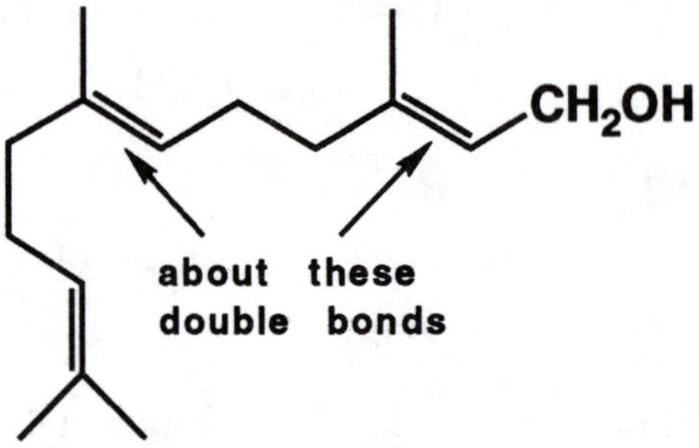

Farnesol
(four *cis-trans* isomers; the *trans-trans* isomer is shown)

(e) zingiberene (Figure 4.7)

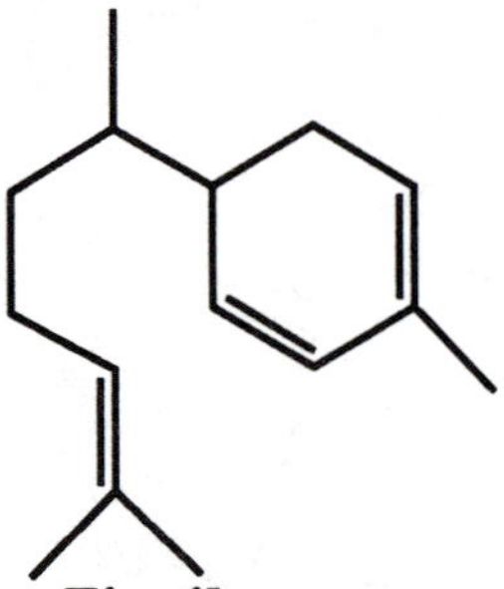

Zingiberene
(no *cis-trans* isomers)

Problem 4.17 Draw structural formulas for all compounds of molecular formula C_5H_{10} that are:
(a) alkenes that do not show *cis-trans* isomerism.

Four alkenes of molecular formula C_5H_{10} do not show *cis-trans* isomerism.

$CH_2{=}CHCH_2CH_2CH_3$
1-pentene

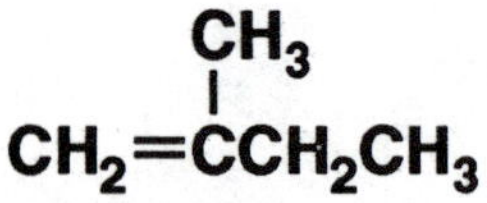

$CH_2{=}C(CH_3)CH_2CH_3$
2-methyl-1-butene

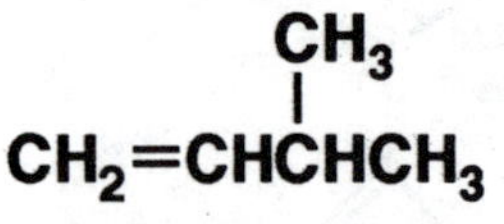

$CH_2{=}CHCH(CH_3)CH_3$
3-methyl-1-butene

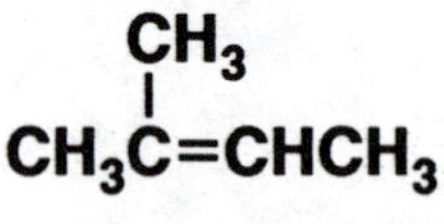

$CH_3C(CH_3){=}CHCH_3$
2-methyl-2-butene

(b) alkenes that do show *cis-trans* isomerism.

One alkene of molecular formula C_5H_{10} shows *cis-trans* isomerism.

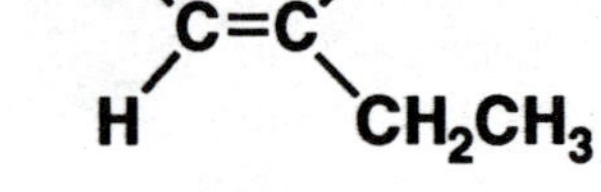

$CH_3(H)C{=}C(H)CH_2CH_3$
***trans*-2-pentene**

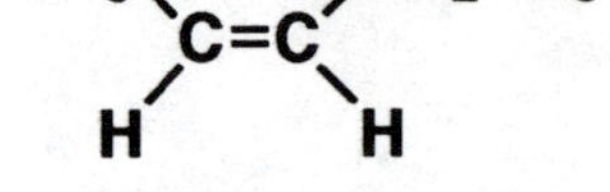

$CH_3(H)C{=}C(H)CH_2CH_3$
***cis*-2-pentene**

(c) cycloalkanes that do not show *cis-trans* isomerism.

Four cycloalkanes of molecular formula C_5H_{10} do not show *cis-trans* isomerism.

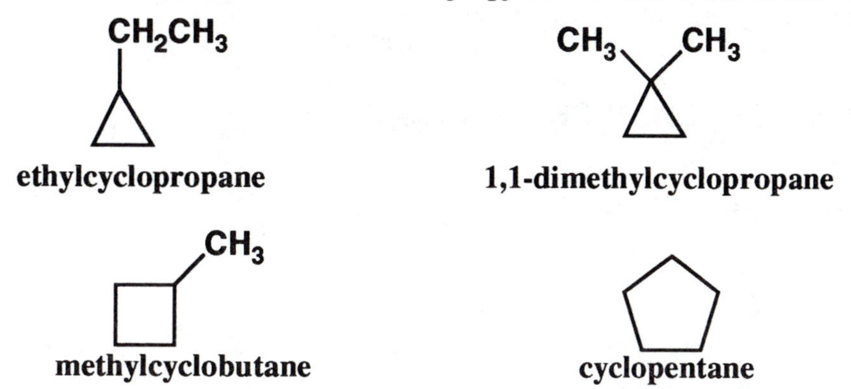

(d) cycloalkanes that do show *cis-trans* isomerism.

Only one cycloalkane of molecular formula C_5H_{10} shows *cis-trans* isomerism.

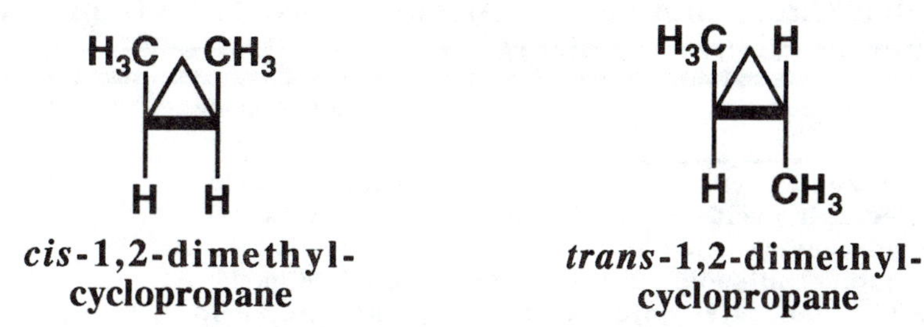

Problem 4.18 Draw structural formulas for the four isomeric chloropropenes, molecular formula C_3H_5Cl.

There are four isomeric chloropropenes.

$CH_2{=}CHCH_2Cl$

3-chloropropene (allyl chloride)

$CH_2{=}C(Cl)CH_3$

2-chloropropene

Cl(H)C=C(H)(CH₃) — Cl and H on one carbon; H and CH_3 on the other, trans

***trans*-1-chloropropene**

Cl(H)C=C(CH₃)(H) — Cl and CH_3 on the same side

***cis*-1-chloropropene**

Problem 4.19 Following are structural formulas for six of the most abundant carboxylic acids found in animal fats, plant oils, and biological membranes. Because many of these carboxylic acids were first isolated from animal fats, they are often referred to as fatty acids.

(1) $CH_3(CH_2)_{14}CO_2H$ Palmitic acid

(2) $CH_3(CH_2)_{16}CO_2H$ Stearic acid

(3) $CH_3(CH_2)_7CH{=}CH(CH_2)_7CO_2H$ Oleic acid

(4) $CH_3(CH_2)_4CH{=}CHCH_2CH{=}CH(CH_2)_7CO_2H$ Linoleic acid

(5) $CH_3CH_2(CH{=}CHCH_2)_3(CH_2)_6CO_2H$ Linolenic acid

(6) $CH_3(CH_2)_3(CH_2CH{=}CH)_4(CH_2)_3CO_2H$ Arachidonic acid

(a) How many *cis-trans* isomers are possible for each?

(b) Give each an IUPAC name.

The following table shows number of carbon atoms, IUPAC name, common name, and number of *cis-trans* isomers.

	Carbon atoms	IUPAC name	Common name	cis-trans isomers
(1)	16	hexadecanoic acid	palmitic acid	0
(2)	18	octadecanoic acid	stearic acid	0
(3)	18	9-octadecenoic acid	oleic acid	2
(4)	18	9,12-octadecadienoic acid	linoleic acid	2x2=4
(5)	18	9,12,15-octadecatrienoic acid	linolenic acid	2x2x2=8
(6)	20	5,8,11,14-dodecatetraenoic acid	arachidonic acid	2x2x2x2=16

(c) For those that show *cis-trans* isomerism, consult a textbook of biochemistry and see if you can determine which of the possible configurations is the most prevalent in the biological world.

The most common *cis-trans* isomers of each in the biological world are all the *cis* isomers.

Nomenclature

Problem 4.20 Arrange the following groups in order of increasing priority.

(a) $-CH_3$ $-H$ $-Br$ $-CH_2CH_3$

$$-H < -CH_3 < -CH_2CH_3 < -Br$$

(b) $-OCH_3$ $-CH(CH_3)_2$ $-B(CH_2CH_3)_2$ $-H$

$$-H < -B(CH_2CH_3)_2 < -CH(CH_3)_2 < -OCH_3$$

(c) $-CH_3$ $-CH_2OH$ $-CH_2NH_2$ $-CH_2Br$

$$-CH_3 < -CH_2NH_2 < -CH_2OH < -CH_2Br$$

Problem 4.21 Assign E-Z configurations to the following alkenes.

(a) 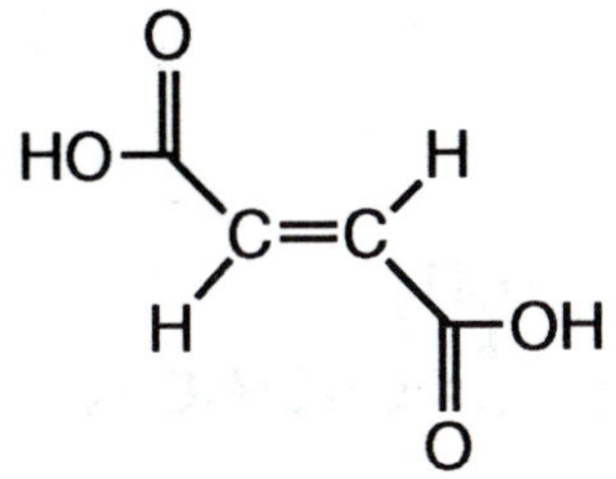 $HOOC-CH=CH-COOH$ (HO–C(=O) and H on one carbon; H and C(=O)–OH on the other)

(b) $HOCH_2(H)C=C(CH_3)Cl$

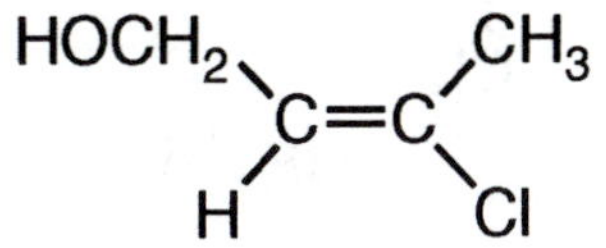

(c) $Cl(H)C=C(CH_3)CH_2Cl$

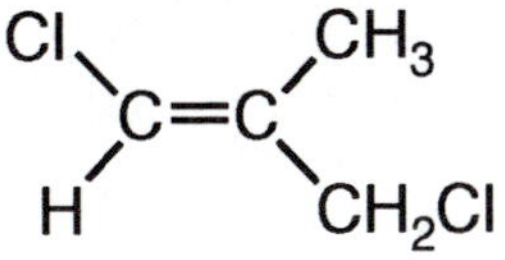

All have the (E)-configuration.

Preparation of Alkenes

Problem 4.22 Draw structural formulas for the alkenes formed on dehydrohalogenation of the following compounds using KOH. Where more than one alkene may be formed, predict which is the major product.

(a) 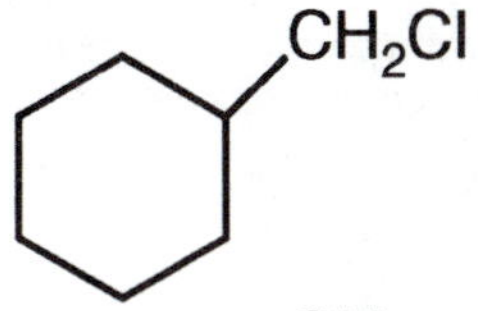 cyclohexyl–CH_2Cl

methylenecyclohexane (cyclohexane=CH_2)

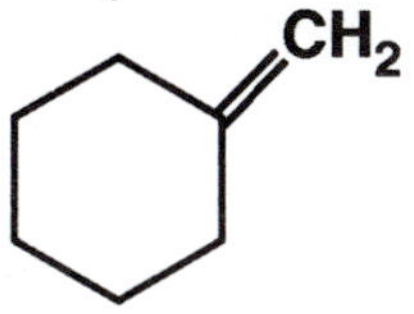

(b) 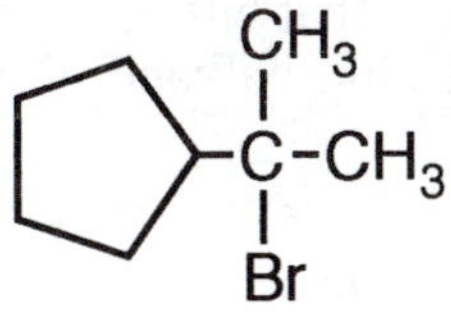 cyclopentyl–$C(CH_3)_2Br$

cyclopentylidene=$C(CH_3)_2$ (major) + cyclopentyl–$C(CH_3)=CH_2$

(c) 2-bromo-1,1-dimethylcyclohexane (Br, CH_3, CH_3)

3,3-dimethylcyclohexene (CH_3, CH_3)

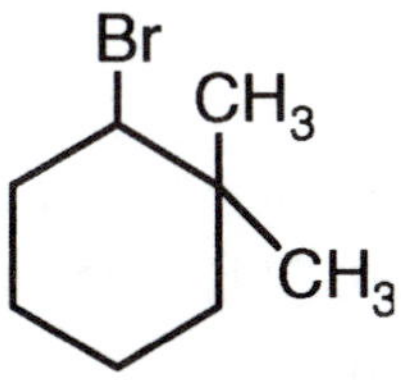

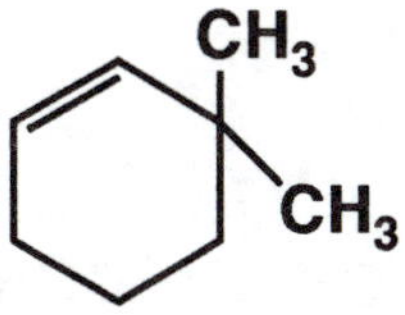

(d) 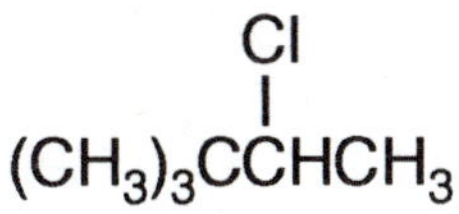 $(CH_3)_3CCHClCH_3$

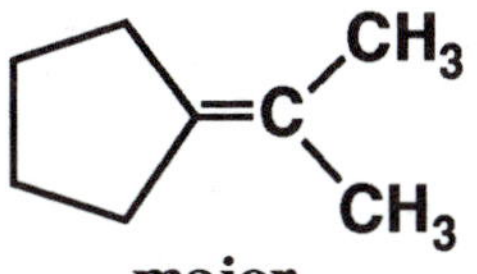

$CH_3C(CH_3)_2CH=CH_2$

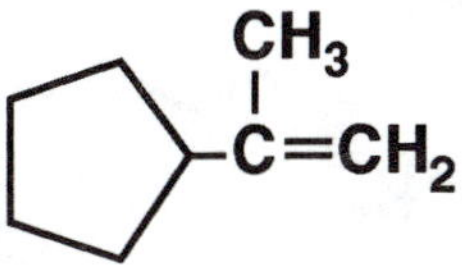

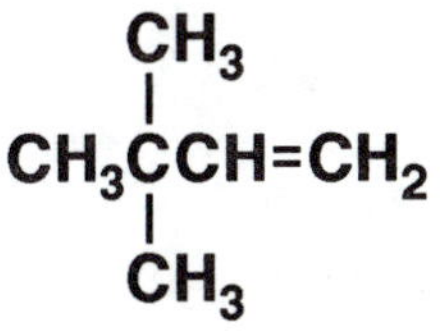

(e) $CH_3CH_2C(CH_3)(Cl)CH_2CH_2CH_2CH_3$

$CH_3CH{=}C(CH_3)CH_2CH_2CH_2CH_3$ + $CH_3CH_2C(CH_3){=}CHCH_2CH_2CH_3$

major

$CH_3CH_2C({=}CH_2)CH_2CH_2CH_2CH_3$

Problem 4.23 Draw the structural formula of all chloroalkanes which undergo dehydrohalogenation in the presence of KOH to give each alkene as the major product. For some parts, there is only one chloroalkane that will give the desired alkene as the major product. For other parts, there may be two or more chloroalkanes that give the desired alkene as the major product.

(a) 1-methylcyclohexene (CH_3 on ring double bond)

2-chloro-1-methylcyclohexane (CH_3, Cl on adjacent ring carbons) or 1-chloro-1-methylcyclohexane (CH_3, Cl on same ring carbon)

(b) methylenecyclohexane (ring $={CH_2}$)

(chloromethyl)cyclohexane (ring–CH_2Cl)

(c) $CH_2{=}C(CH_3)CH(CH_3)CH_2CH_3$

$CH_2(Cl)CH(CH_3)CH(CH_3)CH_2CH_3$

(d) $CH_3{-}C(CH_3){=}C(CH_3)CH_2CH_3$

$CH_3C(CH_3)(Cl){-}CH(CH_3)CH_2CH_3$ or $CH_3CH(CH_3){-}C(Cl)(CH_3)CH_2CH_3$

(e)

$$\begin{array}{c} \quad CH_3 \\ \quad | \\ CH_3CHC{=}CHCH_3 \\ \qquad | \\ \qquad CH_3 \end{array}$$

$$\begin{array}{c} \quad CH_3 \quad Cl \\ \quad | \qquad | \\ CH_3CHCHCHCH_3 \\ \qquad | \\ \qquad CH_3 \end{array}$$

Problem 4.24 Elimination of HBr from 2-bromonorbornane gives only 2-norbornene and no 1-norbornene. How do you account for the regioselectivity of this dehydrohalogenation? In answering this question, you will find it helpful to make molecular models of both 1-norbornene and 2-norbornene and analyze the angle strain in each.

2-bromonorbornane | 2-norbornene | 1-norbornene

As can be seen with molecular models, there is a great deal of angle strain in 1-norbonene. In other words, the bridgehead carbon atom in norbornane must remain roughly tetrahedral and sp^3 to accommodate the bonds to adjacent atoms. 1-Norbonene would have an sp^2 bridgehead atom that would have a large amount of angle strain.

Terpenes

Problem 4.25 Show that the carbon skeleton of farnesol can be coiled and then cross-linked to give the carbon skeleton of caryophyllene (Figure 4.7).

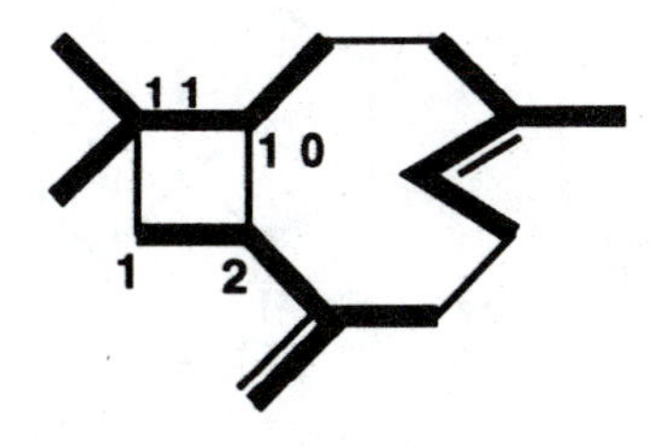

Caryophyllene showing cross-linking of the carbon chain of farnesol

Problem 4.26 Show that the structural formula of Vitamin A (Figure 4.8) can be divided into four isoprene units joined by head-to-tail linkages and cross linked at one point to form the six-member ring.

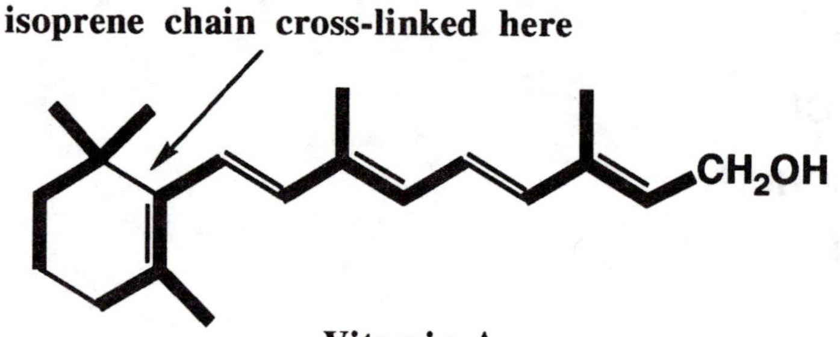

isoprene chain cross-linked here

CH_2OH

Vitamin A

Problem 4.27 Show that the structural formula of β-carotene (Section 23.6A) can be divided into eight isoprene units.

The following structural formula shows the eight isoprene units of β-carotene.

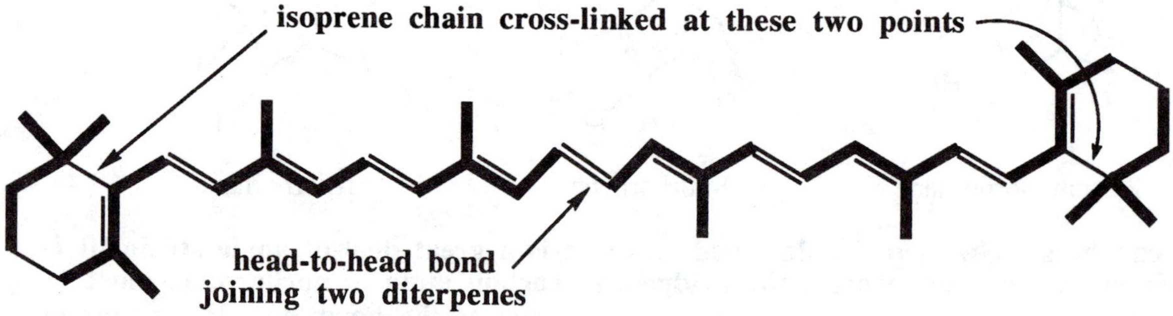

isoprene chain cross-linked at these two points

head-to-head bond joining two diterpenes

Problem 4.28 Santonin, $C_{15}H_{18}O_3$, isolated from the flower heads of certain species of Artemisia, is an anthelmintic, that is, a drug used to rid the body of worms (helminths). It has been estimated that over one-third of the world's population is infested with these parasites. Santonin in oral doses of 60 mg is used as an anthelmintic for roundworms (*Ascaris lumbricoides*).

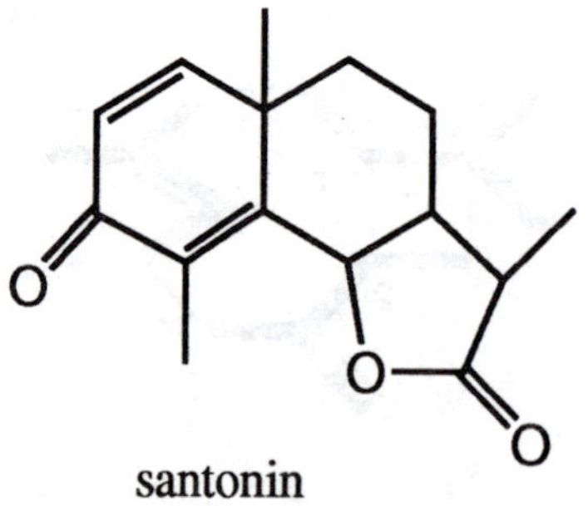

O O O

santonin

Locate the three isoprene units in santonin and show how the carbon skeleton of farnesol might be coiled and then cross-linked to give santonin. Two different coiling patterns of the carbon skeleton of farnesol that could lead to santonin. Try to find them both.

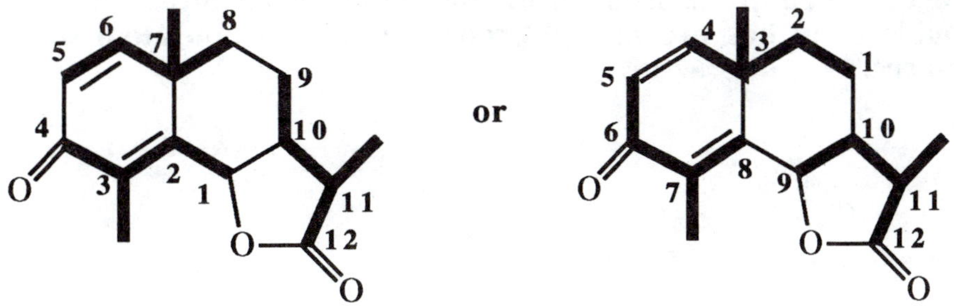

or

Problem 4.29 In many parts of South America, extracts of the leaves and twigs of *Montanoa tomentosa* are used as a contraceptive, to stimulate menstruation, to facilitate labor, or to terminate early pregnancy. Phytochemical investigations of this plant have resulted in isolation of a very potent fertility-regulating compound called zoapatanol.

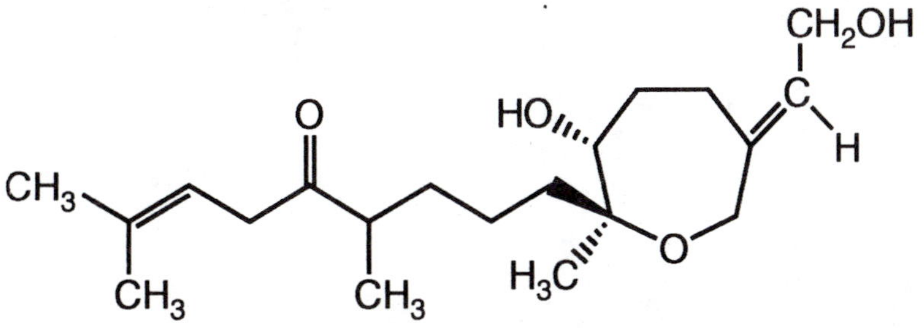

(a) Show that the carbon skeleton can be divided into four isoprene units bonded head-to-tail and then cross-linked in one point along the chain.

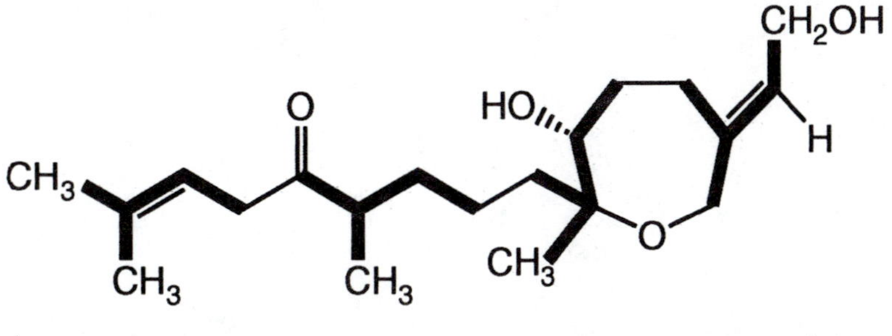

(b) Specify the configuration about the carbon-carbon double bond to the seven-member ring according to the E,Z system.

The double bond in question has the E configuration, because the hydroxymethyl group is on the side of the double bond opposite the higher priority carbon atom that is linked to the ether oxygen.

(c) How many *cis-trans* isomers are possible for zoapatanol? Consider the possibilities for *cis-trans* isomerism in both cyclic compounds and in alkenes.

There are total of 2 x 2 or 4 *cis-trans* isomers. There are two *cis-trans* isomers on the ring, and two *cis-trans* isomers of the double bond attached to the ring. The other double bond has two methyl groups on one carbon atom so it has no *cis-trans* isomers.

CHAPTER 5: ALKENES II

SUMMARY OF REACTIONS

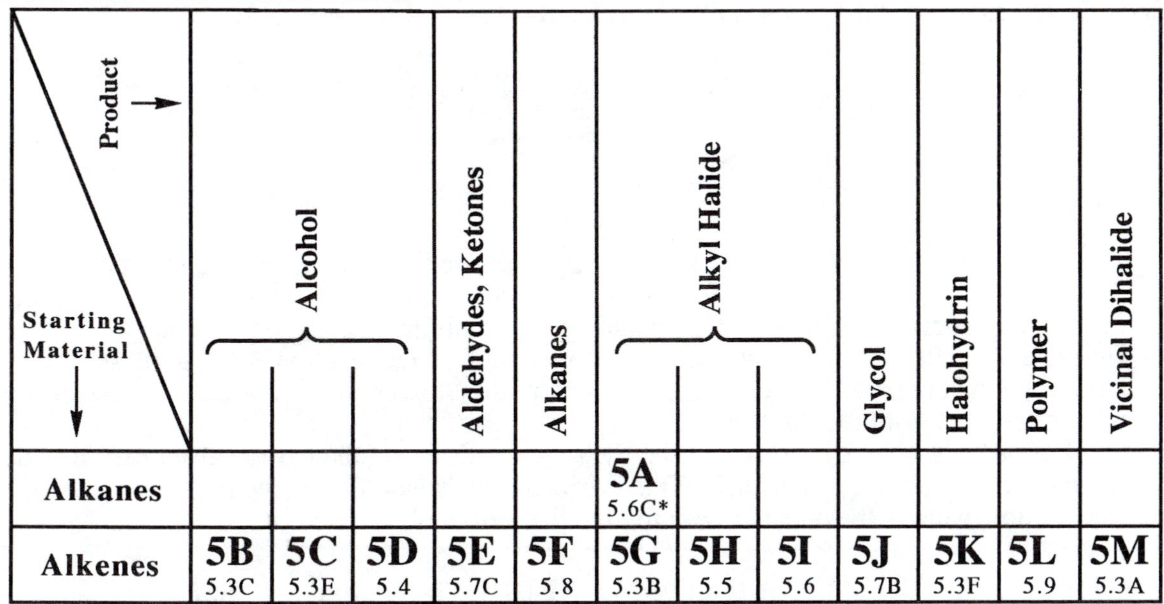

Starting Material ↓ / Product →	Alcohol	Alcohol	Alcohol	Aldehydes, Ketones	Alkanes	Alkyl Halide	Alkyl Halide	Alkyl Halide	Glycol	Halohydrin	Polymer	Vicinal Dihalide
Alkanes						**5A** 5.6C*						
Alkenes	**5B** 5.3C	**5C** 5.3E	**5D** 5.4	**5E** 5.7C	**5F** 5.8	**5G** 5.3B	**5H** 5.5	**5I** 5.6	**5J** 5.7B	**5K** 5.3F	**5L** 5.9	**5M** 5.3A

*Section in book that describes reaction.

REACTION 5A: HALOGENATION OF ALKANES (Section 5.6C)

$$\text{—C—C'(H)—} + X_2 \xrightarrow[\text{or light}]{\text{heat}} \text{—C—C'(X)—} + \text{H-X}$$

- **Halogenation of alkanes** occurs by a radical mechanism. The reaction begins with an initiation step involving the homolytic cleavage of a halogen such as chlorine with heat or light.
- Propagation continues until termination occurs involving two radical species reacting with each other to produce a new covalent bond and thereby quenching both radicals.
- Halogenation of alkanes is **regioselective**. That is, reactions using a variety of different alkanes demonstrate that tertiary hydrogen atoms are replaced in preference to secondary hydrogen atoms which are replaced in preference to primary hydrogen atoms.
- Halogenation using bromine is significantly more regioselective than chlorination. In order to understand the detailed regioselectivity observed with halogenation of alkanes, it is helpful to consider **Hammond's postulate**. According to Hammond's postulate, the *structure* of a transition state more closely resembles the stable species (reactants or products) that is closest in energy to the transition state. In other words, the transition state of an endothermic reaction resembles products while the transition state for an exothermic reaction resembles reactants. ✱ *[This concept is useful in a number of situations, so it is important to make sure it is understood before going on.]*
- Abstraction of hydrogen by chlorine or bromine radicals is the first propagation step of the radical chain reaction with alkanes. This step has a higher energy of activation for bromine relative to chlorine, therefore, according to Hammond's postulate the transition state for the bromine reaction resembles reactants less than the transition state for chlorine. In other

words, the transition state for the bromine reaction has more radical character and, therefore, a higher degree of selectivity is observed with bromine in favor of the more stable radicals such as allyl or tertiary radicals. *[This is a complex argument ; make sure you understand it before moving on.]*

REACTION 5B: ACID-CATALYZED HYDRATION (Section 5.3C)

$$\mathrm{>C{=}C'<} \xrightarrow[H_2O]{H^+} \mathrm{-\overset{H}{\underset{|}{C}}-\underset{OH}{C'}-}$$

- In the presence of an acid catalyst like sulfuric acid, water adds to alkenes to give alcohols.
- The reaction mechanism involves formation of a carbocation from protonation of the pi bond, followed by nucleophilic attack of the resulting electrophilic carbocation by water.
- The water can attack from either side of the carbocation so the reaction is not stereoselective.
- This reaction is **regioselective**, that is one constitutional isomer is produced in preference to other possible constitutional isomers.
- **Markovnikov's rule** is followed so the hydrogen ends up on the carbon atom that already has more hydrogen atoms attached. This is because the water attacks the more stable carbocation, namely the one that is more highly substituted.

REACTION 5C: OXYMERCURATION / REDUCTION (Section 5.3E)

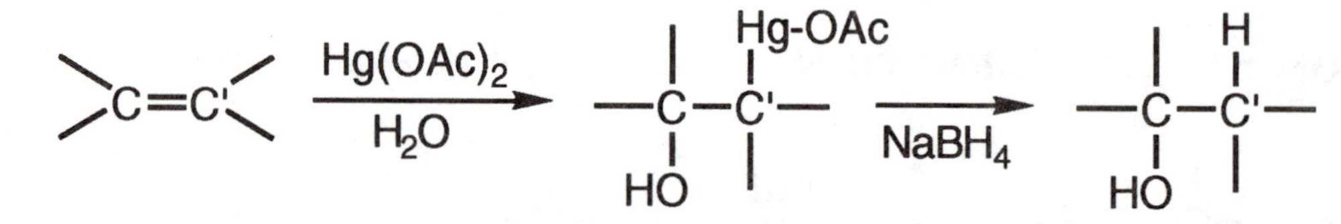

- Hydration of alkenes can be accomplished by treatment of an alkene with mercury(II) acetate in water followed by reduction with $NaBH_4$.
- The reaction mechanism involves formation of a mercurinium ion intermediate that is then attacked by water. The reaction is completed by adding $NaBH_4$ that results in displacement of the mercury atom by hydrogen.
- Oxymercuration is both regioselective and stereoselective because Markovnikov's rule is followed and anti addition predominates. This means the mercurinium ion intermediate has a partially bridged structure that prevents attack of the nucleophilic water from the side with the mercury ion. There is some carbocation character to the intermediate because the carbon atom that forms the more stable carbocation is preferentially attacked by the water. Note that the intermediate is not a full blown carbocation since it does not rearrange.

REACTION 5D: HYDROBORATION (Section 5.4)

$$\mathrm{>C{=}C'<} \xrightarrow[2)\ H_2O_2]{1)\ BH_3} \mathrm{-\overset{H}{C}-\overset{OH}{C'}-}$$

- Addition of borane, BH_3, to an alkene forms a trialkylborane. Hydroboration followed by treatment with peroxide gives an alcohol in which the -OH group is added to the *less-substituted* carbon of the alkene. Thus, the hydroboration reaction is distinct from and

complementary to the acid-catalyzed hydration and oxymercuration reactions in which the OH group adds primarily on the *more-substituted* carbon in agreement with Markovnikov's rule.
- In the first step of the hydroboration reaction mechanism, the pi electrons of the alkene react with the boron atom, a Lewis acid. It is thought that the hydrogen atom is simultaneously transferred to the other atom of the alkene. Reactions in which bond-making and bond-breaking occur simultaneously are called **concerted**.
- Note that addition of BH_3 is stereoselective in that the boron and hydrogen atom are added to the same side of the alkene molecule, a situation referred to as **syn** addition.
- The reaction continues two more times to give a trialkylborane, and oxidation of a trialkylborane with hydrogen peroxide replaces the boron with OH.

REACTION 5E: OZONOLYSIS (Section 5.7C)

C=C'(H) + O_3 → ozonide (C–C' with O–O–O bridge) —$(CH_3)_2S$→ C=O + C'(=O)H

- Ozone reacts with an alkene to form an ozonide intermediate that can be cleaved by the addition of $(CH_3)_2S$ into the two carbonyl species, namely ketones and aldehydes.
- Ozonolysis can produce an aldehyde product if there was a hydrogen on one of the original alkene sp2 carbon atoms.

REACTION 5F: CATALYTIC HYDROGENATION (Section 5.8)

C=C' —H_2 / metal→ H–C–C'–H

- Alkenes react quantitatively with molecular hydrogen (H_2) in the presence of a transition metal catalyst (platinum, palladium, ruthenium, and nickel) to give alkanes.
- Catalytic hydrogenation probably occurs because the H_2 molecule splits apart and makes two metal-hydrogen bonds on the metal surface. The alkene adsorbs onto the metal surface and then two sequential C-H bonds are made.
- The carbon-carbon sigma bond usually does not have a chance to rotate during the reaction so both hydrogen atoms are added to the same face of the alkene. This is referred to as **cis** or **syn addition**.

REACTION 5G: HYDROHALOGENATION (Section 5.3B)

C=C' —H-X→ H–C–C'–X

-HF, HCl, HBr, or HI can add to alkenes to give an alkyl halide.
- Like reaction 5B, protonation of the pi bond results in formation of a carbocation intermediate, then the halide anion is the nucleophile that reacts with the carbocation.
- The reaction follows Markovnikov's rule, so the hydrogen ends up on the carbon atom that has the greater number of hydrogens already attached to it.

- Because there is no bridged intermediate, the halide anion can attack from either side of the trigonal planar (sp^2 hybridized) carbocation. Therefore, addition of hydrogen halides to alkenes is not a stereoselective reaction.

REACTION 5H: RADICAL HYDROBROMINATION (Section 5.5)

$$\rangle C{=}C'\langle \xrightarrow[\text{peroxides}]{\text{H-Br}} -\overset{\text{H}}{\underset{|}{\overset{|}{C}}}-\underset{\text{Br}}{\underset{|}{\overset{|}{C'}}}-$$

- In the presence of radical initiators such as peroxides, HBr adds to alkenes via a radical chain reaction mechanism.
- In the **initiation step**, an alkoxy radical derived from the homolytic cleavage of a peroxide reacts with H-Br to give an alcohol and ·Br.
- In the first **chain propagation step**, the ·Br reacts with one electron of the alkene to give the bromoalkyl radical. The more stable radical predominates, that is, the one with the unpaired electron on the more highly substituted carbon atom.
- In the second propagation step, the bromoalkyl radical reacts with H-Br to give the product bromoalkane and ·Br. The interesting part is that ·Br reacts with another alkene molecule and the process is repeated over and over.
- The chain propagation continues until a **chain termination** step in which two radical species react with one another to make a covalent bond and thus quench the radicals.
- The net result is that the reaction does not follow Markovnikov's rule, because the bromine ends up on the less substituted carbon atom of the alkene.
- H-Cl and H-I do not undergo radical chain addition reactions with alkenes.

REACTION 5I: ALLYLIC HALOGENATION (Section 5.6)

$$\rangle C{=}C'\langle\ \underset{\text{H}}{\underset{|}{C''}}\langle \xrightarrow[\text{light or peroxides}]{\text{NBS}} \rangle C{=}C'\langle\ \underset{\text{Br}}{\underset{|}{C''}}\langle$$

- When different conditions are used, alkenes can react with radical species by a different chain reaction mechanism called **allylic substitution**.
- In allylic substitution, radicals such as ·Br at low concentration react with an alkene to form an **allylic radical**.
- The ·Br could be derived from heat or light causing the homolytic bond dissociation of Br_2, or by using a reagent such as N-bromosuccinimide in the presence of either heat or peroxides.
- The allylic radical then reacts with Br_2 to generate the bromoalkene and ·Br. The ·Br can then continue the chain propagation step by reacting with another molecule of alkene.
- The allylic radical, although still highly reactive, is relatively stable for a radical because the unpaired electron is delocalized. Allylic radicals are even more stable than tertiary radicals. This can be understood in terms of resonance stabilization as depicted in the following two contributing structures.

$$CH_2{=}CH{-}\dot{C}H_2 \longleftrightarrow \dot{C}H_2{-}CH{=}CH_2$$

- The radical delocalization is facilitated by the sp^2 hybridization of the radical carbon, since the 2p orbital containing the unpaired electron can overlap with the pi bond of the alkene.
- For unsymmetrical molecules, the reaction is regioselective. For example, secondary allylic hydrogens are substituted in preference to primary allylic hydrogens, etc.
- Allylic substitution can also occur with Cl_2 and light or heat.

REACTION 5J: OXIDATION TO VICINAL DIOLS (GLYCOLS) (Section 5.7B)

$$>C{=}C'< + KMnO_4 \longrightarrow \text{cyclic manganate ester } (-C(O)-C'(O)-,\ O{-}Mn(=O)(O^-){-}O) \xrightarrow{H_2O} -C(OH)-C'(OH)- + MnO_2$$

vicinal diol

- In these reactions a cyclic intermediate is formed from the alkene and then hydrolyzed to yield a **vicinal diol** (OH groups on adjacent carbon atoms).
- Note how the cyclic intermediate insures that both oxygen atoms are added to the same face of the alkene (syn addition).
- OsO_4 reacts with alkenes to give vicinal diols according to a mechanism that is similar to that shown for $KMnO_4$ above. In the case of OsO_4, the reagent can be used catalytically with the addition of other reagents such as peroxides (ROOR). When using $KMnO_4$, care must be taken to keep the reaction basic to prevent further reaction.

REACTION 5K: HALOHYDRIN FORMATION (Section 5.3F)

$$>C{=}C'< \xrightarrow[H_2O]{X_2} -C(HO)-C'(X)-$$

- Treatment of an alkene with Br_2 or Cl_2 in the presence of water results in addition of HO- and Br-, or HO- and Cl- to the alkene. The resulting compounds are called **halohydrins**, either **bromohydrins** or **chlorohydrins**.
- The reaction involves initial formation of a bridged bromonium or chloronium ion intermediate, followed by nucleophilic attack of water.
- The bridged halonium intermediate is very analogous to the mercurinium ion intermediate, displaying both bridged and partial carbocation characteristics. As a result, the nucleophilic attack of water occurs on the side of the intermediate opposite the halogen, and at the site of the more stable carbocation. Consistent with this, anti stereochemistry is observed, and the HO- ends up on the carbon atom that is more highly substituted (makes the more stable cation).

REACTION 5L: POLYMERS (Section 5.9)

$$CH_2{=}CH_2 \xrightarrow{\text{Catalyst}} -CH_2-CH_2-CH_2-CH_2-CH_2-CH_2-CH_2-CH_2-$$

ethylene polyethylene

- Industrially, the most important reaction of alkenes is **polymerization**, the building together of many small molecules into a very large molecular chain.
- Polymerization of ethylene is initiated by trace cations, anions or free radicals. When radicals are used, the ethylene polymerizes by a radical chain reaction mechanism.

REACTION 5M: FORMATION OF VICINAL DIHALIDES (Section 5.3A)

$$>C{=}C'< \xrightarrow{X_2} -\overset{X}{\underset{|}{\overset{|}{C}}}-\overset{|}{\underset{X}{\underset{|}{C'}}}-$$

- **Bromination** and **chlorination** involve the addition of Br_2 and Cl_2 to an alkene, respectively.
- In these reactions, one of the halogen atoms acts as an electrophile, breaking the halogen-halogen bond. This creates a positively-charged intermediate and a halide anion. The positively-charged intermediate has a unique bridged structure and is referred to as a bridged halonium ion. The halide anion then completes the reaction by creating a bond to the positively-charged species from the side of the molecule opposite the halogen bridge.
- The halogen bridge blocks the top of the structure, so the halide anion *must* attack from the side opposite the bridging group. The net result is that the two halogens end up on opposite faces of the molecule. This orientation is referred to as **anti addition.**

SUMMARY OF IMPORTANT CONCEPTS

5.0 OVERVIEW

- **Reaction mechanisms** describe how chemical bonds are made and broken during the course of a reaction, the order in which the bonds are broken and formed, the rates at which these processes occurs, and the role of solvent or a catalyst if any. Mechanisms provide a theoretical framework within which to organize a great deal of descriptive chemistry.

5.1 REACTIONS OF ALKENES-AN OVERVIEW

- In contrast to alkanes, alkenes react with a variety of compounds in at least three characteristic ways:
  - First, **addition reactions** involve breaking the pi bond of an alkene and replacing it with two sigma bonds.
  - Second, **allylic substitution reactions** can take place in which a hydrogen atom on the carbon atom adjacent to a carbon-carbon double bond can be replaced by a new atom or group of atoms.
  - Third, **polymer addition reactions** involve the formation of polymer chains from monomer alkene molecules.

5.2 REACTION MECHANISMS

- **Transition state theory** provides a model for understanding the relationships between reaction rates, molecular structure, and energetics.
  - The **total energy** of any chemical system is always conserved, and is the sum of the **kinetic energy** and **potential energy**. As molecules collide they convert kinetic energy into potential energy in the form of bond vibrational energy.
  - A **reaction coordinate** is a plot of the motion of atoms associated with changing energy as reactants proceed to products during a reaction.
  - For simple **one step reactions**, reaction occurs if sufficient potential energy becomes concentrated in the proper bonds.

The **transition state** or **activated complex** is the point on the reaction coordinate where the potential energy is a maximum.
An activated complex has essentially zero lifetime because it is a maximum on the energy diagram, yet it does have a definite arrangement of atoms and electrons.
The difference in potential energy between reactants and the activated complex is called the **energy of activation**. A molecule must have more potential energy than the energy of activation to proceed from starting materials to products.
The greater the energy of activation, the slower the **rate of reaction**, and vice versa.
The rate of reaction also depends on the **collisional frequency between molecules**, the **fraction of collisions with proper orientation** for reaction and the **fraction of collisions with energy greater than the energy of activation**.

- In multi-step reactions, each step has its own transition state and energy of activation.
  An **intermediate** is a potential energy minimum between two transition states on a reaction coordinate for a multi-step reaction. Reactive intermediates are never present in appreciable concentrations because the energy of activation for their conversion back to reactants or on to products is so small.
  The slowest step is the one that has the highest energy of activation and is called the **rate-determining step**. *[This is a very important concept in the study of reaction mechanisms. Notice that the overall rate of a multi-step reaction cannot be faster than the rate determining step.]*

5.3 ADDITION TO ALKENES-ELECTROPHILIC ADDITIONS

• The details of **alkene addition reactions** can best be understood by considering the mechanism of the reaction as well as the structure of the alkene.
  - The electrons of the alkene pi bond are located relatively far from the atomic nuclei, so they can act as a type of nucleophile with extremely electron deficient chemical species, referred to as **electrophiles**. *[This is the key idea of the chapter, and all of the following reactions should be thought of as starting with the weakly nucleophilic pi electrons attacking a highly electrophilic species.]*
  - When the pi electrons react with an electrophile, the pi bond is broken and a new sigma bond is formed with the electrophile. This creates a positively charged intermediate that is itself attacked by a nucleophile to form another new sigma bond, thereby completing the reaction.
    The key to understanding the details of these reactions is to keep track of the electrophile, the nucleophile, and the structure of the positively charged intermediate.

$$>C{=}C< \;+\; E^{+} \longrightarrow \text{Positively-Charged Intermediate} \xrightarrow{Nu^{-}} -\overset{E}{\underset{|}{\overset{|}{C}}}-\underset{Nu}{\overset{|}{C}}-$$

  - The bottom line is that the pi bond is replaced by two new sigma bonds, one to an electrophile and one to a nucleophile.

• In a **stereoselective reaction**, one stereoisomer is formed or destroyed in preference to all of the other possible stereoisomers.
  - The bromination and chlorination reactions of alkenes are stereoselective, since only the anti isomer is produced.

• **Markovnikov's rule** can be understood by considering the structure of the carbocation intermediate formed during the addition reaction.
  - The basic idea is that the more stable carbocation intermediate leads to the predominant (Markovnikov) product. The more stable carbocation is the one that has more alkyl groups attached to the positively charged carbon atom.

- In other words, a tertiary (3°) carbocation is more stable than a secondary (2°) carbocation, which is more stable than a primary (1°) carbocation, which is more stable than a methyl carbocation.
- A carbocation is sp^2 hybridized. All three sp^2 hybridized orbitals take part in bonds, but the 2p orbital is empty.

- A characteristic of carbocations is that they can **rearrange** if transfer of a **hydride ion** (a hydrogen atom plus the two bonding electrons) or alkyl group can create a more stable carbocation. Only consider rearrangements if a more stable carbocation can be formed from a less stable one, and if the reaction contains a full carbocation intermediate.

$$(H_3C)(H_3C)(H)C-\overset{+}{C}H-CH_3 \longrightarrow (H_3C)(H_3C)\overset{+}{C}-CH_2-CH_3$$

2° carbocation can rearrange to 3° carbocation

$$H_3C-CH_2-\overset{+}{C}H_2 \longrightarrow H_3C-\overset{+}{C}H-CH_3$$

1° carbocation can rearrange to 2° carbocation

5.7 OXIDATION OF ALKENES

- **Oxidation/reduction reactions** are a very important class of reactions in organic chemistry in which electrons are gained or lost by a reactant during the course of a reaction. Oxidation/reduction reactions can be recognized by writing **balanced half-reactions** that keep track of the organic structures involved.
  - To write a balanced half-reaction, first write a half-reaction showing the organic reactants and products. Complete a material balance using H^+ and H_2O for reactions carried out in acid, or OH^- and H_2O for reactions carried out in base. Finally complete the charge balance by adding electrons on one side or the other.
- An **oxidation** is defined as a reaction in which electrons are lost from a reactant being transformed into products.
- A **reduction** is defined as a reaction in which electrons are gained by a reactant being transformed into products.

5.8 ADDITION OF HYDROGEN-CATALYTIC REDUCTION

- **Heat of hydrogenation** is defined as the change in enthalpy, ΔH, for the reaction between an alkene and hydrogen to form an alkane.
  - Heats of hydrogenation are negative. In other words reduction of an alkene to an alkane is an exothermic process, because the reaction involves the breaking of a pi bond (pi bonds are relatively weak) and a sigma bond (the H-H bond) to from two stronger C-H bonds.
  - An alkene with a lower heat of hydrogenation (less exothermic) is the more stable alkene. This makes sense because you expect to get less energy out of a molecule that is already more stable.
  - From the comparison of heats of hydrogenation of a variety of molecules, the following general conclusions can be reached:

    More highly substituted alkenes are more stable than less highly substituted alkenes. ✱
    Trans alkenes are more stable than *cis* alkenes. This is a **steric affect** in that the cis substituents are actually so close that there is a net repulsion between the electron clouds of each. ✱

5.9 ALKENE SELF ADDITION-POLYMERIZATION

- Several small alkenes are used to make important polymers including vinyl chloride (makes PVC construction tubing), $CF_2{=}CF_2$ (makes Teflon), $CH_2{=}CCl_2$ (makes Saran wrap) and $CH_2{=}CHC_6H_5$ (makes Styrofoam).
- Because of the importance of polymers such as polyethylene, vast amounts of ethylene are produced every year. Ethylene is produced by the thermal cracking of hydrocarbons that come natural gas and petroleum. Ethylene is also converted into other useful compounds such as vinyl chloride and ethylene oxide that are themselves used for a wide variety of industrial applications including polymers.

CHAPTER 5
Solutions to the Problems

Problem 5.1 Name and draw structural formulas for the two possible products of the following alkene addition reactions. Use Markonikov's rule to predict which is the major product.

(a) $CH_3-CH=CH_2$ + HI ⟶ $CH_3CH(I)CH_3$ + $CH_3CH_2CH_2I$

major product ($CH_3CH(I)CH_3$)

(b) methylenecyclohexane ($C_6H_{10}=CH_2$) + HI ⟶ 1-iodo-1-methylcyclohexane + (iodomethyl)cyclohexane ($C_6H_{11}-CH_2I$)

major product (1-iodo-1-methylcyclohexane)

Problem 5.2 Draw structural formulas for the products of the following alkene addition reactions. Predict which is the major product.

(a) $CH_3-C(CH_3)=CH-CH_3$ + H_2O $\xrightarrow{H_2SO_4}$ $CH_3C(CH_3)(OH)CH_2CH_3$ + $CH_3CH(CH_3)CH(OH)CH_3$

major product ($CH_3C(CH_3)(OH)CH_2CH_3$)

(b) $CH_2=C(CH_3)-CH_2-CH_3$ + H_2O $\xrightarrow{H_2SO_4}$ $CH_3C(CH_3)(OH)CH_2CH_3$ + $CH_2(OH)CH(CH_3)CH_2CH_3$

major product ($CH_3C(CH_3)(OH)CH_2CH_3$)

Problem 5.3 Arrange the following carbocations in order of increasing stability.

(a) cyclohexyl cation with + on the ring carbon bearing $-CH_3$ (b) cyclohexyl cation with + on the ring carbon adjacent to the one bearing $-CH_3$ (c) cyclohexyl$-\overset{+}{C}H_2$

The order of increasing stability of carbocations is methyl < primary < secondary < tertiary.

(c) cyclohexyl$-\overset{+}{C}H_2$ (b) ring cation adjacent to $-CH_3$ (a) ring cation bearing $-CH_3$

primary carbocation **secondary carbocation** **tertiary carbocation**

Problem 5.4 Draw structural formulas for the trialkylborane and alkene that give the following alcohols under the reaction conditions shown:

(a) an alkene $\xrightarrow{BH_3}$ a trialkylborane $\xrightarrow[NaOH]{H_2O_2}$ $CH_3-CH(CH_3)-CH_2-CH_2-OH$

$CH_3CH(CH_3)CH=CH_2 \xrightarrow{BH_3} CH_3CH(CH_3)CH_2CH_2BR_2 \xrightarrow[NaOH]{H_2O_2} CH_3-CH(CH_3)-CH_2-CH_2-OH$

(b) an alkene $\xrightarrow{BH_3}$ a trialkylborane $\xrightarrow[NaOH]{H_2O_2}$ cyclopentyl-CH_2OH

methylenecyclopentane (cyclopentane$=CH_2$) $\xrightarrow{BH_3}$ cyclopentyl-CH_2BR_2 $\xrightarrow[NaOH]{H_2O_2}$ cyclopentyl-CH_2OH

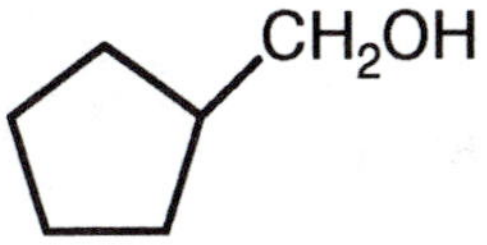

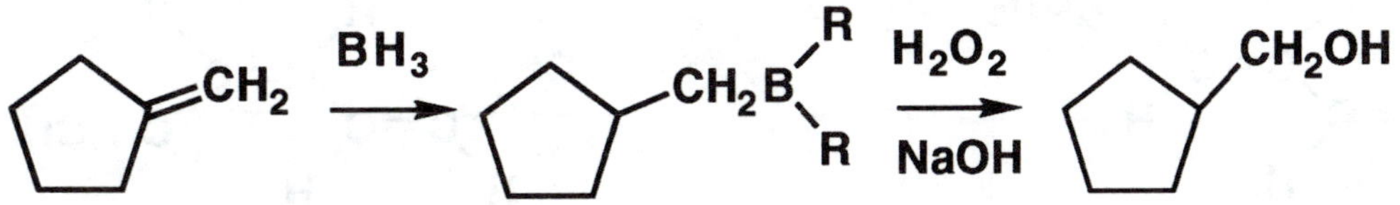

Problem 5.5 What alkene of molecular formula C_6H_{12}, when treated with ozone and then dimethyl sulfide, gives the following product(s)?

(a) $C_6H_{12} \xrightarrow[2.\ (CH_3)_2S]{1.\ O_3} CH_3CH_2CH{=}O$ (only product)

$CH_3CH_2CH=CHCH_2CH_3$

(b) $C_6H_{12} \xrightarrow[2.\ (CH_3)_2S]{1.\ O_3} CH_3CH{=}O + CH_3C({=}O)CH_2CH_3$ (equal moles of each)

$CH_3CH=C(CH_3)CH_2CH_3$

(c) $C_6H_{12} \xrightarrow[2.\ (CH_3)_2S]{1.\ O_3} CH_3C({=}O)CH_3$ (only product)

$CH_3C(CH_3)=C(CH_3)CH_3$

Problem 5.6 Arrange the four *cis-trans* isomers of 2,4-heptadiene in order of decreasing stability.

Following are structural formulas for these four dienes. The most stable has a trans configuration about each double bond. The least stable diene is the all-*cis* isomer. The *trans-cis* and *cis-trans* isomers are intermediate in stability.

most stable

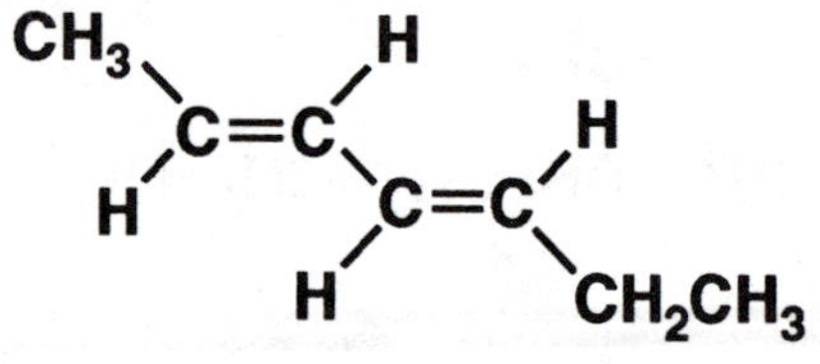

trans--trans

intermediate stability

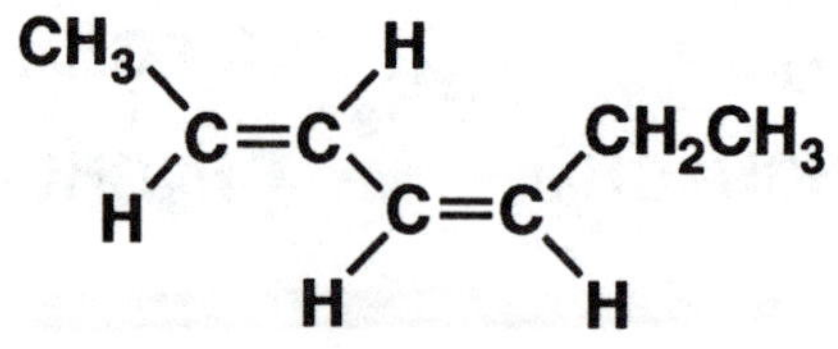

trans--cis

intermediate stability

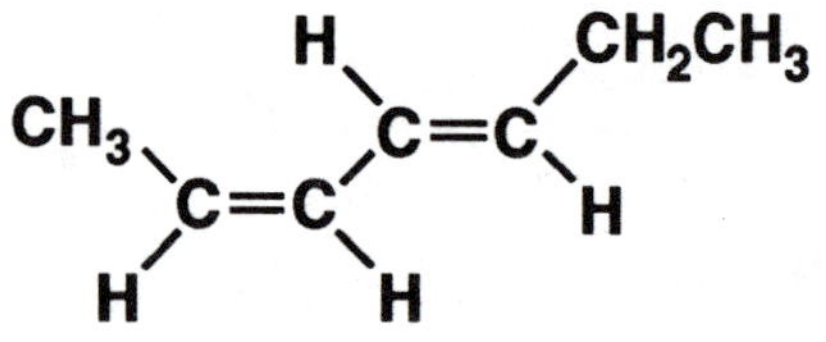

cis--trans

least stable

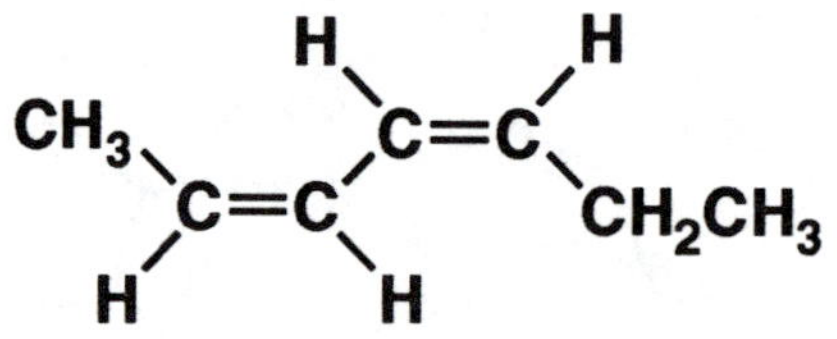

cis--cis

Energetics of Chemical Reactions

Problem 5.7 Following are some bond dissociation energies.

Bond	Bond dissociation energy (kcal/mol)	Bond	Bond dissociation energy (kcal/mol)
H-H	104	C-Si	72
O-H	110.6	C=C	146
C-H	98.7	C=O (aldehyde)	174
N-H	93.4	C=O (CO_2)	192
Si-H	76	C≡O	257
C-C	82.6	N≡N	227
C-N	73	C≡C	200
C-O	85.5	O=O	119
C-I	51		

If a suitable catalyst could be found, which of the following reactions are energetically favorable?

The following reactions can only occur to a significant extent as written if they are exothermic, that is if the bonds that are formed are stronger than the ones that are broken in the reaction. Recall that a catalyst increases the rate, but not the overall thermodynamics of a reaction.

To find out if a reaction is exothermic, the dissociation energy of all the bonds in the molecules on each side of the equation are added together. If the bond dissociation energy total from the right side of the equation is higher than the total from the left side of the equation, then the reaction is exothermic (ΔH for the reaction is negative).

(a) $CH_2{=}CH_2 + H_2 + N_2 \longrightarrow H_2N\text{-}CH_2\text{-}CH_2\text{-}NH_2$

The bond dissociation energies from the left side of the equation:

146 + (4 x 98.7) + 104 + 227 = 871.8 kcal/mol
(C=C) (4 C-H) (H-H) (N≡N)

The bond dissociation energies from the right side of the equation:

(4 x 93.4) + (2 x 73) + 82.6 + (4 x 98.7) = 997 kcal/mol
(4 N-H) (2 C-N) (C-C) (4 C-H)

This reaction is exothermic because 997 is larger than 871.8.

(b) $CH_2{=}CH_2 + CH_4 \longrightarrow H\text{-}CH_2\text{-}CH_2\text{-}CH_3$

The bond dissociation energies from the left side of the equation:

146 + (4 x 98.7) + (4 x 98.7) = 935.6 kcal/mol
(C=C) (4 C-H) (4 C-H)

The bond dissociation energies from the right side of the equation:

(3 x 82.6) + (8 x 98.7) = 1037.4 kcal/mol
(3 C-C) (8 C-H)

This reaction is exothermic because 1037.4 is larger than 934.6.

(c) $CH_2{=}CH_2 + (CH_3)_3SiH \longrightarrow H\text{-}CH_2\text{-}CH_2\text{-}Si(CH_3)_3$

The bond dissociation energies from the left side of the equation:

146 + (4 x 98.7) + (9 x 98.7) + (3 x 72) + 76 = 1721.1 kcal/mol
(C=C) (4 C-H) (9 C-H) (3 C-Si) (Si-H)

The bond dissociation energies from the right side of the equation:

(82.6) + (5 x 98.7) + (9 x 98.7) + (4 x 72) = 1752.4 kcal/mol
(C-C) (5 C-H) (9 C-H) (4 C-Si)

This reaction is exothermic because 1752.4 is larger than 1721.1.

(d) $CH_2{=}CH_2 + CHI_3 \longrightarrow H\text{-}CH_2\text{-}CH_2\text{-}C(I)_3$

The bond dissociation energies from the left side of the equation:

146 + (5 x 98.7) + (3 x 51) = 792.5 kcal/mol
(C=C) (5 C-H) (3 C-I)

The bond dissociation energies from the right side of the equation:

(3 x 82.6) + (5 x 98.7) + (3 x 51) = 894.3 kcal/mol
(3 C-C) (5 C-H) (3 C-I)

This reaction is exothermic because 894.3 is larger than 792.5, so a catalyst might be found for this reaction.

(e) $CH_2{=}CH_2 + CO + H_2 \longrightarrow H{-}CH_2{-}CH_2{-}\overset{\overset{\displaystyle O}{\|}}{C}H$

The bond dissociation energies from the left side of the equation:

146 + (4 x 98.7) + 257 + 104 = 901.8 kcal/mol
(C=C) (4 C-H) (C≡O) (H-H)

The bond dissociation energies from the right side of the equation:

(3 x 82.6) + (6 x 98.7) + 174 = 1014 kcal/mol
(3 C-C) (6 C-H) (C=O)

This reaction is exothermic because 1014 is larger than 901.8.

(f) 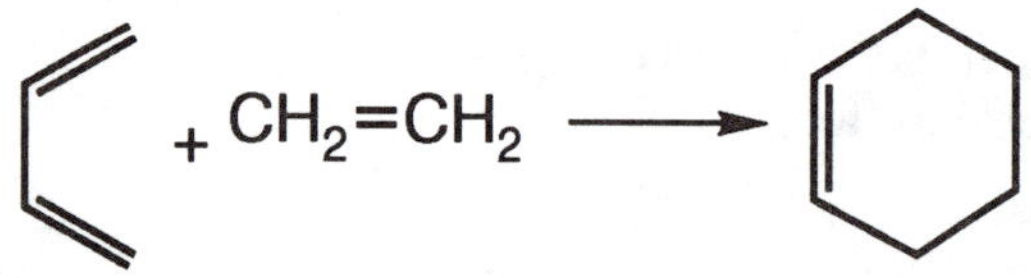 [butadiene] + $CH_2{=}CH_2 \longrightarrow$ [cyclohexene]

The bond dissociation energies from the left side of the equation:

(3 x 146) + 82.6 + (10 x 98.7) = 1507.6 kcal/mol
(3 C=C) (C-C) (10 C-H)

The bond dissociation energies from the right side of the equation:

146 + (5 x 82.6) + (10 x 98.7) = 1546 kcal/mol
C=C (5 C-C) (10 C-H)

This reaction is exothermic because 1546 is larger than 1507.6.

(g) 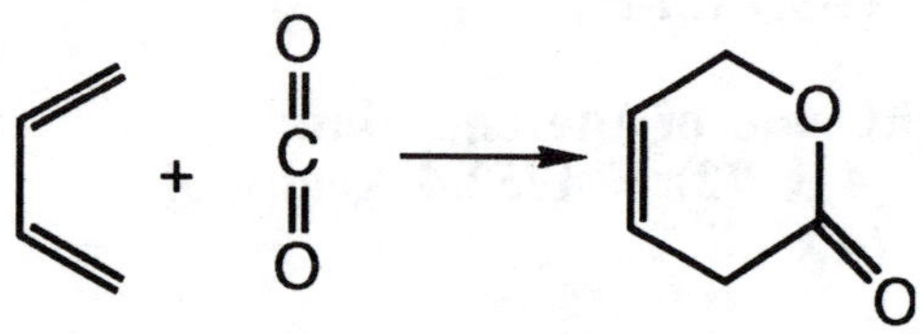 [butadiene] + O=C=O $\longrightarrow$ [dihydropyranone]

The bond dissociation energies from the left side of the equation:

(2 x 146) + 82.6 + (2 x 192) + (6 x 98.7) = 1350.8 kcal/mol
(2 C=C) (C-C) (2 C=O) (6 C-H)

The bond dissociation energies from the right side of the equation:

146 + (3 x 82.6) + (6 x 98.7) + (2 x 85.5) + 174 = 1331 kcal/mol
C=C (3 C-C) (6 C-H) (2 C-O) (C=O)

This reaction is endothermic because 1331 is smaller than 1350.8.

(h) $HC\equiv CH + O_2 \longrightarrow H\text{-}\overset{O}{\overset{\|}{C}}\text{-}\overset{O}{\overset{\|}{C}}\text{-}H$

The bond dissociation energies from the left side of the equation:

200 + (2 x 98.7) + 119 = 516.4 kcal/mol
(C≡C) (2 C-H) (O=O)

The bond dissociation energies from the right side of the equation:

82.6 + (2 x 174) + (2 x 98.7) = 628 kcal/mol
(C-C) (2 C=O) (2 C-H)

This reaction is exothermic because 628 is larger than 516.4.

Electrophilic Additions

Problem 5.8 Using your knowledge of resonance, predict which double-bonded carbon atoms in the following molecules are most reactive toward electrophiles (such as H^+):

(a) $CH_2=CH\text{-}\ddot{\underset{\cdot\cdot}{O}}\text{-}CH_3$

The sp^2 carbon atom on the end is the one that is most reactive with electrophiles, because as shown below, the adjacent oxygen atom helps to stabilize an adjacent positive charge. In the contributing structure on the right, all atoms have a complete octet.

$$CH_3-\overset{+}{C}H\text{-}\ddot{\underset{\cdot\cdot}{O}}\text{-}CH_3 \longleftrightarrow CH_3-CH=\underset{+}{\ddot{O}}\text{-}CH_3$$

(b) $CH_2=CH\text{-}CH=CH_2$

Again, the sp^2 carbon atoms on the end are the most reactive with electrophiles, because the pi electrons from the other double bond can be used to stabilize an adjacent positive charge.

$$CH_3-\overset{+}{C}H\text{-}CH=CH_2 \longleftrightarrow CH_3-CH=CH\text{-}\overset{+}{C}H_2$$

Problem 5.9 Draw structural formulas for the isomeric carbocations formed by addition of H^+ to the following alkenes. Label each carbocation primary, secondary or tertiary, and state which of the isomeric carbocations is formed more readily.

(a) $CH_3-CH_2-\overset{CH_3}{\overset{|}{C}}=CH\text{-}CH_3$

$$CH_3-CH_2-\overset{\overset{CH_3}{|}}{\underset{+}{C}}-CH_2-CH_3 \quad + \quad CH_3-CH_2-\overset{\overset{CH_3}{|}}{CH}-\underset{+}{CH}-CH_3$$

tertiary (more stable) — secondary (less stable)

(b) $CH_3-CH_2-CH=CH-CH_3$

$$CH_3-CH_2-\overset{+}{C}H-CH_2-CH_3 \quad + \quad CH_3-CH_2-CH_2-\overset{+}{C}H-CH_3$$

both secondary carbocations (of equal stability)

(c) 1-methylcyclopentene (CH_3) → tertiary carbocation (more stable) + secondary carbocation (less stable)

tertiary (more stable)

secondary (less stable)

(d) methylenecyclohexane ($=CH_2$) → cyclohexyl-$\overset{+}{C}H_2$ + 1-methylcyclohexyl cation ($-CH_3$)

primary (less stable)

tertiary (more stable)

Problem 5.10 Arrange the following compounds in order of increasing rate of reaction with HI. Draw the structural formula of the major product formed in each case and explain the basis for your ranking.

(a) $CH_3-CH=CH-CH_3$

$$CH_3-CH=CH-CH_3 \longrightarrow CH_3-CH_2-\overset{+}{C}H-CH_3 \longrightarrow CH_3-CH_2-\overset{\overset{I}{|}}{C}H-CH_3$$

a secondary carbocation — 2-iodobutane (sec-butyl iodide)

(b) $CH_3-\overset{\overset{CH_3}{|}}{C}=CH-CH_3$

$$CH_3-\overset{\overset{CH_3}{|}}{C}=CH-CH_3 \longrightarrow CH_3-\overset{\overset{CH_3}{|}}{\underset{+}{C}}-CH_2-CH_3 \longrightarrow CH_3-\overset{\overset{CH_3}{|}}{\underset{\underset{I}{|}}{C}}-CH_2-CH_3$$

a tertiary carbocation — 2-iodo-2-methylbutane

(c) $CH_3-CH=CH_2$

$$CH_3-CH=CH_2 \longrightarrow CH_3-\overset{+}{C}H-CH_3 \longrightarrow CH_3-CH(I)-CH_3$$

a secondary carbocation — 2-iodopropane (isopropyl iodide)

Reaction rates in these electrophilic additions are a function of the stability of the carbocation intermediate formed in the rate-determining step. The more stable the carbocation intermediate (tertiary > secondary > primary) the greater the rate of the reaction. In this problem, (b) reacts most rapidly, followed by (a) and (c) with approximately equal reactivities toward addition of HI.

Problem 5.11 Predict the organic product(s) of the reaction of 2-butene with the following reagent(s).

(a) H_2O (H_2SO_4)

$CH_3-CH(OH)-CH_2-CH_3$

(b) Br_2

$CH_3-CH(Br)-CH(Br)-CH_3$

(c) Cl_2

$CH_3-CH(Cl)-CH(Cl)-CH_3$

(d) Br_2 in H_2O

$CH_3-CH(Br)-CH(OH)-CH_3$

(e) HI

$CH_3-CH(I)-CH_2-CH_3$

(f) Cl_2 in H_2O

$CH_3-CH(Cl)-CH(OH)-CH_3$

(g) $Hg(OAc)_2$, H_2O

$CH_3-CH(OH)-CH(HgOAc)-CH_3$

(h) the product in (g) + $NaBH_4$

$CH_3-CH(OH)-CH_2-CH_3$

Problem 5.12 Reaction of 2-methyl-2-pentene with each of the following shows a high regioselectivity. Draw a structural formula for the major product of each reaction, and account for the observed regioselectivity.

(a) HBr (absence of peroxides)

$CH_3-C(CH_3)(Br)-CH_2-CH_2-CH_3$

(b) HBr (presence of peroxides)

$CH_3-CH(CH_3)-CH(Br)-CH_2-CH_3$

(c) H_2O in the presence of H_2SO_4

$CH_3-C(CH_3)(OH)-CH_2-CH_2-CH_3$

(d) Br_2 in H_2O

$CH_3-C(CH_3)(HO)-CH(Br)-CH_2-CH_3$

(e) $Hg(OAc)_2$ in H_2O

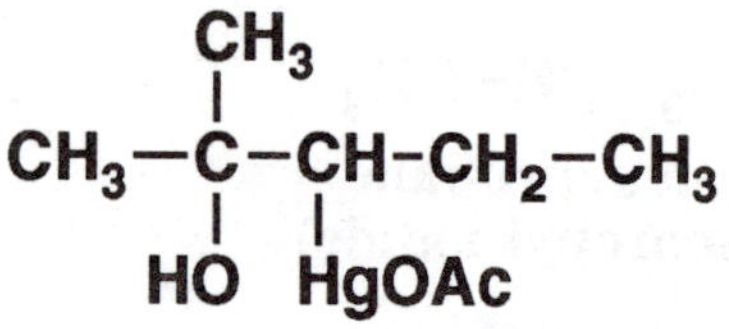

$$CH_3-C(CH_3)(HO)-CH(HgOAc)-CH_2-CH_3$$

Problem 5.13 Reaction of 1-methylcyclopentene with the following reagents shows a high degree of regioselectivity and stereoselectivity. Propose a mechanism for each reaction and account for the observed regioselectivity and stereoselectivity.

(a) Br_2

Stereoselectivity is accounted for by formation of a bridged bromonium ion intermediate that is then attacked by a nucleophile at the carbon of the bridged intermediate most able to accommodate partial positive charge.

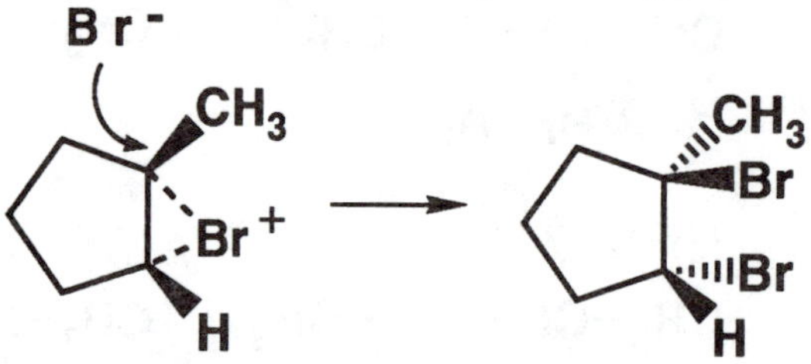

Br⁻ → CH₃, Br⁺, H → CH₃, Br, Br, H

a bridged bromonium ion intermediate

(b) Br_2 in H_2O

Attack by H_2O on the carbon bearing the methyl group followed by loss of a proton gives the trans bromohydrin.

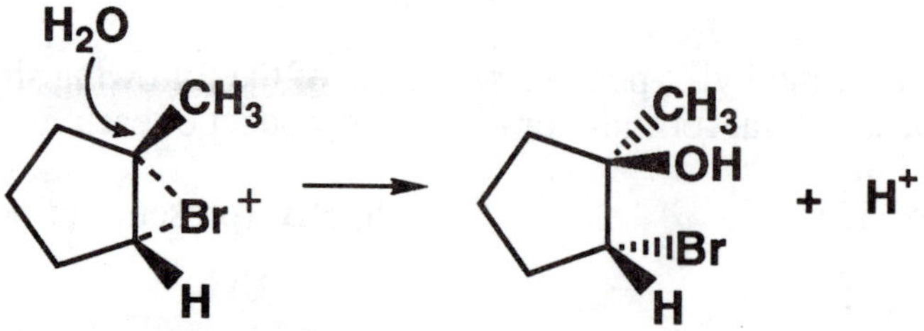

H_2O → CH₃, Br⁺, H → CH₃, OH, Br, H + H^+

a bridged bromonium ion intermediate

(c) $Hg(OAc)_2$ in H_2O

Attack by water on the bridged mercurinium ion followed by loss of a proton results in -OH trans to -HgOAc.

H_2O CH_3 + HgOAc H → CH_3 OH HgOAc H + H^+

a bridged mercurinium ion intermediate

Problem 5.14 Draw the structural formula for an alkene or alkenes with the indicated molecular formula that give the compound shown as the major product. Note that more than one alkene may give the same compound as the major product.

(a) C_5H_{10} + H_2O $\xrightarrow{H_2SO_4}$ $CH_3-C(CH_3)(OH)-CH_2-CH_3$

$CH_2=C(CH_3)-CH_2-CH_3$ or $CH_3-C(CH_3)=CH-CH_3$

(b) C_5H_{10} + Br_2 → $CH_3-CH(CH_3)-CH(Br)-CH_2Br$

$CH_3-CH(CH_3)-CH=CH_2$

(c) C_7H_{12} + HCl → 1-chloro-1-methylcyclohexane (CH_3, Cl)

1-methylcyclohexene (CH_3) or methylenecyclohexane (CH_2)

Problem 5.15 Reaction of the following bicycloalkene with bromine in carbon tetrachloride gives a *trans*-dibromide. In both (a) and (b), the bromine atoms are trans to each other. However, only one of these products shown is formed. Which *trans*-dibromide is formed and how might you account for the fact that it is formed to the exclusion of the other *trans*-dibromide?

CH_3 + Br_2 $\xrightarrow{CCl_4}$ or

(a) (b)

Product (a) is formed. Electrophilic addition of bromine to an alkene occurs via a bridged bromonium ion intermediate and anti addition of the two bromine atoms. In a cyclohexane ring, anti addition corresponds to trans and diaxial addition. Only in formula (a) are the two added bromines trans and diaxial. In (b) they are trans, but diequatorial.

(a) *trans*-diequatorial **(b) *trans*-diaxial**

Problem 5.16 Terpin hydrate is prepared commercially by addition of two mol of water to limonene (Figure 4.6) in the presence of dilute sulfuric acid. Terpin hydrate is used medicinally as an expectorant for coughs. It may be given as terpin hydrate and codeine.

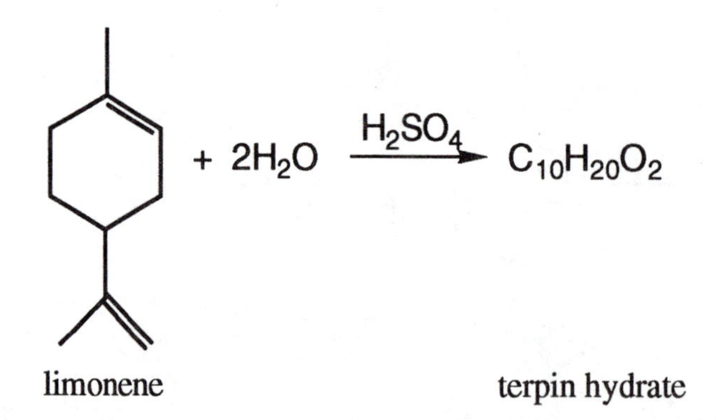

(a) Propose a structural formula for terpin hydrate and propose a mechanism to account for the formation of the product you have proposed.

Add water to each double bond by (1) protonation of each double bond to give a 3° carbocation, (2) reaction of each carbocation with water, and (3) loss of a proton to give terpin hydrate.

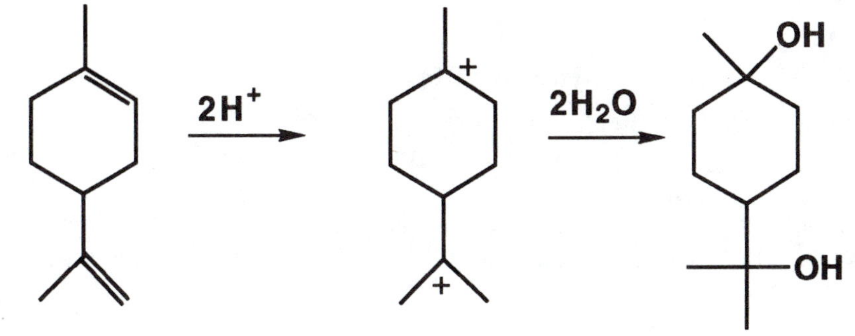

$$\xrightarrow{2H^+} \xrightarrow{2H_2O}$$

limonene **terpin hydrate**

(b) How many *cis-trans* isomers are possible for the structural formula you have proposed?

There are two *cis-trans* isomers, shown here as chair conformations with the $(CH_3)_2COH$- side chain equatorial.

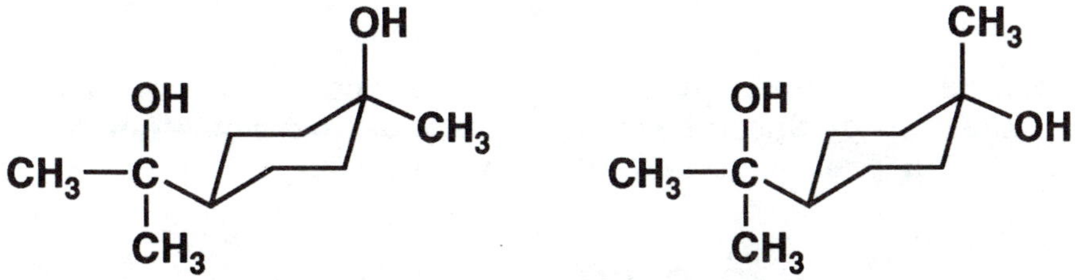

Problem 5.17 Treatment of 2-methylpropene with methanol in the presence of sulfuric acid yields a compound of formula $C_5H_{12}O$.

$$CH_3-\overset{CH_3}{\overset{|}{C}}=CH_2 + CH_3OH \xrightarrow{H_2SO_4} C_5H_{12}O$$

Propose a structural formula for this compound and also a mechanism to account for its formation.

Reaction of the alkene with a proton gives a tertiary carbocation intermediate. Reaction of this intermediate with the oxygen atom (a nucleophile) of methanol followed by loss of a proton gives *tert*-butyl methyl ether.

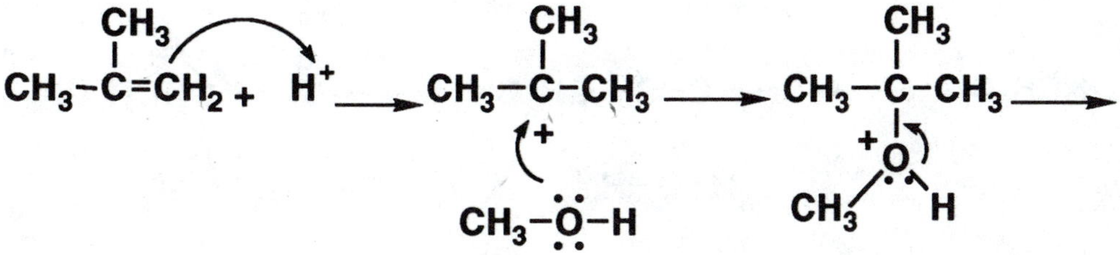

$$CH_3-\overset{CH_3}{\overset{|}{C}}=CH_2 + H^+ \longrightarrow CH_3-\overset{CH_3}{\overset{|}{C^+}}-CH_3 \xrightarrow{CH_3-\ddot{O}-H} CH_3-\overset{CH_3}{\overset{|}{C}}-CH_3 \longrightarrow$$

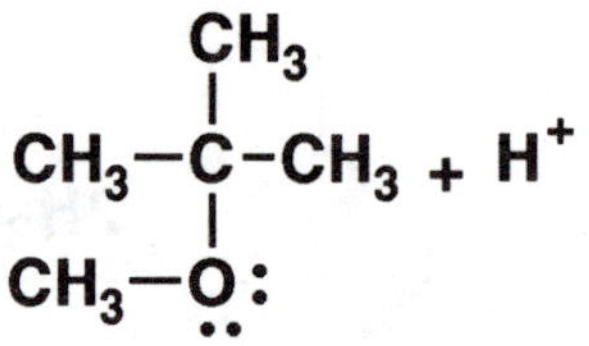

$$CH_3-\overset{\overset{CH_3}{|}}{\underset{\underset{CH_3-\ddot{O}:}{|}}{C}}-CH_3 + H^+$$

Problem 5.18 Treatment of cyclohexene and HBr in the presence of acetic acid gives bromocyclohexane (85%) and cyclohexyl acetate (15%). Propose a mechanism for formation of the latter product.

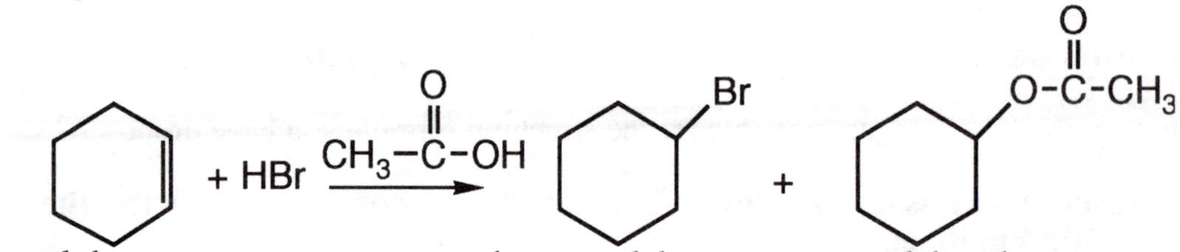

$$\text{cyclohexene} + HBr \xrightarrow{CH_3-\overset{O}{\overset{\|}{C}}-OH} \text{bromocyclohexane (85\%)} + \text{cyclohexyl acetate (15\%)}$$

cyclohexene

bromocyclohexane (85%)

cyclohexyl acetate (15%)

Reaction of cyclohexene with a proton gives a secondary carbocation intermediate. Reaction of this intermediate with the oxygen atom of acetic acid, followed by loss of a proton gives cyclohexyl acetate.

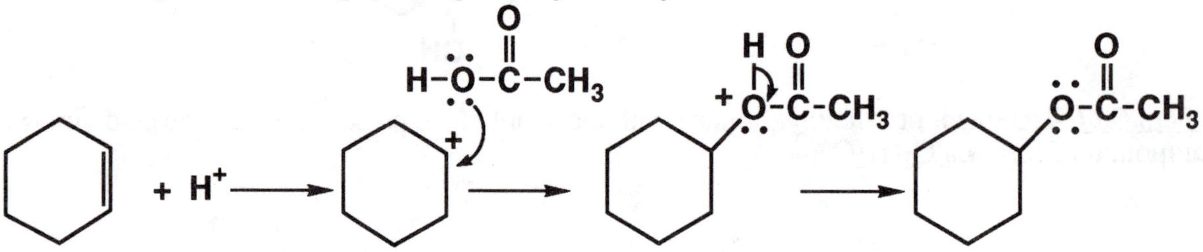

$$\text{cyclohexene} + H^+ \longrightarrow \text{cyclohexyl cation} \xrightarrow{H-\ddot{O}-\overset{O}{\overset{\|}{C}}-CH_3} \text{cyclohexyl}-\overset{+}{O}(H)-\overset{O}{\overset{\|}{C}}-CH_3 \longrightarrow \text{cyclohexyl}-\ddot{O}-\overset{O}{\overset{\|}{C}}-CH_3$$

Problem 5.19 Propose a mechanism for the following reaction.

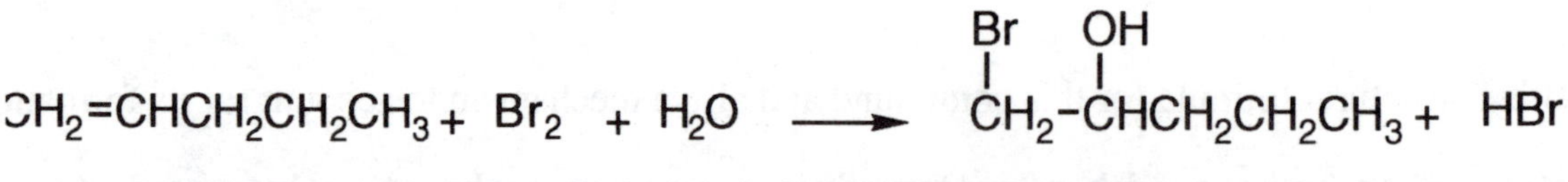

$$CH_2{=}CHCH_2CH_2CH_3 + Br_2 + H_2O \longrightarrow \underset{Br}{CH_2}-\underset{OH}{CH}CH_2CH_2CH_3 + HBr$$

Reaction of 1-pentene with bromine gives a bridged bromonium ion intermediate. Anti attack of water on this intermediate at the more substituted secondary carbon, followed by loss of a proton gives 1-bromo-2-pentanol.

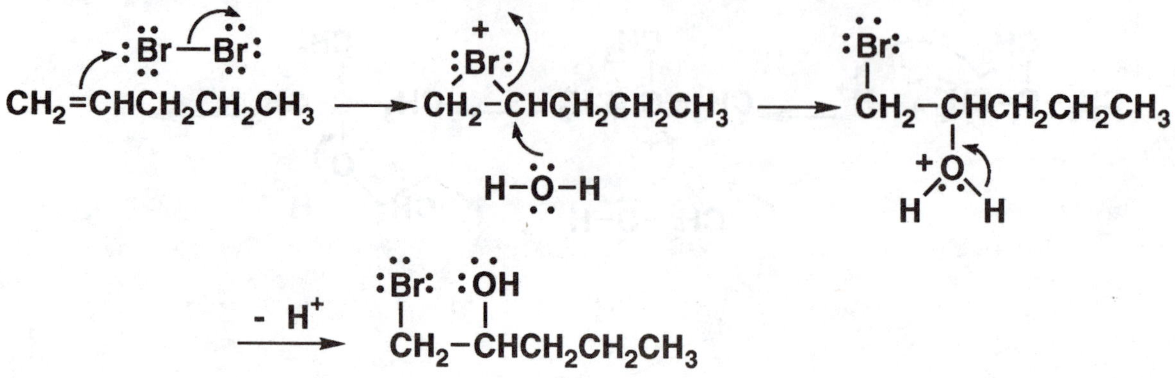

$$CH_2{=}CHCH_2CH_2CH_3 \xrightarrow{:\ddot{Br}-\ddot{Br}:} \text{bromonium ion } (:\overset{+}{Br}: \text{ bridging } CH_2-CHCH_2CH_2CH_3) \xrightarrow{H-\ddot{O}-H} \underset{}{:\ddot{Br}:}-CH_2-CH(\overset{+}{O}H_2)CH_2CH_2CH_3$$

$$\xrightarrow{-H^+} :\ddot{Br}:-CH_2-CH(:\ddot{O}H)CH_2CH_2CH_3$$

Problem 5.20 Reaction of 4-penten-1-ol with bromine in water forms a cyclic bromoether. How do you account for the formation of this product rather than a simple bromohydrin as was formed in problem 5.19?

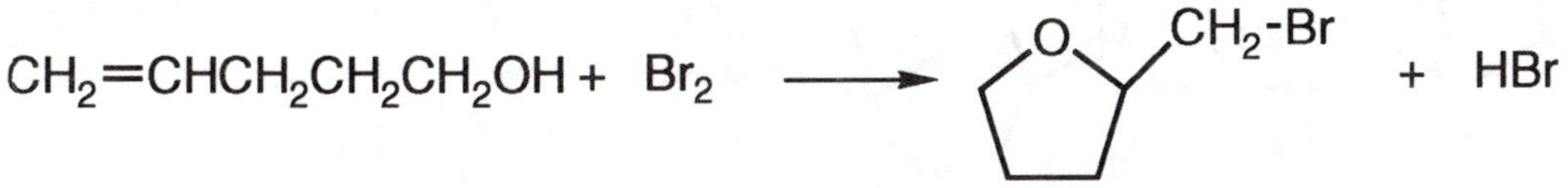

$CH_2{=}CHCH_2CH_2CH_2OH + Br_2 \longrightarrow$ (2-(bromomethyl)tetrahydrofuran, CH_2-Br) $+ HBr$

4-penten-1-ol

Reaction of the alkene with bromine gives a bridged bromonium ion intermediate. Reaction of this intermediate with the oxygen atom of the hydroxyl group followed by loss of a proton gives the observed cyclic ether, a derivative of tetrahydrofuran.

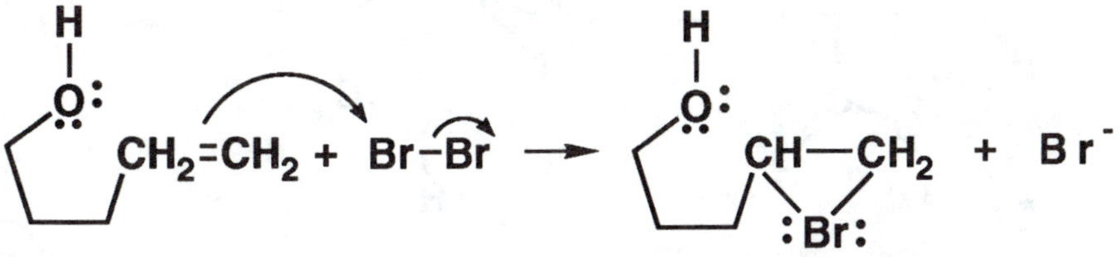

H–O: ... $CH_2{=}CH_2$ + Br–Br ⟶ H–O: ... CH—CH_2 (:Br: +) + Br^-

a bridged bromonium ion intermediate

CH—CH_2 (:Br: +) + Br^- ⟶ +O(H) ... CH—CH_2–Br ⟶ (tetrahydrofuran ring, CH_2–Br) + H^+

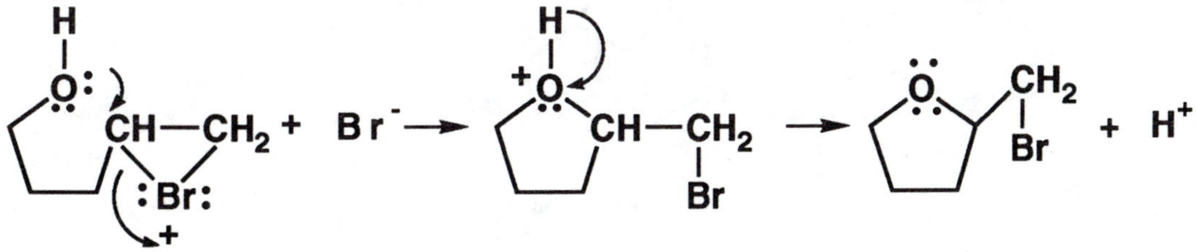

Problem 5.21 Provide a mechanism for each of the following reactions.

(a) (5-hexen-1-ol, OH) $\xrightarrow[H_2O]{H_2SO_4}$ (2-methyltetrahydropyran, O, CH_3)

(alkene ... :OH) + H–O–S(=O)(=O)–O–H ⟶ CH_3–$\overset{+}{C}H$ ... :OH

⟶ H–Ö:–H + (ring +O–H, CH_3) ⟶ (ring O, CH_3) + H_3O^+

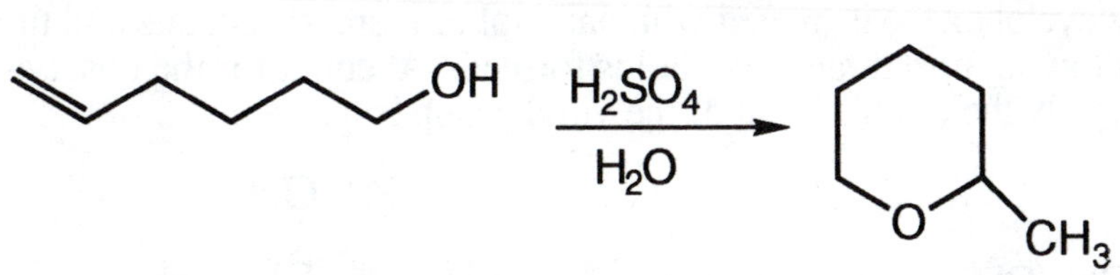

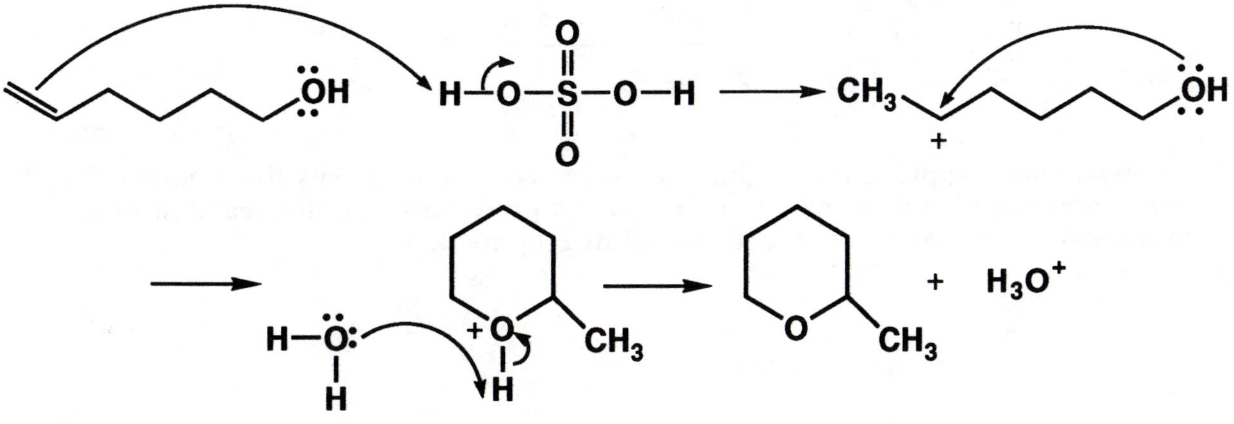

(b)

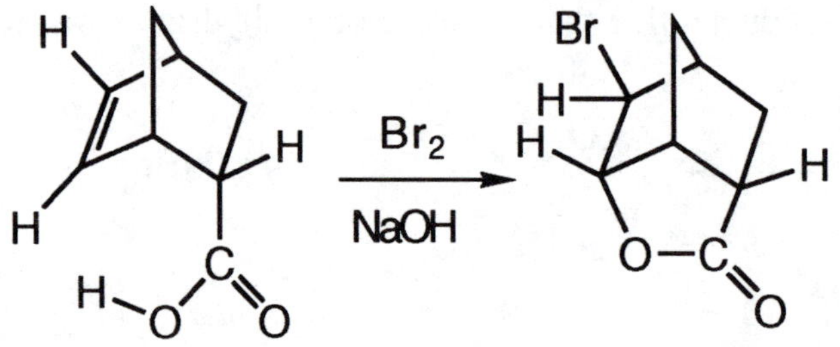

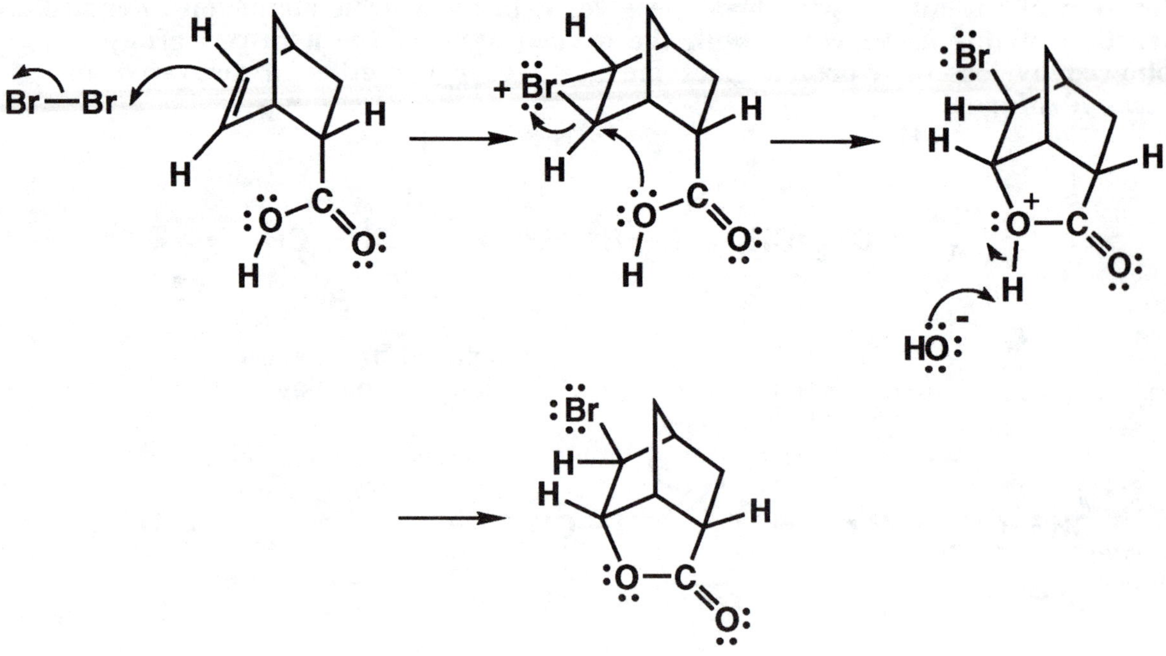

$\xrightarrow[\text{NaOH}]{Br_2}$

Problem 5.22 When 4-vinylcyclohexene is treated with one mol of mercuric acetate and then with sodium borohydride, the following unsaturated alcohol is formed. Account for the observation that under these conditions, only the double bond of the vinyl group is hydrated?

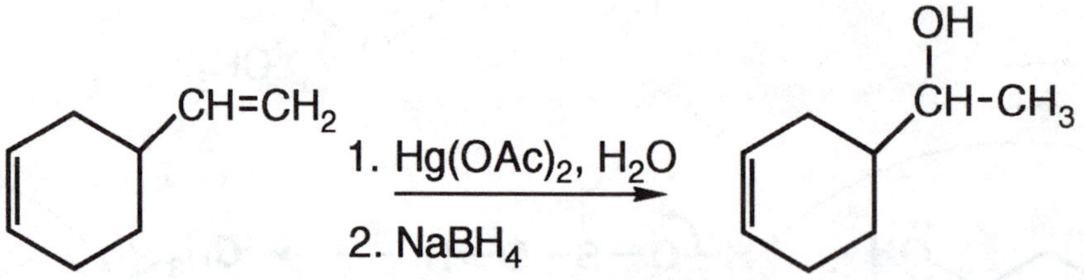

$\xrightarrow[\text{2. } NaBH_4]{\text{1. } Hg(OAc)_2,\ H_2O}$

The most likely explanation is that the oxymercuration is sensitive enough to steric crowding about the alkene that there is preference for the reaction of a monosubstituted alkene over a disubstituted cycloalkene.

Concerted Additions: Hydroboration

Problem 5.23 Each alkene is treated with diborane in tetrahydrofuran (THF) to form a trialkylborane and then with hydrogen peroxide in aqueous sodium hydroxide. Draw a structural formula of the alcohol formed in each case. Specify stereochemistry where appropriate.

(a) CH_2 (b) CH_3

CH_2OH CH_3 OH

(c) $CH_3-\overset{CH_3}{\overset{|}{C}}=CH-CH_2-CH_3$

$CH_3\overset{CH_3}{\overset{|}{C}}H\underset{OH}{\underset{|}{C}}HCH_2CH_3$

(d) $CH_2=CH(CH_2)_5CH_3$

$HOCH_2CH_2(CH_2)_5CH_3$

(e) $(CH_3)_3CCH=CH_2$

$CH_3\overset{CH_3}{\overset{|}{\underset{CH_3}{\underset{|}{C}}}}CH_2CH_2OH$

Problem 5.24 Reaction of α-pinene with diborane followed by treatment of the resulting trialkylborane with alkaline hydrogen peroxide gives an alcohol with the following structural formula.

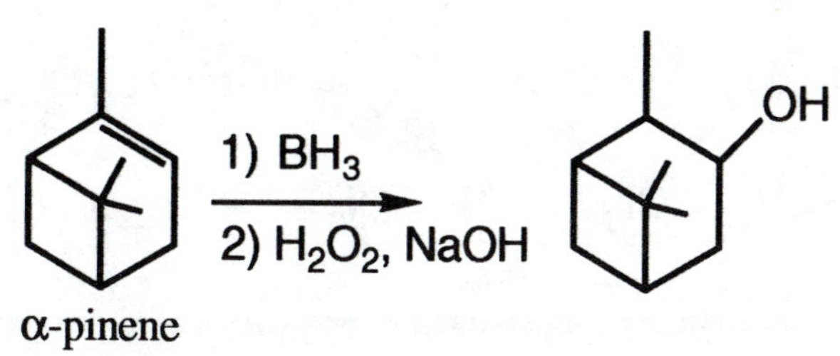

α-pinene

(a) Four *cis-trans* isomers possible for this bicyclic alcohol. Draw formulas for all four.

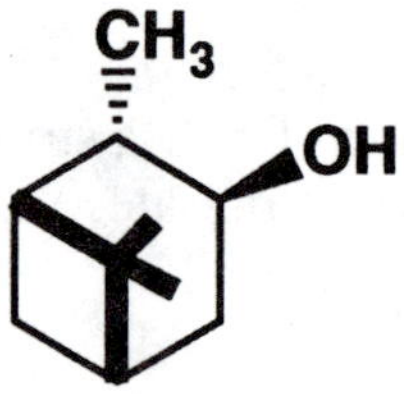

CH_3 OH [1] CH_3 OH [2] CH_3 OH [3] CH_3 OH [4]

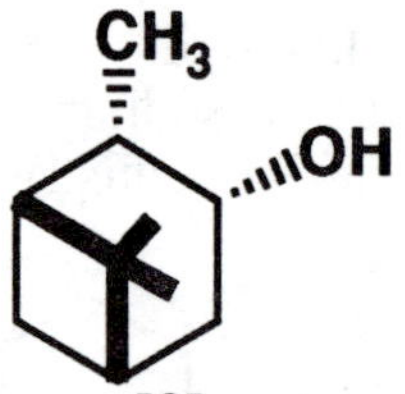

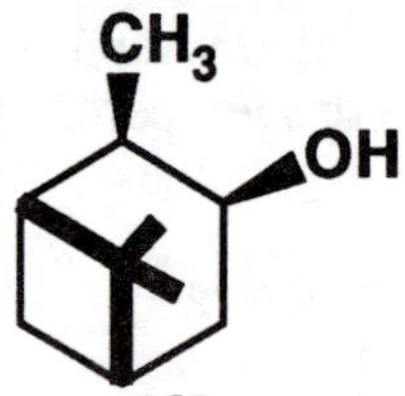

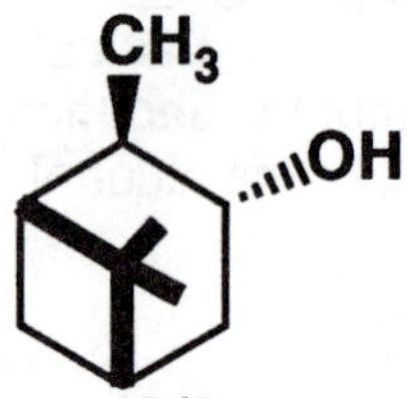

(b) Of the four possible *cis-trans* isomers that could exist for a molecule of this structural formula, only one is formed (in over 85% yield) in the above reaction. Which *cis-trans* isomer is formed and how do you account for the observed stereoselectivity ?

Above are perspective formulas for the four possible *cis-trans* isomers. Hydroboration followed by treatment with alkaline hydrogen peroxide results in syn (*cis*) addition of -H and -OH. Furthermore, boron adds to the less substituted carbon and from the least hindered side. In hydroboration of α-pinene, boron adds to the disubstituted carbon of the double bond and from the side opposite the bulky dimethyl substituted bridge. Compound [4] is the product formed in 85% yield.

Radical Additions

Problem 5.25 In the presence of radical initiators, carbon tetrachloride reacts with alkenes to give a 1:1 adduct as shown in the following example.

$$CH_3(CH_2)_5CH{=}CH_2 + CCl_4 \xrightarrow[\text{heat}]{\text{peroxide}} CH_3(CH_2)_5CH(Cl){-}CH_2{-}CCl_3$$

Propose a radical chain mechanism to account for the formation of the observed product. In addition, demonstrate that your chain propagation steps add up to the observed reaction.

Initiation is by a peroxide radical reacting with tetrachloromethane (carbon tetrachloride) to form a trichloromethyl radical. As shown by the following balanced equation, the proposed chain propagation steps add up to the observed reaction.

Initiation: $Rad\cdot + Cl{-}CCl_3 \longrightarrow Rad{-}Cl + \cdot CCl_3$

Propagation:

$$R{-}CH{=}CH_2 + \cdot CCl_3 \longrightarrow R{-}\dot{C}H{-}CH_2{-}CCl_3$$

$$R{-}\dot{C}H{-}CH_2{-}CCl_3 + Cl{-}CCl_3 \longrightarrow R{-}CH(Cl){-}CH_2{-}CCl_3 + \cdot CCl_3$$

Sum of chain propagation steps:

$$R{-}CH{=}CH_2 + Cl{-}CCl_3 \longrightarrow R{-}CH(Cl){-}CH_2{-}CCl_3$$

Termination: (one of several possible) $R-\dot{C}H-CH_2-CCl_3 + \cdot CCl_3 \longrightarrow R-CH(CCl_3)-CH_2-CCl_3$

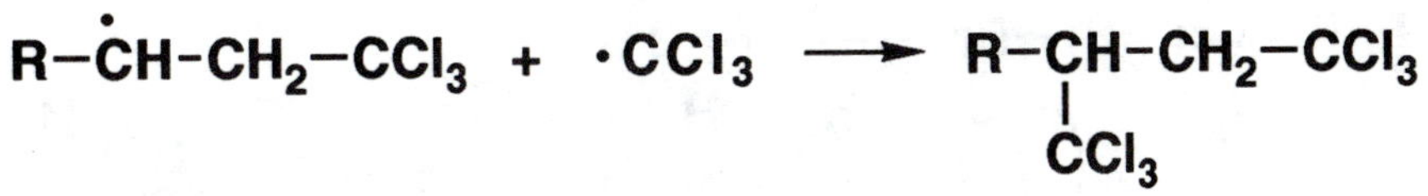

Problem 5.26 Cyclobutane reacts with bromine to give bromocyclobutane, but bicyclo[1.1.0]butane reacts with bromine to give 1,3-dibromocyclobutane. Account for the difference in reaction between the two compounds.

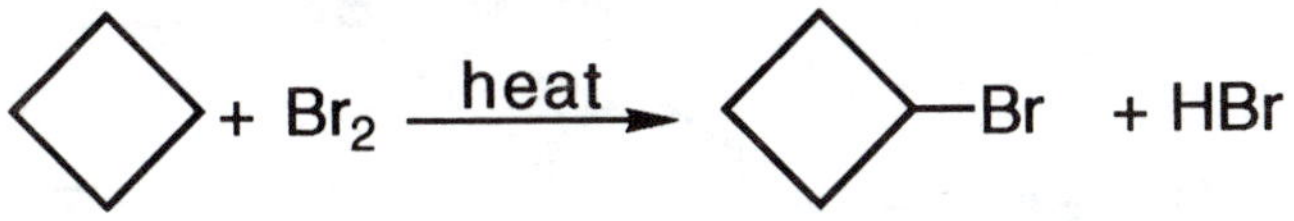

Cyclobutane + Br_2 —heat→ Bromocyclobutane + HBr

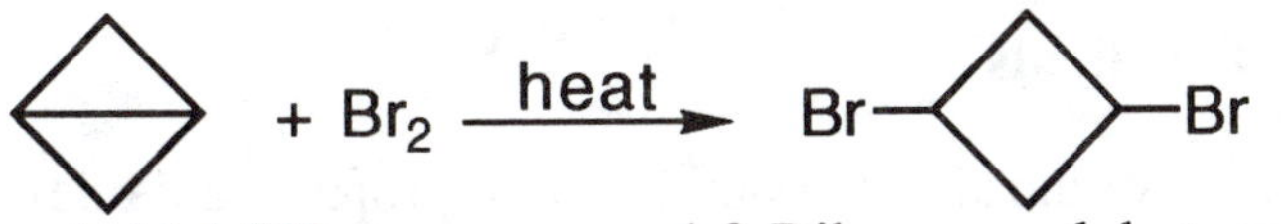

Bicyclo{1.1.0}butane + Br_2 —heat→ 1,3-Dibromocyclobutane

The bridging double bond in bicyclo[1.1.0]butane is highly strained, thus Br• radical reacts with this bond to create a new radical followed by reaction with another Br_2 molecule to yield the final products.

Initiation: $Br—Br \xrightarrow{\text{heat}} 2Br\cdot$

Propagation: Br• + bicyclo[1.1.0]butane ⟶ Br–cyclobutyl•

Br–cyclobutyl• + Br—Br ⟶ Br–cyclobutyl–Br + Br•

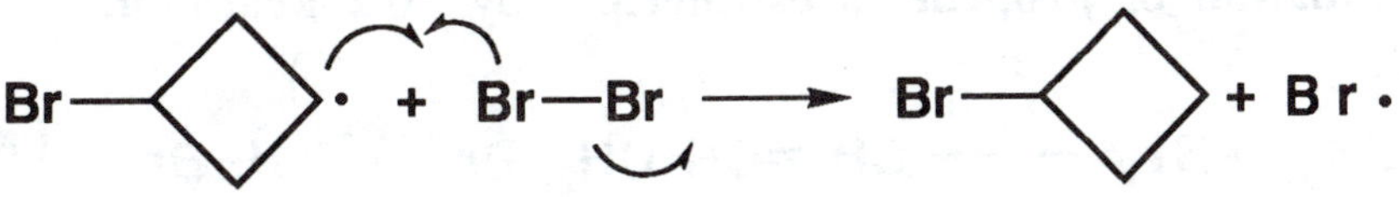

Allylic Halogenation

Problem 5.27 Propose a series of chain initiation, propagation, and termination steps to account for the following reaction.

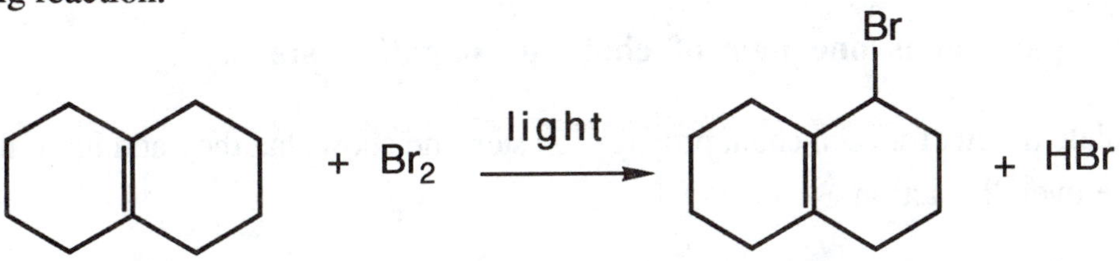

+ Br_2 —light→ (Br) + HBr

Following is a possible radical chain mechanism for this reaction.

Initiation: $Br{-}Br \xrightarrow{light} 2Br\cdot$

Propagation:

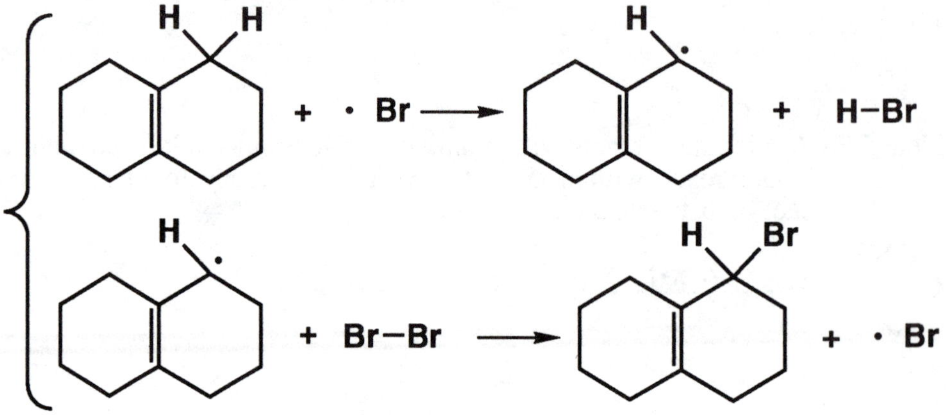

(cyclic allylic H abstraction by $\cdot Br$ giving allylic radical $+ H{-}Br$; allylic radical $+ Br{-}Br \longrightarrow$ allylic bromide $+ \cdot Br$)

Termination: Coupling of any pair of radicals

Problem 5.28 Following is a balanced equation for the allylic bromination of propene and relevant bond dissociation energies.

$$CH_2{=}CH{-}CH_3 + Br_2 \longrightarrow CH_2{=}CH{-}CH_2{-}Br + HBr$$

Bond	Bond dissociation energy (kcal/mol)
$CH_2{=}CHCH_2{-}H$	+87.0
$Br{-}Br$	+46.0
$CH_2{=}CHCH_2{-}Br$	+55.7
$Br{-}H$	+87.5

(a) Calculate the heat of reaction (ΔH) for this conversion.

Allylic bromination of propene is exothermic by 10.2 kcal/mol.

$CH_2{=}CH{-}CH_3$	$+Br_2$	$\longrightarrow$	$CH_2{=}CH{-}CH_2{-}Br$	+	$H{-}Br$	ΔH (kcal/mol
+87.0	+46.0		-55.7		-87.5	-10.2

(b) Propose a pair of chain propagation steps and show that, taken together, your steps add up to the observed reaction.

Following part (c) is one pair of chain propagation steps.

(c) Calculate the ΔH for each chain propagation step and show that they add up to the observed ΔH for the overall reaction.

The sum of enthalpies of each step adds up to the value for the heat of reaction calculated in part (a).

Chain Propagation Steps

	ΔH (kcal/mol)
$CH_2{=}CH{-}CH_3 + \cdot Br \longrightarrow CH_2{=}CH{-}\dot{C}H_2 + H{-}Br$	
+87.0 ... -87.5	-0.5
$CH_2{=}CH{-}\dot{C}H_2 + Br{-}Br \longrightarrow CH_2{=}CH{-}CH_2{-}Br + \cdot Br$	
+46.0 ... -55.7	-9.7
$CH_2{=}CH{-}CH_3 + Br_2 \longrightarrow CH_2{=}CH{-}CH_2{-}Br + H{-}Br$	-10.2

Halogenation of Alkanes

Problem 5.29 Following is a balanced equation for bromination of propane.

$$CH_3{-}CH_2{-}CH_3 + Br_2 \longrightarrow CH_3{-}CH(Br){-}CH_3 + HBr$$

(a) Using the values for bond dissociation energies given in Appendix 1, calculate ΔH for this reaction.

Formation of 2-bromopropane (isopropyl bromide) by radical bromination of propane is exothermic by 14 kcal/mol.

$$CH_3{-}CH_2{-}CH_3 + Br_2 \longrightarrow CH_3{-}CH(Br){-}CH_3 + H{-}Br$$

$CH_3{-}CH_2{-}CH_3$	Br_2	$CH_3{-}CH(Br){-}CH_3$	$H{-}Br$	ΔH (kcal/mol
+96	+46	-68	-88	-14

(b) Propose a pair of chain propagation steps and show that, taken together, yours add up to the observed reaction.

Following is a pair of chain propagation steps for this reaction. Of these steps, the first involving hydrogen abstraction has the higher energy of activation.

	ΔH (kcal/mol)
$CH_3{-}CH_2{-}CH_3 + \cdot Br \longrightarrow CH_3{-}\dot{C}H{-}CH_3 + H{-}Br$	
+96 ... -88	+8
$CH_3{-}\dot{C}H{-}CH_3 + Br_2 \longrightarrow CH_3{-}CH(Br){-}CH_3 + \cdot Br$	
+46 ... -68	-22
sum of ΔH for chain propagation steps:	-14

Following is an alternative pair of chain propagation steps. Because of the considerably higher energy of activation of the first of these steps, the rate of chain propagation by this mechanism is so low that it is not competitive with the chain mechanism first proposed.

Reaction	ΔH (kcal/mol)
$CH_3-CH_2-CH_3 + \cdot Br \longrightarrow CH_3-CH(Br)-CH_3 + \cdot H$	
+96 -68	+28
$\cdot H + Br_2 \longrightarrow H-Br + \cdot Br$	
+46 -88	-42
sum of ΔH for chain propagation steps:	-14

(c) Calculate ΔH for each of your chain propagation steps.

See answer to part (b)

(d) Of your chain propagation steps, which is the slower, i.e., which has the higher energy of activation.

See answer to part (b)

Problem 5.30 Following are balanced equations for fluorination of propane to produce both 1-fluoropropane and 2-fluoropropane.

$$CH_3-CH_2-CH_3 + F_2 \longrightarrow CH_3-CH_2-CH_2-F + HF$$

$$CH_3-CH_2-CH_3 + F_2 \longrightarrow CH_3-\underset{}{CH}(F)-CH_3 + HF$$

Assume that each product is formed by a radical chain reaction.

(a) Calculate ΔH for each reaction.

Formation of 1-fluoropropane (propyl fluoride) and 2-fluoropropane (isopropyl fluoride) are both exothermic.

				ΔH (kcal/mol)
$CH_3{-}CH_2{-}CH_3$ + F_2	$\longrightarrow$	$CH_3{-}CH_2{-}CH_2{-}F$ + HF		
+100	+38	-106	-136	-109

$CH_3{-}CH_2{-}CH_3$ + F_2	$\longrightarrow$	$CH_3{-}CH(F){-}CH_3$ + HF		
+96	+38	-107	-136	-109

(b) Propose a pair of chain propagation steps for each reaction and calculative ΔH for each step.

Following are pairs of chain propagation steps for formation of each product. Both steps in each pair of reactions are highly exothermic and have energies of activation of only a few kcal/mol.

				ΔH (kcal/mol)
$CH_3{-}CH_2{-}CH_3$ + ·F	$\longrightarrow$	$CH_3{-}CH_2{-}\dot{C}H_2$	+ HF	
+100			-136	-36
$CH_3{-}CH_2{-}\dot{C}H_2$ + F_2	$\longrightarrow$	$CH_3{-}CH_2{-}CH_2{-}F$	+ ·F	
+38		-106		-68
				-104

				ΔH (kcal/mol)
$CH_3{-}CH_2{-}CH_3$ + ·F	$\longrightarrow$	$CH_3{-}\dot{C}H{-}CH_3$	+ HF	
+96			-136	-40
$CH_3{-}\dot{C}H{-}CH_3$ + F_2	$\longrightarrow$	$CH_3{-}CH(F){-}CH_3$	+ ·F	
+38		-107		-69
				-109

(c) Reasoning from the Hammond postulate, predict the regioselectivity of radical fluorination relative to that of chlorination and bromination.

Because the hydrogen abstraction step in each sequence is highly exothermic, the transition state is reached very early in hydrogen abstraction and the intermediate in this step has very little radical character. Therefore, the relative stabilities of primary versus secondary radicals is of little importance in determination of product. Accordingly predict very low regioselectivity for fluorination of hydrocarbons.

Oxidation

Problem 5.31 Write structural formulas for the major organic product(s) formed by the reaction of 1-methylcyclohexene with the following oxidizing agents.

(a) H_2O_2/OsO_4

(b) $KMnO_4$ (cold, dilute)

For both (a) and (b):

(c) O_3 followed by $(CH_3)_2S$

Problem 5.32 Each of the following alkenes is reacted with ozone to form an ozonide and then with dimethyl sulfide. Draw the structural formula of the organic product(s) formed from each alkene.

(a) $CH_3-C(CH_3)=CH-CH_2-CH(CH_3)-CH_3$

$CH_3-C(CH_3)=O$ + $H-C(=O)-CH_2-CH(CH_3)-CH_3$

(b) $CH_3-C(CH_3)=CH-CH_2-C(CH_3)=CHCH_2CH_3$

$CH_3-C(CH_3)=O$ + $H-C(=O)-CH_2-C(CH_3)=O$ + $H-C(=O)-CH_2CH_3$

(c) α-pinene (Fig. 4.6)

$H-C(=O)-CH_2-$ [cyclobutane with gem-dimethyl] $-C(=O)-CH_3$

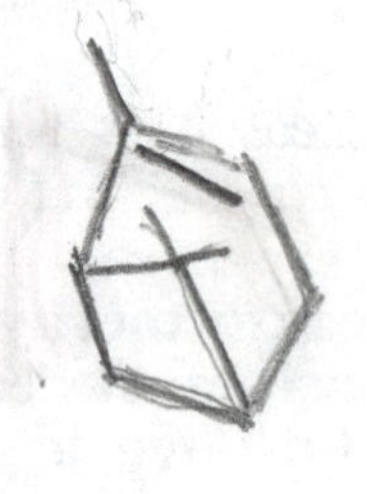

(d) limonene (Fig. 4.6)

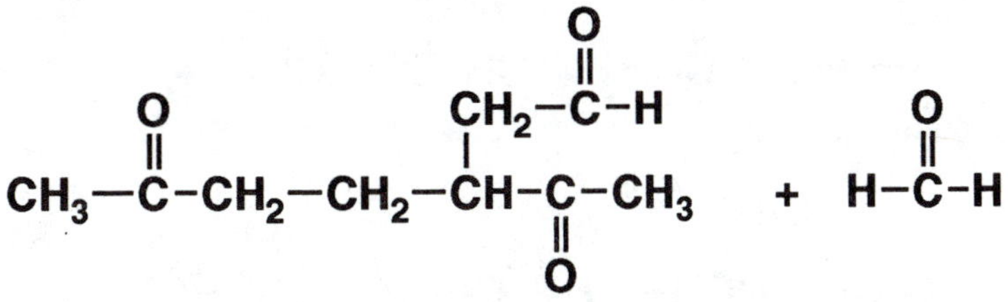

$$CH_3-\overset{O}{\overset{\|}{C}}-CH_2-CH_2-\underset{CH_2-\overset{O}{\overset{\|}{C}}-H}{\overset{|}{CH}}-\underset{\underset{O}{\|}}{C}-CH_3 \quad + \quad H-\overset{O}{\overset{\|}{C}}-H$$

(e) zingiberene (Fig. 4.7)

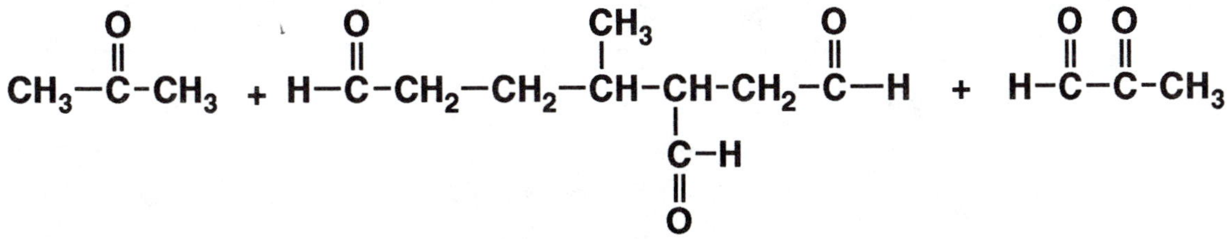

$$CH_3-\overset{O}{\overset{\|}{C}}-CH_3 \; + \; H-\overset{O}{\overset{\|}{C}}-CH_2-CH_2-\overset{CH_3}{\overset{|}{CH}}-\underset{\underset{\underset{O}{\|}}{C-H}}{\underset{|}{CH}}-CH_2-\overset{O}{\overset{\|}{C}}-H \; + \; H-\overset{O}{\overset{\|}{C}}-\overset{O}{\overset{\|}{C}}-CH_3$$

(f) caryophyllene (Fig. 4.7)

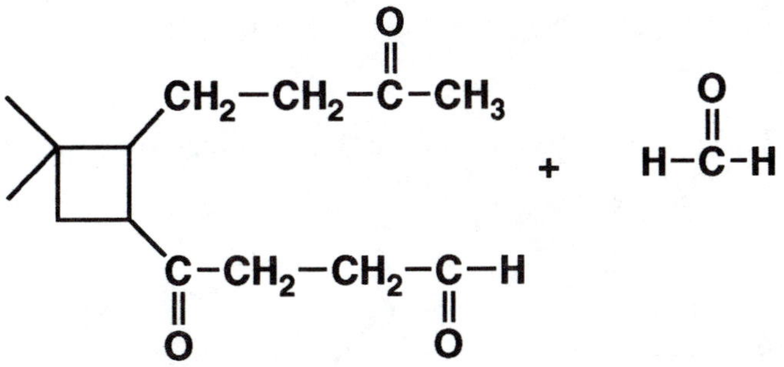

Cyclobutane ring with gem-dimethyl, substituted with $CH_2-CH_2-\overset{O}{\overset{\|}{C}}-CH_3$ and $\underset{\underset{O}{\|}}{C}-CH_2-CH_2-\underset{\underset{O}{\|}}{C}-H$ $\quad + \quad H-\overset{O}{\overset{\|}{C}}-H$

Problem 5.33 Draw the structural formula of the alkene that reacts with ozone followed by dimethyl sulfide to give the following products.

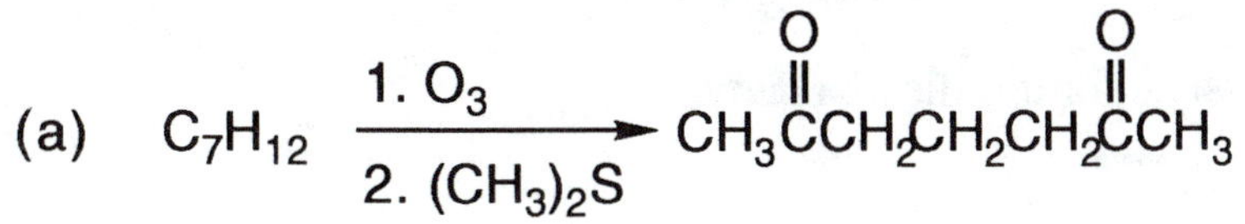

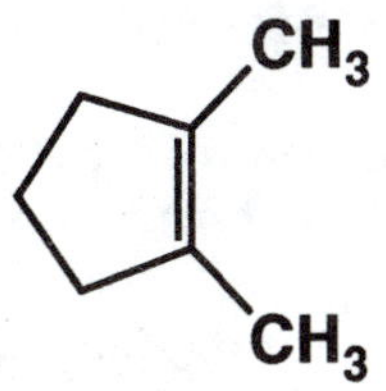

(a) $C_7H_{12} \xrightarrow[2.\ (CH_3)_2S]{1.\ O_3} CH_3\overset{O}{\overset{\|}{C}}CH_2CH_2CH_2\overset{O}{\overset{\|}{C}}CH_3$

1,2-dimethylcyclopentene (CH_3, CH_3 on the ring double bond)

(b) $C_{10}H_{18} \xrightarrow[2.\ (CH_3)_2S]{1.\ O_3} CH_3\overset{O}{\overset{\|}{C}}CH_3 + CH_3\overset{O}{\overset{\|}{C}}CH_2CH_3 + H\overset{O}{\overset{\|}{C}}CH_2\overset{O}{\overset{\|}{C}}H$

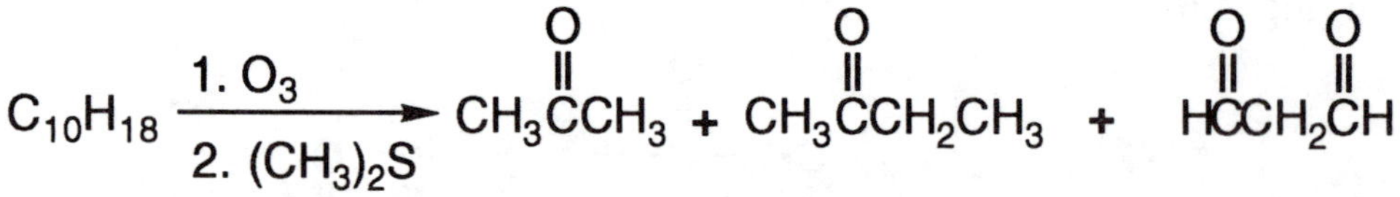

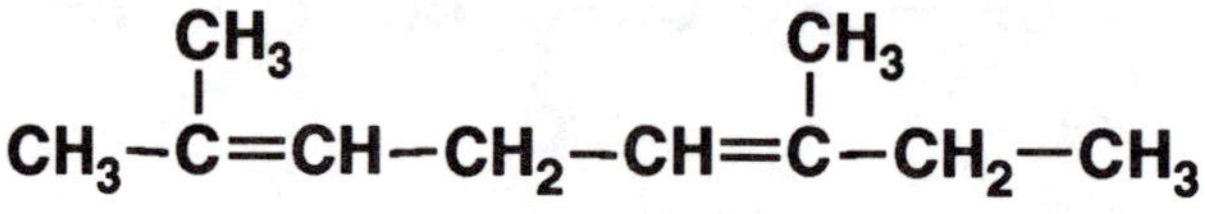

$$CH_3-\underset{}{\overset{CH_3}{\overset{|}{C}}}=CH-CH_2-CH=\overset{CH_3}{\overset{|}{C}}-CH_2-CH_3$$

(c) $C_{10}H_{18} \xrightarrow[2.\ (CH_3)_2S]{1.\ O_3} CH_3\overset{CH_3}{\overset{|}{C}}HCH_2\overset{O}{\overset{\|}{C}}CH_2CH_2CH_2CH_2\overset{O}{\overset{\|}{C}}H$

$CH_2\overset{CH_3}{\overset{|}{C}}HCH_3$ (substituent on cyclohexene ring)

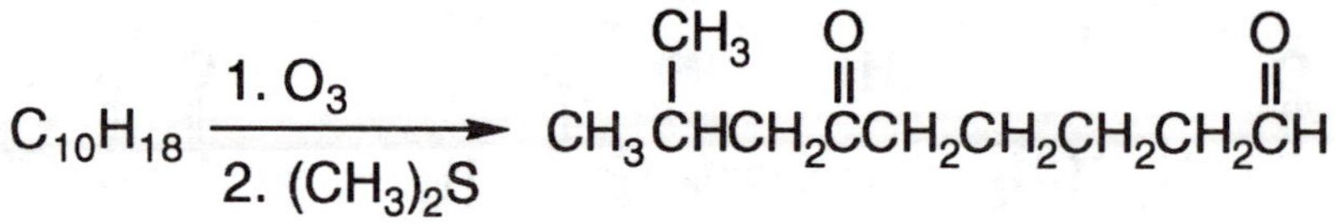

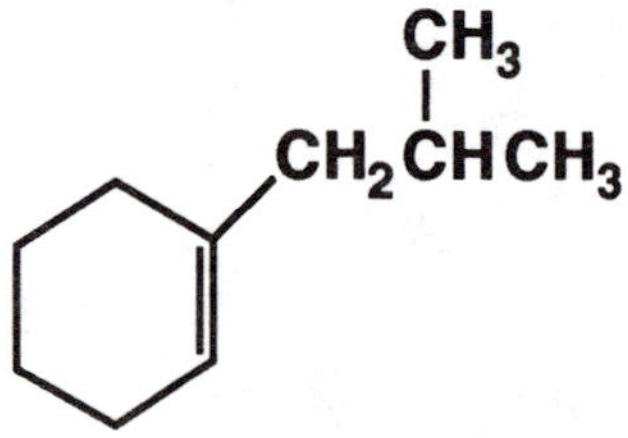

Problem 5.34 Bicyclo[2.2.1]-2-heptene (norbornene) is oxidized by ozone/dimethyl sulfide to cyclopentane-1,3-dicarbaldehyde.

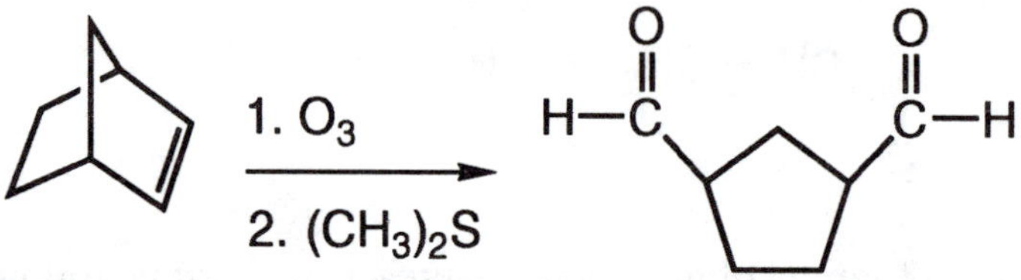

1. O_3
2. $(CH_3)_2S$

H—C(=O) ... C(=O)—H

bicyclo[2.2.1]-2-heptene (norbornene)

cyclopentane-1,3-dicarbaldehyde

(a) How many *cis-trans* isomers are possible for this dicarbaldehyde?

One pair of *cis-trans* isomers is possible.

(b) Which of the possible *cis-trans* isomers is formed by ozonolysis of norbornene?

Only the *cis* isomer is formed. Because of the geometry of the bicycloalkene, the two carbon atoms of the alkene double bond must be fused *cis* to each other.

Problem 5.35 (a) Draw a structural formula for the bicycloalkene of molecular formula C_8H_{12} that, on ozonolysis followed by dimethyl sulfide, gives cyclohexane-1,4-dicarboxaldehyde.

Following are two stereorepresentations for the bicycloalkene.

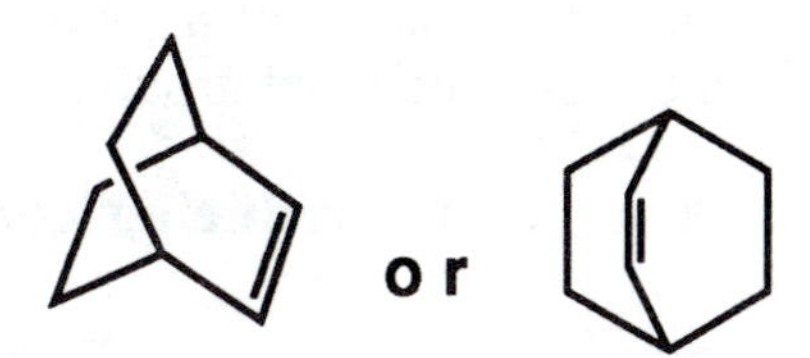

bicyclo[2.2.2]-2-heptene

(b) Do you predict the product to be the *cis* isomer, the *trans* isomer, or a mixture of *cis* and *trans* isomers? Explain.

The product is the *cis* isomer. In either of the alternative chair conformations of the product, one carbaldehyde group is axial and the other is equatorial.

(c) Draw a suitable stereorepresentation for the more stable chair conformation of the dicarbaldehyde formed in this oxidation.

C_8H_{12} $\xrightarrow[2.\ (CH_3)_2S]{1.\ O_3}$ H-C(=O)–C_6H_{10}–C(=O)-H

cyclohexane-1,4-dicarbaldehyde

O=C(H)– (chair) –C(H)=O

***cis*-cyclohexane-1,4-dicarbaldehyde**

Problem 5.36 Natural rubber is a polymer of 2-methyl-1,3-butadiene (isoprene).

$$\left(\!-CH_2-\overset{CH_3}{\overset{|}{C}}=CH-CH_2-\!\right)_n$$

Poly(2-methyl-1,3-butadiene)
(Polyisoprene)

(a) Draw the structural formula of a section of natural rubber showing three repeating isoprene units.

$$-CH_2-\overset{CH_3}{\overset{|}{C}}=CH-CH_2-CH_2-\overset{CH_3}{\overset{|}{C}}=CH-CH_2-CH_2-\overset{CH_3}{\overset{|}{C}}=CH-CH_2-$$

a section of three repeating units of the natural rubber polymer

(b) Draw the structural formula of the product of oxidation of natural rubber by ozone followed by work up in the presence of $(CH_3)_2S$. Name each functional group present in this product.

The product of ozonolysis and work up under reducing conditions is a ketoaldehyde.

$$-CH_2-\overset{\overset{CH_3}{|}}{C}=CH-CH_2-CH_2-\overset{\overset{CH_3}{|}}{C}=CH-CH_2-CH_2-\overset{\overset{CH_3}{|}}{C}=CH-CH_2-$$

$$\xrightarrow[\text{2. }(CH_3)_2S]{\text{1. }O_3} H-\overset{\overset{O}{||}}{C}-CH_2-CH_2-\overset{\overset{CH_3}{|}}{C}=O$$

(c) The smog prevalent in Los Angeles contains oxidizing agents. Account for the fact that this type of smog attacks natural rubber (automobile tires, etc.) but does not attack polyethylene or polyvinyl chloride?

Polyethylene and polyvinyl chloride do not contain carbon-carbon double bonds.

Reduction

Problem 5.37 Predict the major organic product(s) of the following reactions. Show stereochemistry where appropriate.

(a) geraniol $+ 2H_2 \xrightarrow{Pt}$

CH₂OH

geraniol
(rose and other flowers)

Reduction of geraniol adds hydrogen atoms to each carbon-carbon double bond. There is no possibility for cis-trans isomerism in the product.

CH₂OH

3,7-dimethyl-1-octanol

(b) + H_2 $\xrightarrow{Pt}$

α-pinene
(turpentine)

Reduction of α-pinene adds hydrogen atoms preferentially from the less hindered side of the double bond, namely the side opposite the one-carbon bridge bearing the two methyl groups. Predict, therefore, that the major isomer formed is the first one shown.

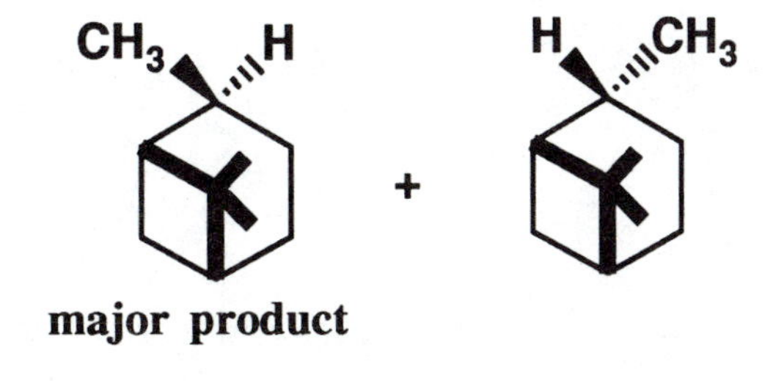

Polymerization

Problem 5.38 Following is the structural formula of a section of polypropylene derived from three units of propylene monomer.

$$-CH_2\underset{|}{\overset{CH_3}{CH}}CH_2\overset{CH_3}{CH}CH_2\overset{CH_3}{CH}-$$

Draw structural formulas for comparable sections of:
(a) poly (vinyl chloride)

$$-CH_2-\overset{Cl}{CH}-CH_2-\overset{Cl}{CH}-CH_2-\overset{Cl}{CH}-$$

(b) polytetrafluoroethylene (Teflon)

$$-CF_2-CF_2-CF_2-CF_2-CF_2-CF_2-$$

(c) poly (methyl methacrylate) (Plexiglass)

$$-CH_2-\underset{CH_3}{\overset{CH_3-O-C=O}{C}}-CH_2-\underset{CH_3}{\overset{CH_3-O-C=O}{C}}-CH_2-\underset{CH_3}{\overset{CH_3-O-C=O}{C}}-$$

(d) poly (1,1-dichloroethylene)

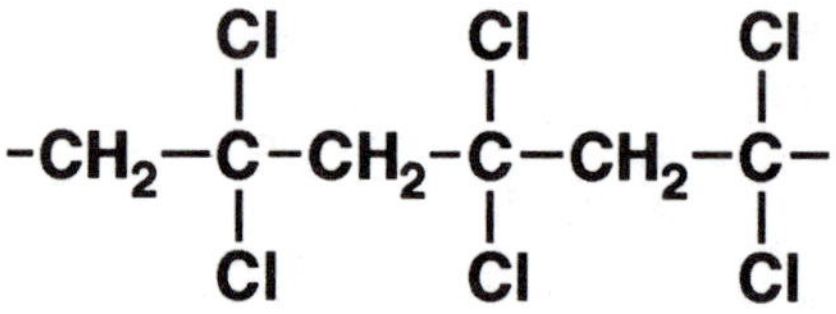

$$-CH_2-\underset{Cl}{\overset{Cl}{C}}-CH_2-\underset{Cl}{\overset{Cl}{C}}-CH_2-\underset{Cl}{\overset{Cl}{C}}-$$

Synthesis

Problem 5.39 Show how to convert cyclopentene into these compounds. Following each structural formula is the number of different reactions described in this chapter for bringing about each synthesis.

(a) 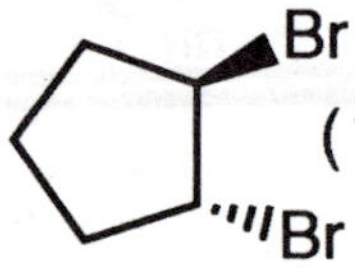 trans-1,2-dibromocyclopentane (1)

Reaction with Br_2 dissolved in an inert solvent such as CCl_4.

(b) 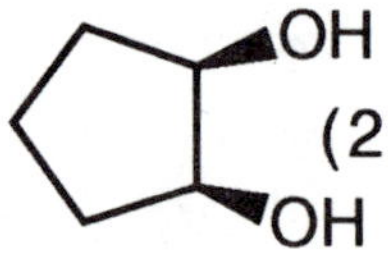 cis-1,2-cyclopentanediol (2)

(1) Formation of a *cis*-glycol by oxidation with OsO_4 in the presence of H_2O_2, or (2) Oxidation with $KMnO_4$ at pH 11.8.

(c) 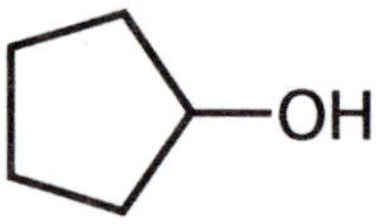 cyclopentanol (3)

(1) Acid-catalyzed hydration of cyclopentene using H_2O in the presence of H_2SO_4, or (2) hydroboration followed by oxidation of the trialkylborane with H_2O_2 in aqueous NaOH, or (3) oxymercuration using mercury(II) acetate followed by $NaBH_4$ reduction of the organomercury intermediate.

(d) 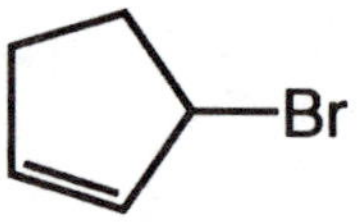 3-bromocyclopentene (1)

Reaction with Br_2 at 350°C by a radical chain mechanism. The most reactive position is at the allylic carbon (the carbon adjacent to the double bond).

(e)  iodocyclopentane (1)

Addition of HI to cyclopentene.

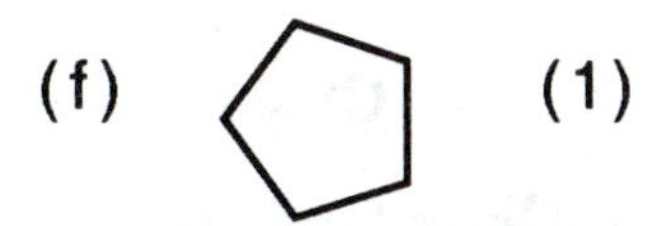

(f) (1)

Catalytic reduction of cyclopentene using H_2 in the presence of a Pd, Pt, or Ni catalyst.

(g) $H-\overset{\overset{O}{\|}}{C}-(CH_2)_3-\overset{\overset{O}{\|}}{C}-H$ (1)

Reaction with ozone, O_3, followed by treatment of the ozonide with dimethyl sulfide, $(CH_3)_2S$.

Problem 5.40 Show how to convert the given starting material into the desired product. Note that some of these syntheses require only one step, whereas others may require two or more steps.

(a) $CH_3-\overset{\overset{CH_3}{|}}{CH}-CH_2-Cl \longrightarrow CH_3-\overset{\overset{CH_3}{|}}{C}=CH_2$

Reaction of 1-chloro-2-methylpropane (isobutyl chloride) with NaOH or KOH in ethanol brings about a β-elimination reaction (Section 4.5), or more specifically dehydrohalogenation, to give 2-methylpropene (isobutylene).

$$CH_3-\overset{\overset{CH_3}{|}}{CH}-CH_2-Cl \xrightarrow[CH_3CH_2OH]{NaOH} CH_3-\overset{\overset{CH_3}{|}}{C}=CH_2$$

(b) $CH_3-\overset{\overset{CH_3}{|}}{C}=CH_2 \longrightarrow CH_3-\underset{\underset{Br}{|}}{\overset{\overset{CH_3}{|}}{C}}-CH_3$

Addition of HBr in the absence of peroxides.

$$CH_3-\overset{\overset{CH_3}{|}}{C}=CH_2 \xrightarrow{HBr} CH_3-\underset{\underset{Br}{|}}{\overset{\overset{CH_3}{|}}{C}}-CH_3$$

(c) $CH_3-\overset{\overset{CH_3}{|}}{CH}-CH_2-Cl \longrightarrow CH_3-\underset{\underset{OH}{|}}{\overset{\overset{CH_3}{|}}{C}}-CH_3$

Dehydrohalogenation of 1-chloro-2-methylpropane to 2-methylpropene as in part (a) followed by acid-catalyzed hydration of the alkene to give 2-methyl-2-propanol.

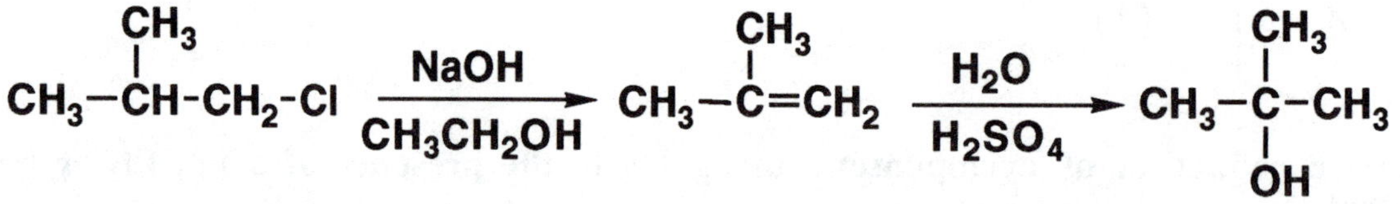

$$CH_3-\underset{|}{\overset{CH_3}{CH}}-CH_2-Cl \xrightarrow[CH_3CH_2OH]{NaOH} CH_3-\overset{CH_3}{C}=CH_2 \xrightarrow[H_2SO_4]{H_2O} CH_3-\underset{OH}{\overset{CH_3}{C}}-CH_3$$

Hydration of the alkene to the tertiary alcohol can also be brought about by oxymercuration of the alkene followed by reduction with $NaBH_4$.

(d)

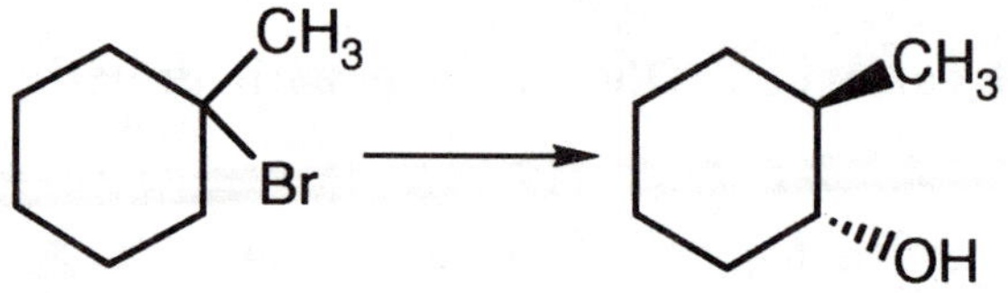

CH3, Br → CH3, OH

Dehydrohalogenation of 1-bromo-1-methylcyclohexane using KOH or NaOH in ethanol gives 1-methylcyclohexene. Hydroboration followed by oxidation of the trialkylborane gives *trans*-2-methylcyclohexanol.

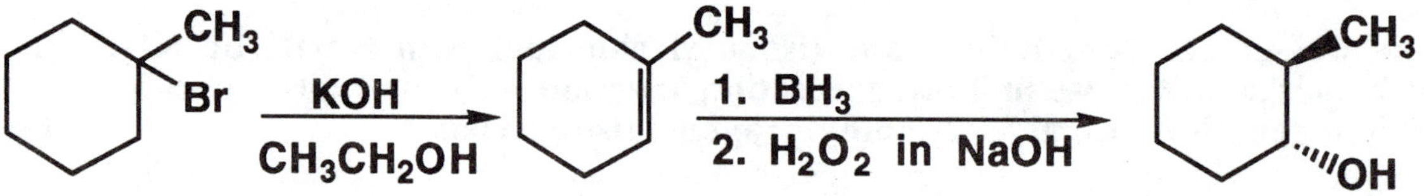

CH3, Br $\xrightarrow[CH_3CH_2OH]{KOH}$ 1-methylcyclohexene $\xrightarrow[2.\ H_2O_2\ in\ NaOH]{1.\ BH_3}$ CH3, OH

(e)

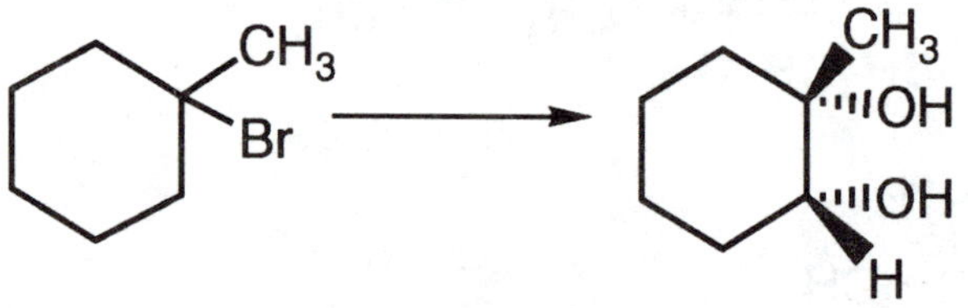

CH3, Br → CH3, OH, OH, H

Dehydrohalogenation of 1-bromo-1-methylcyclohexane as in part (d) gives 1-methylcyclohexene. Oxidation to the *cis*-glycol can be brought about using either OsO_4 in the presence of H_2O_2 or using KMnO4 at pH 11.8.

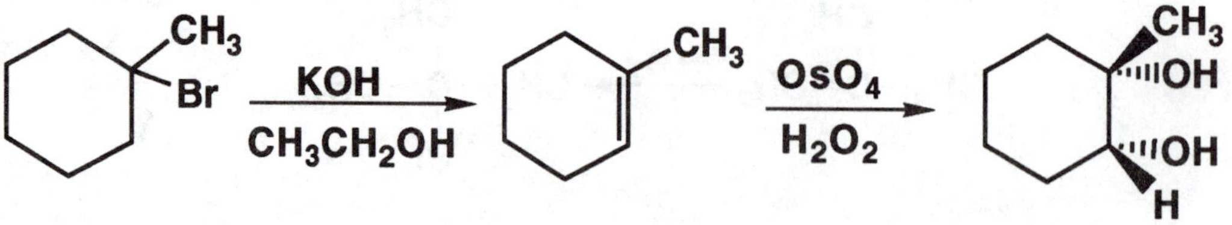

CH3, Br $\xrightarrow[CH_3CH_2OH]{KOH}$ CH3 (1-methylcyclohexene) $\xrightarrow[H_2O_2]{OsO_4}$ CH3, OH, OH, H

(f)

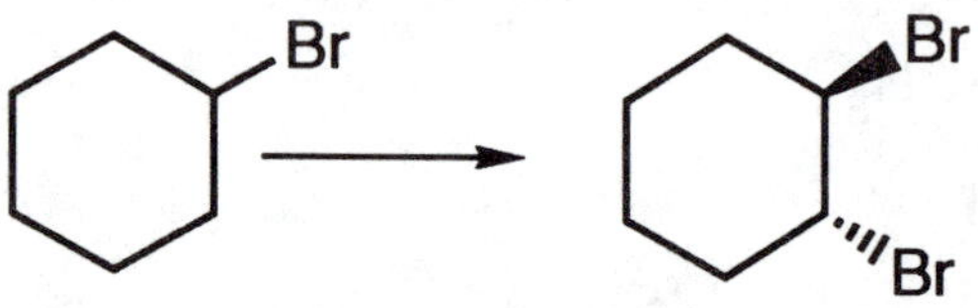

Br → Br, Br

Dehydrohalogenation of bromocyclohexane followed by addition of Br_2.

$$\text{C}_6\text{H}_{11}\text{Br} \xrightarrow[\text{CH}_3\text{CH}_2\text{OH}]{\text{KOH}} \text{cyclohexene} \xrightarrow[\text{CCl}_4]{\text{Br}_2} \text{trans-1,2-dibromocyclohexane}$$

(g) cyclohexane ⟶ bromocyclohexane

Substitution of bromine for hydrogen by a radical chain mechanism.

$$\text{cyclohexane} \xrightarrow[350^\circ\text{C}]{\text{Br}_2} \text{bromocyclohexane}$$

(h) cyclohexane ⟶ cyclohexene

Conversion of cyclohexane to bromocyclohexane as in part (g) followed by dehydrohalogenation using KOH in ethanol.

$$\text{cyclohexane} \xrightarrow[350^\circ\text{C}]{\text{Br}_2} \text{bromocyclohexane} \xrightarrow[\text{CH}_3\text{CH}_2\text{OH}]{\text{KOH}} \text{cyclohexene}$$

(i) bromocyclohexane ⟶ $\text{H-C(=O)-(CH}_2)_4\text{-C(=O)-H}$

Dehydrohalogenation of bromocyclohexane as in part (h); then oxidation with ozone followed by work-up in the presence of dimethyl sulfide.

$$\text{bromocyclohexane} \xrightarrow[\text{CH}_3\text{CH}_2\text{OH}]{\text{KOH}} \text{cyclohexene} \xrightarrow[\text{2) (CH}_3)_2\text{S}]{\text{1) O}_3} \text{H-C(=O)-(CH}_2)_4\text{-C(=O)-H}$$

(j) cyclohexane ⟶ 3-bromocyclohexene

Conversion of cyclohexane to cyclohexene as in part (g) followed by allylic bromination by a radical chain mechanism.

cyclohexane $\xrightarrow[350^\circ C]{Br_2}$ bromocyclohexane $\xrightarrow[CH_3CH_2OH]{KOH}$ cyclohexene $\xrightarrow[350^\circ C]{Br_2}$ 3-bromocyclohexene

Qualitative Organic Analysis

Problem 5.41 Describe how you would distinguish between the members of each pair of compounds by a simple chemical test. For each pair, tell what test you would perform, what you expect to observe, and write an equation for each positive test. For example, to distinguish between cyclohexane and 1-hexene in part (a), you might consider the reaction of each with Br_2 and CCl_4 or reaction with $KMnO_4$ in basic solution.

(a) cyclohexane and 1-cyclohexene

Add a solution of Br_2 in CCl_4 to each hydrocarbon. No reaction with cyclohexane. Bromine adds to 1-hexene to form 1,2-dibromohexane. Observations are that a solution of bromine in carbon tetrachloride is deep purple. Addition of this solution to 1-hexene results in loss of the purple color and formation of a colorless solution.

$$CH_2{=}CHCH_2CH_2CH_2CH_3 + Br_2 \xrightarrow{CCl_4} CH_2(Br){-}CH(Br)CH_2CH_2CH_2CH_3$$

colorless   **purple**   **colorless**

Alternatively, treat each with alkaline $KMnO_4$, pH 11.8. No reaction with cyclohexane. 1-Hexene reacts to form a glycol and a brown precipitate of manganese dioxide, MnO_2.

$$CH_2{=}CH(CH_2)_3CH_3 + MnO_4^- \xrightarrow{pH\ 11.8} CH_2(OH){-}CH(OH)(CH_2)_3CH_3 + MnO_2$$

colorless   **purple**   **colorless**   **brown precipitate**

(b) 1-hexene and 2-chlorohexane

Apply the same tests as in part (a). No reactions are seen with 2-chlorohexane.

(c) 1,1-dimethylcyclopentane and 2,3-dimethyl-2-butene

Apply the same tests as in part (a). No reactions are seen with 1,1-dimethyl-cyclopentane.

$$CH_3{-}C(CH_3){=}C(CH_3){-}CH_3 + Br_2 \xrightarrow{CCl_4} CH_3{-}C(CH_3)(Br){-}C(CH_3)(Br){-}CH_3$$

colorless   **purple**   **colorless**

$$CH_3-\underset{}{\overset{H_3C}{\overset{|}{C}}}=\overset{CH_3}{\overset{|}{C}}-CH_3 + MnO_4^- \xrightarrow[\text{pH 11.8}]{} CH_3-\overset{H_3C}{\overset{|}{\underset{OH}{\underset{|}{C}}}}-\overset{CH_3}{\overset{|}{\underset{OH}{\underset{|}{C}}}}-CH_3 + MnO_2$$

colorless purple colorless brown precipitate

CHAPTER 6: ALKYNES

SUMMARY OF REACTIONS

Starting Material ↓ / Product →	Alkanes	Alkenes (cis)		Alkenes (trans)	Alkynes	Alkyne Anions	Dihaloalkanes	Dihaloalkenes	Haloalkenes	Ketones		Tetrahaloalkanes	Vinyl Esters
Alkynes	6A 6.5A*	6B 6.5A	6C 6.5B	6D 6.5A			6E 6.5C	6F 6.5C	6G 6.5C	6H 6.5B	6I 6.5C	6J 6.5C	6K 6.5C
Haloalkenes					6L 6.4								
Terminal Alkynes						6M 6.5D							
Vicinal Dihalides					6N 6.4								

*Section in book that describes reaction.

REACTION 6A: CATALYTIC HYDROGENATION (Section 6.5A)

$$-C{\equiv}C'- \xrightarrow[\text{Pd, Pt, Ni}]{2H_2} -CH_2\text{-}C'H_2-$$

- **Catalytic reduction** of an alkyne with hydrogen in the presence of a palladium, platinum, or nickel catalyst results in addition of two mol of hydrogen to the alkyne to produce an alkane.

REACTION 6B: CATALYTIC HYDROGENATION, LINDLAR CATALYST (Section 6.5A)

$$-C{\equiv}C'- \xrightarrow[\text{Pd/CaCO}_3\text{ (Lindlar catalyst)}]{H_2} \text{(cis) } C{=}C' \text{ with H on each carbon, same side}$$

- If a special catalyst called the **Lindlar catalyst** is used, the reduction stops after the addition of one mol of hydrogen to give the ***cis* alkene** in high yield. The *cis* stereochemistry is observed because the reaction apparently involves the simultaneous or nearly simultaneous transfer of two hydrogen atoms from the surface of the metal catalyst to the alkyne.

REACTION 6C: ADDITION OF DIBORANE: HYDROBORATION (Section 6.5B)

$$-C\equiv C- \xrightarrow[2)\ CH_3CO_2H]{1)\ BH_3} \text{(cis) } C{=}C \text{ with H, H on same side}$$

- Borane adds to internal alkynes to give a trialkenylborane. The hydrogen and boron end up on the same side of the double bonds in the trialkenylborane (syn addition).
- Treatment of the trialkenylborane with acetic acid replaces the boron with hydrogen to produce the ***cis*** **alkene.**
- For terminal alkynes, a special sterically hindered borane reagent such as **(sia)$_2$BH** is used instead of regular borane to prevent the unwanted addition of a second borane to make a dihydroborated alkane. The steric bulk of the (sia)$_2$BH prevents the second borane from reacting. Note that for terminal alkynes, the boron ends up on the less substituted carbon atom.

REACTION 6D: CHEMICAL REDUCTION (Section 6.5A)

$$-C\equiv C'- \xrightarrow[NH_3\ (l)]{Li\ or\ Na} \text{(trans) } C{=}C' \text{ with H on opposite sides}$$

- **Chemical reduction** of an alkyne with sodium or lithium metal in liquid ammonia results in formation of a ***trans*** **alkene** product. This is complementary to the use of the Lindlar catalyst that produces the ***cis*** **alkene** product.

REACTION 6E: ADDITION OF 2 MOL OF HYDROGEN HALIDES (Section 6.5C)

$$-C\equiv C'- \xrightarrow[\text{HX is HBr or HCl}]{2\ eq.\ HX} -CX_2-C'H_2-$$

- Alkynes add two moles of **HBr** and **HCl**.
- For terminal alkynes, both the first and second mol of the hydrogen halide follow Markovnikov's rule.
- Addition of **HCl** to acetylene yields **chloroethylene**, a compound of considerable industrial importance for the manufacturing of plastics such as **polyvinyl chloride** that is used for pipes and other fittings. Today, cheaper routes are used for the production of chloroethylene involving ethylene and chlorine followed by heating.

REACTION 6F: ADDITION OF 1 MOL OF BROMINE OR CHLORINE (Section 6.5C)

$$-C\equiv C'- \xrightarrow[X_2\text{ is }Br_2\text{ or }Cl_2]{1\ eq.\ X_2} \text{(trans) } C{=}C' \text{ with X on opposite sides}$$

- Alkynes react with 1 mol of Cl_2 or Br_2 to give **anti addition** and thus the *trans* dihaloalkene. This can be isolated or reacted with another mol of the halogen to give the tetrahaloalkane (reaction 6J).

REACTION 6G: ADDITION OF 1 MOL OF HYDROGEN HALIDES (Section 6.5C)

$$-C{\equiv}C'- \xrightarrow[\text{HX is HBr or HCl}]{\text{1 eq. HX}} \text{(X)C=C'(H)}$$

- Alkynes add one mol of **HBr** and **HCl**. The addition is stereoselective and results in **anti addition**.
- For terminal alkynes, the addition of both the first and second mol of hydrogen halide follows Markovnikov's rule.

REACTION 6H: REACTION WITH DIBORANE AND PEROXIDE (Section 6.5B)

$$-C{\equiv}C'- \xrightarrow[\text{2) } H_2O_2\text{, NaOH}]{\text{1) } BH_3} -CH_2-C'(=O)-$$

- Treatment with diborane results in formation of an alkenylborane. Note that for terminal alkynes, the boron ends up on the less substituted carbon atom. Reaction of this alkenyl borane with hydrogen peroxide in aqueous sodium hydroxide replaces the boron atom with -OH. The alcohol that is formed is called an **enol** , and it is not stable because it can rearrange to the more stable **ketone** or **aldehyde**. This process of rearrangement is called **tautomerization.**
- The position of equilibrium for most keto-enol tautomer pairs lies very far to the side of the keto form.

REACTION 6I: ADDITION OF WATER: HYDRATION (Section 6.5C)

$$-C{\equiv}C'- \xrightarrow[H_2SO_4,\ HgSO_4]{H_2O} -CH{=}C'(OH)- \xrightarrow{\text{tautomerization}} -CH_2-C'(=O)-$$

- Addition of **water** occurs in the presence of strong acids and Hg(II) salts, the resulting enol then undergoes keto-enol tautomerism to give a ketone. For terminal alkynes, the -OH group adds to the carbon atom with the alkyl group (according to Markovnikov's rule).

REACTION 6J: ADDITION OF 2 EQUIVALENTS BROMINE OR CHLORINE (Section 6.5C)

$$-C\equiv C'- \xrightarrow[X_2 \text{ is } Br_2 \text{ or } Cl_2]{2 \text{ eq. } X_2} -CX_2-C'X_2-$$

- Alkynes react with 1 mol of Cl_2 or Br_2 to give **anti addition** and thus the trans dihaloalkene (reaction 6F). This can be isolated or reacted with another mol of the halogen to give the tetrahaloalkane.

REACTION 6K: ADDITION OF CARBOXYLIC ACIDS - FORMATION OF VINYL ESTERS (Section 6.5C)

$$-C\equiv C- \;+\; -C'(=O)OH \xrightarrow[HgSO_4]{H_2SO_4} -C'(=O)-O-C(-)=CH-$$

- Addition of **carboxylic acids** occurs in the presence of strong acids and Hg(II) salts to give **vinyl esters**. An important example involves the addition of acetic acid to acetylene to give vinyl acetate. Polymerization gives **poly(vinyl acetate)**, which is used in adhesives.

REACTION 6L: DEHYDROHALOGENATION OF HALOALKENES (Section 6.4)

$$X-C=C' \xrightarrow{NaNH_2} -C\equiv C'-$$

- Reaction of haloalkenes with strong base such as $NaNH_2$ results in formation of an alkyne.
- Dehydrohalogenation of a haloalkene with at least one hydrogen on each adjacent carbon atom can also form a side product with cumulated double bonds called an **allene**.
- Allenes are compounds that have two carbon-carbon double bonds adjacent to each other.

$$H_2C{=}C{=}CH_2$$

allene

- Allenes are less stable than isomeric alkynes. For this reason, they are only minor side products during the dehydrohalogenation reactions used to produce alkynes.

REACTION 6M: DEPROTONATION OF TERMINAL ALKYNES (Section 6.5D)

$$-C\equiv C'-H \xrightarrow{NaNH_2} -C\equiv C'^{-}\,Na^{+} \;+\; NH_3$$

- One of the major differences between the chemistry of alkynes and alkenes is that the **hydrogen attached** to a **carbon-carbon triple bond** is sufficiently **acidic** that it can be conveniently removed by a strong base, usually sodamide ($NaNH_2$), sodium hydride (NaH) or lithium diisopropylamide (LDA). ✱

- The pK_a for a terminal alkyne is a relatively acidic 25, compared with nearly 44 for a normal alkene and 48 for an alkane. This low pK_a for the terminal alkyne is due to the fact that the lone pair on carbon produced upon deprotonation resides in an sp orbital for the alkyne which has 50% s character, and is thus more stable.

REACTION 6N: DEHYDROHALOGENATION OF VICINAL DIHALIDES (Section 6.4)

$$-\underset{H}{\overset{X}{C}}-\underset{H}{\overset{X}{C'}}- \xrightarrow{NaNH_2} \overset{X}{\ }C{=}C'\underset{H}{\ } \xrightarrow{NaNH_2} -C{\equiv}C'-$$

- Alkynes can be synthesized from vicinal dihalides by **double dehydrohalogenation** with a very strong base such as sodium amide. The dehydrohalogenation occurs in two steps. The first dehydrohalogenation converts the vicinal dihalide into a haloalkene, that is then converted into the alkyne. ✱
- Alkenes can be converted into alkynes by a very useful sequence of reactions involving initial reaction with a halogen to form a vicinal dihalide, that is then converted to the alkyne by addition of two mol of strong base like $NaNH_2$.

$$R{-}CH{=}C'H{-}R \xrightarrow{Br_2} R{-}\underset{Br}{\overset{H}{C}}{-}\underset{Br}{\overset{H}{C'}}{-}R \xrightarrow{2\ NaNH_2} R{-}C{\equiv}C'{-}R$$

vicinal dibromide

SUMMARY OF IMPORTANT CONCEPTS

6.0 OVERVIEW

- **Alkynes** are molecules with a carbon-carbon triple bond composed of one sigma bond and two pi bonds. As with alkenes, the chemistry of alkynes is largely determined by the pi bonds, so in many ways the chemistry of alkynes is similar to that of alkenes. ✱

6.1 STRUCTURE

- Because alkynes have a triple bond composed of one sigma bond and two pi bond, the carbon-carbon bond length (0.121 nm) is shorter than for alkenes (0.134 nm) or alkanes (0.154 nm). The bond dissociation energy is also considerably higher for alkynes. ✱
  - In valence bond terms, a triple bond consists of one sigma bond formed from sp-sp overlap, one pi bond formed from $2p_y$-$2p_y$ overlap, and a second pi bond perpendicular to the first that is formed from $2p_z$-$2p_z$ overlap. See Figure 1.22 in the text for a diagram.
  - The shorter carbon-carbon bond length in alkynes is due to the sp-sp overlap of the sigma bond. The sp orbitals are 50% s in character and the larger the percent s character the shorter the bond.

6.2 NOMENCLATURE

- According to the **IUPAC system**, the infix **yn** is used to show the presence of a carbon-carbon triple bond. ✱

- The longest chain that contains the triple bond is numbered from the end that gives the triply bonded carbons the lower numbers. The location of the triple bond is indicated by the number of the first carbon of the triple bond.
- If more than one triple bond is present in the chain then the infixes **adiyn**, **atriyn**, etc. are used.
- If a double and triple bond are found in the same molecule, then **en** and **yn** are both used. The triple bond takes precedence, so it is written last and the numbering scheme is chosen that gives the triple bond the lower number. For example, 4-hexen-1-yne is a correct name, not 2-hexen-5-yne for  .
- The IUPAC system retains the name acetylene for ethyne (HC≡CH).

• **Common names** for simple alkynes are derived from that of acetylene by prefixing the names of substituents on the carbon-carbon triple bonds to the name **acetylene**. For example, dimethylacetylene is the common name of 2-butyne.

6.3 PHYSICAL PROPERTIES

• The **physical properties** of alkynes are similar to those of alkanes and alkenes with similar carbon skeletons. For example, alkynes with four or less carbon atoms are gases at room temperature while those with five or more carbon atoms are liquids at room temperature.

• When acetylene is passed through a solution of certain catalysts, it polymerizes to make a shiny polymer of **polyacetylene** that contains alternating double and single bonds. This polymer is very interesting because it can be doped (an electron can be removed or added to the pi bonding system) and then it becomes a **conductor** of electricity. When stretched, the polymer chains become more ordered and the conductivity is greater along the chains than perpendicular to them. This and other properties of polyacetylene may lead to some important applications in the future.

6.4 PREPARATION OF ALKYNES

• **Mechanism-based enzyme inhibitors** (also called **suicide substrates**) are compounds that irreversibly inactivate enzymes. They are designed based on an understanding of the mechanism of catalysis by the enzyme. The enzyme reacts with the inhibitor to create a species that then irreversibly inactivates the enzyme by forming a covalent bond to the enzyme.
- In an important example of a mechanism-based inhibitor, an alkyne was used to inhibit an enzyme that usually catalyzes the dehydration of a secondary alcohol to give an alkene. With the inhibitor, the alkyne is first converted into an allene by the enzyme. The reactive allene then reacts with the enzyme, thereby causing irreversible deactivation.

6.5 REACTIONS OF ALKYNES

• Alkenes and alkynes have a combination of sigma and pi bonds, so they show clear similarities in the way they react. This is because it is the pi electrons that dictate the reactions. The similarities are especially apparent with addition reactions, although with alkynes it is often possible to stop addition after reaction with just one of the pi bonds. ✳ *[This powerful idea accurately predicts a large number of reaction of alkynes.]*

• Two types of **reduction** reactions are used to convert alkynes into alkenes and alkanes, namely **catalytic reduction** and **chemical reduction**.

• Alkynes undergo many of the same electrophilic additions as alkenes. ✳

CHAPTER 6
Solutions to the Problems

Problem 6.1 Give each compound an IUPAC name.

(a) $CH_3(CH_2)_5C{\equiv}CH$

1-octyne

(b) 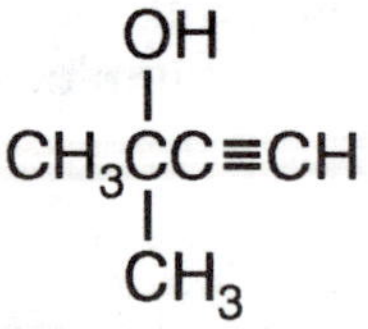 $CH_3C(OH)(CH_3)C{\equiv}CH$

2-methyl-3-butyn-2-ol

(c) 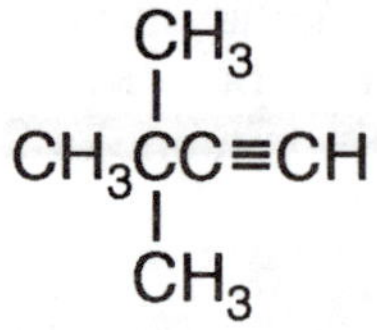 $CH_3C(CH_3)_2C{\equiv}CH$

3,3-dimethyl-1-butyne

Problem 6.2 Give each compound a common name.

(a) 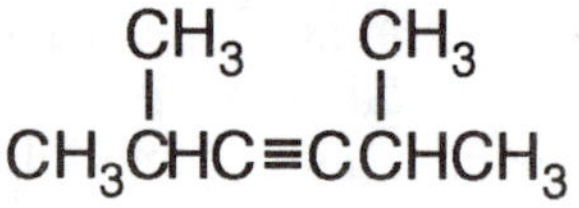 $CH_3CH(CH_3)C{\equiv}CCH(CH_3)CH_3$

diisopropylacetylene

(b) 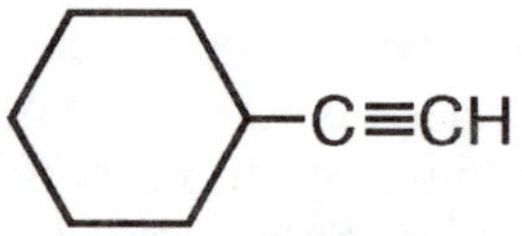 cyclohexyl–$C{\equiv}CH$

cyclohexylacetylene

(c) $HC{\equiv}CCH_2CH_2CH_2CH_3$

butylacetylene

Problem 6.3 Draw a structural formula for the alkene and dichloroalkane of given molecular formula that will yield the indicated alkyne from each reaction sequence.

The key to this problem is to realize that wherever the triple bond is in the carbon chain, that is where the double bond was.

(a) $C_6H_{12} \xrightarrow{Cl_2} C_6H_{12}Cl_2 \xrightarrow{2NaNH_2} CH_3CH_2C{\equiv}CCH_2CH_3$

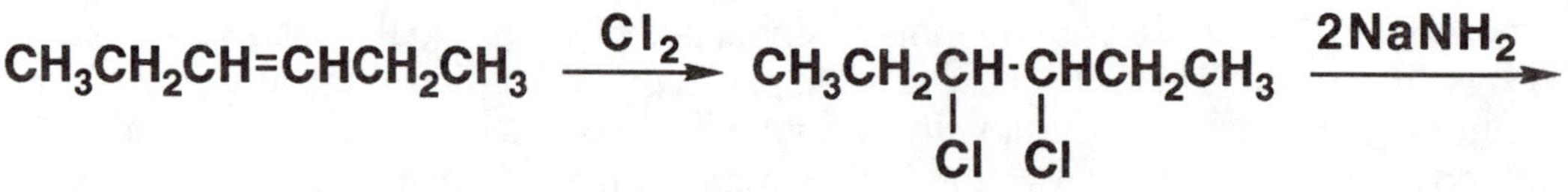

$$\mathbf{CH_3CH_2CH{=}CHCH_2CH_3 \xrightarrow{Cl_2} CH_3CH_2CH(Cl){\cdot}CH(Cl)CH_2CH_3 \xrightarrow{2NaNH_2}}$$

(b) 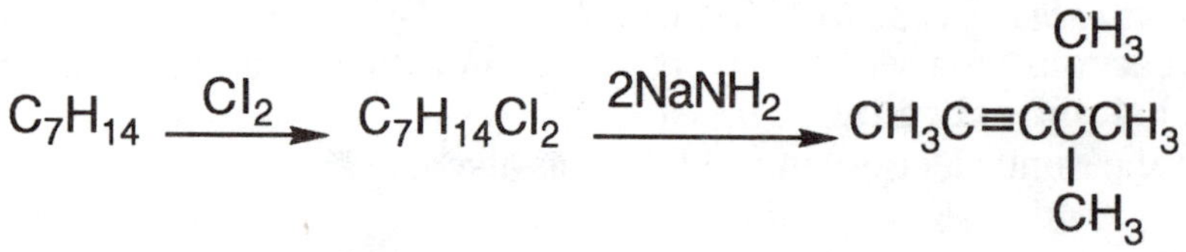 $C_7H_{14} \xrightarrow{Cl_2} C_7H_{14}Cl_2 \xrightarrow{2NaNH_2} CH_3C{\equiv}CC(CH_3)_2CH_3$

$$\mathbf{CH_3{-}CH{=}CH{-}C(CH_3)_2{-}CH_3 \xrightarrow{Cl_2} CH_3{-}CH(Cl){-}CH(Cl){-}C(CH_3)_2{-}CH_3 \xrightarrow{2NaNH_2}}$$

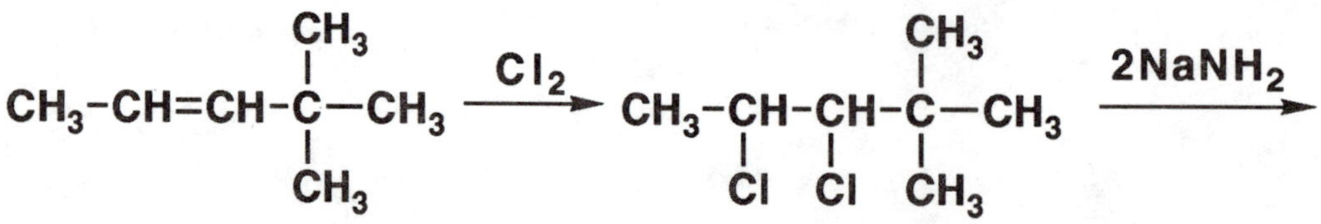

Problem 6.4 Draw the structural formula for the hydrocarbon of given molecular formula that will undergo hydroboration / oxidation to give the following compounds.

(a) 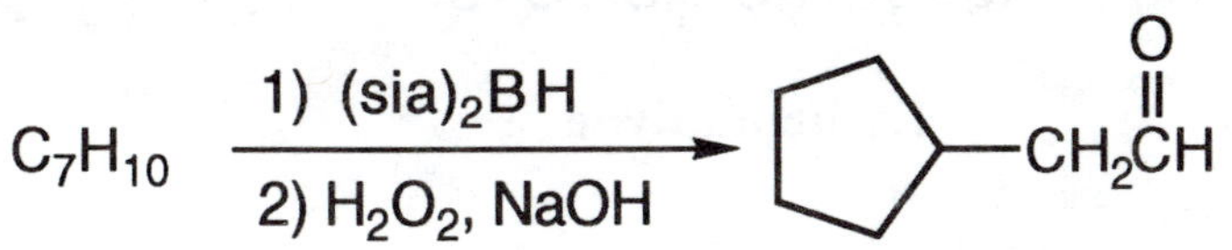 C_7H_{10} $\xrightarrow[\text{2) } H_2O_2\text{, NaOH}]{\text{1) (sia)}_2BH}$ cyclopentyl–CH_2CH(=O)

cyclopentyl–C≡CH

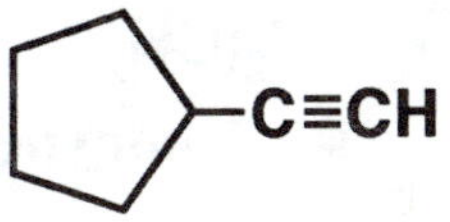

(b) 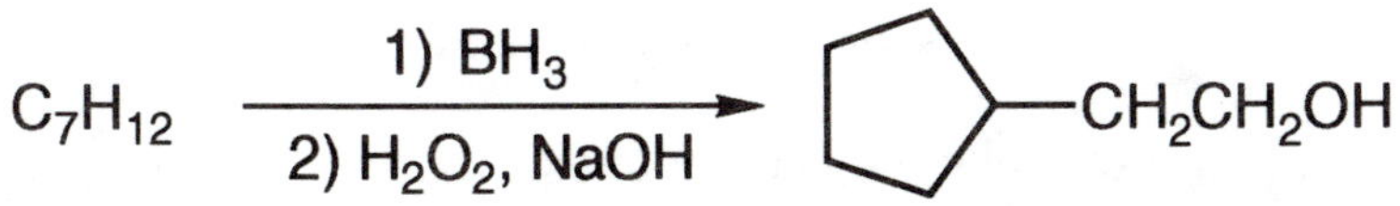 C_7H_{12} $\xrightarrow[\text{2) } H_2O_2\text{, NaOH}]{\text{1) } BH_3}$ cyclopentyl–CH_2CH_2OH

cyclopentyl–$CH=CH_2$

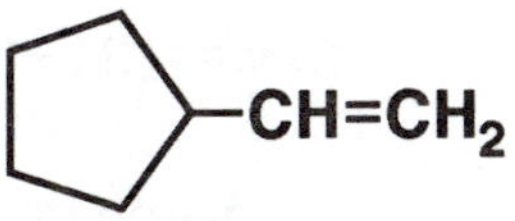

Problem 6.5 Acid-catalyzed hydration of 2-pentyne gives a mixture of two ketones, each of molecular formula $C_5H_{10}O$. Propose structural formulas for these two ketones and for the enol from which each is derived.

$$CH_3CH_2C{\equiv}CCH_3 + H_2O \xrightarrow[HgSO_4]{H_2SO_4} C_5H_{10}O$$

The two ketones are 2-pentanone and 3-pentanone.

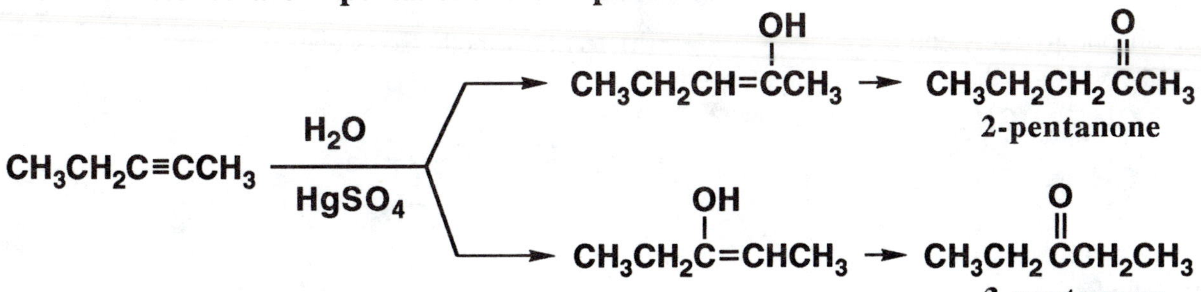

$CH_3CH_2C{\equiv}CCH_3$ $\xrightarrow[HgSO_4]{H_2O}$

- $CH_3CH_2CH{=}C(OH)CH_3$ → $CH_3CH_2CH_2C(=O)CH_3$ **2-pentanone**
- $CH_3CH_2C(OH){=}CHCH_3$ → $CH_3CH_2C(=O)CH_2CH_3$ **3-pentanone**

Structure and Nomenclature

Problem 6.6 Write IUPAC names for the following compounds.

(a) 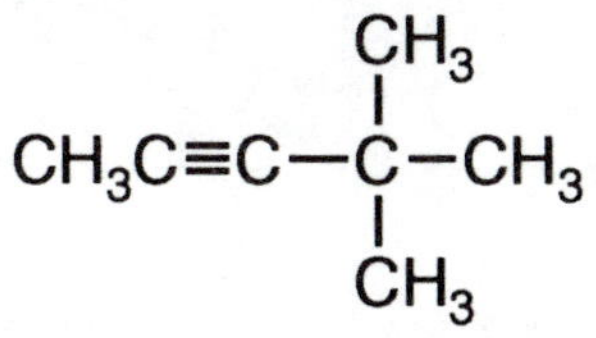 $CH_3C{\equiv}C{-}C(CH_3)_2{-}CH_3$

4,4-dimethyl-2-pentyne
(*tert*-butylmethylacetylene)

(b) $HC{\equiv}CCH_2Br$

3-bromopropyne

(c) 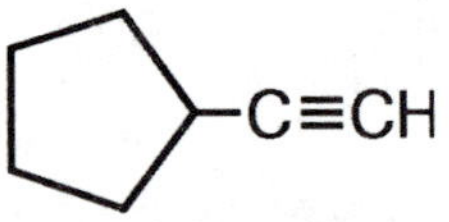 cyclopentyl–$C\equiv CH$

ethynylcyclopentane (cyclopentylacetylene)

(d) $HC\equiv CCH_2CH_2CH_2C\equiv CH$

1,6-heptadiyne

(e) $CH_3(CH_2)_5C\equiv CCH_2OH$

2-nonyn-1-ol

(f) $CH_3(CH_2)_6C\equiv CH$

1-nonyne

(g) $CH_3C\equiv CCH_2OH$

2-butyn-1-ol

(h) $CH_3(CH_2)_7C\equiv C(CH_2)_7CO_2H$

9-octadecynoic acid

Problem 6.7 Draw structural formulas for the following compounds.

(a) 3-hexyne

$CH_3CH_2C\equiv CCH_2CH_3$

(b) vinylacetylene

$CH_2{=}CH{-}C\equiv CH$

(c) 3-chloro-1-butyne

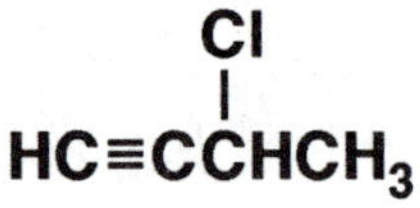

$HC\equiv CCH(Cl)CH_3$

(d) 5-isopropyl-3-octyne

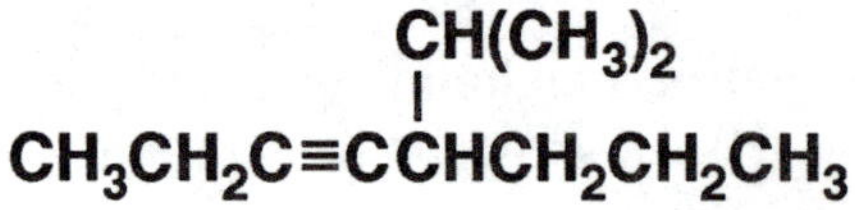

$CH_3CH_2C\equiv CCH(CH(CH_3)_2)CH_2CH_2CH_3$

(e) 3-pentyn-2-ol

$CH_3C\equiv CCH(OH)CH_3$

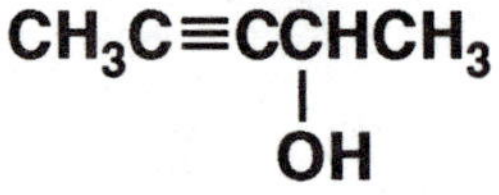

(f) 2-butyne-1,4-diol

$HOCH_2C\equiv CCH_2OH$

(g) diisopropylacetylene

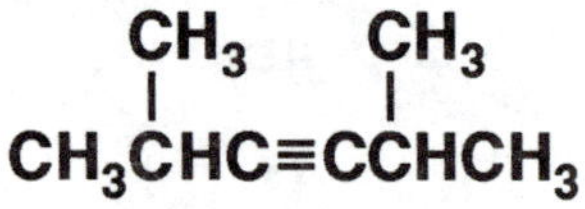

$CH_3CH(CH_3)C\equiv CCH(CH_3)CH_3$

(h) *tert*-butylmethylacetylene

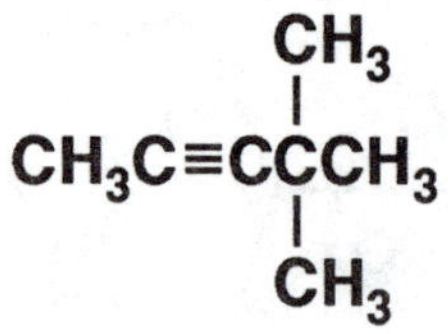

$CH_3C\equiv CC(CH_3)_2CH_3$

(i) cyclodecyne

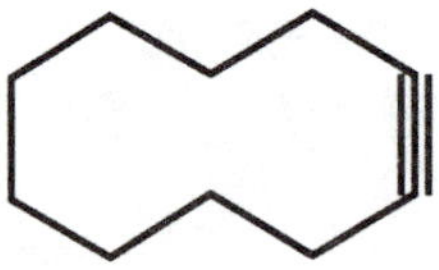

Problem 6.8 Predict all bond angles about each circled atom.

(a) $CH_3-C\equiv C-CH_3$

180°

$CH_3-C\equiv C-CH_3$

(b) $CH_2=CH-C\equiv CH$

120°

$CH_2=CH-C\equiv CH$

(c) $CH_2=C=CH-CH_3$

180°

$CH_2=C=CH-CH_3$

(d) $O=C=O$

180°

$O=C=O$

Problem 6.9 Following are Lewis structures for several small molecules. State the orbital hybridization of each circled atom.

(a) $CH_3-C\equiv C-CH_3$

sp

$CH_3-C\equiv C-CH_3$

(b) $CH_2=CH-C\equiv CH$

sp^2

$CH_2=CH-C\equiv CH$

(c) $CH_2=C=CH-CH_3$

sp

$CH_2=C=CH-CH_3$

(d) $O=C=O$

sp

$O=C=O$

Problem 6.10 Describe each circled carbon-carbon bond in terms of the overlap of atomic orbitals.

(a) $CH_3-C\equiv C-CH_3$

σ_{sp-sp^3}

$CH_3-C\equiv C-CH_3$

(b) $CH_2=CH-C\equiv CH$

σ_{sp^2-sp}

$CH_2=CH-C\equiv CH$

(c) $CH_2=C=CH-CH_3$

σ_{sp-sp^2}

$CH_2=C=CH-CH_3$

π_{2p-2p}

(d) $CH_2=CH-CH=CH_2$

$\sigma_{sp^2-sp^2}$

$CH_2=CH-CH=CH_2$

Preparation of Alkynes

Problem 6.11 Show how to prepare each alkyne from the given starting material.

(a) $CH_3CH_2CH_2CH{=}CH_2 \longrightarrow CH_3CH_2CH_2C{\equiv}CH$

First treat 1-pentene with either bromine (Br_2) or chlorine (Cl_2) to form a 1,2-dihalopentane. Then carry out a double dehydrohalogenation with two moles of sodium amide ($NaNH_2$) to form 1-pentyne.

$$CH_3(CH_2)_2CH{=}CH_2 \xrightarrow{Br_2} CH_3(CH_2)_2CH(Br){-}CH_2Br \xrightarrow{2NaNH_2} CH_3CH_2CH_2C{\equiv}CH$$

(b) $CH_3(CH_2)_5CH(Cl)CH_3 \longrightarrow CH_3(CH_2)_4C{\equiv}CCH_3$

First convert 2-chlorooctane to 2-octene by dehydrohalogenation with sodium amide. This is an example of a β-elimination reaction. Then follow the procedure in (a) to convert an alkene to an alkyne, namely addition of Br_2 or Cl_2 followed by a double dehydrohalogenation.

$$CH_3(CH_2)_5CH(Cl)CH_3 \xrightarrow{NaNH_2} CH_3(CH_2)_4CH{=}CHCH_3 \xrightarrow[2)\ 2NaNH_2]{1)\ Br_2} CH_3(CH_2)_4C{\equiv}CCH_3$$

(c) $CH_3CH_2CH_2C{\equiv}CH \longrightarrow CH_3CH_2CH_2C{\equiv}CD$

First form the alkyne anion with sodium amide or sodium hydride (NaH) and then react the anion with a deuterium donor such as D_2O or CH_3CH_2OD.

$$CH_3CH_2CH_2C{\equiv}CH \xrightarrow{NaNH_2} CH_3CH_2CH_2C{\equiv}C{:}^-\ Na^+ \xrightarrow{D_2O} CH_3CH_2CH_2C{\equiv}CD$$

Reactions of Alkynes

Problem 6.12 Draw structural formulas for the major product(s) formed by reaction of 3-hexyne with each of the following reagents. Where you predict no reaction, write NR.

(a) H_2(excess) / Pt

$CH_3CH_2CH_2CH_2CH_2CH_3$

(b) H_2(excess) / Lindlar catalyst

cis-$CH_3CH_2CH{=}CHCH_2CH_3$ (both H on the same side: CH_3CH_2 and CH_2CH_3 on top, H and H below)

(c) Na in NH_3 (liquid)

trans-$CH_3CH_2CH{=}CHCH_2CH_3$ (CH_3CH_2 and H on top, H and CH_2CH_3 below)

(d) BH_3 followed by H_2O_2 / NaOH

$CH_3CH_2C({=}O)CH_2CH_2CH_3$

(e) BH_3 followed by CH_3CO_2H

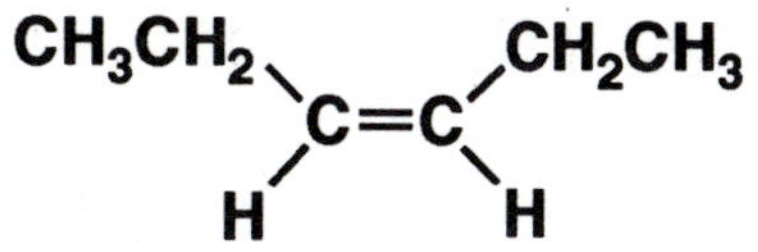

$CH_3CH_2(H)C=C(H)CH_2CH_3$ (cis)

(f) BH_3 followed by CH_3CO_2D

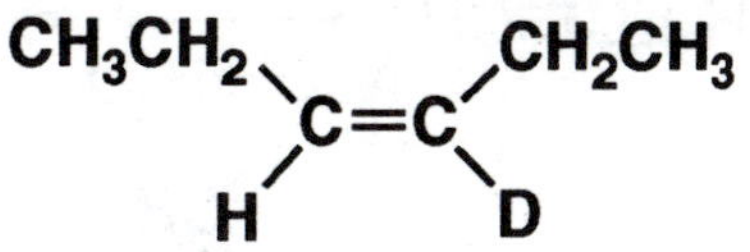

$CH_3CH_2(H)C=C(D)CH_2CH_3$ (cis)

(g) HCl (one mol) / LiCl in CH_3CO_2H

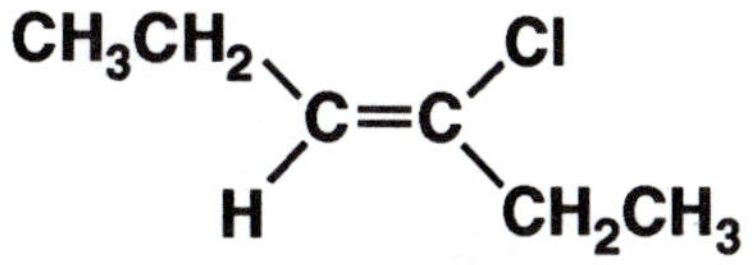

$CH_3CH_2(H)C=C(Cl)CH_2CH_3$

(h) $NaNH_2$ in NH_3 (liquid)

No Reaction

(i) HBr (one mol)

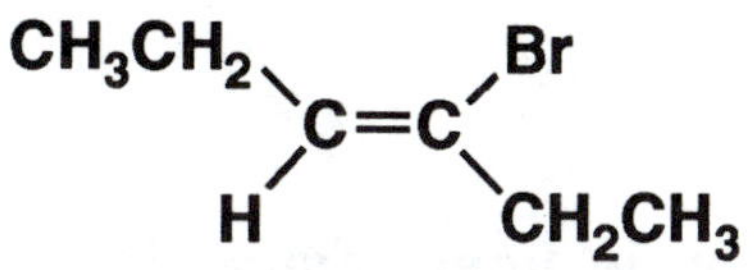

$CH_3CH_2(H)C=C(Br)CH_2CH_3$

(j) HBr (two mol)

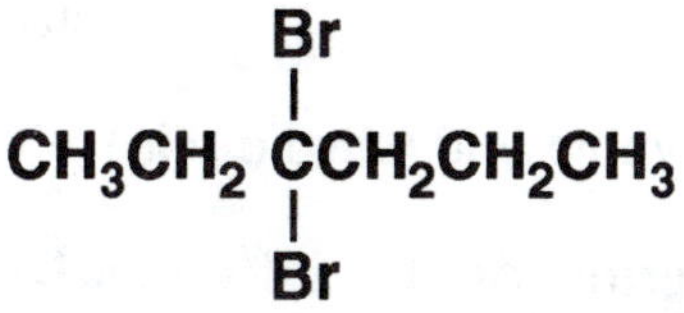

$CH_3CH_2CBr_2CH_2CH_2CH_3$

(k) H_2O in H_2SO_4 / $HgSO_4$

$CH_3CH_2C(=O)CH_2CH_2CH_3$

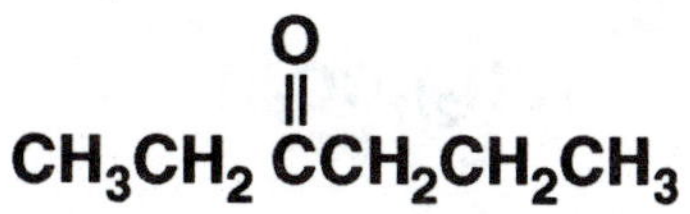

(l) CH_3CO_2H in H_2SO_4 / $HgSO_4$

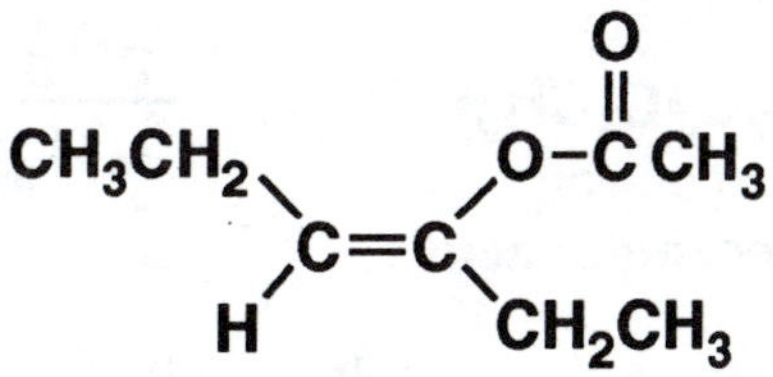

$CH_3CH_2(H)C=C(OC(=O)CH_3)CH_2CH_3$

Problem 6.13 Draw the structural formula of the bracketed enol formed in each alkyne hydration reaction and then draw the structural formula of the carbonyl compound with which each enol is in equilibrium.

An enol contains a hydroxyl group on a carbon-carbon double bond and is in equilibrium, by keto-enol tautomerism, with the isomeric aldehyde or ketone.

(a) $CH_3(CH_2)_5C\equiv CH + H_2O \xrightarrow[H_2SO_4]{HgSO_4}$ (an enol) $\longrightarrow$

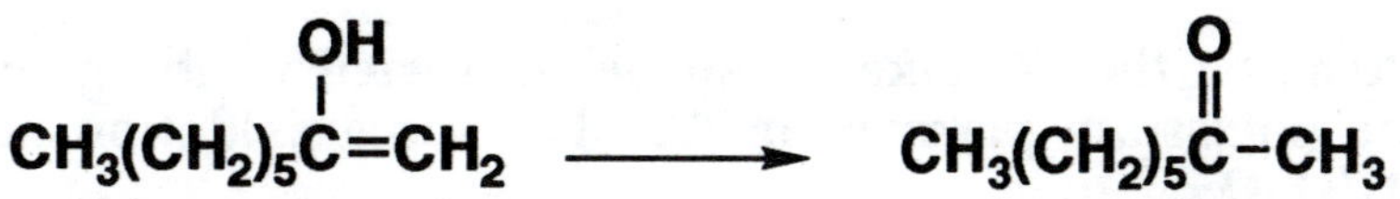

$CH_3(CH_2)_5C(OH)=CH_2 \longrightarrow CH_3(CH_2)_5C(=O)-CH_3$

1-octen-2-ol (an enol) → **2-octanone (a ketone)**

(b) $CH_3(CH_2)_5C{\equiv}CH \xrightarrow[NaOH/H_2O_2]{(sia)_2BH}$ (an enol) $\longrightarrow$

$$\underset{\substack{\textbf{1-octen-1-ol}\\ \textbf{(an enol)}}}{CH_3(CH_2)_5CH{=}\overset{OH}{\overset{|}{C}}H} \longrightarrow \underset{\substack{\textbf{octanal}\\ \textbf{(an aldehyde)}}}{CH_3(CH_2)_5CH_2\overset{O}{\overset{\|}{C}}{-}H}$$

Syntheses

Problem 6.14 Show how you might convert 9-octadecynoic acid to the following:

$$CH_3(CH_2)_7C{\equiv}C(CH_2)_7CO_2H$$

9-octadecynoic acid

(a) (E)-9-octadecynoic acid (eliadic acid)

Chemical reduction of the alkyne with two moles of sodium in liquid ammonia converts the alkyne to an (E)-alkene. Note that the carboxyl group is unaffected by this condition.

$$CH_3(CH_2)_7C{\equiv}C(CH_2)_7CO_2H \xrightarrow[NH_3(l)]{2Na} \begin{matrix} CH_3(CH_2)_7 & & H \\ & C{=}C & \\ H & & (CH_2)_7CO_2H \end{matrix}$$

9-octadecenoic acid

(b) (Z)-9-octadecenoic acid (oleic acid)

Reduction of the alkyne with one mole of hydrogen with the specially prepared Lindlar's catalyst gives a (Z)-alkene. The carboxyl group is unaffected by these conditions.

$$CH_3(CH_2)_7C{\equiv}C(CH_2)_7CO_2H \xrightarrow[Pd/CaCO_3]{H_2} \begin{matrix} CH_3(CH_2)_7 & & (CH_2)_7CO_2H \\ & C{=}C & \\ H & & H \end{matrix}$$

9-octadecynoic acid

(c) 9,10-dihydroxydecanoic acid

Either the (E)-alkene or the (Z)-alkene can be converted to the glycol by oxidation with cold, alkaline potassium permanganate. The same oxidation can be brought about using OsO_4/H_2O_2.

$$CH_3(CH_2)_7CH{=}CH(CH_2)_7CO_2H \xrightarrow[pH\ 11.8]{KMnO_4} CH_3(CH_2)_7\underset{HO}{\underset{|}{C}}H{-}\underset{OH}{\underset{|}{C}}H(CH_2)_7CO_2H$$

9-octadecenoic acid
(*cis* or *trans*)

(d) 9,10-dibromooctadecanoic acid

Addition of Br_2 to either the (E)-alkene or the (Z)-alkene forms the desired product.

$$CH_3(CH_2)_7CH{=}CH(CH_2)_7CO_2H \xrightarrow{Br_2} CH_3(CH_2)_7\underset{Br}{CH}{-}\underset{Br}{CH}(CH_2)_7CO_2H$$

9-octadecenoic acid
(*cis* or *trans*)

(e) octadecanoic acid

Reduction of either the (E)-alkene or the (Z)-alkene with one mole of H_2 in the presence of a Ni, Pd, or Pt catalyst, or reduction of the alkyne with two moles of H_2 gives the desired product.

$$CH_3(CH_2)_7CH{=}CH(CH_2)_7CO_2H \xrightarrow[Ni]{H_2} CH_3(CH_2)_7CH_2{-}CH_2(CH_2)_7CO_2H$$

9-octadecenoic acid
(*cis* or *trans*)

$$CH_3(CH_2)_7C{\equiv}C(CH_2)_7CO_2H \xrightarrow[Ni]{2H_2} CH_3(CH_2)_7CH_2{-}CH_2(CH_2)_7CO_2H$$

9-octadecynoic acid

Problem 6.15 For small scale and consumer welding applications, many hardware stores sell cylinders of MAAP gas, which is a mixture of propyne (methylacetylene) and 1,2-propadiene gases, with other hydrocarbons. How would you prepare this gas mixture in the laboratory?

This gas mixture could be prepared from a double dehydrohalogenation of 1,2-dibromopropane. As described in section 6.4, the 1,2-propadiene (allene) is a side product of the reaction, being derived from β-elimination from the intermediate 2-bromopropene.

$$CH_3{-}CHBr{-}CH_2Br \xrightarrow{NaNH_2} CH_3{-}C{\equiv}CH + CH_2{=}C{=}CH_2$$

1,2-dibromopropane → **propyne** + **1,2-propadiene (allene)**

Problem 6.16 Show reagents and experimental conditions you might use to convert propyne into each product. Some of these syntheses can be done in one step. Others require two or more steps.

$$\text{(a)}\quad CH_3{-}C{\equiv}CH \longrightarrow CH_3{-}\overset{Br}{\underset{Br}{C}}{-}\overset{Br}{\underset{Br}{CH}}$$

Addition of two moles of Br_2 to propyne gives 1,1,2,2-tetrabromopropane.

$$CH_3-C\equiv CH \xrightarrow{Br_2} CH_3\overset{Br}{\overset{|}{C}}=\overset{Br}{\overset{|}{C}}H \xrightarrow{Br_2} CH_3-\underset{Br}{\underset{|}{\overset{Br}{\overset{|}{C}}}}-\underset{Br}{\underset{|}{\overset{Br}{\overset{|}{C}}}}H$$

(b) $$CH_3-C\equiv CH \longrightarrow CH_3-\underset{Br}{\underset{|}{\overset{Br}{\overset{|}{C}}}}-CH_3$$

Addition of two moles of HBr in the absence of peroxides occurs by electrophilic addition and gives first 2-bromopropene and then 2,2-dibromopropane.

$$CH_3-C\equiv CH \xrightarrow{HBr} CH_3\overset{Br}{\overset{|}{C}}=CH_2 \xrightarrow{HBr} CH_3-\underset{Br}{\underset{|}{\overset{Br}{\overset{|}{C}}}}-CH_3$$

(c) $$CH_3-C\equiv CH \longrightarrow CH_3-\overset{O}{\overset{\|}{C}}-CH_3$$

Acid-catalyzed hydration of the alkyne gives an enol which is in equilibrium, by keto-enol tautomerism, with the isomeric ketone, in this case propanone (acetone).

$$CH_3-C\equiv CH + H_2O \xrightarrow[HgSO_4]{H_2SO_4} \left[CH_3-\overset{OH}{\overset{|}{C}}=CH_3 \right] \longrightarrow CH_3-\overset{O}{\overset{\|}{C}}-CH_3$$

(d) $$CH_3-C\equiv CH \longrightarrow CH_3CH_2-\overset{O}{\overset{\|}{C}}-H$$

Hydroboration with (sia)2BH or other hindered derivative of borane followed by oxidation with alkaline hydrogen peroxide gives an enol which is in equilibrium, by keto-enol tautomerism, with the isomeric aldehyde, in this case propanal.

$$CH_3-C\equiv CH \xrightarrow[2)\ NaOH/H_2O_2]{1)\ (sia)_2BH} \left[CH_3CH=\overset{OH}{\overset{|}{C}}H \right] \longrightarrow CH_3-CH_2-\overset{O}{\overset{\|}{C}}H$$

Problem 6.17 Show reagents and experimental conditions you might use to convert each starting material into the desired product. Some of these syntheses can be done in one step. Others require two or more steps.

(a) $CH_3CH_2CH_2C{\equiv}CCH_3 \longrightarrow$ (E)-$CH_3CH_2CH_2CH{=}CHCH_3$ ($CH_3CH_2CH_2$ and H on one carbon; H and CH_3 on the other, trans)

Chemical reduction of the alkyne with sodium in liquid ammonia gives (E)-2-hexene.

$$CH_3CH_2CH_2C{\equiv}CCH_3 \xrightarrow[NH_3(l)]{2Na} \text{(E)-}CH_3CH_2CH_2CH{=}CHCH_3$$

(b) $CH_3CH_2CH_2C{\equiv}CCH_3 \longrightarrow$ (Z)-$CH_3CH_2CH_2CH{=}CHCH_3$ ($CH_3CH_2CH_2$ and CH_3 cis; H and H cis)

Hydroboration of the internal alkyne followed by reaction of the organoborane with acetic acid gives (Z)-2-hexene.

$$CH_3CH_2CH_2C{\equiv}CCH_3 \xrightarrow[2)\ CH_3CO_2H]{1)\ BH_3} \text{(Z)-}CH_3CH_2CH_2CH{=}CHCH_3$$

Alternatively, catalytic reduction with hydrogen in the presence of Lindlar catalyst gives the (Z)-alkene.

$$CH_3CH_2CH_2C{\equiv}CCH_3 + H_2 \xrightarrow[\text{Lindlar catalyst}]{Pd/CaCO_3} \text{(Z)-}CH_3CH_2CH_2CH{=}CHCH_3$$

(c) $CH_3(CH_2)_4C{\equiv}CH \longrightarrow CH_3(CH_2)_4C{\equiv}C{:}^-\ Na^+$

The anion can be formed using sodium amide, $NaNH_2$, or sodium hydride, NaH.

$$CH_3(CH_2)_4C{\equiv}CH + NaNH_2 \longrightarrow CH_3(CH_2)_4C{\equiv}C{:}^-\ Na^+ + NH_3$$

(d) $CH_3CH_2C{\equiv}CH \longrightarrow CH_3CH_2C{\equiv}CD$

Formation of the terminal acetylide anion followed by reaction with a deuterium donor such as D_2O gives 1-deutero-1-butyne.

$$CH_3CH_2C{\equiv}CH \xrightarrow{NaNH_2} CH_3CH_2C{\equiv}C{:}^-\ Na^+ \xrightarrow{D_2O} CH_3CH_2C{\equiv}CD$$

(e) $CH_3CH_2C{\equiv}CH \longrightarrow$ $\begin{matrix} CH_3CH_2 & & H \\ & C{=}C & \\ H & & D \end{matrix}$

Hydroboration with this disubstituted derivative of diborane followed by reaction of the organoborane with deuteroacetic acid gives the desired 1-deutero-1-butene.

$$CH_3CH_2C{\equiv}CH \xrightarrow[2.\ CH_3CO_2D]{1.\ (sia)_2BH} \begin{matrix} CH_3CH_2 & & H \\ & C{=}C & \\ H & & D \end{matrix}$$

(f) $CH_3CH_2C{\equiv}CH \longrightarrow$ $\begin{matrix} CH_3CH_2 & & H \\ & C{=}C & \\ D & & H \end{matrix}$

Hydroboration of the terminal alkyne with deuteroborane adds deuterium to the more substituted carbon of the alkyne. Reaction of the deuterated organoborane with acetic acid gives the 2-deutero-1-butene.

$$CH_3CH_2C{\equiv}CH \xrightarrow[2.\ CH_3CO_2H]{1.\ (sia)_2BD} \begin{matrix} CH_3CH_2 & & H \\ & C{=}C & \\ D & & H \end{matrix}$$

(g) $HOCH_2CH{=}CHCH_2OH \longrightarrow HOCH_2CH_2{-}CH_2CH_2OH$

Catalytic reduction of the carbon-carbon double bond with one mol of hydrogen in the presence of a transition metal catalyst gives 1,4-butanediol.

$$HOCH_2CH{=}CHCH_2OH + H_2 \xrightarrow{Ni} HOCH_2CH_2{-}CH_2CH_2OH$$

(h) 1-ethynylcyclohexanol (HO, C≡CH) $\longrightarrow$ 1-acetylcyclohexanol (HO, $C({=}O){-}CH_3$)

Acid-catalyzed hydration of the carbon-carbon triple bond followed by keto-enol tautomerism of the resulting enol gives the desired ketone. Note that the tertiary alcohol in the starting material is unaffected by these conditions.

HO C≡CH (cyclohexane) + H_2O $\xrightarrow[HgSO_4]{H_2SO_4}$ [HO, OH, C=CH_2 (cyclohexane)] ⟶ HO, O, C–CH_3 (cyclohexane)

<u>Problem 6.18</u> Rimantadine is effective in preventing naturally occurring infections caused by the influenza A virus and in treating established illness. It is thought to exert its antiviral effect by blocking a late stage in the assembly of the virus. Rimantadine is synthesized from adamantane in the following sequence. We have covered the types of chemistry involved in Steps 1-4. We will discuss the chemistry of Step 5 in Chapter 17.

Adamantane $\xrightarrow[(1)]{Br_2}$ (adamantyl) Br $\xrightarrow[(2)]{CH_2=CHBr,\ AlBr_3}$ (adamantyl) CH_2CHBr_2

$\xrightarrow[(3)]{}$ (adamantyl) C≡CH $\xrightarrow[(4)]{}$ (adamantyl) C(=O)–CH_3 $\xrightarrow[(5)]{}$ (adamantyl) $CH(NH_2)CH_3$ (NH₂, CHCH₃)

(a) Describe experimental conditions to bring about Step 1. By what type of mechanism does this reaction occur? Account for the regioselectivity of bromination in Step 1.

This reaction occurs via a radical bromination, with the bromine ending up on the tertiary carbon atom.

(b) Propose a mechanism for Step 2. Hint: As we shall see in Chapter 16, reaction of an alkyl bromide such as adamantyl bromide with aluminum bromide (a Lewis acid, Section 3.4) results in formation of a carbocation and $AlBr_4^-$. Assume that adamantyl cation is formed in Step 2 and proceed from there to describe a mechanism.

The bromoethene pi electrons attack the admantyl cation, to create a new cation that captures a bromide from $AlBr_4^-$ to yield dibromoethyl-adamantane.

$AlBr_4^-$ $(Br—AlBr_3)^-$

Br $AlBr_3$ CH_2CHBr $CH_2{=}CHBr$ CH_2CHBr_2

(c) Account for the regioselectivity of carbon-carbon bond formation in Step 2.

The carbon atom without the halogen ends up attached to the adamantane group, because this allows formation of the more stable intermediate with the cation adjacent to the halogen atom. This formation of this cation has a lower energy of activation because of resonance stabilization of the cation provided by the halogen.

(d) Describe experimental conditions to bring about Step 3.

This reaction occurs via double dehydrohalogenation using a strong base such as $NaNH_2$.

(e) Describe experimental conditions to bring about Step 4.

This transformation occurs via an acid-catalyzed hydration of the alkyne using H_2O, H_2SO_4, and $HgSO_4$. The initially formed enol equilibrates to the more stable keto form.

CHAPTER 7: CONJUGATED DIENES

SUMMARY OF REACTIONS

Starting Material ↓ / Product →	Cyclohexenes	Dihaloalkenes (1,2 addition)	Dihaloalkenes (1,4 addition)	Haloalkenes (1,2 addition)	Haloalkenes (1,4 addition)
Conjugated Dienes	7A 7.5	7B 7.4A		7C 7.4A	

REACTION 7A: THE DIELS-ALDER REACTION (Section 7.5)

- Conjugated dienes undergo an interesting reaction with an alkene or alkyne called a **cycloaddition** reaction that results in production of a six-membered ring through the formation of two carbon-carbon bonds. This reaction is named for its discoverers, and hence is called the **Diels-Alder reaction.** ✱
- The Diels-Alder reaction is a very important and useful synthetic reaction because (1) it is one of the few reactions that creates six-membered rings (2), it is one of the few reactions that forms two carbon-carbon bonds simultaneously (3), it is stereoselective.

REACTION 7B: CONJUGATE ADDITION OF X_2 (Section 7.4A)

1,2-addition

1,4-addition or conjugate addition

- Conjugated dienes undergo the same kind of addition reactions with X_2 as simple alkenes, but the reaction usually gives mixtures of 1,2-addition and 1,4-addition. The numbers do not refer to IUPAC nomenclature, but to the sp^2 atoms of the conjugated diene. Therefore, 1,2-addition indicates that addition takes place at adjacent atoms at one end of the conjugated diene, and 1,4-addition (also referred to as **conjugate addition**) indicates that addition takes place at atoms 1 and 4 of the conjugated diene. ✳

REACTION 7C: CONJUGATE ADDITION OF H-X (Section 7.4A)

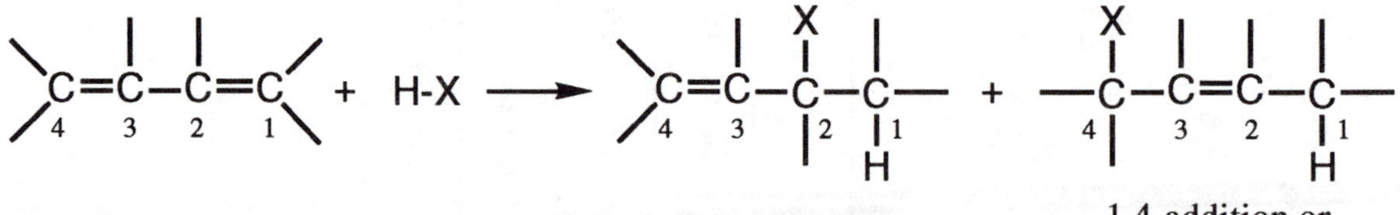

>C=C–C=C< (4 3 2 1) + H-X ⟶ >C=C–C(X)–C(H)– (4 3 2 1) + –C(X)–C=C–C(H)– (4 3 2 1)

1,2-addition

1,4-addition or conjugate addition

- Conjugated dienes undergo the same kind of addition reactions with H-X as normal alkenes, but the reaction usually gives mixtures of 1,2-addition and 1,4-addition.

SUMMARY OF IMPORTANT CONCEPTS

7.0 OVERVIEW

• This chapter involves the study of molecules that have two or more conjugated dienes. Conjugated dienes undergo many of the same reactions characteristic of normal, unconjugated alkenes, but they also undergo their own unique set of reactions.

7.1 STRUCTURE AND NOMENCLATURE

• **Dienes** are compounds that contain two carbon-carbon double bonds, and they can be divided into three groups:
  - **Unconjugated dienes** are ones in which there is at least one atom between the double bonds.

    For example: or

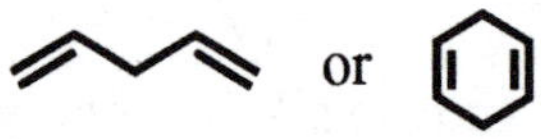

  - **Cumulated dienes** are ones in which the two double bonds share a common carbon atom.

    For example: $H_2C{=}C{=}CH_2$

  - **Conjugated dienes** are ones in which the two double bonds are adjacent to each other.

    For example: or ✳

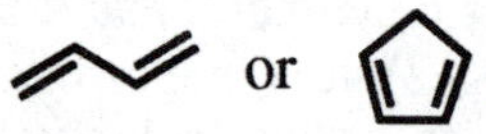

    Conjugated dienes are more stable than would be predicted from the stabilities of simple alkenes. For example, judging from heats of hydrogenation, 1,3-butadiene is 4.1 kcal/mol more stable than predicted from twice the heat of hydrogenation of 1-butene. This stabilization is the result of the pi electron delocalization of conjugated dienes described below.

7.3 STRUCTURE OF CONJUGATED DIENES

• According to the **valence bond model**, as shown in Figure 7.2, there is partial overlap between the adjacent 2p orbitals such as those on carbon-2 and carbon-3 in butadiene. This partial overlap allows **delocalization** of the pi electrons into all four parallel 2p orbitals. According to this model, the extra stability of conjugated dienes is the result of this delocalization. ✳

-Note that the delocalization is only possible if all four of the 2p orbitals of the conjugated diene are parallel, a situation that requires all four of the sp^2 hybridized carbon atoms to be in the same plane. In other words, the carbon-carbon single bond in conjugated dienes has partial double bond character, and thus cannot rotate freely. ✱ *[This would be a good time to review valence bond theory (Section 1.8) if this section is confusing.]*

- As shown in Figure 7.3, the **molecular orbital model** describes conjugated dienes as a set of two filled pi bonding molecular orbitals with zero and one node, respectively, and two unfilled antibonding orbitals with two and three nodes, respectively.
  - The lowest energy molecular orbitals are only possible if there is overlap between all four 2p orbitals. Thus, like the valence bond model, the molecular orbital model explains why all four sp^2 atoms of the conjugated dienes must be in the same plane to allow for the stabilization provided by electron delocalization.

7.4 ELECTROPHILIC ADDITION TO CONJUGATED DIENES

- The two isomeric products (1,2 and 1,4 addition) observed with addition reactions to conjugated dienes can be explained by a two-step mechanism that involves rate limiting formation of an **allylic cation intermediate**, that then reacts with a nucleophile at the 2 or 4 position to yield products.

$$\underset{4}{CH_2}=\underset{3}{CH}-\underset{2}{CH}=\underset{1}{CH_2} + H-Br \longrightarrow \left(\underset{4}{CH_2}=\underset{3}{CH}-\overset{+}{\underset{2}{CH}}-\overset{H}{\underset{1}{CH_2}} \longleftrightarrow \overset{+}{\underset{4}{CH_2}}-\underset{3}{CH}=\underset{2}{CH}-\overset{H}{\underset{1}{CH_2}}\right)$$

allylic cation intermediate

Br^- ↙ ↘ Br^-

$$\underset{4}{CH_2}=\underset{3}{CH}-\overset{Br}{\underset{2}{CH}}-\overset{H}{\underset{1}{CH_2}} \qquad \overset{Br}{\underset{4}{CH_2}}-\underset{3}{CH}=\underset{2}{CH}-\overset{H}{\underset{1}{CH_2}}$$

1,2-addition　　　　1,4-addition or conjugate addition

- The **allylic cation intermediate** is considerably more stable than a comparably substituted alkyl carbocation. This extra stability is the result of delocalization of the pi electrons and positive charge as described by the following resonance forms. ✱ *[Distribution of a charge over more than one atom almost always leads to net stabilization. This principle can be used in a wide variety of situations to predict relative stabilities of charged species.]*

$$CH_2{=}CH-\overset{+}{CH_2} \longleftrightarrow \overset{+}{CH_2}-CH{=}CH_2$$

- The relative stabilities of different carbocations are as follows:

$$\text{methyl} < 1^\circ < 2^\circ = 1^\circ \text{ allylic} < 3^\circ < 2^\circ \text{ allylic} < 3^\circ \text{ allylic}$$

- For conjugated dienes reacting with molecules such as HBr or Br_2, 1,2-addition products predominate over 1,4-addition products at low temperature, while at high temperature the 1,4-addition products predominate. These facts are explained by the concepts of **kinetic** and **thermodynamic control** of reaction product distributions. ✱
- **Kinetic or rate control** occurs at low temperature when the distribution of products is determined by the relative **rates** of formation of each product. **Thermodynamic or**

equilibrium control occurs at high temperature so the products equilibrate, thus the distribution of products is determined by the relative **thermodynamic stabilities** of each product. ✷

- As can be seen in Figure 7.4, the energy of activation for 1,2-addition is lower than that for 1,4-addition. As a result, under **kinetic control** at low temperature, the 1,2-addition reaction predominates because it has the **lower reaction barrier** and thus a **faster rate** of formation.

  As predicted by the two most important resonance forms, the allylic cation intermediate has the positive charge concentrated at carbon atoms number 2 and 4. However, the **greater concentration of positive charge** is at carbon number 2, because this is a **secondary carbon atom. Less positive charge** is concentrated at the **primary carbon atom 4**. As a result, there is a lower reaction barrier for reaction at the atom with the greater concentration of positive charge, namely carbon atom two. *[Recall that carbocations with more alkyl substituents are more stable, thus a secondary carbocation is more stable than a primary carbocation.]*

- On the other hand, the 1,4-addition product contains a disubstituted carbon-carbon double bond, so it is thermodynamically more stable than the 1,2-addition product that has only one alkyl group attached to the double bond. Thus, at equilibrium (high temperature) the reaction is under **thermodynamic control**, and the **more stable** 1,4-addition product predominates. ✷

  In general, alkenes with more alkyl groups attached are thermodynamically more stable than alkenes with fewer alkyl groups. *[This is a useful concept that will also be used to explain the product distribution of β-elimination reactions in chapter 10.]*

  The two products can **equilibrate at higher temperature** because the product molecules that have been formed **collide with enough energy to convert them back into the allylic cation intermediate** that can then form products again. In this way, the thermodynamically less stable 1,2-addition product can be converted into the thermodynamically more stable 1,4-addition product. At lower temperature, the product molecules do not collide with enough energy to convert them back into the allylic cation, so the 1,2-addition products that form the fastest remain as products.

7.5 THE DIELS-ALDER REACTION

- In the Diels-Alder reaction, the electrons in the pi bonds move to create the new bonds as shown below. The arrows are not meant to indicate a mechanism, but rather to help keep track of the electrons during the reaction.

diene dienophile

  - The alkene or alkyne that reacts with the diene is referred to as the **dienophile**.
  - The Diels-Alder reaction is usually favorable (the product is more stable than the reactants) because the overall transformation involves breaking two carbon-carbon pi bonds in order to make two carbon-carbon sigma bonds. Recall that generally carbon-carbon sigma bonds are stronger than pi bonds, so Diels-Alder reactions generally have negative values for ΔH. *[It is always helpful to understand why a reaction occurs as well as simply how it occurs.]*
- Studies of the mechanism of Diels-Alder reactions have shown that the pi bonds are broken and the sigma bonds are made in a **single, concerted step**. In other words, the cyclic redistribution of bonding electrons occurs such that bond-breaking and bond-making occur simultaneously. Reactions of this type involving a single step and a cyclic rearrangement of bonding electrons are called **pericyclic reactions**.

- During the reaction, the diene and the dienophile approach each other in parallel planes, allowing interaction between the pi bonds of both molecules. The reaction is depicted in Figure 7.5.
- This **pericyclic reaction** mechanism is very different than the ionic and radical reaction mechanisms we have discussed previously. *[This would be an excellent time to review what is meant by ionic and radical mechanisms. Ionic mechanisms involve species with full or partial charges like electrophilic addition reactions to alkenes (Section 5.3). Radical mechanisms involve radical species, such as the radical addition reactions described in Section 5.5.]*

• The above mechanism explains some important details of the Diels-Alder reaction.
- The diene must adopt the **s-*cis* conformation** in order to react with the dienophile. This conformation is required to allow both sigma bonds to be formed simultaneously with the dienophile.

  As stated earlier, in order to allow for maximum overlap of the 2p orbitals, the four carbon atoms of conjugated dienes prefer to lie in the same plane. This means that there are two preferred conformations of conjugated dienes, the s-*cis* and s-*trans* conformations. The **"s"** indicates that it is the **single bond** between the two alkenes that is referred to with the *cis* or *trans* designation. ✱ *[This is an important point that is best understood using models.]*

s-*trans*-1,3-butadiene cannot take part in Diels-Alder reaction

s-*cis*-1,3-butadiene can take part in Diels-Alder reaction

- Conjugated dienes that cannot adopt the s-*trans* conformation cannot take part in Diels-Alder reactions. On the other hand, conjugated dienes such as cyclopentadiene ( ) that can only adopt the s-*cis* geometry are very good reactants in Diels-Alder reactions.
- The stereochemistry of the dienophile is retained in the product of the Diels-Alder reaction, since the two sigma bonds are made simultaneously. For example, the *cis* stereochemistry of dimethyl maleate is retained in the product. ✱

dimethyl maleate (*cis* dienophile)

cis product

• When **cyclic conjugated dienes** are used for a Diels-Alder reaction, the product is a **bicyclic** structure. As a result, substituents on the dienophile can end up in the **endo** or **exo** positions. In practice, the substituents end up predominantly in endo position.
- The **exo** position is the one that is on the same side as the shortest bridge of the bicyclic structure, while the **endo** position is the one that is on the side opposite the shortest bridge. ✱

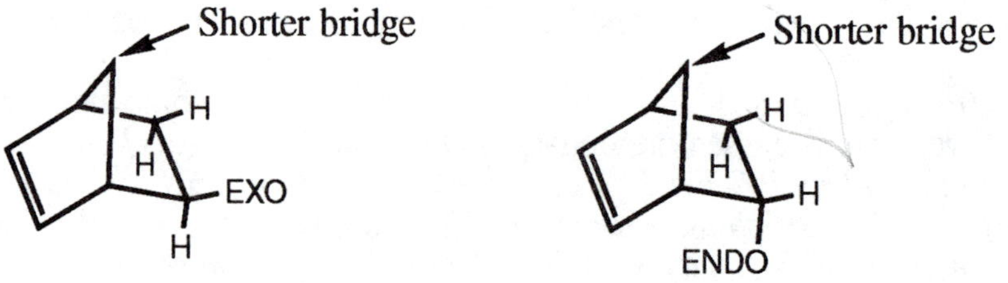

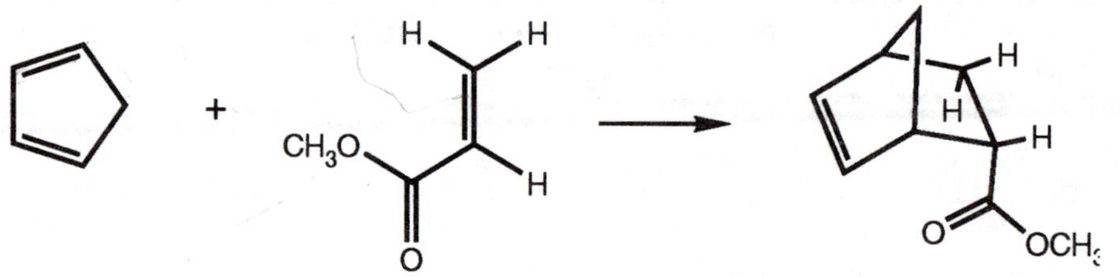

Shorter bridge
EXO
Shorter bridge
ENDO

CH_3O OCH_3

the endo product is formed predominantly

- Substituents on the reactants can greatly alter the rates of Diels-Alder reactions. The reaction is **facilitated** by a combination of **electron-withdrawing substituents** on one of the reactants and **electron-releasing substituents** on the other. Usually, the electron-withdrawing groups are found on the dienophile. ✱
  - Examples of **electron-withdrawing groups** usually found on dienophiles include ketones, carboxylic esters, nitro groups, and nitriles. These groups pull electron density out of the pi bond, and make it easier for the reaction to take place.
- It is important to point out that cycloaddition reactions have specific electronic requirements, so alkenes do not dimerize to form a four-membered ring product, and dienes do not dimerize to form an eight-membered ring product.
- Certain Diels-Alder reactions can be catalyzed by Lewis acids such as $AlCl_3$. The Lewis acid is thought to operate by coordinating to the dienophile to help polarize it, thus facilitating reaction with the diene.

CHAPTER 7
Solutions to the Problems

Problem 7.1 Which of these terpenes (Figure 4.6) contain conjugated dienes?

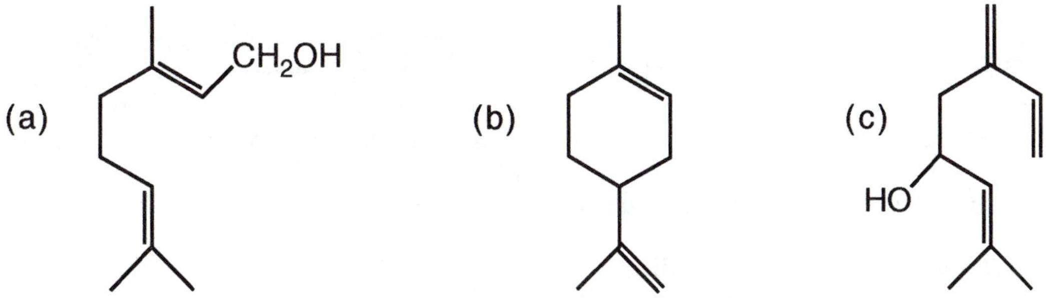

(a) CH_2OH geraniol

(b) limonene

(c) HO an aggregating pheromone of bark beetles

Of the terpenes shown in Figure 4.6, only the aggregating pheromone of the bark beetle contains a conjugated diene. Both geraniol and limonene are dienes, but in them the double bonds are not conjugated.

Problem 7.2 Estimate the stabilization gained due to conjugation when 1,4-pentadiene is converted to *trans*-1,3-pentadiene. Note that the answer is not as simple as comparing the heats of hydrogenation of 1,4-pentadiene and *trans*-1,3-pentadiene because, although the double bonds are moved from unconjugated to conjugated, the degree of substitution of one of the double bonds is also changed, in this case from a monosubstituted double bond to a disubstituted double bond. To answer this question you must separate the effect due to conjugation from that due to change in degree of substitution.

On the assumption that there is no interaction between the double bonds in 1,3-pentadiene, estimate that its heat of hydrogenation is -30.1 kcal/mol (based on that of the terminal double bond in 1-pentene) plus -27.6 kcal/mol (based on that for the internal double bond in *trans*-2-pentene). The calculated heat of hydrogenation is -57.6 kcal/mol. The experimental heat of hydrogenation is -54.1 kcal/mol. Thus, the stabilization due to conjugation of the double bonds is 3.5 kcal/mol. These values are shown graphically on the energy diagram.

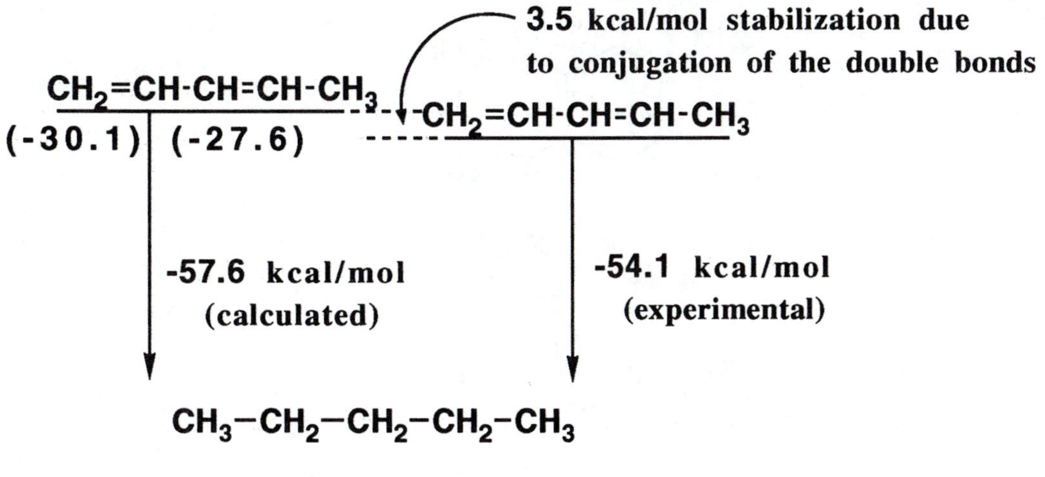

3.5 kcal/mol stabilization due to conjugation of the double bonds

CH_2=CH-CH=CH-CH_3
(-30.1) (-27.6)

CH_2=CH-CH=CH-CH_3

-57.6 kcal/mol (calculated)

-54.1 kcal/mol (experimental)

$CH_3-CH_2-CH_2-CH_2-CH_3$

Problem 7.3 Write an additional contributing structure for each carbocation and state which of the two makes the greater contribution to the resonance hybrid.

(a) (b)

Greater contribution **Equal contribution**

Problem 7.4 Predict the product(s) formed by addition of one mole of Br_2 to 2,4-hexadiene.

Predict both 1,2-addition and 1,4-addition.

$$CH_3-CH=CH-CH=CH-CH_3 \xrightarrow{Br_2}$$

$$CH_3-CH(Br)-CH(Br)-CH=CH-CH_3 + CH_3-CH(Br)-CH=CH-CH(Br)-CH_3$$

Problem 7.5 What combination of diene and dienophile will undergo Diels-Alder reaction to give each adduct.

(a) CH=CH₂ substituted cyclohexene → 1,3-butadiene + $CH_2=CH-CH=CH_2$

(b) cyclohexene with CO_2CH_3, CO_2CH_3 → 1,3-butadiene + $(CH_3O_2C)_2C=CH_2$

(c) cyclohexadiene with CO_2CH_3, CO_2CH_3 → 1,3-butadiene + $CH_3O_2C-C\equiv C-CO_2CH_3$

In each part of this problem, the diene is 1,3-butadiene. In (a), the dienophile is one of the double bonds of a second molecule of 1,3-butadiene. In part (b), the dienophile is a 1,1-disubstituted alkene and in part (c), it is a disubstituted alkyne.

Problem 7.6 Which molecules can function as dienes in Diels-Alder reactions? Explain your reasoning.

(a)  (b) 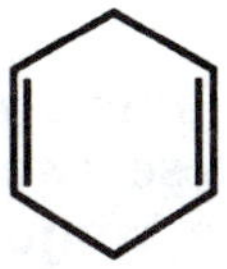 (c)

To function as a diene, the double bonds must be conjugated and able to assume an s-*cis* conformation. Both (a) 1,3-cyclohexadiene and (c) 1,3-cyclopentadiene can function as dienes. The double bonds in 1,4-cyclohexadiene are not conjugated and, therefore, this molecule cannot function as a diene.

Problem 7.7 What diene and dienophile might you use to prepare the following Diels-Alder adduct?

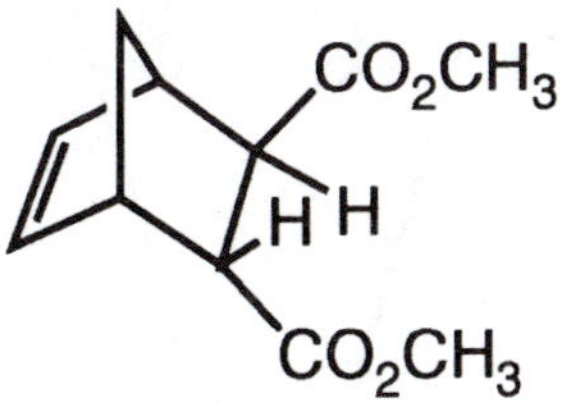

CO_2CH_3 H H CO_2CH_3

Use cyclopentadiene and the E-alkene.

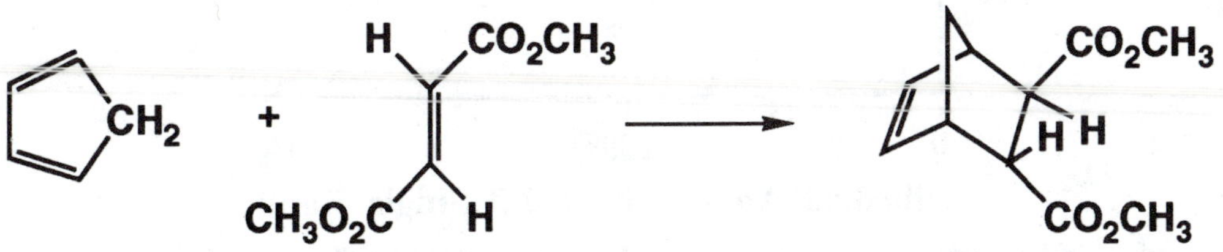

CH_2 + H CO_2CH_3 CH_3O_2C H ⟶ CO_2CH_3 H H CO_2CH_3

Structure and Stability

Problem 7.8 If an electron is added to 1,3-butadiene, into which molecular orbital does it go? If an electron is removed from 1,3-butadiene, from which molecular orbital is it taken?

If an electron is added to 1,3-butadiene, it will be placed into the lowest unoccupied molecular orbital. As shown in Figure 7.3, this means the $\pi_3{}^*$ antibonding molecular orbital that has two nodes.

If an electron is removed from 1,3-butadiene, it will be removed from the highest occupied molecular orbital. As shown in Figure 7.3, this means the π_2 bonding molecular orbital that has one node.

Problem 7.9 The heat of hydrogenation of allene (1,2-propadiene) to propene is -35.3 kcal/mol. Compare this to the heat of hydrogenation of 1,3-butadiene and 1-butene. Does allene have the characteristics of a conjugated or a non-conjugated diene?

Using the values given in the table in Section 7.2, the heat of hydrogenation for the reaction of 1,3-butadiene going to 1-butene is:

$$-56.5 \text{ kcal/mol} - (-30.3 \text{ kcal/mol}) = -26.5 \text{ kcal/mol}$$

On the other hand, the heat of hydrogenation of allene going to propene is -35.5 kcal/mol. Thus, as can be seen by comparing these two values, allene does not have the characteristic heat of hydrogenation of a conjugated diene (-26.5 kcal/mol), but is closer to that of a simple pi bond found in a non-conjugated diene (-30 kcal/mol).

Problem 7.10 Draw a potential energy diagram (potential energy vs. dihedral angle 0° - 360°, see Figure 2.8) for rotation about the 2,3 single bond in 1,3-butadiene.

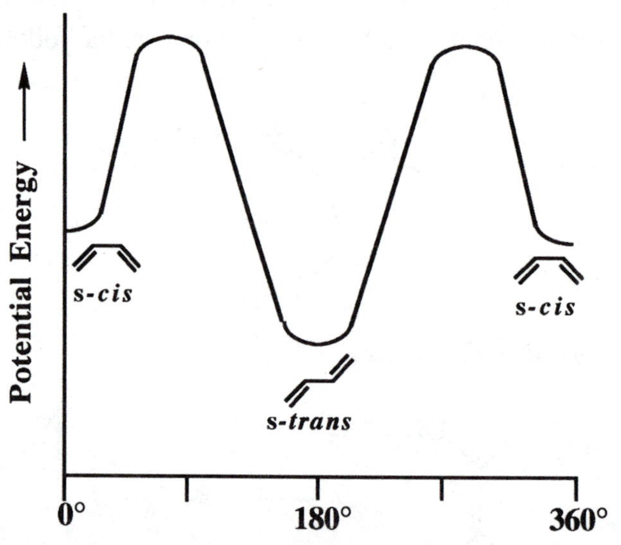

Dihedral Angle About 2,3 Single Bond

The s-*cis* and s-*trans* conformations are the most stable for 1,3-butadiene, because these provide for an overall planar geometry. This maximizes overlap of the 2p orbitals and thus conjugation. The s-*trans* geometry is more stable than the s-*cis* geometry because of steric interactions. That is, there is a small steric interaction between the hydrogen atoms on atoms 1 and 4 in the s-*cis* conformation, which is absent in the s-*trans* conformation.

Problem 7.11 Draw all important contributing structures for the following allylic cations and then rank the structures you have drawn in order of relative contributions to the resonance hybrid.

(a) $\overset{+}{CH_2}$

The secondary allylic cation makes the greater contribution.

$\overset{+}{CH_2}$ (cyclohexene ring) ⟷ CH_2 (=, exocyclic) (cyclohexane ring, +)

primary allylic (lesser contribution) **secondary allylic (greater contribution)**

(b) $CH_2{=}CH{-}CH{=}CH{-}CH_2^+$

The secondary allylic cation makes the greater contribution.

$$CH_2{=}CH{-}CH{=}CH{-}CH_2^+ \longleftrightarrow CH_2{=}CH{-}\overset{+}{C}H{-}CH{=}CH_2 \longleftrightarrow \overset{+}{C}H_2{-}CH{=}CH{-}CH{=}CH_2$$

primary allylic (lesser contribution) **secondary allylic (greater contribution)** **primary allylic (lesser contribution)**

(c) $CH_3{-}\overset{+}{C}(CH_3){-}CH{=}CH_2$

The tertiary allylic cation makes the greater contribution.

$$CH_3{-}\overset{+}{C}(CH_3){-}CH{=}CH_2 \longleftrightarrow CH_3{-}C(CH_3){=}CH{-}\overset{+}{C}H_2$$

tertiary allylic (greater contribution) **primary allylic (lesser contribution)**

Electrophilic Addition

Problem 7.12 Predict the structure of the major product formed by 1,2-addition of HCl to 2-methyl-1,3-butadiene (isoprene). To arrive at a prediction, first consider proton transfer to carbon 1 of this diene. Second, consider proton transfer to carbon 4 of this diene. Then compare the relative stabilities of the two allylic carbocation intermediates.

A tertiary allylic cation is more stable than a secondary allylic cation. Assume that because the tertiary allylic cation is the more stable of the two, the energy of activation is lower for formation and accordingly it is formed at a greater rate than the secondary allylic cation. Therefore, the major product of 1,2-addition is 3-chloro-3-methyl-1-butene.

$$CH_2{=}C(CH_3){-}CH{=}CH_2 \xrightarrow{H^+} CH_2{=}C(CH_3){-}\overset{+}{C}H{-}CH_2(H) \longrightarrow CH_2{=}C(CH_3){-}CH(Cl){-}CH_3$$

secondary allylic cation

$$CH_2{=}C(CH_3){-}CH{=}CH_2 \xrightarrow{H^+} (H)CH_2{-}\overset{+}{C}(CH_3){-}CH{=}CH_2 \longrightarrow CH_3{-}C(CH_3)(Cl){-}CH{=}CH_2$$

tertiary allylic cation — major product

Problem 7.13 Predict the major product formed by 1,4-addition (conjugate addition) of HCl to isoprene. Follow the reasoning suggested in the previous problem.

The major product of 1,4-addition is 1-chloro-3-methyl-2-butene.

$$CH_2{=}C(CH_3){-}CH{=}CH_2$$

$$\xrightarrow{H^+} CH_2{=}C(CH_3){-}\overset{+}{C}H{-}CH_2(H) \longleftrightarrow \overset{+}{C}H_2{-}C(CH_3){=}CH{-}CH_2(H) \longrightarrow CH_2(Cl){-}C(CH_3){=}CH{-}CH_2(H)$$

less stable allylic cation

$$\xrightarrow{H^+} (H)CH_2{-}\overset{+}{C}(CH_3){-}CH{=}CH_2 \longleftrightarrow (H)CH_2{-}C(CH_3){=}CH{-}\overset{+}{C}H_2 \longrightarrow (H)CH_2{-}C(CH_3){=}CH{-}CH_2(Cl)$$

more stable allylic cation — major product

Problem 7.14 Predict the structure of the major 1,2-addition product formed by reaction of one mol of Br_2 with isoprene. Also predict the structure of the major 1,4-addition product formed under these conditions.

Following the reasoning developed in the previous problems, predict that the major product of 1,2-addition to isoprene is 3,4-dibromo-3-methyl-1-butene.

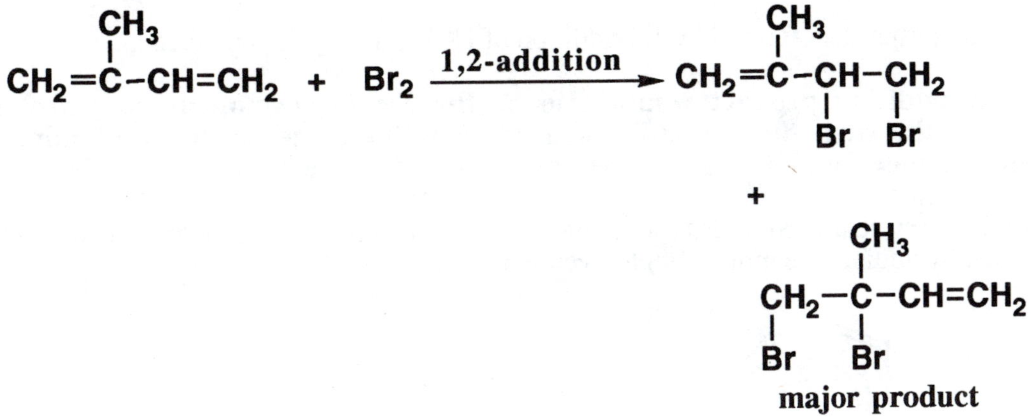

$$CH_2=C(CH_3)-CH=CH_2 + Br_2 \xrightarrow{\text{1,2-addition}} CH_2=C(CH_3)-CH(Br)-CH_2Br$$

$$+ \; CH_2(Br)-C(CH_3)(Br)-CH=CH_2$$

major product

There is only one 1,4-addition product possible from addition of bromine.

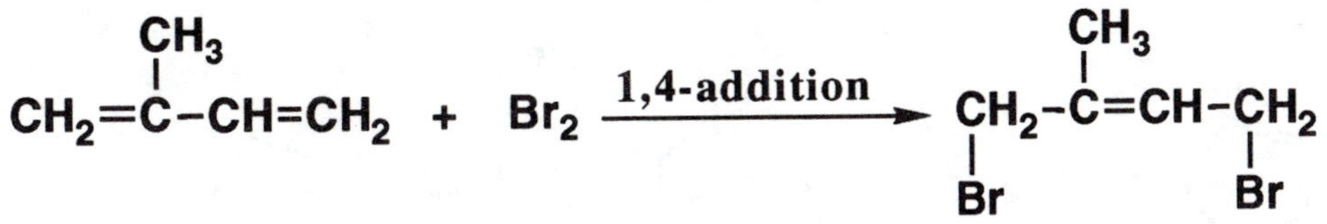

$$CH_2=C(CH_3)-CH=CH_2 + Br_2 \xrightarrow{\text{1,4-addition}} CH_2(Br)-C(CH_3)=CH-CH_2Br$$

Problem 7.15 (a) Which of the two molecules shown do you expect to be the major product formed by 1,2-addition of HCl to cyclopentadiene? Explain.

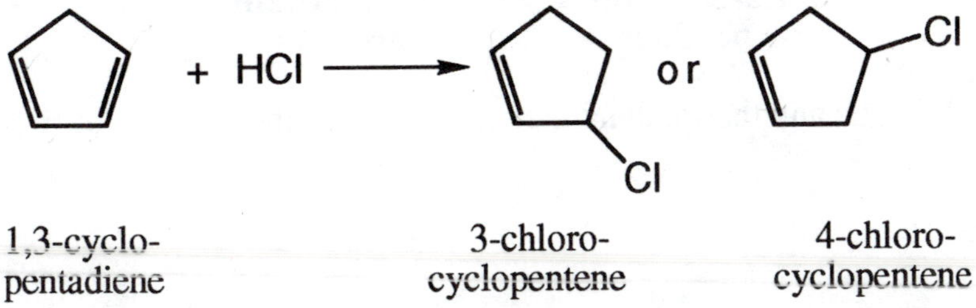

1,3-cyclopentadiene + HCl ⟶ 3-chlorocyclopentene or 4-chlorocyclopentene

The major product of 1,2-addition will be 3-chlorocyclopentene. This is because 3-chlorocyclopentene is the only product that will be derived from the more stable allylic cation intermediate. The 4-chlorocyclopentene derives from a less stable secondary cation intermediate, so it is formed in lesser amounts.

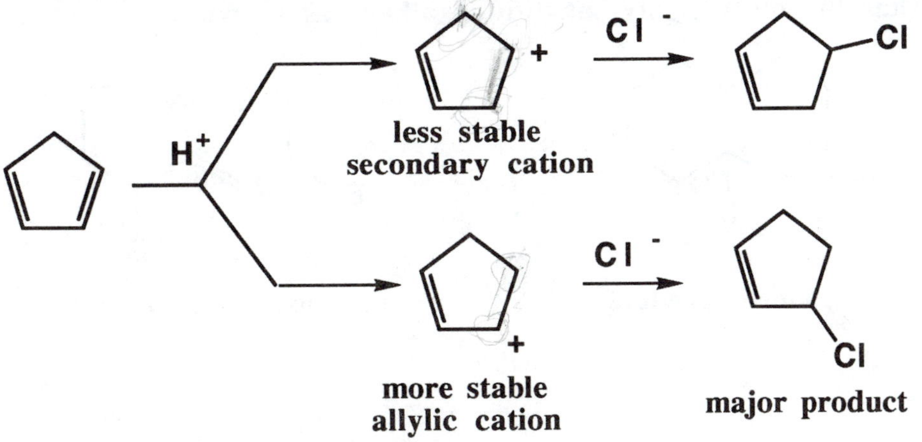

Cyclopentadiene + H^+ ⟶ less stable secondary cation $\xrightarrow{Cl^-}$ 4-chlorocyclopentene

Cyclopentadiene + H^+ ⟶ more stable allylic cation $\xrightarrow{Cl^-}$ 3-chlorocyclopentene (major product)

(b) Predict the major product formed by 1,4-addition of HCl to 1,3-cyclopentadiene.

The conjugate addition product would also be the 3-chlorocyclopentene. Because of symmetry in the ring, the 1,4-addition product is the same as the predominant 1,2-addition product, both being derived from the allylic cation.

Problem 7.16 Draw structural formulas for the two constitutional isomers of molecular formula $C_5H_6Br_2$ formed by adding one mol of Br_2 to cyclopentadiene.

+ Br_2 ⟶ $C_5H_6Br_2$

The two constitutional isomers are the 1,2 and 1,4-addition products.

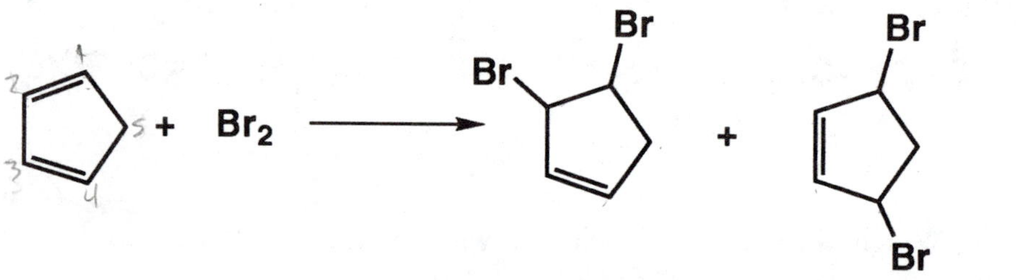

+ Br_2 ⟶ Br Br + Br Br

1,2-addition product **1,4-addition product**

Problem 7.17 What are the expected kinetic and thermodynamic products from the addition of Br_2 to the following dienes?

(a)

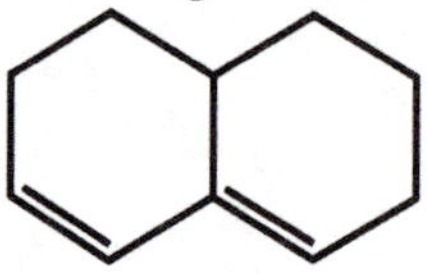

The kinetic product will be the 1,2-addition product that is derived from the more highly substituted and thus more stable bridged bromonium ion intermediate. The thermodynamic product is the most stable one, namely the 1,4-addition product, since this has the most highly substituted alkene as shown.

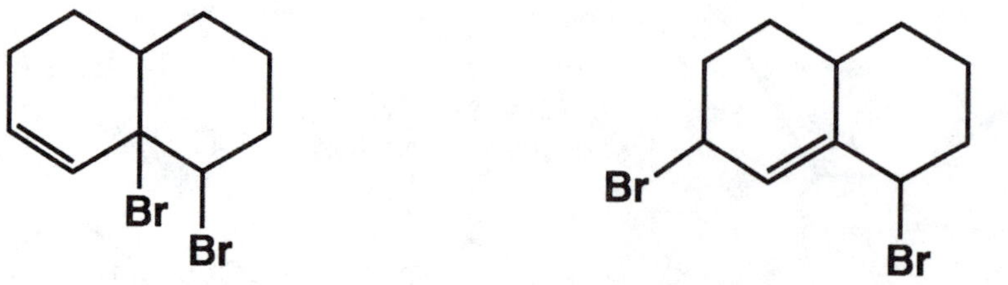

Br Br Br Br

kinetic product **thermodynamic product**

(b)

Because of symmetry in the molecule, both possible bromonium ion intermediates are the same. Thus there is only one possible 1,2-addition product and this is the kinetic product. The thermodynamic product is the more stable one, namely the 1,4-addition product, because this has the most highly substituted alkene as shown.

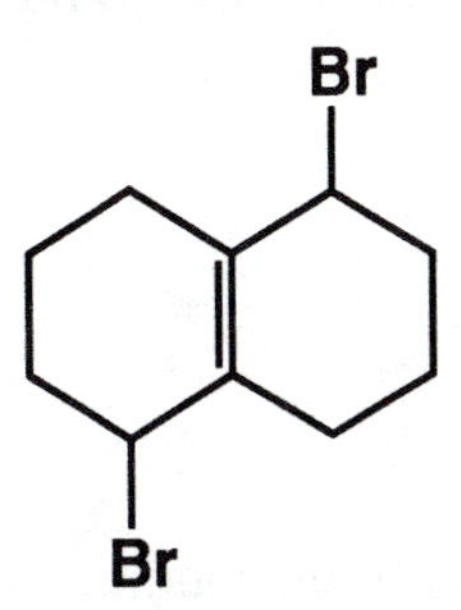

kinetic product **thermodynamic product**

Diels-Alder Reactions

Problem 7.18 Draw structural formulas for the products of the reaction of cyclopentadiene with each of the following dienophiles.

Underneath each dienophile is the corresponding Diels-Alder adduct.

(a) $CH_2{=}CHCl$

(b) $CH_2{=}CHCOCH_3$ (C=O)

(c) $HC{\equiv}CH$

(d) $CH_3OCC{\equiv}CCOCH_3$ (each C=O)

Problem 7.19 Propose structural formulas for compounds (A) and (B) and specify the configuration of compound (B).

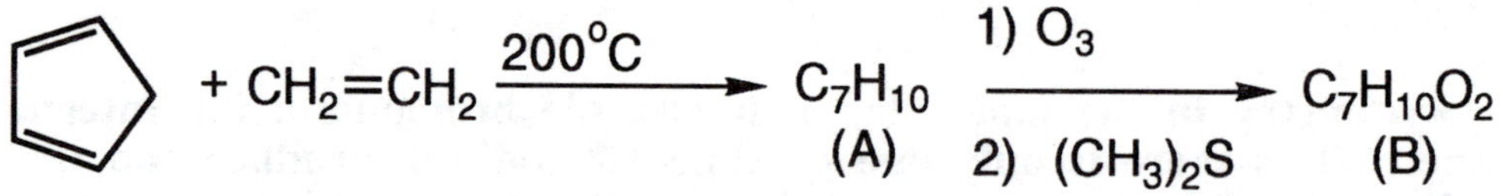

$$\text{cyclopentadiene} + CH_2{=}CH_2 \xrightarrow{200^\circ C} \underset{(A)}{C_7H_{10}} \xrightarrow[2)\ (CH_3)_2S]{1)\ O_3} \underset{(B)}{C_7H_{10}O_2}$$

The Diels-Alder adduct is bicyclo[2.2.1]-2-heptene, more commonly known as norbornene. The oxidation product is *cis*-1,3-cyclopentanedicarbaldehyde.

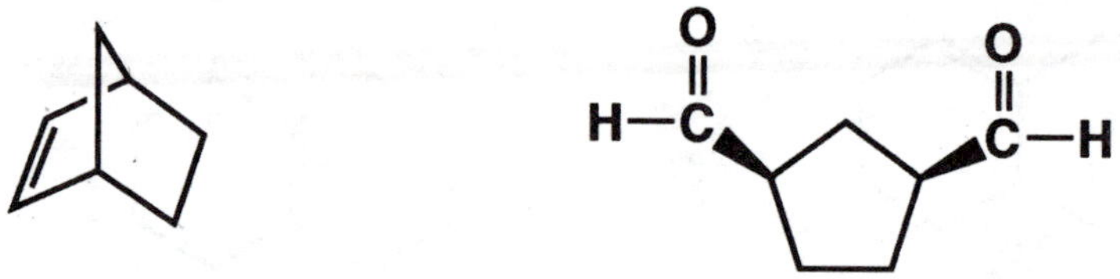

O O
|| ||
H—C C—H

Problem 7.20 Under certain conditions, 1,3-butadiene can function both as a diene and a dienophile. Draw a structural formula for the Diels-Alder adduct formed by reaction of 1,3-butadiene with itself.

In the formulas below, one molecule of butadiene is shown as the diene and the other is shown as the dienophile.

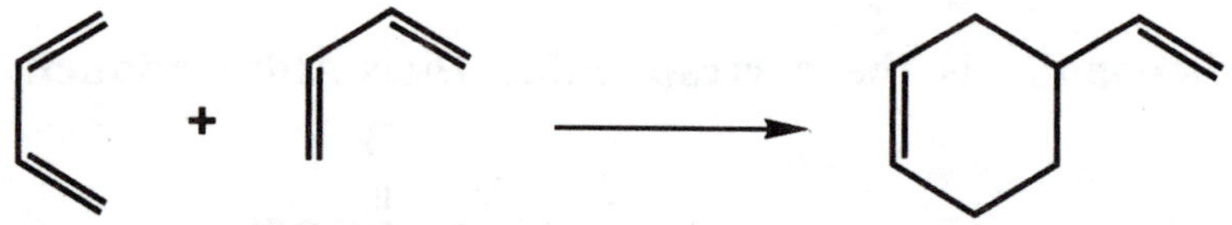

+

butadiene **butadiene** **4-vinylcyclohexene**

Note that an electron pushing mechanism can be drawn for cyclization of two molecules of butadiene to 1,5-cyclooctadiene. This reaction does not take place. Rather, the Diels-Alder reaction shown above takes place instead.

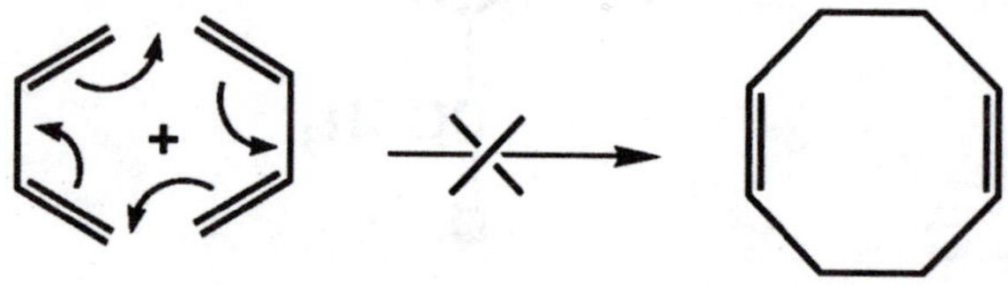

+

1,5-cyclooctadiene

Problem 7.21 Butadiene is a gas at room temperature and requires gas-handling apparatus to use it in a Diels-Alder reaction. Butadiene sulfone is a convenient substitute for gaseous butadiene. This sulfone is a solid at room temperature (mp 66°C) and when heated above its boiling point of 110°C, decomposes by a reverse Diels-Alder reaction to give s-*cis*-butadiene and sulfur dioxide. Draw a Lewis structure for butadiene sulfone and show by curved arrows the path of this reverse Diels-Alder reaction.

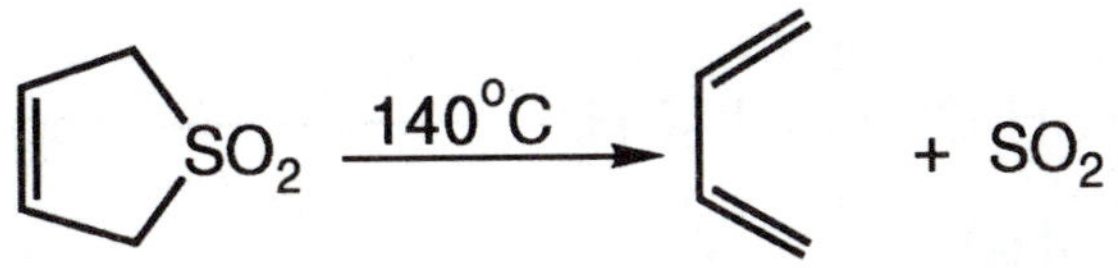

$\ce{SO2}$ —140°C→ + SO_2

butadiene sulfone ("3-sulfolene") butadiene sulfur dioxide

The electron flow in this decomposition is the reverse of that observed in a Diels-Alder reaction.

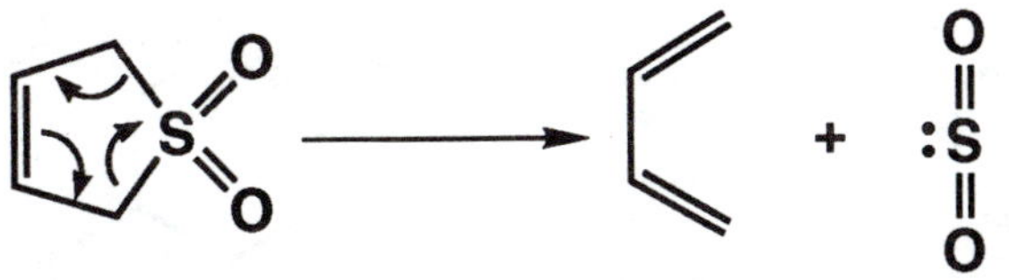

butadiene sulfone **butadiene** **sulfur dioxide**

Problem 7.22 The following triene undergoes an intramolecular Diels-Alder reaction to give the product shown. Show how the carbon skeleton of the triene must be coiled to give this product and show by curved arrows the reorganization of electron pairs that takes place to give the product.

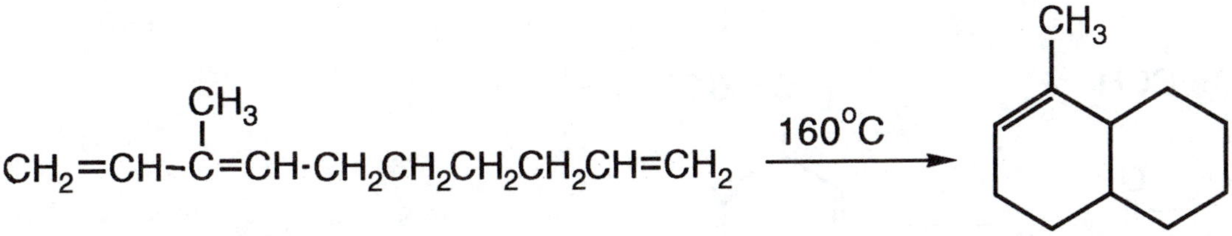

$CH_2{=}CH{-}C(CH_3){=}CH{-}CH_2CH_2CH_2CH_2CH{=}CH_2$ —160°C→ (CH_3)

A clue to this problem is to locate the position of the carbon-carbon double bond in the product and to realize that the two carbons of this double bond were carbons 2 and 3 of the conjugated diene. The other two carbon atoms making up this six-membered ring were from the dienophile.

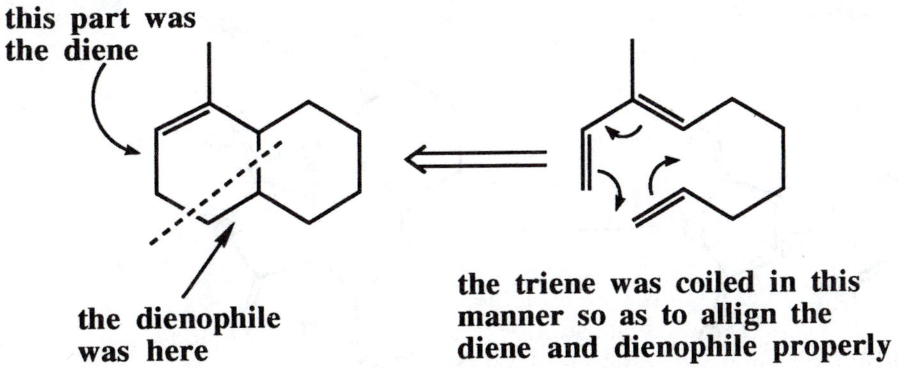

this part was the diene

the dienophile was here

the triene was coiled in this manner so as to allign the diene and dienophile properly

Problem 7.23 The following triene undergoes an intramolecular Diels-Alder reaction to give a bicyclic product. Propose a structural formula for the product. Account for the observation that the Diels-Alder reaction given in this problem takes place under milder conditions (at lower temperature) than the analogous Diels-Alder reaction shown in the previous problem.

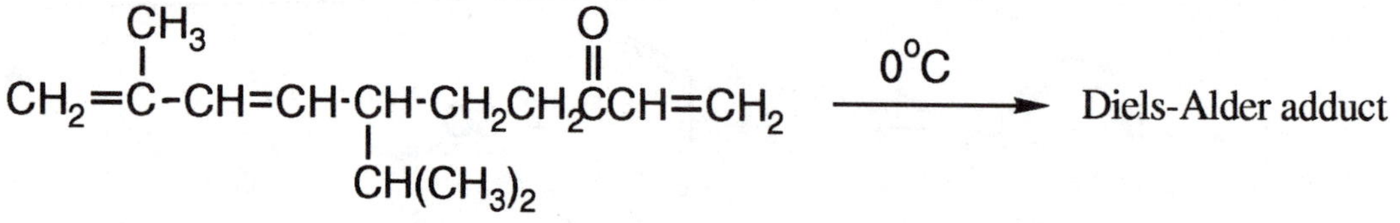

$CH_2{=}C(CH_3){-}CH{=}CH{-}CH(CH(CH_3)_2){-}CH_2CH_2C(=O)CH{=}CH_2 \xrightarrow{0^\circ C}$ Diels-Alder adduct

Follow the arrangement of diene and dienophile illustrated in the previous problem.

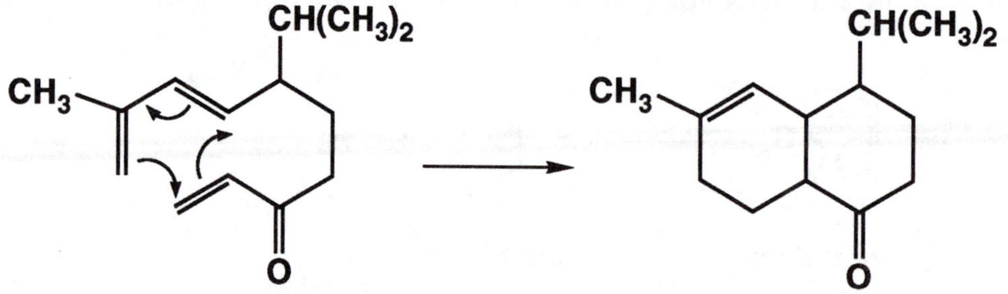

CH_3 $CH(CH_3)_2$ O → CH_3 $CH(CH_3)_2$ O

Problem 7.24 The Diels-Alder reaction is not limited to making six membered rings with only carbon atoms. Predict the products of the following reactions which produce heterocycles, rings with atoms other than carbon in them.

(a) $O{-}\overset{O}{\overset{\|}{C}}CH_3$ H CH_3 → $O{-}\overset{O}{\overset{\|}{C}}CH_3$ O CH_3

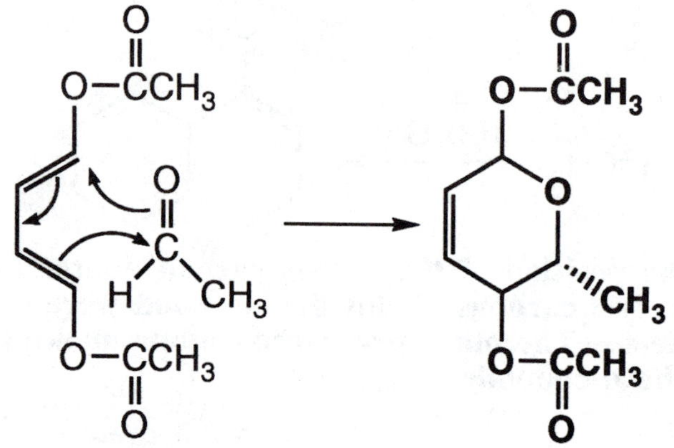

(b) N O O O $(CH_3)_3C$ → N H O O O H $(CH_3)_3C$

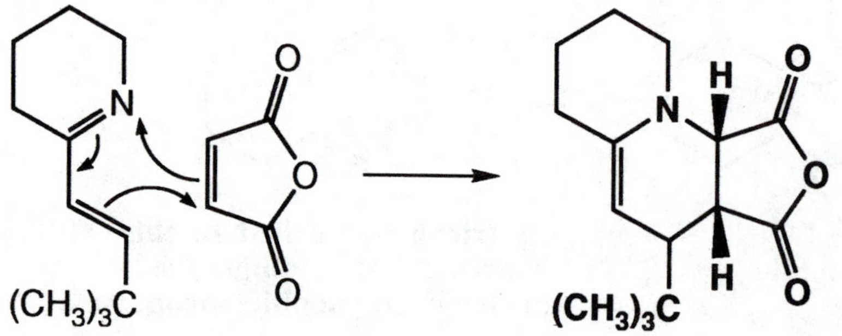

(c)

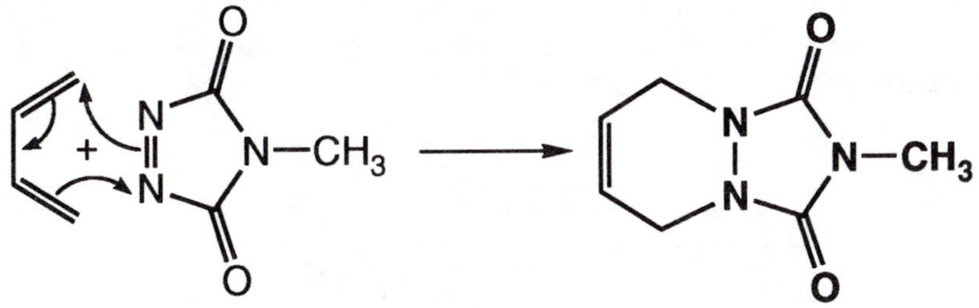

(d)

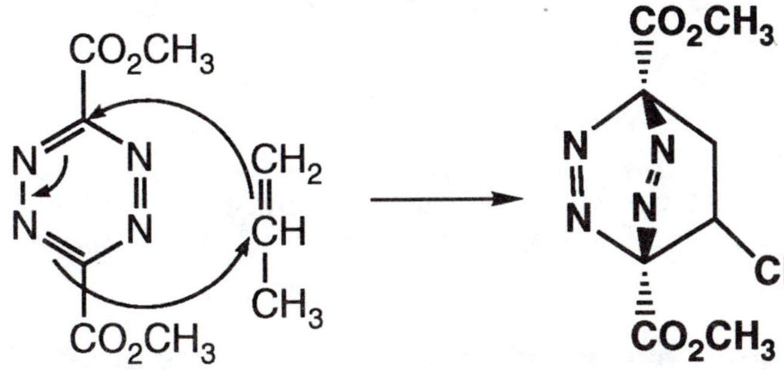

(e)

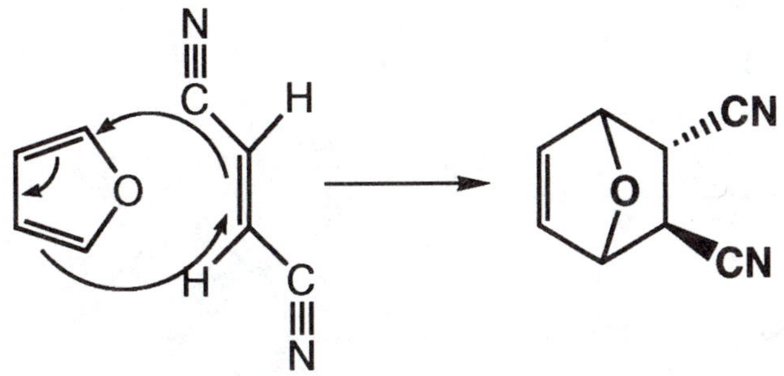

Problem 7.25 The Diels-Alder reaction is often used by organic chemists in their syntheses of naturally occuring molecules. It is seldom used, however, by plants and animals to synthesize the molecules of nature. One exception is a class of naturally occurring molecules isolated from the Australian plant *Endiandra introrsa* (Lauraceae), which include endiandric acid B and endiandric acid C. Studies strongly suggest that the plant synthesizes precursors, which then cyclize to the observed acids. Propose the structural formula of a precursor that will undergo an intramolecular Diels-Alder reaction to produce endriandric acid C?

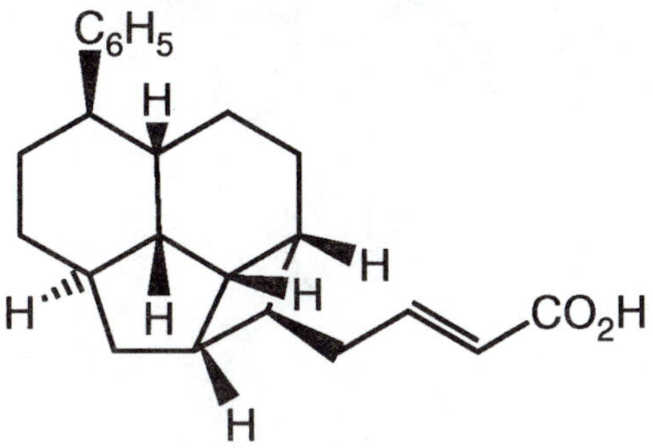

endiandric acid B

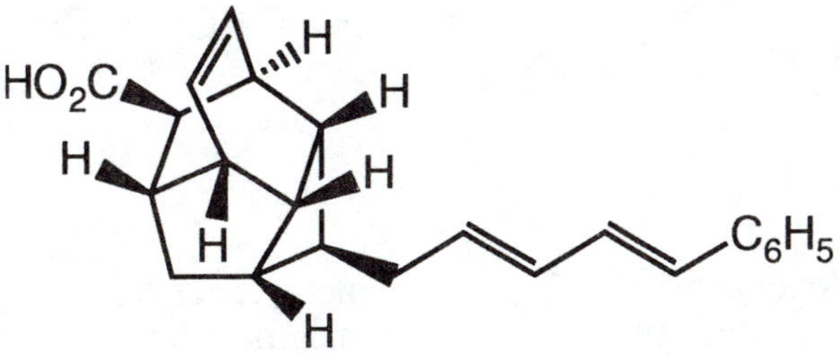

endiandric acid C

The appropriate precursor to endiandric acid C can be deduced from the usual bond disconnections. Note how the electron-withdrawing carboxylic acid group is on the dienophile portion of the molecule.

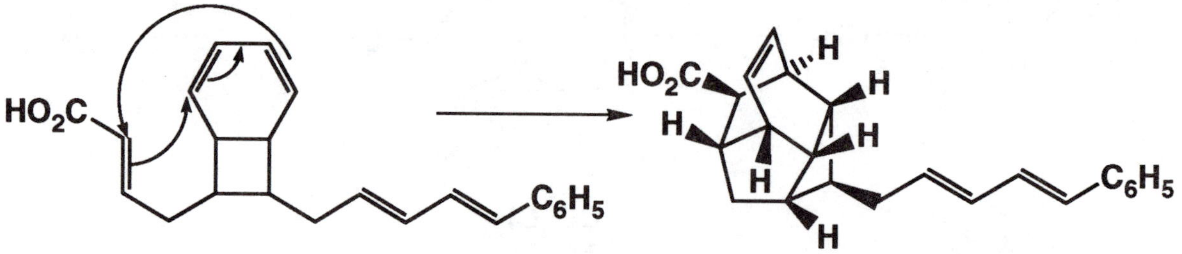

Problem 7.26 Because endiandric acid B has two cyclohexene rings, there are two possible intramolecular Diels-Alder reactions that could give rise to this natural product. Keeping in mind your answer to the previous question what is the more likely structure to undergo an intramolecular Diels-Alder reaction to give endiandric acid B? See K.C. Nicolaou et al., *J. Am. Chem. Soc.*, **104**, 5555-5562 (1982)

A molecule very similar to that shown above can be used to produce endiandric acid B. In this case, the diene is located adjacent to the C_6H_5 group.

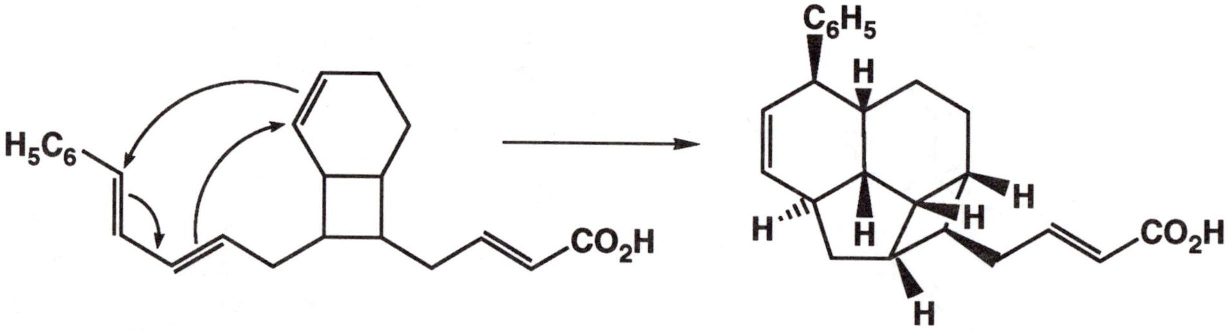

Problem 7.27 The first step in a synthesis of dodecahedrane involves a Diels-Alder reaction between the cyclopentadiene derivative (1) and dimethyl acetylenedicarboxylate (2). Show how these two molecules react to form the dodecahedrane synthetic intermediate (3). (L.A. Paquette, R.J. Ternansky, D.W. Balogh, and G. Kentgen, *J. Am. Chem. Soc.*, **105**, 5446 (1983)

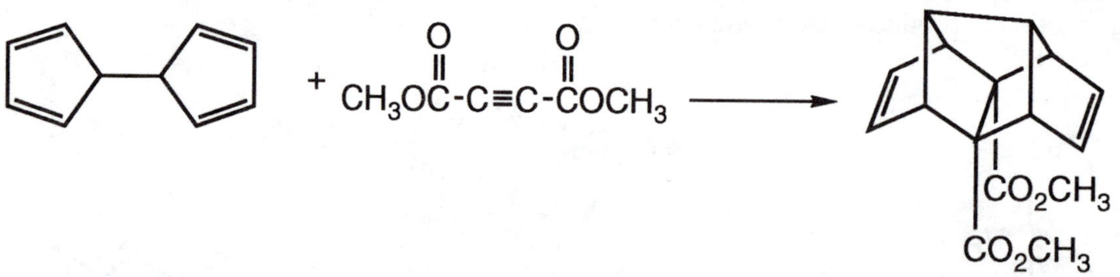

+ $CH_3OC(=O)-C\equiv C-C(=O)OCH_3$ →

cyclopentadienyl-cyclopentadiene (1)

dimethyl acetylene-dicarboxylate (2)

(3)

This reaction is best understood as two successive Diels-Alder reactions. The second one is an intramolecular reaction that takes place in the product of the first reaction.

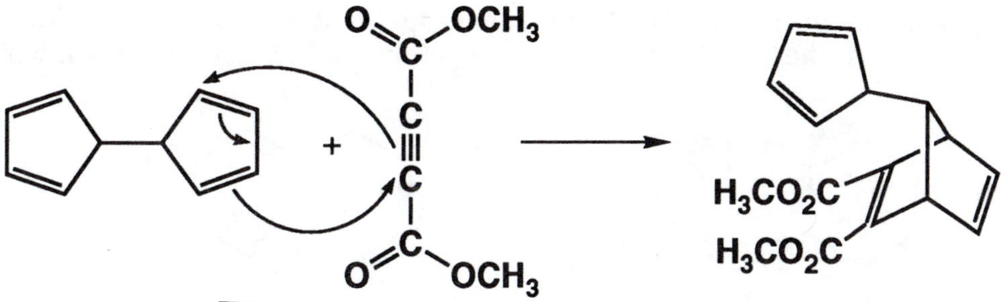

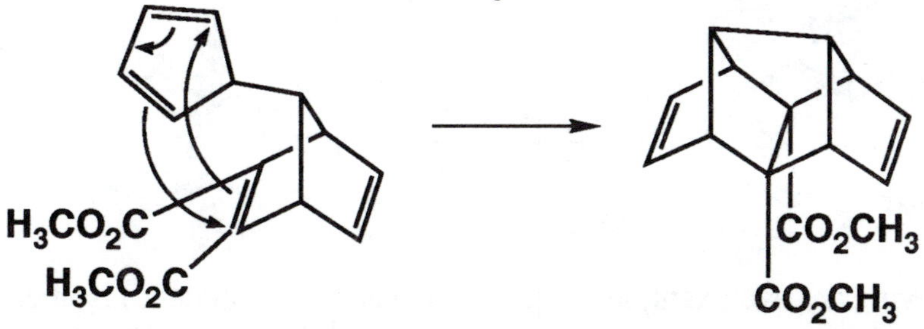

$O=C-OCH_3$, $C \equiv C$, $O=C-OCH_3$ → H_3CO_2C, H_3CO_2C

H_3CO_2C, H_3CO_2C → CO_2CH_3, CO_2CH_3

Problem 7.28 The Diels-Alder reaction proceeds through a cyclic, six-electron transition state. Another example of a reaction that takes place via such a transition state is the Cope rearrangement of 1,5-dienes.

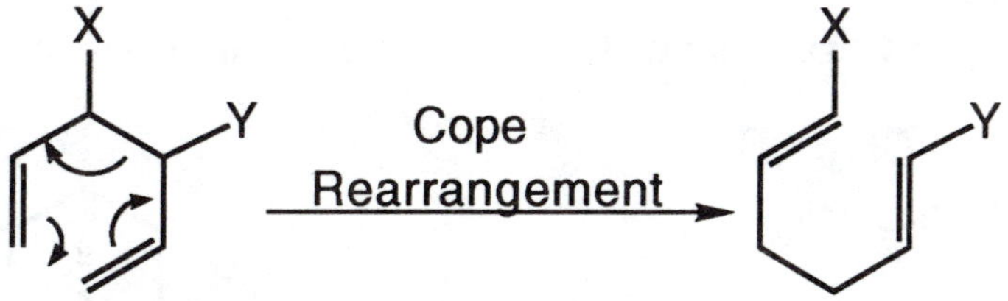

X, Y — Cope Rearrangement → X, Y

Show that the following rearrangements can be explained as Cope rearrangements.

(a) CH_2CH_2 — 350°C → CH_2-CH_2

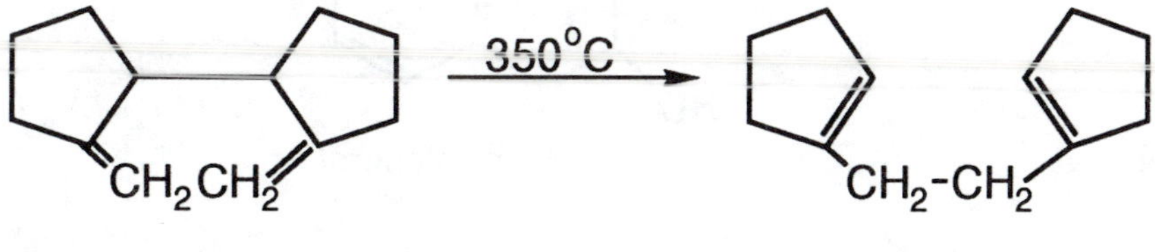

The electron flow is indicated directly on the structure.

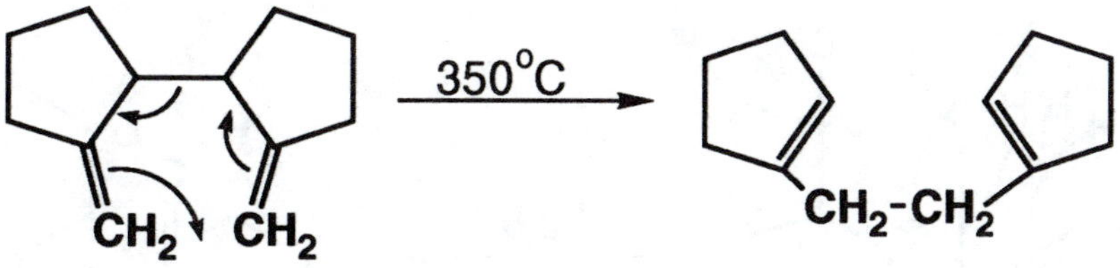

CH_2 CH_2 — 350°C → CH_2-CH_2

(b) OH, CH=CH₂, CH=CH₂ — 320°C → O

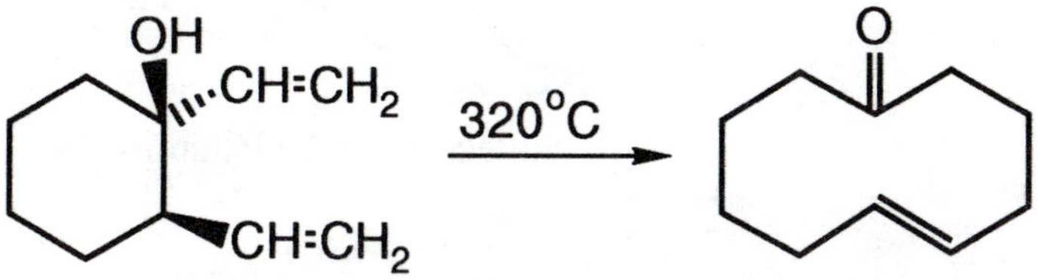

The key to this problem is that the Cope rearrangement is immediately followed by tautomerization of the resulting enol to the more stable ketone as shown.

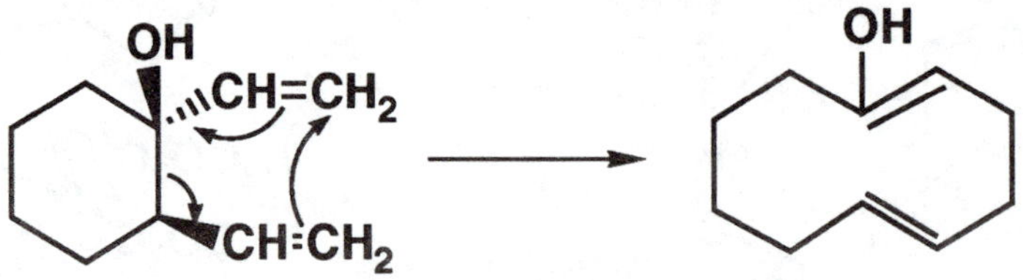

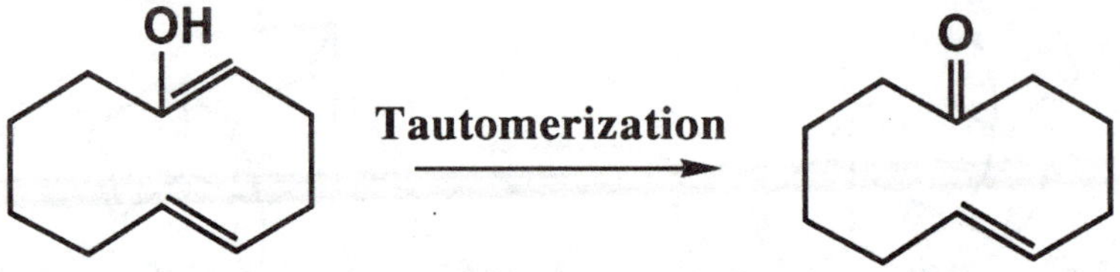

OH, CH=CH$_2$, CH=CH$_2$ → OH

OH —Tautomerization→ O

Problem 7.29 Following are two examples of photoinduced (light-induced) isomerizations. Vitamin D3 (cholecalciferol) is produced by the action of sunlight on 7-dehydrocholesterol in the skin. First is formed precalciferol and then cholecalciferol. Use curved arrows to show the flow of electrons in these photoisomerizations. Cholecalciferol is shown here in an s-*cis* conformation. After its formation it assumes an s-*trans* conformation.

The appropriate arrows are drawn directly on the structures to indicate the flow of electrons.

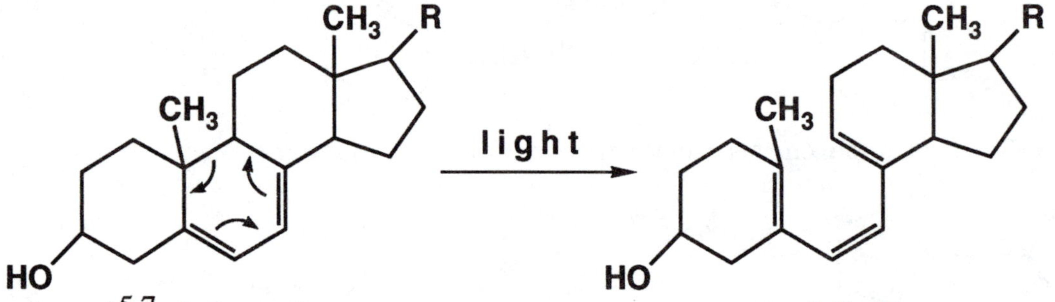

CH$_3$ R, CH$_3$, HO — light → CH$_3$ R, CH$_3$, HO

$\Delta^{5,7}$-cholesterol precalciferol

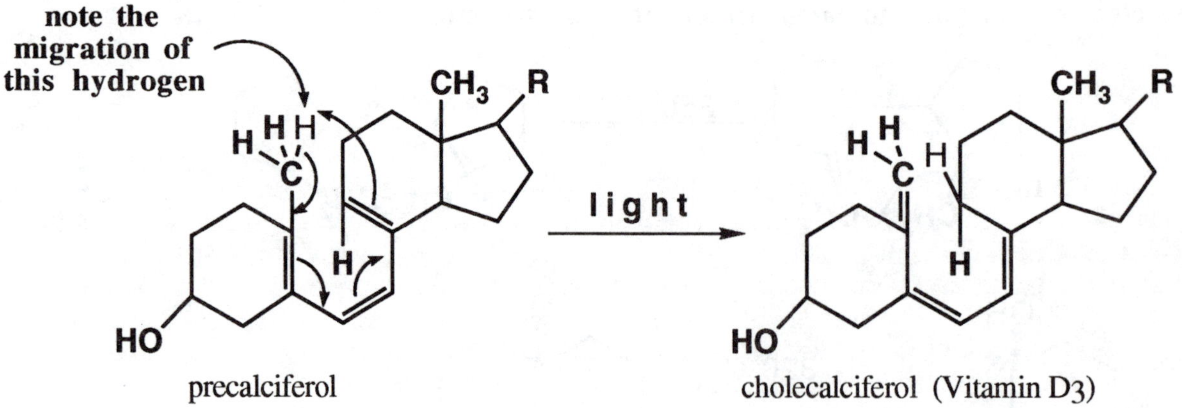

note the migration of this hydrogen

H H H C, CH$_3$ R, H, HO — light → H H C H, CH$_3$ R, H, HO

precalciferol cholecalciferol (Vitamin D3)

CHAPTER 8: CHIRALITY

8.0 OVERVIEW

- Molecules are three-dimensional, and this chapter describes very important consequences of that three-dimensionality, namely stereochemistry and chirality.

8.1 ISOMERISM

- **Isomers** are *different* molecules that have the same molecular formula. ✱ *[This is the first of several very important definitions that should be learned now to avoid confusion later. They may sound relatively simple, but learning to apply them correctly in cases of complex molecules can be quite challenging.]*
- **Constitutional isomers** are different molecules with the same molecular formula, but with a different order of attachment of atoms. In other words, the atoms are connected to each other differently. For example, pentane (pentane structure) and 2-methylbutane (2-methylbutane structure) are constitutional isomers of molecular formula C_5H_{12}.
- **Stereoisomers** are different molecules with the same molecular formula, the same order of attachment of atoms, but a different **orientation** of those atoms or groups in space. ✱ *[This concept is subtle and may be confusing at first, but it is actually a very powerful concept that should be understood before moving on.]*
  - Previously, two types of **stereoisomers** have been presented; E,Z stereoisomers of alkenes and cis-trans stereoisomers of cyclic compounds. In general, two different stereoisomers will have the same systematic name, but an additional element is added to the name such as "Z" or "cis" to specify the specific stereoisomer being named. For example, the following two molecules are the same constitutional isomer (2-butene), but they differ in the relative **orientation** of their atoms is space. For the molecule on the left (E-2-butene), the two methyl groups are **oriented** away from each other, while for the molecule on the right the two methyl groups are **oriented** together.

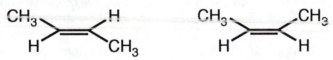

E-2-butene Z-2-butene

  - E,Z isomers of alkenes and cis,trans isomers of cyclic molecules are examples of **diastereomers**, that is stereoisomers that are *not* mirror images of each other. This definition is best understood in the context of the definition of enantiomers.
- **Enantiomers** are stereoisomers that are mirror images of each other. ✱
  - A **mirror image** is what you see as the reflection of an object in a mirror.

8.2 CHIRALITY

- **Chirality** is a property of three-dimensional objects that is very important in chemistry. An object is **chiral** if it is not superposable on its mirror image. That is, a chiral object and its mirror image cannot be oriented in space so that all of their features (corners, edges, points, etc.) correspond exactly to each other. ✱ *[This is a very difficult concept, but one that is absolutely central to the rest of this chapter and the study of stereochemistry. Chirality should be understood before proceeding.]*
  - Great examples of chiral objects are your hands. They are mirror images of each other (if you hold your left hand up to a mirror you see an image that looks like your right hand in the mirror), yet you cannot orient your two hands in space so that they are superposable. Try

this for yourself if you are having trouble understanding chirality. Because your hands are chiral, they have "handedness", that is the left hand is different in three-dimensional space than your right hand even though they are mirror images of each other and both have four fingers and a thumb. For this reason, a glove that fits your left hand well will not fit your right hand and you must buy one left-handed and one right-handed glove.

- Other examples of chiral objects include your feet, an airplane propeller or ceiling fan, a wood screw or a drill bit. In fact, the vast majority of objects around you are chiral.

• An object is **achiral** if it is superposable on its mirror image. Examples of objects that are achiral include a perfect cube or a perfect sphere. To be achiral, the molecule must have symmetry.

-Achiral objects can be identified because they possess at least one **plane** or **center of symmetry.**

A **plane of symmetry** is an imaginary plane passing through an object dividing it such that one half is the mirror reflection of the other half. Only highly symmetric objects such as a perfect sphere have a plane of symmetry. ✳ *[This is an important concept. It may be helpful to carefully examine Figure 8.2 to make sure you understand when an object has a plane of symmetry.]*

A **point of symmetry** is a point situated such that identical components of the object are located equidistant from the point along any axis passing through the point of symmetry. ✳ *[This is another difficult concept, and again it may be helpful to examine Figure 8.3 to make sure you understand when an object has a point of symmetry.]*

• **Molecules can be chiral**. To be chiral, a molecule cannot have a plane of symmetry or a point of symmetry. Thus, **the best way to check for chirality in a molecule is to look for a plane or point of symmetry**. If no symmetry element is found, then the molecule is chiral. ✳

• Often, molecules are chiral if they contain a tetrahedral carbon atom with four different substituents. This is because a tetrahedral carbon atom that has four different substituents is **not** superposable on its mirror image. ✳ *[This important concept is best understood using models. Try making a model of a tetrahedral carbon atom with four different substituents attached. Next, make a model of its mirror image, and try to superpose the two models. The fact that the two models cannot be superposed confirms that the molecules display chirality. Since the molecules are non-superposable mirror images of each other, they are **enantiomers**.]*

- A **stereocenter** is a point in a molecule at which interchange of two atoms or groups of atoms bonded to that point produces a different stereoisomer. Note that the presence of a stereocenter does not guarantee that a molecule is chiral. It is worth repeating that **the best way to see if a molecule is chiral is to make sure that there are no planes or points of symmetry present.** ✳

A carbon atom with four different substituents is a **tetrahedral stereocenter**. For example, the central carbon atom in lactic acid (2-hydroxypropanoic acid) is a tetrahedral stereocenter. ✳

A properly substituted trigonal carbon atom can also be a **trigonal stereocenter**. For example, carbons 2 and 3 of 2-butene are trigonal stereocenters.

Atoms other than carbon can be a stereocenter including silicon, phosphorus, nitrogen, and germanium.

• Although most chiral molecules have a tetrahedral carbon stereocenter, some organic molecules are chiral even though they do not have a stereocenter. These molecules are relatively rare, but important examples include allenes with four different substituents or appropriately substituted biphenyl derivatives.

8.3 NAMING ENANTIOMERS : THE R-S SYSTEM

- The absolute configuration of a tetrahedral stereocenter is assigned using the **R-S convention** first introduced by Cahn, Ingold, and Prelog. This convention is based on assigning **priorities** to the four different substituents around the stereocenter. The priority rules are as follows:
  - 1. Each atom attached directly to the stereocenter is assigned a priority based on atomic number. The higher the atomic number, the higher the priority. For example, -Br is assigned a higher priority than -Cl, which is assigned a higher priority than -OH, etc.
  -2. For isotopes, the higher the atomic mass of the isotope, the higher its priority.
  -3. If a priority based on atomic number cannot be assigned to the atoms directly bonded to the stereocenter, then look at the next set of atoms and continue until a priority can be assigned. **It is the first point of difference** that matters here.
  -4. Atoms of double or triple bonds are considered as if they are bonded to an equivalent number of similar atoms by single bonds. *[This is the hardest priority rule, and practice is usually needed to fully understand it.]*
- In order to assign the absolute configuration of a tetrahedral stereocenter according to the R-S convention use the following procedure:
  -1. Locate the tetrahedral stereocenter and identify its four substituents.
  -2. Assign a priority 1,2,3, or 4 to each substituent using the rules listed above.
  -3. Orient the molecule in space such that the group with the lowest priority (4) is directed away from you. In other words, orient the molecule so that when you look at it, the lowest priority group is directly behind the carbon stereocenter. The three remaining groups (1-3) will be arranged like the legs of a tripod directed toward you.
  -4. Read the remaining three groups in order from highest (1) to lowest priority (3).
  -5. If reading the groups proceeds in a clockwise direction, the absolute configuration is designated as **R**, if reading proceeds in a counterclockwise direction, the absolute configuration is **S.** ✱ *[It is essential that you become very good at assigning absolute stereochemistry, and practice is the best way to become very good.]*

8.4 NONCYCLIC MOLECULES WITH TWO OR MORE STEREOCENTERS

- For a molecule with **n** tetrahedral stereocenters, the maximum number of stereoisomers possible is $\mathbf{2^n}$. This is because each stereocenter can be either R or S.
- A good way to learn about the stereochemical consequences of having more than one stereocenter in the same molecule is to examine all four possible stereoisomers that arise when there are two stereocenters in the same molecule. For example, consider the four stereoisomers of 2,3,4-trihydroxybutanal drawn below:

a	b	c	d
CHO	CHO	CHO	CHO
H–C(R)–OH	HO–C(S)–H	H–C(R)–OH	HO–C(S)–H
H–C(R)–OH	HO–C(S)–H	HO–C(S)–H	H–C(R)–OH
CH_2OH	CH_2OH	CH_2OH	CH_2OH

- From right to left the four isomers are the R,R (**a**); S,S (**b**); R,S (**c**) and S,R (**d**) stereoisomers. When examined in pairs, the following relationships become apparent:
  a,b and **c,d** are non-superposable mirror images of each other, thus they are pairs of **enantiomers**. *[This would be a good time to review the definition of enantiomers if necessary.]*
  a,c and **a,d** and **b,c** and **b,d** are all pairs of stereoisomers that are not mirror images of each other, thus these are all pairs of **diastereomers**.

- Hint: when asked to identify the stereochemical relationships between pairs of stereoisomers, it is helpful to first assign the absolute configuration (R or S) to each stereoisomer then compare these, instead of trying to compare the molecules directly. ✳
- Certain molecules have special symmetry properties that reduce the number of stereoisomers to fewer than the predicted 2^n. In these cases, some of the stereoisomers contain a plane of symmetry so the molecules are achiral. Molecules or ions that contain two or more tetrahedral stereocenters but are achiral because of a plane of symmetry are called **meso compounds**. ✳
  - **Meso compounds** will always be the R,S/S,R isomer of a molecule. For example, the R,S/S,R isomer of tartaric acid is a meso compound.

plane of symmetry ⟹ [CO_2H / H–C(R)–OH / H–C(S)–OH / CO_2H] = [CO_2H / HO–C(S)–H / HO–C(R)–H / CO_2H]

meso-tartaric acid

8.5 CYCLOALKANES WITH TWO OR MORE STEREOCENTERS

- Understanding stereochemical relationships between cycloalkane stereoisomers can be very difficult. The key is being able to identify planes of symmetry in the molecules. If there is a plane of symmetry present then the molecule is achiral, it there is no plane of symmetry then the molecule is chiral. ✳
  - For example, *cis*-3-methylcyclopentanol is chiral because there is no plane of symmetry in the molecule as can be seen by the planar representation. On the other hand, *cis*-1,3-cyclopentanediol is a achiral because there is a plane of symmetry.

HO CH_3 H H

cis-3-methylcyclopentanol
No plane of symmetry; chiral

Plane of Symmetry
HO OH H H

cis-1,3-cyclopentanediol
Plane of symmetry; achiral

  - With disubstituted cyclohexane derivatives, it is especially important to keep track of the substitution pattern. For example, *trans*-1,4-cyclohexanediol has a plane of symmetry and is thus achiral, while *trans*-1,3-cyclohexanediol does not have a plane of symmetry so it is achiral.

HO H H OH

trans-1,4-cyclohexanediol
Plane of symmetry, it extends right through both OH groups perpendicular to ring; chiral.

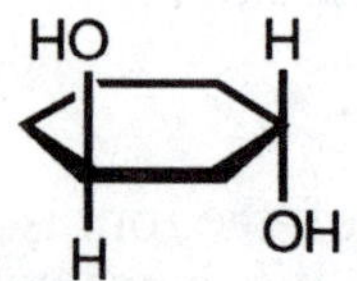

trans-1,3-cyclohexanediol
No plane of symmetry; achiral.

  - Note that sometimes the cyclohexane chair conformations must be considered when looking for planes of symmetry. This is explained further in the text.

8.6 PHYSICAL PROPERTIES OF ENANTIOMERS

- In general, enantiomers have identical physical properties when those properties are measured in an achiral environment. For example, two enantiomers have the same boiling points, melting points, solubilities in achiral solvents, the same pK_a values, etc. On the other hand, two diastereomers have different physical properties like melting points or boiling points. ✱

8.7 HOW CHIRALITY IS DETECTED IN THE LABORATORY

- Although enantiomers have identical physical properties when those properties are measured in an achiral way, they are different compounds. Therefore, they have different physical properties when the measurements are made in a chiral way such as interactions with plane polarized light. Each member of a pair of enantiomers rotates the plane of plane polarized light, so they are called **optically active**.
  - Normal light consists of waves vibrating in all planes perpendicular to its path. Certain materials such as Polaroid sheets only allow waves vibrating in a single plane to pass through. The resulting light is thus called **plane polarized light**, because all of the resulting light is vibrating in the same plane.
  - Samples of enantiomers rotate the plane of plane polarized light. A **polarimeter** is the instrument used in the laboratory to measure the direction and magnitude of rotation of plane polarized light.

    If a sample of an enantiomer rotates plane polarized light in a **clockwise** direction, it is called **dextrorotary**. If a sample rotates plane polarized light in a **counterclockwise** direction, it is called **levorotary**. ✱

    In order to standardize optical rotation data, **specific rotation** has been defined according to the following equation:

$$[\alpha]_{\lambda}^{T} = \frac{\text{observed rotation (degrees)}}{\text{cell length (dm) x concentration (g/mL)}}$$

    Note that the length of the cell in which the sample is placed is measured in unusual units, namely decimeters. 1 dm = 10 cm. The T stands for the measurement temperature and the λ stands for the wavelength of light used to make the measurement (usually the sodium D line at 589 nm).

    By convention, a dextrorotary compound is designated with a plus sign (+) and a levorotary compound is designated with a minus sign (-).
- Two enantiomers rotate plane polarized light by the same number of degrees, but with opposite sign. Meso compounds and all other achiral molecules do not rotate plane polarized light. ✱
- An equimolar mixture of two enantiomers is called a **racemic mixture**. For racemic mixtures, the rotations of plane polarized light exactly cancel and there is no overall rotation. If two enantiomers in a mixture are in unequal amounts, the entire mixture will have a net rotation, and the magnitude and direction of that rotation can be used to determine the exact ratio of the two enantiomers in the mixture according to the following equation:

$$\%\ \text{optical purity} = \frac{[\alpha]_{obs}}{[\alpha]_{\text{pure enantiomer}}} \times 100$$

  - **Enantiomeric excess**, abbreviated **ee**, is equal to the percent optical purity, and is often referred to when two different enantiomers are made in different amounts in the same reaction. **Enantiomeric excess** is defined according to the following equation:

$$\%\ \text{enantiomeric excess (ee)} = \left(\frac{\text{moles of one enantiomer - moles of other enantiomer}}{\text{moles of both enantiomers}}\right) \text{X } 100$$

- There is **no** necessary **relationship between** an **absolute configuration** (R and S) and the **sign of rotation** (+ and -). For some pairs of stereoisomers, the "R" enantiomer has a (+) rotation, while the "S" enantiomer has a (-) rotation, yet for other pairs of stereoisomers, the "S" enantiomer has a (+) rotation and the "R" enantiomer has a (-) rotation. This makes sense since the R and S designations are based on artificial nomenclature rules, while the (+) and (-) rotations are the result of actual measurements. ✳

8.8 SEPARATION OF ENANTIOMERS-RESOLUTION

- **Resolution** is the process whereby a racemic mixture is separated into the two enantiomers. A pair of enantiomers are difficult to separate, because they both have identical physical properties such as melting points, boiling points, etc. A common method of resolution involves combining the racemic mixture with another compound that is a single enantiomer. The combination results in the production of two diastereomers that can usually be separated because diastereomers have different physical properties. Following resolution, the enantiomers are recovered. ✳
  - A reaction that lends itself to resolution is salt formation. For example, one enantiomer of a chiral compound such as a positively-charged amine like (+)-cinchonine can be used to form diastereomeric salts with a racemic mixture of negatively-charged species like deprotonated form of chiral carboxylic acids. Following resolution, the pure enantiomers of the carboxylic acid are recovered by acid precipitation.

8.9 THE SIGNIFICANCE OF CHIRALITY IN THE BIOLOGICAL WORLD

- Almost all of the molecules of living systems are chiral. Remarkably, even though all of these chiral molecules could in theory exist as a mixture of stereoisomers, almost invariably only one stereoisomer is found in nature.
  - Enzymes, Mother Nature's molecular machines, are composed of chiral amino acids. As a result, the enzymes are themselves chiral. For that reason, they are able to distinguish substrate enantiomers. A pair of enantiomers can be distinguished in a chiral environment like the active site of an enzyme.

8.10 MOLECULES CONTAINING STEREOCENTERS AS REACTANTS OR PRODUCTS

- During a reaction, one or more stereocenters in a molecule may be created or destroyed. It is therefore important to keep track of stereocenters during the entire reaction mechanism in order to predict accurately the stereochemistry of the final product. ✳
- In general, optical activity is never produced from optically inactive starting materials, even though the products may be chiral. In other words, if a stereocenter is created from an achiral starting material, then a racemic mixture of the two possible enantiomers is formed (or a meso compound if applicable).
  - Examples of this include the addition of Br_2 to *cis*-2-butene to create a racemic mixture of the two enantiomers of 2,3-dibromobutane, and the $KMnO_4$ oxidation of 1-butene to give a racemic mixture of the two enantiomers of 1,2-butanediol.
- Alternatively, optical activity is generated in a reaction only if at least one of the reactants itself is chiral, or if the reaction is carried out in the presence of a catalyst that is itself chiral.
- A **regioselective reaction** is one in which one direction of bond-making or bond-breaking occurs preferentially over all other possible directions.
- A **stereoselective reaction** is one in which one stereoisomer or an enantiomeric pair of stereoisomers is formed or destroyed preferentially over all others that may be formed or destroyed.
- A **stereospecific reaction** is one in which different stereoisomeric starting materials give stereoisomeric products.

CHAPTER 8
Solutions to the Problems

Problem 8.1 Which of the following molecules have a tetrahedral stereocenter? For each that does, label the stereocenter and draw stereorepresentations for pairs of enantiomers.

Each part has a tetrahedral stereocenter. The stereocenters are labeled with an asterisk.

(a) cyclopentyl–CH(OH)–CH_3 — enantiomers: OH, *C, H, CH_3 | OH, H, *C, H_3C

(b) CH_3–CH(Br)–CH_2Cl — enantiomers: Br, *C, CH_3, H, CH_2Cl | Br, *C, CH_3, ClH_2C, H

(c) CH_3–CH(OH)–CH(CH_3)–CH_3 — enantiomers: $CH(CH_3)_2$, H–*C–OH, CH_3 | $CH(CH_3)_2$, HO–*C–H, CH_3

Problem 8.2 Assign an R or S configuration to each stereocenter and give each molecule an IUPAC name.

The drawings underneath each molecule show the order of priority, the perspective from which to view the molecule, and the R,S designation for the absolute configuration.

(a)

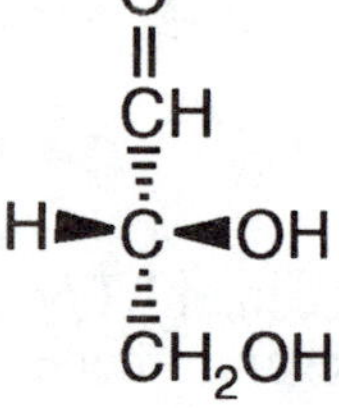

O
||
CH
H—C—OH
CH$_2$OH

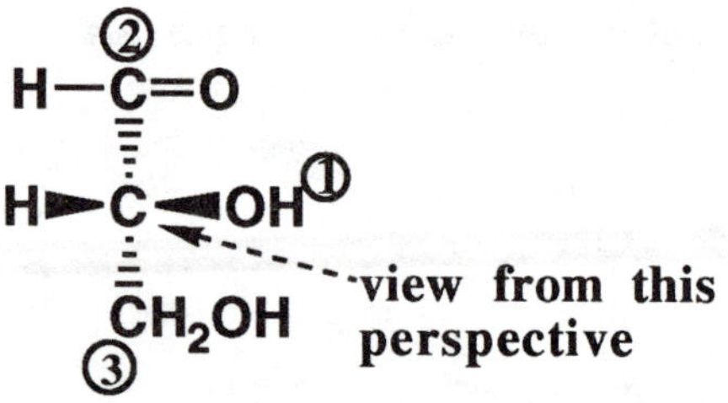

② H—C=O
H—C—OH ①
CH$_2$OH ③
view from this perspective

(2R)-2,3-dihydroxypropanal

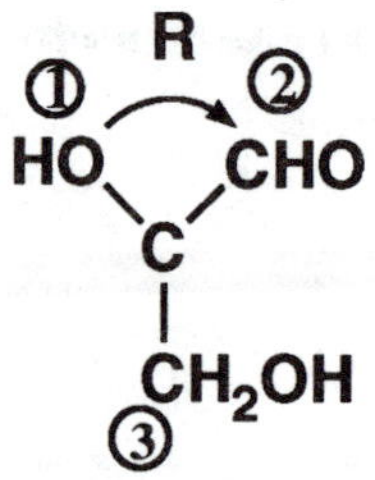

R
① HO ② CHO
C
CH$_2$OH ③

If you view from the perspective shown, this is what you see

(b)

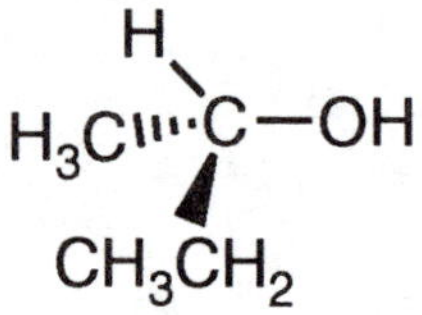

H
H_3C—C—OH
CH_3CH_2

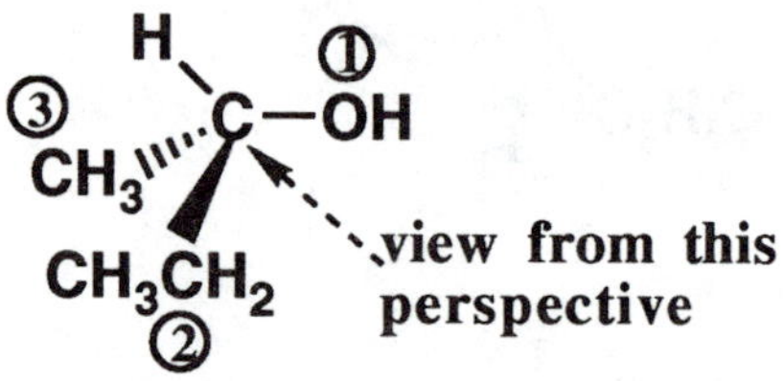

H ① OH
③ CH_3 C
CH_3CH_2 ②
view from this perspective

(S)-2-butanol

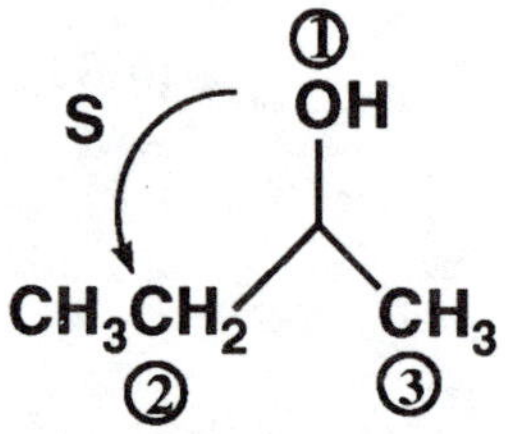

S
① OH
CH_3CH_2 ② CH_3 ③

If you view from the perspective shown, this is what you see

(c)

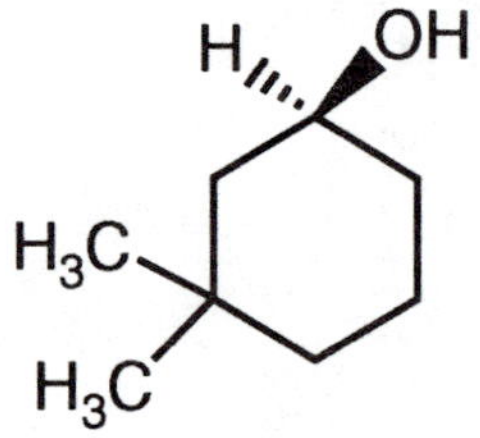

H OH
H_3C
H_3C

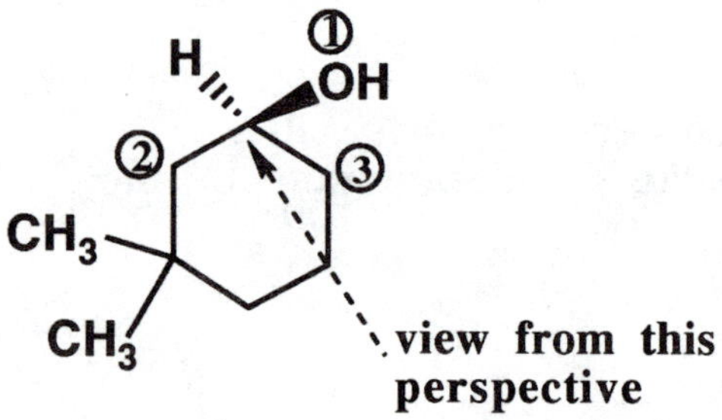

H ① OH
② ③
CH_3
CH_3
view from this perspective

(1S)-3,3-dimethylcyclohexanol

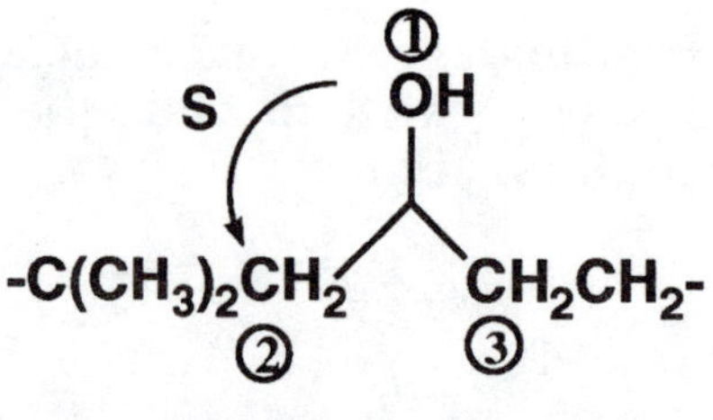

S
① OH
-C(CH_3)$_2$$CH_2$ ② CH_2CH_2- ③

If you view from the perspective shown, this is what you see

Problem 8.3 Convert the following three-dimensional formulas to Fischer projections. Note that there is more than one correct Fischer projection corresponding to each three-dimensional formula.

(a) (b) (c)

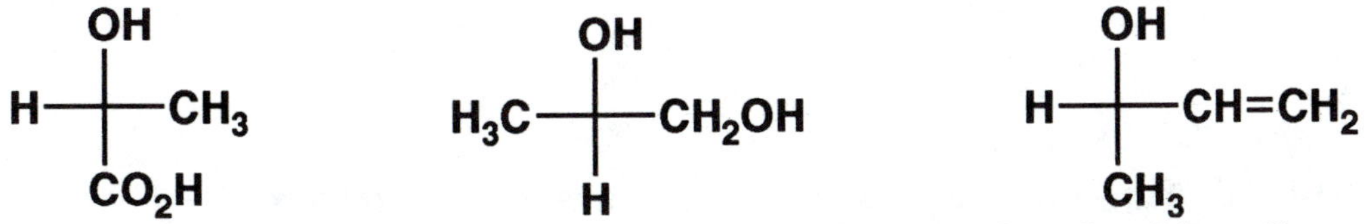

OH
H—+—CH_3
CO_2H

OH
H_3C—+—CH_2OH
H

OH
H—+—CH=CH_2
CH_3

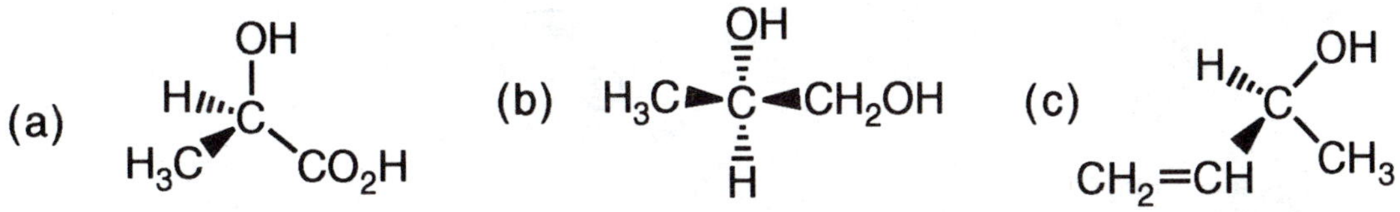

The Fischer projections were drawn with the OH group placed arbitrarily at the top.

Problem 8.4 Barry Sharpless, currently Professor of Chemistry at the Research Institute of Scripps Clinic, La Jolla, California, has devised a procedure for oxidation of alkenes in which the oxidizing agent, osmium tetroxide, is bound to a chiral organic base. Thus, oxidation of the alkene to a glycol takes place in a chiral environment. Under these conditions, vinylcyclooctane is oxidized to an (R)-glycol in 93% enantiomeric excess.

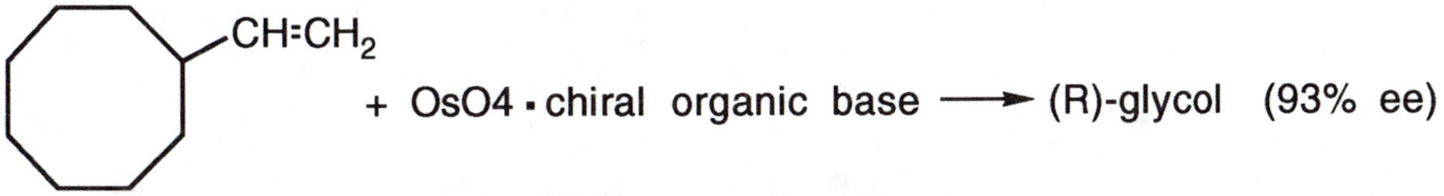

CH=CH_2 + OsO4 • chiral organic base ⟶ (R)-glycol (93% ee)

Draw the structural formula for the product of this oxidation, showing its configuration, and calculate the percentages of (R)-glycol and (S)-glycol in the product.

Following is a stereorepresentation of the (R)-glycol. The composition of the product is 96.5% (R)-glycol and 3.5% (S)-glycol, as derived from the equation for % ee given in Section 8.7D.

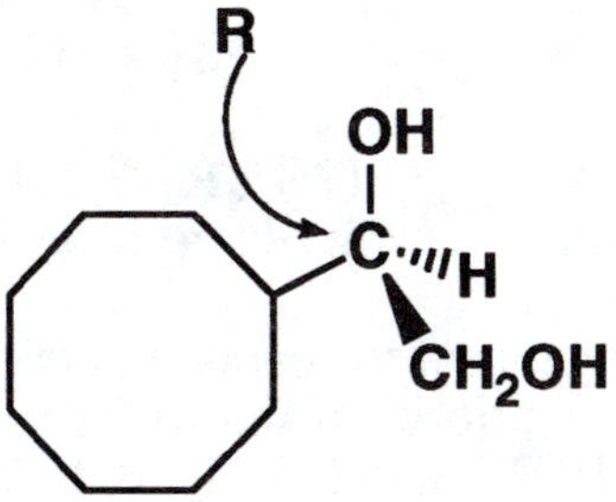

Problem 8.5 Assign an (R) or (S) configuration to (+)-2-aminopropanoic acid, more commonly known as (+)-alanine.

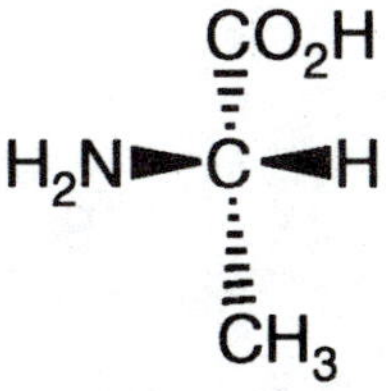

(+)-2-Aminopropanoic acid
(+)-Alanine

The tetrahedral stereocenter has an (S)-configuration.

(S)-(+)-2-Aminopropanoic acid
(S)-(+)-Alanine

when viewed from the proper perspective, this is what you see

Problem 8.6 Assign R or S configuration to each stereocenter in the enantiomers of threose.

(2R,3S)-2,3,4-trihydroxybutanal

(2S,3R)-2,3,4-trihydroxybutanal

Problem 8.7 (a) Think about the configurations of the enantiomers of *trans*-1,2-cyclopentanediol, but do not go through the work of assigning priorities, and so forth. Do you predict configurations of the enantiomers to be (1R,2S) and (1S,2R) or (1R,2R) and (1S,2S)?

If the enantiomers of *cis*-2-methylcyclopentanol are (1R,2S) and (1S,2R) combination, then the diastereomers must be (1S,2S) and (1R,2R) as shown on the following stereorepresentations.

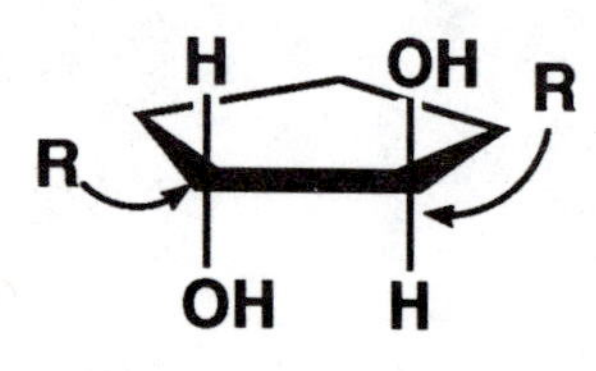

(1R,2R)-1,2-cyclopentanediol

(1S,2S)-1,2-cyclopentanediol

(b) Now do the work of assigning configurations. Did you predict correctly in (a), and if so well done. If not, why not and what did you overlook?

As another way to think about this problem, the two tetrahedral stereocenters in this molecule cannot be reflections (an R,S combination) of each other; if they were, then the molecule would have a plane of symmetry so it would be meso. A meso compound cannot have an enantiomer, so it would not be a correct answer to the question.

Chirality

Problem 8.8 Think about the helical coil of a telephone cord or a spiral binding and suppose that you view the spiral from one end and find that it is a left-handed twist. If you view the same spiral from the other end, do you find it to be a right-handed twist, or is it a left-handed twist from that end as well?

A helical coil has the same handedness viewed from either end.

Problem 8.9 Next time you have the opportunity to view a collection of whelks, augers, or other sea shells that have a helical twist, study the chirality of their twists. Do you find an equal number of left-handed and right-handed whelks or are they all or mostly all of one chirality? What about the chirality of whelks compared to augers and other spiral shells?

This question was just meant to make you think about chirality in nature, but if you do know the answer please share it with your class.

Problem 8.10 One reason we can be sure that sp^3-hybridized carbon atoms are tetrahedral is the number of stereoisomers that can exist for different organic compounds.
(a) How many stereoisomers are possible for $CHCl_3$, CH_2Cl_2, and CHClBrF if the four bonds to carbon have a tetrahedral geometry?

Both tetrahedral $CHCl_3$ and tetrahedral CH_2Cl_2 are achiral, so there is only one stereoisomer of each.

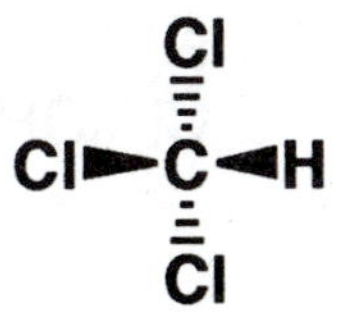

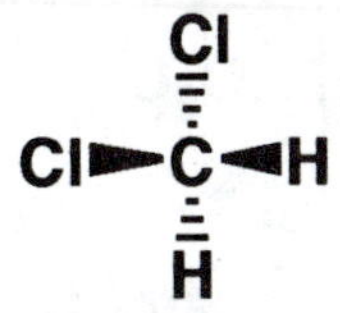

On the other hand, tetrahedral CHBrClF is chiral so there are two isomers possible.

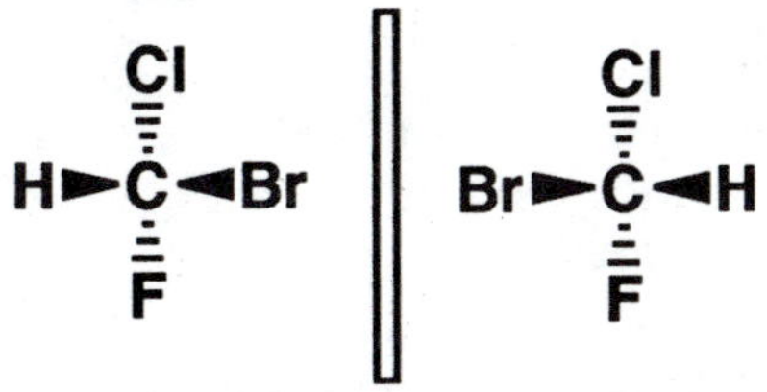

(b) How many stereoisomers are possible for each of these compounds if the four bonds to carbon have a square planar geometry?

Even as a square planar complex (the H and three Cl atoms are in the same plane as the C atom) there is only one stereoisomer possible.

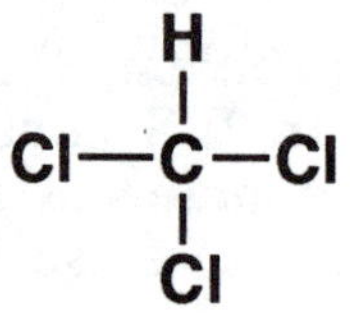

H
|
Cl—C—Cl
|
Cl

There are two possible stereoisomers of CH_2Cl_2, one with the Cl atoms adjacent to each other, and another with the Cl atoms opposite to each other.

H | H—C—Cl | Cl   H | Cl—C—Cl | H

There are three possible stereoisomers of a square planar CHBrClF as shown.

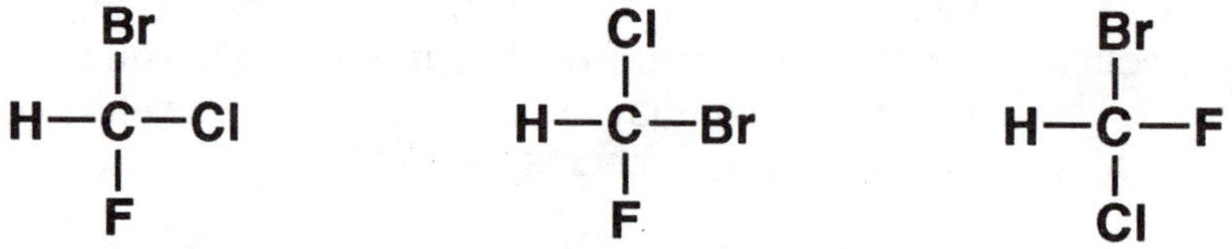

Br | H—C—Cl | F   Cl | H—C—Br | F   Br | H—C—F | Cl

Enantiomers

Problem 8.11 Draw mirror images for these molecules.

The mirror images are shown in bold.

(a) 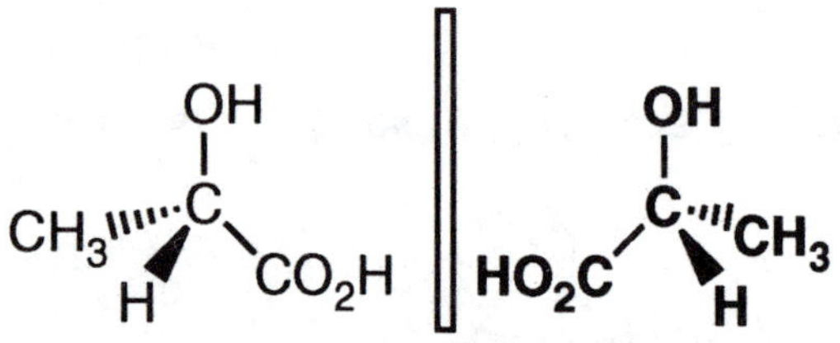 OH, CH_3, C, H, CO_2H | **OH, C, CH_3, HO_2C, H**

(b) CHO, H►C◄OH, CH_2OH | **CHO, HO►C◄H, CH_2OH**

(c) CO_2H, H_2N►C◄H, CH_3 | **CO_2H, H►C◄NH_2, CH_3**

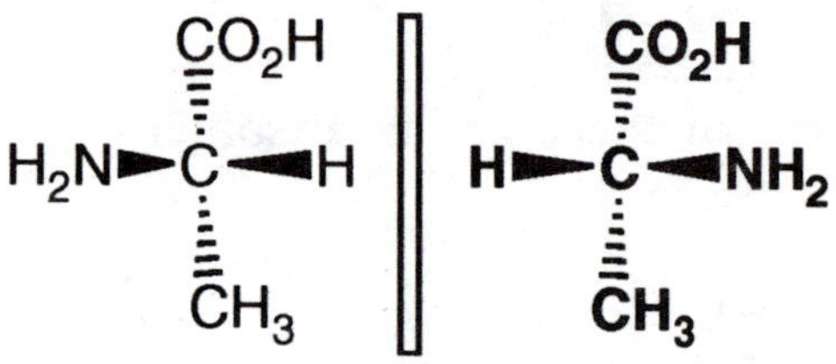

(d) 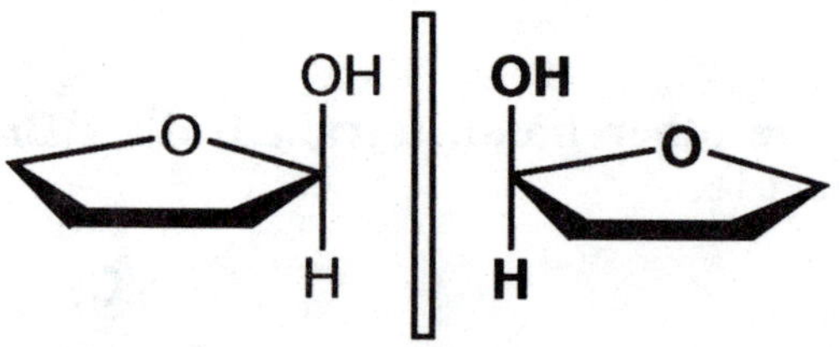 O, OH, H | **OH, O, H**

(e) 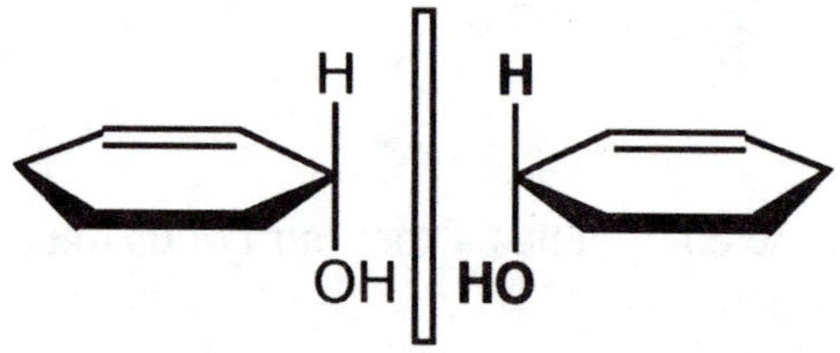 H, OH | **H, HO**

(f) 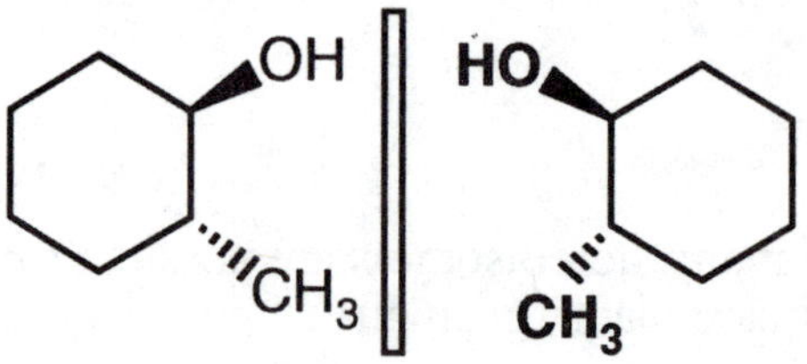 OH, CH_3 | **HO, CH_3**

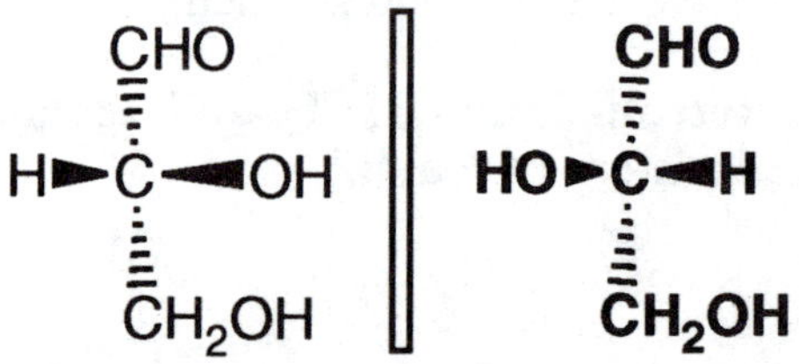

(g) 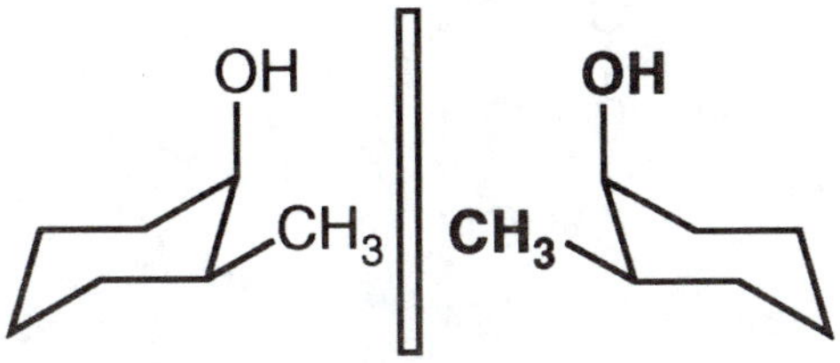 [cyclohexane structures with OH and CH_3, mirror images] (h) 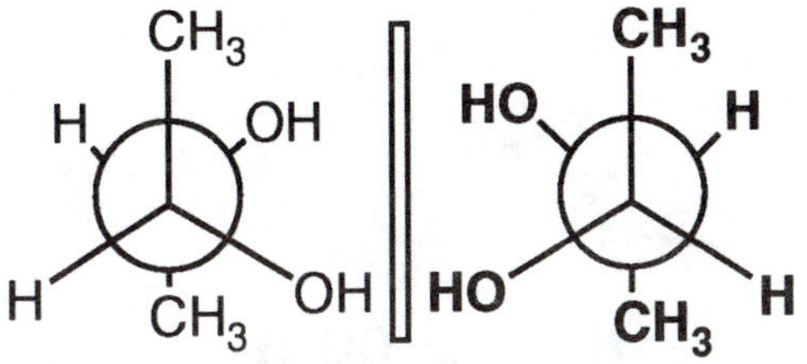 [Newman projections with CH_3, H, OH, mirror images]

Problem 8.12 Following are several stereorepresentations for lactic acid. Take (a) as a reference structure. Which of the stereorepresentations are identical with (a) and which are mirror images of (a)?

(a) 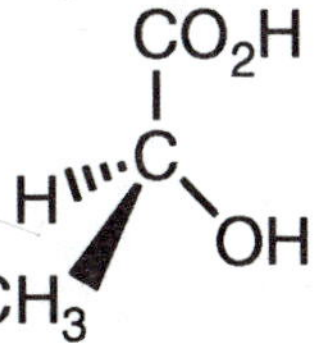 CO_2H, H, OH, CH_3 (b) 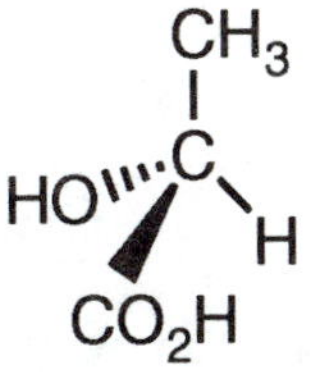 CH_3, HO, H, CO_2H (c) 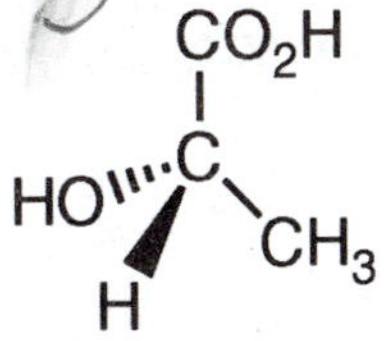 CO_2H, HO, CH_3, H (d) 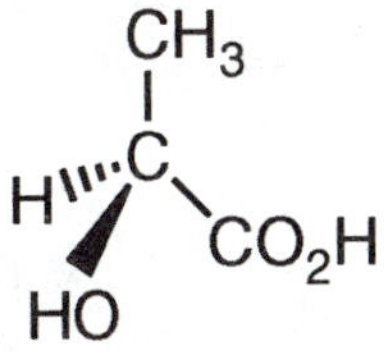 CH_3, H, CO_2H, HO

All stereorepresentations have the (S)-configuration so they are identical.

Problem 8.13 Mark each stereocenter in the following molecules with an asterisk. How many stereoisomers are possible for each molecule?

(a) 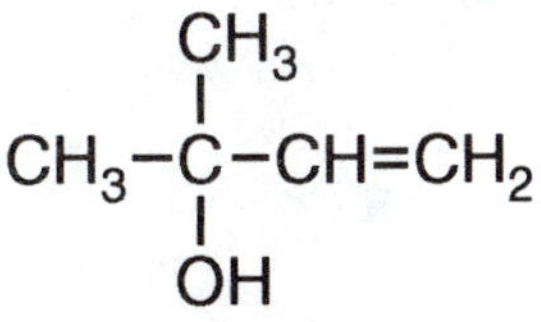 $CH_3-C(CH_3)(OH)-CH=CH_2$

No stereocenters

(b) 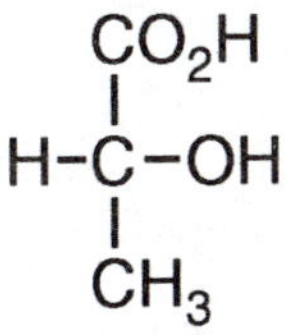 $H-C(CO_2H)(CH_3)-OH$

$H-\overset{*}{C}(CO_2H)(CH_3)-OH$

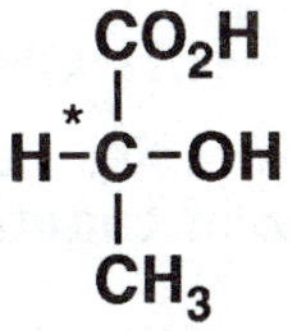

2 stereoisomers (a pair of enantiomers)

(c) 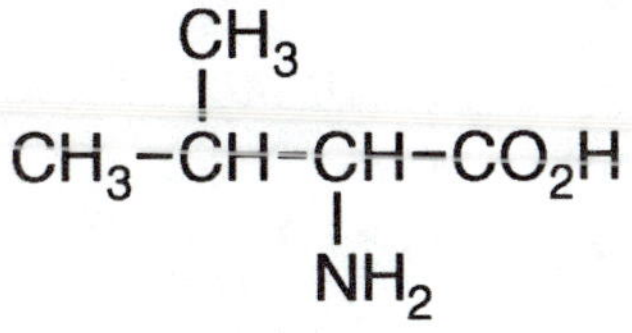 $CH_3-CH(CH_3)-CH(NH_2)-CO_2H$

$CH_3-CH(CH_3)-\overset{*}{C}H(NH_2)-CO_2H$

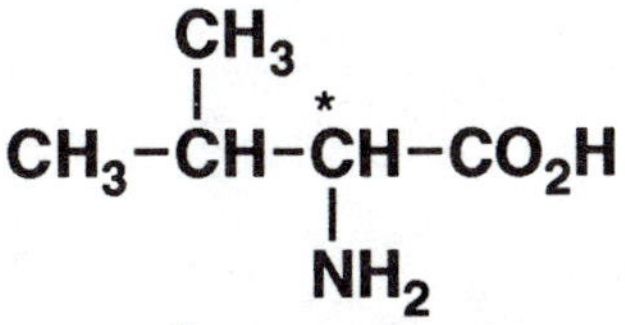

2 stereoisomers (a pair of enantiomers)

(d) 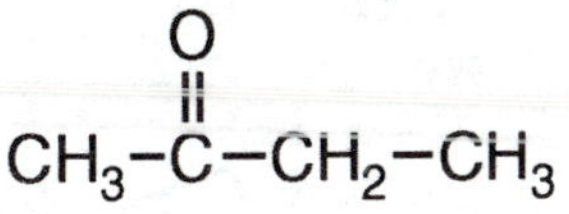 $CH_3-C(=O)-CH_2-CH_3$

No stereocenters

(e)

H, CO_2H C=C CO_2H, H

H * * CO_2H C=C HO_2C H

2 stereoisomers
(a pair of diastereomers: E,Z-isomers)

(f) CH_2OH H—C—OH CH_2OH

No stereocenters

(g) $CH_3-CH_2-CH(OH)-CH=CH_2$

$CH_3-CH_2-\overset{*}{C}H(OH)-CH=CH_2$

2 stereoisomers
(a pair of enantiomers)

(h) $CH_2\text{-}CO_2H$ H-C-CO_2H $CH_2\text{-}CO_2H$

No Stereocenters

Problem 8.14 Following are structural formulas of two natural products recently synthesized in the laboratory by Theodore Cohen of the University of Pittsburgh. How many stereoisomers are possible for each compound? Explain how you arrived at your answer.

(a)

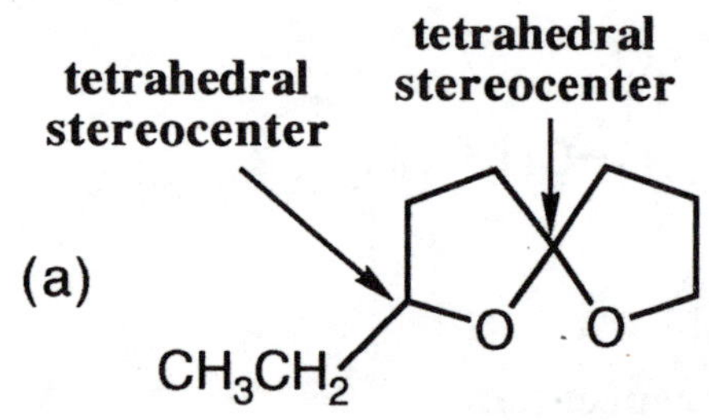

a pheromone of the Norway spruce beetle

There are two tetrahedral stereocenters in the molecule. As a result, there are 2^2 or 4 possible stereoisomers.

(b)

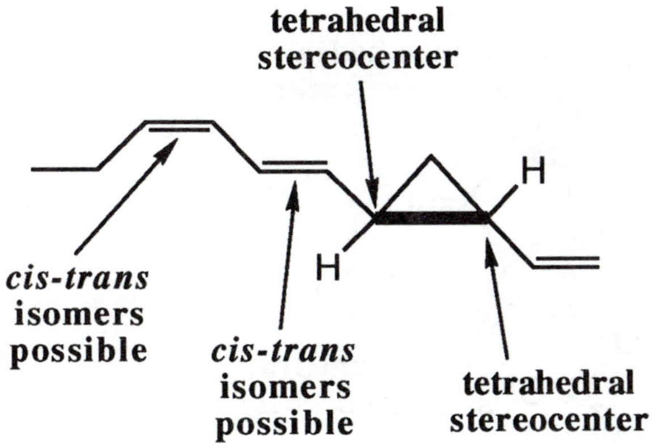

tetrahedral stereocenter

H

H

cis-trans isomers possible

cis-trans isomers possible

tetrahedral stereocenter

Dictyopterene B (a marine gamete attractant)

There are two alkenes with possible *cis-trans* isomers and two tetrahedral stereocenters in this molecule. Therefore there are $2^2 \times 2^2 = 16$ possible stereoisomers.

Problem 8.15 Following are structural formulas for two natural products isolated from plants of the chrysanthemum family. Pyrethrin II is a natural insecticide and is marketed as such.
(a) Label all tetrahedral stereocenters in each molecule and all carbon-carbon double bonds about which there is the possibility for *cis-trans* isomerism.

The tetrahedral stereocenters and carbon-carbon double bonds with possible *cis-trans* isomers are indicated on the structures.

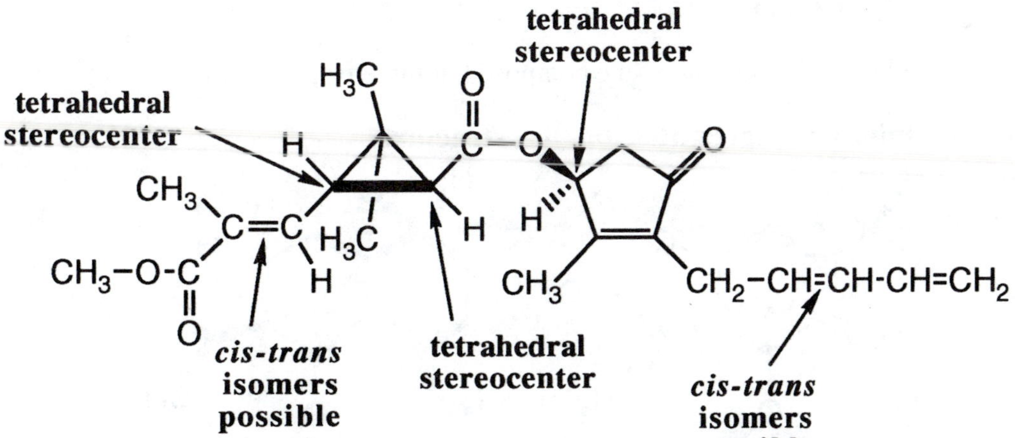

tetrahedral stereocenter

tetrahedral stereocenter

H_3C O

H C—O O

CH_3

C=C H_3C H H

CH_3-O-C H CH_3 CH_2-CH=CH-CH=CH_2

O

cis-trans isomers possible

tetrahedral stereocenter

cis-trans isomers possible

pyrethrin II

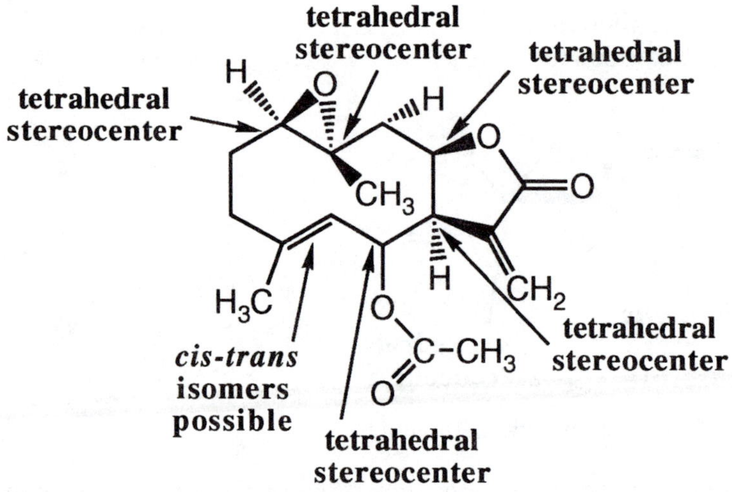

pyrethrosin

(b) State the number of stereoisomers possible for each molecule.

For pyrethrin II there are 2 alkenes with possible *cis-trans* isomers and three tetrahedral stereocenters. Thus, there $2^2 \times 2^3 = 32$ possible stereoisomers of this molecule.

For pyrethrosin, there are five tetrahedral stereocenters and one alkene with possible *cis-trans* isomers. Recall that a ten-member ring is large enough to allow the *trans* alkene configuration. There are a total of $2^5 \times 2 = 64$ possible stereoisomers of this molecule.

(c)Show that pyrethrosin is a sesquiterpene composed of three isoprene units.

The isoprene units are highlighted on the structure.

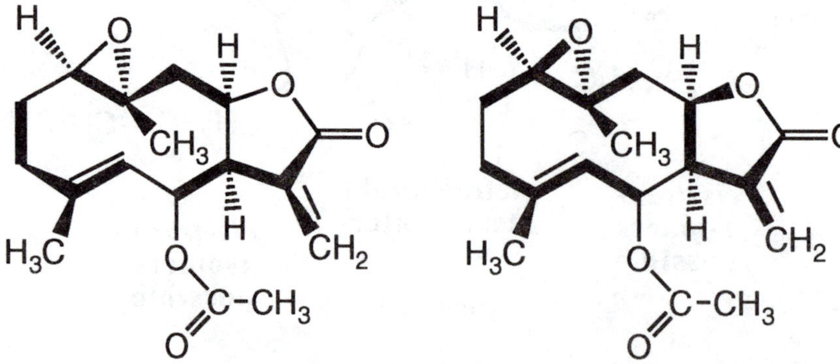

Problem 8.16 *Trans*-cyclooctene has been resolved, and its enantiomers are stable at room temperature. *Trans*-cyclononene has also been resolved, but it racemizes with a half-life of 4 min at 0°C. How can racemization of these cycloalkenes take place without breaking any bonds? Why does *trans*-cyclononene racemize under these conditions but *trans*-cyclooctene does not?

Before trying to answer this question, it may be helpful to make models to help understand the structure of the enantiomers and the dynamics of the ring systems.

The enantiomers of *trans*-cyclooctene and *trans*-cyclononene are based upon two different *conformations* of the rings that are mirror images of each other. The *trans*-alkene places sufficient strain into the ring that interconversion between these conformations is difficult, especially for the smaller *trans*-cyclooctene. Thus, at room temperature it cannot interconvert into both enantiomers. *Trans*-cyclononene is slightly larger, so it is less strained and can interconvert more rapidly. Thus, low temperature is required to "freeze out" the two enantiomers of *trans*-cyclononene.

Problem 8.17 Show that butane in a gauche conformation is chiral. Do you expect that resolution of butane at room temperature is possible?

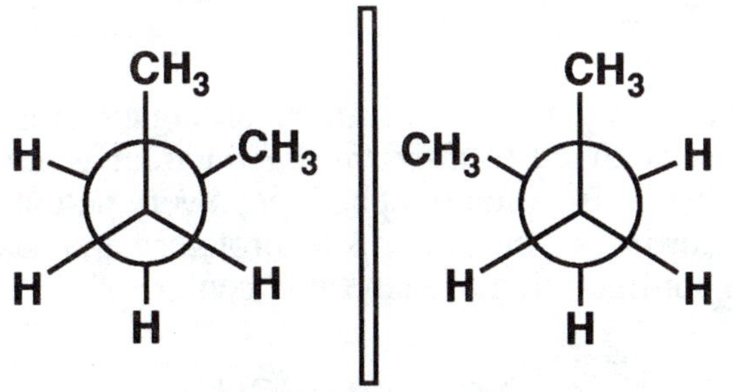

CH_3 H CH_3 H   H   H | CH_3 CH_3 H   H   H   H

As can be seen from the above Newman projections, the two gauche conformations are non-superposable mirror images of each other. However, these conformations rapidly interconvert at room temperature via rotation around the central C-C bond, so they cannot be resolved.

Absolute Configuration: the R-S Convention

Problem 8.18 Following are structural formulas for the enantiomers of carvone. Each has a distinctive odor characteristic of the source from which it can be isolated. Assign an R-S configuration to each enantiomer.

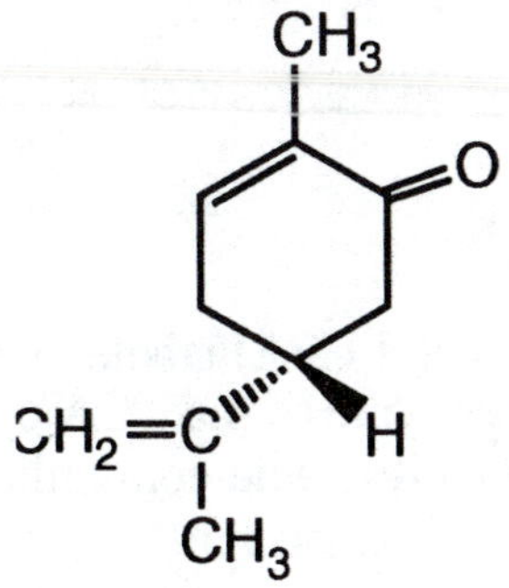

CH_3 O $CH_2{=}C$ H CH_3

(-)-Carvone $[\alpha]_D^{20} = -62.5^o$
(spearmint oil)

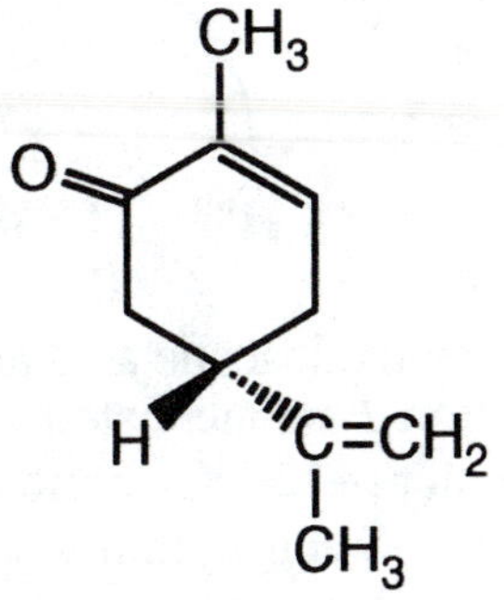

CH_3 O   H $C{=}CH_2$ CH_3

(+)-carvone $[\alpha]_D^{20} = +62.5^o$
(caraway oil)

Following are R-S designations for each enantiomer.

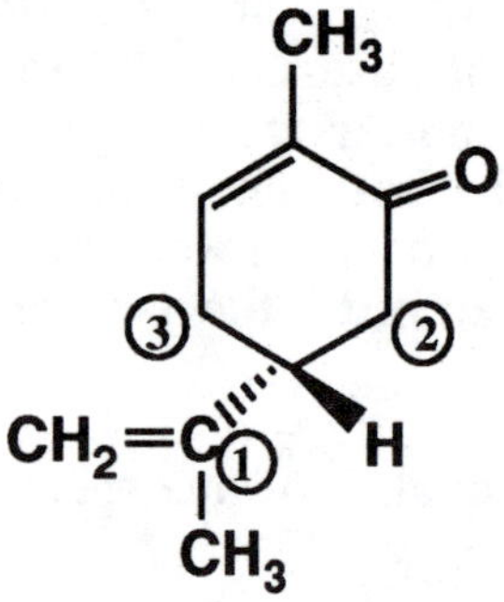

CH_3 O ③ ② $CH_2=C$① H CH_3

(R)-(-)-carvone

CH_3 O ② ③ H ①$C=CH_2$ CH_3

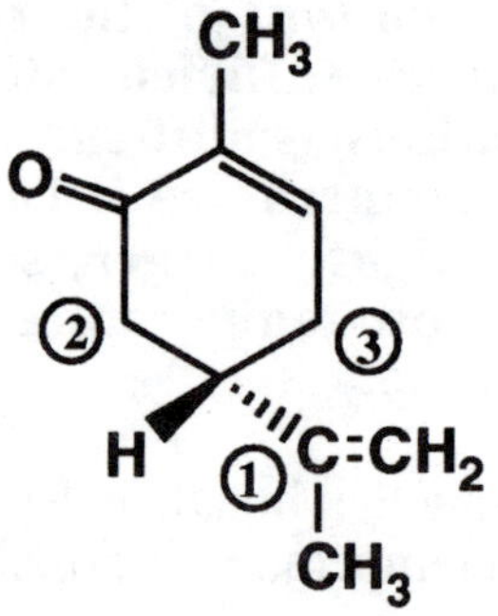

(S)-(+)-carvone

Problem 8.19 An organic base widely used as a chiral resolving agent is ephedrine. For centuries, Chinese herbal medicine has used extracts of *Ephedra sinica* totreat asthma. Phytochemical investigation of this plant resulted in isolation of ephedrine, a very potent dilator of the air passages of the lungs. The naturally occurring stereoisomer is levorotatory and has the following configuration. Assign an R-S configuration to each stereocenter.

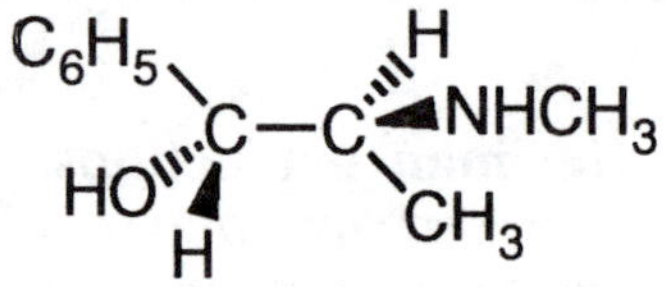

C_6H_5 H C—C $NHCH_3$ HO H CH_3

Ephedrine

$[\alpha]_D^{21} = -41^o$

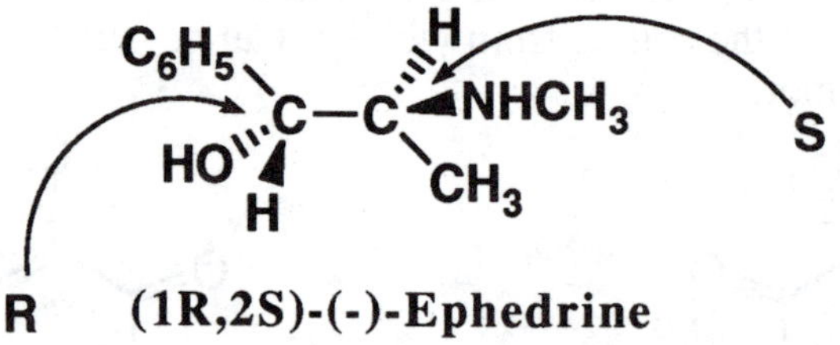

C_6H_5 H C—C $NHCH_3$ S HO H CH_3

R **(1R,2S)-(-)-Ephedrine**

Problem 8.20 When oxaloacetic acid and acetyl coenzyme A (acetyl-CoA) labeled with radioactive carbon-14 in position 2 are incubated with citrate synthase, an enzyme of the TCA cycle, only the following enantiomer of [2-^{14}C]-citric acid is formed. Note that citric acid containing only ^{12}C is achiral. Assign an R-S configuration to this enantiomer of [2-^{14}C] citric acid.

$O=C(-CO_2H)(-CH_2-CO_2H)$ + $^{14}CH_3-C(=O)-S-CoA$ $\xrightarrow{\text{citrate synthase}}$ $HO_2C-C(-OH)(^{14}CH_2-CO_2H)(CH_2-CO_2H)$

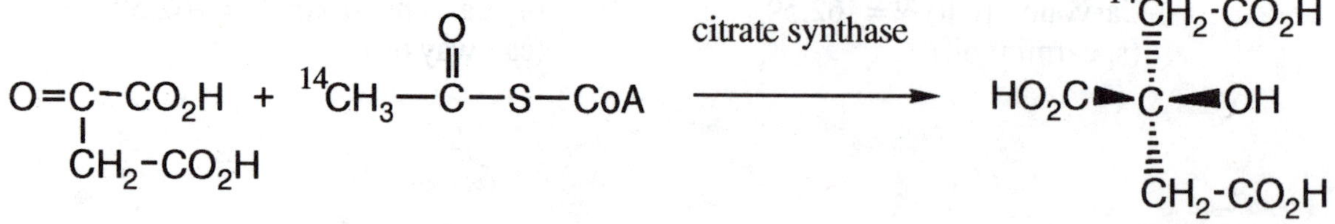

Oxaloacetic acid Acetyl-CoA [2-^{14}C]citric acid

This enantiomer is S-(2-^{14}C) citric acid.

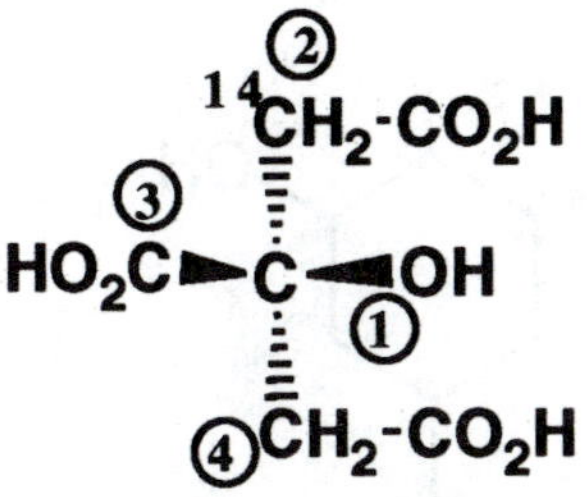

$^{14}CH_2$-CO_2H (2)
(3) HO_2C—C—OH (1)
(4) CH_2-CO_2H

S-[2-^{14}C]Citric acid

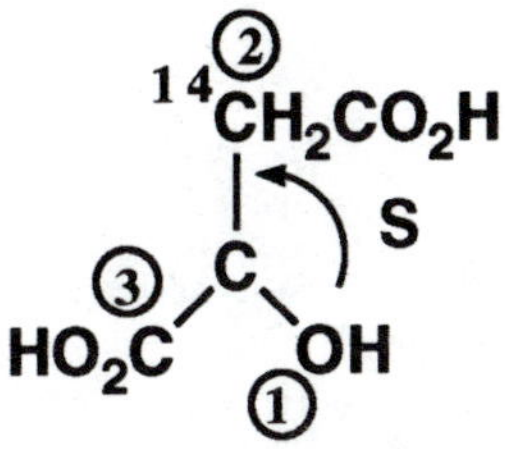

$^{14}CH_2CO_2H$ (2)
(3) HO_2C—C—OH (1)
S

If you view from the proper perspective, this is what you see

Molecules With Two or More Stereocenters

Problem 8.21 Mark each stereocenter in the following molecules with an asterisk. How many stereoisomers are possible for each molecule?

(a) 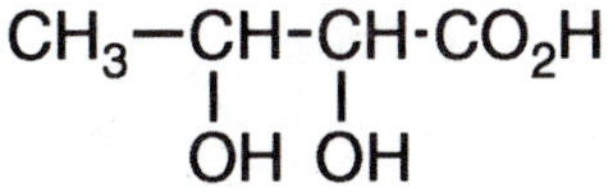 CH_3—CH(OH)-CH(OH)-CO_2H

CH_3—C*H(OH)-C*H(OH)-CO_2H

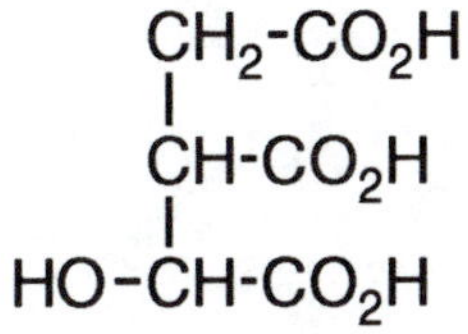

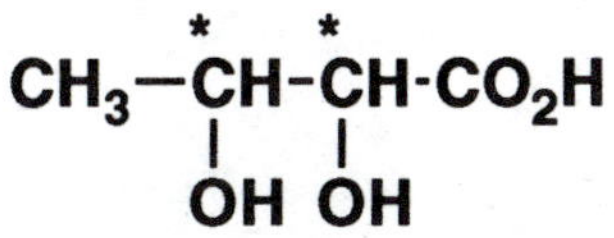

4 stereoisomers (two pairs of enantiomers)

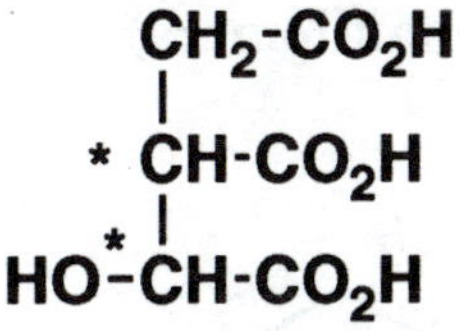

(b) CH_2-CO_2H / CH-CO_2H / HO-CH-CO_2H

CH_2-CO_2H / * CH-CO_2H / HO-*CH-CO_2H

4 stereoisomers (two pairs of enantiomers)

(c) 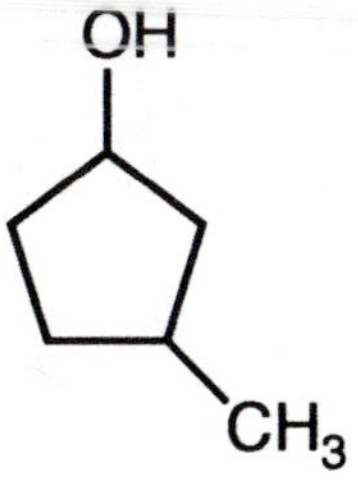 OH, CH_3 (3-methylcyclopentanol)

OH, *, *, CH_3

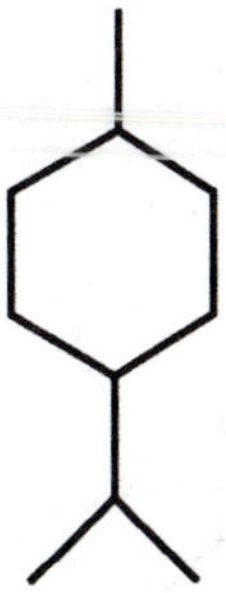

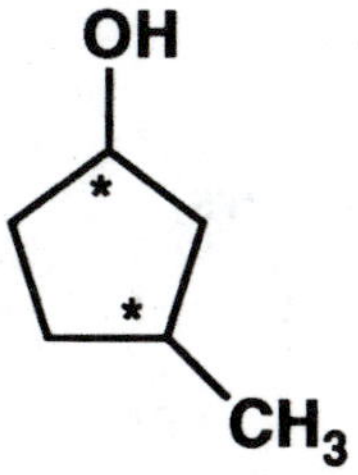

4 stereoisomers (two pairs of enantiomers)

(d)

*, *

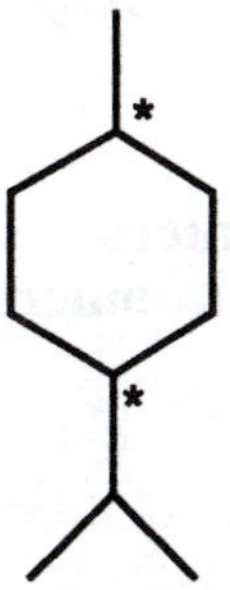

2 stereoisomers (two diastereomers, *cis* and *trans*)

(e)

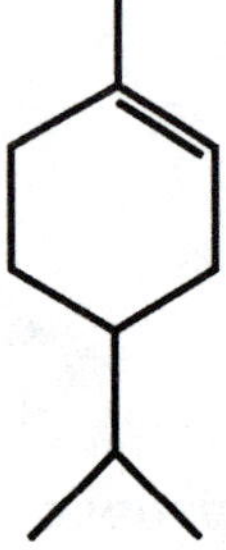

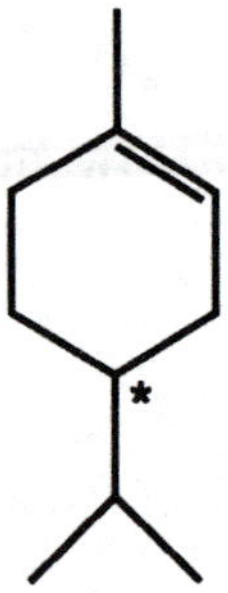

OH

(f)

*

*

OH

*

*

2 stereoisomers
(one pair of enantiomers)

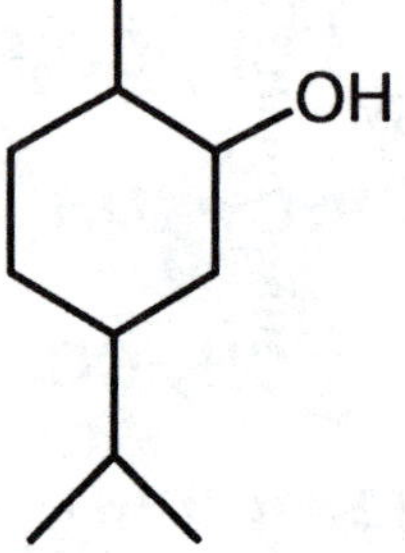

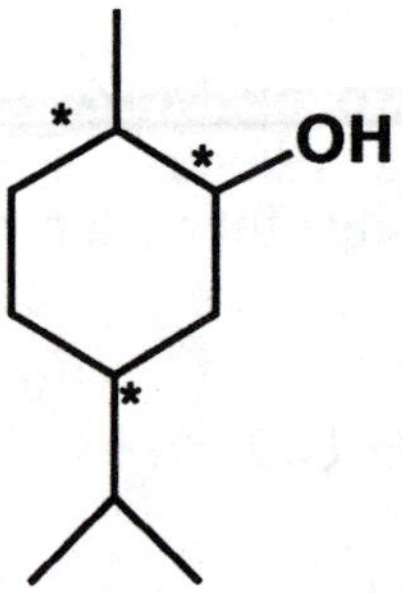

8 stereoisomers
(four pairs of enantiomers)

(g)

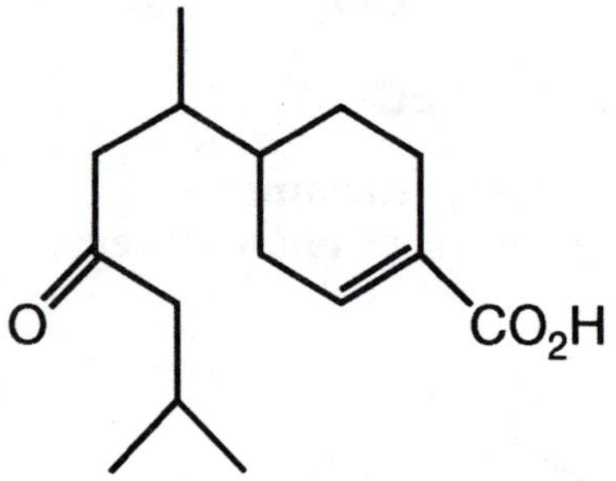

O

CO_2H

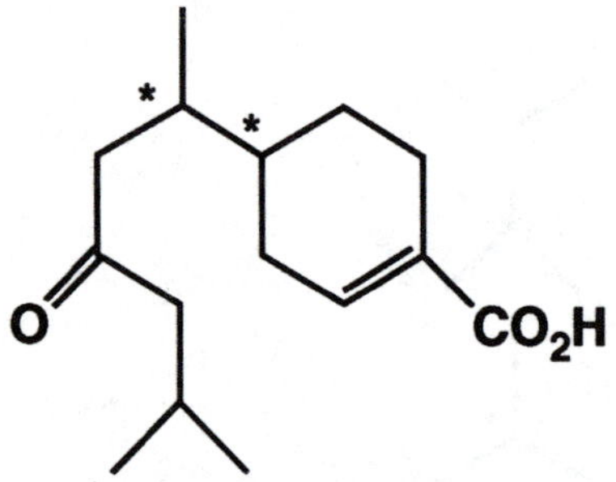

(h)

O

OH

OH

*

*

O

CO_2H

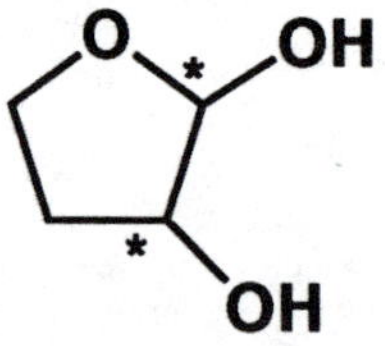

O

*

OH

*

OH

4 stereoisomers
(two pairs of enantiomers)

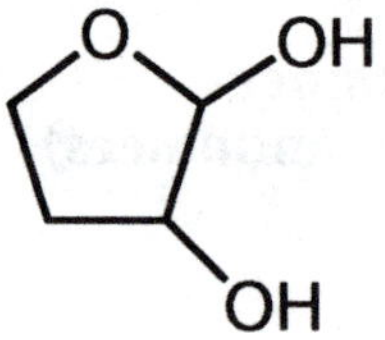

4 stereoisomers
(two pairs of enantiomers)

(i)

4 stereoisomers
(2 pairs of enantiomers)

Problem 8.22 Draw the structural formula for at least one bromoalkene of molecular formula C_5H_9Br that shows:

(a) Neither E-Z-isomerism nor enantiomerism.

Draw any structural formula in which there is no tetrahedral stereocenter and in which at least one carbon of the double bond has two identical atoms or groups of atoms on it. For example:

2-bromo-3-methyl-2-butene

(b) E-Z-isomerism but not enantiomerism.

Each carbon of the double bond must have two different atoms or groups of atoms on it, and there can be no tetrahedral stereocenter. For example:

(Z)-1-bromo-2-methyl-2-butene

(c) Enantiomerism but not E-Z-isomerism.

One carbon of the double bond must have two identical atoms or groups of atoms on it, and there must be a tetrahedral stereocenter. For example:

(S)-4-bromo-1-pentene

(d) Both enantiomerism and E-Z-isomerism.

There is only one constitutional isomer that shows E-Z isomerism and has a tetrahedral stereocenter. Drawn is the *cis*-isomer:

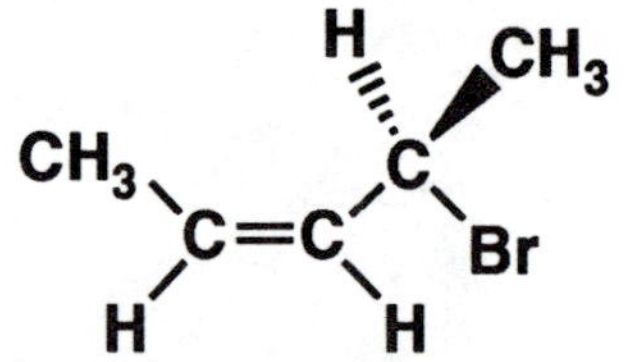

(4R)-4-bromo-(Z)-2-pentene

Problem 8.23 Following are several molecules isolated from natural systems. Each is isolated as a single stereoisomer with the specific rotation given. Mark all stereocenters and state the maximum number of stereoisomers possible for each molecule.

Each stereocenter is marked with an asterisk and the maximum number of stereoisomers possible is given under the structural formula.

(a)

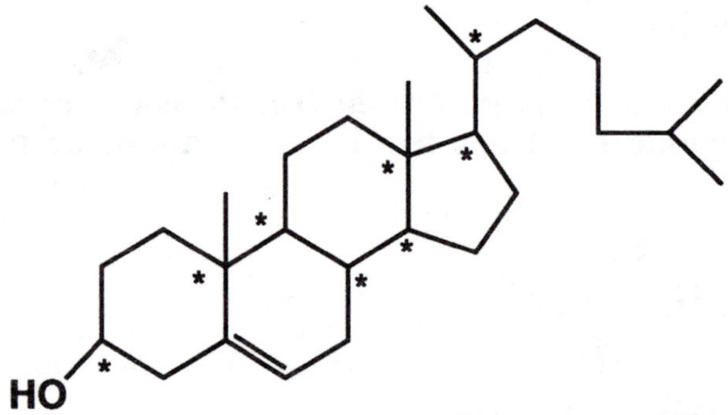

cholesterol

$[\alpha]_D^{24} = +15°$

256 stereoisomers (128 pairs of enantiomers)

(b)

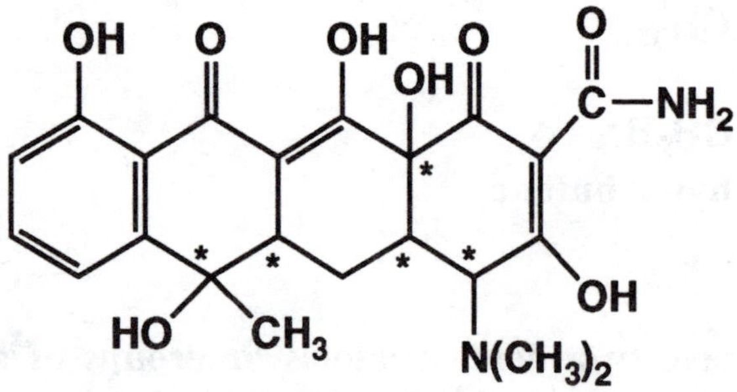

tetracycline

$[\alpha]_D^{25} = +225°$

32 stereoisomers (16 pairs of enantiomers)

(c) HO—(3,4-dihydroxyphenyl)—CH_2—C*H—CO_2H with NH_2

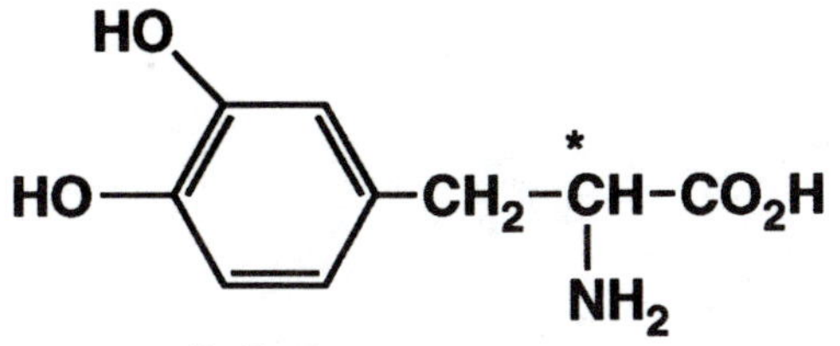

L-Dopa

3-(3,4-dihydroxyphenyl)alanine

$[\alpha]_D^{27} = -11.5^o$

2 stereoisomers (1 pair of enantiomers)

(d) α-pinene

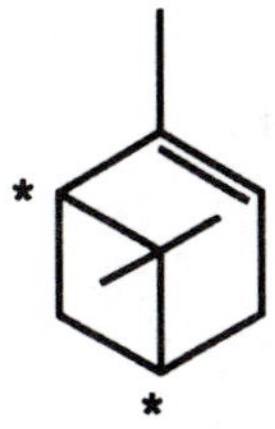

$[\alpha]_D^{21} = +50.7^o$

4 stereoisomers (2 pairs of enantiomers)

(e) $CH_3-\overset{*}{C}H(OH)-\overset{*}{C}H(NH_2)-CO_2H$

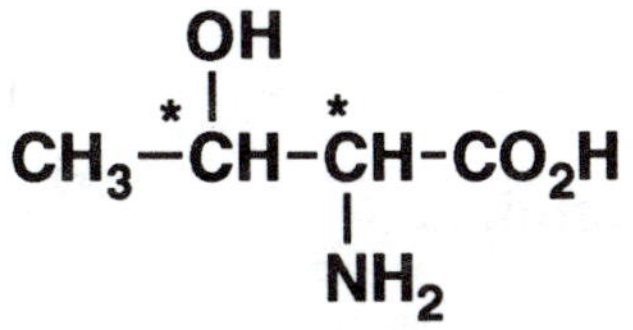

threonine

$[\alpha]_D^{20} = -27.4^o$

4 stereoisomers (2 pairs of enantiomers)

Problem 8.24 If the optical rotation of a new compound is measured and found to have a specific rotation of 40°, how can you tell if the actual rotation is not really 40° plus some multiple of 360° (that is the rotation is not actually 40 + (n x 360)°, where n has only integer values. That is, how can you tell if the rotation is not actually a value such as 400° or 760°?

You could dilute the solution by a factor of two and then remeasure the rotation. If the new rotation is 20°, then the original rotation was 40°. If the new rotation is 200°, then the original rotation was 400°.

Problem 8.25 Are the formulas within each set identical, enantiomers, or diastereomers?

(a) CH_3—C(H)—Cl / CH_3—C(H)—OH and CH_3—C(H)—OH / Cl—C(H)—CH_3

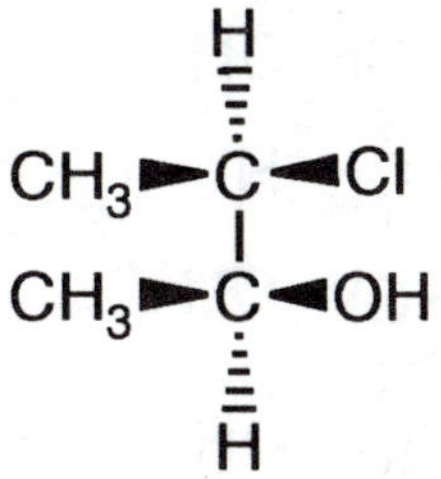

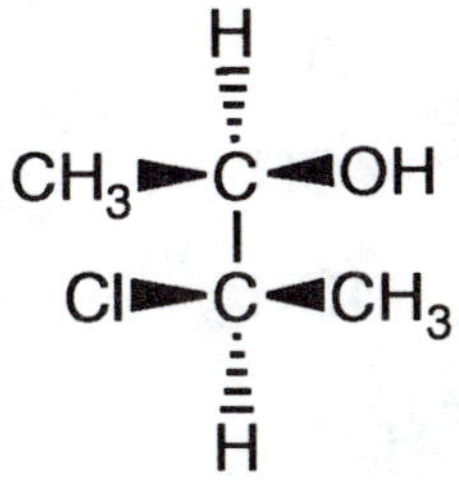

Diastereomers. Configurations are (2S,3R)-3-chloro-2-butanol and (2R,3R)-3-chloro-2-butanol.

(b) and

Identical. They are both (2R,3S)-2,3-butanediol, a meso compound.

(c) and

Identical. They are both (1S,2R)-*cis*-2-methylcyclohexanol.

(d) and

Diastereomers. *Cis* and *trans* isomers.

Problem 8.26 Which of the following are meso compounds?

(a) (b) (c)

(d) (e) (f)

(g) (h)

The meso compounds are (a), (c), (d), and (e).

Problem 8.27 Vigorous oxidation of the following bicycloalkene of known configuration gives 2,2-dimethylcyclopentane-1,3-dicarboxylic acid. Assume that the conditions of oxidation have no effect on the stereocenters of either the starting bicycloalkene or the resulting dicarboxylic acid. Is the dicarboxylic acid one enantiomer, a racemic mixture, or a meso compound?

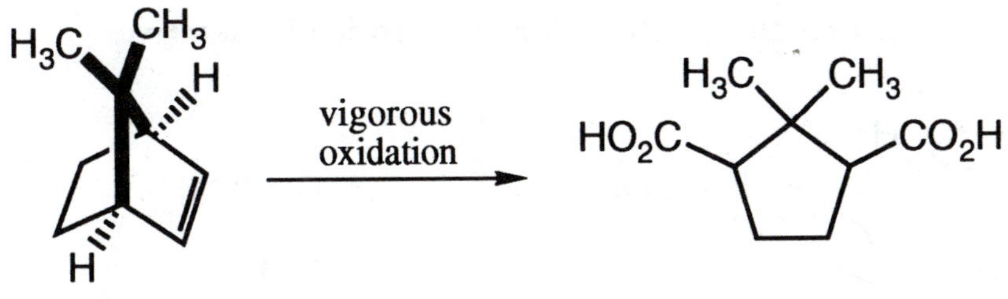

vigorous oxidation →

(1R,4S)-7,7-Dimethyl-bicyclo[2.2.1]hept-2-ene

2,2-Dimethylcyclopentane-1,3-dicarboxylic acid

The two carboxyl groups must be *cis* to each other and, therefore, the compound is meso with the following configuration.

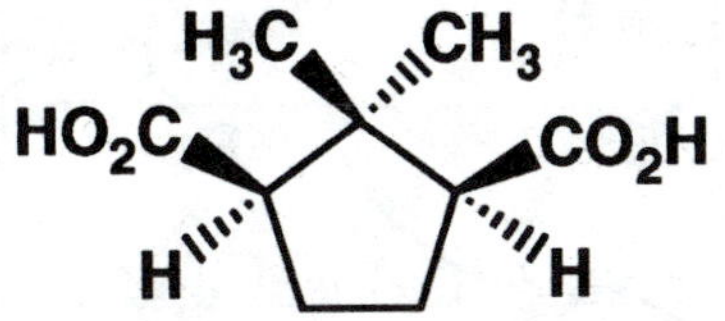

meso-2,2-dimethylcyclopentane-1,3-dicarboxylic acid

Reactions That Produce Chiral Compounds

Problem 8.28 In each of the following reactions, the organic starting material is achiral. The structural formula of the product is given. For each product state:

(1) How many stereoisomers are possible for the products?

(2) Which of the possible stereoisomers is/are formed in the reaction shown?

(3) Determine whether the product is optically active or optically inactive.

(a) cyclopentyl-$CH{=}CH_2$ → 1) $Hg(OAc)_2$, H_2O; 2) $NaBH_4$ → cyclopentyl-CH(OH)-CH_3

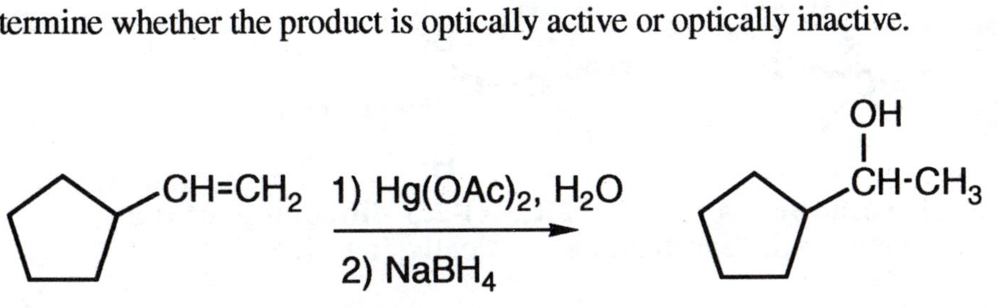

The alcohol contains one stereocenter; two stereoisomers (one pair of enantiomers) are possible. The product will be a racemic mixture and optically inactive.

a racemic mixture formed; optically inactive

(b) 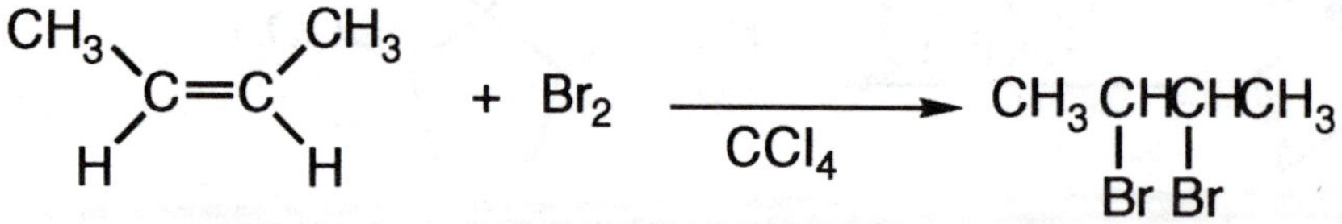 $CH_3CH{=}CHCH_3$ (cis) + Br_2 $\xrightarrow{CCl_4}$ $CH_3CH(Br)CH(Br)CH_3$

Three stereoisomers are possible; one pair of enantiomers and one meso compound. Given the anti stereoselectivity of the addition of bromine to a double bond, the product is the pair of enantiomers and is optically inactive.

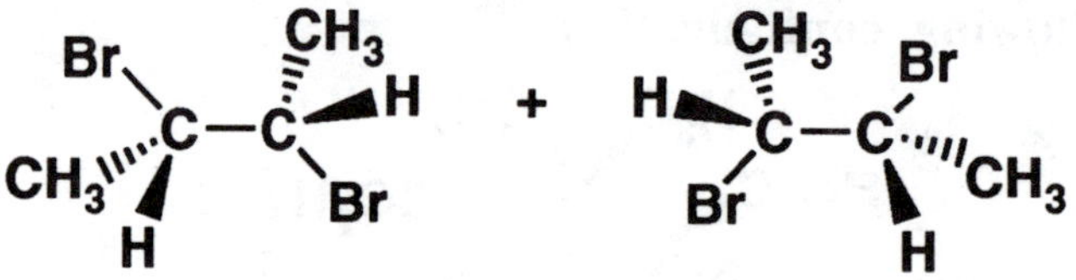

a racemic mixture formed; optically inactive

(c) 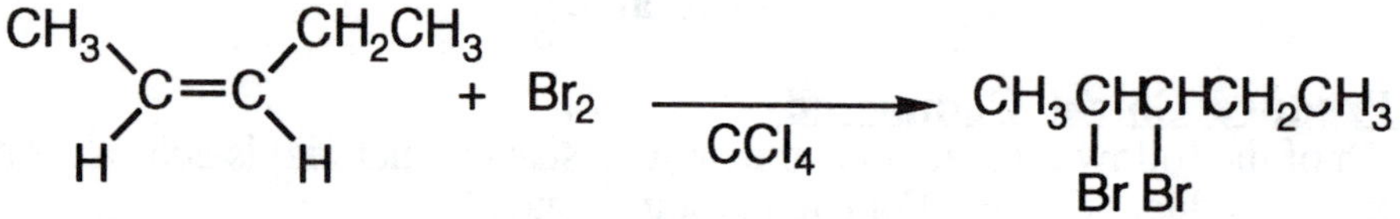 $CH_3CH{=}CHCH_2CH_3$ (cis) + Br_2 $\xrightarrow{CCl_4}$ $CH_3CH(Br)CH(Br)CH_2CH_3$

There are two stereocenters and four possible stereoisomers; two pairs of enantiomers. Given the anti stereoselectivity of addition of bromine to this *cis* alkene, the product is the following pair of enantiomers and is optically inactive.

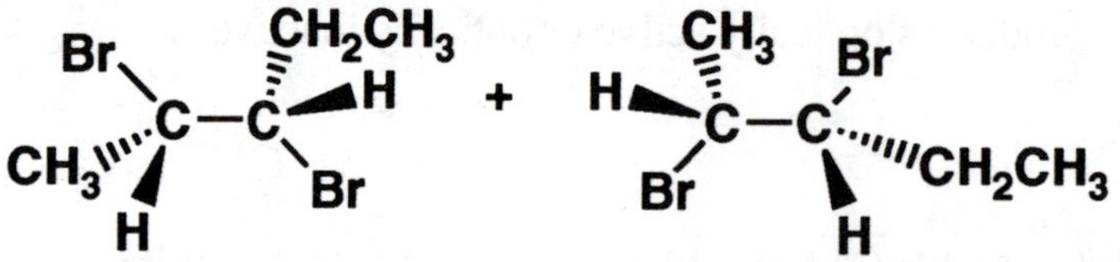

(2S,3S)-2,3-dibromopentane (2R,3R)-2,3-dibromopentane

a racemic mixture formed; optically inactive

(d) 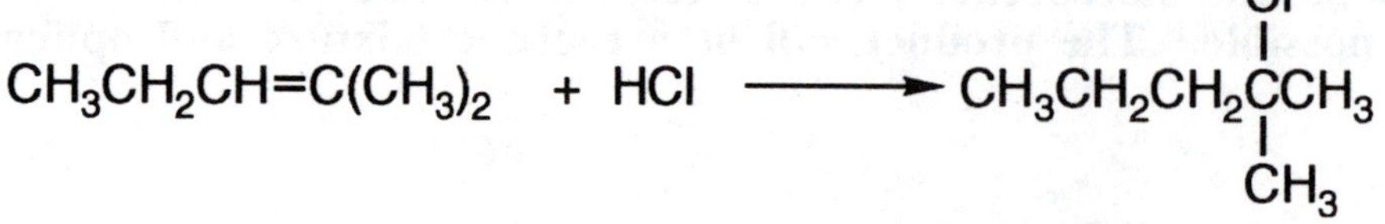 $CH_3CH_2CH{=}C(CH_3)_2$ + HCl $\longrightarrow$ $CH_3CH_2CH_2CCl(CH_3)CH_3$

The product is achiral and, therefore, optically inactive.

(e) 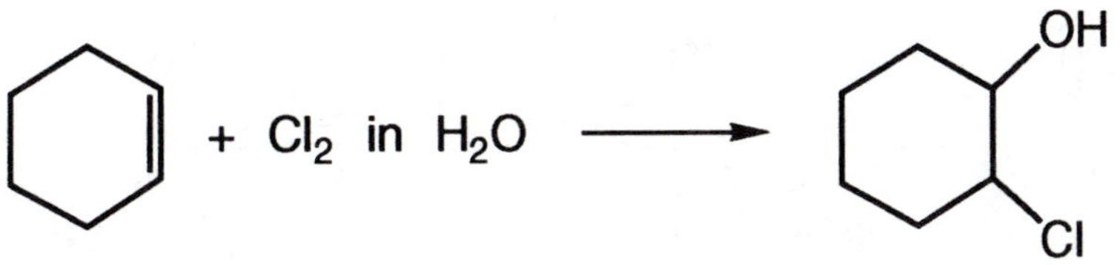 cyclohexene + Cl_2 in H_2O ⟶ 2-chlorocyclohexanol (OH, Cl)

This molecule contains two stereocenters and four stereoisomers (two pairs of enantiomers) are possible. The *cis* isomer is one pair of enantiomers, the *trans* isomer is a second pair of enantiomers. Given the known stereoselectivity (anti) of addition of HOCl to an alkene, the product is the following pair of enantiomers. Because the product is a racemic mixture, it is optically inactive.

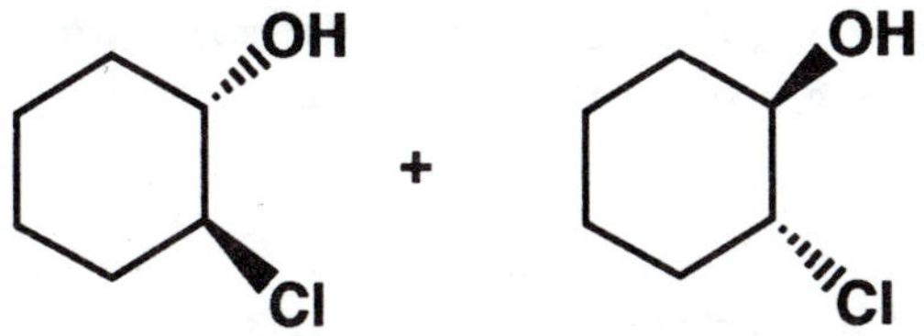

OH, Cl + OH, Cl

a racemic mixture formed; optically inactive

(f) 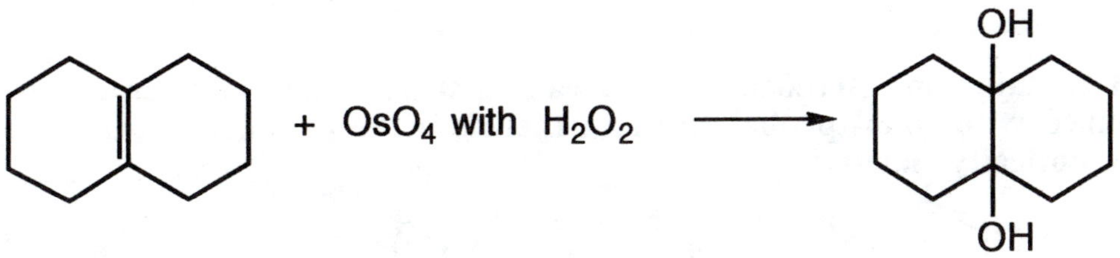 + OsO_4 with H_2O_2 ⟶ OH, OH

In theory, there could be a *cis*-diol and a *trans*-diol. When the cyclohexane rings are drawn as planar hexagons, each stereoisomer has an apparent plane of symmetry. It is more accurate, however, to draw the cyclohexane rings as puckered conformations as shown below. The *trans*-diol is achiral; it has both a plane of symmetry and a point of symmetry.

OH, OH

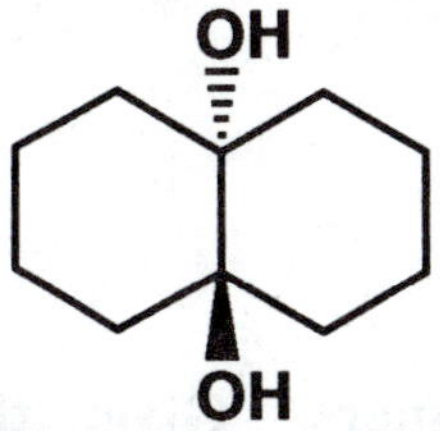

the *trans*-diol with cyclohexane rings drawn as planar hexagons

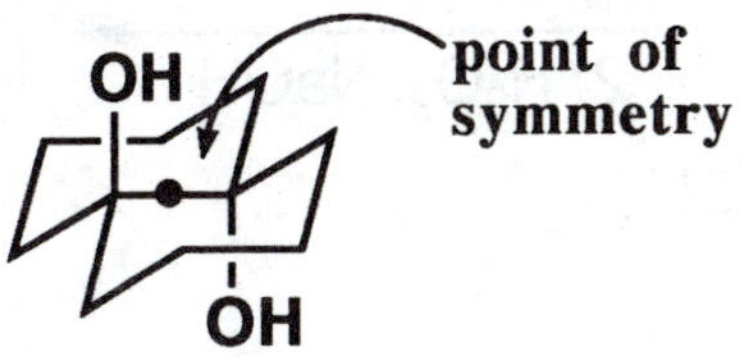

OH, OH, point of symmetry

the *trans*-diol is achiral; it has a plane of symmetry and a point of symmetry

The *cis*-diol is chiral; it has neither a plane of symmetry nor a point of symmetry.

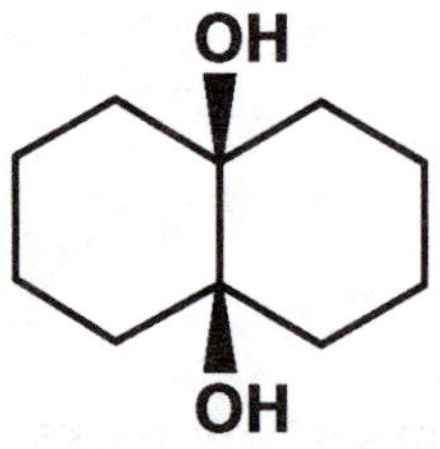

OH

OH

HO

HO

OH

OH

the *cis*-diol with cyclohexane rings drawn as planar hexagons

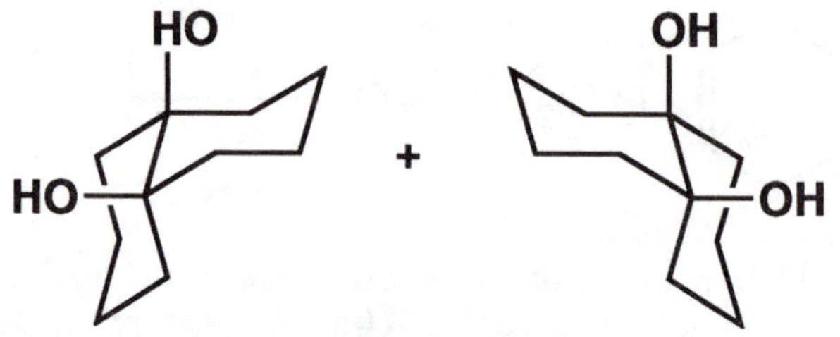

a racemic mixture formed; optically inactive

Because of the *cis* stereoselectivity of oxidative hydroxylation by OsO_4, the product is the *cis*-diol formed as a racemic mixture and therefore optically inactive.

(g) 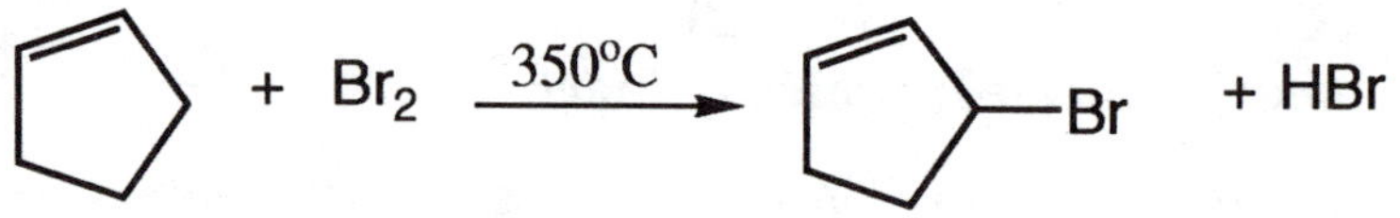 + Br_2 $\xrightarrow{350°C}$ Br + HBr

The product contains one stereocenter and can exist as a pair of enantiomers. Both are formed with equal probability and therefore the product is a racemic mixture and optically inactive.

H Br + H Br

a racemic mixture formed; optically inactive

(h) 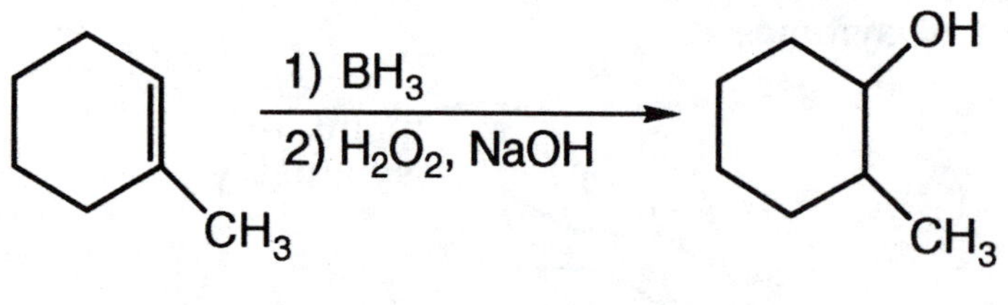 CH_3 1) BH_3 2) H_2O_2, NaOH → OH CH_3

There are two stereocenters and four possible stereoisomers. Given the stereoselectivity of hydroboration/oxidation, only the *trans* isomer is formed as a racemic mixture. The product is optically inactive.

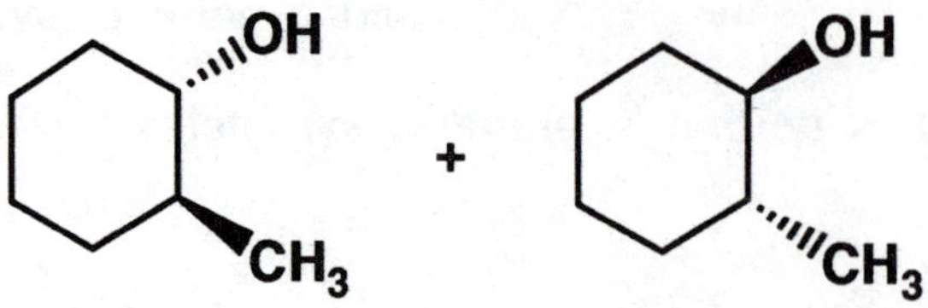

OH CH_3 + OH CH_3

a racemic mixture formed; optically inactive

(i) 1-methylcyclopentene CH_3 + HBr $\xrightarrow{\text{absence of peroxides}}$ 1-bromo-1-methylcyclopentane (CH_3, Br)

The product is achiral, since it has a plane of symmetry; optically inactive.

(j) 1-methylcyclopentene CH_3 + HBr $\xrightarrow{\text{peroxides}}$ 2-bromo-1-methylcyclopentane (CH_3, Br)

There are two stereocenters created so a total of 4 stereoisomers are possible. The reaction is not stereoselective so all four stereoisomers are produced; optically inactive.

Problem 8.29 A long polymer chain, such as polyethylene ($-CH_2CH_2-)_n$, can potentially exist in solution as a chiral object. Give two examples of chiral secondary structures that a polyethylene chain could adopt.

The polymer chain could wind itself into a chiral helix with either left-handed or right-handed twist. This would be analogous to right-handed and left-handed spiral staircases.

Problem 8.30 State the number and kind of stereoisomers formed when (R)-3-methyl-1-pentene is treated with the following reagents.

the tetrahedral stereocenter is (R)

$$CH_3-CH_2-\underset{}{\overset{CH_3}{\overset{|}{C}}}H-CH=CH_2$$

(R)-3-methyl-1-pentene

(a) $Hg(OAc)_2$, H_2O followed by $NaBH_4$

These reagents would place a hydroxyl group at the more highly substituted carbon of the alkene, namely the carbon in the two position. This creates a new stereocenter at the two position, thus there will be two products; (2R,3R)-3-methyl-2-pentanol and (2S,3R)-3-methyl-2-pentanol; a pair of diastereomers.

(b) H_2/Pt

These reagents will reduce the double bond to give a single achiral product.

(c) BH_3 followed by H_2O_2 in NaOH

These reagents will place a hydroxyl group at the less substituted carbon of the alkane, namely the carbon in the one position. This does not create a stereocenter so there will be only one product, (R)-3-methyl-1-pentanol.

(d) Br_2 in CCl_4

These reagents will place bromine atoms at the two carbon atoms of the double bond. This will create a stereocenter at the two position. This will give two products (2R,3R)-1,2-dibromo-3-methylpentane and (2S,3R)-1,2-dibromo-3-methylpentane; a pair of diastereomers.

CHAPTER 9: ALCOHOLS AND THIOLS

SUMMARY OF REACTIONS

Starting Material ↓ / Product →	Aldehydes	Alkenes	Alkyl Halides		Carboxylic Acids	Disulfides	Ketones	Ketones/Aldehydes	Metal Alkoxides
Alcohols		**9A** 9.5E*	**9B** 9.5D1	**9C** 9.5D2					**9D** 9.5A
Alcohols (Primary)	**9E** 9.5G				**9F** 9.5G				
Alcohols (Secondary)							**9G** 9.5G		
Thiols						**9H** 9.6B			
Vicinal Diols							**9I** 9.5F	**9J** 9.5H	

*Section of book that describes reaction.

REACTION 9A: ACID-CATALYZED DEHYDRATION (Section 9.5E)

$$-\overset{|}{\underset{|}{C}}-\overset{OH}{\underset{|}{C'}}- \xrightarrow[H_3PO_4]{H_2SO_4 \text{ or}} >C=C'< + H_2O$$

- Alcohols can be heated with H_3PO_4 or H_2SO_4 to generate an alkene. The net result of this process is the removal of H_2O from the alcohol, thus the process is called **dehydration.** ✱
- As with dehydrohalogenation discussed in Section 4.5, when more than one alkene can be formed from dehydration, the more stable alkene is formed in larger amounts. This generalization is known as Zaitsev's rule. In general, the more stable alkene is the one that is more highly substituted, that is, the one with more alkyl groups on the sp^2 carbon atoms.
- The mechanism of dehydration involves protonation of the oxygen atom of the -OH group, followed by loss of water to form a carbocation. Since there is now no good nucleophile to react with the carbocation, a different reaction takes place, namely loss of H^+ to give the alkene.

REACTION 9B: REACTION WITH H-X (Section 9.5D)

$$\text{—}\overset{|}{\underset{|}{C}}\text{—}\overset{OH}{\underset{|}{\overset{|}{C'}}}\text{—} \xrightarrow{\text{H-X}} \text{—}\overset{|}{\underset{|}{C}}\text{—}\overset{X}{\underset{|}{\overset{|}{C'}}}\text{—}$$

- Tertiary alcohols can be converted to alkyl halides by treatment with H-X, where X = Cl, Br, or I. The mechanism involves initial protonation of the oxygen atom of the -OH group followed by departure of the resulting H_2O to create a carbocation that reacts with X^- to give the final product.
  The step involving loss of water is the slowest, and this is called the **rate-determining step**.
  This reaction is considered to follow an **S_N1 mechanism**, because it involves Substitution of X for -OH via a Nucleophilic attack by X^- onto the carbocation and there is only **1** reactant in the rate-determining step (the protonated oxonium species that simply loses water without reacting with any other molecules).
- Primary alcohols react much more slowly than tertiary alcohols with H-X reagents. The Lucas reagent (ZnX_2) is used to speed up the reaction of secondary and primary alcohols. The mechanism of reaction with Lucas reagents involves initial complexation of the Zn^{2+} with the oxygen atom of the -OH group, followed by displacement of the Zn-O-H by X^- to give the alkyl halide product.
  This reaction is classified as an **S_N2** reaction, because the reaction involves Substitution of the X for -OH via a Nucleophilic attack by X^- onto the backside of the C-O bond and the rate determining step involves **2** reagents, the X^- and the Zn complex of the alcohol.

REACTION 9C: REACTION WITH PX_3 OR $SOCl_2$ (Section 9.5D)

$$\text{—}\overset{|}{\underset{|}{C}}\text{—}\overset{OH}{\underset{|}{\overset{|}{C'}}}\text{—} \xrightarrow{PX_3 \text{ or } SOCl_2} \text{—}\overset{|}{\underset{|}{C}}\text{—}\overset{X}{\underset{|}{\overset{|}{C'}}}\text{—}$$

- The -OH group of alcohols can also be replaced with a halide using PCl_3, PBr_3, or $SOCl_2$. In these cases, the products of the reaction are the alkyl halide and phosphorus acid (H_3PO_3) or SO_2 and HCl, respectively.

REACTION 9D: REACTION WITH ACTIVE METALS (Section 9.5A)

$$\text{—}\overset{|}{\underset{|}{C}}\text{—OH} \xrightarrow{M} \text{—}\overset{|}{\underset{|}{C}}\text{—O}^- M^+ + H_2$$

- Alcohols react with active metals such as Li, Na, and K to produce hydrogen gas (H_2) and a metal alkoxide. ✱
- The product alkoxides are slightly more basic than hydroxide, HO^-. This means that alcohols are slightly weaker acids than water *[Section 3.3]*.
- The sodium alkoxides can also be produced using sodium hydride (NaH).

REACTION 9E: OXIDATION OF PRIMARY ALCOHOLS TO ALDEHYDES (Section 9.5G)

$$\text{R-CH}_2\text{-OH} \xrightarrow{\text{PCC}} \text{R-CHO}$$

- A special reagent called **pyridinium chlorochromate** reacts with **primary alcohols** and **stops at the aldehyde**, without reacting further to the carboxylic acid.

REACTION 9F: OXIDATION OF PRIMARY ALCOHOLS TO CARBOXYLIC ACIDS (Section 9.5G)

$$\text{R-CH}_2\text{-OH} \xrightarrow{[O]} \text{R-COOH}$$

- **Primary alcohols are oxidized** all the way **to carboxylic acids** when aqueous solutions of various forms of chromium(VI) such as CrO_3, $K_2Cr_2O_7$, and especially H_2CrO_4. The H_2CrO_4 is prepared from CrO_3 or $K_2Cr_2O_7$ and H_2SO_4. In this case, an aldehyde is initially formed, but is oxidized further to a carboxylic acid before it can be isolated.

REACTION 9G: OXIDATION OF SECONDARY ALCOHOLS TO KETONES (Section 9.5G)

$$\text{C-CH(OH)-C''} \xrightarrow{H_2CrO_4} \text{C-C'(=O)-C''}$$

- **Secondary alcohols** can be **oxidized to ketones**. Oxidizing agents such as PCC can be used, but a reagent called the **Jones reagent** is also common. The Jones reagent is prepared by dissolving $K_2Cr_2O_7$ in sulfuric acid.

REACTION 9H: OXIDATION OF THIOLS TO DISULFIDES (Section 9.6B)

$$\text{-C-SH} \xrightarrow{I_2 \text{ or } 1/2\ O_2} \text{-C-S-S-C-}$$

- Even relatively mild oxidizing agents such as I_2 or O_2 react with thiols to produce disulfides. In fact, thiols are so susceptible to oxidation that they must be protected from contact with air (the O_2) to avoid spontaneous disulfide formation.
- Disulfide bonds are especially important in proteins, where they help stabilize three-dimensional structure.

REACTION 9I: THE PINACOL REARRANGEMENT OF GLYCOLS (Section 9.5F)

$$-\overset{|}{\underset{|}{C}}-\overset{HO}{\underset{|}{C'}}-\overset{OH}{\underset{|}{C''}}-\overset{|}{\underset{|}{C'''}}- \xrightarrow{H_2SO_4} O{=}C'(-)-\overset{-C-}{\underset{|}{C''}}-\overset{|}{\underset{|}{C'''}}- + H_2O$$

- Glycols such as pinacol (2,3-dimethyl-2,3-butanediol) undergo a unique reaction under acid catalyzed dehydration conditions called a **pinacol rearrangement**. In this reaction, there is loss of water like a normal dehydration, yet the intermediate carbocation rearranges via migration of an alkyl group, leading to formation of the ketone or aldehyde product.

REACTION 9J: CLEAVAGE OF GLYCOLS WITH PERIODIC ACID (Section 9.5H)

$$-\overset{HO}{\underset{|}{C}}-\overset{OH}{\underset{|}{C'}}- \xrightarrow{HIO_4} -C{=}O + O{=}C'-$$

- **Periodic acid (H_5IO_6 or $HIO_4 \cdot 2H_2O$)** can cleave **glycols** to give carbonyl compounds, either aldehydes or ketones depending on the starting material. Recall that **Glycols** are compounds with -OH groups on adjacent carbon atoms. The mechanism of the reaction involves a cyclic periodate ester that decomposes to give the two carbonyl compounds and HIO_3. Thus, the net reaction is a two-electron oxidation of the glycol, and a corresponding two-electron reduction of periodic acid.
- The reagent $Pb(OAc)_4$ can also be used for the same cleavage reaction with glycols, and the mechanism is usually analogous including a cyclic intermediate.
- Alkenes can be cleaved to carbonyl compounds directly using O_3. Sometimes the yields for these reactions are low. In those cases, higher yields can be obtained by reacting the alkene with OsO_4/H_2O_2 to give the glycol in high yield, that is then cleaved to the carbonyl compounds with H_5IO_6. *[This is a good illustration of how sometimes the synthetic route with more steps (in this case two steps) actually gives higher overall yields than the route with fewer steps.]*

SUMMARY OF IMPORTANT CONCEPTS

9.0 OVERVIEW

- Alcohols and thiols are important functional groups that are involved in a number of characteristic reactions, and are very common in nature.

9.1 STRUCTURE OF ALCOHOLS AND THIOLS

- An **alcohol** is a molecule that contains an -OH (**hydroxyl**) group attached to an sp^3 hybridized carbon atom. ✳
- A **thiol** is analogous to an alcohol, except that thiols have an -SH (**sulfhydryl**) group attached to an sp^3 hybridized carbon atom. ✳
- Both oxygen and sulfur are in Column 6A of the periodic table so each has **6 valence electrons**. The **sulfur atom** is significantly **larger** than the oxygen atom (0.104 nm vs.

0.066 nm), because the valence electrons in the sulfur reside in the third principle energy level compared with the second principle energy level for the valence electrons of oxygen. For the same reason, the **bonding orbitals of sulfur** have **more p character** than the bonding orbitals of oxygen.

9.2 NOMENCLATURE

- All alcohols and thiols can be named according to IUPAC rules, but numerous common names are still used for the simpler ones so these must be learned as well.
- In the IUPAC system, an alcohol is named by selecting the parent chain as the longest continuous chain that contains the -OH group. The suffix **e** is changed to **ol** and a number is added to designate the position of the -OH group.
  - When there is a choice, the chain is numbered to give the -OH group the lowest number. This includes cyclic and bicyclic alcohols. Examples of IUPAC names for alcohols include 2-pentanol (not 4-pentanol!) and cyclohexanol.
  - A molecule that has more than one -OH group is called a **diol** if it has two, or a **triol** if it has three, etc. These molecules are named by adding the suffix **diol** or **triol**, etc., after the suffix **e** and then providing a number for each carbon atom having an -OH group attached. Again the chain is numbered to give the lowest possible numbers to -OH groups.
    Examples of IUPAC names for diols or triols include 1,2-pentanediol or 2,3,5-octanetriol.
  - Diols that contain -OH groups on adjacent carbon atoms are still referred to with the common name of **glycols**. For example ethylene glycol is really 1,2-ethanediol, the major component of antifreeze.
  - Compounds that contain an -OH group and a C=C bond are named as an alcohol, with the parent chain numbered so that the -OH group is assigned the lowest number. The infix **en** is used in place of **an** and the suffix **ol** is used in place of the suffix **e**. Numbers are assigned as normal.
    For example, 3-penten-2-ol is the correct IUPAC name for CH_3-CH=CH-CHOH-CH_3.
- According to the IUPAC rules, a thiol is named the same as an alcohol, except the suffix **thiol** is placed *after* the suffix **e**. For example 2-butanethiol is a correct IUPAC name. In the common nomenclature rules, the -SH group is referred to as a mercaptan. For example, butyl mercaptan is the common name for 1-butanethiol.
- The alcohol group has higher priority than the thiol group, so molecules that contain both an -OH group and an -SH group are named as alcohols. In this case, the -SH group is referred to as **mercapto**. For example, 4-mercapto-2-butanol is the correct IUPAC name for CH_2SH-CH_2-CHOH-CH_3. *[This concept of functional group priority for nomenclature will become more important as we learn about more functional groups.]*
- Alcohols are characterized according to how many alkyl groups are attached to the carbon atom bearing the -OH group. A **primary alcohol (1°)** is one in which the -OH group is attached to a primary carbon atom, for example 1-propanol. A **secondary alcohol (2°)** is one in which the -OH group is attached to a secondary carbon atom, for example 2-propanol . A **tertiary alcohol (3°)** is one in which the -OH group is attached to a tertiary carbon atom, for example 2-methyl-2-propanol. ✱ *[Characterizing an alcohol as primary, secondary, or tertiary will be the key to understanding some of the reactions of alcohols.]*

OH OH OH

1-propanol (1°) 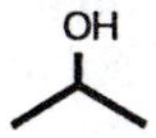 2-propanol (2°) 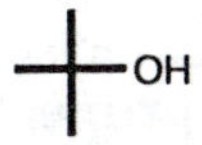 2-methyl-2-propanol (3°)

9.3 PHYSICAL PROPERTIES

- Alcohols and thiols have very different physical properties, because the O-H bond is much more polar than the S-H bond.

- Because oxygen is more electronegative than either carbon or hydrogen, there is increased electron density and thus a partial negative charge on the oxygen atom of an alcohol. Similarly, since hydrogen is so much less electronegative than oxygen, the hydrogen atom has relatively little electron density and thus a partial positive charge. ✱ *[This picture of the alcohol group having partial positive charge on hydrogen and partial negative charge on oxygen is very important and explains most of the reactions of alcohols. Better yet, thinking of the alcohol group in this way will allow you to* <u>*predict*</u> *successfully these reactions.]*

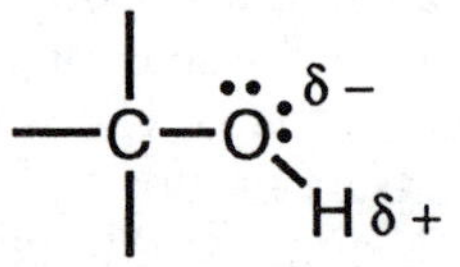

- As the result of the polarization of the O-H bond, there is a permanent dipole moment in the molecule. Two molecules with permanent dipole moments are attracted to each other since the negative end of one dipole moment is attracted to the positive end of the other dipole moment, and *vice versa*. This attraction is called **dipole-dipole interaction**, an example of a weak intermolecular force that holds different molecules together. ✱
- A **hydrogen bond** is a special type of dipole-dipole interaction that occurs when the positive end of one of the dipoles is a hydrogen that is bonded to a very electronegative element (O, N, F). For example, hydrogen bonds that occur between different alcohol molecules involve the hydrogens of one -OH group interacting with the lone pair of electrons on the oxygen atom of another -OH group.

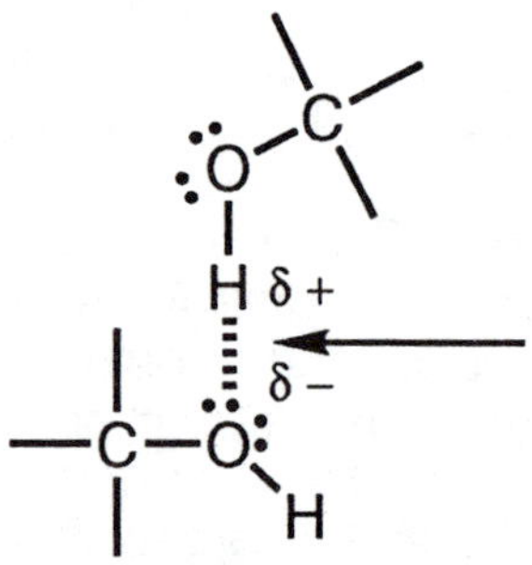

Hydrogen bond

  - A hydrogen bond is worth about 5 kcal/mol in water. This is small compared to the strength of covalent bonds that are about 100 kcal/mol, but when several hydrogen bonds are working together, they are strong enough to hold large linear chains in precise, three dimensional structures such as those found in proteins and nucleic acids.
  - The alcohol group can take part in hydrogen bonding, so molecules with -OH groups can stick to each other. As a result, alcohols have boiling points and melting points that are significantly higher than the corresponding alkanes or alkenes of similar molecular weight.
  - Because water (H_2O) molecules are hydrogen bonded to each other in solution, only molecules that can themselves take part in dipole-dipole interactions like hydrogen bonds can break up the water molecules and thus dissolve. For that reason, the -OH group on small alcohols allows them to dissolve in water.

    The longer the non-polar alkyl chain on an alcohol, the lower the solubility in water and the higher the solubility in non-polar solvents like hexane and benzene.
- The **physical properties of thiols** are completely **different** than those of alcohols, because the **S-H bond** is **not** as **polar** as the O-H bond. This is because sulfur and hydrogen atoms are similar in electronegativity, so these atoms do not have significant partial charges in the -SH group. As a result, thiols show little association by hydrogen bonding so they have lower boiling/melting points and are less soluble in water compared with analogous alcohols. ✱ *[This correlation between hydrogen bonding ability, water solubility and boiling/melting points is a*

good illustration of how understanding molecular structure can lead to accurate predictions of physical properties.]

9.4 PREPARATION OF ALCOHOLS

- Three methods of producing alcohols have already been discussed; 1) Acid-catalyzed hydrolysis of alkenes (Section 5.3C), 2) Oxymercuration of alkenes (Section 5.3E), and 3) Hydroboration of alkenes followed by oxidation (Section 5.4). *[This is a good time to review these reactions if necessary.]*

9.5 REACTIONS OF ALCOHOLS

- Alcohols are valuable starting materials because they can take part in a number of different reactions including formation of alkoxide salts, dehydration, and oxidation. *
- **Alcohols** are **weakly acidic** in aqueous solution, the relative acidities of which are determined by a combination of the degree of solvation of the respective alkoxides and/or inductive effects.
  - The bulkier the alkyl group, the lesser the ability of water to solvate the alkoxide, so the less acidic the alcohol. Thus 2-methyl-2-propanol is less acidic than methanol.
  - The more electron withdrawing groups near the oxygen atom, the greater the acidity of the alcohol. This is because the electron withdrawing groups accommodate some of the negative charge of the alkoxide species, and this charge distribution stabilizes the alkoxide species. The more stable the alkoxide, the more acidic the alcohol. Thus trifluoroethanol is substantially more acidic than ethanol.
- The oxygen atom of an alcohol reacts as a weak Lewis base and can be protonated by extremely strong acids to generate oxonium ions.
- Alcohols react with H-X to form alkyl halides (Reaction 9B, Section 9.5D).
  - In general, tertiary alcohols react faster by the S_N1 mechanism because carbocation formation is involved in the rate determining step, and as explained in Section 5.3, tertiary carbocations are the most stable. Primary alcohols form the least stable carbocations, so they are the least likely to undergo an S_N1 reaction. Secondary alcohols fall somewhere in-between in both carbocation stability and S_N1 reactivity.
  - On the other hand, primary alcohols are most likely to react by the S_N2 mechanism, because primary alcohols have less steric hindrance to interfere with nucleophilic attack. Tertiary alcohols have more alkyl groups that can act to screen out nucleophiles trying to react. Secondary alcohols fall somewhere in-between.
  - Secondary and primary alcohols with branching at the β carbon can rearrange because a carbocation intermediate is involved. Carbocations are notorious for rearrangements, which involve the migration of an alkyl group to create a new carbocation of equal or greater stability.

$$CH_3-C(CH_3)_2-CH_2-\ddot{O}H + H-X \rightleftharpoons CH_3-C(CH_3)_2-CH_2-\overset{+}{O}H_2 \;(+\,X^-) \longrightarrow CH_3-\overset{+}{C}(CH_3)-CH_2CH_3 \;(+\,H_2O) \longrightarrow CH_3-C(X)(CH_3)-CH_2CH_3$$

Rearrangement

- An alcohol is dehydrated to an alkene in anhydrous acid (reaction 9A, section 9.5E). Note how this dehydration reaction is just the exact reverse of acid-catalyzed hydration of an alkene.

$$CH_3CH(\ddot{O}H)CH_3 \overset{H^+}{\rightleftharpoons} H_2C{=}CH-CH_3 + H_2O$$

The reaction conditions determine the position of this equilibrium. Large amounts of water favor formation of the alcohol, removing all traces of water favors formation of the alkene.

The mechanism of forming the alkene from the alcohol with acid catalysis is exactly the reverse of the mechanism of forming an alcohol from the alkene under acid catalysis. This is a good illustration of the important concept of **microscopic reversibility**; which says that for any equilibrium reaction, the transition states and intermediates for the forward reaction are exactly the same as the transition states and intermediates of the backward reaction. In other words, the reactions proceed via the same mechanism in both directions.

- **Primary alcohols** can be **oxidized to aldehydes** (reaction 9.5E) or **carboxylic acids** (reaction 9.5F) and **secondary alcohols** can be **oxidized to ketones** (reaction 9G). In order to be oxidized, an alcohol must have a hydrogen attached to the carbon atom containing the -OH group, thus tertiary alcohols are extremely resistant to oxidization (because they do not have any such hydrogens). ✻

9.6 REACTIONS OF THIOLS

- **Thiols** are **slightly more acidic than** the corresponding **alcohols**. In other words, they are easier to deprotonate in base.
- **Thiols** readily **form insoluble salts with** most **heavy metal ions**, especially mercury and lead. ✻
- **Thiols** are **easily oxidized** to a number of higher oxidation states. As a result, thiols can be considered as mild reducing agents. ✻ *[Recall that when a molecule is oxidized, it gives up electrons. Thus an easily oxidized substance can be considered a reducing agent since it likes to give away electrons.]*
  - Thiols (R-SH) can be oxidized to disulfides (R-S-S-R), sulfinic acids ($R\text{-}SO_2H$), and sulfonic acids ($R\text{-}SO_3H$).

9.7 METHANOL-A KEY INDUSTRIAL ALCOHOL

- Methanol is produced in vast amounts (10.5 billion pounds in 1993), and is used for a variety of applications including conversion to formaldehyde (used in adhesives, etc.), *tert*-butyl methyl ether (an antiknock additive to gasoline) and liquid hydrocarbon fuels.

CHAPTER 9
Solutions to the Problems

Problem 9.1 Give IUPAC names for the following alcohols.

(a) $CH_3-CH_2-CH(CH_2CH_3)-CH_2-OH$ (b) 1-hydroxy-1-methylcyclopentane structure (HO, CH_3) (c) bicyclic structure with OH

2-Ethyl-1-butanol **1-Methylcyclopentanol** **Bicyclo[2.2.1]-7-heptanol**

Problem 9.2 Classify each alcohol as primary, secondary, or tertiary.

(a) $CH_3C(CH_3)_2CH_2OH$ (b) cyclopropyl–OH (c) $CH_2=CH-CH_2-OH$ (d) cyclopentane with CH_3 and OH

Primary **Secondary** **Primary** **Tertiary**

Problem 9.3 Write IUPAC names for the following unsaturated alcohols:

(a) $CH_3-CH=CH-CH_2OH$ (b) cyclohexene ring –OH

2-Buten-1-ol **2-Cyclohexenol**

Problem 9.4 Arrange the following compounds in order of increasing boiling point. Explain the basis for your answer.

$CH_3OCH_2CH_2OCH_3$ $HOCH_2CH_2OH$ $CH_3OCH_2CH_2OH$

In order of increasing boiling point they are:

$CH_3OCH_2CH_2OCH_3$	$CH_3OCH_2CH_2OH$	$HOCH_2CH_2OH$
1,2-dimethoxyethane	**2-methoxyethanol (methyl cellosolve)**	**1,2-ethanediol (ethylene glycol)**
bp 84°C	**bp 125°C**	**bp 198°C**

Hydrogen bonding, or lack of it, is the key. Although 1,2-dimethoxyethane is a polar molecule, there is little intermolecular association between molecules in the

pure liquid because centers of positive and negative charge on adjacent molecules cannot approach each other close enough to develop appreciable dipole-dipole interactions. However, the fact that its boiling point is higher than that of hexane (bp 69°C), a nonpolar hydrocarbon of comparable molecular weight, indicates that there is some dipole-dipole interaction. Both 2-ethoxyethanol and 1,2-ethanediol can associate by hydrogen bonding. Because 1,2-ethanediol has more sites for hydrogen bonding, it has a higher boiling point than 2-methoxyethanol.

Problem 9.5 Arrange the following in order of increasing solubility in water.

$ClCH_2CH_2Cl$ $CH_3CH_2CH_2OH$ $CH_3CH_2CH_2CH_2OH$

In order of increasing solubility in water, they are:

$ClCH_2CH_2Cl$	$CH_3CH_2CH_2CH_2OH$	$CH_3CH_2CH_2OH$
1,2-Dichloroethane	**1-Butanol**	**1-Propanol**
slightly soluble	**8 g/100 g H_2O**	**soluble in all proportions**

Problem 9.6 Write balanced equations for the following reactions:

(a) $$CH_3C(CH_3)_2OH + K \longrightarrow$$

$$2\,CH_3C(CH_3)_2OH + 2K \longrightarrow 2\,CH_3C(CH_3)_2O^-\,K^+ + H_2$$

(b) $$C_6H_{11}\text{-}OH + Na \longrightarrow$$

$$2\,C_6H_{11}\text{-}OH + 2Na \longrightarrow 2\,C_6H_{11}\text{-}O^-\,Na^+ + H_2$$

Problem 9.7 Draw structural formulas for the alkenes formed on acid-catalyzed dehydration of the following alcohols. For each, predict which is the major product.

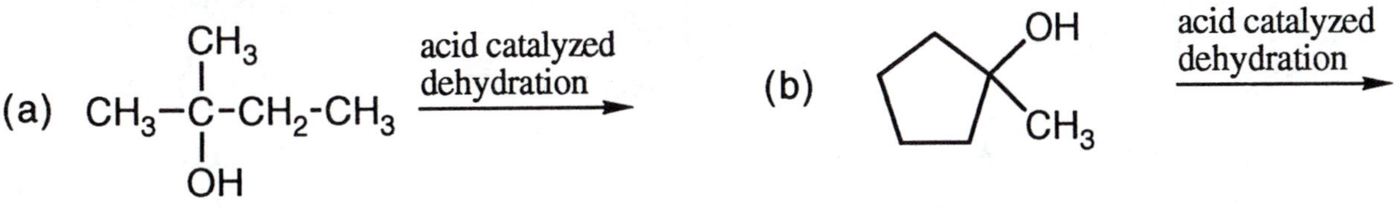

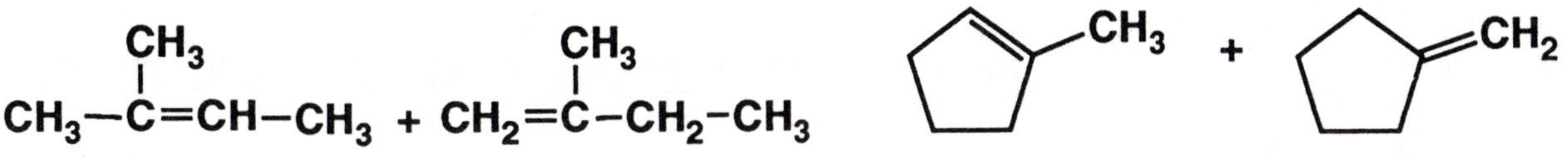

(a) $CH_3-C(CH_3)(OH)-CH_2-CH_3$ —acid catalyzed dehydration→

(b) 1-methylcyclopentanol (OH, CH_3) —acid catalyzed dehydration→

(a) $CH_3-C(CH_3)=CH-CH_3$ + $CH_2=C(CH_3)-CH_2-CH_3$

major product (the first)

(b) 1-methylcyclopentene (CH_3) + methylenecyclopentane ($=CH_2$)

major product (the first)

Problem 9.8 Propose a mechanism to account for the following transformation:

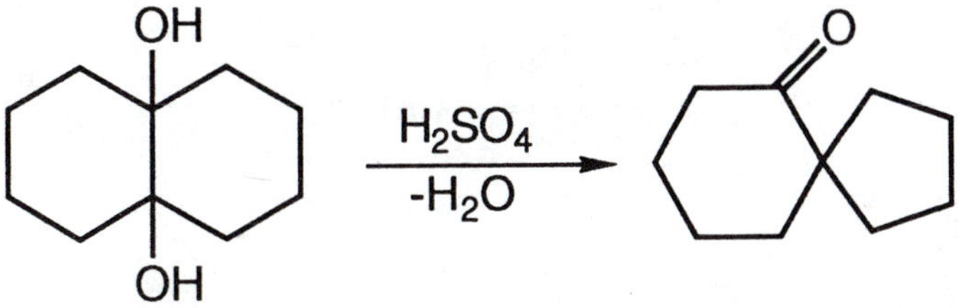

decalin-4a,8a-diol (OH, OH) —H_2SO_4, $-H_2O$→ spiro[4.5]decan-6-one (O)

Protonation of either hydroxyl group followed by departure of water gives a tertiary carbocation. Migration of a pair of electrons from an adjacent bond and loss of a proton to give a ketone gives the observed product.

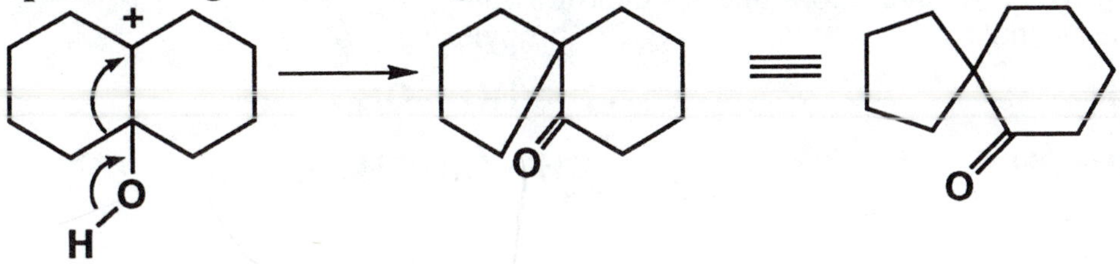

Structure and Nomenclature

Problem 9.9 Name each of the following:

(a) $CH_3CH_2CH_2CH_2CH_2OH$ — **1-pentanol**

(b) $HOCH_2CH_2CH_2CH_2OH$ — **1,4-butanediol**

(c) $CH_2=C(CH_3)CH_2CH_2OH$ — **3-methyl-3-buten-1-ol**

(d) $CH_3CH(OH)CH_2Cl$ — **1-chloro-2-propanol**

(e) $HOCH_2CH=CHCH_2OH$ — **2-buten-1,4-diol**

(f) $HOCH_2CH_2CH(CH_3)CH_3$ — **3-methyl-1-butanol (isopentyl alcohol)**

(g) 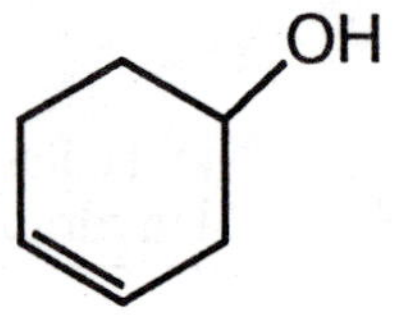 OH

3-cyclohexenol

(h) 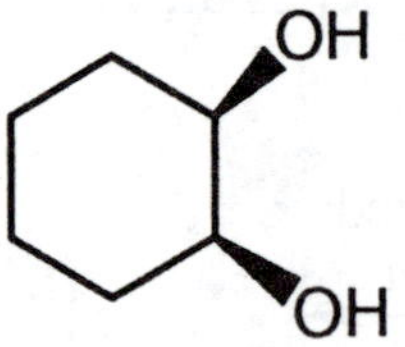 OH OH

***cis*-1,2-cyclohexanediol**

(i) $CH_3CH(OH)CH(OH)CH_3$

2,3-butanediol

(j) 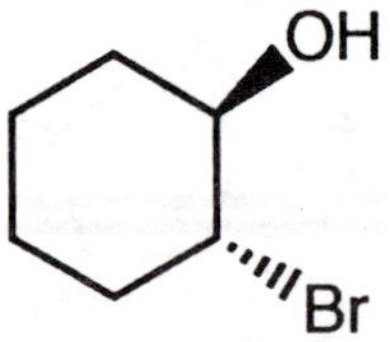 OH Br

***trans*-2-bromocyclohexanol**

(k) $CH_3CH_2CH_2CH_2SH$

1-butanethiol

(l) $CH_3CH{=}CHCH_2SH$

2-buten-1-thiol

(m) $HSCH_2CH_2SH$

1,2-ethanedithiol

(n) $CH_3(CH_2)_4CH(OH)C{\equiv}CH$

1-octyn-3-ol

(o) $CH_3C(CH_3){=}CHCH_2CH_2C(CH_3){=}CHCH_2OH$

3,7-dimethyl-2,6-octadien-1-ol

Problem 9.10 Write structural formulas for the following:

(a) isopropyl alcohol

$CH_3{-}CH(CH_3){-}OH$

(b) propylene glycol

$CH_3{-}CH(OH){-}CH_2(OH)$

(c) 5-methyl-2-hexanol

$CH_3{-}CH(OH){-}CH_2{-}CH_2{-}CH(CH_3){-}CH_3$

(d) 2-methyl-2-propyl-1,3-propanediol

$HO{-}CH_2{-}C(CH_3)(CH_2{-}CH_2{-}CH_3){-}CH_2{-}OH$

(e) 1-chloro-2-hexanol

$Cl{-}CH_2{-}CH(OH){-}CH_2{-}CH_2{-}CH_2{-}CH_3$

(f) *cis*-3-isobutylcyclohexanol

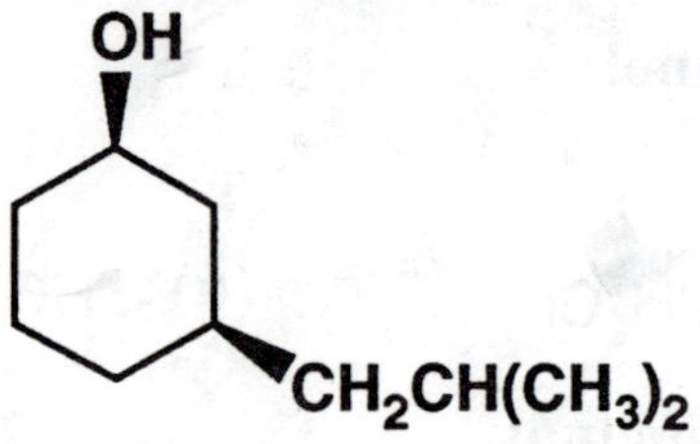

OH

$CH_2CH(CH_3)_2$

(g) 2,2-dimethyl-1-propanol

$$CH_3-C(CH_3)_2-CH_2-OH$$

(h) 2-mercaptoethanol

$$HS-CH_2-CH_2-OH$$

(i) allyl alcohol

$$CH_2=CH-CH_2-OH$$

(j) *trans*-2-vinylcyclohexanol

OH, CH=CH_2 (trans-substituted cyclohexane ring)

(k) (Z)-5-methyl-2-hexen-1-ol

$$HO-CH_2-CH=CH-CH_2-CH(CH_3)-CH_3$$

(cis: H and H on the same side)

(l) 2-propyn-1-ol

$$HC\equiv C-CH_2OH$$

(m) 3-chloro-1,2-propanediol

$$HO-CH_2-CH(OH)-CH_2-Cl$$

(n) *cis*-3-pentene-1-ol

$$HO-CH_2-CH_2-CH=CH-CH_3$$

(H and H on the same side)

(o) bicyclo[2.2.1]-heptan-7-ol

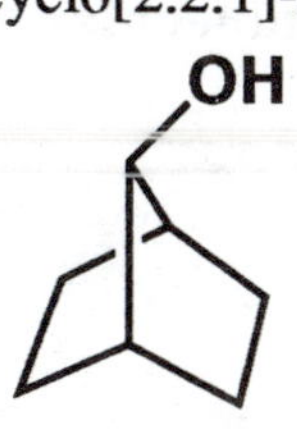

Problem 9.11 Name and draw structural formulas for the eight isomeric alcohols of molecular formula $C_5H_{12}O$. In addition, classify each as primary, secondary, or tertiary.

The eight isomeric alcohols of molecular formulas $C_5H_{12}O$ are grouped by carbon skeleton. First, the three alcohols derived from pentane, then the four derived from 2-methylbutane (isopentane), and then the single alcohol derived from 2,2-dimethylpropane (neopentane). Each is given an IUPAC name. Common names, where appropriate, are given in parentheses.

$$CH_3CH_2CH_2CH_2CH_2OH$$

1-pentanol (pentyl alcohol) (primary)

$$CH_3CH_2CH_2CH(OH)CH_3$$

2-pentanol (secondary)

$$CH_3CH_2CH(OH)CH_2CH_3$$

3-pentanol (secondary)

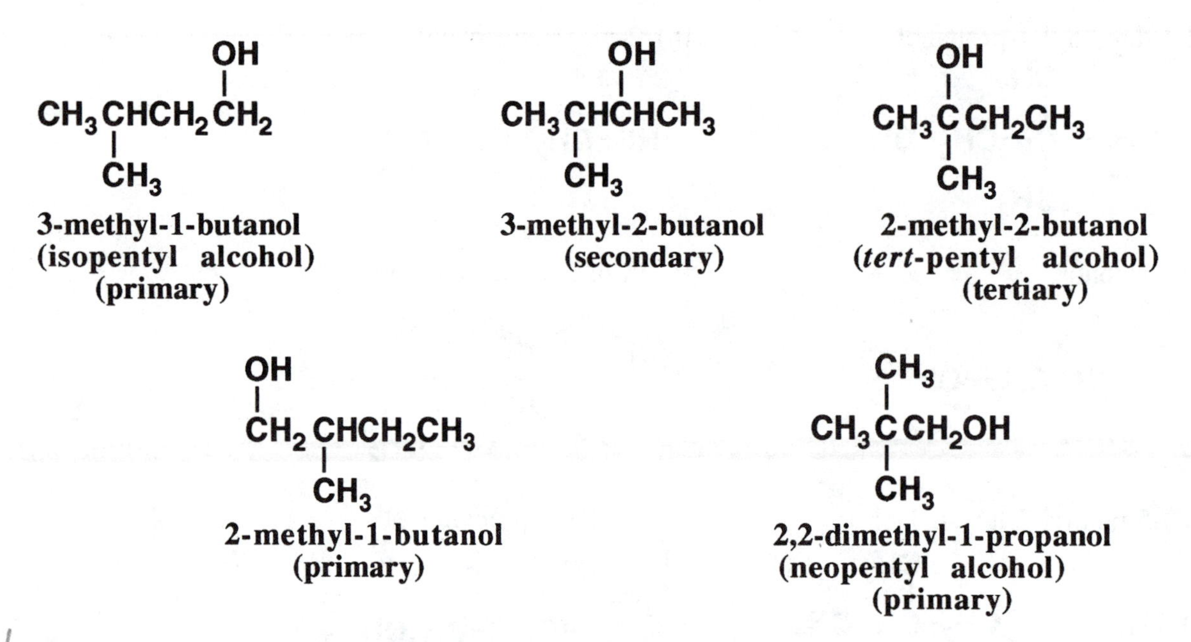

Physical Properties of Alcohols and Thiols

Problem 9.12 Arrange the compounds in each set in order of decreasing boiling point (highest to lowest). Explain the basis for your answers.

(a) $CH_3CH_2CH_3$ $CH_3CH_2CH_2CH_2CH_2CH_2CH_3$ $CH_3CH_2CH_2CH_2CH_3$

C_7H_{16}	C_5H_{12}	C_3H_8
heptane	pentane	propane
(bp 98°C)	(bp 36°C)	(bp -42°C)

These are all unbranched hydrocarbons, so size is important. The largest structure will have the highest boiling point because of increased molecular weights and increased dispersion forces.

(b) N_2H_4 H_2O_2 CH_3CH_3

H_2O_2	NH_2NH_2	CH_3CH_3
hydrogen peroxide	hydrazine	ethane
(bp 151°C)	(bp 113°C)	(bp -89°C)

Ethane clearly has the lowest boiling point, because it has no groups that could take part in hydrogen bonding. Each molecule of both hydrogen peroxide and hydrazine can take part in the same number of hydrogen bonds, but the hydrogen bonds in hydrazine are weaker. This is because nitrogen is not as electronegative as oxygen, so N-H bonds do not have as large of a dipole moment as O-H bonds. (See problem 9.14).

(c) CH_3CHCH_3 (OH on C2) CH_3CHCH_2 (HO, OH) CH_2CHCH_2 (HO HO OH)

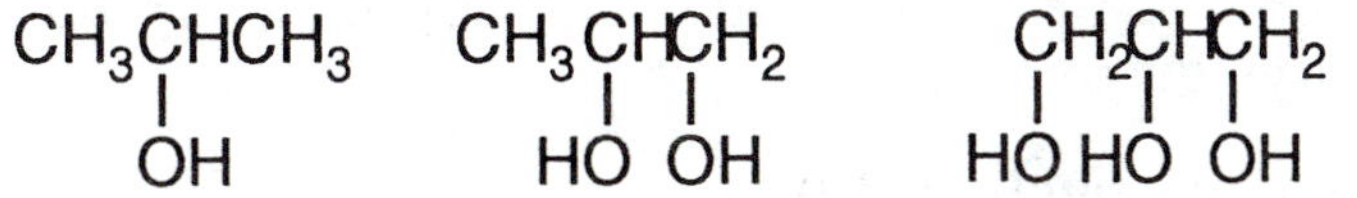

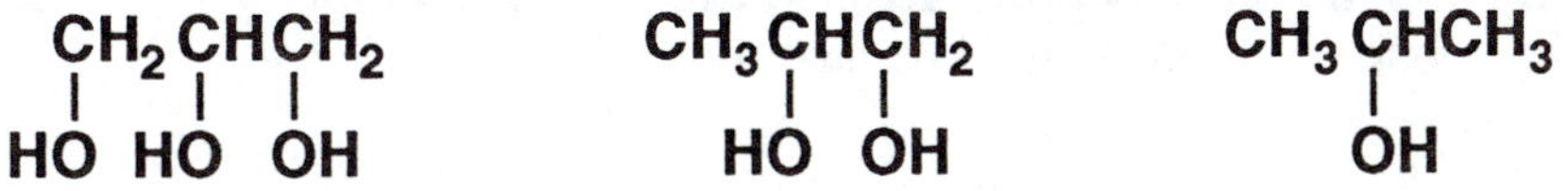

CH_2CHCH_2 (HO HO OH)	CH_3CHCH_2 (HO OH)	CH_3CHCH_3 (OH)
1,2,3-propanetriol (glycerol, glycerin) (bp 290°C)	1,2-propanediol (propylene glycol) (bp 189°C)	2-propanol (isopropyl alcohol) (bp 83°C)

The molecules with the most -OH groups can take part in the most hydrogen bonds, thus increasing intermolecular attractions and raising the boiling points.

Problem 9.13 Arrange the following compounds in each set in order of decreasing boiling point (highest to lowest). Explain the basis for your answers.

(a) $CH_3CH_2CH_2CH_2CH_3$ $CH_3CH_2CH_2CH_3$ $CH_3CH_2CH_2CH_2OH$

$CH_3CH_2CH_2CH_2OH$	$CH_3CH_2CH_2CH_2CH_3$	$CH_3CH_2CH_2CH_3$
bp 117°C	bp 36°C	bp -0.5°C

1-Butanol, $CH_3CH_2CH_2CH_2OH$, has the highest boiling point (117°C), because of the presence of its -OH group that can function as both a hydrogen bond donor and hydrogen bond acceptor. The pentane has a higher boiling point than butane because of its higher molecular weight.

(b) CH_3CH_2SH $CH_3CH_2CH_2OH$ $CH_3CH_2CH_2CH_3$

$CH_3CH_2CH_2OH$	CH_3CH_2SH	$CH_3CH_2CH_2CH_3$
bp 91°C	bp 61°C	bp -0.5°C

1-Propanol, $CH_3CH_2CH_2OH$, has the highest boiling point for the same reason given in part (a). The thiol group of ethanethiol, CH_3CH_2SH, provides for dipole-dipole interactions that raise its boiling point above that of butane.

(c) CH_3CH_2F CH_3OCH_3 CH_3CH_2OH

CH_3CH_2OH	CH_3CH_2F	CH_3OCH_3
bp 78.5°C	bp 37.7°C	bp -25°C

All three are polar molecules. In fact, the dipole moment of fluoroethane (1.94 Debye units) is larger than that of ethanol (1.69 Debye units) indicating that fluoroethane has a higher molecular polarity than ethanol. However, molecular polarity is not the entire story. A second factor is how close the oppositely-charged dipoles can approach each other. In the case of fluoroethane, the

positively-charged carbon of one molecule is shielded from approach by its surrounding hydrogens and methyl group.

Because of the large internuclear distance, there is no appreciable interaction between the positive and negative dipoles

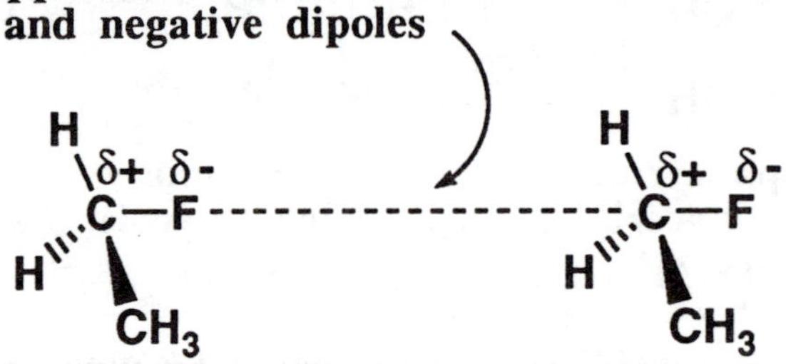

H $\delta+$ $\delta-$ C—F - - - - - - - - - - C—F (each carbon bearing H, H, CH_3; H $\delta+$ $\delta-$)

In the case of ethanol, the positively-charged hydrogen of one molecule has easy access to the negatively-charged oxygen of another molecule of ethanol and dipole-dipole interaction is appreciable in the form of a hydrogen bond.

Problem 9.14 Compounds that contain an N-H group show evidence of association by hydrogen bonding. Do you expect this association to be stronger or weaker than that of compounds containing O-H groups? Explain.

Weaker. The O-H bond is more polar, because the difference in electronegativity between N and H is less than the difference between O and H. Thus, the degree of intermolecular interaction between compounds containing an N-H group is less than that between compounds containing an -OH group.

Bond	Difference in electronegativity	% Partial ionic character
N-H	3.0 - 2.1	30%
O-H	3.5 - 2.1	40%

Problem 9.15 Which of the following compounds can participate in hydrogen bonding with water. For each that can, indicate which site(s) can function as a hydrogen bond acceptor; which can function as a hydrogen bond donor.

The molecules in bold can function as hydrogen bond donors and/or acceptors. Each site is labeled "donor" or "acceptor" as appropriate.

(a) $CH_3CH_2CH_2CH_2CH_2OH$

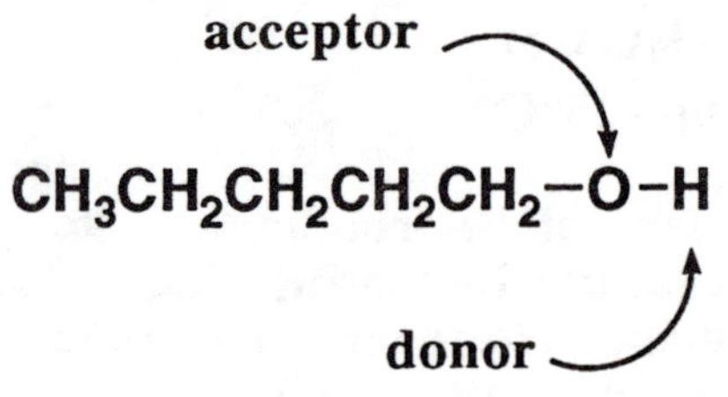

acceptor (O)

$CH_3CH_2CH_2CH_2CH_2-O-H$

donor (H)

(b) $CH_3CH_2CH_2OCH_2CH_2OH$

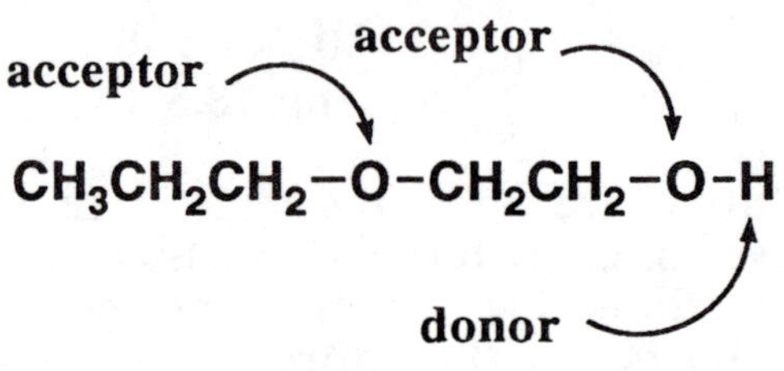

acceptor (ether O) acceptor (hydroxyl O)

$CH_3CH_2CH_2-O-CH_2CH_2-O-H$

donor (H)

(c) $(CH_3CH_2)_2NH$

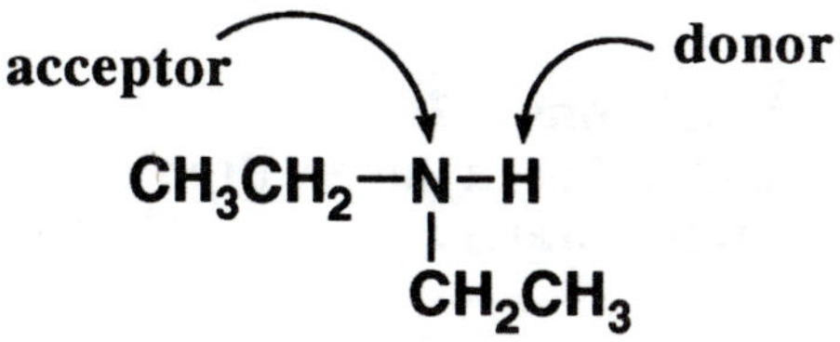

acceptor donor

$CH_3CH_2-N(CH_2CH_3)-H$

(d) $CH_3C(=O)CH_3$

acceptor

$CH_3-C(=O)-CH_3$

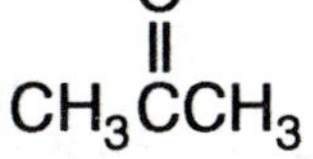

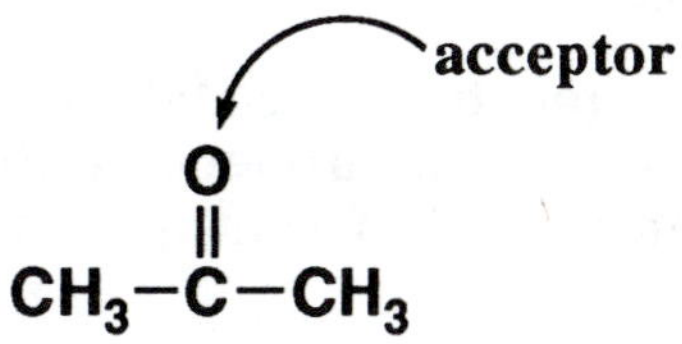

(e) $CH_3C{\equiv}CH$

None

(f) $CH_3(H)C{=}C(H)CH_3$

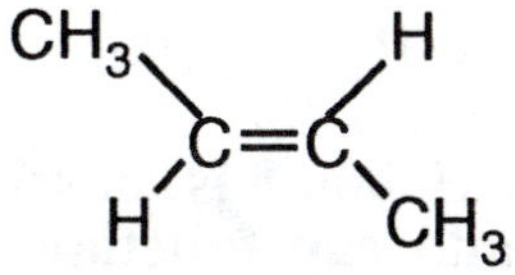

None

(g) $HSCH_2(H)C{=}C(H)CH_3$

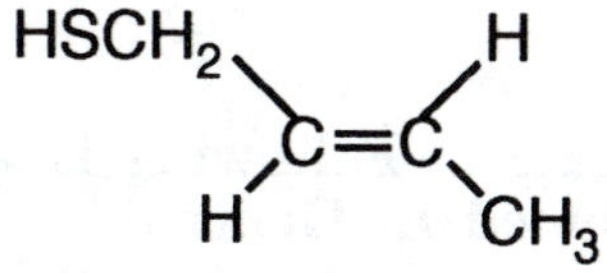

None

(h) $CH_3CH_2C(=O)OH$

acceptor acceptor

$CH_3CH_2C(=O)-O-H$

donor

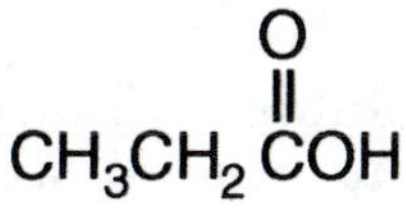

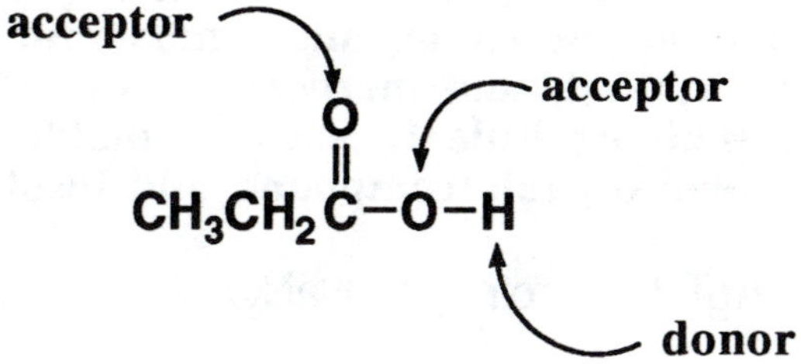

(i) $CH_3CH_2S(=O)CH_3$

acceptor

$CH_3CH_2S(=O)CH_3$

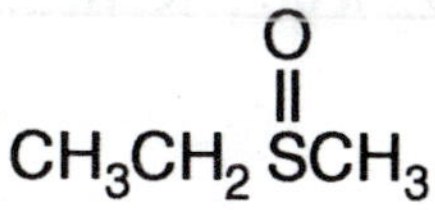

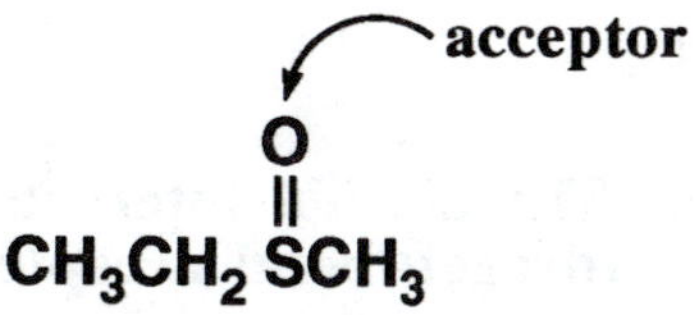

Problem 9.16 From each pair of compounds, select the more soluble in water; explain the basis for your reasoning.

(a) CH_2Cl_2 or CH_3OH

Methanol, CH_3OH, is soluble in all proportions in water. Dichloromethane, CH_2Cl_2, is insoluble. The highly polar -OH group of methanol is capable of participating both as a hydrogen bond donor and hydrogen bond acceptor with water and, therefore, interacts strongly with water by intermolecular association. No such interaction is possible with dichloromethane.

(b) $CH_3\overset{\overset{\displaystyle O}{\|}}{C}CH_3$ or $CH_3\overset{\overset{\displaystyle CH_2}{\|}}{C}CH_3$

Propanone (acetone), CH_3COCH_3, is soluble in all proportions. 2-Methylpropene (isobutylene) is insoluble in water. Acetone has a large dipole moment and can function as a hydrogen bond acceptor from water.

(c) $CH_3CH_2CH_2SH$ or $CH_3CH_2CH_2OH$

1-Propanol, $CH_3CH_2CH_2OH$, is more soluble than 1-propanethiol (propyl mercaptan). Because the difference in electronegativity between S and H is small (2.5 - 2.1), an S-H bond is classified as nonpolar. An -OH bond has 40% partial ionic character and can function as both a hydrogen bond donor and hydrogen bond acceptor with water.

(d) CH_3CH_2Cl or NaCl

NaCl is the more soluble. Chloroethane is insoluble in water. Following is a review of some of the general water solubility rules developed in General Chemistry. For these rules, soluble is defined as dissolving greater than 0.10 mol/L. Slightly soluble is dissolving between 0.01 mol/L and 0.10 mol/L.
1. Sodium, potassium, and ammonium salts of halogens or nitrates are soluble.
2. Silver, lead, and mercury(I) salts of halogens are insoluble.
Thus, applying Rule 1, NaCl is soluble in water. Chloroethane (ethyl chloride) is a nonpolar organic compound and insoluble in water.

(e) AgCl or $NaNO_3$

Applying Rule 1, $NaNO_3$ is soluble in water. Applying Rule 2, AgCl is insoluble in water.

(f) $CH_3CH_2\overset{\overset{\displaystyle OH}{|}}{C}HCH_2CH_3$ or $CH_3CH_2\overset{\overset{\displaystyle O}{\|}}{C}CH_2CH_3$

3-Pentanol is more soluble in water than 3-pentanone. The alcohol interacts more strongly with water because it can function both as a hydrogen bond acceptor and hydrogen bond donor.

Problem 9.17 Arrange the compounds in each set in order of decreasing solubility in water; explain the basis for your answers.
(a) ethanol; butane; diethyl ether

CH_3CH_2OH	$CH_3CH_2OCH_2CH_3$	$CH_3CH_2CH_2CH_3$
soluble in all proportions	**8 g/100 mL water**	**insoluble in water**

In general, the greater a molecule can take part in hydrogen bonding with water, the greater the molecule will be able to interact with the water molecules and dissolve. Only the ethanol can act both as a donor and acceptor of hydrogen bonds with water. Diethyl ether can act as an acceptor of hydrogen bonds. Butane can act as neither a donor nor an acceptor of hydrogen bonds.

(b) 1-hexanol; 1,2-hexanediol; hexane

$CH_2CHCH_2CH_2CH_2CH_3$ (with OH on the first and second carbons) $\quad$ $CH_3CHCH_2CH_2CH_2CH_3$ (with OH on the second carbon) $\quad$ $CH_3CH_2CH_2CH_2CH_2CH_3$

OH OH $\quad\quad$ OH

The 1,2-hexanediol molecules can take part in more hydrogen bonds with water than the 1-hexanol, since the diol has two -OH groups. The hexane has no polar bonds and thus cannot take part in any dipole-dipole interactions with water molecules.

Problem 9.18 Diethyl ether and 1-butanol have about the same solubility in water; they differ in boiling points by approximately 82°C.

$CH_3CH_2CH_2CH_2OH$	$CH_3CH_2OCH_2CH_3$
1-butanol	diethyl ether
8 g/100 mL water	8 g/100 mL water
bp 117°C	bp 35.6°C

How do you account for the similar solubilities of these two compounds but their widely differing boiling points?

(a) While there is both bond and molecular polarity in diethyl ether, steric hindrance prevents close approach of the positively charged carbons and the negatively charged oxygen of adjacent molecules and, therefore, resulting dipole-dipole interactions between molecules of diethyl ether in the pure liquid are small. 1-Butanol in the liquid state can associate by hydrogen bonding and, therefore, the energy required to change 1-butanol from the liquid state to the gaseous state is greater than that required for a comparable change of state for diethyl ether. Hence, 1-butanol has the higher boiling point.

(b) There are two factors to take into considerationin determining water solubility; (1) the degree of dipole-dipole interaction in the pure liquid (no water) and (2) the degree of dipole-dipole interaction of each compound with water. A high degree of dipole-dipole interaction in the pure liquid favors a lower solubility in water; conversely, a high degree of dipole-dipole interaction with water favors greater solubility in water. In the case of 1-butanol and diethyl ether, these two factors are closely balanced. 1-Butanol has a high degree of dipole-dipole interaction with molecules of water in aqueous solution. Diethyl ether has a lesser degree of dipole-dipole interaction in the pure liquid and also a lesser degree of dipole-dipole interaction with molecules of water in aqueous solution.

Problem 9.19 Propanoic acid and methyl acetate are constitutional isomers and both liquids at room temperature. One of these compounds has a boiling point of 141°C, the other of 57°C.

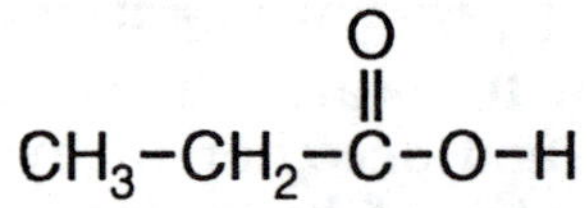

$CH_3-CH_2-\overset{O}{\overset{\|}{C}}-O-H$
propanoic acid

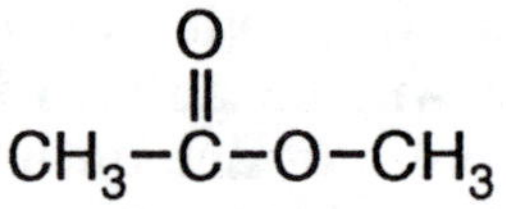

$CH_3-\overset{O}{\overset{\|}{C}}-O-CH_3$
methyl acetate

Which compound has which boiling point? Explain your reasoning.

Propanoic acid has the higher boiling point. Because the carboxyl group of propanic acid can function as both a hydrogen bond donor (through the O-H group) and a hydrogen bond acceptor (through the C=O and C-O groups) there is a high degree of intermolecular association between its molecules of propanoic acid in the liquid state. Methyl acetate, in the pure liquid state, cannot associate by hydrogen bonding.

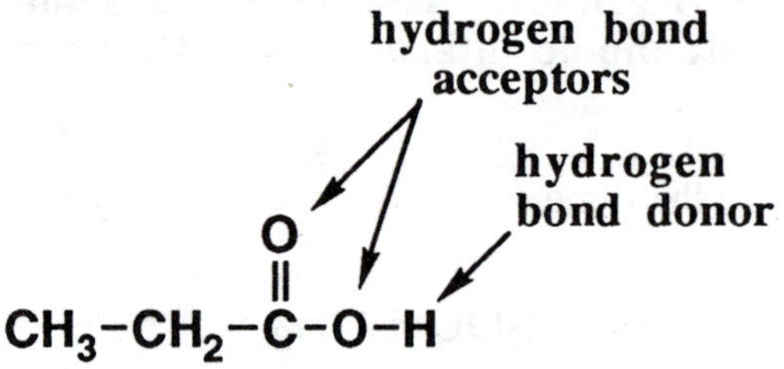

hydrogen bond acceptors
hydrogen bond donor

$CH_3-CH_2-\overset{O}{\overset{\|}{C}}-O-H$
propanoic acid
bp 141°C
soluble in water in all proportions

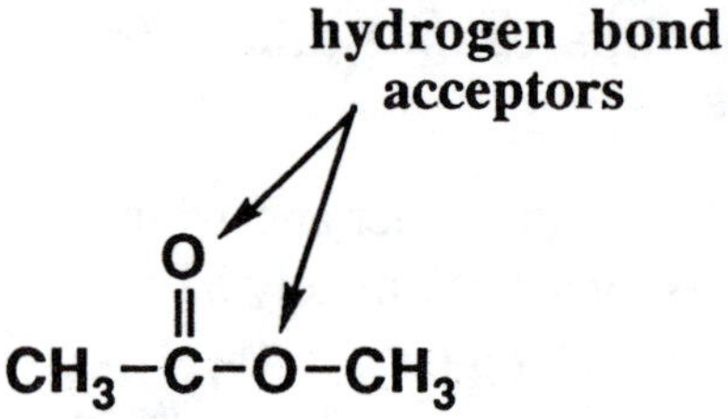

hydrogen bond acceptors

$CH_3-\overset{O}{\overset{\|}{C}}-O-CH_3$
methyl acetate
bp 57°C
33 g/100 mL water

Both molecules can interact with water as hydrogen bond acceptors from O-H of water. Only propanoic acid, however, can interact with water as a hydrogen bond donor as well. Therefore, the solubility of propanoic acid in water is greater than that of methyl acetate.

Problem 9.20 Following are structural formulas along with boiling points and solubilities in water of 1-butanol and 2-methyl-2-propanol, a pair of constitutional isomers.

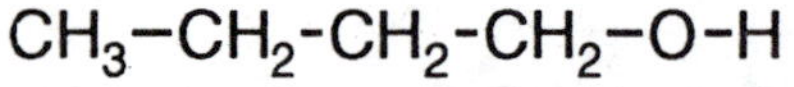

$CH_3-CH_2-CH_2-CH_2-O-H$
1-butanol
bp 117°C
8 g/100 mL H_2O

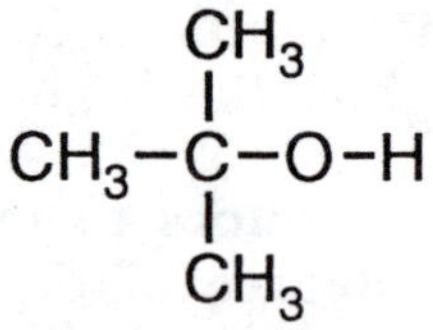

$CH_3-\overset{CH_3}{\overset{|}{\underset{CH_3}{\underset{|}{C}}}}-O-H$
2-methyl-2-propanol
bp 82°C
soluble in all proportions

(a) How might you account for the observation that the boiling point of 2-methyl-2-propanol is lower than that of 1-butanol?

While both have polar -OH groups, that in 2-methyl-2-propanol (*tert*-butyl alcohol) is more hindered and, therefore, not as accessible for hydrogen bonding in the pure liquid. 1-Butanol, with the less sterically hindered -OH group has the greater degree of intermolecular association by hydrogen bonding and, therefore, the higher boiling point.

(b) How might you account for the observation that the solubility of 2-methyl-2-butanol in water is greater than that of 1-butanol?

The fact that 2-methyl-2-propanol (*tert*-butyl alcohol) is more soluble in water than 1-butanol is another demonstration that the shape of the hydrocarbon portion of a molecule is important in determining its physical properties. A *tert*-butyl group is more compact than a butyl group and causes less disruption of hydrogen bonding in water when *tert*-butyl alcohol.

Synthesis of Alcohols

Problem 9.21 Draw the alcohol formed when 1,1-dimethyl-4-methylenecyclohexane is treated with each set of reagents.

(a) CH_3, CH_3 ... $=CH_2$ — (1) BH_3; (2) H_2O_2, NaOH → CH_3, CH_3 ... CH_2OH

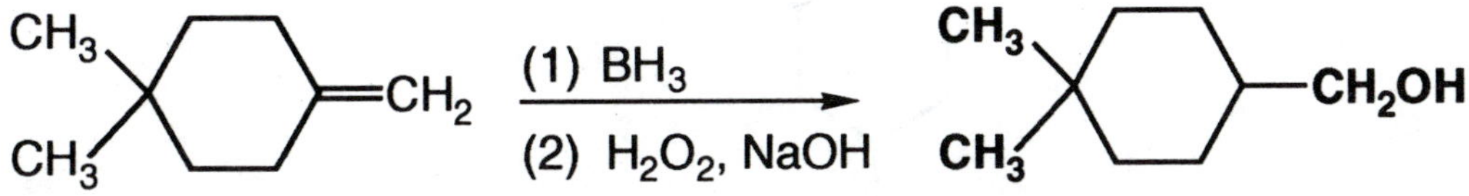

(b) CH_3, CH_3 ... $=CH_2$ — (1) $Hg(OAc)_2$, H_2O; (2) $NaBH_4$ → CH_3, CH_3 ... CH_3, OH

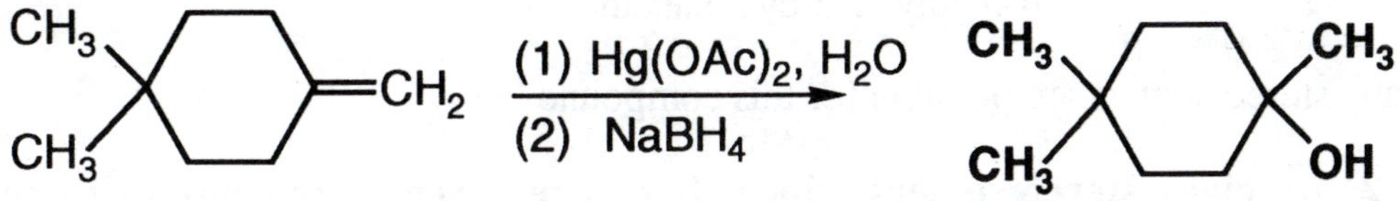

(c) CH_3, CH_3 ... $=CH_2$ — H_2O, H_2SO_4 → CH_3, CH_3 ... CH_3, OH

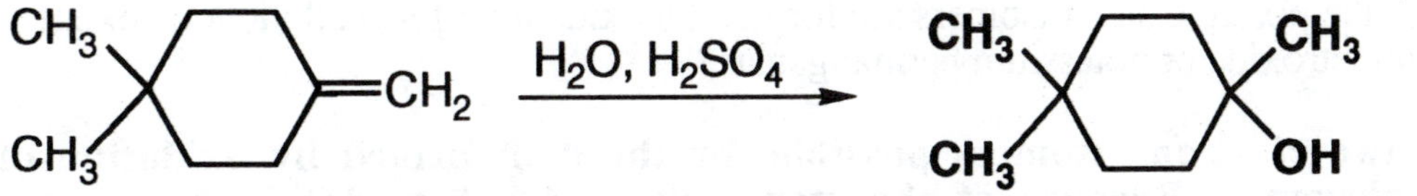

Problem 9.22 Oxymercuration of an alkene followed by reduction with sodium borohydride is both regioselective and stereoselective. Oxymercuration of the following bicycloalkene followed by reduction gives a single alcohol in better than 95% yield. Propose a structural formula for the alcohol formed and account for the stereoselectivity of this reaction sequence.

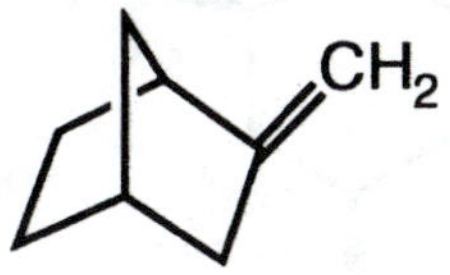

CH_2

Following is the structural formula of the bicyclic alcohol formed by oxymercuration/reduction.

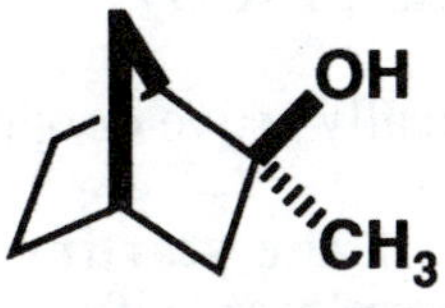

OH

CH_3

First, given the known regioselectivity of oxymercuration, predict that -OH adds to the more substituted carbon of the double bond. To account for the stereoselectivity, predict that a molecule of water will approach the bridged mercurinium ion intermediate from the same side as the one-carbon bridge (from the top of the molecule as it is drawn in the problem) rather than on the same side as the two-carbon bridge (from the bottom side as the molecule is drawn in the problem). Approach of a molecule of water on the same side as the one-carbon bridge is less hindered.

Problem 9.23 Following is the structural formula of 4-methyl-1,2-cyclohexanediol.

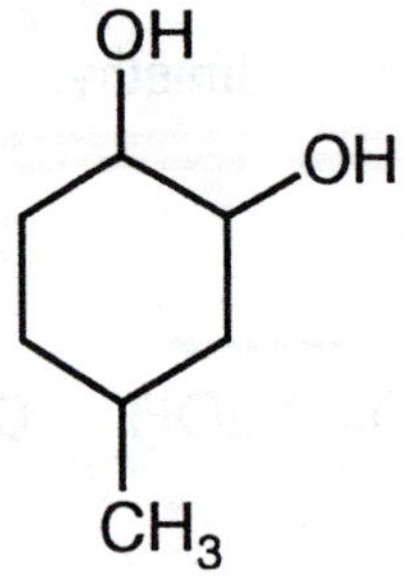

OH

OH

CH_3

4-methyl-1,2-cyclohexanediol

(a) How many stereoisomers are possible for this compound?

There are 2^3 or eight stereoisomers, since there are three stereocenters in this molecule and no planes of symmetry.

(b) Which of the possible stereoisomers are formed by oxidation of 4-methylcyclohexene with either osmium tetroxide or potassium permanganate?

There are two *cis-trans* isomers possible for the diol formed by oxidation of 4-methylcyclohexene. Because of the stereoselectivity of oxidation by osmium tetroxide and potassium permanganate under these conditions, the two -OH groups are *cis* to each other, and both are either *cis* or *trans* to the methyl group.

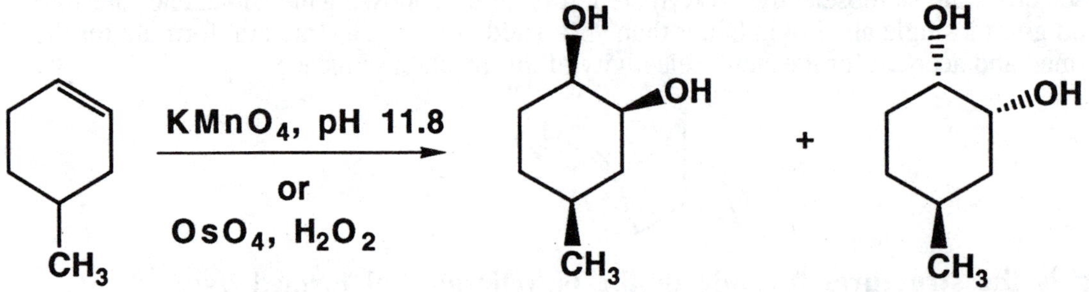

CH_3 — $KMnO_4$, pH 11.8 or OsO_4, H_2O_2 → (OH, OH, CH_3) + (OH, OH, CH_3)

both -OH groups cis to -CH_3 | **both -OH groups trans to -CH_3**

(c) Are the products formed in part (b) optically active or optically inactive?

The products will be optically active if the starting material is either pure 4R or 4S enantiomer, but the problem doesn't specify configuration so assume 4R,4S (racemic) and therefore inactive product.

Acidity of Alcohols

Problem 9.24 Complete the following acid-base reactions. In addition, show all valence electrons on the interacting atoms and show by the use of curved arrows the flow of electrons in each reaction.

(a) $CH_3CH_2{-}O{-}H + H{-}O^{+}(H){-}H \longrightarrow$

$CH_3CH_2{-}\ddot{O}{-}H + H{-}\ddot{O}^{+}(H){-}H \longrightarrow CH_3CH_2{-}\ddot{O}^{+}(H){-}H + :\ddot{O}(H){-}H$

B-L acid

(b) $CH_3CH_2{-}O{-}CH_2CH_3 + H{-}O{-}S(=O)_2{-}O{-}H \longrightarrow$

$CH_3CH_2{-}\ddot{O}{-}CH_2CH_3 + H{-}\ddot{O}{-}S(=O)_2{-}O{-}H \longrightarrow CH_3CH_2{-}\ddot{O}^{+}(H){-}CH_2CH_3 + {}^{-}:\ddot{O}{-}S(=O)_2{-}O{-}H$

B-L acid

(c) $CH_3CH_2CH_2CH_2CH_2{-}O{-}H + H{-}I \longrightarrow$

$CH_3CH_2CH_2CH_2CH_2{-}\ddot{O}{-}H + H{-}\ddot{I}: \longrightarrow CH_3CH_2CH_2CH_2CH_2{-}\ddot{O}^{+}(H){-}H + :\ddot{I}:^{-}$

B-L acid

(d) $CH_3CH_2CH_2C(=O){-}O{-}H + H{-}O{-}S(=O)_2{-}O{-}H \longrightarrow$

$CH_3CH_2CH_2C(=\ddot{O}:){-}O{-}H + H{-}\ddot{O}{-}S(=O)_2{-}O{-}H \longrightarrow CH_3CH_2CH_2C(=:O^{+}{-}H){-}O{-}H + {}^{-}:\ddot{O}{-}S(=O)_2{-}O{-}H$

B-L acid

(e) Cyclohexyl–O-H + BF_3 ⟶

Cyclohexyl–$\ddot{O}$-H + BF_3 ⟶ Cyclohexyl–$\overset{+}{O}(H)$–$\overset{-}{B}F_3$

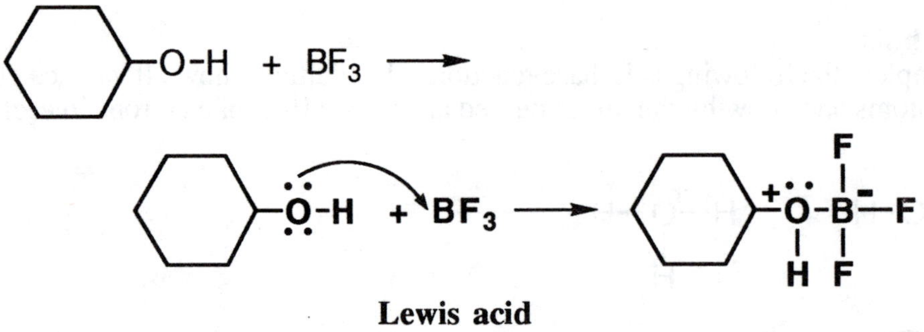

Lewis acid

(f) $CH_3CH{=}CH\overset{+}{C}HCH_3$ + H–O–H ⟶

$CH_3CH{=}CH\overset{+}{C}HCH_3$ + H–$\ddot{O}$–H ⟶ $CH_3CH{=}CHCH(\overset{+}{O}H_2)CH_3$

Lewis acid

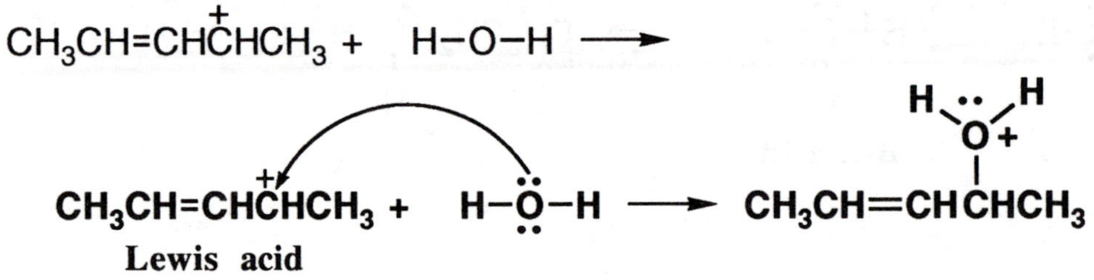

(g) Cyclohexyl–O-H + H-O-P(=O)(OH)-O-H ⟶

Cyclohexyl–$\ddot{O}$-H + H-$\ddot{O}$-P(=O)(OH)-O-H ⟶ Cyclohexyl–$\overset{+}{O}H_2$ + $^-$:$\ddot{O}$-P(=O)(OH)-O-H

B-L acid

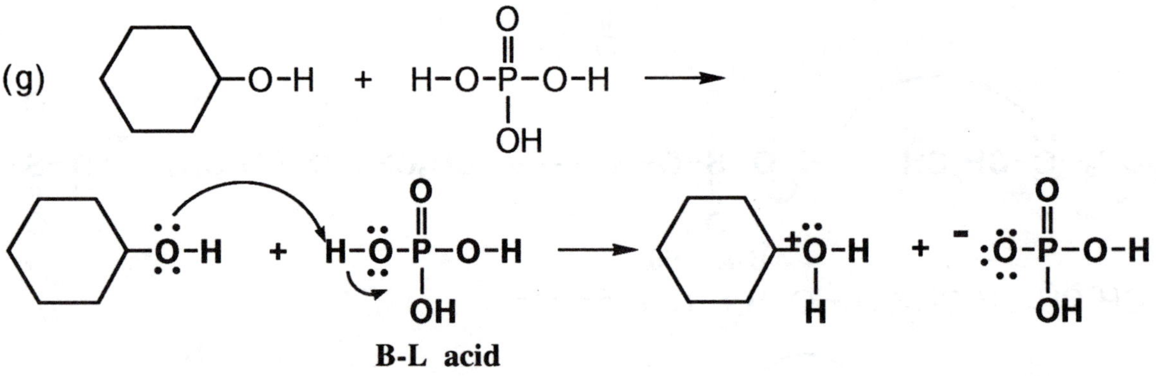

Problem 9.25 Select the stronger acid from each pair and explain your reasoning. For each stronger acid, write a structural formula for its conjugate base.

(a) H_2O or H_2CO_3 (b) CH_3OH or CH_3CO_2H

(c) CH_3CH_2OH or $CH_3C{\equiv}CH$ (d) CH_3CH_2OH or CH_3CH_2SH

Under each acid is given its pK_a. The stronger acid has the smaller value of pK_a.

	weaker acid	stronger acid	conjugate base of stronger acid
(a)	H_2O pK_a 15.7	H_2CO_3 pK_a 6.36	HCO_3^-
(b)	CH_3OH pK_a 15.5	CH_3CO_2H pK_a 4.76	$CH_3CO_2^-$

(c) $CH_3C{\equiv}CH$ pKa 25 | CH_3CH_2OH pKa 15.9 | $CH_3CH_2O^-$

(d) CH_3CH_2OH pKa 15.9 | CH_3CH_2SH pKa 8.5 | $CH_3CH_2S^-$

Problem 9.26 From each pair, select the stronger base. For each stronger base, write the structural formula of its conjugate acid.

(a) OH^- or CH_3O^- (each in H_2O)

$$HO^- \longrightarrow HOH$$

stronger base → weaker acid

(b) $CH_3CH_2O^-$ or $CH_3C{\equiv}C^-$

$$CH_3C{\equiv}C^- \longrightarrow CH_3C{\equiv}CH$$

stronger base → weaker acid

(c) $CH_3CH_2S^-$ or $CH_3CH_2O^-$

$$CH_3CH_2O^- \longrightarrow CH_3CH_2OH$$

stronger base → weaker acid

(d) $CH_3CH_2O^-$ or NH_2^-

$$NH_2^- \longrightarrow NH_3$$

stronger base → weaker acid

Problem 9.27 For the following equilibria, label the stronger acid, stronger base, weaker acid, and weaker base. Also predict the position of each equilibria.

(a) $CH_3CH_2O^- + CH_3C{\equiv}CH \rightleftharpoons CH_3CH_2OH + CH_3C{\equiv}C^-$

$$CH_3CH_2O^- + CH_3C{\equiv}CH \longleftarrow CH_3CH_2OH + CH_3C{\equiv}C^-$$

$CH_3CH_2O^-$	$CH_3C{\equiv}CH$	CH_3CH_2OH	$CH_3C{\equiv}C^-$
	pKa 25	pKa 15.9	
weaker base	weaker acid	stronger acid	stronger base

(b) $CH_3CH_2O^- + HCl \rightleftharpoons CH_3CH_2OH + Cl^-$

$$CH_3CH_2O^- + HCl \longrightarrow CH_3CH_2OH + Cl^-$$

$CH_3CH_2O^-$	HCl	CH_3CH_2OH	Cl^-
	pKa -7	pKa 15.5	
stronger base	stronger acid	weaker acid	weaker base

(c) $CH_3C(=O)\text{-}OH + CH_3CH_2O^- \rightleftharpoons CH_3C(=O)\text{-}O^- + CH_3CH_2OH$

$$CH_3C(=O)\text{-}OH + CH_3CH_2O^- \longrightarrow CH_3C(=O)\text{-}O^- + CH_3CH_2OH$$

$CH_3C(=O)\text{-}OH$	$CH_3CH_2O^-$	$CH_3C(=O)\text{-}O^-$	CH_3CH_2OH
pKa 4.76			pKa 15.9
stronger acid	stronger base	weaker base	weaker acid

Reactions of Alcohols

Problem 9.28 Write equations for the reaction of 1-butanol, a primary alcohol, with the following reagents. While you should show formulas for all products, you need not balance equations. Where you predict no reaction, write NR.

(a) Na metal

$$CH_3CH_2CH_2CH_2OH + Na \longrightarrow CH_3CH_2CH_2CH_2O^- Na^+ + H_2$$

(b) PCl_3

$$CH_3CH_2CH_2CH_2OH + PCl_3 \longrightarrow CH_3CH_2CH_2CH_2Cl + P(OH)_3$$

(c) HCl, cold

No reaction

(d) HCl, $ZnCl_2$, heat

$$CH_3CH_2CH_2CH_2OH + HCl \xrightarrow[\text{heat}]{ZnCl_2} CH_3CH_2CH_2CH_2Cl + H_2O$$

(e) PBr_3

$$CH_3CH_2CH_2CH_2OH + PBr_3 \longrightarrow CH_3CH_2CH_2CH_2Br + P(OH)_3$$

(f) $K_2Cr_2O_7$, H_2SO_4, heat

$$CH_3CH_2CH_2CH_2OH + K_2Cr_2O_7 \xrightarrow[\text{heat}]{H_2SO_4} CH_3CH_2CH_2\overset{\overset{\large O}{\|}}{C}\text{-}OH + Cr^{3+}$$

(g) HIO_4

No reaction

(h) $SOCl_2$

$$CH_3CH_2CH_2CH_2OH + SOCl_2 \xrightarrow{\text{pyridine}} CH_3CH_2CH_2CH_2\text{-}Cl + SO_2 + HCl$$

(i) pyridinium chlorochromate (PCC)

$$CH_3CH_2CH_2CH_2OH + PCC \longrightarrow CH_3CH_2CH_2\overset{\overset{\large O}{\|}}{C}\text{-}H + Cr^{3+}$$

Problem 9.29 Write equations for the reaction of 2-butanol, a secondary alcohol, with the following reagents. While you should show formulas for all products, you need not balance equations. Where you predict no reaction, write NR.

(a) Na metal

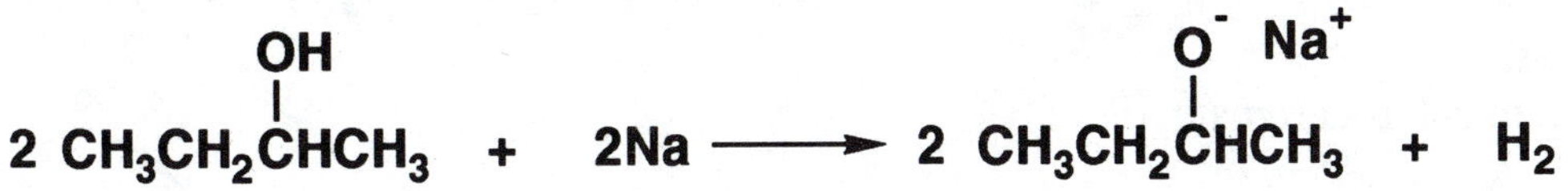

$$2\ CH_3CH_2CH(OH)CH_3 + 2Na \longrightarrow 2\ CH_3CH_2CH(O^-Na^+)CH_3 + H_2$$

(b) H_2SO_4, heat

$$CH_3CH_2CH(OH)CH_3 \xrightarrow[\text{heat}]{H_2SO_4} CH_3CH{=}CHCH_3 + H_2O$$

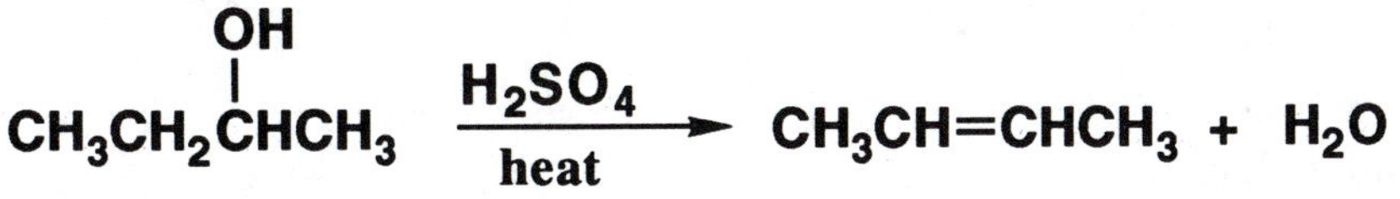

(c) HBr, cold

No reaction

(d) HBr, $ZnBr_2$, heat

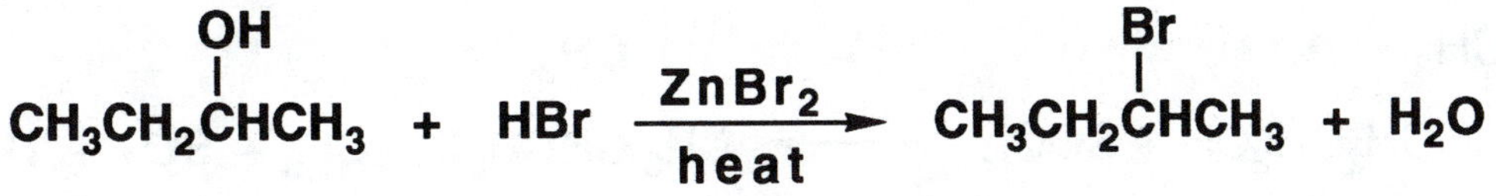

$$CH_3CH_2CH(OH)CH_3 + HBr \xrightarrow[\text{heat}]{ZnBr_2} CH_3CH_2CH(Br)CH_3 + H_2O$$

(e) PBr_3

$$CH_3CH_2CH(OH)CH_3 + PBr_3 \longrightarrow CH_3CH_2CH(Br)CH_3 + P(OH)_3$$

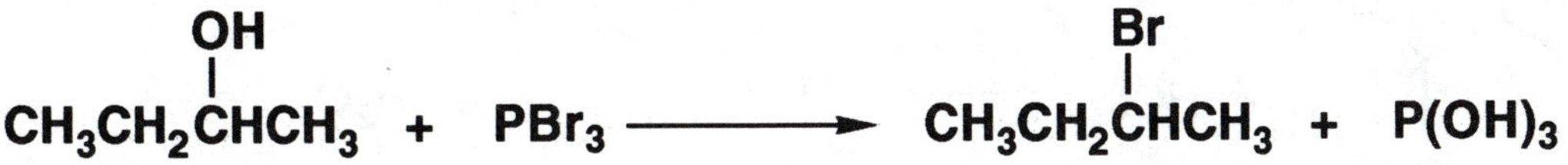

(f) $K_2Cr_2O_7$, H_2SO_4, heat

$$CH_3CH_2CH(OH)CH_3 + K_2Cr_2O_7 \xrightarrow[\text{heat}]{H_2SO_4} CH_3CH_2C(=O)CH_3 + Cr^{3+}$$

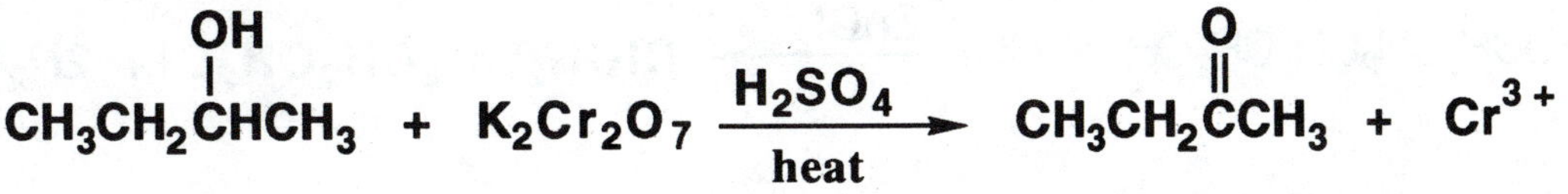

(g) HIO_4

No reaction

(h) $SOCl_2$

$$CH_3CH_2CH(OH)CH_3 + SOCl_2 \xrightarrow{\text{pyridine}} CH_3CH_2CH(Cl)CH_3 + SO_2 + HCl$$

(i) pyridinium chlorochromate (PCC)

$$CH_3CH_2CH(OH)CH_3 + PCC \longrightarrow CH_3CH_2C(=O)CH_3 + Cr^{3+}$$

Problem 9.30 Complete the following equations. Show structural formulas for the major products, but do not balance.

(a) $CH_3CH_2CH_2OH + H_2CrO_4 \longrightarrow CH_3CH_2C(=O)OH + Cr^{3+}$

(b) $CH_3CH(CH_3)CH_2CH_2OH + SOCl_2 \longrightarrow CH_3CH(CH_3)CH_2CH_2Cl + SO_2 + HCl$

(c) 1-methylcyclohexanol (ring with CH_3 and OH) + HCl ⟶ 1-chloro-1-methylcyclohexane (ring with CH_3 and Cl) + H_2O

(d) $HOCH_2CH_2CH_2CH_2OH + HCl \xrightarrow{ZnCl_2} ClCH_2CH_2CH_2CH_2Cl + 2H_2O$

(e) cyclooctanol (ring with OH) + H_2CrO_4 ⟶ cyclooctanone (ring with O) + Cr^{3+}

(f) bicyclic diol (OH, OH) + $Pb(OAc)_4$ ⟶ cyclodecane-1,6-dione (O, O) + $Pb(OAc)_2$ + $2CH_3CO_2H$

(g) cyclohexene $\xrightarrow[\text{2) } H_5IO_6]{\text{1) } OsO_4/H_2O_2}$

cyclohexene $\xrightarrow{OsO_4/H_2O_2}$ cis-1,2-cyclohexanediol (OH, OH) $\xrightarrow{H_5IO_6}$ hexanedial (CHO, CHO)

(h) 4,4-dimethylcyclohexanol (CH_3, CH_3, OH) + $SOCl_2$ $\longrightarrow$ 4-chloro-1,1-dimethylcyclohexane (CH_3, CH_3, Cl) + SO_2 + HCl

Problem 9.31 Show how you might bring about the conversion of 4-methylcyclohexanol to the following.

(a) O^- K^+ on cyclohexane ring, CH_3 at position 4

(b) 4-methylcyclohexene, CH_3

(c) Cl on cyclohexane ring, CH_3 at position 4

(d) methylcyclohexane, CH_3

(e) 4-methylcyclohexanone (O, CH_3)

(f) $HCCH_2CH_2CHCH_2CH$ (with O= on both terminal carbons; CH_3 on CH)

(g) $HOCCH_2CH_2CHCH_2COH$ (with O= on both carbonyl carbons; CH_3 on CH)

(a) Reaction of 4-methylcyclohexanol with potassium metal to form the potassium alkoxide and hydrogen gas.
(b) Acid-catalyzed dehydration of 4-methylcyclohexanol by heating in the presence of concentrated H_2SO_4 or 85% H_3PO_4.
(c) Reaction of 4-methylcyclohexanol with either $SOCl_2$, PCl_3, or HCl in the presence of $ZnCl_2$.
(d) Catalytic reduction of 4-methylcyclohexene from part (b) using hydrogen gas in the presence of a Ni, Pd, or Pt catalyst. Reduction of the carbon-carbon double

double bond can also be brought about by hydroboration followed by reaction of the trialkylborane with acetic acid.
(e) Oxidation of 4-methylcyclohexanol using chromic acid in acetone or pyridinium chlorochromate.
(f) Oxidative cleavage of the alkene from part (b) using OsO_4/H_2O_2 to form the glycol followed by reaction of the glycol with either periodic acid, H_5IO_6, or lead tetraacetate, $Pb(OAc)_4$. Alternatively, oxidative cleavage of the alkene from part (b) using ozone followed by work up in the presence of dimethyl sulfide.
(g) Oxidation of the dialdehyde from part (f) with a reagent such as $KMnO_4$ to produce the dicarboxylic acid.
These transformations are summarized in the following flow chart.

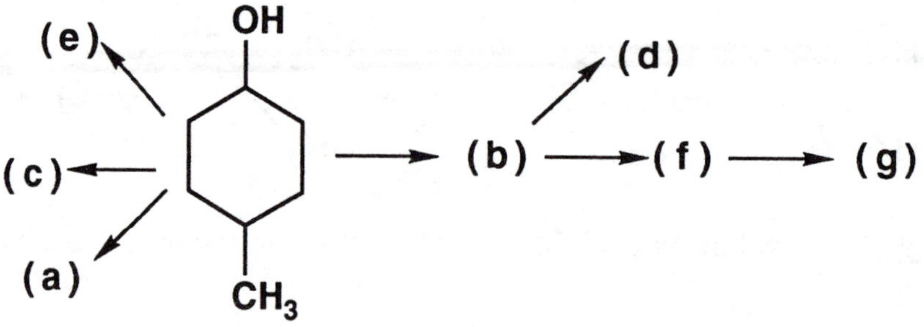

(4-methylcyclohexanol, OH at top, CH_3 at bottom) → (a), (c), (e); → (b) → (d); (b) → (f) → (g)

Problem 9.32 Show how you might bring about the following conversions. For any conversion involving more than one step, show each intermediate compound formed.

(a) 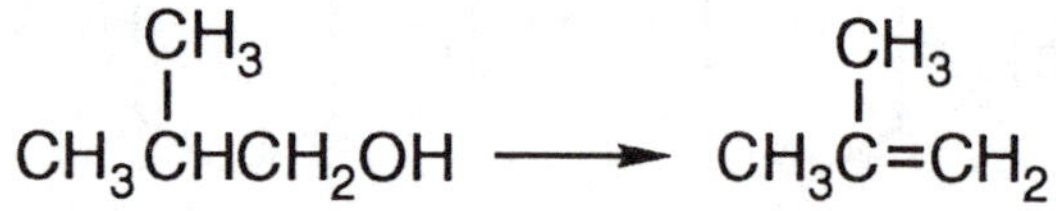 $CH_3CH(CH_3)CH_2OH \longrightarrow CH_3C(CH_3)=CH_2$

The most common laboratory methods for dehydration of an alcohol to an alkene involve heating the alcohol with either 85% phosphoric acid or concentrated sulfuric acid.

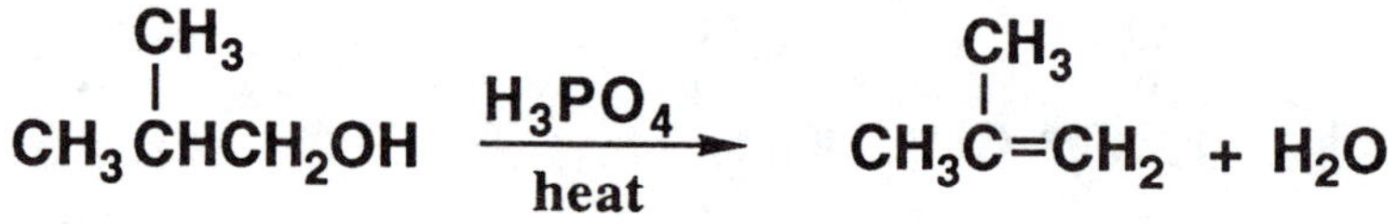

$$CH_3CH(CH_3)CH_2OH \xrightarrow[\text{heat}]{H_3PO_4} CH_3C(CH_3)=CH_2 + H_2O$$

(b) 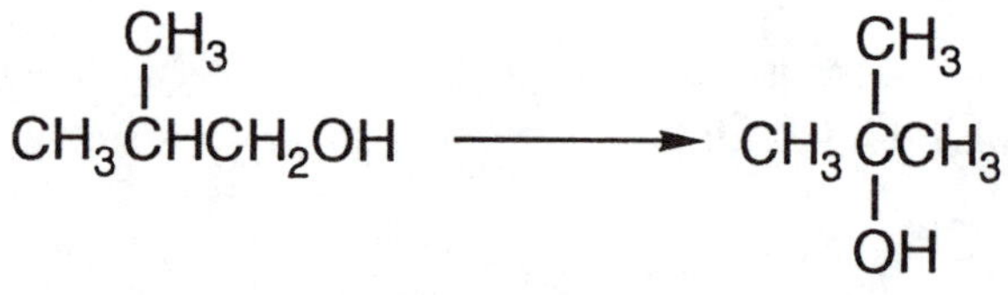 $CH_3CH(CH_3)CH_2OH \longrightarrow CH_3C(CH_3)(OH)CH_3$

We have not seen any reaction that can switch an -OH group from one carbon atom to an adjacent carbon atom. We have seen, however, reactions by which we can (1) bring about dehydration of an alcohol to an alkene and then (2) hydrate the alkene to an alcohol in the following way.

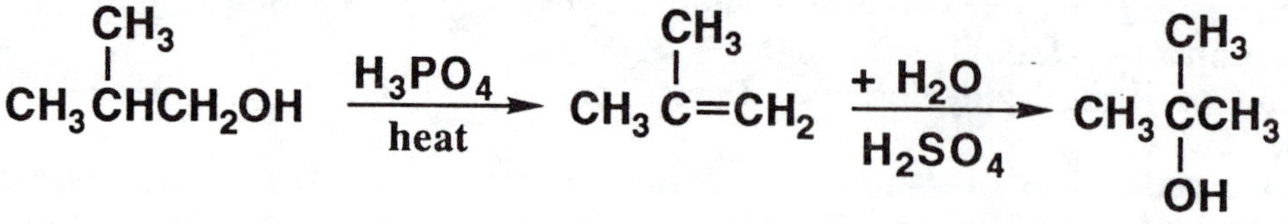

$$CH_3CH(CH_3)CH_2OH \xrightarrow[\text{heat}]{H_3PO_4} CH_3C(CH_3)=CH_2 \xrightarrow[H_2SO_4]{+ H_2O} CH_3C(CH_3)(OH)CH_3$$

(c) 1-methylcyclohexanol ⟶ *trans*-2-methylcyclohexanol

Acid-catalyzed dehydration of this tertiary alcohol to 1-methylcyclohexene followed by hydroboration/oxidation to form *trans*-2-methylcyclohexanol.

1-methylcyclohexanol $\xrightarrow[\text{heat}]{H_3PO_4}$ 1-methylcyclohexene $\xrightarrow[\text{2) } H_2O_2/NaOH]{\text{1) } B_2H_6}$ *trans*-2-methylcyclohexanol

(d) 1-methylcyclohexanol ⟶ $CH_3CCH_2CH_2CH_2CH_2CH$ (with C=O at C2 and C7)

Acid-catalyzed dehydration of the tertiary alcohol as in part (c) followed by oxidative cleavage of the carbon-carbon double bond using osmium tetroxide in the presence of periodic acid, or ozonolysis followed by work-up in the presence of dimethyl sulfide.

1-methylcyclohexanol $\xrightarrow[\text{heat}]{H_3PO_4}$ 1-methylcyclohexene $\xrightarrow[\text{2) } H_5IO_6]{\text{1) } OsO_4}$ $CH_3C\ H_2CH_2CH_2CH_2CH$ (with C=O at C2 and C7)

(e) methylenecyclohexane ($=CH_2$) ⟶ cyclohexyl–CH_2Br

Hydroboration of the alkene followed by oxidation of the resulting organoborane with alkaline hydrogen peroxide gives a primary alcohol. Reaction of this alcohol with PBr_3 gives the desired primary bromide.

methylenecyclohexane ($=CH_2$) $\xrightarrow[\text{2) } H_2O_2/NaOH]{\text{1) } B_2H_6}$ cyclohexyl–CH_2OH $\xrightarrow{PBr_3}$ cyclohexyl–CH_2Br

Alternatively, addition of HBr to the alkene in the presence of a peroxide or other radical source so that addition to the carbon-carbon double bond is by radical mechanism rather than by an electrophilic addition mechanism.

$$\text{C}_6\text{H}_{10}\text{=CH}_2 \xrightarrow[\text{peroxide}]{\text{HBr}} \text{C}_6\text{H}_{11}\text{–CH}_2\text{Br}$$

(f)

$$\text{C}_6\text{H}_{10}\text{=CHCH}_3 \longrightarrow \text{C}_6\text{H}_{11}\text{–}\overset{\text{O}}{\overset{\|}{\text{C}}}\text{CH}_3$$

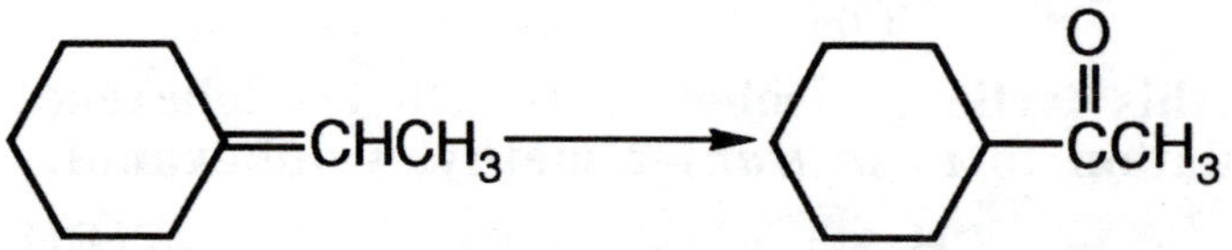

Hydroboration of the alkene followed by oxidation in alkaline hydrogen peroxide gives a secondary alcohol. Oxidation of this alcohol with chromic acid in aqueous sulfuric acid, by chromic acid in pyridine, or by pyridinium chlorochromate gives the desired ketone.

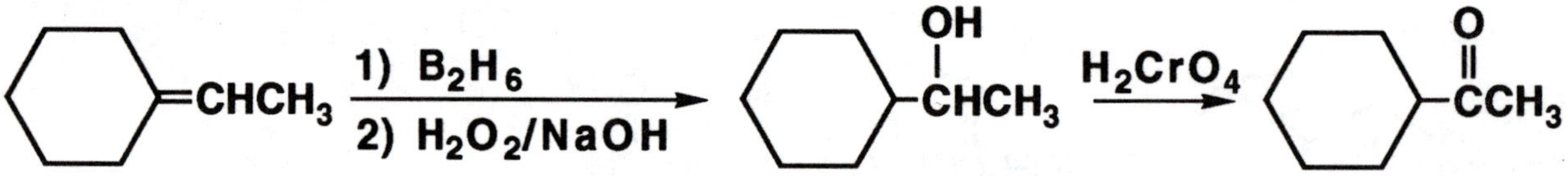

$$\text{C}_6\text{H}_{10}\text{=CHCH}_3 \xrightarrow[\text{2) } H_2O_2/NaOH]{\text{1) } B_2H_6} \text{C}_6\text{H}_{11}\text{–}\overset{\text{OH}}{\overset{|}{\text{C}}}\text{HCH}_3 \xrightarrow{H_2CrO_4} \text{C}_6\text{H}_{11}\text{–}\overset{\text{O}}{\overset{\|}{\text{C}}}\text{CH}_3$$

(g)

$$(CH_3)_2C\text{=}CHCH_2CH_2C(CH_3)\text{=}CHCH_2OH \longrightarrow (CH_3)_2C\text{=}CHCH_2CH_2C(CH_3)\text{=}CH\overset{\text{O}}{\overset{\|}{\text{C}}}\text{H}$$

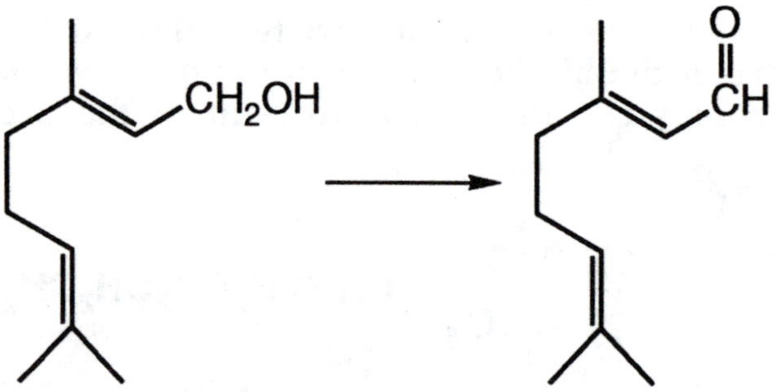

Transformation of the primary alcohol to an aldehyde can be brought about by oxidation with chromium trioxide in pyridine or pyridinium chlorochromate.

$$(CH_3)_2C\text{=}CHCH_2CH_2C(CH_3)\text{=}CHCH_2OH \xrightarrow[CH_2Cl_2]{CrO_3\text{, pyridine}} (CH_3)_2C\text{=}CHCH_2CH_2C(CH_3)\text{=}CH\overset{\text{O}}{\overset{\|}{\text{C}}}\text{H}$$

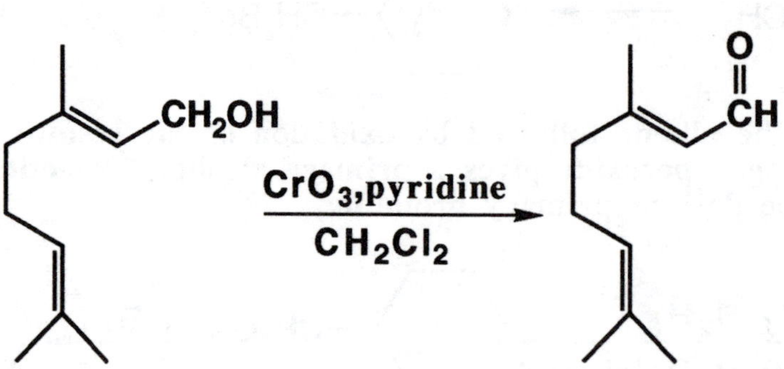

(h) cyclooctanol → cis-1,2-cyclooctanediol

Acid-catalyzed dehydration of the secondary alcohol to an alkene followed by oxidation to the *cis*-glycol using either osmium tetroxide in the presence of hydrogen peroxide, or using cold potassium permanganate at pH 11.8.

$$\text{cyclooctanol} \xrightarrow[\text{heat}]{H_3PO_4} \text{cyclooctene} \xrightarrow[H_2O_2]{OsO_4} \text{cis-1,2-cyclooctanediol}$$

(i) $CH_3(CH_2)_4C{\equiv}CH \longrightarrow CH_3(CH_2)_6OH$

Catalytic reduction of the terminal alkyne to an alkene using hydrogen gas in the presence of Lindlar's catalyst. Then hydroboration/oxidation of the alkene to a primary alcohol.

$$CH_3(CH_2)_4C{\equiv}CH \xrightarrow[\text{Lindlar catalyst}]{H_2} CH_3(CH_2)_4CH{=}CH_2 \xrightarrow[\text{2) } H_2O_2/NaOH]{\text{1) } B_2H_6} CH_3(CH_2)_6OH$$

(j) $CH_3(CH_2)_4C{\equiv}CH \longrightarrow CH_3(CH_2)_4\overset{O}{\overset{\|}{C}}CH_3$

Mercuric ion catalyzed hydration of the terminal alkyne in the presence of sulfuric acid/mercury(II) sulfate. The immediate product of hydration is an enol which is in equilibrium with the desired ketone by keto-enol tautomerism.

$$CH_3(CH_2)_4C{\equiv}CH \xrightarrow[HgSO_4,\ H_2SO_4]{H_2O} CH_3(CH_2)_4\overset{OH}{\overset{|}{C}}{=}CH_2 \longrightarrow CH_3(CH_2)_4\overset{O}{\overset{\|}{C}}CH_3$$

an enol

(k) $CH_3(CH_2)_4C{\equiv}CH \longrightarrow CH_3(CH_2)_4CH_2\overset{O}{\overset{\|}{C}}H$

Hydroboration of the terminal alkyne using a sterically hindered dialkylborane, followed by alkaline hydrogen peroxide oxidation of the resulting trialkylborane. The intermediate product of hydration is an enol which is in equilibrium with the desired aldehyde by keto-enol tautomerism.

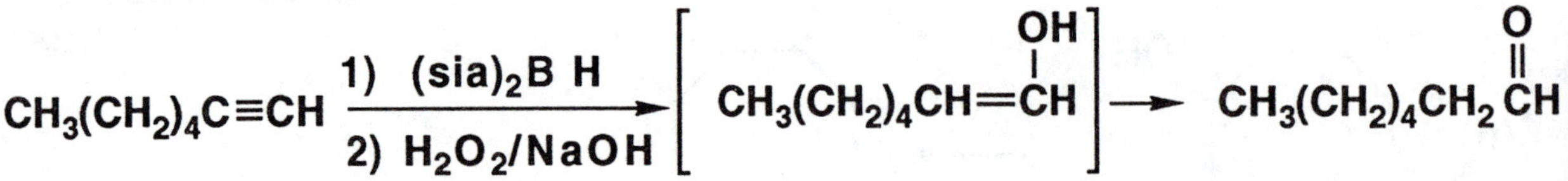

$$CH_3(CH_2)_4C{\equiv}CH \xrightarrow[\text{2) } H_2O_2/NaOH]{\text{1) (sia)}_2BH} \left[CH_3(CH_2)_4CH{=}\overset{OH}{\overset{|}{C}H} \right] \longrightarrow CH_3(CH_2)_4CH_2\overset{O}{\overset{\|}{C}H}$$

(l) cyclopentyl–CH_2OH ⟶ cyclopentyl–CH_2I

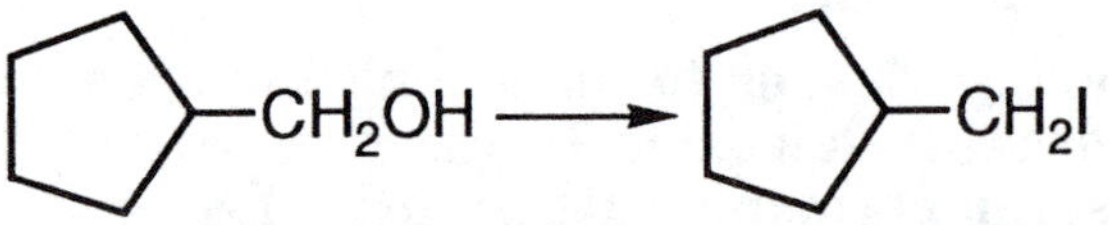

Reaction of the primary alcohol with concentrated HI.

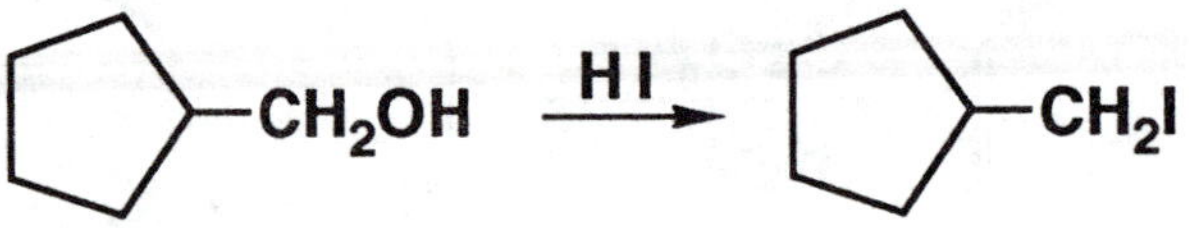

$$\text{cyclopentyl–}CH_2OH \xrightarrow{HI} \text{cyclopentyl–}CH_2I$$

Problem 9.33 Following are structural formulas for two constitutional isomers of molecular formula $C_{15}H_{26}O$.

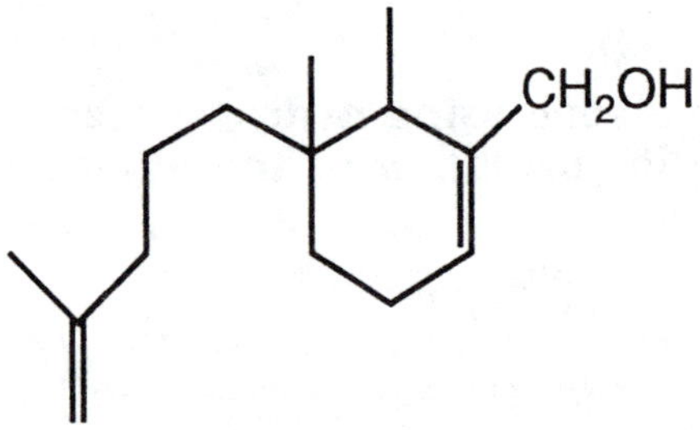

A: CH_2OH

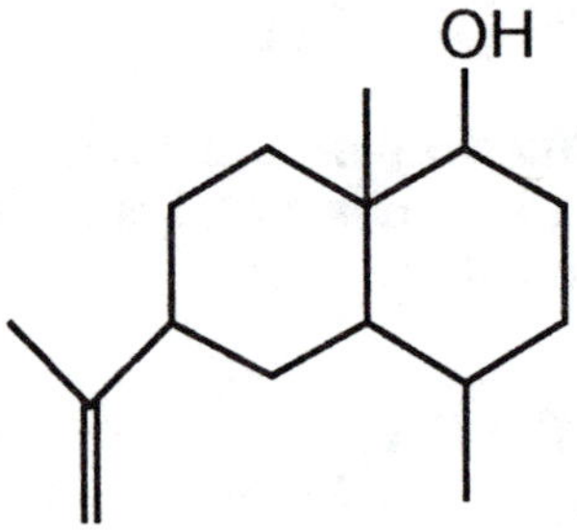

B: OH

(a) How might you oxidize the primary alcohol group of (A) to an aldehyde without affecting either of the carbon-carbon double bonds?

(b) How might you oxidize the secondary alcohol group of (B) to a ketone without affecting the carbon-carbon double bond?

For (a and b) oxidize the alcohol using either chromium trioxide in pyridine or pyridinium chlorochromate.

(c) How many *cis-trans* isomers are possible for each compound?

There are four *cis-trans* isomers possible for (A) and 32 possible for (B). The four *cis-trans* isomers for (A) arise because the carbons marked by dark circles are stereocenters so there are 2^2 or 4 isomers. For (B), there are five groups which may have a *cis-trans* relationship to each other. These possibilities give 2^5 or 32 possible isomers.

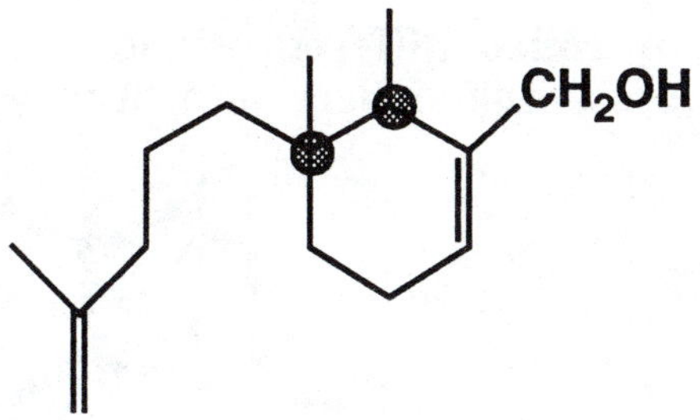

CH$_2$OH

OH

4 isomers

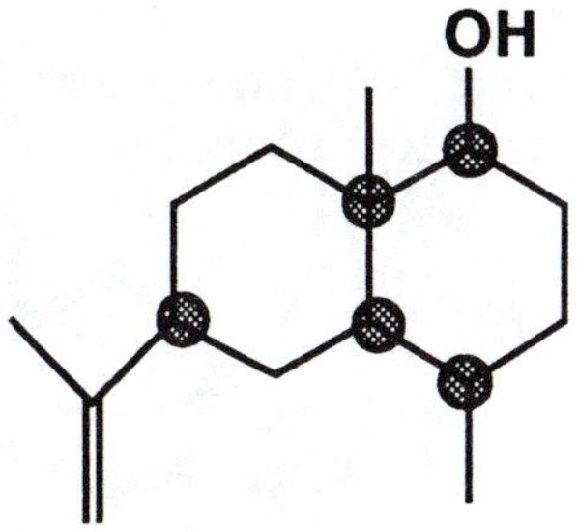

2x2x2x2x2 = 32 isomers

(d) Draw the most stable *cis-trans* isomer of alcohol (B), showing each cyclohexane ring in a chair conformation. *Hint:* Review the relative stabilities of axial and equatorial substituents (Section 2.6B), and of *cis*-decalin and *trans*-decalin (Section 2.7B).

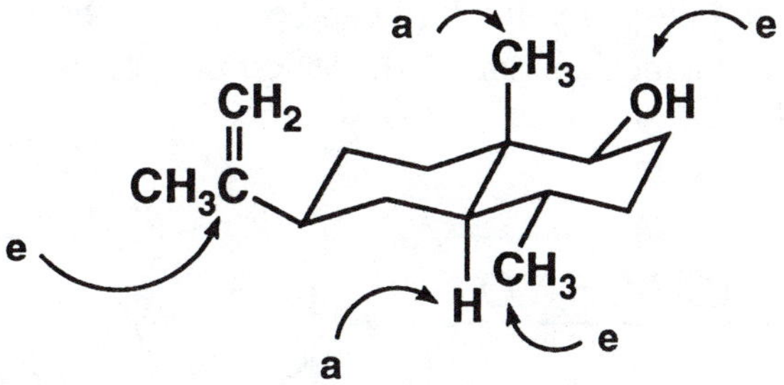

a CH$_3$ e
CH$_2$ OH
CH$_3$C
e
CH$_3$
H
a e

The ring junction is *trans*; -CH$_3$ and -H groups at the junction are axial. The remaining -OH, -CH$_3$, and isopropenyl groups are all equatorial.

(e) Show that the carbon skeleton of each can be divided into three isoprene units joined head-to-tail and then cross-linked at one or more places.

The carbon skeleton of (A) may be divided into three isoprene units lined head-to-tail in a chain and then cross-linked at one place.

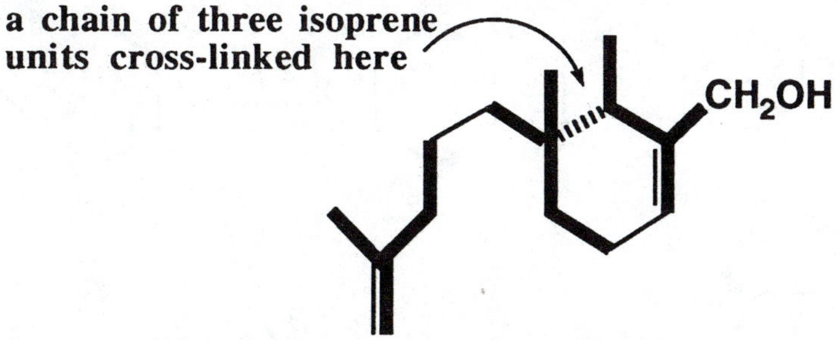

a chain of three isoprene units cross-linked here

CH$_2$OH

A

The carbon skeleton of (B) may be divided into three isoprene units, coiled in two different fashions, and then cross-linked in two places.

a chain of three isoprene units cross-linked in two places

or coiled differently and cross-linked here and here

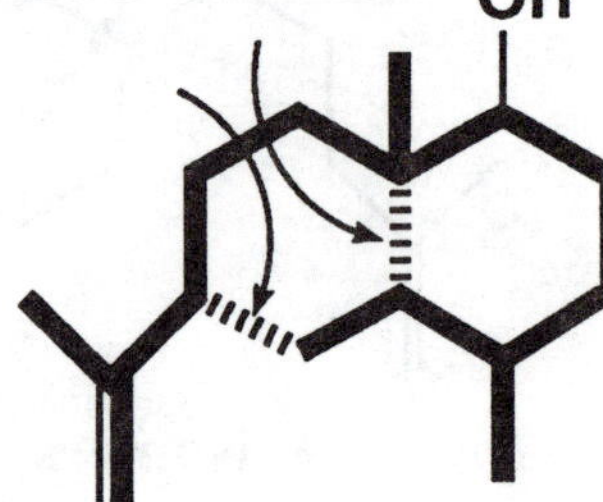

Ignore

Pinacol Rearrangement

Problem 9.34 Heating six-member and larger cyclic diols with concentrated sulfuric acid gives mainly dienes and only small amounts of spiroketones by way of a pinacol rearrangement. Pinacol rearrangement of a six-member and larger cyclic diols can be affected, however, by use of boron trifluoride etherate at 0°C. Propose a mechanism for the following boron trifluoride etherate catalyzed pinacol rearrangement.

HO OH

$BF_3 \bullet (CH_3CH_2)_2O$ → + H_2O

(7) (1) (6) (5)

spiro[5.6]dodecan-7-one

BF_3 HO :ÖH (1) → H O $^-BF_3$ + :OH (2) → H$^+$ O $^-BF_3$:OH

≡ H^+ + $HO{-}\bar{B}F_3$ (3) → HF + $HO{-}BF_2$

Problem 9.35 It is one of the postulates of internal strain theory of rings that (1) a six-member ring with all sp^3 carbons is less strained than a five-member ring with all sp^3 hybridized carbons and (2) a five-member ring with an sp^2 carbon is less strained than a six-member ring with an sp^2 carbon. The following cyclic diol can, in principal, undergo boron trifluoride etherate-catalyzed pinacol rearrangement to give isomeric spiroketones. In practice, only one of the possible spiroketones is formed. Using the above postulates, predict which of the two is formed.

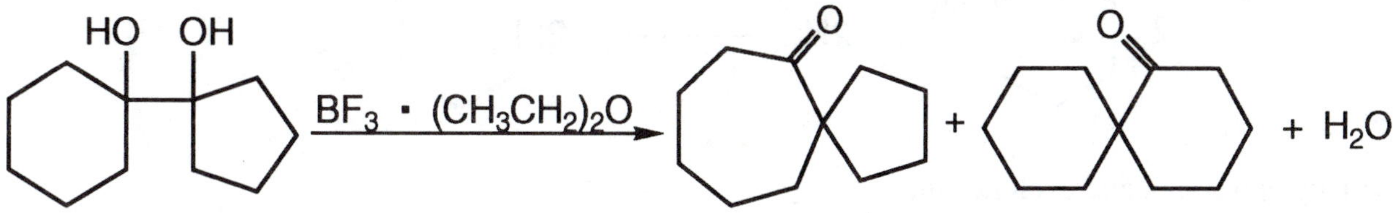

HO OH

$BF_3 \cdot (CH_3CH_2)_2O$ → + + H_2O

spiro[4.6]undecan-6-one spiro[5.5]undecan-1-one

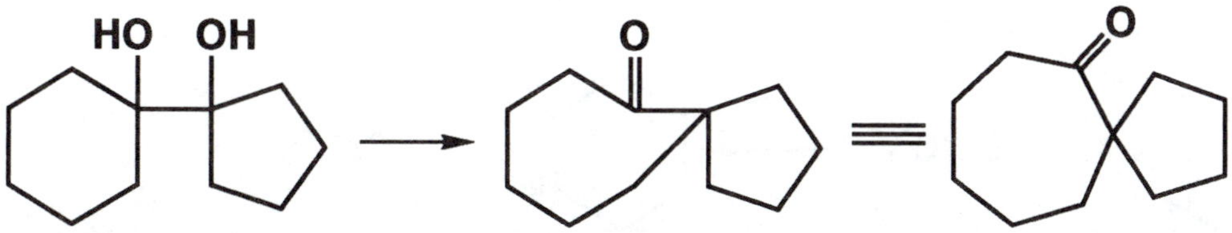

HO OH

Reaction with boron trifluoride at -OH of the five-member ring gives the more stable of the two possible secondary carbocations.

<u>Qualitative Organic Analysis</u>

<u>Problem 9.36</u> Show how you might distinguish between the members of each pair of compounds by a simple chemical test. In each case, tell what test you would perform, what you would expect to observe, and write an equation for each positive test.

(a) cyclohexanol and cyclohexene

Cyclohexene discharges the color of bromine in carbon tetrachloride and also of cold, alkaline potassium permanganate to produce a brown precipitate of manganese dioxide. Cyclohexanol does not react with either of these reagents under these conditions.

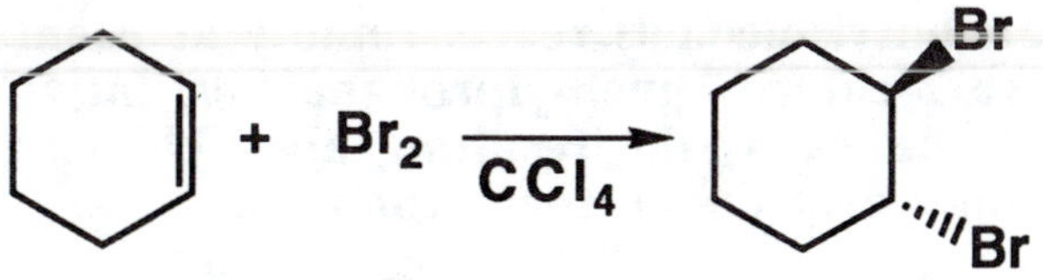

+ Br_2 $\xrightarrow{CCl_4}$ Br, Br

(colorless) (red) (colorless)

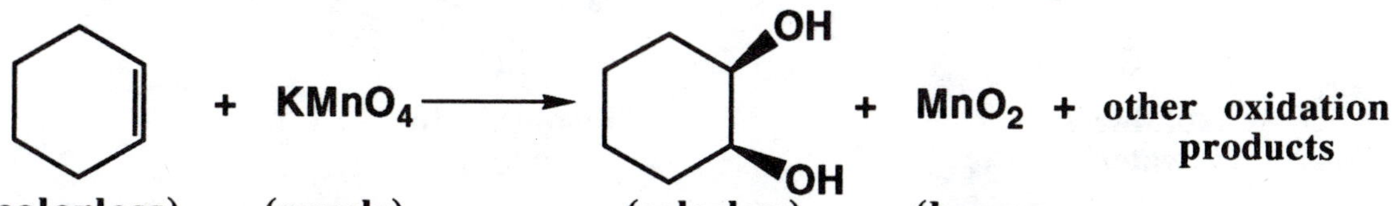

+ $KMnO_4$ → OH, OH + MnO_2 + other oxidation products

(colorless) (purple) (colorless) (brown precipitate)

Alternatively, only cyclohexanol reacts with active metals such as sodium and potassium to liberate hydrogen gas which then bubbles from the test solution.

$$2\ \text{cyclohexanol} + 2Na \longrightarrow 2\ \text{cyclohexyl-}O^- Na^+ + H_2$$

(b) cyclohexanol and cyclohexane

Only cyclohexanol is oxidized by potassium dichromate. An aqueous solution of dichromate ion is orange; a solution of Cr^{3+} is yellow.

$$\text{cyclohexanol} + Cr_2O_7^- \longrightarrow \text{cyclohexanone} + Cr^{3+}$$

(orange) **(yellow)**

Alternatively, only cyclohexanol reacts with sodium to liberate hydrogen gas. See part (a) for an equation for this reaction.

(c) ethanol and 1,2-dimethoxyethane

Only ethanol reacts with sodium metal to liberate hydrogen gas.

$$2CH_3CH_2OH + 2Na \longrightarrow 2CH_3CH_2O^-\ Na^+ + H_2$$

(d) 1-butanol and 2-methyl-2-propanol

2-Methyl-2-propanol (*tert*-butyl alcohol) reacts rapidly at room temperature with concentrated HCl to give 2-chloro-2-methylpropane (*tert*-butyl chloride). The starting alcohol is soluble in water, the resulting alkyl chloride is insoluble in water. 1-Butanol does not react under these conditions.

$$(CH_3)_3C{-}OH + HCl \longrightarrow (CH_3)_3C{-}Cl + H_2O$$

(soluble in water) **(insoluble in water)**

(e) 1-pentanol and 1-pentanethiol

When mixed with an aqueous solution of lead(II) acetate or mercury(II) acetate, 1-pentanethiol forms a water insoluble salt. 1-Pentanol does not react under these conditions.

$$2CH_3(CH_2)_4SH + Pb(OAc)_2 \longrightarrow [CH_3(CH_2)_4S]_2Pb + 2CH_3CO_2H$$

insoluble in water

CHAPTER 10: ALKYL HALIDES

SUMMARY OF REACTIONS

Starting Material ↓ / Product →	Alcohols	Alkenes	Alkyl Halides			Alkynes	Amines	Ammonium Ions	Ethers	Nitriles	Thioethers	Thiols
Alcohols			**9B** 9.5D*	**9C** 9.5D								
Alkanes			**2B** 2.9B									
Alkenes			**5G** 5.3B	**5H** 5.5	**5I** 5.6							
Alkyl Halides	**10A** 10.6	**10B** 10.10	**10C** 10.6			**10D** 10.6	**10E** 10.6	**10F** 10.6	**10G** 10.6	**10H** 10.6	**10I** 10.6	**10J** 10.6

*Section of book that describes reaction.

REACTION 2B: HALOGENATION OF ALKANES (Section 2.9B)

$$-\overset{|}{\underset{|}{C}}-H \xrightarrow[h\nu]{X_2} -\overset{|}{\underset{|}{C}}-X$$

REACTION 5G: HYDROHALOGENATION OF ALKENES (Section 5.3B)

$$>C=C'< \xrightarrow{H\text{-}X} -\overset{H}{\underset{|}{C}}-\overset{|}{\underset{X}{C'}}-$$

REACTION 5H: RADICAL HYDROBROMINATION (Section 5.5)

$$>C=C'< \xrightarrow[\text{peroxides}]{H\text{-}Br} -\overset{H}{\underset{|}{C}}-\overset{|}{\underset{Br}{C'}}-$$

REACTION 5I: ALLYLIC HALOGENATION (Section 5.6)

NBS; light or peroxides: C=C'–C''–H → C=C'–C''–Br

REACTION 9B: REACTION OF ALCOHOLS WITH H-X (Section 9.5D1)

H-X: –C–C'(OH)– → –C–C'(X)–

REACTION 9C: REACTION OF ALCOHOLS WITH PX_3 OR $SOCl_2$ (Section 9.5D2)

PX_3 or $SOCl_2$: –C–C'(OH)– → –C–C'(X)–

REACTION 10A: FORMATION OF ALCOHOLS: REACTION WITH HYDROXIDE ION AND WATER (Section 10.6)

HO^- or H_2O: –C–X → –C–OH

- Alkyl halides can be converted into alcohols by treatment with either hydroxide or water.
- For HO^-, substitution occurs via an S_N2 mechanism with primary alkyl halides, but with secondary and especially tertiary alkyl halides elimination (reaction 10B) can become important because HO^- is also a relatively strong base.
- The H_2O can react predominantly via an S_N2 mechanism with primary alkyl halides, but with secondary and especially tertiary alkyl halides the S_N1 mechanism can become important.

REACTION 10B: FORMATION OF ALKENES: β-ELIMINATION (Section 10.10)

base: H–C–C'–X → C=C'

- Alkyl halides undergo β-elimination in the presence of base to produce alkenes.
- Primary alkyl halides will only undergo appreciable elimination (E2) with a very strong base, for example RO^-.

- Secondary alkyl halides may undergo some E2 elimination with strong bases, or E1 elimination in solvolysis reactions.
- Tertiary alkyl halides readily undergo E2 elimination with base, or E1 in solvolysis reactions.

REACTION 10C: HALOGEN EXCHANGE (Section 10.6)

$$-\overset{|}{\underset{|}{C}}-X \xrightarrow{X'^-} -\overset{|}{\underset{|}{C}}-X'$$

- The halogen of an alkyl halide can be exchanged by using the halide ion as a nucleophile in a substitution reaction.
- Like most of the non-basic nucleophiles, the reaction will take place via an S_N2 mechanism for methyl, primary, and secondary alkyl halides, but the S_N1 mechanism is important for tertiary alkyl halides.

REACTION 10D: ALKYLATION OF ALKYNES: REACTION WITH TERMINAL ALKYNE ANIONS (Section 10.6)

$$-\overset{|}{\underset{|}{C}}-X \xrightarrow{^-C'\equiv C''-} -\overset{|}{\underset{|}{C}}-C'\equiv C''-$$

- Alkyne anions react with certain alkyl halides to generate other alkynes. The alkyne anions are produced by deprotonation of terminal alkynes (reaction 6N, section 6.5D in the book).
- Alkyne anions are such strong bases, that substitution is important only for methyl or primary alkyl halides. Elimination is an important side reaction for secondary alkyl halides and the only reaction tertiary alkyl halides.

REACTION 10E: ALKYLATION OF DEPROTONATED AMINES (Section 10.6)

$$-\overset{|}{\underset{|}{C}}-X \xrightarrow{\diagdown N^- \diagup} -\overset{|}{\underset{|}{C}}-\overset{|}{\underset{|}{N}}:$$

- Deprotonated amines react with alkyl halides to produce alkylated amines.
- This reaction is most important for methyl and primary alkyl halides, because the deprotonated amines are such strong bases that elimination is a competeing reaction with secondary and tertiary alkyl halides.

REACTION 10F: ALKYLATION OF AMINES (Section 10.6)

$$-\overset{|}{\underset{|}{C}}-X \xrightarrow{-\overset{|}{\underset{|}{N}}:} -\overset{|}{\underset{|}{C}}-\overset{|}{\underset{|}{N^+}}-$$

- Amines also react with alkyl halides to produce alkylated amines.

REACTION 10G: FORMATION OF ETHERS: REACTION WITH ALKOXIDE ANIONS (Section 10.6)

$$-\overset{|}{\underset{|}{C}}-X \xrightarrow{-\overset{|}{\underset{|}{C'}}-O^-} -\overset{|}{\underset{|}{C}}-O-\overset{|}{\underset{|}{C'}}-$$

- Alkyl halides react with alkoxide anions to produce ethers.
- This reaction is most important for methyl and primary alkyl halides. The alkoxide anions are such strong bases that β-elimination is a competing reaction with secondary and tertiary alkyl halides.

REACTION 10H: FORMATION OF NITRILES: REACTION WITH CYANIDE ANION (Section 10.6)

$$-\overset{|}{\underset{|}{C}}-X \xrightarrow{N{\equiv}C^-} -\overset{|}{\underset{|}{C}}-C{\equiv}N$$

- Alkyl halides react with the cyanide anion to produce nitriles.

REACTION 10I: FORMATION OF THIOETHERS: REACTION WITH THIOLATE ANIONS (Section 10.6)

$$-\overset{|}{\underset{|}{C}}-X \xrightarrow{-\overset{|}{\underset{|}{C'}}-S^-} -\overset{|}{\underset{|}{C}}-S-\overset{|}{\underset{|}{C'}}-$$

- Alkyl halides react with the thiolate anions to produce thioethers.

REACTION 10J: FORMATION OF THIOLS: REACTION WITH HS^- (Section 10.6)

$$-\overset{|}{\underset{|}{C}}-X \xrightarrow{HS^-} -\overset{|}{\underset{|}{C}}-SH$$

- Alkyl halides react with HS^- to produce thiols.

SUMMARY OF IMPORTANT CONCEPTS

10.0 OVERVIEW

- **Haloalkanes**, also known as **alkyl halides** in the common nomenclature, are compounds that contain a **halogen atom** attached to an **sp^3 carbon atom**. The terms haloalkanes and alkyl halides are used interchangeably for the rest of the chapter. They can be prepared from a variety of different compounds, and they can be converted into a variety of different compounds. Thus, alkyl halides are an important class of molecules for organic synthesis. ✳

10.1 STRUCTURE

- The general symbol for haloalkanes is **R-X**, where **R** is **any alkyl group** (the carbon attached to the halogen must be sp3 hybridized) and **X** can be any of the **halogens**, namely -F, -Cl, -Br, or -I.

10.2 NOMENCLATURE

- IUPAC names are derived for haloalkanes by naming the parent hydrocarbon according to normal rules, and treating the halogen atom as a substituent to be listed in alphabetical order like the other substituents. For example, 2-bromobutane or 3-fluoro-4-methylnonane are acceptable IUPAC names.
- Common names of haloalkanes consist of the name of the alkyl group followed by the name of the halide as a separate word. For example, propyl iodide is the common name for the compound called 1-iodopropane in the IUPAC nomenclature.
  - Haloalkanes in which all of the hydrogens of a hydrocarbon are replaced by halogen atoms are called perhaloalkanes. For example, perchloropropane is the common name for the compound of molecular formula C_3Cl_8.

10.3 PHYSICAL PROPERTIES

- Since the halogens are more electronegative than carbon, the C-X bond is polar. Electronegativity increases in the order I<Br<Cl<F, but bond length increases in the opposite order, namely F<Cl<Br<I. The maximal combination of electronegativity and bond length occurs with C-Cl bonds, so chloroalkanes are the most polar, having the largest bond dipole moment. ✳
  - As liquids, haloalkane molecules are attracted to each other by a combination of dipole-dipole interactions and dispersion forces. These attractive forces are eventually balanced by repulsive forces when the molecules approach each other too closely.
- The **boiling points** of haloalkanes are generally higher than comparable alkanes of similar size and shape. This is because of the dipole-dipole interactions that are possible with the haloalkanes, as well as the increased **polarizability** of halogen atoms compared with hydrocarbons. Large atoms such as bromine or iodine with lone pairs of electrons are highly polarizable, that is their electron density can be temporarily "moved around," which increases the strength of induced dipole interactions between molecules.
- The **densities** of haloalkanes are higher than hydrocarbons because of the relatively high atomic mass of the halogens, especially bromine and iodine.
- As shown in Table 10.6 in the text, the larger the halogen, the longer and weaker the C-X bond. Only C-F bonds are stronger than C-H bonds.

10.4 PREPARATION OF ALKYL HALIDES

- Several methods have already been discussed for the preparation of haloalkanes. These include preparation from alcohols (Section 9.5D), radical halogenation of alkanes (Section 2.9B), addition of HX to alkenes (Section 5.3B), and allylic halogenation (Section 5.6). *[This would be a great time to review these methods if necessary. It is often helpful to study reactions from different chapters in groups like this to help reinforce the fact that they all have something in common, namely they all produce alkyl halides]*

10.5 REACTIONS OF ALKYL HALIDES WITH BASES AND NUCLEOPHILES

- Alkyl halides react with electron rich reagents that are both **Lewis bases** and **nucleophiles**. A **nucleophile** is any species capable of **donating a pair of electrons** to form a new covalent bond. Because nucleophiles can also function as bases, this must always be considered when nucleophiles react with alkyl halides.

10.6 NUCLEOPHILIC ALIPHATIC SUBSTITUTION

- **Nucleophilic substitution reactions** are reactions in which one nucleophile is substituted for another. They are very important reactions for alkyl halides, because a wide variety of different functional groups can be prepared in this way. In these reactions, the **halogen atom** is **replaced by** the **nucleophile.** ✱
  - For example, Table 10.6 in the text lists a number of negatively-charged and neutral nucleophiles that can react with alkyl halides such as methyl bromide to produce numerous types of molecules.
  - The solvent can have a strong influence on nucleophilic aliphatic substitution reactions. Two different types of solvents are used in substitution reactions, **protic solvents** and **aprotic solvents. Protic solvents** are solvents that contain a functional group such as -OH that can act a hydrogen bond donor. **Aprotic solvents** do not have any functional groups that can act as hydrogen bond donors.
  - Solvents are further classified as **polar** and **nonpolar. Polar** solvents interact strongly with ions and polar molecules, while **nonpolar** solvents do not interact with ions and polar molecules. **Dielectric constant** is a common measure of solvent polarity and is defined as the amount of electrostatic insulation provided by molecules placed between two charges.
  - Water, formic acid, methanol, and ethanol are considered to be **polar protic solvents.** Important **polar aprotic** solvents include dimethyl sulfoxide, acetonitrile, dimethylformamide, and acetone while **nonpolar protic** solvents include dichloromethane, diethyl ether, and benzene. It is helpful to categorize solvents this way, because solvents in a given category influence reactions between nucleophiles and alkyl halides in the same ways.

10.7 MECHANISM OF NUCLEOPHILIC ALIPHATIC SUBSTITUTION

- There are two different limiting mechanisms for nucleophilic aliphatic substitution reactions. These are called **S_N2 and S_N1** mechanisms, and they will now be discussed individually. ✱
- The term **S_N2** stands for Substitution reaction, Nucleophilic, 2nd order (also called bimolecular). According to the **S_N2** mechanism, bond-breaking and bond-making occur at the same time. Thus, both the nucleophile and alkyl halide are involved in the rate determining step, hence this is a **bimolecular reaction.** The reaction takes place in a single step, so there is only a single transition state, not any intermediates. In this case, the departing halogen atom is called the **leaving group.** ✱

$$Nu:^- + \text{C—Br} \longrightarrow \left[\overset{\delta^-}{Nu}\text{----C----}\overset{\delta^-}{Br}\right] \longrightarrow \overset{\delta^+}{Nu}\text{—C} + Br^-$$

Transition state in which
Nu-C bond is formed as
C-Br bond is broken.

  - A **bimolecular reaction** is one in which two reactants take part in the transition state of the slow or rate-determining step of a reaction. Thus the rates of bimolecular reactions such as **S_N2** reactions are proportional to the concentration of both the alkyl halide and nucleophile.
  - Since the nucleophile is involved in the rate-determining-step of the **S_N2** reaction, stronger nucleophiles react with a faster rate. Stronger nucleophiles are said to have increased **nucleophilicity.**

    In the gas phase, there is a correlation between nucleophilicity and basicity; the stronger the base, the higher the nucleophilicity. In general, within a period of the periodic table, nucleophilicity increases from right to left. Furthermore, for different reagents with the same nucleophilic atom, an anion is a better nucleophile than a neutral species. Solvents have a dramatic effect on nucleophilicity. Since **polar aprotic** solvents are good at solvating cations but not anions, anionic nucleophiles participate readily in nucleophilic substitution reactions. In **polar aprotic** solvents, the relative nucleophilicities of halide ions are the same as observed in the gas phase. **Polar protic** solvents strongly solvate anions. In **polar protic** solvents, the relative nucleophilicities of halide ions are reversed compared to what is observed in the gas phase due to differential solvation.

- Polar protic solvents greatly inhibit S_N2 reactions with negatively charged nucleophiles, because the nucleophile is so highly solvated and thus unreactive. As a result, S_N2 reactions are dramatically faster in polar aprotic solvents such as acetonitrile (CH_3CN) compared with polar protic solvents like water.
- If the halide leaving group is attached to a stereocenter, then the **configuration** of the stereocenter is **inverted** during an **S_N2 reaction.** This is because the nucleophile enters from the **opposite side** of the molecule **as the departing leaving group**, thus the molecule inverts analogous to how an umbrella is inverted in the wind.
- **S_N2** reactions are particularly **sensitive** to **steric factors**, since they are greatly retarded by steric hindrance (crowding) at the site of reaction.
- Since there is no carbocation produced in S_N2 reactions, there is no skeletal rearrangement observed.

• The term **S_N1** stands for Substitution reaction, Nucleophilic, 1st order (also called unimolecular). According to the **S_N1** mechanism, there are two steps. The carbon-halide bond breaks in the rate-determining first step, creating a carbocation intermediate that then makes a new bond to the nucleophile in the second step. Only the alkyl halide is involved with the rate-determining step, thus the reaction is **unimolecular**. Since the reaction involves two steps, there are two transition states and one intermediate. ✻

C—Br → (rate-determining step, 1st step) → [carbocation intermediate +C] (+ Br⁻) → (:Nu, 2nd step) → C—Nu + Nu—C

- A **unimolecular reaction** is one in which only one reactant takes part in the transition state of the rate-determining-step. Thus the rates of unimolecular reactions such as **S_N1** reactions are proportional to the concentration of the alkyl halide.
- Since nucleophiles are not involved in the rate-determining step, stronger nucleophiles do not react faster in S_N1 reactions.
- Because the S_N1 mechanism involves creation and separation of unlike charges in order to form the carbocation intermediate, polar solvents that can stabilize these charges by solvation greatly accelerate S_N1 reactions. For example, S_N1 reactions are much faster in water than in ethanol.
- If the leaving group is attached to a stereocenter, then the **configuration** of the stereocenter is **racemized** during an **S_N1 reaction.** This is because the carbocation intermediate is achiral and the nucleophile can approach from either side, leading to both possible

enantiomers as products. In theory, an S_N1 reaction will result in complete racemization, but in fact only partial racemization is observed; with the inversion product predominating. This is accounted for by proposing that while bond-breaking between carbon and the leaving group is complete, the leaving group remains associated for some period of time with the carbocation as an ion pair. To the extent that the leaving group remains associated as an ion pair, it hinders approach of the nucleophile from that face, favoring attack from the opposite face resulting in an excess of inversion.

- **S_N1** reactions are greatly accelerated by electronic factors that stabilize carbocations.
- Since there is a carbocation intermediate in S_N1 reactions, **skeletal rearrangements** are observed if they produce another carbocation of equal or greater stability.

• The **structure** of the **alkyl halide** greatly **influences** which **mechanism** will be followed. The order of reactivity for the S_N2 mechanism increases in the order: 3° < 2° < 1° allylic = 1° < methyl, since steric hindrance is highest for 3° alkyl halides and least for methyl halides. On the other hand, the order of reactivity for the S_N1 mechanism increases in the order: methyl < 1° < 2° < 1° allylic < 3°, since 3° carbocations are most stable and methyl carbocations are least stable.

✱

- The net result is that when nucleophilic substitution reactions occur, **methyl** and **1° alkyl** halides react **exclusively** by the **S_N2** mechanism. **1° allylic** and **2° alkyl halides** can react by **either** the **S_N2** or **S_N1** mechanism. **3° alkyl halides** react **exclusively** by the **S_N1** mechanism. ✱
- Other reaction mechanisms such as β-elimination reactions can take place when a nucleophile reacts with an alkyl halide, and the structure of the alkyl halide greatly influences which of these reactions occurs.

• The leaving group develops a partial negative charge as it is departing by either the S_N1 or S_N2 mechanism. Thus, the lower the basicity, the better a halide is able to function as a leaving group. I^- is the best leaving group, and leaving group ability increases in the order: $F^- < Cl^- < Br^- < I^-$.

10.8 CONVERSION OF ALCOHOLS TO ALKYL HALIDES-A CLOSER LOOK

• Conversions of alcohols into alkyl halides through the action of H-X can take place according to S_N1 or S_N2 mechanisms, depending on the structure of the alcohol and other experimental conditions.

- **1° alcohols** without extensive ß-branching react via an **S_N2** mechanism, but first the -OH group is turned into a better leaving group (H_2O) via protonation. The water is displaced by the halide ion nucleophile in an S_N2 reaction.

$$R{-}CH_2{-}\ddot{O}{-}H \xrightleftharpoons{H_3O^+} R{-}CH_2{-}\overset{+}{\underset{..}{O}}(H){-}H \xrightarrow{S_N2} R{-}CH_2{-}X$$

$$(+\ H_2O) \qquad\qquad (+\ H_2O)$$

- **3° alcohols** react via an **S_N1** mechanism. Again, the -OH group is turned into a better leaving group (H_2O) via protonation. Since the water cannot be displaced by the halide ion due to steric hindrance, the water departs to create a carbocation, that then reacts with X^-.

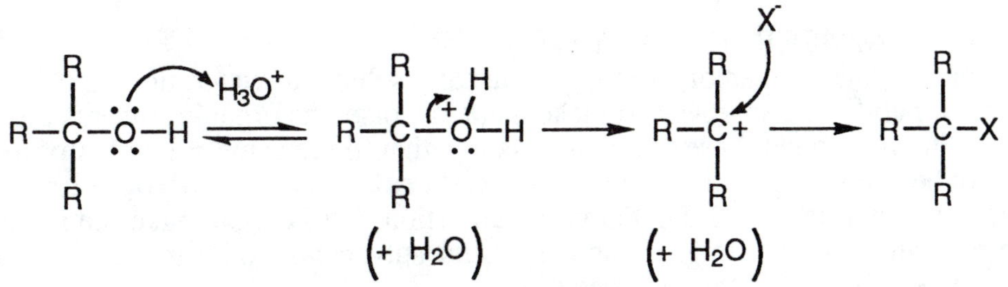

- **2° alcohols** can react via a combination of $\mathbf{S_N1}$ and $\mathbf{S_N2}$ mechanisms.
- **Rearrangements** can occur with **1° or 2° alcohols** having extensive ß-branching, and in this case the alkyl group that migrates can be thought of as an internal nucleophile that displaces the water molecule.

• Reaction of alcohols with thionyl chloride, $\mathbf{SOCl_2}$, are interesting because experimental conditions can be chosen that provide for an $\mathbf{S_N2}$ **or** $\mathbf{S_Ni}$ **mechanism**. The key difference is how much Cl^- is available in the reaction to act as a nucleophile to take part in an S_N2 reaction.

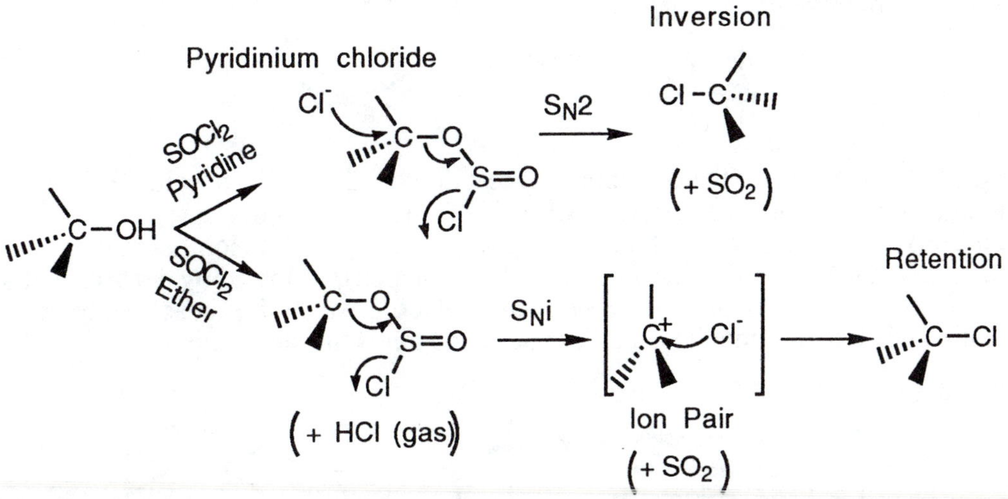

- With bases like pyridine, Cl^- is kept in solution as pyridinium chloride, so there is enough chloride present to act as a nucleophile in an S_N2 reaction. As a result, inversion of configuration is observed.
- In solvents like ether, Cl^- leaves as HCl gas. With only small amounts of Cl^- in solution, the SO_2 has a chance to leave, generating a carbocation. The carbocation forms an ion pair with chloride, that facilitates formation of the Cl-C bond with retention of configuration. This front side displacement reaction is called an S_Ni reaction.

• **Alcohols** react with **sulfonyl chlorides** such as *p*-toluenesulfonyl chloride or methanesulfonyl chloride to create a **sulfonate ester**. These molecules are important, because the sulfonates are good leaving groups, analogous to halides, that can depart during S_N2 reactions. Thus, the -OH group is turned into a good leaving group. Good nucleophiles such as I^- or HO^- can be used to displace the sulfonate with inversion of configuration.

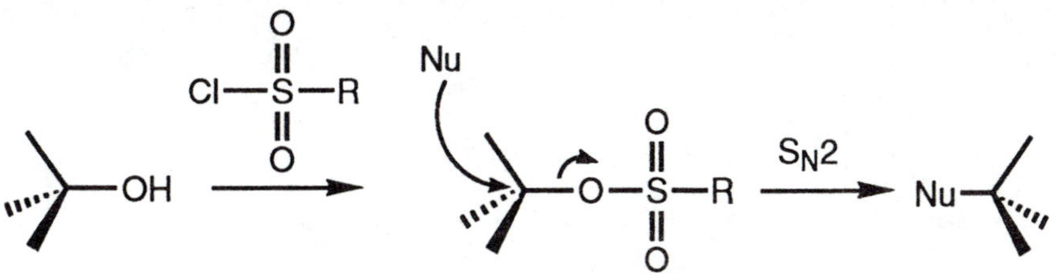

10.9 PHASE TRANSFER CATALYSIS

- Numerous nucleophiles are anions so they are usually soluble in water, not the organic solvents in which alkyl halides are soluble. To make matters worse, the organic solvents are not miscible with water. In order to get these two reactants together in the same solvent, a **phase transfer catalyst** can be used. A phase transfer catalyst is a **cation** that is **hydrophobic** (dissolves in organic solvent) such as the tetrabuytlammonium cation. This cation makes an ionic bond with the anionic nucleophile and brings it into the organic solvent where it can react with the alkyl halide in a nucleophilic displacement reaction.

10.10 ß-ELIMINATION

- Most nucleophiles are also bases and alkyl halides are predisposed to ß-elimination, so this must always be considered as a competing reaction. **ß-Elimination** involves **loss of the halide ion** and a **proton from a ß-carbon atom** (the carbon adjacent to the one with the halide). The stronger the base, the higher the percentage of ß-elimination product formed in a reaction. Note that when there is more than one ß carbon atom with a hydrogen atom attached, multiple alkene products are possible.

X, H, C, β, α, β-Elimination, C=C

- There are two mechanisms for the ß-elimination reaction of alkyl halides, called **E1** and **E2**. These are analogous in some ways to the S_N1 and S_N2 mechanisms discussed above.
  - **E1 reactions** involve departure of a leaving group such as a halide ion to create a carbocation (analogous to the first step of the S_N1 reaction), followed by departure of a hydrogen atom on a ß-carbon to yield the final product. Like the S_N1 reaction, the carbocation is a true intermediate, and loss of halide ion is the slow step.

H—C—C—X, Slow, Base, H—C—C+, (+ X⁻), Fast, (+ Base-H)

The rate-determining step in an E1 reaction is loss of the halide to generate the carbocation. Thus, the **E1 reaction is unimolecular** (first order) since the rate only depends on the concentration of alkyl halide.

E1 reactions give predominantly the **Zaitsev elimination product**, namely the **more highly substituted alkene**. This is because the product determining step has partial double bond character, the transition state with lower energy is the one with the more stable partial double bond.

  - **E2 reactions** are concerted in that the base removes the ß-hydrogen at the same time the C-X bond is broken.

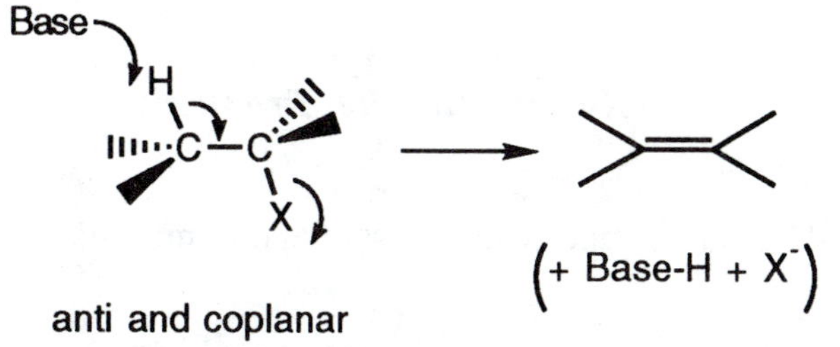

Base → H ... C—C ... X → alkene (+ Base-H + X^-)

anti and coplanar geometry of H and X

The only step in the E2 reaction involves both the base and the alkyl halide. Thus, the **E2 reaction is bimolecular** (second order), since the rate depends on both the concentration of base and the concentration alkyl halide.
E2 reactions also give predominantly the **Zaitsev elimination product**, since there is significant partial double bond character in the transition state.
E2 reactions proceed preferentially when the **ß-hydrogen atom** removed by the base and the **departing X atom** are oriented **anti** and **coplanar** to one another. This is particularly important for cyclohexane derivatives, where the ß-hydrogen and X atom must both be axial to satisfy the anti and coplanar arrangement.

10.11 SUBSTITUTION VERSUS ELIMINATION

- In the absence of any base, tertiary alkyl halides in polar solvents undergo unimolecular reactions to give a combination of substitution (S_N1) and elimination (E1). Although the exact ratios are hard to predict, the amount of substitution can be increased by increasing the concentration of non-basic nucleophile.
- In general, for bimolecular reactions, increased steric hindrance increases the ratio of elimination to substitution products. This is because steric hindrance interferes with the approach of the nucleophile to the backside of the C-X bond, thus impeding the substitution reaction.
  - **Tertiary halides** react with all basic reagents to give **elimination products**. There is too much steric hindrance for substitution to compete effectively with elimination.
  - **Secondary alkyl halides** have an intermediate amount of steric hindrance and are **borderline**. Substitution or elimination may predominate depending on the particular nucleophile/base, solvent, and temperature of the reaction. Strongly basic nucleophiles such as alkoxides favor E2 reactions, but weakly basic strong nucleophiles favor substitution.
  - **Primary alkyl halides** and **methyl halides** have very little steric hindrance, so they react with all nucleophiles, even strongly basic nucleophiles like hydroxide ions and alkoxides ions, to give predominantly **substitution products**.

CHAPTER 10
Solutions to the Problems

Problem 10.1 Give IUPAC names and where possible, common names for each compound.

(a) 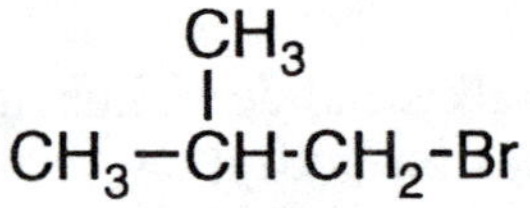 $CH_3-CH(CH_3)-CH_2-Br$

1-bromo-2-methylpropane (isobutyl bromide)

(b)

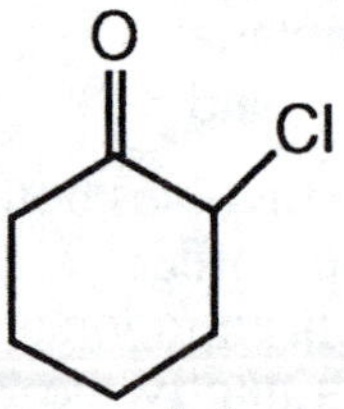

2-chlorocyclohexanone

(c)

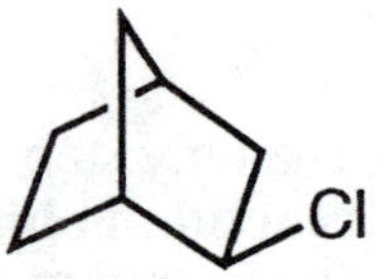

2-chlorobicyclo[2.2.1] heptane

(d) 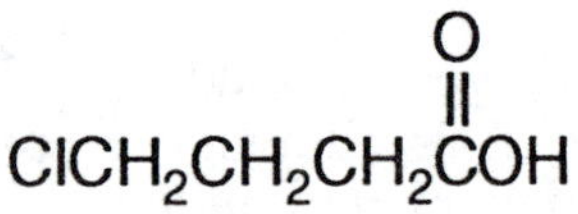 $ClCH_2CH_2CH_2COH$ (C=O)

4-chlorobutanoic acid

(e) 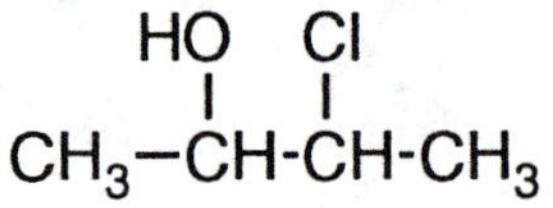 $CH_3-CH(OH)-CH(Cl)-CH_3$

3-chloro-2-butanol

(f)

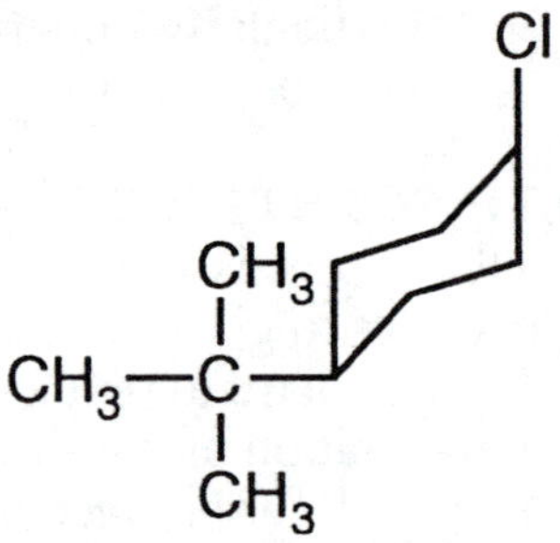

***cis*-1-*tert*-butyl-4-chlorocyclohexane**

Problem 10.2 What alcohol and reaction conditions will give each compound in good yield?

(a)

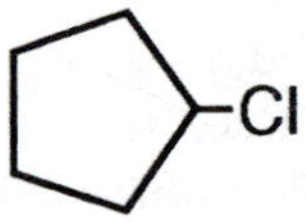

Cyclopentanol can be converted to the desired product by reaction with either HCl, $SOCl_2$, or PCl_3.

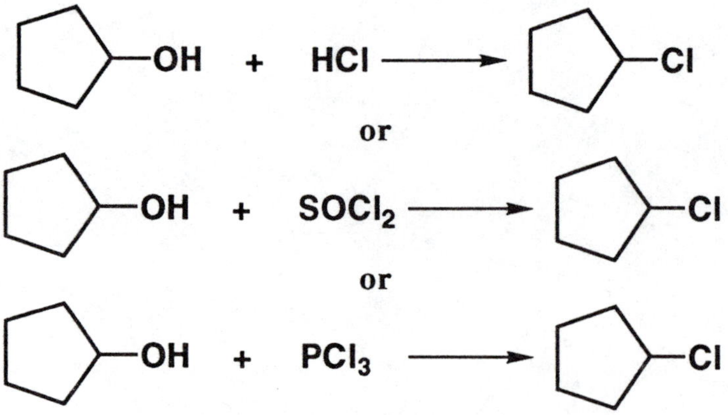

Cyclopentanol-OH + HCl ⟶ Cyclopentyl-Cl

or

Cyclopentanol-OH + $SOCl_2$ ⟶ Cyclopentyl-Cl

or

Cyclopentanol-OH + PCl_3 ⟶ Cyclopentyl-Cl

(b) 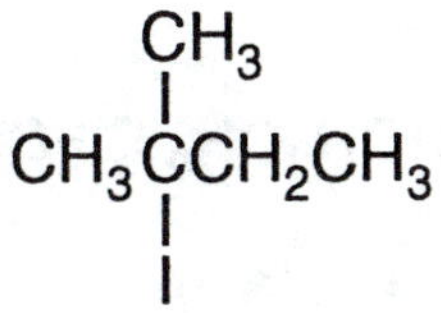 $CH_3C(CH_3)(I)CH_2CH_3$

This reaction utilizes a 3° alcohol so it procedes via an S_N1 mechanism.

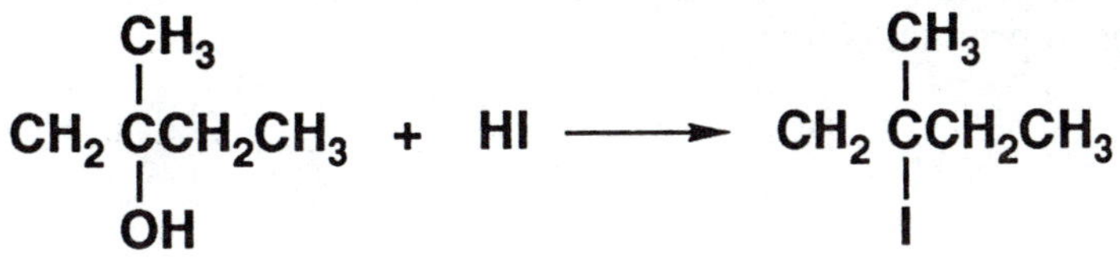

$$CH_3C(CH_3)(OH)CH_2CH_3 + HI \longrightarrow CH_3C(CH_3)(I)CH_2CH_3$$

(c) 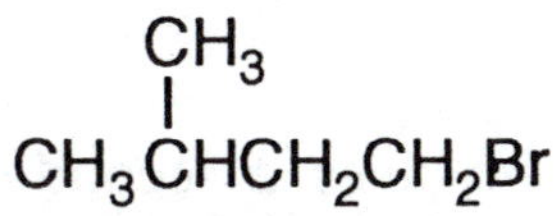 $CH_3CH(CH_3)CH_2CH_2Br$

Either HBr or PBr_3 can be used to carry out the conversion of 3-methyl-1-butanol to the desired product.

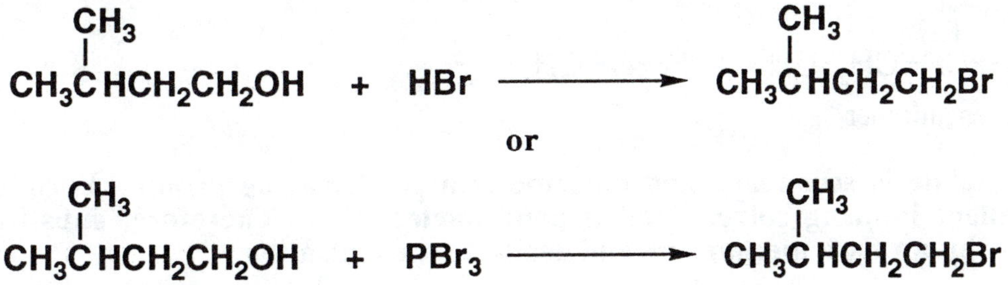

$$CH_3CH(CH_3)CH_2CH_2OH + HBr \longrightarrow CH_3CH(CH_3)CH_2CH_2Br$$

or

$$CH_3CH(CH_3)CH_2CH_2OH + PBr_3 \longrightarrow CH_3CH(CH_3)CH_2CH_2Br$$

Problem 10.3 Draw structural formulas for the products of the following nucleophilic aliphatic substitution reactions.

(a) 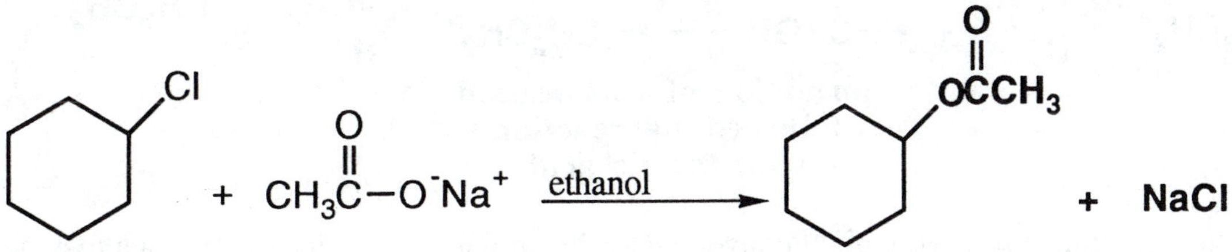 $$C_6H_{11}Cl + CH_3C(=O){-}O^-Na^+ \xrightarrow{\text{ethanol}} C_6H_{11}OC(=O)CH_3 + NaCl$$

(b) 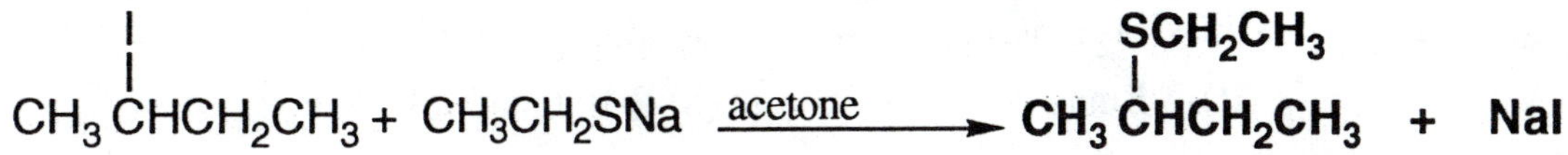 $$CH_3CH(I)CH_2CH_3 + CH_3CH_2SNa \xrightarrow{\text{acetone}} CH_3CH(SCH_2CH_3)CH_2CH_3 + NaI$$

(c) $CH_3CH(CH_3)CH_2CH_2Br + CH_3C{\equiv}C^-Na^+ \xrightarrow{\text{dimethyl sulfoxide}} CH_3CH(CH_3)CH_2CH_2C{\equiv}CCH_3 + NaBr$

Problem 10.4 Write the expected substitution product(s) for each reaction and predict the mechanism by which each product is formed.

(a) $(CH_3)C$–cyclohexyl–Br + NaSH $\xrightarrow{\text{acetone}}$ $(CH_3)C$–cyclohexyl–SH

The SH^- is a very good nucleophile and since the reaction involves a secondary alkyl halide with a good leaving group, the reaction mechanism is S_N2.

(b) $CH_3-CH(Cl)-CH_2-CH_3 + H-C(=O)-OH \longrightarrow$

R enantiomer

The alkyl halide is secondary and chloride is a good leaving group. Formic acid is an excellent ionizing solvent and a poor nucleophile. Therefore, substitution takes place by an S_N1 mechanism and leads to racemization.

$CH_3CH_2C(Cl)(H)CH_3 + H-C(=O)-OH \xrightarrow{S_N1} CH_3CH_2C(OCH{=}O)(H)CH_3 + CH_3CH_2C(HCO_2)(H)CH_3$

formation of carbocation followed by reaction with formic acid

Poblem 10.5 Show how you might convert (R)-2-butanol to (S)-2-butanethiol via a tosylate.

$CH_3CH_2CH(OH)CH_3 \xrightarrow{?} CH_3CH_2CH(SH)CH_3$

(R)-2-Butanol (S)-2-Butanethiol

$CH_3CH_2C(OH)(H)CH_3 \xrightarrow{Ts\text{-}Cl} CH_3CH_2C(OTs)(H)CH_3 \xrightarrow{NaSH} CH_3CH_2C(CH_3)(H)SH$

(R)-2-Butanol **(S)-2-Butanethiol**

Problem 10.6 Complete the indicated reactions.

(a) $CH_3CH(CH_3)CH_2CH_2CH_2Br + CH_3C{\equiv}C^-Na^+ \xrightarrow[DMSO]{substitution}$

$CH_3CH(CH_3)CH_2CH_2CH_2C{\equiv}CCH_3 + NaBr$

(b) $CH_3CH(CH_3)CH_2CH_2CH_2Br + CH_3C{\equiv}C^-Na^+ \xrightarrow[DMSO]{\beta\text{-elimination}}$

$CH_3CH(CH_3)CH_2CH{=}CH_2 + CH_3C{\equiv}CH + NaBr$

Problem 10.7 1-*tert*-Butyl-4-chlorocyclohexane exists as two stereoisomers; one *cis* isomer and one *trans* isomer. Reaction of either isomer with sodium ethoxide in ethanol by an E2 reaction gives 4-*tert*-butylcyclohexene.

Cl–C_6H_{10}–$C(CH_3)_3$ + CH_3CH_2ONa $\xrightarrow[ethanol]{E2}$ C_6H_9–$C(CH_3)_3$ + CH_3CH_2OH + NaCl

1-*tert*-butyl-4-chlorocyclohexane 4-tert-butylcyclohexene

While each isomer gives 4-*tert*-butylcyclohexene, the *cis* isomer undergoes E2 reaction several orders of magnitude faster than the *trans* isomer. How might you account for this experimental observation?

The *tert*-butyl group is the largest substituent on the cyclohexane ring. In the most stable chair conformation of both the *cis* and *trans* isomers, it will be in an equatorial position. In the most stable chair conformation of the *cis* isomer, -Cl is axial and coplanar to -H on adjacent carbons. This chair conformation undergoes β-elimination by an E2 mechanism.

t-Bu (chair, axial Cl, axial H) + $CH_3CH_2O^-$ $\xrightarrow{E2}$ t-Bu–cyclohexene + CH_3CH_2OH + Cl^-

In the most stable chair conformation of the *trans* isomer, chlorine is equatorial and not coplanar to either -H on an adjacent carbon. Interconversion from this chair to the less stable chair results in the -Cl and -H becoming axial and coplanar. It is this conformation that undergoes E2 elimination to give the cycloalkene. The bottom line is that the *cis* isomer undergoes E2 reaction more slowly because of the energy required to convert the more stable chair, but E2-unreactive chair, to the less stable, but E2-reactive, chair.

t-Bu–(chair)–Cl ⇌ (chair with axial t-Bu, axial H and axial Cl; $^{-}OCH_2CH_3$ attacks H) $\xrightarrow{E2}$ t-Bu–cyclohexene $+ CH_3CH_2OH + Cl^-$

more stable chair conformation — less stable chair conformation

Problem 10.8 Predict whether each reaction proceeds predominantly by substitution, elimination, or whether the two compete. Write structural formulas for the major organic product(s).

All will proceed predominantly by substitution.

(a) $CH_3CH_2\text{-}CH(I)\text{-}CH_2\text{-}CH_3 + CH_3O^-Na^+ \xrightarrow{\text{methanol}} CH_3CH_2\text{-}CH(OCH_3)\text{-}CH_2\text{-}CH_3 + Na^+I^-$

(b) cyclohexane with CH_3 (wedge) and Cl (hashed), 1,4 $+ Na^+I^- \xrightarrow{\text{acetone}}$ cyclohexane with CH_3 (wedge) and I (wedge), 1,4 $+ Na^+Cl^-$

(c) $C_6H_5CH_2CH_2Br + Na^+CN^- \xrightarrow{\text{methanol}} C_6H_5CH_2CH_2CN + Na^+Br^-$

Nomenclature

Problem 10.9 Give IUPAC names for the following compounds. Where stereochemistry is shown, include a designation of configuration in your answer.

(a) **Z-2-bromo-2-hexene (*trans*-2-bromo-2-hexene)**

(b) **(R)-3-bromo-3-methylcyclohexene**

(c) ***trans*-1-bromo-4-methylcyclohexane**

(d) $ClCH_2CH_2CH_2CH_2Cl$

1,4-dichlorobutane

(e) **(S)-2-iodooctane**

(f) **(S)-2-bromopentane**

(g) **3-fluorocycloheptene**

(h) $CH_3CH(CH_3)CH_2Cl$

2-methyl-1-chloropropane

(i) **1-chlorobicyclo [2.2.1]heptane**

Problem 10.10 Draw structural formulas for the following compounds:

(a) Allyl iodide

$CH_2{=}CHCH_2I$

(b) (R)-2-Chlorobutane

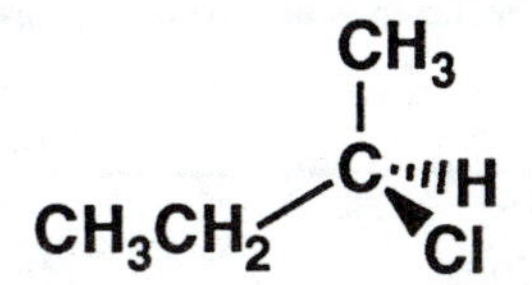

(c) Meso-2,3-dibromobutane

(d) *trans*-1-Bromo-3-isopropylcyclohexane

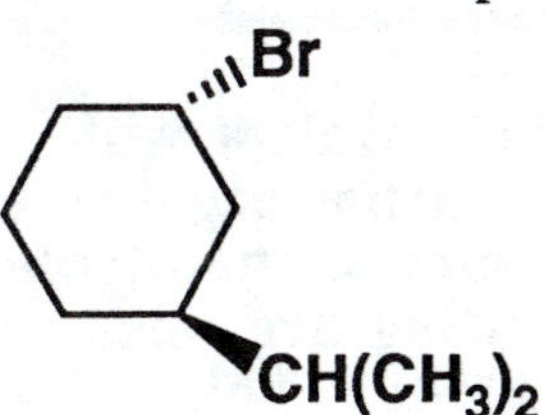

(e) Neopentyl iodide

$CH_3C(CH_3)_2CH_2I$

(e) Cyclobutane bromide

Physical properties

Problem 10.11 Water and methylene chloride are insoluble in each other. When each is added to a test tube, two layers form. Which layer is water and which layer is methylene chloride? Explain.

The upper layer is water and the lower layer is methylene chloride. The heavy chlorine atoms provide for a relatively high formula weight for such a small molecule, and thus a density that is higher than that of water.

Problem 10.12 The boiling point of methylcyclohexane (C_7H_{14}, MW 98.2) is 101°C. The boiling point of perfluoromethylcyclohexane (C_7F_{14}, MW 350) is 76°C. Account for the fact that although the molecular weight of perfluoromethylcyclohexane is over three times that of methylcyclohexane, its boiling point is lower than that of methylcyclohexane.

Name	Molecular formula	Molecular weight (g/mol)	Boiling point (C)
methylcyclohexane	C_7H_{14}	98.2	100.9
perfluoromethylcyclohexane	C_7F_{14}	350.1	76.1

The most significant intermolecular forces of attraction between molecules of each pure substance are dispersion (induced dipole-dipole) forces. Because of the extremely low polarizability of fluorine, perfluoroalkanes are unique among the halocarbons in having boiling points close to that of their parent alkanes. In the case of the molecules considered in this problem, the boiling point of the perfluoro derivative is lower than that of the parent cycloalkane.

Problem 10.13 Account for the fact that among the chlorinated derivatives of methane, chloromethane has the largest dipole moment and tetrachloromethane has the smallest dipole moment.

Name	Molecular formula	Dipole moment (Debye units)
chloromethane	CH_3Cl	1.87
dichloromethane	CH_2Cl_2	1.60
trichloromethane	$CHCl_3$	1.01
tetrachloromethane	CCl_4	0

Each C-Cl bond is polar covalent with carbon bearing a partial positive charge and chlorine bearing a partial negative charge. The dipole moment of each molecule is the vector sum of the dipole moments of each bond. The lengths of the arrows on the following drawings represent the relative magnitudes of the vector sums of the individual bond dipole moments for chloromethane, dichloromethane, and trichloromethane.

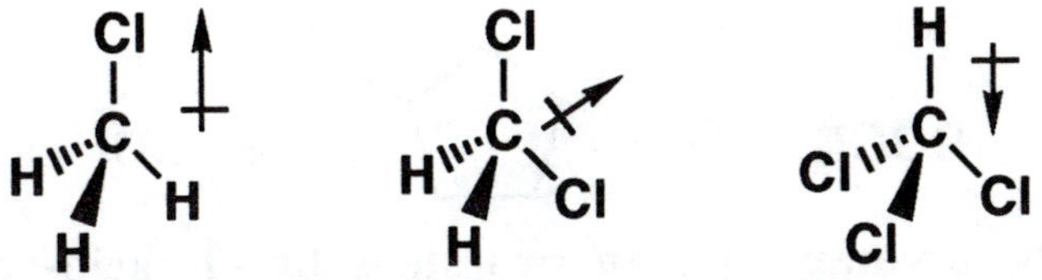

The product of charge times distance is greater for chloromethane than for dichloromethane or trichloromethane. Because tetrachloromethane is tetrahedral, the vector sum of the individual bond dipoles is zero; that is, the center of positive charge in the molecule as well as the center of negative charge is on carbon.

Synthesis of Alkyl Halides

Problem 10.14 Show reagents and conditions to bring about the following conversions.

(a)

Addition of HCl to 2-methyl-2-pentene.

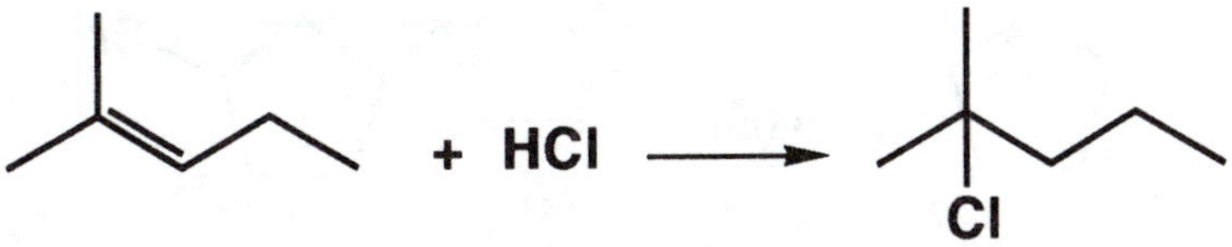

+ HCl →

(b) $CH_3CH{=}CHCH_3 \longrightarrow CH_3CH{=}CHCH_2Br$

Radical bromination using NBS at room temperature.

$$CH_3CH{=}CHCH_3 + NBS \xrightarrow[\text{peroxides}]{\text{light or}} CH_3CH{=}CHCH_2Br + HBr$$

(c) $CH_3CH{=}CHCH_3 \longrightarrow CH_3CHBrCHBrCH_3$

One step bromination with Br_2.

$$CH_3CH{=}CHCH_3 + Br_2 \longrightarrow CH_3CHBrCHBrCH_3$$

(d)

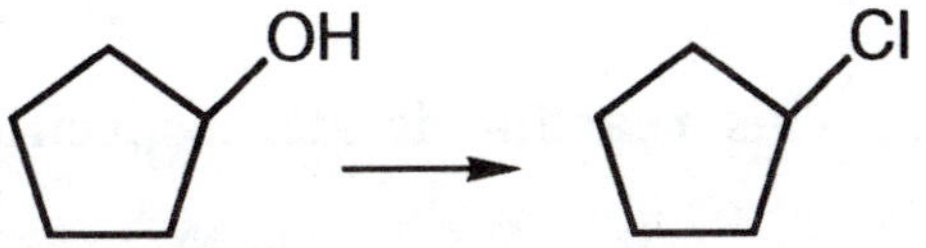

This transformation can be accomplished in one step by reaction of cyclopentanol with concentrated HCl or with thionyl chloride, $SOCl_2$.

+ HCl → + H_2O

$$\text{cyclopentanol} + SOCl_2 \longrightarrow \text{chlorocyclopentane} + HCl + SO_2$$

Alternatively, it can be accomplished in two steps by (1) acid-catalyzed dehydration of cyclopentanol to cyclopentene followed by (2) addition of HCl to the cycloalkene.

$$\text{cyclopentanol} \xrightarrow[-H_2O]{H_2SO_4} \text{cyclopentene} \xrightarrow{HCl} \text{chlorocyclopentane}$$

(e) cyclopentene $\longrightarrow$ bromocyclopentane

Electrophilic addition of HBr to cyclopentene, or free radical addition of HBr in the presence of peroxides or other radical initiators.

$$\text{cyclopentene} + HBr \longrightarrow \text{bromocyclopentane}$$

(f) $CH_3{-}CH_2{-}C{\equiv}CH \longrightarrow CH_3{-}CH_2{-}CBr_2{-}CH_3$

Electrophilic addition of two HBr to 1-butyne.

$$CH_3{-}CH_2{-}C{\equiv}CH \xrightarrow{HBr} CH_3{-}CH_2{-}CBr{=}CH_2 \xrightarrow{HBr} CH_3{-}CH_2{-}CBr_2{-}CH_3$$

(g) cyclohexene $\longrightarrow$ trans-1,2-dichlorocyclohexane

Electrophilic addition of Cl_2 to cyclohexene. This reaction is stereospecific and results in anti addition.

$$\text{cyclohexene} + Cl_2 \longrightarrow \text{trans-1,2-dichlorocyclohexane}$$

(h) 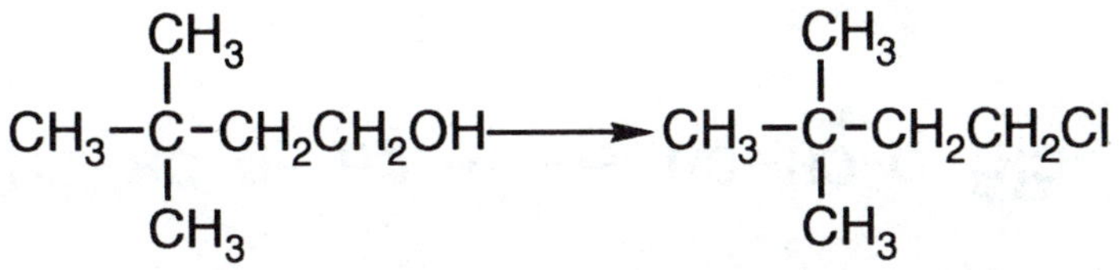 $CH_3-C(CH_3)_2-CH_2CH_2OH \longrightarrow CH_3-C(CH_3)_2-CH_2CH_2Cl$

Reaction of this primary alcohol with two branches on the β-carbon with concentrated HCl gives almost exclusively the rearranged alkyl chloride (Section 10.7G). The reagent most commonly used to avoid this type of rearrangement is thionyl chloride, $SOCl_2$.

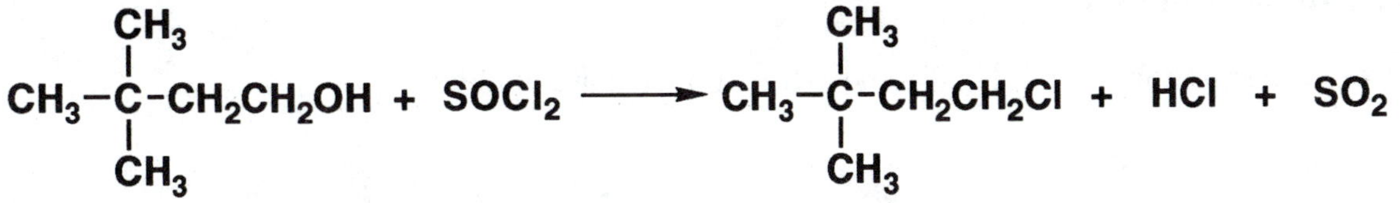

$$CH_3-C(CH_3)_2-CH_2CH_2OH + SOCl_2 \longrightarrow CH_3-C(CH_3)_2-CH_2CH_2Cl + HCl + SO_2$$

<u>Nucleophilic Aliphatic Substitution</u>

<u>Problem 10.15</u> Draw a structural formula for the most stable carbocation of each molecular formula and indicate how each might be formed.

(a) $C_4H_9^+$

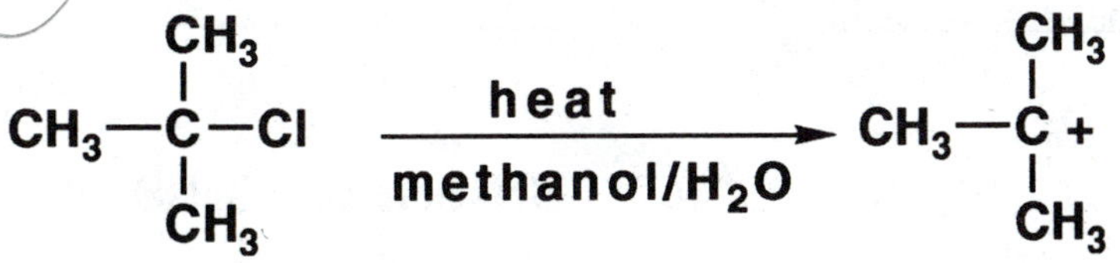

$$CH_3-C(CH_3)_2-Cl \xrightarrow[\text{methanol}/H_2O]{\text{heat}} CH_3-\overset{+}{C}(CH_3)_2$$

(b) $C_3H_7^+$

$$CH_3-CH(CH_3)-Cl \xrightarrow[\text{methanol}/H_2O]{\text{heat}} CH_3-\overset{+}{C}H(CH_3)$$

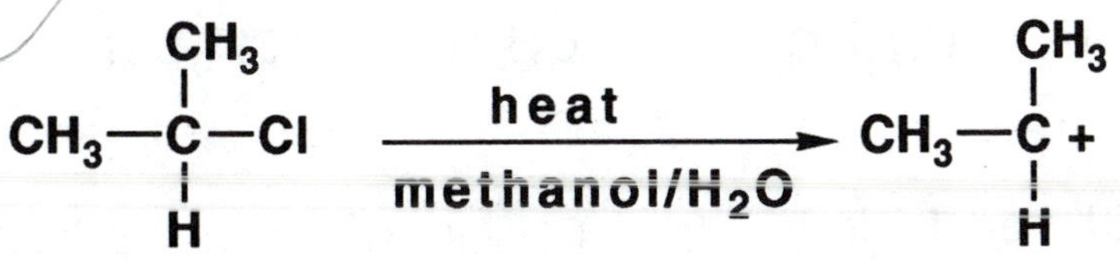

(c) $C_8H_{15}^+$

$$(CH_3)_2C(OH)-C(CH_3)=C(CH_3)_2 \xrightarrow{H^+} (CH_3)_2C=C(CH_3)-\overset{+}{C}(CH_3)_2 \longleftrightarrow (CH_3)_2\overset{+}{C}-C(CH_3)=C(CH_3)_2$$

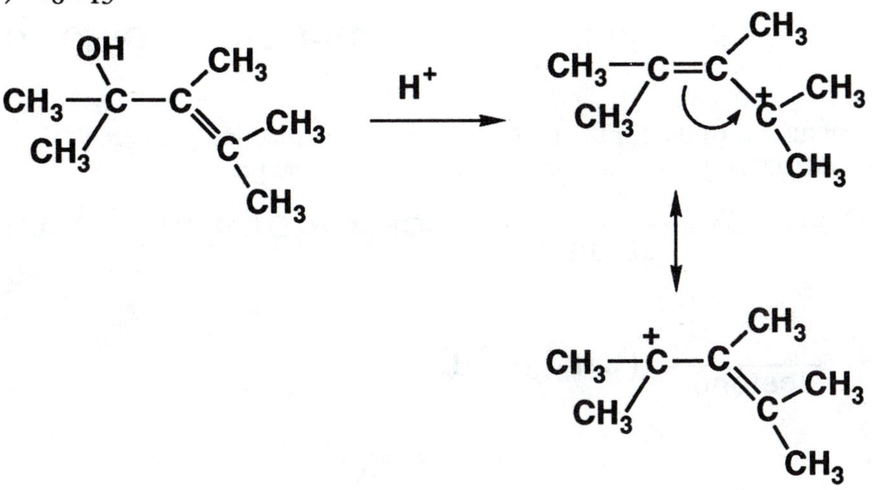

(d) $C_3H_7O^+$

$$CH_3{-}O{-}CH{=}CH_2 \xrightarrow{H^+} CH_3{-}O{-}\overset{+}{C}H{-}CH_3 \longleftrightarrow CH_3{-}\overset{+}{O}{=}CH{-}CH_3$$

Problem 10.16 Reaction of 1-bromopropane and sodium hydroxide in ethanol follows an S_N2 mechanism. What will happen to the rate of this reaction if:
(a) the concentration of NaOH is doubled?

The rate will double.

(b) the concentration of both NaOH and 1-bromobutane are doubled?

The rate will quadruple.

(c) the volume of the solution in which the reaction is carried out is doubled?

The rate will decrease by a factor of four because the concentration of each reactant is halved.

Problem 10.17 From each pair, select the stronger acid.

(a) H_2O or OH^- (b) $CH_3\overset{O}{\overset{\|}{C}}O^-$ or OH^- (c) CH_3SH or CH_3S^-

$OH^- > H_2O$ **$OH^- > CH_3\overset{O}{\overset{\|}{C}}O^-$** **$CH_3S^- > CH_3SH$**

(d) Cl^- or I^- in DMSO (e) Cl^- or I^- in methanol (f) CH_3OCH_3 or CH_3SCH_3

$Cl^- > I^-$ **$I^- > Cl^-$** **$CH_3SCH_3 > CH_3OCH_3$**

Problem 10.18 Draw the structural formula for the product of each S_N2 reaction. Where configuration of the stating material is given, show the configuration of the product.

(a) $CH_3CH_2CH_2Cl + CH_3CH_2ONa \xrightarrow[\text{ethanol}]{} \mathbf{CH_3CH_2CH_2OCH_2CH_3 + NaCl}$

(b) $(CH_3)_3N\text{:} + CH_3I \xrightarrow[\text{acetone}]{} \mathbf{(CH_3)_4N^+\ I^-}$

(c) $C_6H_5CH_2Br + NaCN \xrightarrow{\text{acetone}} C_6H_5CH_2CN + NaBr$

(d) CH_3-cyclohexyl-Cl $+ CH_3SNa \xrightarrow{\text{ethanol}} CH_3$-cyclohexyl-$SCH_3$ $+ NaCl$

(e) $CH_3CH_2CH_2Cl + CH_3C{\equiv}C{:}^{-}\ Na^{+} \longrightarrow CH_3CH_2CH_2C{\equiv}CCH_3 + NaCl$

(f) cyclopentyl-$CH_2Cl + {:}NH_3 \xrightarrow{\text{ethanol}}$ cyclopentyl-$CH_2\overset{+}{N}H_3Cl^{-}$

(g) morpholine (O, :NH) $+ CH_3(CH_2)_6CH_2Cl \xrightarrow{\text{ethanol}}$ morpholinium (O, $^{+}N(H)-CH_2(CH_2)_6CH_3$) Cl^{-}

(h) $CH_3CH_2CH_2OSO_2CH_3 + NaCN \xrightarrow{\text{acetone}} CH_3CH_2CH_2CN + CH_3SO_3Na$

(i) cyclopentyl-$CH_2OH + Cl{-}SO_2{-}C_6H_4{-}CH_3 \xrightarrow{\text{pyridine}}$

cyclopentyl-$CH_2O{-}SO_2{-}C_6H_4{-}CH_3$

Problem 10.19 You were told that each reaction in the previous problem proceeds by an S_N2 mechanism. Suppose you were not told the mechanism. Describe how you could conclude from the structure of the alkyl halide, the nucleophile, and the solvent that each reaction is in fact an S_N2 reaction.

(a) A primary halide, strong nucleophile/strong base in ethanol, a moderately ionizing solvent all favor S_N2.
(b) Trimethylamine is a moderate nucleophile. A methyl halide in acetone, a weakly ionizing solvent, all work together to favor S_N2.
(c) Cyanide is a moderate nucleophile. A primary halide in acetone, a weakly ionizing solvent, all work together to favor S_N2.
(d) The alkyl chloride is secondary, so either an S_N1 or S_N2 mechanism is possible. Ethylsulfide ion is a strong nucleophile, but weak base. It therefore reacts by an S_N2 pathway.
(e) The sodium salt of the terminal alkyne is a moderate nucleophile, but also a strong base. Because the halide is primary, an S_N2 pathway is favored.
(f) Ammonia is a weak base and good nucleophile, and the halide is primary. Therefore S_N2 is favored.
(g) The major factor here favoring an S_N2 pathway is that the leaving group is halide and on a primary carbon.
(h) The cyanide anion is a strong nucleophile and mesylate is a good leaving group on a primary carbon. Therefore S_N2 is favored.
(i) The chlorine on sulfur is a good leaving group and the alcohol is a strong enough nucleophile to favor an S_N2 reaction.

Problem 10.20 Treatment of 1-aminoadamantane with methyl 2,4-dibromobutanoate involves two successive S_N2 reactions and gives compound A which is an intermediate in the synthesis of carmantidine. Propose a structural formula for this intermediate. Carmantidine has been used in treating the spasms associated with Parkinson's disease.

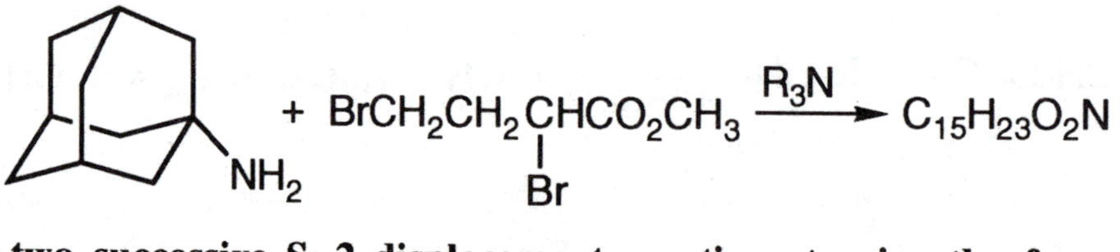

NH_2 + $BrCH_2CH_2CHCO_2CH_3$ (Br on C-2) $\xrightarrow{R_3N}$ $C_{15}H_{23}O_2N$

There are two successive S_N2 displacement reactions to give the four-member ring.

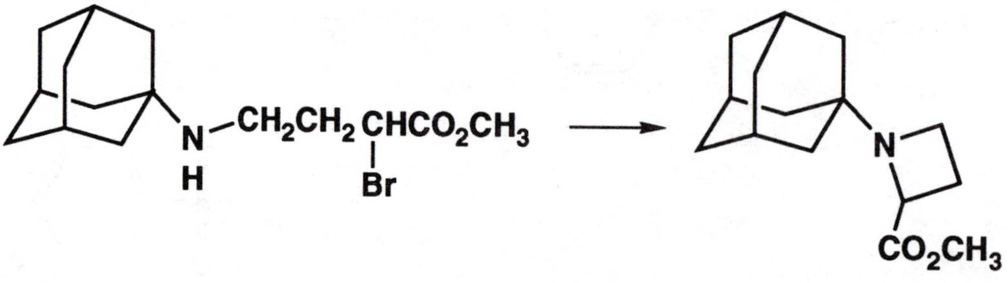

N(H)–$CH_2CH_2CHCO_2CH_3$ (Br on C-2) ⟶ N (four-membered ring) with CO_2CH_3

Problem 10.21 Select the member from each pair that shows the faster rate of S_N2 reaction with KI in acetone.

The relative rates of S_N2 reactions for pairs of molecules in this problem depend on two factors: (1) bromine is a better leaving group than chlorine and (2) a

primary carbon without β-branching is less hindered and more reactive toward S_N2 substitution than a primary carbon with one, two or three branches on the β-carbon.

(a) $CH_3CH_2CH_2CH_2Cl$ or $CH_3\underset{}{\overset{CH_3}{C}}HCH_2Cl$

$CH_3CH_2CH_2CH_2Cl$ faster than $CH_3\overset{CH_3}{C}HCH_2Cl$

(b) $CH_3CH_2CH_2CH_2Cl$ or $CH_3CH_2CH_2CH_2Br$

$CH_3CH_2CH_2CH_2Br$ faster than $CH_3CH_2CH_2CH_2Cl$

(c) $CH_3\overset{CH_3}{C}HCH_2CH_2Cl$ or $CH_3\underset{CH_3}{\overset{CH_3}{C}}CH_2Cl$

$CH_3\overset{CH_3}{C}HCH_2CH_2Cl$ faster than $CH_3\underset{CH_3}{\overset{CH_3}{C}}CH_2Cl$

(d) $CH_3CH_2CH_2\overset{Br}{C}HCH_3$ or $CH_3\underset{CH_3}{C}H\overset{Br}{C}HCH_3$

$CH_3CH_2CH_2\overset{Br}{C}HCH_3$ faster than $CH_3\underset{CH_3}{C}H\overset{Br}{C}HCH_3$

Problem 10.22 What hybridization best describes the reacting carbon in the S_N2 transition state? Would electron withdrawing groups or electron donating groups be expected to better stabilize the transition state?

The hybridization state of the reacting carbon is best described as sp^2 in the S_N2 transition state. Electron withdrawing groups would be expected to stabilize the transition state.

Problem 10.23 Attempts to prepare optically active iodides by nucleophilic displacement on optically active compounds with I^- normally produce racemic alkyl iodides. Why are the product alkyl iodides racemic?

Iodide is a good nucleophile as well as a good leaving group. The alkyl iodide that is formed will therefore react with other iodide nucleophiles under the conditions of the reaction, thus leading to racemization.

Problem 10.24 Draw the structural formula for the product of each S_N1 reaction. Where configuration of the starting material is given, show the configuration of the product.

(a) (S)-Ph-CH(Cl)CH$_2$CH$_3$ + CH$_3$CH$_2$OH $\xrightarrow[\text{ethanol}]{}$ **(R,S)-Ph-CH(OCH$_2$CH$_3$)CH$_2$CH$_3$ + HCl**

(b) 1-chloro-1-methylcyclopentane (CH_3, Cl) + CH_3OH $\xrightarrow[\text{methanol}]{}$ **1-methoxy-1-methylcyclopentane (CH_3, OCH_3) + HCl**

(c) $CH_3-C(CH_3)_2-Cl$ + CH_3COH (O) $\xrightarrow[\text{acetic acid}]{}$ **$CH_3-C(CH_3)_2-O-C(=O)-CH_3$ + HCl**

(d) 3-bromocyclohexene (Br) + CH_3OH $\xrightarrow[\text{methanol}]{}$ **3-methoxycyclohexene (OCH_3) + HBr**

Problem 10.25 You were told that each reaction in the previous problem proceeds by an S_N1 mechanism. Suppose you were not told the mechanism. Describe how you could conclude from the structure of the alkyl halide, the nucleophile, and the solvent that each reaction is in fact an S_N1 reaction.

(a) Chlorine is a good leaving group and the resulting secondary carbocation is a very stable carbocation intermediate. Ethanol is a moderately ionizing solvent and a poor nucleophile.
(b) Methanol is a moderately ionizing solvent and a poor nucleophile. Chlorine is a good leaving group and the resulting carbocation is tertiary.
(c) Acetic acid is a strongly ionizing solvent and a poor nucleophile. Chlorine is a good leaving group and the resulting carbocation is tertiary.
(d) Methanol is a moderately ionizing solvent and a poor nucleophile. Bromine is a good leaving group, and the resulting carbocation is both secondary and allylic.

Problem 10.26 Vinylic halides such as vinyl bromide, CH_2=CHBr, do not undergo S_N1 reactions. Nor do they undergo S_N2 reactions. What factors account for this lack of reactivity of vinylic halides?

In vinyl bromide, the bromine atom is bonded to an sp^2 hybridized carbon atom. An S_N1 reaction would give a vinyl carbocation, but such a carbocation is high in energy and very difficult to generate. In order to undergo an S_N2 reaction, the nucleophile must approach in a direction opposite the C-X bond. This trajectory is not possible for a vinyl halide.

Problem 10.27 Select the member of each pair that will undergo S_N1 solvolysis in aqueous ethanol more rapidly.

Relative rates for each pair of compounds listed in this problem depend on a combination of two factors: (1) bromine is a better leaving group than chlorine and (2) the stability of the resulting carbocation.

(a) $CH_3CH_2CH_2CH_2Cl$ or $CH_3-C(CH_3)_2-Cl$

$CH_3-C(CH_3)_2-Cl$ **faster than** $CH_3CH_2CH_2CH_2Cl$

The energy of activation for formation of a tertiary carbocation is lower than that for formation of a primary carbocation.

(b) $CH_3-C(CH_3)_2-Cl$ or $CH_3-C(CH_3)_2-Br$

$CH_3-C(CH_3)_2-Br$ **faster than** $CH_3-C(CH_3)_2-Cl$

Bromine is a better leaving group than chlorine.

(c) CH_2=$CHCH_2Cl$ or $CH_3CH_2CH_2Cl$

CH_2=$CHCH_2Cl$ **faster than** $CH_3CH_2CH_2Cl$

The energy of activation for formation of a resonance-stabilized allylic carbocation is lower than that for formation of a primary carbocation.

(d) 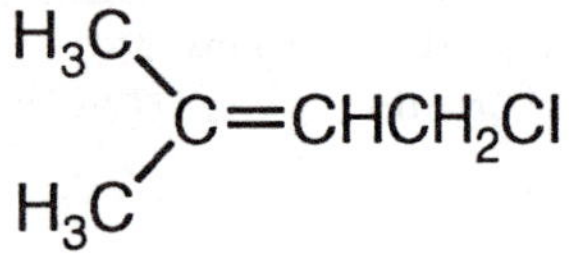 $(H_3C)_2C{=}CHCH_2Cl$ or $H_2C{=}CHCH_2Cl$

$(H_3C)_2C{=}CHCH_2Cl$ faster than $H_2C{=}CHCH_2Cl$

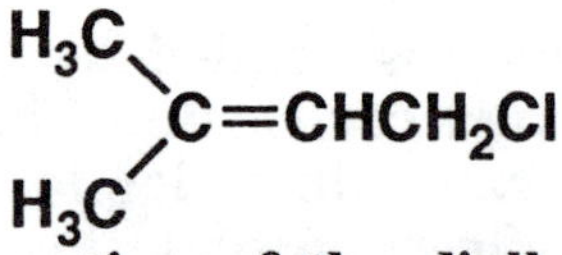

The energy of activation for formation of the dialkyl allylic carbocation is lower than that for formation of an unsubstituted allylic carbocation.

(e) $CH_3(CH_2)_3CH_2Cl$ or $CH_3(CH_2)_2CHClCH_3$

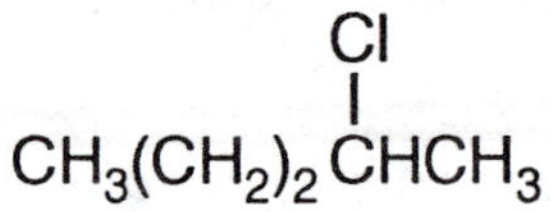

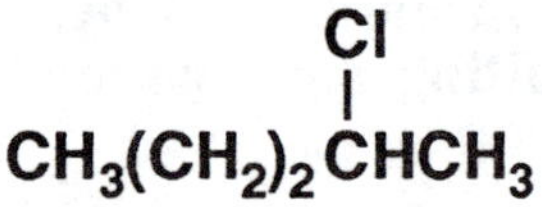

$CH_3(CH_2)_2CHClCH_3$ faster than $CH_3(CH_2)_3CH_2Cl$

Energy of activation for formation of a secondary carbocation is lower than that for formation of a primary carbocation.

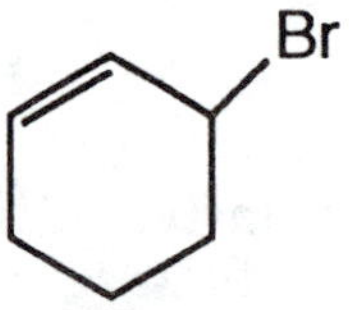

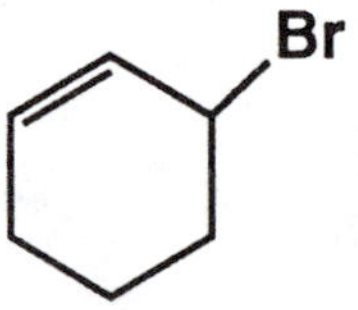

(f) 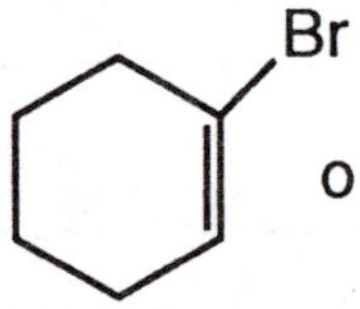 1-bromocyclohexene or 3-bromocyclohexene

3-bromocyclohexene faster than 1-bromocyclohexene

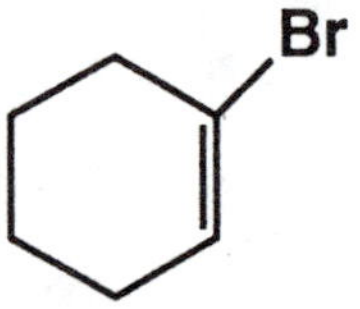

Energy of activation for formation of an allylic carbocation is lower than that for formation of a vinyl carbocation.

Problem 10.28 Account for the following rates of solvolysis under experimental conditions favoring S_N1 reaction.

	$CH_3\ddot{O}CH_2CH_2Cl$	$CH_3CH_2CH_2CH_2Cl$	$CH_3CH_2\ddot{O}CH_2Cl$
relative rate of solvolysis (S_N1)	0.2	1	10^9

1-Chloro-1-ethoxyethane (chloromethyl ethyl ether) reacts the fastest by far in a solvolysis reaction because the carbocation produced by loss of the chlorine atom is stabilized by the adjacent ether oxygen atom. Thus, the energy of activation for this reaction is significantly lower than for the other two molecules. The most important resonance forms are shown below for the stabilized carbocation.

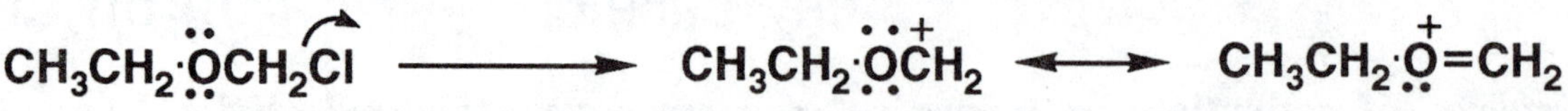

$$CH_3CH_2\ddot{O}CH_2Cl \longrightarrow CH_3CH_2\ddot{O}\overset{+}{C}H_2 \longleftrightarrow CH_3CH_2\overset{+}{O}{=}CH_2$$

1-Chloro-2-methoxyethane (chloroethyl methyl ether) reacts the slowest because the carbocation produced during the reaction is somewhat destabilized by the ether oxygen atom that is two carbon atoms away. This is because oxygen is more electronegative than carbon, so there is a partial positive charge on the carbon atoms bonded to the oxygen. Thus, the carbocation produced by departure of the chlorine atom is adjacent to this partially positive carbon atom; a destabilizing arrangement.

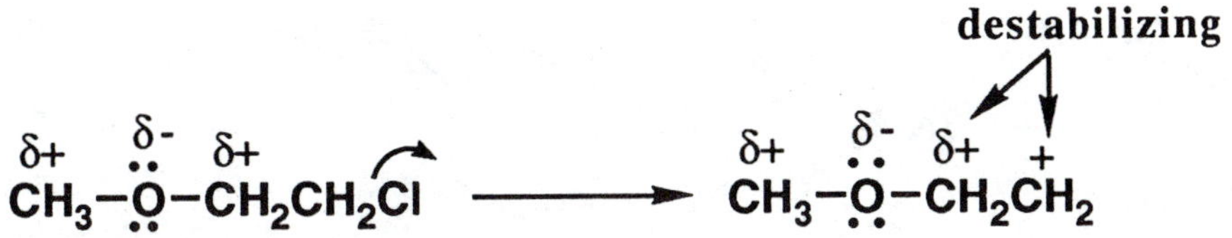

destabilizing

$$\overset{\delta+}{CH_3}-\overset{\delta-}{\ddot{\underset{\cdot\cdot}{O}}}-\overset{\delta+}{CH_2}CH_2Cl \longrightarrow \overset{\delta+}{CH_3}-\overset{\delta-}{\ddot{\underset{\cdot\cdot}{O}}}-\overset{\delta+}{CH_2}\overset{+}{CH_2}$$

Please note that in the case of 1-chloro-2-methoxyethane, there is no way to produce resonance forms with any positive charge on the oxygen atom like that shown for 1-chloro-1-ethoxymethane above.

Problem 10.29 Not all tertiary halides are prone to undergo S_N1 reactions. For example, 1-iodobicyclo[2.2.2]octane is very unreactive toward S_N1 chemistry. What feature of this molecule is responsible for such lack of reactivity?

bridgehead carbon atom

I

1-iodobicyclo[2.2.2]octane

In order to form a cation, great geometrical strain would have to be produced in the molecule. This is because carbocations prefer to be trigonal planar (sp^2 hybridized), and loss of iodine would place a carbocation at the bridgehead position. However, the bicyclic structure of the molecule enforces a tetrahedral geometry at the bridgehead position (109.5° bond angles), thus preventing formation of the carbocation.

Problem 10.30 Show how you might synthesize the following compounds from an alkyl halide and a nucleophile.

(a) CN

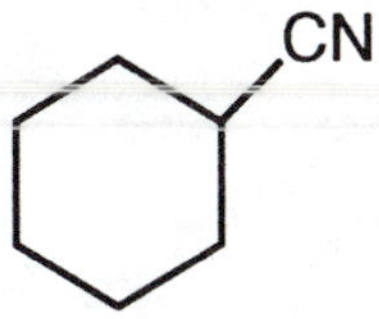

Treatment of a halocyclohexane with cyanide.

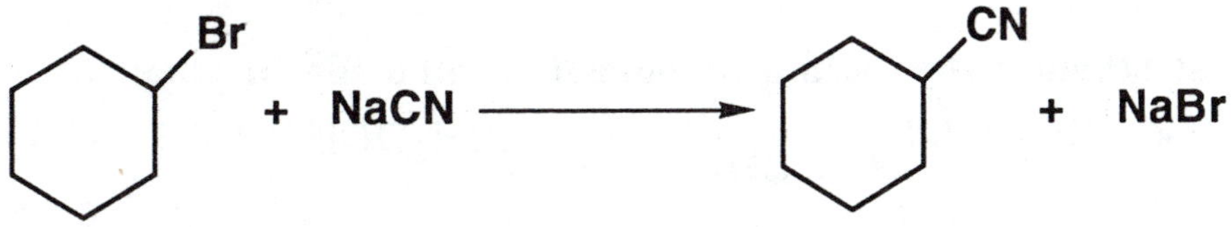

Br + NaCN ⟶ CN + NaBr

(b) CH_2NH_2

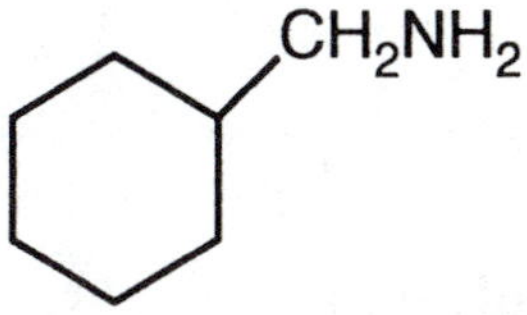

Treatment of chloromethylcyclohexane with two moles of ammonia. The first mole of ammonia is for displacement of chlorine. The second mole of ammonia is to neutralize the HCl formed in the substitution reaction.

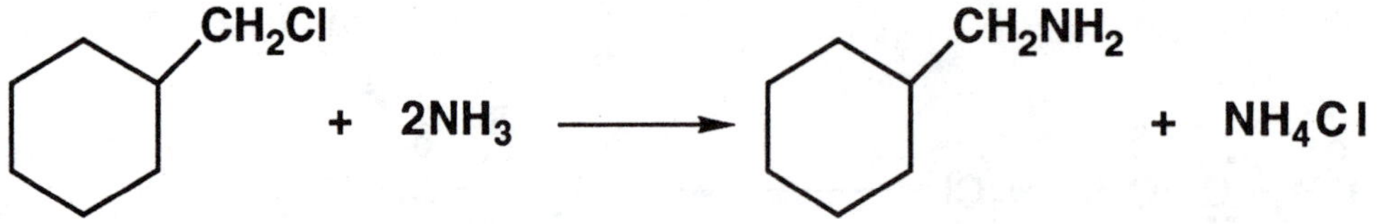

$C_6H_{11}CH_2Cl + 2NH_3 \longrightarrow C_6H_{11}CH_2NH_2 + NH_4Cl$

(c) $C_6H_{11}O\text{-}C(=O)\text{-}CH_3$

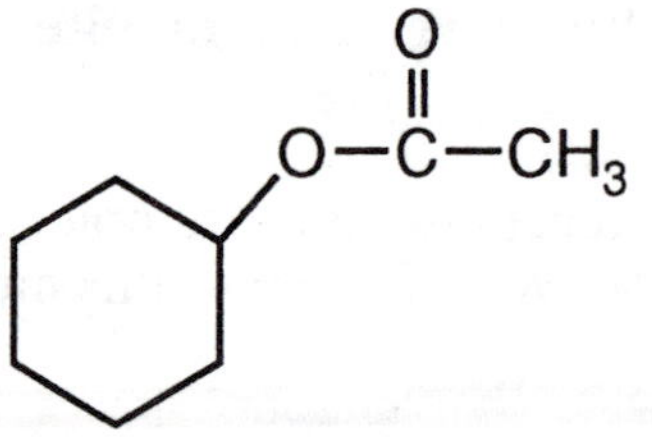

Treatment of a halocyclohexane with the sodium salt of acetic acid.

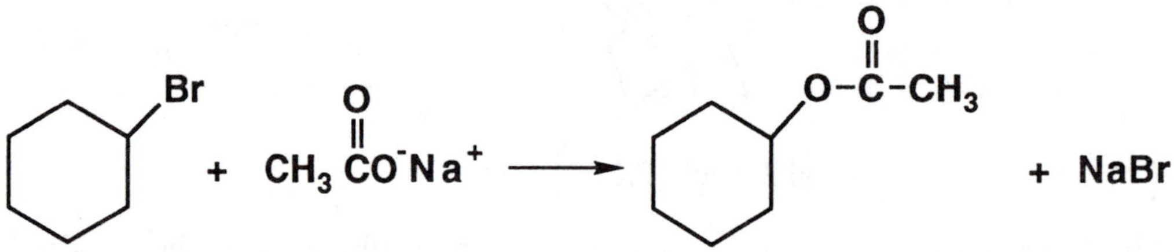

$C_6H_{11}Br + CH_3CO^-Na^+ \longrightarrow C_6H_{11}O\text{-}C(=O)\text{-}CH_3 + NaBr$

(d) $CH_3(CH_2)_3CH_2SH$

Treatment of a 1-halopentane with sodium hydrosulfide.

$CH_3(CH_2)_3CH_2Br + HS^-Na^+ \longrightarrow CH_3(CH_2)_3CH_2SH + NaBr$

(e) $CH_3(CH_2)_5C{\equiv}CH$

Treatment of a 1-halohexane with the sodium salt of acetylene.

$CH_3(CH_2)_4CH_2Br + HC{\equiv}C^-Na^+ \longrightarrow CH_3(CH_2)_4C{\equiv}CH + NaCl$

(f) $CH_3CH_2OCH_2CH_3$

Treatment of a haloethane with sodium or potassium ethoxide in ethanol.

$$CH_3CH_2O^-Na^+ + CH_3CH_2I \xrightarrow[CH_3CH_2OH]{} CH_3CH_2OCH_2CH_3 + NaI$$

(g) cis-3-methylcyclopentanethiol: H_3C and SH (both wedged) on cyclopentane

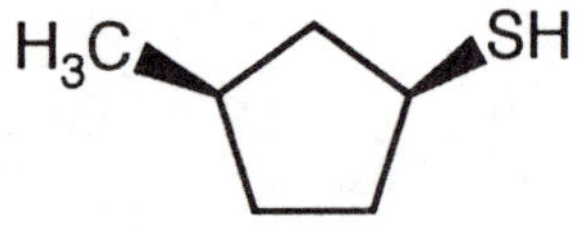

Treatment of the appropriate halocyclopentane with the thiol anion.

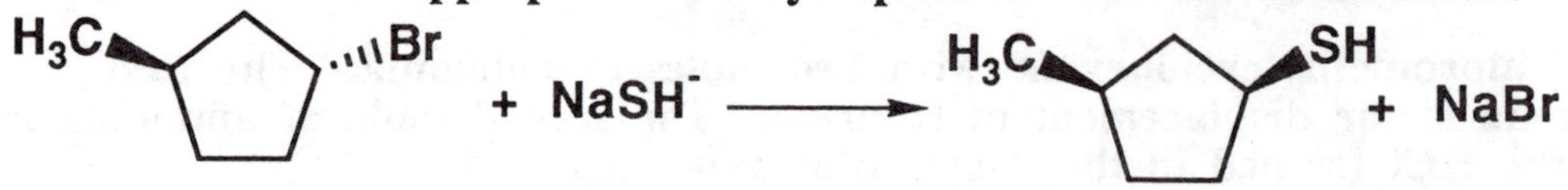

H_3C(wedge)-cyclopentane-(hash)Br $+ NaSH^- \longrightarrow H_3C$(wedge)-cyclopentane-(wedge)SH $+ NaBr$

Problem 10.31 Propose a synthesis for (Z)-9-tricosene (muscalure), the sex pheromone for the common house fly (*Musca domestica*) starting with acetylene and alkyl halides as sources for carbon atoms.

$CH_3(CH_2)_7CH=CH(CH_2)_{12}CH_3$ (cis; both H on the same side)

Muscalure

$$H-C\equiv C-H \xrightarrow[\substack{2)\ CH_3(CH_2)_6CH_2Br \\ 3)\ H_2O}]{1)\ NaNH_2} CH_3(CH_2)_7-C\equiv C-H$$

$$\xrightarrow[\substack{2)\ CH_3(CH_2)_{11}CH_2Br \\ 3)\ H_2O}]{1)\ NaNH_2} CH_3(CH_2)_7-C\equiv C-(CH_2)_{12}CH_3 \xrightarrow[\substack{Pd/CaCO_3 \\ \text{(Lindlar Catalyst)}}]{H_2}$$

$CH_3(CH_2)_7(H)C=C(H)(CH_2)_{12}CH_3$ (cis)

Problem 10.32 Show reagents and experimental conditions required to bring about the following transformations:

$CH_3CH_2CH_3 \xrightarrow{(1)} CH_3CHBrCH_3 \xrightarrow{(2)} CH_3CH=CH_2 \xrightarrow{(3)} CH_3CHClCH_2Cl$

$CH_3CHClCH_2Cl \xrightarrow{(4)} CH_3C\equiv CH \xrightarrow{(5)} CH_3CCl=CH_2 \xrightarrow{(6)} CH_3CCl_2CH_3$

$CH_3C\equiv CH \xrightarrow{(7)} CH_3C\equiv CCH_3$

$CH_3C\equiv CCH_3 \xrightarrow{(8)} CH_3COCH_2CH_3$

$CH_3C\equiv CCH_3 \xrightarrow{(9)} (H_3C)(Br)C=C(Br)(CH_3)$ (trans)

$CH_3C\equiv CCH_3 \xrightarrow{(10)} (H_3C)(H)C=C(H)(CH_3)$ (trans)

$CH_3C\equiv CCH_3 \xrightarrow{(11)} (H_3C)(H)C=C(H)(CH_3)$ (cis)

(1) Br_2 and hv
(2) $tBuO^-K^+$ (E2 elimination)
(3) Cl_2
(4) 1) $NaNH_2$ 2) H_2O
(5) HCl
(6) HCl
(7) 1) $NaNH_2$ 2) CH_3I
(8) H_2O, H_2SO_4, $HgSO_4$
(9) Br_2
(10) Li, NH_3 (*l*)
(11) H_2, Pd/$CaCO_3$ (Lindlar Catalyst)

Problem 10.33 When the phase transfer catalyst tetrabutylammonium hydrogen sulfate is used at elevated temperatures, some decomposition of the catalyst occurs. For example, when the nucleophile is NaCN, the catalyst decomposition products are tributylamine (Bu_3N) and $CH_3CH_2CH_2CH_2CN$. How are these products formed?

This product is formed as the result of an S_N2 reaction, in which the CN^- is acting as the nucleophile and tributyl amine is acting as a leaving group.

$$CN^- + CH_3CH_2CH_2CH_2{-}N^+(CH_2CH_2CH_2CH_3)_3 \xrightarrow{S_N2} CH_3CH_2CH_2CH_2CN + N{-}(CH_2CH_2CH_2CH_3)_3$$

β-Elimination Reactions

Problem 10.34 Draw structural formulas for the alkene(s) formed by treatment of each haloalkane with sodium ethoxide in ethanol. Assume that elimination is by an E2 mechanism.

The major and minor products for each E2 reaction are shown. Where there is more than one combination of leaving groups anti and coplanar, the major product is the more substituted alkene (the so-called Zaitzev product).

(a) $CH_3CH(Br)C(CH_3)_2CH_3$ (Br on C2; two CH_3 on C3)

$CH_2{=}CHC(CH_3)_2CH_3$

(b) 1-chloro-1-methylcyclohexane (ring carbon bearing CH_3 and Cl)

1-methylcyclohexene (CH_3) **(major)** + methylenecyclohexane (CH_2)

(c) 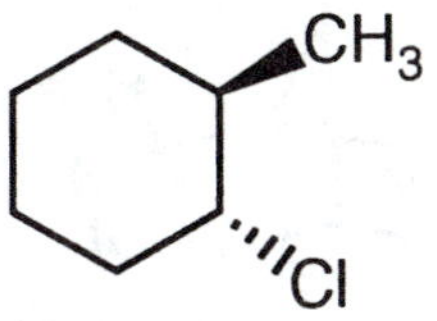 CH_3 Cl

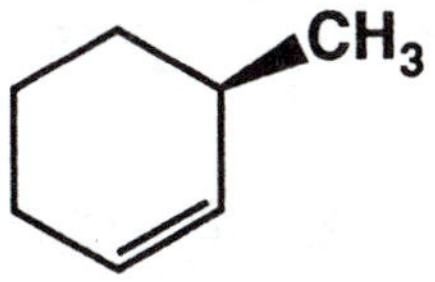

CH_3

(the other elimination cannot occur because the required anti co-planar geometry is not possible)

(d) 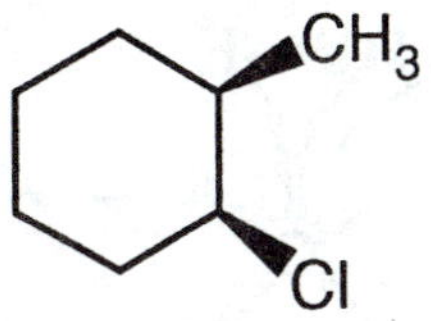 CH_3 Cl

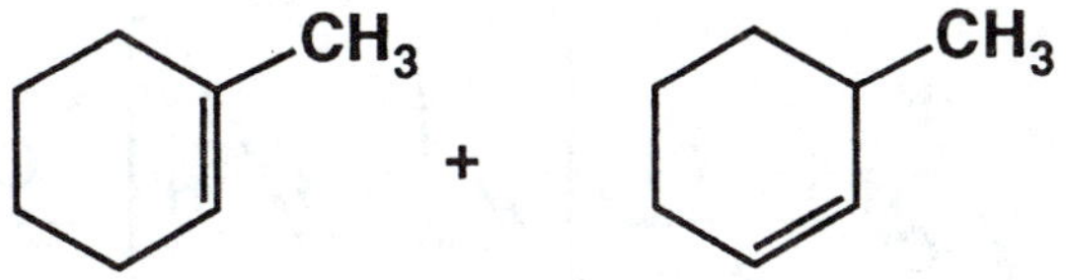

CH_3 + CH_3

(major)

(e) 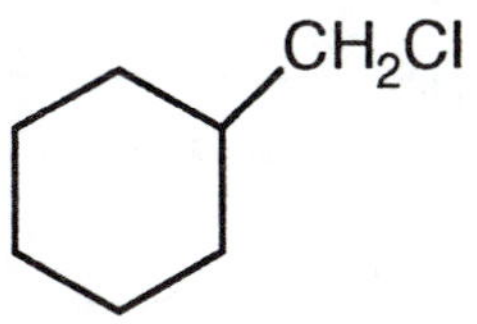 CH_2Cl

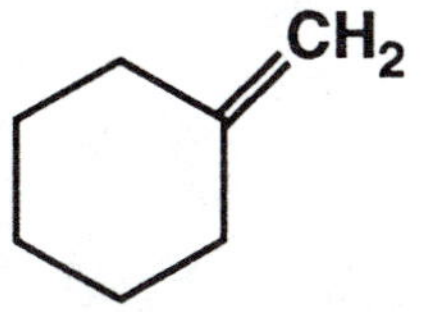

CH_2

(f) 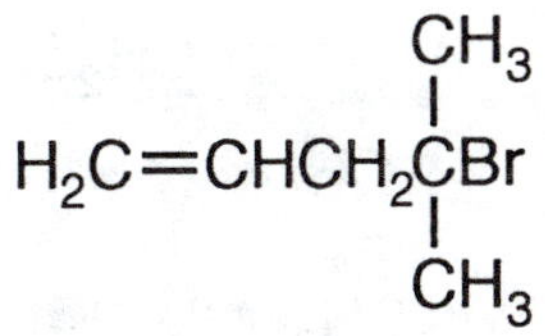 $H_2C{=}CHCH_2CBr(CH_3)_2$

$H_2C{=}CH{-}CH{=}C(CH_3)_2$

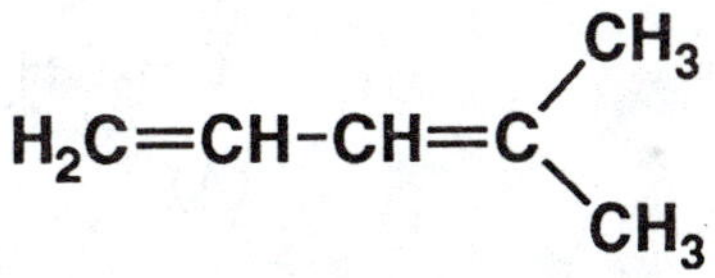

Problem 10.35 Bicyclo[2.2.1]-2,5-heptadiene can be prepared in two steps from cyclopentadiene and vinyl chloride. Provide a mechanism for each step. *Hint:* Review the Diels-Alder reaction (Section 7.5).

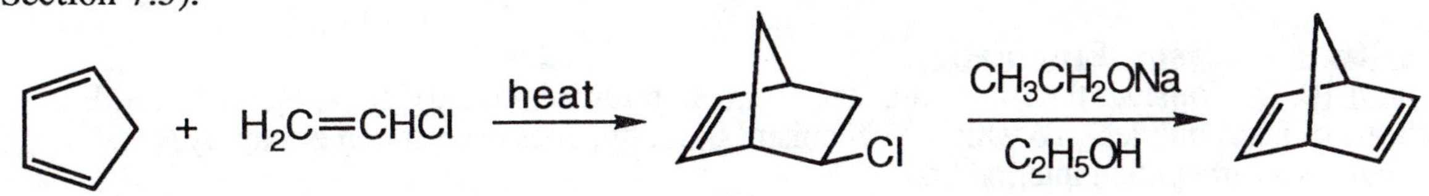

+ $H_2C{=}CHCl$ —heat→ Cl —CH_3CH_2ONa / C_2H_5OH→

The standard Diels-Alder reaction is followed by an E1 elimination reaction. Note that the elimination step cannot follow an E2 mechanism, because the bicyclic structure prevents an anti and coplanar geometry for any of the H atoms on the carbon atom adjacent to the chlorine atom.

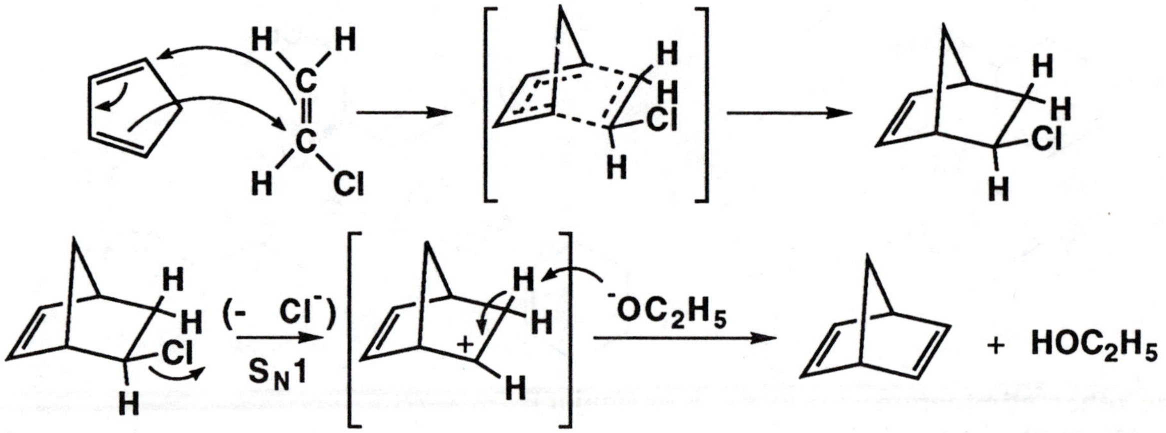

(A) (B)

Problem 10.36 Following are diastereomers (A) and (B) of 3-bromo-3,4-dimethylhexane. On reaction with sodium ethoxide in ethanol, each gives 3,4-dimethyl-3-hexene as the major product. One of these diastereomers gives the (E)-alkene, and the other gives the (Z)-alkene. Which diastereomer gives which alkene? Account for the stereospecificity of each β-elimination.

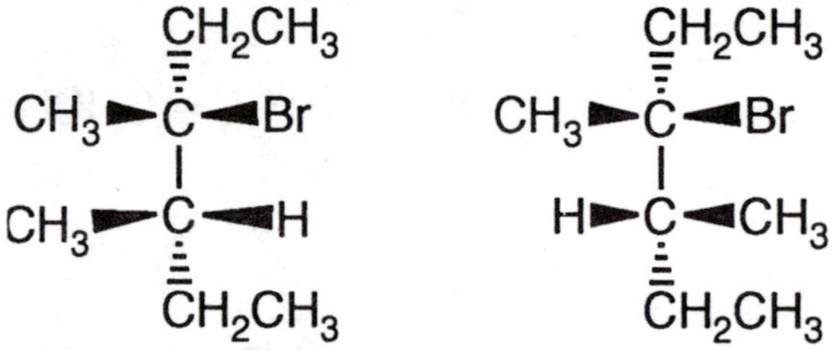

(A) (B)

Rotate the given conformation of each stereoisomer into a conformation in which Br and H are anti and coplanar and then undergo E2 elimination. You will find that diastereomer (A) gives the (E)-isomer and diastereomer (B) gives the (Z)-isomer.

Substitution versus Elimination

Problem 10.37 Consider the following statements in reference to S_N1, S_N2, S_Ni, E1, and E2 reactions of alkyl halides. To which mechanism(s), if any, does each statement apply?

(a) Involves a carbocation intermediate
S_N1, E1

(b) Is first-order in alkyl halide and first-order in nucleophile
S_N2

(c) Involves inversion of configuration at the site of substitution
S_N2

(d) Involves retention of configuration at the site of substitution
S_Ni

(e) Substitution at a stereocenter gives predominantly a racemic product
S_N1

(f) Is first-order in alkyl halide and zero-order in base
E1

(g) Is first-order in alkyl halide and first-order in base
E2

(h) Is greatly accelerated in protic solvents of increasing polarity
S_N1, E1

(i) Rearrangements are common
S_N1, E1

(j) Order of reactivity is 3° > 2° > 1° > methyl
S_N1, E2, E1

(k) Order of reactivity is methyl > 1° > 2° > 3°
S_N2

Problem 10.38 Draw a structural formula for the major product of each reaction and specify the most likely mechanism for formation of the product you have drawn.

The substitution and elimination products for the reactions are given in bold. In each case, the different parameters discussed in the chapter are considered including the type of alkyl halide (primary, secondary, tertiary, etc.) and the relative strength of the nucleophile/base.

(a) 3-bromocyclohexene-Br + CH_3OH $\xrightarrow{\text{methanol}}$

1,3-cyclohexadiene (E1) + 3-methoxycyclohexene-OCH_3 (S_N1)

(b) $CH_3C(CH_3)(Cl)CH_2CH_3$ + NaOH $\xrightarrow[H_2O]{80°}$ $(CH_3)_2C{=}CHCH_3$ (E2)

(c) (R)-$CH_3CH(Cl)CH_2CH_2CH_3$ + $CH_3CO^-Na^+$ (with C=O) $\xrightarrow{\text{DMSO}}$

(S)-$CH_3CH(O{-}C(=O){-}CH_3)CH_2CH_2CH_3$ (S_N2)

(d) [cyclohexane with $C(CH_3)_3$ and Cl] + $CH_3O^-Na^+$ —methanol→ [cyclohexene with $C(CH_3)_3$] (E2)

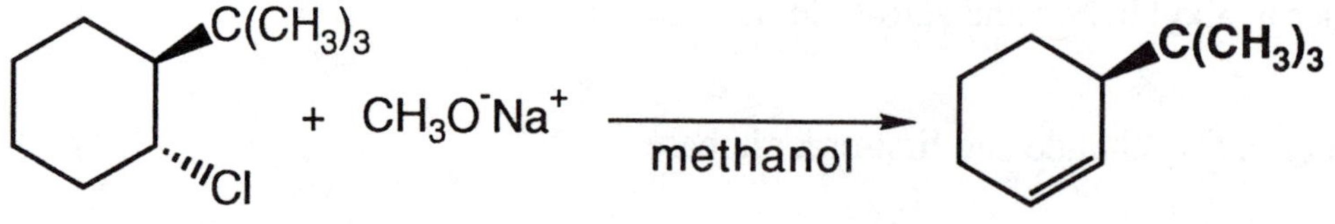

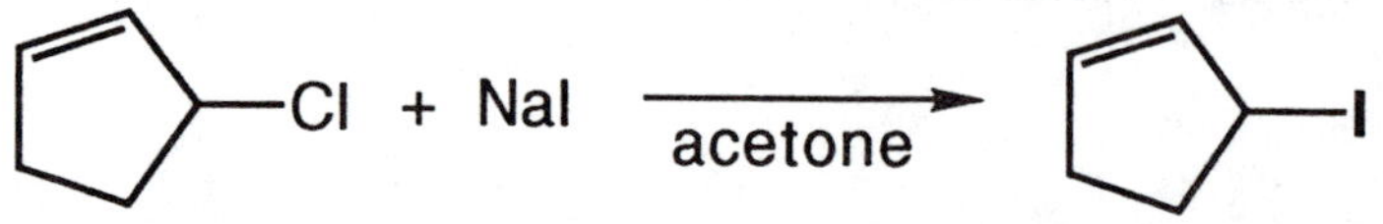

(e) [cyclopentene–Cl] + NaI —acetone→ [cyclopentene–I] (S_N2)

R Isomer S Isomer

(f) $CH_3CH(Cl)CH_2CH_3$ + $HC(=O)OH$ → $CH_3CH(OC(=O)H)CH_2CH_3$ (S_N1)

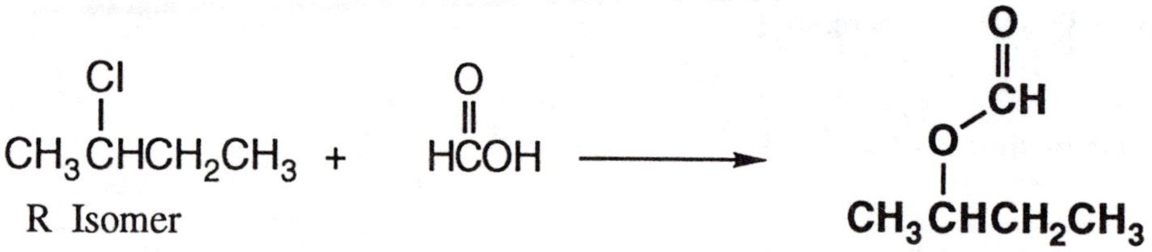

R Isomer R,S Isomer

(g) CH_3CH_2ONa + $CH_2{=}CHCH_2Cl$ —ethanol→

$CH_3CH_2OCH_2CH{=}CH_2$ (S_N2)

Problem 10.39 When *cis*-4-chlorocyclohexanol is treated with sodium hydroxide in ethanol, it gives only the substitution product *trans*-1,4-cyclohexanediol (1). Under the same reaction conditions, *trans*-4-chlorocyclohexanol gives 3-cyclohexenol (2) and the bicyclic ether (3).

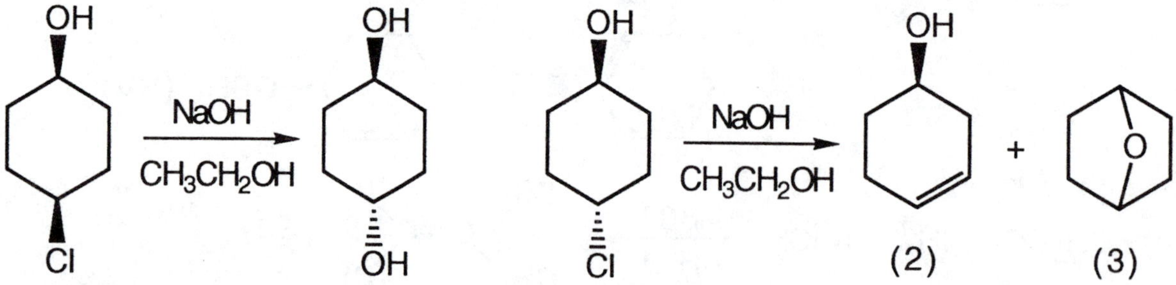

cis-4-Chloro-cyclohexanol —NaOH, CH_3CH_2OH→ (1)

trans-4-Chloro-cyclohexanol —NaOH, CH_3CH_2OH→ (2) + (3)

(a) Propose a mechanism for formation of product (1), and account for its configuration.

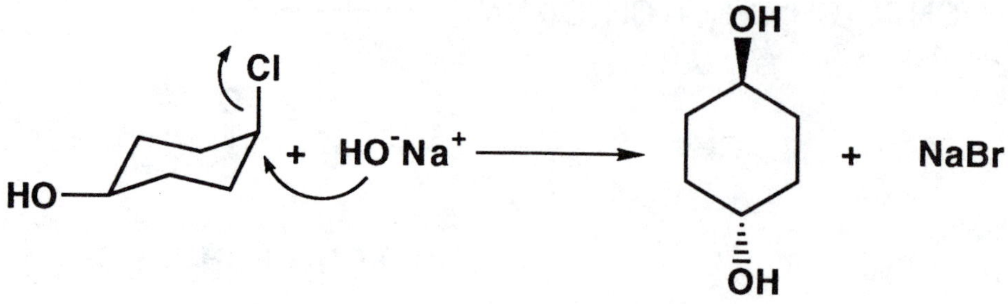

[HO-cyclohexane-Cl chair] + HO^-Na^+ → [trans-1,4-cyclohexanediol] + NaBr

Inversion of configuration is observed because of an S_N2 mechanism.

(b) Propose a mechanism for formation of product (2).

The reaction takes place by an E2 mechanism. The molecule must adopt the chair conformation that places both the HO- and Cl- groups in the axial position in order for the reaction to occur.

HO–(chair)–Cl ⇌ (chair, axial Cl and axial HO, H) + HO^-Na^+ ⟶ (cyclohexenol, OH) + H_2O

(c) Account for the fact that the bicyclic ether (3) is formed from the *trans* isomer but not from the *cis* isomer.

The bicyclic ether product (3) is formed from an intramolecular backside attack of the deprotonated axial hydroxyl group upon an axial chlorine atom. Only the *trans* isomer can adopt the diaxial orientation necessary for this process.

(chair, axial Cl, axial HO) + HO^-Na^+ ⇌ (chair, axial Cl, axial ^-O) ⟶ (bicyclic ether, O) + NaBr

(chair, axial Cl, equatorial HO) + HO^-Na^+ ⇌ (chair, axial Cl, equatorial ^-O) ⟶✕ (bicyclic ether, O) + NaBr

Problem 10.40 The Williamson ether synthesis involves reaction of an alkyl halide with a metal alkoxide. Following are two reactions intended to give *tert*-butyl ethyl ether. One reaction gives the ether in good yield, the other reaction does not. Which reaction gives the ether? What is the major product of the other reaction, and how do you account for its formation?

(a) $CH_3C(CH_3)_2O^-K^+$ + CH_3CH_2Cl $\xrightarrow{\text{2-methyl-2-propanol}}$ $CH_3C(CH_3)_2OCH_2CH_3$ + KCl

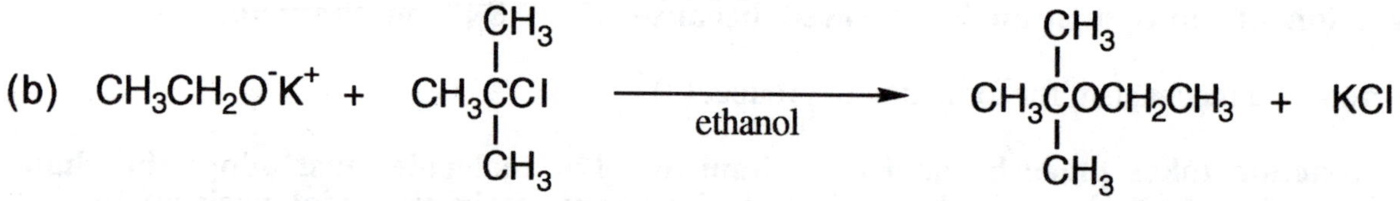

(b) $CH_3CH_2O^-K^+ + CH_3C(CH_3)_2Cl \xrightarrow{\text{ethanol}} CH_3C(CH_3)_2OCH_2CH_3 + KCl$

The only reaction that will give the desired ether product in good yield is the one shown in (a). In (b), the major product will be the elimination product isobutylene, $CH_2{=}C(CH_3)_2$, because the halide is on a tertiary carbon atom and ethoxide is a strong base.

Problem 10.41 The following ethers can, in principle, be synthesized by two different combinations of alkyl halide and alkoxide. Show one combination of alkyl halide and alkoxide that forms ether bond (1) and another that forms ether bond (2). Which combination gives the higher yield of ether?

(a) 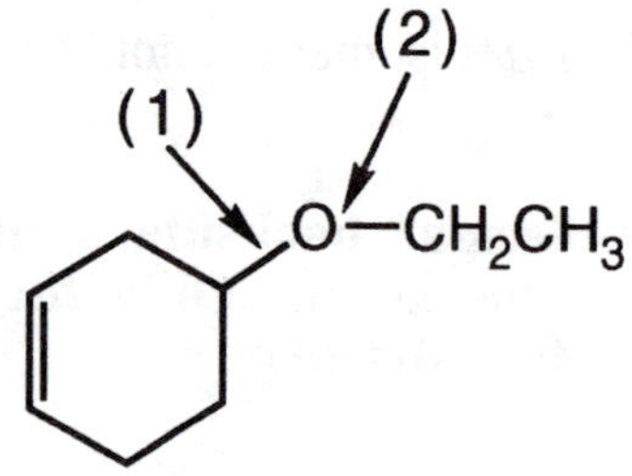 [cyclohexenyl]–O–CH_2CH_3, with bond (1) between the ring carbon and O, and bond (2) between O and CH_2CH_3

As the better combination, choose (2) which involves reaction of an alkoxide with a primary halide and will give substitution as the major product. Scheme (1) involves a strong base/strong nucleophile and secondary halide, conditions that will give both substitution and elimination products.

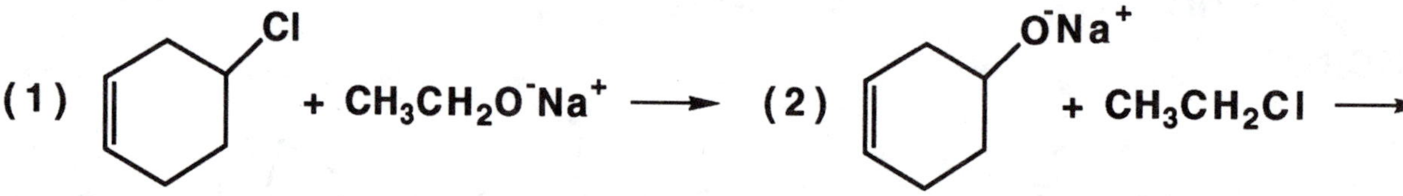

(1) [3-chlorocyclohexene] $+ CH_3CH_2O^-Na^+ \longrightarrow$ (2) [cyclohexenyl]–O^-Na^+ $+ CH_3CH_2Cl \longrightarrow$

(b) 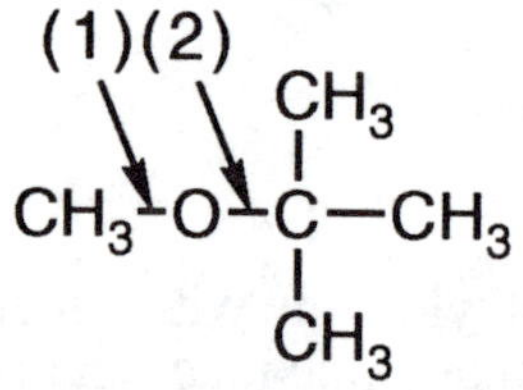 CH_3–O–$C(CH_3)_2$–CH_3, with bond (1) between CH_3 and O, and bond (2) between O and C

Because of the high degree of branching in the haloalkane in (2), S_N2 substitution by this pathway is virtually impossible. Therefore, choose (1) as the only reasonable alternative.

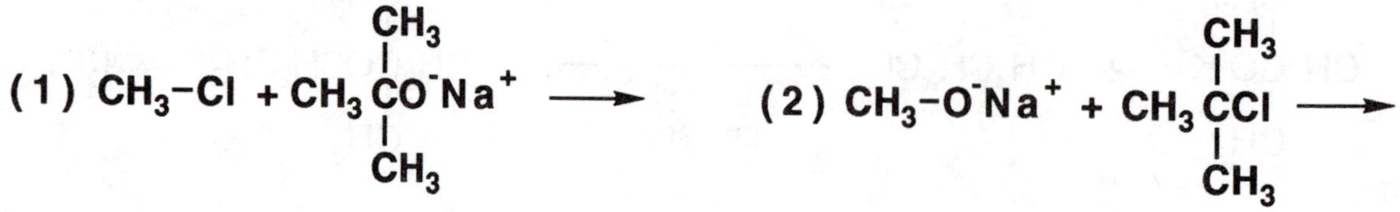

(1) $CH_3{-}Cl + CH_3C(CH_3)_2O^-Na^+ \longrightarrow$ (2) $CH_3{-}O^-Na^+ + CH_3C(CH_3)_2Cl \longrightarrow$

(1) (2)

$$\text{(c)}\ H_2C{=}CHCH_2\overset{(1)}{-}O\overset{(2)}{-}CH_2\underset{CH_3}{\overset{CH_3}{C}}CH_3$$

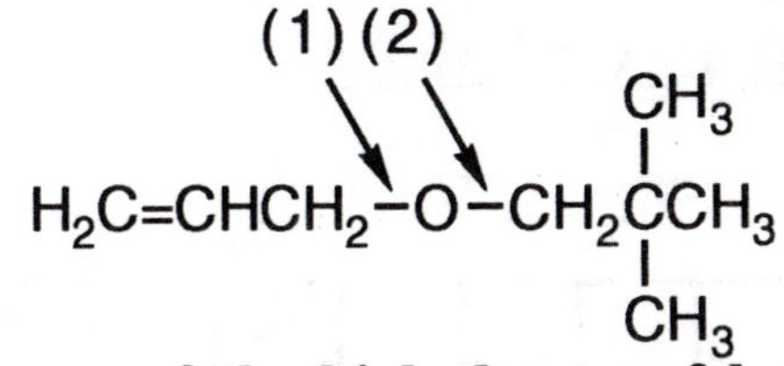

Because of the high degree of branching on the β-carbon in the haloalkane in (2), S_N2 substitution by this pathway is virtually impossible. Therefore, choose (1) as the only reasonable alternative.

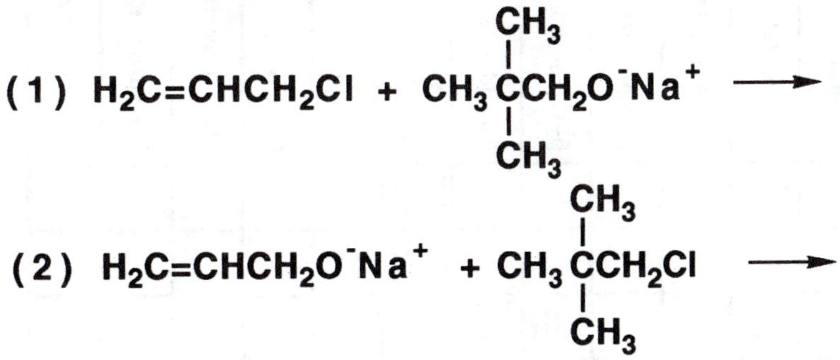

$$\text{(1)}\ H_2C{=}CHCH_2Cl + CH_3\underset{CH_3}{\overset{CH_3}{C}}CH_2O^-Na^+ \longrightarrow$$

$$\text{(2)}\ H_2C{=}CHCH_2O^-Na^+ + CH_3\underset{CH_3}{\overset{CH_3}{C}}CH_2Cl \longrightarrow$$

CHAPTER 11: ETHERS AND EPOXIDES

SUMMARY OF REACTIONS

Starting Material ↓ / Product →	Alcohols	Alkenes	Alkyl Halides	β–Alkynyl Alcohols	β–Amino Alcohols	β–Cyano Alcohols	Epoxides	Ethers		Glycols	Hydroperoxides	β–Hydroxy Ethers	β–Mercapto Alcohols	Sulfones	Sulfoxides	Thioethers	Thiols
Alcohols								**11A** 11.4D*									
Alcohols Alkenes								**11B** 11.4B	**11C** 11.4C								
Alkenes							**11D** 11.7B										
Alkyl Halides																**11E** 11.5	**11F** 11.5
Alkyl Halides Alkoxides								**11G** 11.4A									
Alkoxides Epoxides												**11H** 11.8B					
Alkyne Anions Epoxides				**11I** 11.8B													
Amines Epoxides					**11J** 11.8B												
Epoxides	**11K** 11.8B	**11L** 11.8B				**11M** 11.8B				**11N** 11.8A			**11O** 11.8B				
Ethers			**11P** 11.6A								**11Q** 11.6B						
Halohydrins							**11R** 11.7B										
Sulfoxides														**11S** 11.6C			
Thioethers														**11T** 11.6C	**11U** 11.6C		

*Section in book that describes reaction.

REACTION 11A: ACID-CATALYZED DEHYDRATION OF ALCOHOLS (Section 11.4D)

$$2\ -\overset{|}{\underset{|}{C}}-OH \xrightarrow{H_2SO_4} -\overset{|}{\underset{|}{C}}-O-\overset{|}{\underset{|}{C}}-$$

- Ethers can be produced via acid-catalyzed dehydration of primary alcohols. In this reaction, the -OH group is protonated to give a good leaving group (H_2O) that is displaced in an S_N2 process by another alcohol molecule. ✱

REACTION 11B: ALKOXYMERCURATION / REDUCTION (Section 11.4B)

$$-\overset{|}{\underset{|}{C}}-OH + \;\rangle C'=C''\langle \;\xrightarrow[2)\ NaBH_4]{1)\ Hg(OAc)_2}\; -\overset{|}{\underset{|}{C}}-O-\overset{|}{\underset{|}{C'}}-\overset{|}{\underset{|}{C''}}-H$$

- A second method for synthesizing ethers is **alkoxymercuration/reduction**. In this reaction, an alkene is reacted with $Hg(OAc)_2$ in the presence of an alcohol. The resulting mercury derivative is converted to the ether via reduction with $NaBH_4$. This is analogous to oxymercuration (Section 5.3 E) used to convert alkenes into alcohols. ✱
- This reaction is particularly useful in the cases where the Williamson ether synthesis has difficulty, namely when the alkyl groups are both secondary and/or tertiary.

REACTION 11C: ACID-CATALYZED ADDITION OF ALCOHOLS TO ALKENES (Section 11.4C)

$$-\overset{|}{\underset{|}{C}}-OH + \;\rangle C'=C''\langle \;\xrightarrow{\text{Acid Catalyst}}\; -\overset{|}{\underset{|}{C}}-O-\overset{|}{\underset{|}{C'}}-\overset{|}{\underset{|}{C''}}-H$$

- Ethers can also be prepared through an **acid catalyzed addition of alcohols to alkenes**. In these reactions, the acid reacts with the alkene to create a carbocation that is attacked by the nucleophilic oxygen atom of the alcohol. ✱
- Since a carbocation is involved, the reaction only works with alkenes such as isobutylene in which a tertiary carbocation is produced upon protonation, and when the alcohol is primary and thus resistance to dehydration.

REACTION 11D: OXIDATION OF ALKENES WITH PEROXYACIDS (PERACIDS) (Section 11.7B)

$$\rangle C=C'\langle \;\xrightarrow[\text{(Peroxyacid)}]{RCO_3H}\; -\underset{|}{C}\overset{O}{-}\underset{|}{C'}-$$

- The most common laboratory **synthesis of epoxides** is **from alkenes, using peroxyacids** as oxidizing agents. Commonly used reagents include 3-chloroperoxybenzoic acid and the magnesium salt of monoperoxyphthalic acid. The mechanism of this reaction is thought to be concerted. ✱ *[This is a very interesting mechanism that is worth a close look because it can be trickier than it might first appear.]*

REACTION 11E: PREPARATION OF THIOETHERS FROM ALKYL HALIDES (Section 11.5)

$$-\overset{|}{\underset{|}{C}}-S^- + -\overset{|}{\underset{|}{C'}}-X \longrightarrow -\overset{|}{\underset{|}{C}}-S-\overset{|}{\underset{|}{C'}}-$$

- Thioethers can be prepared from an S_N2 reaction between thiolate anions and alkyl halides.
- Symmetrical thioethers can be prepared from 2 equivalents of primary alkyl halides and one equivalent of sulfide ion (Na_2S).
- The same reaction can be used to prepare cyclic thioethers when an alkyl dihalide such as 1,5-dichloropentane or 1,4-dichlorobutane is used.

REACTION 11F: PREPARATION OF THIOLS FROM ALKYL HALIDES (Section 11.5)

$$-\overset{|}{\underset{|}{C}}-X \xrightarrow{HS^-} -\overset{|}{\underset{|}{C}}-SH$$

- Certain **thiols can be prepared** using an S_N2 reaction between an **alkyl halide and the hydrosulfide ion**. Elimination is a concern here since the hydrosulfide ion is basic, so primary alkyl halides give the highest yields. Yields are lower with secondary halides, and elimination predominates with tertiary alkyl halides. ✱

REACTION 11G: THE WILLIAMSON ETHER SYNTHESIS (Section 11.4A)

$$-\overset{|}{\underset{|}{C}}-O^- + -\overset{|}{\underset{|}{C'}}-X \longrightarrow -\overset{|}{\underset{|}{C}}-O-\overset{|}{\underset{|}{C'}}-$$

- The most common method used to synthesize ethers is the **Williamson ether synthesis**, which is nothing more than an S_N2 reaction between a metal alkoxide and alkyl halide. ✱
- Because the metal alkoxide is basic, an E2 reaction is always a concern with a Williamson ether synthesis. As a result, it is important to choose carefully which piece will be the alkyl halide. In other words, a primary alkyl halide is used if possible. Secondary alkyl halides give lower yields of ethers and tertiary alkyl halides give almost exclusively the elimination product.
- Alkyl tosylates may be used in place of alkyl halides.

REACTION 11H: NUCLEOPHILIC RING OPENING OF EPOXIDES WITH ALKOXIDES (Section 11.8B)

$$-\overset{|}{\underset{|}{C}}-O^- + -\overset{O}{\overbrace{\underset{|}{C'}-\underset{|}{C''}}}- \longrightarrow -\overset{|}{\underset{|}{C}}-O-\overset{|}{\underset{|}{C'}}-\overset{OH}{\underset{|}{C''}}-$$

- **Strong nucleophiles** such as alkoxides can add directly to epoxides via an S_N2 **mechanism.** ✱
- Epoxide reactions are versatile because of the large number of different nucleophiles that can be used including alkyne anions, amines, hydride reagents, cyanide anion, and thiolates. These reactions are described as reactions 11I, 11J, 11K, 11M, and 11O, respectively. ✱
- As expected for an S_N2 reaction, in each case, the **nucleophile adds preferentially** to the **less hindered** epoxide **carbon atom.** ✱

REACTION 11I: NUCLEOPHILIC RING OPENING OF EPOXIDES WITH ALKYNE ANIONS (Section 11.8B)

—C≡C'⁻ + epoxide —C"–C"'— ⟶ —C≡C'–C"–C"'(OH)—

- **Alkyne anions** react with epoxides via an S_N2 mechanism to produce hydroxy alkynes.

REACTION 11J: NUCLEOPHILIC RING OPENING OF EPOXIDES WITH AMINES (Section 11.8B)

—N(H): + epoxide —C–C'— ⟶ —N–C–C'(OH)—

- **Amines** react with epoxides via an S_N2 mechanism to produce β-**amino alcohols**.

REACTION 11K: REDUCTION OF EPOXIDES TO CREATE ALCOHOLS (Section 11.8B)

epoxide —C–C'— ⟶ (1) $LiAlH_4$; 2) H_2O) H–C–C'(OH)—

- **Hydride reagents** such as $LiAlH_4$ react with epoxides to produce **alcohols**.
- The hydrogen ends up on the less hindered carbon atom, consistent with an S_N2 mechanism.

REACTION 11L: REACTION OF EPOXIDES TO CREATE ALKENES (Section 11.8B)

epoxide —C–C'— ⟶ (PPh_3) C=C' + $Ph_3P{=}O$

- **Triphenylphosphine reacts** with epoxides to give **alkenes**.
- The mechanism involves an initial ring opening by the nucleophilic triphenylphosphine to generate a betaine, that then eliminates triphenylphosphine oxide to give the product alkene. This elimination is unusual because the stereochemistry is syn.

REACTION 11M: NUCLEOPHILIC RING OPENING OF EPOXIDES WITH CYANIDE (Section 11.8B)

epoxide —C–C'— ⟶ (1) N≡C"⁻; 2) H_2O) N≡C"–C–C'(OH)—

- The **cyanide anion** reacts with epoxides via an S_N2 mechanism to produce β-**hydroxy nitriles**.

REACTION 11N: ACID-CATALYZED RING OPENING OF EPOXIDES (Section 11.8A)

$$-\overset{O}{C}-C'- \; + \; H_2O \xrightarrow{H^+} HO-C-C'(OH)-$$

- In the presence of an **acid catalyst** (usually perchloric acid), the oxygen of the epoxide is protonated to form an oxonium ion intermediate, that is then susceptible to nucleophilic attack to generate the product.
- The oxonium ion intermediate is analogous to the positively-charged halonium and mercurinium ion intermediates discussed in previous chapters. As a result, the nucleophile attacks the epoxide carbon atom that corresponds to the more stable carbocation, usually the most hindered epoxide carbon atom. Note that this is in distinct contrast to simple nucleophilic attack of epoxides by an S_N2 mechanism (reactions 11H, 11I, 11J, 11K, 11M, and 11O) in which the nucleophile preferentially attacks the less-hindered epoxide carbon atom.
- The nucleophile attacks anti and coplanar to the oxygen atom of the oxonium atom. For epoxides derived from cyclic alkenes, this means the addition results in formation of a *trans* product.

REACTION 11O: NUCLEOPHILIC RING OPENING OF EPOXIDES WITH THIOLATES (Section 11.8B)

$$-\overset{O}{C}-C'- \xrightarrow{HS^-} HS-C-C'(OH)-$$

- The **HS⁻ ion** reacts with epoxides via an S_N2 mechanism to produce **β-hydroxy thiols.**

REACTION 11P: ACID-CATALYZED CLEAVAGE OF ETHERS WITH H-X (Section 11.6A)

$$-C-O-C'- \xrightarrow{2\ H\text{-}X} -C-X \; + \; -C'-X$$

- **Ethers are very robust functional groups**, failing to react with reagents such as oxidizing agents or strong bases. On the other hand, ethers can be cleaved by concentrated aqueous HI or HBr. The mechanism of this reaction involves protonation of the ether oxygen atom to create a positively charged intermediate. If both of the attached alkyl groups are primary, then cleavage occurs by an S_N2 mechanism with the halide ion acting as the nucleophile. If one of the alkyl groups is tertiary, then cleavage is by an S_N1 mechanism to create a carbocation that can react with halide, rearrange, or undergo elimination depending on the details of the reaction. ✱

REACTION 11Q: FORMATION OF HYDROPEROXIDES FROM ETHERS (Section 11.6B)

$$-\overset{|}{\underset{|}{C}}-O-\overset{H}{\underset{|}{\overset{|}{C'}}}- \xrightarrow{O_2} -\overset{|}{\underset{|}{C}}-O-\overset{OOH}{\underset{|}{\overset{|}{C'}}}-$$

- Two **hazards** exist when working with low-molecular-weight ethers. First, they are extremely **flammable** and care should be taken to avoid flames and sparks when working with them. Second, ethers **react slowly** with molecular **oxygen to form explosive hydroperoxides** via a radical process, so they must be stored carefully and disposed of before a problem arises.

REACTION 11R: INTERNAL NUCLEOPHILIC SUBSTITUTION IN HALOHYDRINS TO CREATE EPOXIDES (Section 11.7B)

$$X-\overset{|}{\underset{|}{C}}-\overset{OH}{\underset{|}{\overset{|}{C'}}}- \xrightarrow{NaOH,\ H_2O} -\underset{|}{C}\overset{O}{-}\underset{|}{C'}-$$

- **Epoxides** can also be **produced** by treatment of a **halohydrin with base**. The mechanism of the reaction involves an internal nucleophilic attack and loss of the halogen leaving group. This is a useful reaction because halohydrins can be produced from reaction of an alkene with aqueous Cl_2 or Br_2 (reaction 5K, described in section 5.3F).

REACTION 11S: OXIDATION OF SULFOXIDES TO SULFONES (Section 11.6C)

$$-\overset{|}{\underset{|}{C}}-\overset{O}{\overset{\|}{\ddot{S}}}-\overset{|}{\underset{|}{C'}}- \xrightarrow{H_2O_2} -\overset{|}{\underset{|}{C}}-\overset{O}{\underset{O}{\overset{\|}{\underset{\|}{S}}}}-\overset{|}{\underset{|}{C'}}-$$

- **Sulfoxides** can be **easily oxidized** to **sulfones.**

REACTION 11T: OXIDATION OF THIOETHERS TO SULFONES (Section 11.6C)

$$-\overset{|}{\underset{|}{C}}-\ddot{\underset{..}{S}}-\overset{|}{\underset{|}{C'}}- \xrightarrow{2\ eq.\ H_2O_2} -\overset{|}{\underset{|}{C}}-\overset{O}{\underset{O}{\overset{\|}{\underset{\|}{S}}}}-\overset{|}{\underset{|}{C'}}-$$

- **Thioethers** can be **easily oxidized** to two higher oxidation states, sulfoxides, and sulfones. Several oxidizing agents can be used to carry out these transformations including aqueous hydrogen peroxide (H_2O_2), sodium metaperiodate ($NaIO_4$), and air oxidation in the presence of oxides of nitrogen.

- The extent of oxidation depends on the amount of oxidizing agent added. Two equivalents of an oxidizing agent such as H_2O_2 results in formation of a sulfone.

REACTION 11U: OXIDATION OF THIOETHERS TO SULFOXIDES (Section 11.6C)

$$-\overset{|}{\underset{|}{C}}-\ddot{\underset{..}{S}}-\overset{|}{\underset{|}{C'}}- \xrightarrow{1\ eq.\ H_2O_2} -\overset{|}{\underset{|}{C}}-\overset{O}{\overset{\|}{\underset{..}{S}}}-\overset{|}{\underset{|}{C'}}-$$

- Reaction of thioethers with one equivalent of H_2O_2 results in formation of a sulfoxide.

SUMMARY OF IMPORTANT CONCEPTS

11.1 STRUCTURE OF ETHERS

- The **ether functional group** is composed of an **sp^3 hybridized oxygen atom** bonded to **two carbon atoms**. There are two lone pairs of electrons on the oxygen atom, and these provide many of the reactivity characteristic of ethers. ✱

11.2 NOMENCLATURE OF ETHERS

- In the IUPAC system, ethers are named by choosing the longest alkyl chain as the parent chain. The remaining -OR group is named as an alkoxy substituent. For example, ethoxyethane and 3-methoxyheptane are acceptable IUPAC names.

ethoxyethane 3-methoxyheptane

- Low-molecular-weight ethers have common names that are often used. The common names are constructed by listing the two alkyl groups in alphabetical order followed by the word ether. For example, diethyl ether or methyl propyl ether are acceptable common names.
- Some important ethers are cyclic, and they are given special names such as tetrahydrofuran or 1,4-dioxane.
- Several small ethers are useful solvents, such as diethyl ether, 2-methoxyethanol, 2-ethoxyethanol, and diethylene glycol dimethyl ether
- Sulfur analogs of ethers are referred to as thioethers or sulfides, and are named in common nomenclature by using the word sulfide in place of ether. For example,diethyl sulfide is the sulfur analog of diethyl ether.

11.3 PHYSICAL PROPERTIES OF ETHERS

- Ethers are polar molecules,since the oxygen atom possesses a partial negative charge and each attached carbon atom possesses a partial positive charge. However, there is only limited dipole-dipole interaction between molecules because there is too much steric hindrance around the carbon atoms with the partial positive charges. As a result, the boiling points of ethers are not that much higher than those of similar hydrocarbons. On the other hand, the oxygen atom of ethers can accept hydrogen bonds, so ethers have higher solubilities in water than the corresponding hydrocarbons. ✱

11.7 EPOXIDES

- An epoxide is a cyclic ether contained in a three-membered ring. Epoxides can be named in a number of ways.

- In the IUPAC system, the epoxide is named by listing the substituents on the ring as prefixes to the parent name **oxirane**. For example, ethyloxirane is an acceptable IUPAC name.
- Two different systems are used to assign common names. In the first, the two atoms of the parent chain attached to the oxygen atoms of the epoxide are listed along with the prefix **epoxy**. For example, 1,2-epoxybutane is an acceptable common name. In another system, the alkene from which the epoxide could have been derived is named followed by the word **oxide**. For example, 1-butene oxide is an acceptable common name.

11.8 REACTIONS OF EPOXIDES

- **Epoxides** are **more reactive than normal ethers** because of the strain present in the three-member ring. In particular, the epoxide is susceptible to nucleophilic attack at the less-hindered carbon atom of the epoxide ring with the oxygen atom acting as a leaving group (S_N2 mechanism; reactions 11H, 11I, 11J, 11K, 11M, and 11O), or via an oxonium intermediate with acid catalysis (reaction 11N). ✳

11.9 CROWN ETHERS

- **Crown ethers** are cyclic polyethers derived from ethylene glycol, or a related glycol, that have four or more ether linkages and twelve or more total atoms in the ring. ✳
  - Crown ethers are usually named by a short-hand nomenclature utilizing **crown** as the parent name. This is preceded by the total number of atoms in the ring and followed by the number of oxygen atoms in the ring. For example, 18-crown-6 is a common crown ether.
  - The cavity on the inside of certain crown ethers is exactly the correct size to place an alkali metal inside. Since there are lone pairs of electrons on the oxygen atoms, there is a strong electrostatic attraction between the positively-charged ion and the crown ether. In general, the better the fit between crown ether interior size and ionic radius of the ion, the tighter the binding. For example, 18-crown-6 binds K^+ very tightly, but larger or smaller cations bind less tightly.

  Because the exteriors of crown ether molecules, such as 18-crown-6, are relatively hydrophobic, they can solubilize ions in organic solvents. The bound ion remains in the interior cavity of the crown ether molecule, away from the hydrophobic solvent. This is useful because crown ethers can be used to increase dramatically the solubility of ionic compounds in organic solvents.

CHAPTER 11
Solutions to the Problems

Problem 11.1 Give common names for the following ethers.

(a) $CH_3CH(CH_3)CH_2OCH_2CH_3$

1-ethoxy-2-methylpropane
(ethyl isobutyl ether)

(b) $CH_3OCH_2CH_2OCH_3$

1,2-dimethoxyethane
(ethylene glycol dimethyl ether)

(c) cyclohexane ring with two OCH_2CH_3 groups (wedged, cis)

***cis*-1,2-diethoxycyclohexane**

Problem 11.2 Arrange the following compounds in order of increasing boiling point.

$CH_3OCH_2CH_2OCH_3$ $\quad$ $HOCH_2CH_2OH$ $\quad$ $CH_3OCH_2CH_2OH$

In order of increasing boiling point, they are:

$CH_3OCH_2CH_2OCH_3$	$CH_3OCH_2CH_2OH$	$HOCH_2CH_2OH$
84°C	125°C	198°C

Problem 11.3 Show how you might use the Williamson ether synthesis to prepare the following ethers.

(a) $C_6H_5-CH_2-O-C(CH_3)_2-CH_3$

There is only one combination of alkyl halide and metal alkoxide that gives benzyl *tert*-butyl ether in good yield.

$$C_6H_5-CH_2-Br + CH_3-C(CH_3)_2-O^-K^+ \longrightarrow C_6H_5-CH_2-O-C(CH_3)_2-CH_3 + K^+Br^-$$

If this synthesis were attempted with the alkoxide derived from benzyl alcohol and a *tert*-butyl halide, the reaction would produce an alkene by an E2 pathway.

$$C_6H_5-CH_2-O^-Na^+ + CH_3-\underset{CH_3}{\overset{CH_3}{C}}-Br \longrightarrow C_6H_5-CH_2-OH + CH_2=\overset{CH_3}{C}-CH_3 + Na^+Br^-$$

(b) $(CH_3CH_2CH_2CH_2)_2O$

Treatment of 1-bromobutane with sodium butoxide gives dibutyl ether.

$$CH_3CH_2CH_2CH_2O^-Na^+ + CH_3CH_2CH_2CH_2Br \longrightarrow (CH_3CH_2CH_2CH_2)_2O + Na^+Br^-$$

Problem 11.4 Propose a synthesis for the following ethers using alkoxymercuration followed by sodium borohydride reduction.

(a) $$CH_3CH_2CH_2\overset{CH_3}{C}=CH_2 \xrightarrow{?} CH_3CH_2CH_2\underset{CH_3}{\overset{CH_3}{C}}-OCH_2CH_3$$

Treatment of 2-methyl-1-pentene with mercuric acetate in the presence of ethanol followed by reduction of the organomercury intermediate with sodium borohydride yields the desired product.

$$CH_3CH_2CH_2\overset{CH_3}{C}=CH_2 \xrightarrow{Hg(OAc)_2,\ CH_3CH_2OH} CH_3CH_2CH_2\underset{CH_2HgOAc}{\overset{CH_3}{C}}-OCH_2CH_3$$

$$\xrightarrow{NaBH_4} CH_3CH_2CH_2\underset{CH_3}{\overset{CH_3}{C}}-OCH_2CH_3$$

(b) $$\text{cyclopentene} \xrightarrow{?} \text{cyclopentyl}-O-\overset{CH_3}{CH}CH_3$$

Treatment of cyclopentene with mercuric acetate in the presence of 2-propanol (isopropanol) followed by reduction of the organomercury intermediate with sodium borohydride yields the desired product.

$$\text{cyclopentene} \xrightarrow[CH_3CHOHCH_3]{Hg(OAc)_2,} \text{2-(HgOAc)cyclopentyl}-O-CH(CH_3)_2 \xrightarrow{NaBH_4} \text{cyclopentyl}-O-CH(CH_3)_2$$

Problem 11.5 Show how ethyl hexyl ether might be synthesized by a Williamson ether synthesis, and by alkoxymercuration/reduction.

Using the Williamson ether synthesis, ethyl hexyl ether could be synthesized by either of the two following routes:

$$CH_3CH_2Br + CH_3(CH_2)_4CH_2O^-Na^+$$

or

$$CH_3(CH_2)_4CH_2Br + CH_3CH_2O^-Na^+$$

$$\longrightarrow CH_3CH_2OCH_2(CH_2)_4CH_3$$

Using the alkoxymercuration reaction, ethyl hexyl ether could be synthesized by the following reaction:

$$H_2C{=}CH_2 \xrightarrow[CH_3(CH_2)_4CH_2OH]{Hg(OAc)_2} \underset{|}{\overset{HgOAc}{}}CH_2CH_2OCH_2(CH_2)_4CH_3$$

$$\xrightarrow{NaBH_4} CH_3CH_2OCH_2(CH_2)_4CH_3$$

Problem 11.6 Account for the fact that reaction of *tert*-butyl methyl ether with a limited amount of concentrated HI gives methanol and *tert*-butyl iodide rather than methyl iodide and *tert*-butyl alcohol.

The first step in the reaction involves protonation of the ether oxygen to give an oxonium ion. Cleavage of the oxonium ion on one side gives methanol and a *tert*-butyl cation. Cleavage on the other side gives a methyl cation and *tert*-butyl alcohol. Because of the greater stability of the tertiary carbocation, cleavage to give methanol and *tert*-butyl cation is favored. Reaction of the tertiary cation with the iodide completes the reaction.

Oxonium ion $(CH_3)_3C{-}\overset{+}{O}(H){-}CH_3$:

- cleavage to form a tertiary carbocation $\longrightarrow$ $CH_3{-}\ddot{O}{:}{-}H + {}^+C(CH_3)_3$ (more stable) $\xrightarrow{I^-}$ $I{-}C(CH_3)_3$
- cleavage to form a primary carbocation $\longrightarrow$ $:\ddot{O}(H){-}C(CH_3)_3 + {}^+CH_3$ (less stable)

There is an alternative pathway for formation of product, namely reaction of the oxonium ion with iodide ion by an S_N2 pathway on the less hindered methyl carbon to give iodomethane and the 2-methyl-2-propanol (*tert*-butyl alcohol). The fact is that the S_N1 pathway by way of a tertiary carbocation has a lower energy of activation (a faster rate) than reaction of the oxonium ion with iodide ion by an S_N2 pathway so the S_N1 reaction predominates.

Problem 11.7 Draw structural formulas for the major products of the following reactions.

(a) $H_3C-C(CH_3)_2-O-CH_3$ + HBr (excess) ⟶ $H_3C-C(CH_3)_2-Br$ + CH_3Br

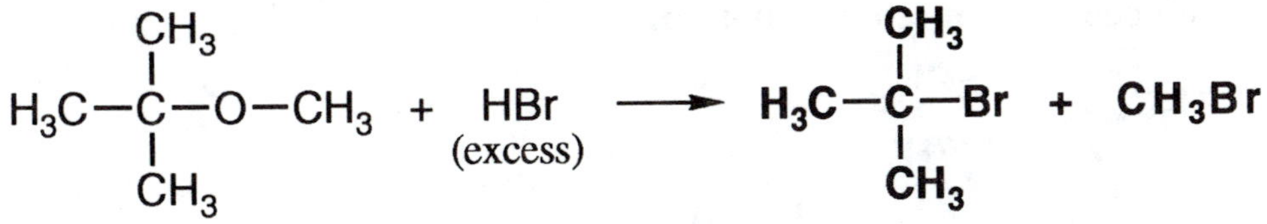

(b) (tetrahydropyran) + HBr (excess) ⟶ $BrCH_2CH_2CH_2CH_2CH_2Br$

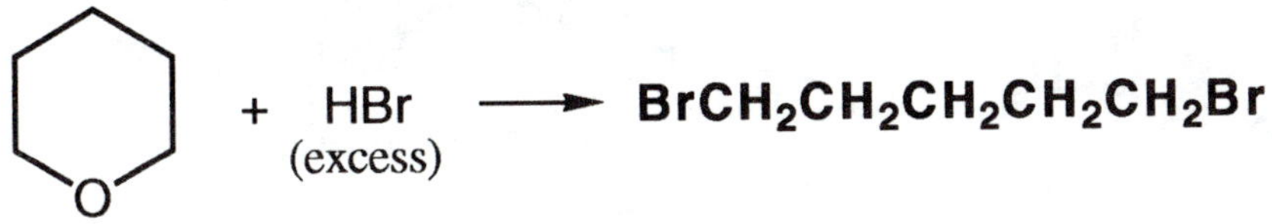

Problem 11.8 Consider the possibilities for stereoisomerism in the halohydrin and epoxide formed in Example 11.8.

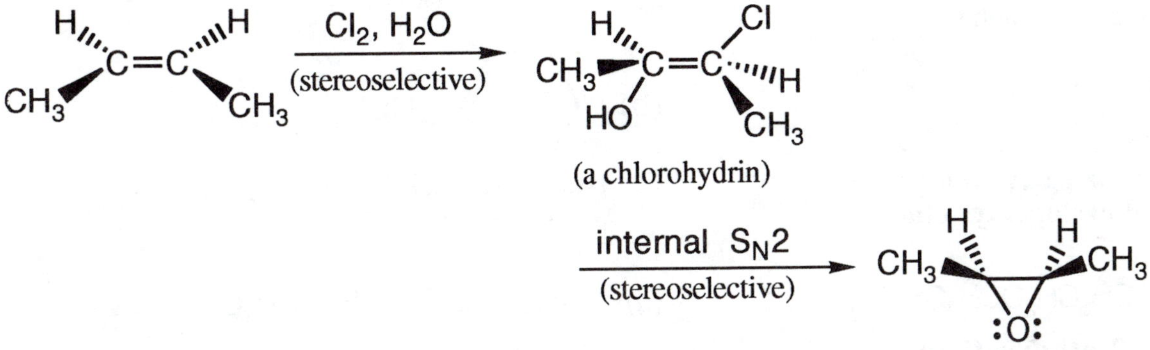

cis-2-butene —Cl_2, H_2O (stereoselective)⟶ (a chlorohydrin) —internal S_N2 (stereoselective)⟶ (an epoxide)

(a) How many stereoisomers are possible for the chlorohydrin? Which of the possible chlorohydrins are formed by reaction of *cis*-2-butene with HOCl?

This chlorohydrin has two stereocenters and, by the 2^n rule, there are four possible stereoisomers - two pairs of enantiomers.

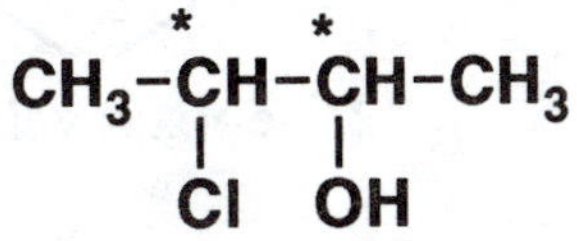

$CH_3-\overset{*}{C}H(Cl)-\overset{*}{C}H(OH)-CH_3$

However, given the stereoselectivity (anti) of halohydrin formation, only one pair of these enantiomers is formed from *cis*-2-butene.

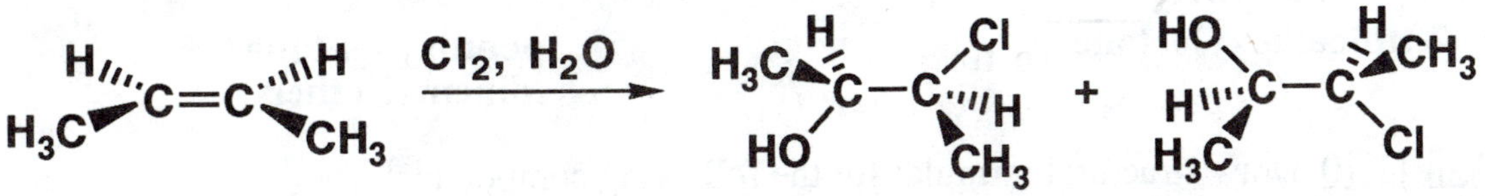

cis-2-butene —Cl_2, H_2O⟶ (chlorohydrin) + (its enantiomer)

(b) How many stereoisomers are possible for this epoxide? Which of the possible stereoisomers is/are formed in this reaction sequence?

There are three stereoisomers possible for this epoxide; a pair of enantiomers and a meso compound. However, because of the stereoselectivity (requirement for anti, planar geometry) of the β-elimination reaction, there is only one stereoisomer produced; the meso compound.

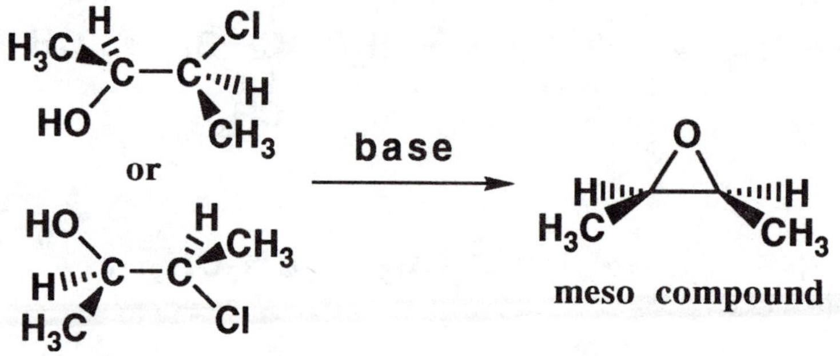

H, H_3C, C—C, Cl, H, HO, CH_3 or HO, H, CH_3, H C—C, H_3C, Cl **base** → O, H, H, H_3C, CH_3

meso compound

Structure and Nomenclature

Problem 11.9 Write names for the following compounds. Where possible, write both IUPAC names and common names.

(a) 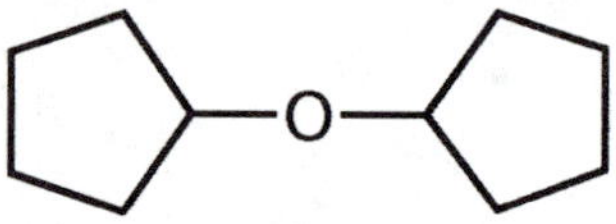 (cyclopentyl)–O–(cyclopentyl)

cyclopentoxycyclopentane
dicyclopentyl ether

(b) 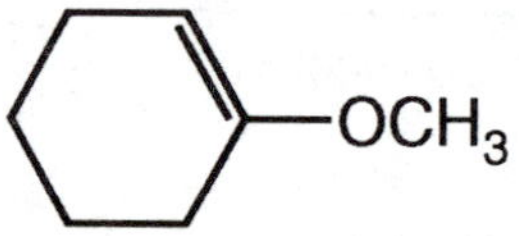 (cyclohexenyl)–OCH_3

1-methoxycyclohexene
1-cyclohexenyl methyl ether

(c) $CH_3CH_2OCH_2CH_2OH$

2-ethoxyethanol

(d) $CH_3CH_2OCH_2CH_2OCH_2CH_3$

2-ethoxyethoxyethane
ethylene glycol diethyl ether

(e) 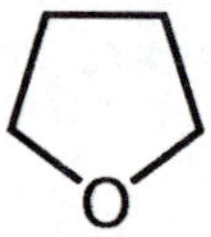 (five-membered ring with O)

tetrahydrofuran

(f) 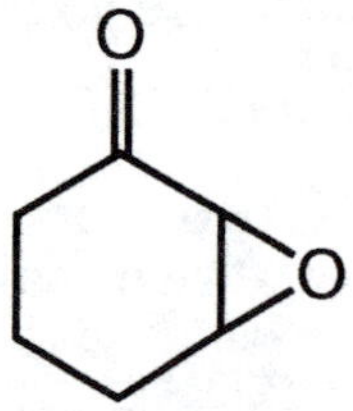 (cyclohexanone ring with epoxide O)

2,3-epoxycyclohexanone

(g) 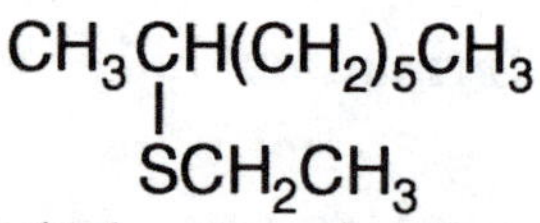 $CH_3CH(CH_2)_5CH_3$ with SCH_2CH_3 on C-2

2-(thioethoxy)octane

(h) $[CH_3(CH_2)_4]_2O$

pentoxypentane
dipentyl ether

Problem 11.10 Draw structural formulas for the following compounds.

(a) diisopropyl ether

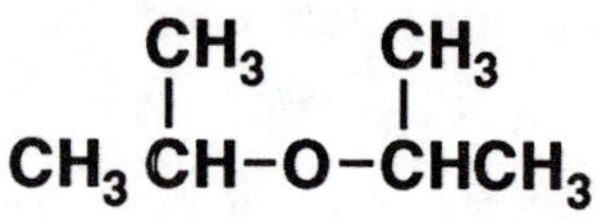

CH_3 CH_3
$CH_3CH–O–CHCH_3$

(b) *trans*-2,3-diethyloxirane

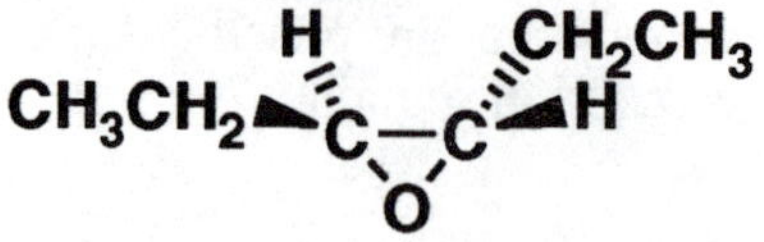

H, CH_2CH_3
CH_3CH_2–C–C–H (O bridging)

(c) *trans*-2-ethoxycyclopentanol

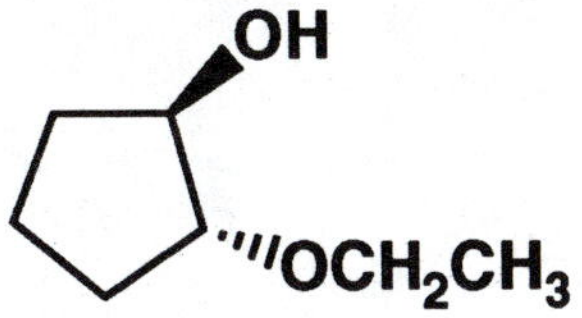

OH

OCH_2CH_3

(d) divinyl ether

$CH_2=CH-O-CH=CH_2$

(e) cyclohexene oxide

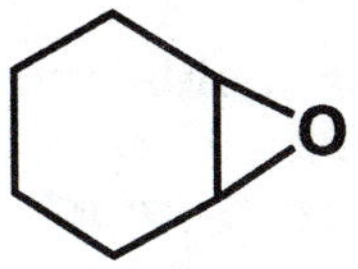

O

(f) allyl cyclopropyl ether

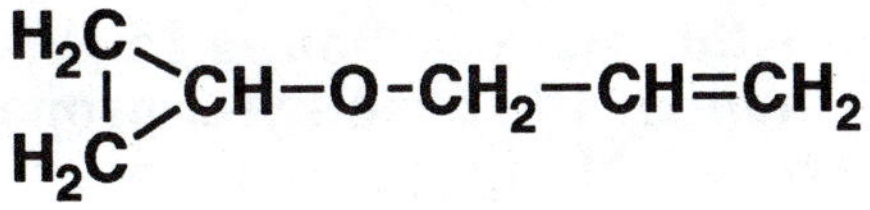

H_2C $CH-O-CH_2-CH=CH_2$ H_2C

(g) (R)-methyloxirane

CH_3 H C CH_2 O

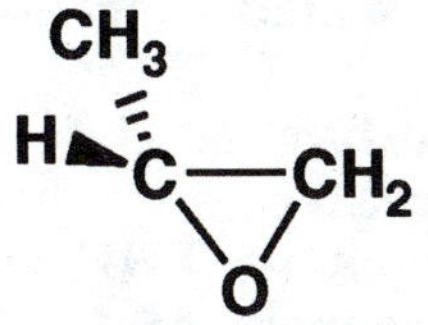

(h) 1,1-dimethoxycyclohexane

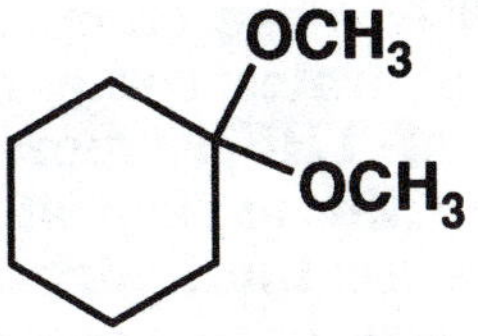

OCH_3 OCH_3

Physical Properties

Problem 11.11 Each compound given in this problem is a common organic solvent. From each pair of compounds, select the solvent with the greater solubility in water.

In each case select the molecule that can make better hydrogen bonds with water or is more polar.

(a) CH_2Cl_2 and CH_3CH_2OH **CH_3CH_2OH**

(b) $CH_3CH_2OCH_2CH_3$ and CH_3CH_2OH **CH_3CH_2OH**

(c) $CH_3C(=O)CH_3$ and $CH_3CH_2OCH_2CH_3$ **$CH_3C(=O)CH_3$**

(d) $CH_3CH_2OCH_2CH_3$ and $CH_3(CH_2)_3CH_3$ **$CH_3CH_2OCH_2CH_3$**

Problem 11.12 Following are structural formulas, boiling points, and solubilities in water for diethyl ether and tetrahydrofuran. Account for the fact that tetrahydrofuran is so much more soluble in water than diethyl ether.

$CH_3CH_2-O-CH_2CH_3$

diethyl ether
bp 35°C
8 g/100 mL water

H_2C-CH_2
$H_2C \quad CH_2$
O

tetrahydrofuran
bp 67°C
very soluble in water

There are two factors to be considered:
(1) The shape of the hydrocarbon is important in determining water solubility. As we saw in Chapter 10, 2-methyl-2-propanol (*tert*-butyl alcohol) is considerably more soluble in water than 1-butanol. A *tert*-butyl group is much more compact than a butyl group and, consequently, there is less disruption of water hydrogen bonding when *tert*-butyl alcohol is dissolved in water than when 1-butanol is dissolved in water. Similarly, the hydrocarbon portion of tetrahydrofuran (THF) is more compact than that of diethyl ether which increases the solubility of THF in water compared to diethyl ether.
(2) The oxygen atom of THF is more accessible for hydrogen bonding and solvation by water than the oxygen atom of diethyl ether. This greater accessibility arises because the hydrocarbon chains bonded to oxygen in THF are "tied back" whereas those on oxygen of diethyl ether have more degrees of freedom and consequently present more steric hindrance to solvation.

Problem 11.13 Because of the Lewis base properties of ether oxygens, crown ethers are excellent complexing agents for Na^+, K^+, and NH_4^+. What kind of molecule might serve as a complexing agent for Cl^- or Br^-?

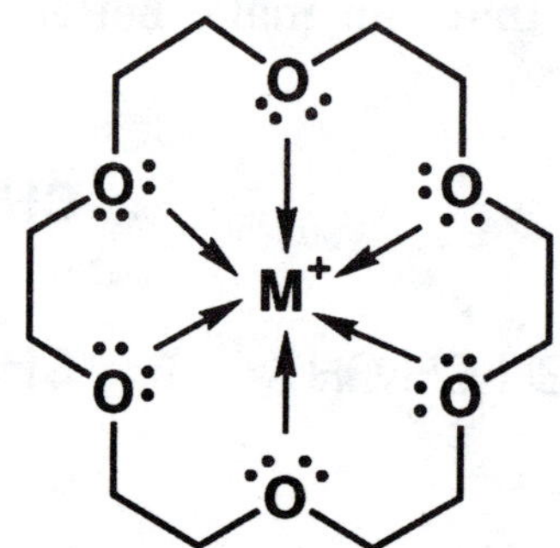

18-crown-6 with a generic metal M^+

To build an analogous system for complexation of Cl^- it is necessary to first realize that the chloride ion is a Lewis base itself. Instead of a cavity with electron pair donors (as in crown ethers) we need a system with electron pair acceptors (or Lewis acid sites), such as the hydrogen atoms of protonated amines.

$$\text{cyclic hexaprotonated hexaaza macrocycle: six } {}^{+}NH \text{ groups, each N–H pointing inward to a central } Cl^{-}$$

The size of the cavity will depend on the diameter of the ion complexed.

Preparation of Ethers

Problem 11.14 Write equations to show a combination of reactants to prepare each ether in good yield by (1) a Williamson ether synthesis and alkoxymercuration/reduction. Which, if either method, gives the better yield? Explain your reasoning.

(a) $CH_3CH_2OCH(CH_3)CH_3$

$$Na^{+}\ {}^{-}OCH(CH_3)CH_3 + CH_3CH_2Br \longrightarrow CH_3CH_2OCH(CH_3)CH_3 + Na^{+}Br^{-}$$

$$CH_2{=}CH{-}CH_3 + Hg(OAc)_2 \xrightarrow[\text{2) } NaBH_4]{\text{1) } CH_3CH_2OH} CH_3CH_2OCH(CH_3)CH_3$$

Either method gives a satisfactory yield.

(b) $CH_3C(CH_3)_2OCH_2CH_2CH_3$

$$CH_3C(CH_3)_2O^{-}K^{+} + CH_3CH_2CH_2Br \longrightarrow CH_3C(CH_3)_2OCH_2CH_2CH_3 + K^{+}Br^{-}$$

$$CH_2{=}CH{-}CH_3 + Hg(OAc)_2 \xrightarrow[\text{2) } NaBH_4]{\text{1) } CH_3C(CH_3)_2OH} CH_3C(CH_3)_2OCH_2CH_2CH_3$$

Either method gives a satisfactory yield.

(c) 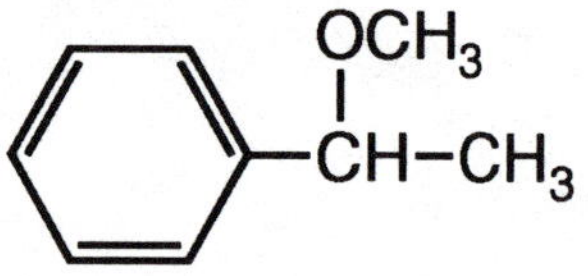 $C_6H_5-CH(OCH_3)-CH_3$

$$C_6H_5-CH(O^-K^+)-CH_3 + CH_3I \longrightarrow C_6H_5-CH(OCH_3)-CH_3 + K^+I^-$$

$$C_6H_5-CH{=}CH_2 + Hg(OAc)_2 \xrightarrow[\text{2) } NaBH_4]{\text{1) } CH_3OH} C_6H_5-CH(OCH_3)-CH_3$$

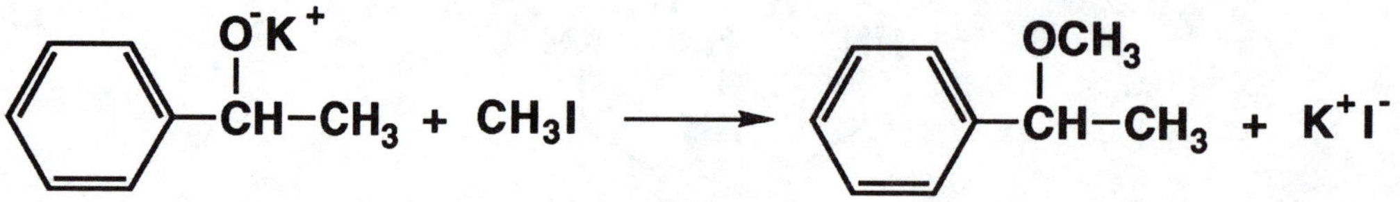

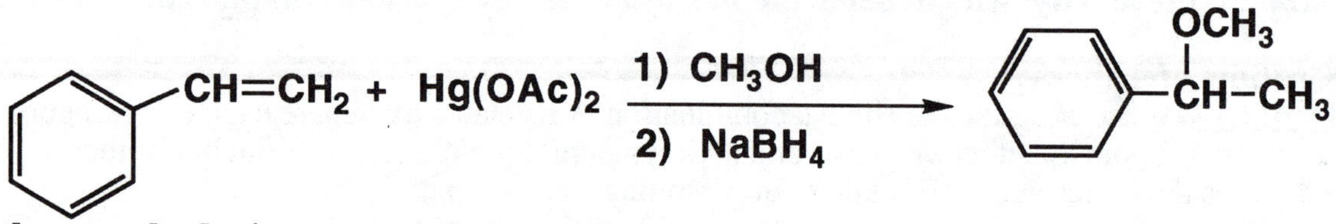

Either method gives a satisfactory yield.

(d) 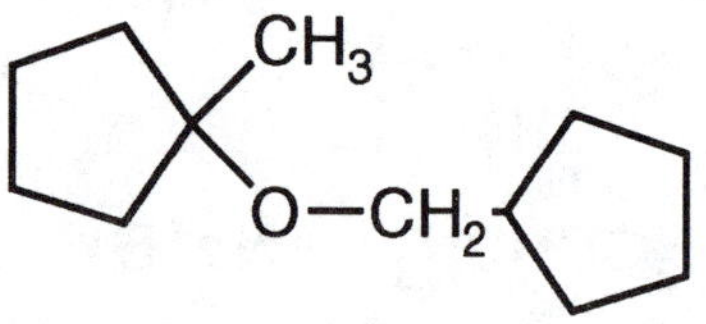 1-methylcyclopentyl cyclopentylmethyl ether: CH_3 and $O-CH_2$-(cyclopentyl) on the same cyclopentane carbon

$$\text{1-methylcyclopentoxide } (O^-K^+) + Br-CH_2-C_5H_9 \longrightarrow \text{1-methylcyclopentyl}-O-CH_2-C_5H_9 + K^+Br^-$$

$$\text{1-methylcyclopentene} + Hg(OAc)_2 \xrightarrow[\text{2) } NaBH_4]{\text{1) } C_5H_9CH_2OH} \text{1-methylcyclopentyl}-O-CH_2-C_5H_9$$

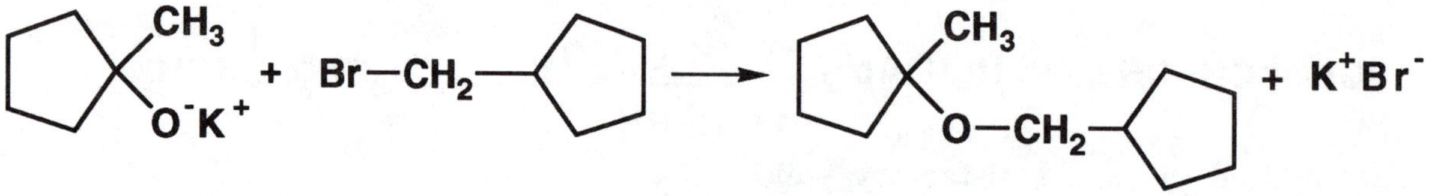

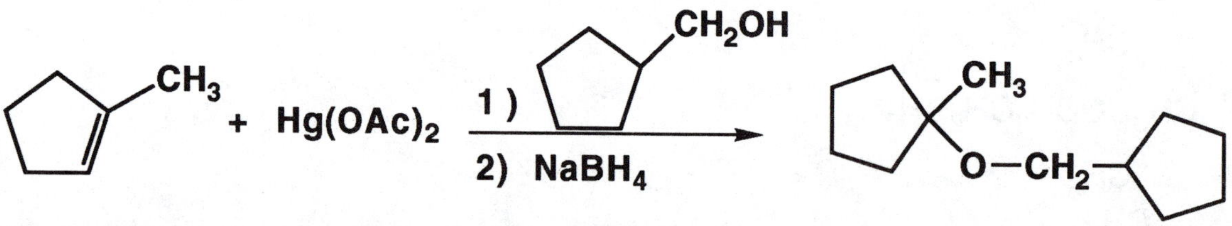

Either method gives a satisfactory yield.

(e) 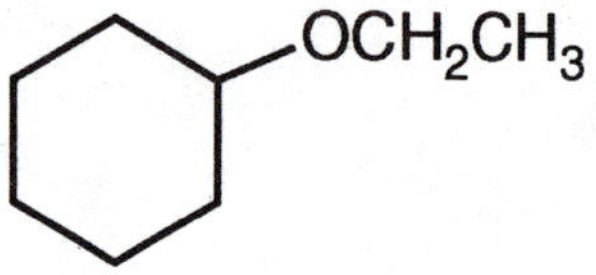 $C_6H_{11}-OCH_2CH_3$

$$C_6H_{11}-O^-K^+ + CH_3CH_2Br \longrightarrow C_6H_{11}-OCH_2CH_3 + K^+Br^-$$

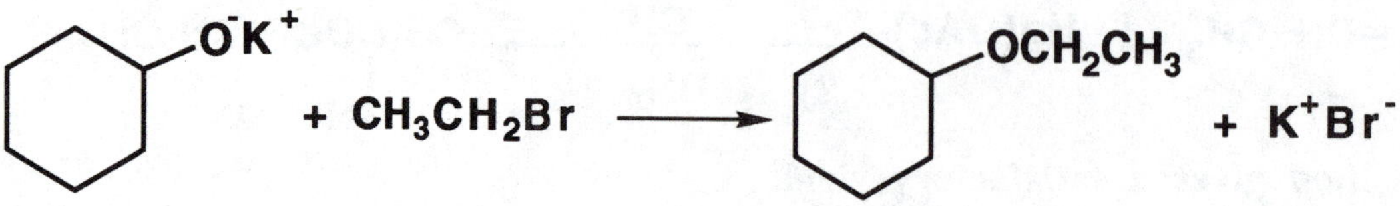

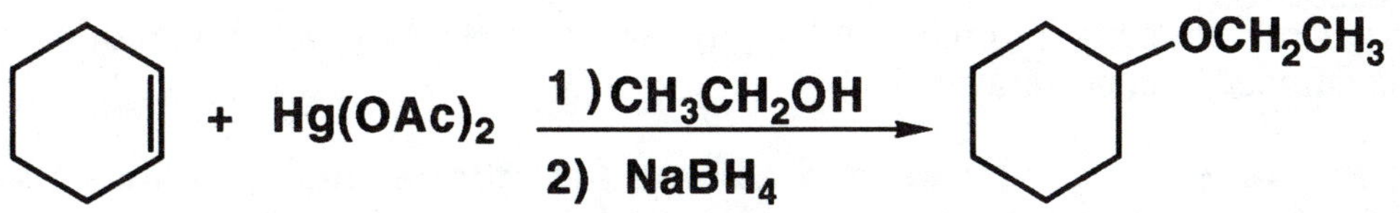

Cyclohexene + $Hg(OAc)_2$ $\xrightarrow[\text{2) } NaBH_4]{\text{1) } CH_3CH_2OH}$ cyclohexyl–OCH_2CH_3

Either method gives a satisfactory yield.

(f) cyclohexyl–$OC(CH_3)_3$

Iodocyclohexane + $KOC(CH_3)_3$ $\longrightarrow$ cyclohexyl–$OC(CH_3)_3$

Cyclohexene + $Hg(OAc)_2$ $\xrightarrow[\text{2) } NaBH_4]{\text{1) } (CH_3)_3COH}$ cyclohexyl–$OC(CH_3)_3$

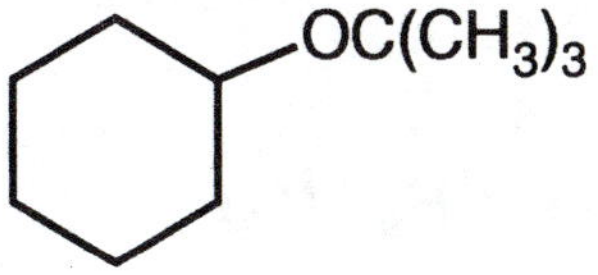

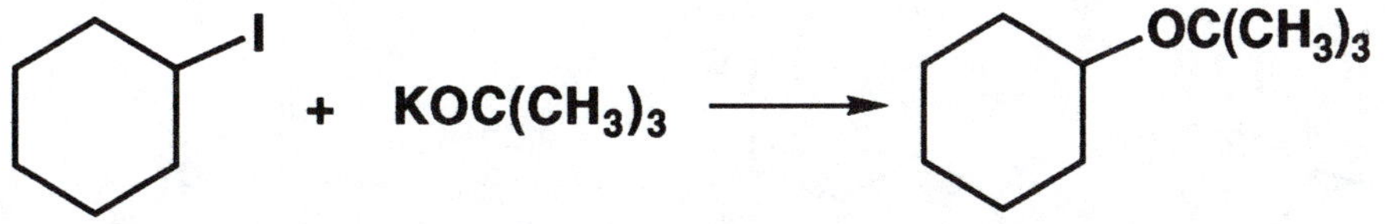

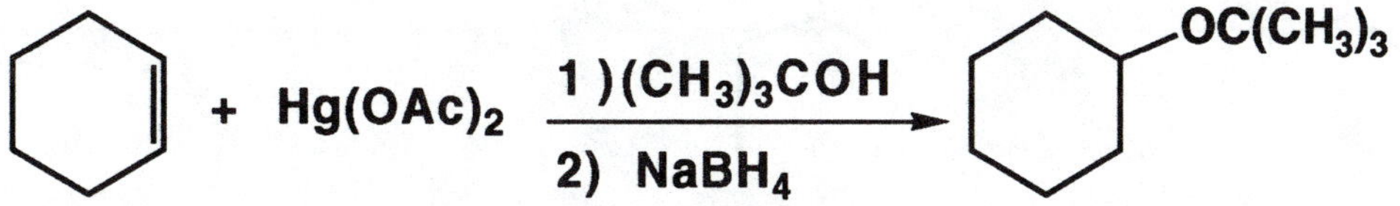

Alkoxymercuration is the method of choice since the Williamson ether synthesis gives poor yields for secondary alkyl halides.

Problem 11.15 Propose a mechanism for the following reaction.

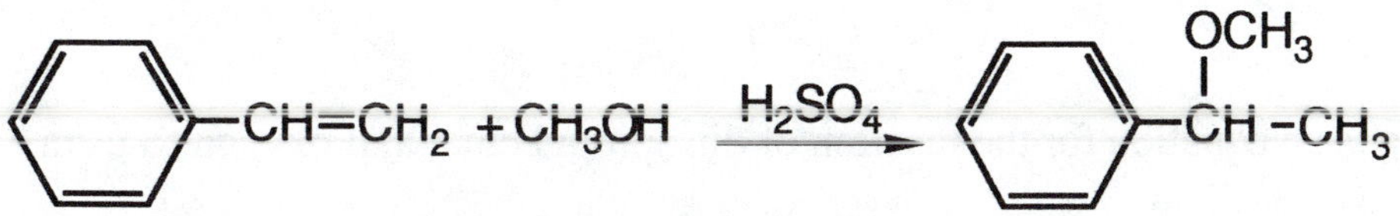

C_6H_5–$CH=CH_2$ + CH_3OH $\xrightarrow{H_2SO_4}$ C_6H_5–$CH(OCH_3)$–CH_3

The following mechanism is in three steps: (1) protonation of the alkene to form a resonance-stabilized carbocation, (2) reaction of the carbocation with methanol to give an oxonium ion, and (3) loss of a proton to give the ether.

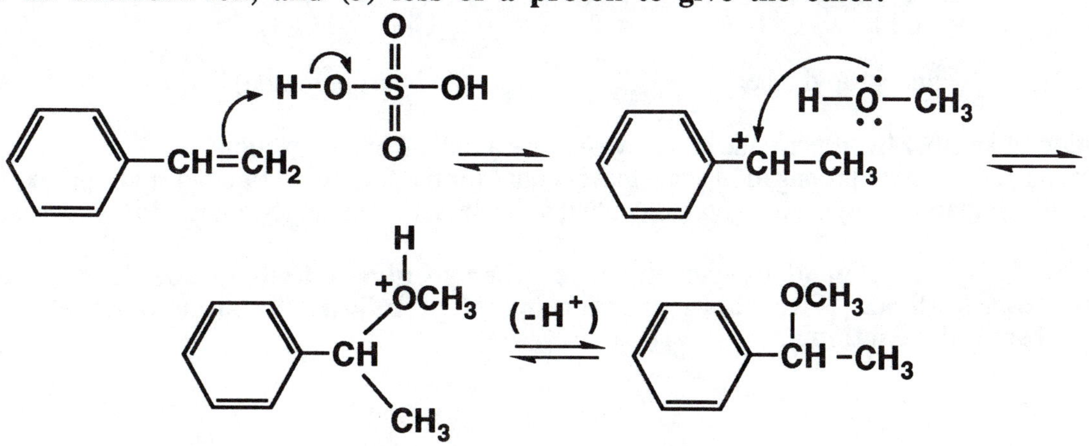

C_6H_5–$CH=CH_2$ + H–O–SO_2–OH $\rightleftharpoons$ C_6H_5–$\overset{+}{C}H$–CH_3 + H–$\ddot{O}$–CH_3 $\rightleftharpoons$

C_6H_5–CH($\overset{+}{O}$(H)CH_3)–CH_3 $\underset{}{\overset{(-H^+)}{\rightleftharpoons}}$ C_6H_5–$CH(OCH_3)$–CH_3

Reactions of Ethers

Problem 11.16 Draw structural formulas for the products formed when each of the following compounds is refluxed in concentrated HI.

Since an excess of HI is used, you can assume that the alcohol products initially produced will be converted to alkyl iodides under these conditions.

(a) $CH_3CH_2OCH_2CH_2CH_3 \xrightarrow{HI} CH_3CH_2CH_2I + CH_3CH_2I$

(b) $C_6H_{11}–CH_2OCH_2CH_3 \xrightarrow{HI} C_6H_{11}–CH_2I + CH_3CH_2I$

(c) [bicyclic cyclohexane-fused tetrahydrofuran] $\xrightarrow{HI}$ 1,2-bis(iodomethyl)cyclohexane

(d) [1,4-dioxane] $\xrightarrow{HI}$ ICH_2CH_2I

Problem 11.17 Following is an equation for the reaction of diisopropyl ether and oxygen to form a hydroperoxide.

$$\underset{\text{diisopropyl ether}}{(CH_3)_2CH\text{-}O\text{-}CH(CH_3)_2} + O_2 \longrightarrow \underset{\text{(a hydroperoxide)}}{(CH_3)_2CH\text{-}O\text{-}C(OOH)(CH_3)_2}$$

Formation of an ether hydroperoxide can be written as a radical chain reaction.
(a) Write a pair of chain propagation steps that account for the formation of ether hydroperoxides. Assume that initiation is by an oxygen molecule that itself is a diradical (Section 1.9B).

Initiation is by reaction of oxygen and the ether to give a hydroperoxide radical and an ether radical. Two chain propagation steps follow, the sum of which adds to the observed reaction.

Initiation:

$$CH_3CH(CH_3)\text{-}O\text{-}CH(CH_3)CH_3 + O_2 \longrightarrow CH_3CH(CH_3)\text{-}O\text{-}\dot{C}(CH_3)CH_3 + H\text{-}O\text{-}O\cdot$$

Propagation:

$$CH_3CH(CH_3)\text{-}O\text{-}\dot{C}(CH_3)CH_3 + O_2 \longrightarrow CH_3CH(CH_3)\text{-}O\text{-}C(O\text{-}O\cdot)(CH_3)CH_3$$

$$CH_3CH(CH_3)\text{-}O\text{-}C(O\text{-}O\cdot)(CH_3)CH_3 + CH_3CH(CH_3)\text{-}O\text{-}CH(CH_3)CH_3 \longrightarrow CH_3CH(CH_3)\text{-}O\text{-}C(OOH)(CH_3)CH_3 + CH_3CH(CH_3)\text{-}O\text{-}\dot{C}(CH_3)CH_3$$

(b) Account for the fact that hydroperoxidation of ethers is regioselective, that it occurs preferentially at a carbon adjacent to the ether oxygen.

The radical formed is secondary rather than primary and it is stabilized by resonance interaction with the oxygen atom.

Synthesis and Reactions of Epoxides

Problem 11.18 Ethylene oxide is the starting material for the synthesis of both Methyl Cellosolve and Ethyl Cellosolve, two important industrial solvents. Propose a mechanism for each of these reactions.

$$CH_2\text{—}CH_2\ (\text{epoxide, O}) + CH_3OH \xrightarrow{H_2SO_4} CH_3OCH_2CH_2OH$$

oxirane (ethylene oxide) → 2-methoxyethanol (Methyl Cellosolve)

$$CH_2\text{—}CH_2\ (\text{epoxide, O}) + CH_3CH_2OH \xrightarrow{H_2SO_4} CH_3CH_2OCH_2CH_2OH$$

2-ethoxyethanol (Ethyl Cellosolve)

For each reaction, protonation of the epoxide gives a cyclic oxonium ion. Reaction of this oxonium ion with the alcohol in an S_N2 reaction opens the epoxide ring. Loss of a proton completes the reaction.

$$CH_2{-}CH_2\ (\text{epoxide}) + H^+ \longrightarrow \text{protonated epoxide} \xrightarrow{H{-}\ddot{O}{-}R} HO{-}CH_2{-}CH_2{-}\overset{+}{O}(H)R \longrightarrow R\ddot{O}CH_2CH_2\ddot{O}H + H^+$$

Problem 11.19 Ethylene oxide is the starting material for the synthesis of 1,4-dioxane. Propose a mechanism for each step of this synthesis.

$$H_2C{-}CH_2\ (\text{ethylene oxide}) + HOCH_2CH_2OH \xrightarrow{H^+} HOCH_2CH_2OCH_2CH_2OH \xrightarrow{H^+} \text{1,4-dioxane}$$

1,4-dioxane

The mechanism for this reaction involves an acid catalyzed attack of the diol on the epoxide followed by protonation of one of the terminal hydroxyl groups, and displacement of water by the other alcohol.

$$H_2C{-}CH_2\ (\text{epoxide}) + H^+ \rightleftharpoons H_2C{-}CH_2\ (\overset{+}{O}{-}H) + H\ddot{O}CH_2CH_2\ddot{O}H \longrightarrow$$

$$HOCH_2CH_2\overset{+}{O}(H)CH_2CH_2\ddot{O}H \underset{}{\overset{(-H^+)}{\rightleftharpoons}} H\ddot{O}CH_2CH_2\ddot{O}CH_2CH_2\ddot{O}H + H^+ \rightleftharpoons$$

$$H_2\overset{+}{O}{-}CH_2CH_2OCH_2CH_2{-}\ddot{O}H \underset{}{\overset{(-H_2O)}{\rightleftharpoons}} \text{protonated 1,4-dioxane} \underset{}{\overset{(-H^+)}{\rightleftharpoons}} \text{1,4-dioxane}$$

Problem 11.20 Propose a synthesis for each of the following ethers, starting with ethylene oxide and any readily available alcohols.

(a) $CH_3OCH_2CH_2OCH_3$

$$H_2C{-}CH_2\ (\text{ethylene oxide}) \xrightarrow{H^+ / CH_3OH} CH_3OCH_2CH_2OCH_3$$

(b) $CH_3OCH_2CH_2OCH_2CH_2OCH_3$

$H_2C{-}CH_2$ (epoxide) + $CH_3O^-Na^+$ $\xrightarrow{CH_3OH}$ $CH_3OCH_2CH_2O^-Na^+$ $\xrightarrow{H_2C{-}CH_2 \text{ (epoxide)}}$

$CH_3OCH_2CH_2OCH_2CH_2O^-Na^+$ $\xrightarrow{H^+ / CH_3OH}$ $CH_3OCH_2CH_2OCH_2CH_2OCH_3$

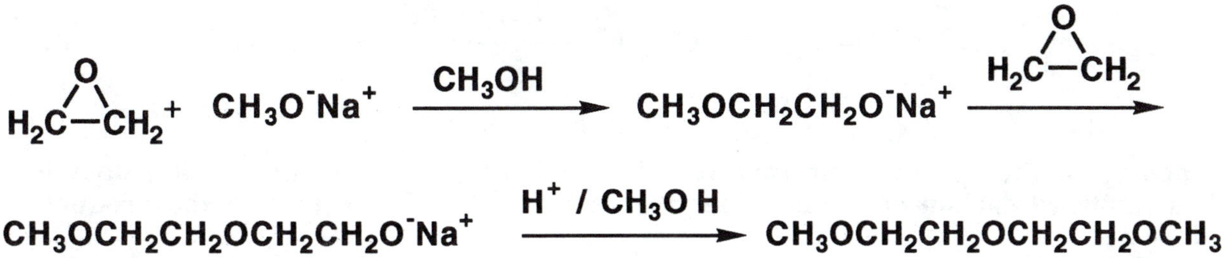

Problem 11.21 Propose a synthesis for 18-crown-6? If a base is used in your synthesis, does it make a difference if it is a lithium salt or a potassium salt?

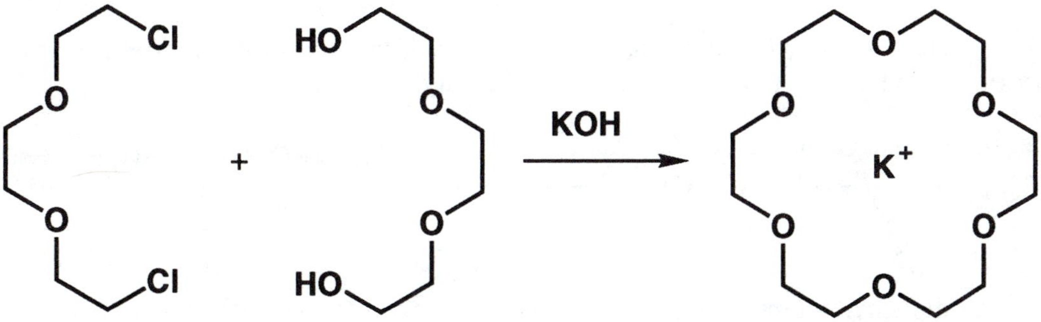

Cl–CH₂CH₂–O–CH₂CH₂–O–CH₂CH₂–Cl + HO–CH₂CH₂–O–CH₂CH₂–O–CH₂CH₂–OH $\xrightarrow{KOH}$ 18-crown-6 (K^+ complex)

18-crown-6

Since 18-crown-6 binds K^+ the best, KOH should be used so that the crown ether can form around the K^+ ion. This type of approach is called the template approach and is used in a variety of similar situations.

The above two pieces can be synthesized as follows:

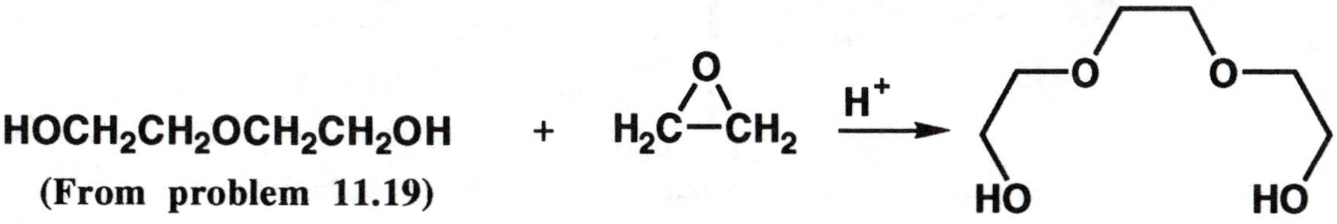

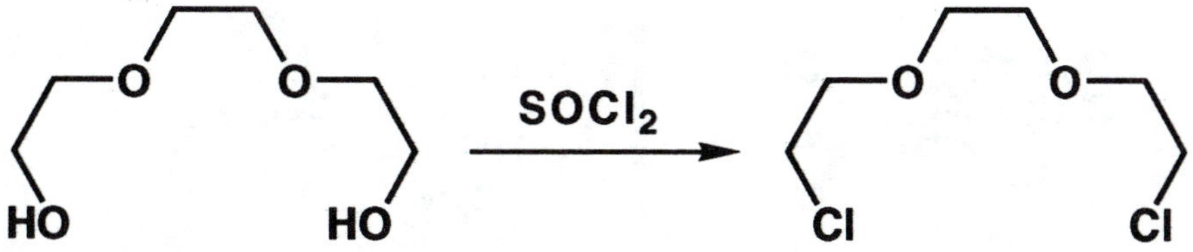

$HOCH_2CH_2OCH_2CH_2OH$ + $H_2C{-}CH_2$ (epoxide) $\xrightarrow{H^+}$ HO–CH₂CH₂–O–CH₂CH₂–O–CH₂CH₂–OH

(From problem 11.19)

HO–CH₂CH₂–O–CH₂CH₂–O–CH₂CH₂–OH $\xrightarrow{SOCl_2}$ Cl–CH₂CH₂–O–CH₂CH₂–O–CH₂CH₂–Cl

Problem 11.22 Predict the structural formula of the major product of the reactions of 2,2,3-trimethyloxirane with methanol in the presence of sodium methoxide.

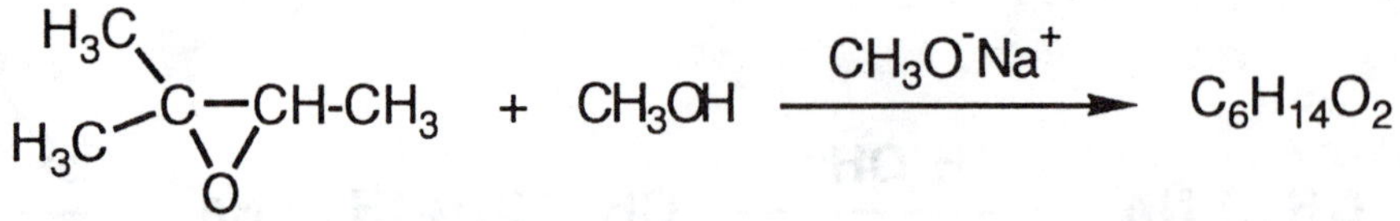

$$(H_3C)_2C\overset{O}{—}CH\text{-}CH_3 + CH_3OH \xrightarrow{CH_3O^-Na^+} C_6H_{14}O_2$$

Opening of the epoxide ring involves attack of methoxide (a nucleophile) on the less hindered carbon of the epoxide ring by an S_N2 pathway to give the product shown.

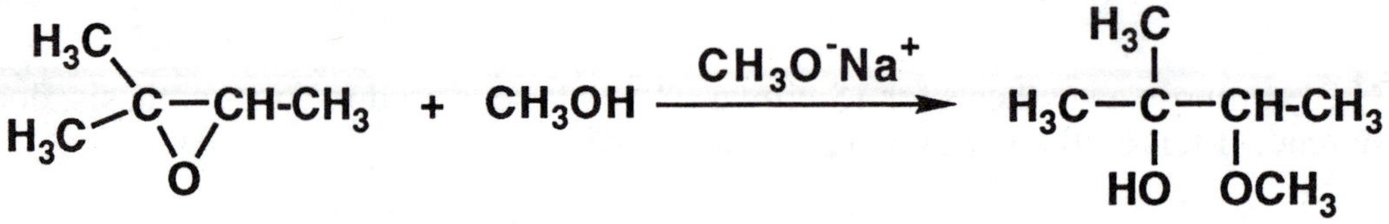

$$(H_3C)_2C\overset{O}{—}CH\text{-}CH_3 + CH_3OH \xrightarrow{CH_3O^-Na^+} H_3C—C(CH_3)(HO)—CH(OCH_3)\text{-}CH_3$$

Problem 11.23 The following equation shows the reaction of *trans*-2,3-diphenyloxirane with hydrogen chloride in benzene to form 2-chloro-1,2-diphenylethanol.

$$C_6H_5(H)C\overset{O}{—}C(H)C_6H_5 + HCl \longrightarrow C_6H_5—CH(HO)—CH(Cl)—C_6H_5$$

trans-2,3-diphenyloxirane 2-chloro-1,2-diphenylethanol

(a) How many stereoisomers are possible for 2-chloro-1,2-diphenylethanol? Draw suitable stereorepresentations for each.

There are two stereocenters in the molecule so there are 2^2 or 4 possible stereoisomers.

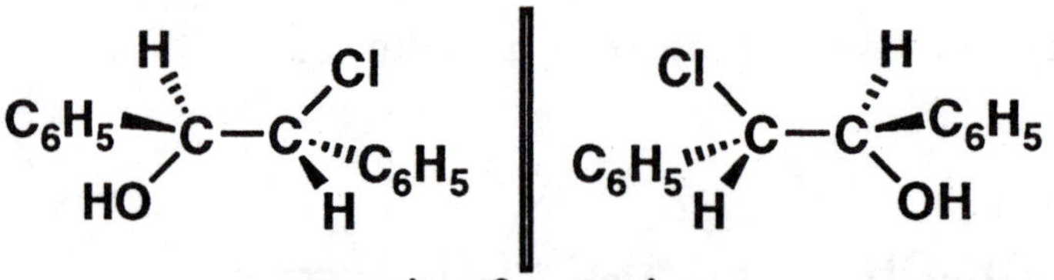

a pair of enantiomers
(from addition of HCl to *trans*-2,3-diphenyloxirane)

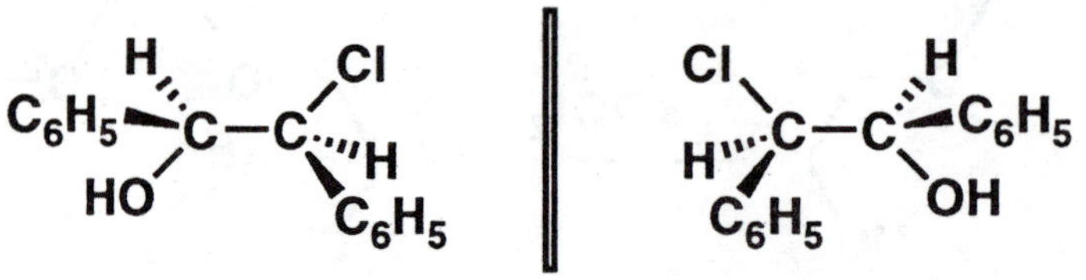

a pair of enantiomers
(from addition of HCl to *cis*-2,3-diphenyloxirane)

(b) Given that opening of the epoxide ring in this reaction is stereoselective, predict which of the possible stereoisomers of 2-chloro-1,2-diphenylethanol is (are) formed in the reaction. Explain.

Opening of the oxirane ring is stereoselective with the nucleophile, chloride ion, displacing the protonated epoxide oxygen in an anti coplanar manner to form the first pair of enantiomers shown in part (a).

Problem 11.24 Propose a mechanism to account for the observation that when tetramethyloxirane is treated with boron trifluoride, it is isomerized to 3,3-dimethyl-2-butanone.

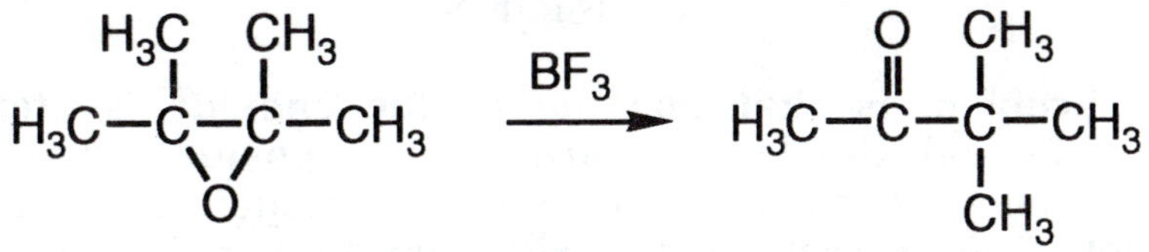

$$H_3C\text{—}C(CH_3)\text{—}C(CH_3)\text{—}CH_3 \ (\text{O bridging}) \xrightarrow{BF_3} H_3C\text{—}C(=O)\text{—}C(CH_3)_2\text{—}CH_3$$

Reaction between boron trifluoride, a Lewis acid, and the epoxide oxygen, a Lewis base, forms an oxonium ion. The ring then opens followed by rearrangement of the resulting secondary carbocation and subsequent loss of boron trifluoride to give the observed ketone.

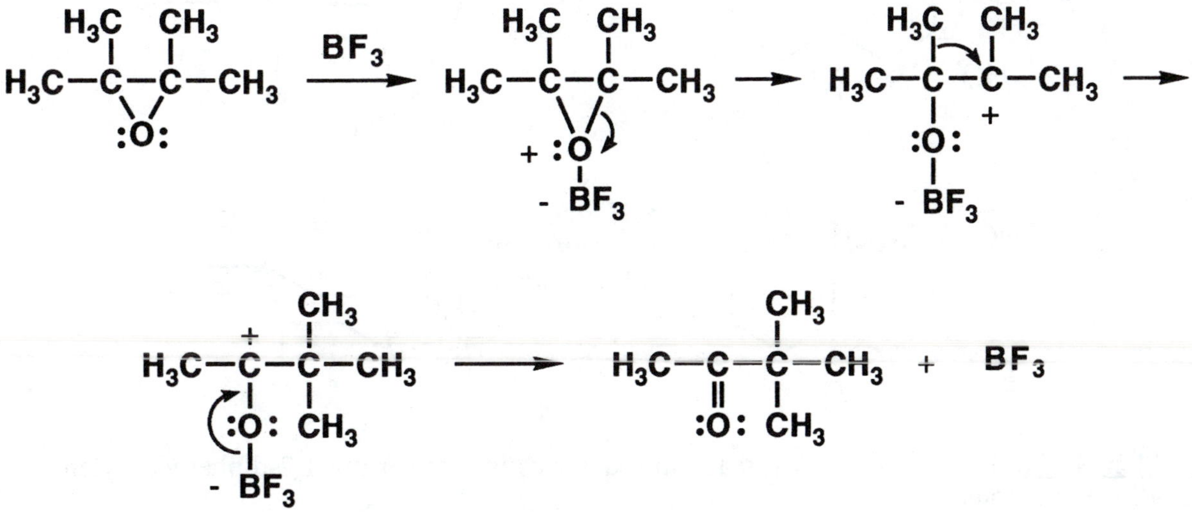

$$H_3C\text{—}C(CH_3)\text{—}C(CH_3)\text{—}CH_3 \ (:\!O\!:) \xrightarrow{BF_3} H_3C\text{—}C(CH_3)\text{—}C(CH_3)\text{—}CH_3 \ (+:\!O\text{—}^{-}BF_3) \longrightarrow H_3C\text{—}C(CH_3)(:\!O\!:\text{—}^{-}BF_3)\text{—}\overset{+}{C}(CH_3)\text{—}CH_3 \longrightarrow$$

$$H_3C\text{—}\overset{+}{C}(:\!O\!:\text{—}^{-}BF_3)\text{—}C(CH_3)_2\text{—}CH_3 \longrightarrow H_3C\text{—}C(=\!:\!O\!:)\text{—}C(CH_3)_2\text{—}CH_3 + BF_3$$

Problem 11.25 Following is the structural formula for an epoxide derived from 9-methyldecalin. Acid-catalyzed hydrolysis of this epoxide gives a *trans*-diol. Of these two possible *trans*-diols, only one is formed. How might you account for this stereoselectivity? *Hint*: Begin by drawing *trans*-decalin with each six-member ring in the more stable chair conformation (Section 2.7B), and then determine whether each substituent in the isomeric glycols is axial or equatorial.

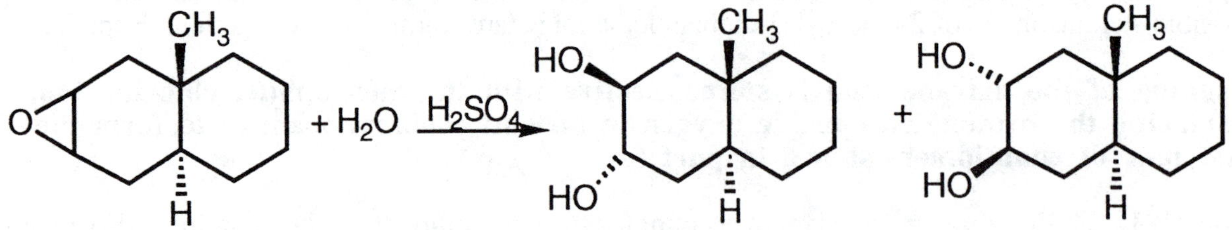

$$\text{epoxide} + H_2O \xrightarrow{H_2SO_4} \text{glycol A} + \text{glycol B}$$

only this glycol is formed

this glycol is not formed

The key to this problem is that opening of the epoxide is stereospecific; the incoming nucleophile and the leaving protonated epoxide oxygen must be anti coplanar. In a cyclohexane ring, anti coplanar corresponds to *trans* and diaxial. An accurate model of the glycol formed will show that in it, the two -OH groups are diaxial and, therefore, *trans* and coplanar. In the alternative glycol, the -OH groups are also *trans*, but because they are diequatorial, they are not coplanar.

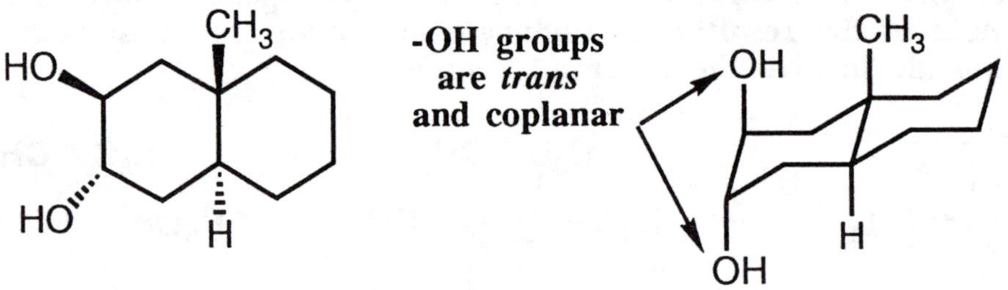

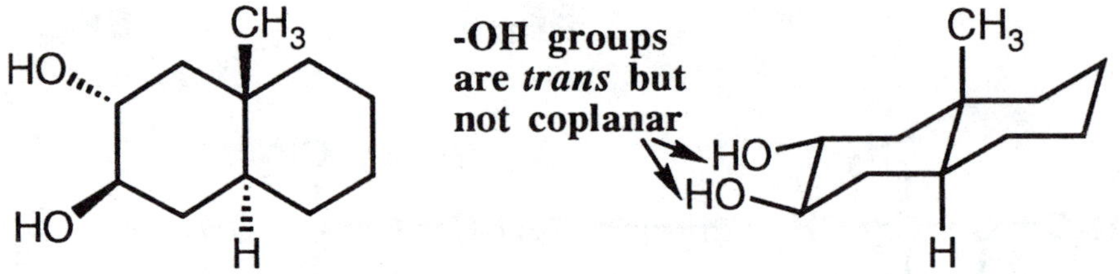

-OH groups are *trans* and coplanar

-OH groups are *trans* but not coplanar

Problem 11.26 Following are two reaction sequences for converting 1,2-diphenylethylene into 2,3-diphenyloxirane.

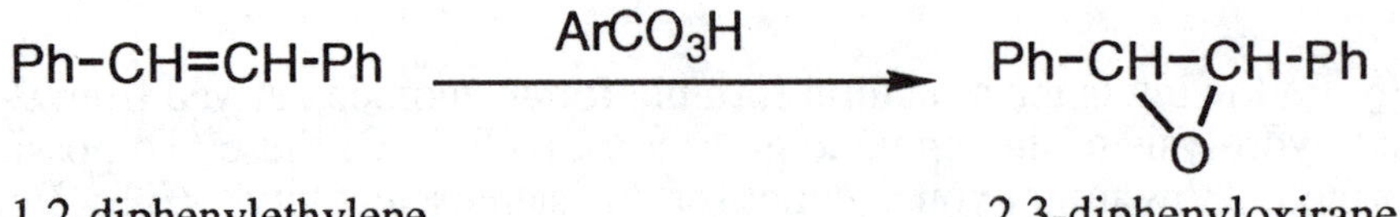

$$\text{Ph-CH=CH-Ph} \xrightarrow{ArCO_3H} \text{Ph-CH}\!-\!\text{CH-Ph (epoxide, O)}$$

1,2-diphenylethylene 2,3-diphenyloxirane

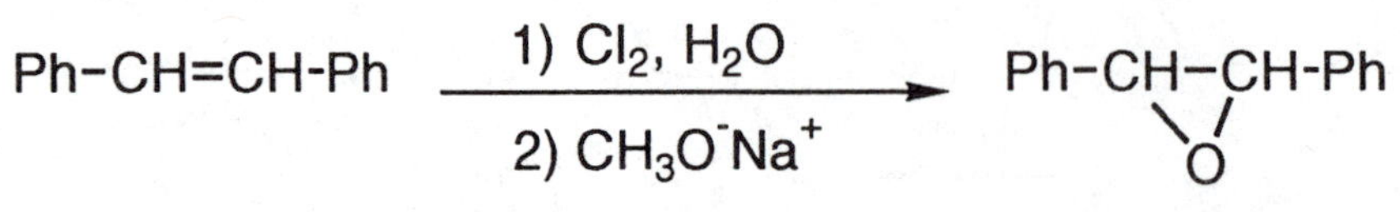

$$\text{Ph-CH=CH-Ph} \xrightarrow[\text{2) } CH_3O^-Na^+]{\text{1) } Cl_2,\ H_2O} \text{Ph-CH}\!-\!\text{CH-Ph (epoxide, O)}$$

1,2-diphenylethylene 2,3-diphenyloxirane

Suppose that the starting alkene is *trans*-1,2-diphenylethylene.
(a) What is the configuration of the oxirane formed in each sequence?

In each case, the oxirane has the *trans* configuration.

(b) Does the oxirane formed in either sequence rotate the plane of polarized light? Explain.

Neither product will rotate the plane of polarized light. The *trans* isomer can exist as a pair of enantiomers. In each reaction, the product formed is a racemic mixture and will not be optically active.

Problem 11.27 In each pair, select the molecule with the greater indicated property.

(a) stronger base: $(CH_3)_3N$: or $(CH_3)_3P$:

(b) better nucleophile: $(CH_3)_3N$: or $(CH_3)_3P$:

Trimethylphosphine is both the stronger base and the better nucleophile.

Problem 11.28 Write equations for the reaction of triphenylphosphine with the following.

(a) CH_3CH_2I (b) 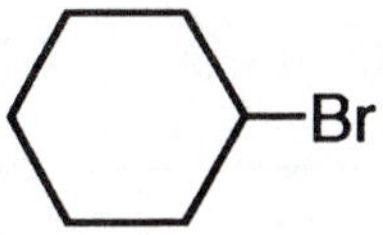 cyclohexyl–Br (c) 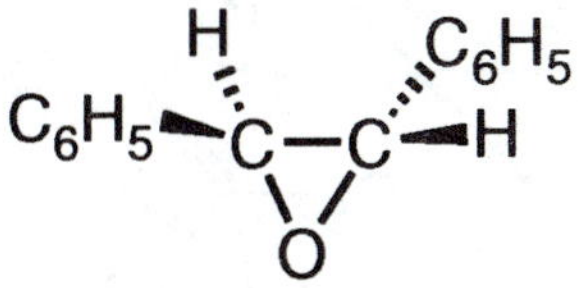 trans-2,3-diphenyloxirane (C_6H_5, H on each carbon)

$CH_3CH_2\overset{+}{P}Ph_3\ I^-$ cyclohexyl–$\overset{+}{P}Ph_3\ Br^-$ H(C_6H_5)C=C(H)(C_6H_5)

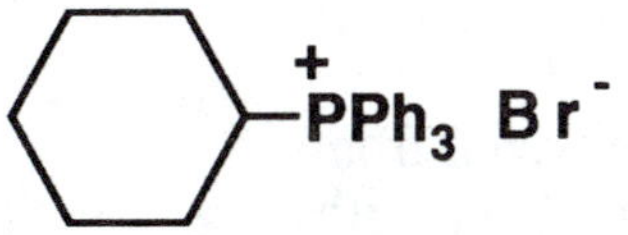

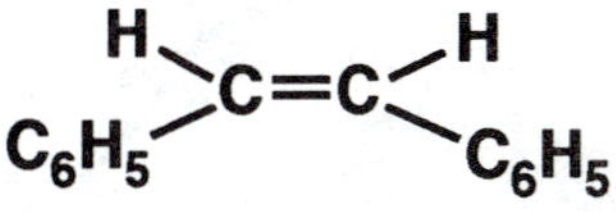

Given above are formulas for the products formed. Note how the stereochemistry is changed in the product in part (c) due to the syn elimination step that occured following heating of the intermediate betaine.

Problem 11.29 Write a series of reactions by which you might convert *cis*-cyclooctene into *trans*-cyclooctene.

Oxidation of *cis*-cyclooctene with a peroxycarboxylic acid gives the *cis*-epoxide. Treatment of this epoxide with triphenylphosphine gives a *trans* betaine. Heating the intermediate betaine results in syn elimination and formation of *trans*-cyclooctene.

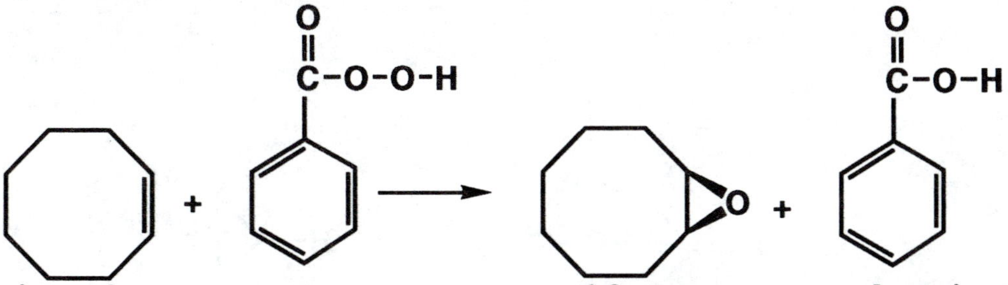

cis-cyclooctene + peroxybenzoic acid (C_6H_5–C(=O)–O–O–H) → 1,2-epoxycyclooctane + benzoic acid (C_6H_5–C(=O)–O–H)

cis-cyclo-octene **peroxy-benzoic acid** **1,2-epoxy-cyclooctane** **benzoic acid**

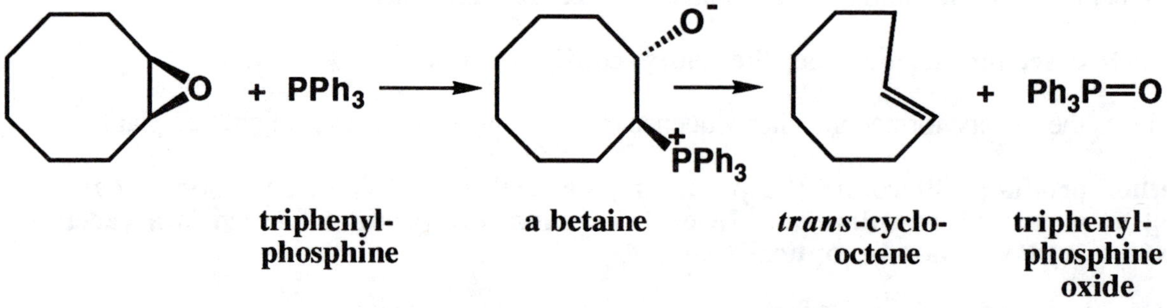

+ PPh_3 → a betaine → *trans*-cyclooctene + $Ph_3P{=}O$

triphenylphosphine | a betaine | *trans*-cyclooctene | triphenylphosphine oxide

Problem 11.30 One of the most useful organic reactions discovered in the last 15 years is the titanium-catalyzed asymmetric epoxidation of allylic alcohols developed by Professor Barry Sharpless and coworkers (see K.B. Sharpless et al., *Pure and Appl. Chem.*, **55**, 589 (1983)). The reagent combination uses Ti(O-iPr)$_4$, a hydroperoxide, and a chiral molecule such as (+)-diethyl tartrate.

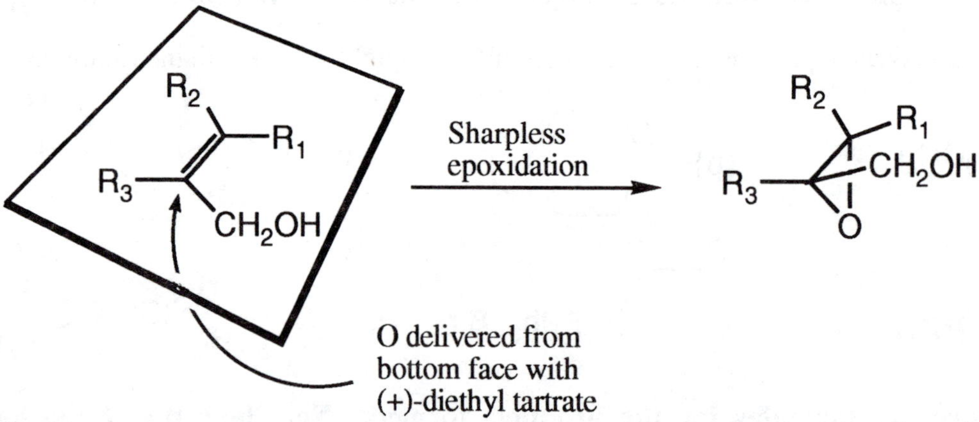

Sharpless epoxidation

O delivered from bottom face with (+)-diethyl tartrate

Two new stereocenters are created in the product. If (+)-diethyl tartrate is used in the reaction, the product arises by delivery of "O" from the hydroperoxide to the bottom face of the molecule and the product epoxide is the stereoisomer shown. If (-)-diethyl tartrate is used, the product is the enantiomer of the stereoisomer shown. Draw the expected products of Sharpless epoxidation of the following allylic alcohols using (+)-diethyl tartrate.

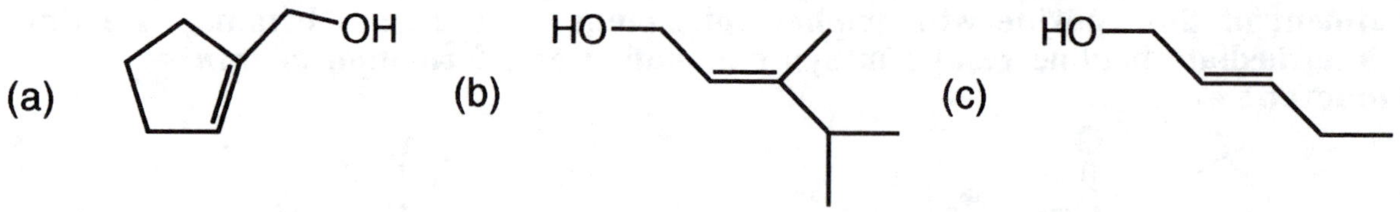

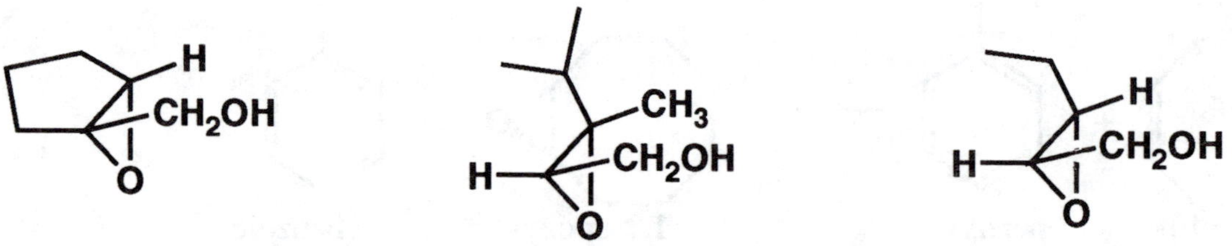

(a) (b) (c)

(d)

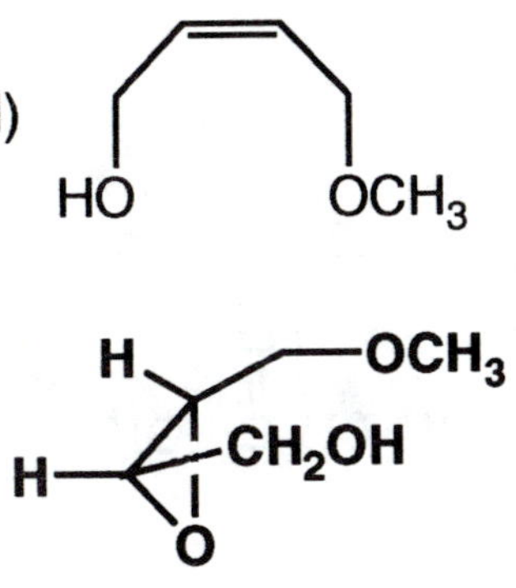

HO OCH$_3$

H OCH$_3$
H CH$_2$OH
O

Problem 11.31 The following chiral epoxide is an intermediate in the synthesis of the insect pheromone, frontalin. How can this epoxide be prepared from an allylic alcohol precursor, using the Sharpless epoxidation reaction described in question 11.30.?

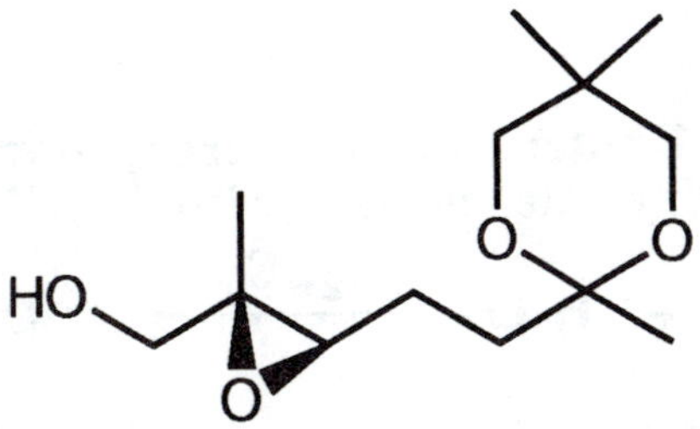

HO O O O

A close inspection indicates that this epoxide has the oxygen atom on the opposite face compared to the ones in problem 11.30 that utilized the (+)-diethyl tartrate. Thus the desired product is produced from the appropriate alkene using (-)-diethyl tartrate along with the rest of the Sharpless epoxidation reagents.

O O H H$_3$C CH$_2$OH **Sharpless Epoxidation** → **(-)-diethyl tartrate** O O O H$_3$C * * H CH$_2$OH

Synthesis

Problem 11.32 Show reagents and experimental conditions to synthesize the following compounds from 1-butanol. Any derivative of 1-butanol already prepared in an earlier part of this problem may then be used for a later synthesis.

(a) butanal

Oxidation of 1-butanol with pyridinium chlorochromate (PCC) or chromic acid, H_2CrO_4, in aqueous acetone.

$$CH_3CH_2CH_2CH_2OH \xrightarrow{PCC} CH_3CH_2CH_2\overset{O}{\overset{\|}{C}}H$$

(b) butanoic acid

Oxidation of 1-butanol with chromic acid in aqueous acid.

$$CH_3CH_2CH_2CH_2OH + H_2CrO_4 \longrightarrow CH_3CH_2CH_2\overset{O}{\overset{\|}{C}}{-}OH$$

(c) 1-butene

Acid-catalyzed dehydration of 1-butanol using H_2SO_4 or H_3PO_4.

$$CH_3CH_2CH_2CH_2OH \xrightarrow{H_2SO_4} CH_3CH_2CH{=}CH_2 \quad + \quad H_2O$$

(d) 2-butanol (two ways)

Acid-catalyzed hydration of 1-butene or oxymercuration of 1-butene followed by reduction of the organomercury intermediate with $NaBH_4$.

$$CH_3CH_2CH{=}CH_2 \ + \ H_2O \xrightarrow{H_2SO_4} CH_3CH_2\underset{OH}{\underset{|}{C}}HCH_3$$

$$CH_3CH_2CH{=}CH_2 \xrightarrow[\text{2) } NaBH_4]{\text{1) } Hg(OAc)_2,\ H_2O} CH_3CH_2\underset{OH}{\underset{|}{C}}HCH_3$$

(e) 2-bromobutane

Addition of HBr (no peroxides) to 1-butene. Another acceptable approach is the reaction of 2-butanol with PBr_3.

$$CH_3CH_2CH{=}CH_2 \ + \ HBr \longrightarrow CH_3CH_2\underset{Br}{\underset{|}{C}}HCH_3$$

(f) 1-chlorobutane

Treatment of 1-butanol with a chlorinating agent such as $SOCl_2$, PCl_3, or HCl / $ZnCl_2$.

$$CH_3CH_2CH_2CH_2OH + SOCl_2 \longrightarrow CH_3CH_2CH_2CH_2OH + SO_2 + HCl$$

(g) 1,2-dibromobutane

Addition of Br_2 to 1-butene.

$$CH_3CH_2CH{=}CH_2 \ + \ Br_2 \longrightarrow CH_3CH_2\underset{Br}{\underset{|}{C}}H\underset{Br}{\underset{|}{C}}H_2$$

(h) 1-butyne

Dehydrohalogenation of 1,2-dibromobutane with two moles of $NaNH_2$.

$$CH_3CH_2CH(Br)CH_2Br + 2\ NaNH_2 \longrightarrow CH_3CH_2C{\equiv}CH + 2\ NH_3 + 2\ NaBr$$

(i) 2-butanone

Oxidation of 2-butanol using $K_2Cr_2O_7$ in aqueous acid, and hydration of 1-butyne in the presence of H_2SO_4 and $HgSO_4$.

$$CH_3CH_2CH(OH)CH_3 + H_2CrO_4 \longrightarrow CH_3CH_2C(=O)CH_3 + Cr^{3+}$$

$$CH_3CH_2C{\equiv}CH + H_2O \xrightarrow{H_2SO_4} \left[CH_3CH_2C(OH){=}CH_2\right] \longrightarrow CH_3CH_2C(=O)CH_3$$

enol

(j) 1-chloro-2-butanol

Addition of chlorine in water (HOCl) to 1-butene.

$$CH_3CH_2CH{=}CH_2 + Cl_2,\ H_2O \longrightarrow CH_3CH_2CH(OH)CH_2Cl + HCl$$

(k) ethyloxirane

Treatment of 1-chloro-2-butanol with base to give the epoxide by an internal S_N2 reaction. Another acceptable method would be epoxidation of 1-butene using 3-chloroperoxybenzoic acid.

$$CH_3CH_2CH(OH)CH_2Cl + KOH \longrightarrow CH_3CH_2\overset{\diagdown\ \diagup}{CHCH_2}\ (\text{bridged by O})$$

(l) dibutyl ether (two ways)

Treatment of 1-butanol with sodium metal to form sodium butoxide followed by treatment of this metal alkoxide with 1-chlorobutane in a Williamson synthesis.

$$2\ CH_3CH_2CH_2CH_2OH + 2Na \longrightarrow 2\ CH_3CH_2CH_2CH_2O^-Na^+ + H_2$$

$$CH_3CH_2CH_2CH_2O^-Na^+ + CH_3CH_2CH_2CH_2Cl \longrightarrow (CH_3CH_2CH_2CH_2)_2O + Na^+Cl^-$$

Alternatively, this reaction could be accomplished by treatment of 1-butanol with concentrated sulfuric acid at a temperature that favors ether formation.

$$2\ CH_3CH_2CH_2CH_2OH \xrightarrow{H_2SO_4} (CH_3CH_2CH_2CH_2)_2O + H_2O$$

(m) butyl *sec*-butyl ether

Treatment of 2-butanol with sodium metal to form sodium 2-butoxide and then treatment of this metal alkoxide with 1-chlorobutane.

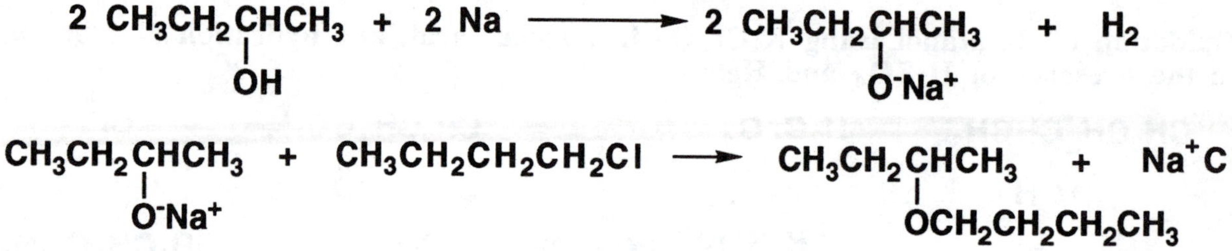

$$2\ CH_3CH_2CH(OH)CH_3 + 2\ Na \longrightarrow 2\ CH_3CH_2CH(O^-Na^+)CH_3 + H_2$$

$$CH_3CH_2CH(O^-Na^+)CH_3 + CH_3CH_2CH_2CH_2Cl \longrightarrow CH_3CH_2CH(OCH_2CH_2CH_2CH_3)CH_3 + Na^+Cl^-$$

(n) 2-butoxy-1-butanol

Treatment of ethyloxirane with 1-butanol in the presence of an acid catalyst.

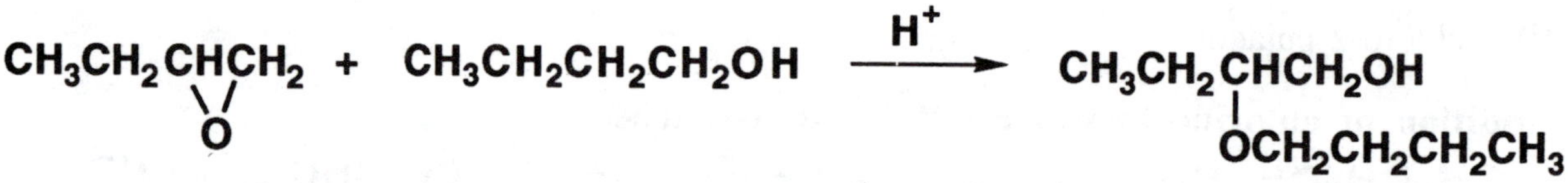

$$CH_3CH_2\underset{\diagdown O \diagup}{CHCH_2} + CH_3CH_2CH_2CH_2OH \xrightarrow{H^+} CH_3CH_2CH(OCH_2CH_2CH_2CH_3)CH_2OH$$

(o) 1-butoxy-2-butanol

Treatment of ethyloxirane with 1-butanol in the presence of sodium butoxide.

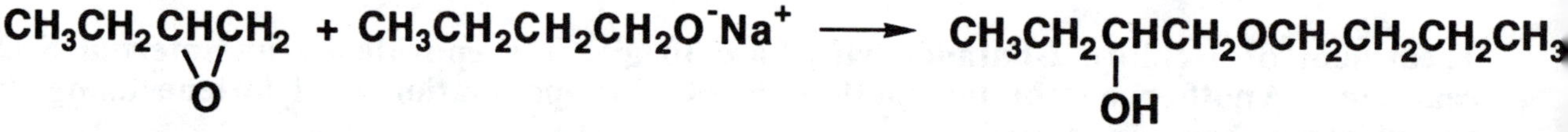

$$CH_3CH_2\underset{\diagdown O \diagup}{CHCH_2} + CH_3CH_2CH_2CH_2O^-Na^+ \longrightarrow CH_3CH_2CH(OH)CH_2OCH_2CH_2CH_2CH_3$$

(p) 1,2-butanediol (two ways)

Acid-catalyzed hydrolysis of ethyloxirane.

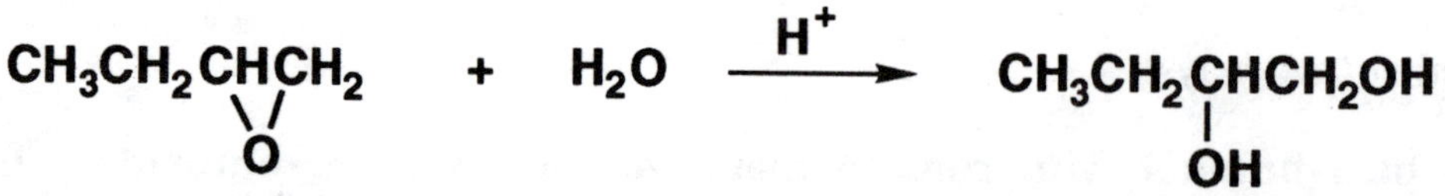

$$CH_3CH_2\underset{\diagdown O \diagup}{CHCH_2} + H_2O \xrightarrow{H^+} CH_3CH_2CH(OH)CH_2OH$$

Alternatively, oxidation of 1-butene using potassium permanganate in cold aqueous solution at high pH (pH=11).

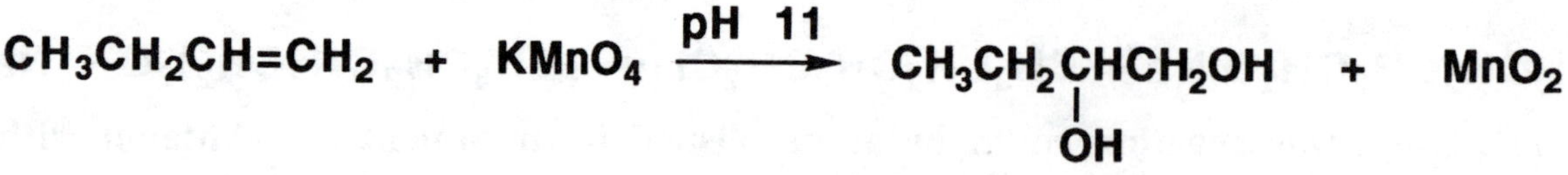

$$CH_3CH_2CH{=}CH_2 + KMnO_4 \xrightarrow{pH\ 11} CH_3CH_2CH(OH)CH_2OH + MnO_2$$

Problem 11.33 Show how to bring about the following conversions.

(a) 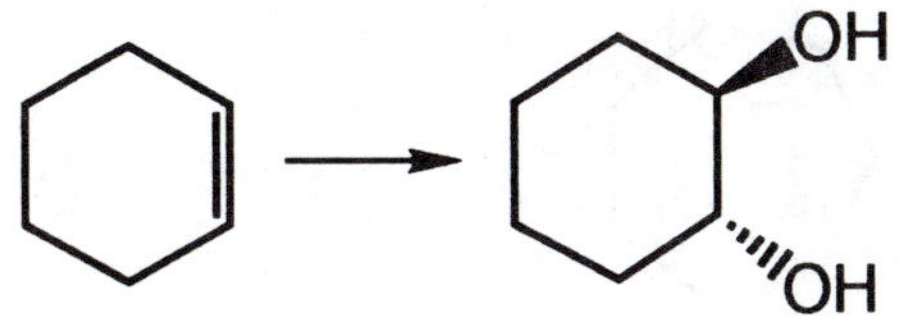 cyclohexene → *trans*-1,2-cyclohexanediol (b) 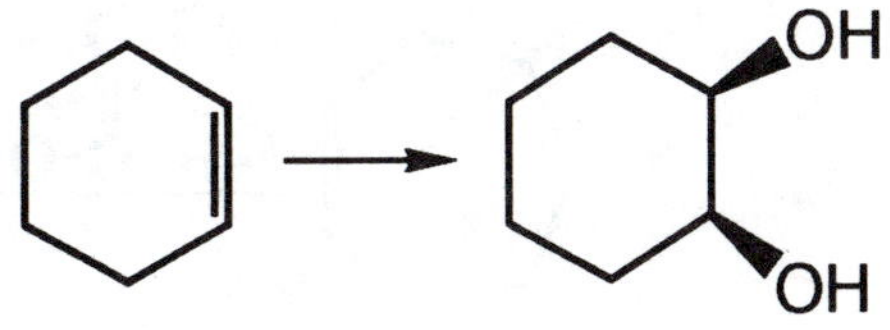 cyclohexene → *cis*-1,2-cyclohexanediol

The *trans*-glycol is prepared by acid-catalyzed hydrolysis of cyclohexene oxide, which in turn may be synthesized from cyclohexene by oxidation with a peroxycarboxylic acid.

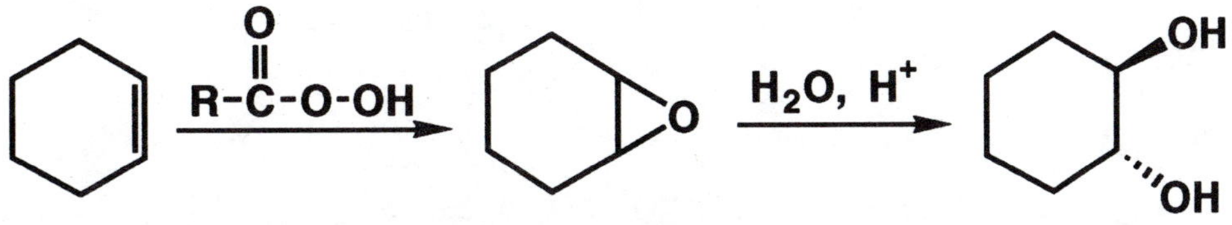

R-C(=O)-O-OH; H_2O, H^+

Alternatively, the epoxide could be prepared by reaction with chlorine in water follwed by treatment with base.

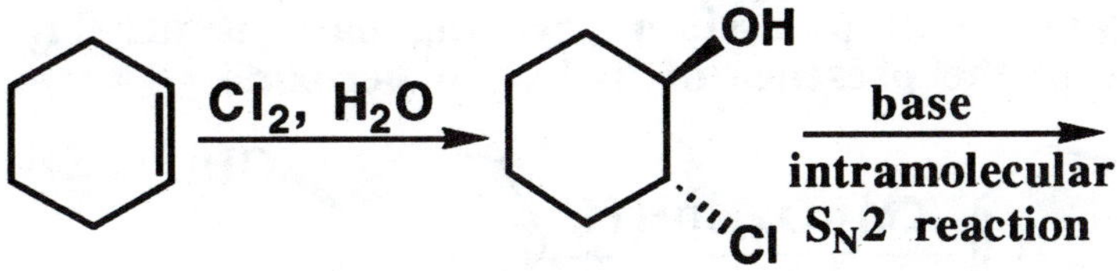

Cl_2, H_2O; base, intramolecular S_N2 reaction

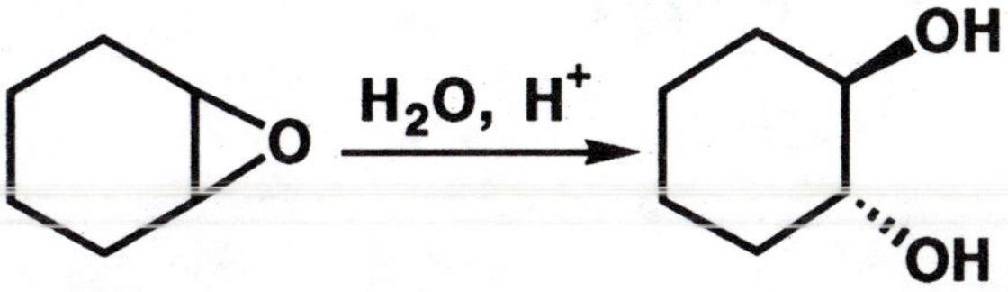

H_2O, H^+

The *cis*-glycol is prepared by oxidation of cyclohexene using either potassium permanganate at pH 11 or using osmium tetroxide in the presence of hydrogen peroxide.

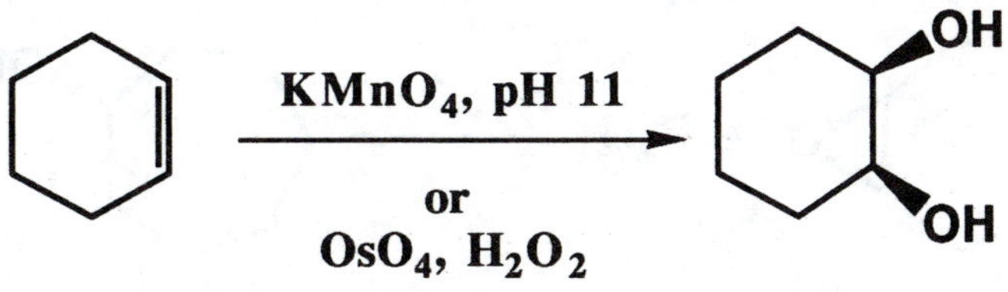

$KMnO_4$, pH 11 or OsO_4, H_2O_2

Problem 11.34 Show reagents and experimental conditions to convert cycloheptene to the following. Any compound made in an earlier part of this problem may be used as an intermediate in any following conversion.

(a)

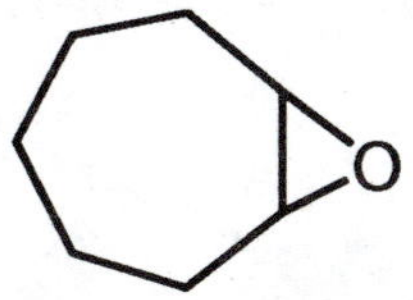

Oxidation of cycloheptene with a peroxyacid such as trifluoroperacetic acid.

Oxidation of cycloheptene with a peroxyacid such as trifluoroperacetic acid.

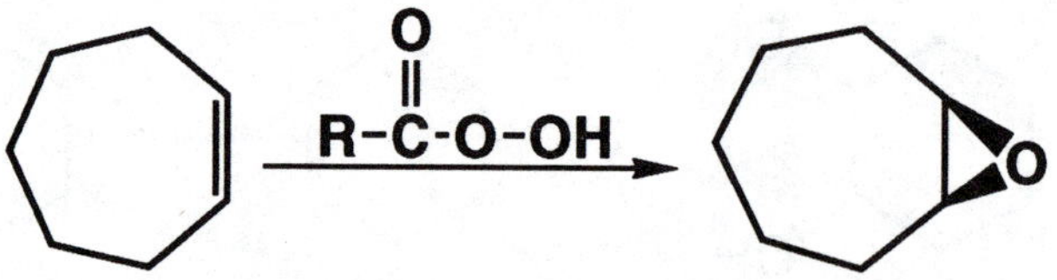

R-C(=O)-O-OH

As an alternative, addition of HOCl to cycloheptene to form a chlorohydrin followed by treatment of the chlorohydrin with strong base.

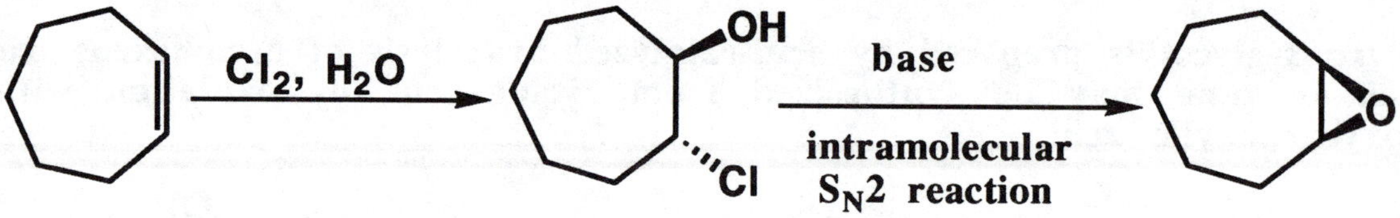

Cl_2, H_2O OH Cl base intramolecular S_N2 reaction O

(b) 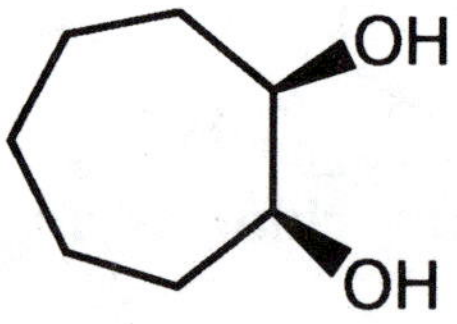 OH OH

Oxidation of cycloheptene with potassium permanganate at pH 11, or oxidation with osmium tetroxide in the presence of hydrogen peroxide.

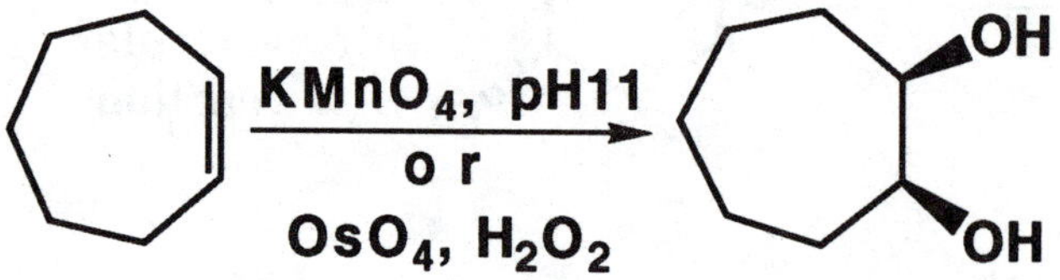

$KMnO_4$, pH11 or OsO_4, H_2O_2 OH OH

(c) 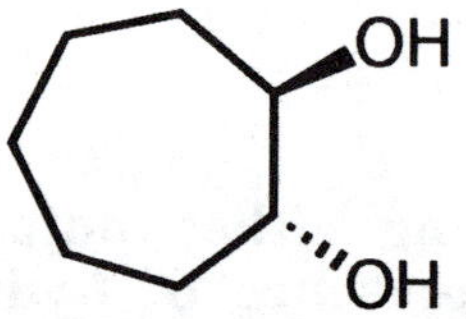 OH OH

Acid-catalyzed hydrolysis of the epoxide from part (a).

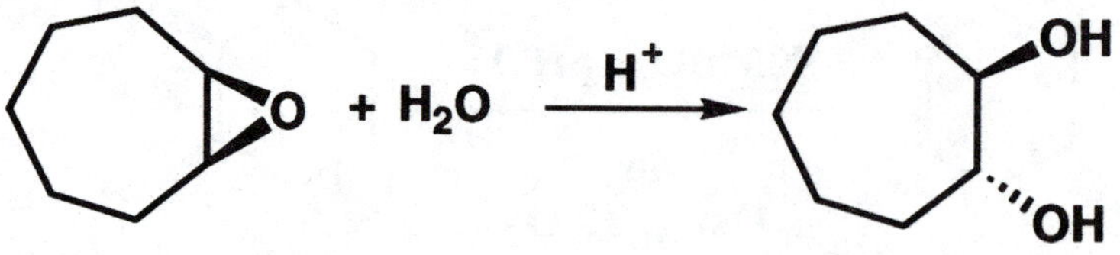

O + H_2O H^+ OH OH

(d) 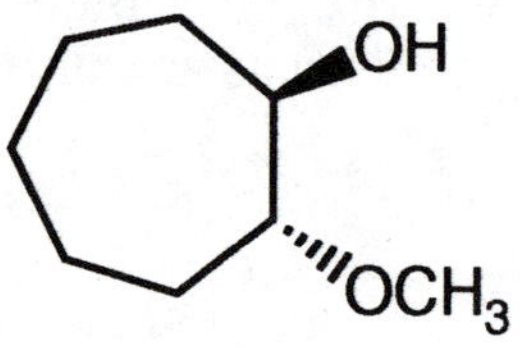 OH OCH_3

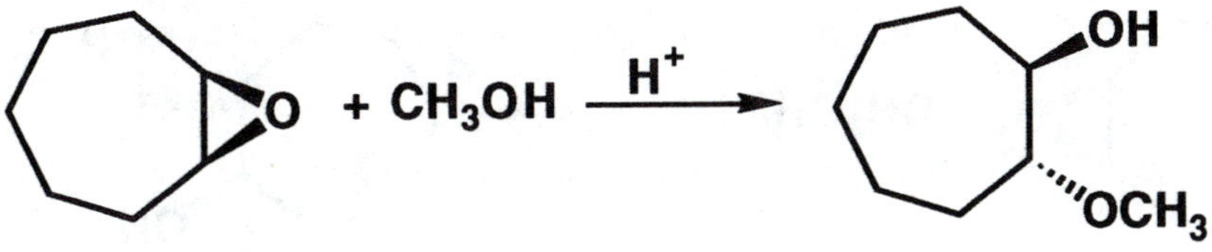

+ CH_3OH $\xrightarrow{H^+}$ OH, OCH_3

(e) 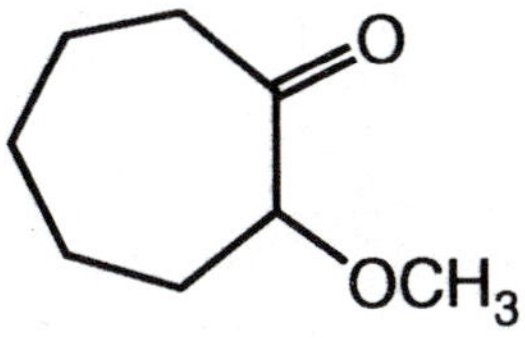 O, OCH_3

Oxidation of the secondary alcohol of part (d) using chromium trioxide in pyridine or as pyridinium chlorochromate.

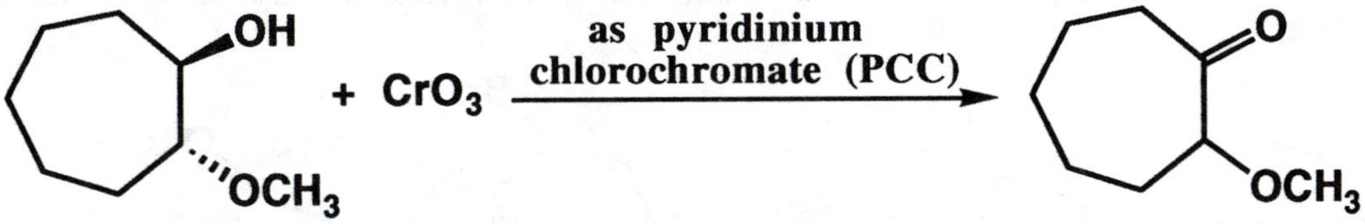

OH, OCH_3 + CrO_3 $\xrightarrow{\text{as pyridinium chlorochromate (PCC)}}$ O, OCH_3

(f) 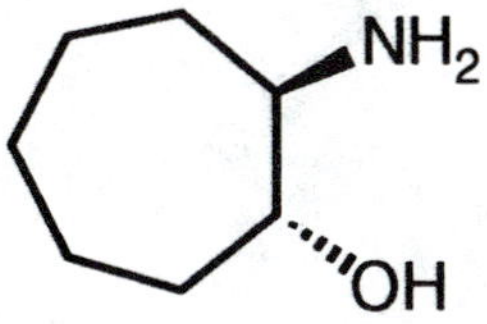 NH_2, OH

Treatment of the epoxide from part (a) with ammonia.

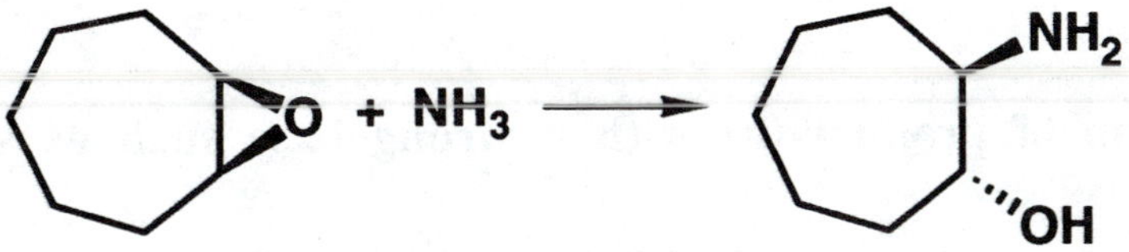

O + NH_3 $\longrightarrow$ NH_2, OH

(g) 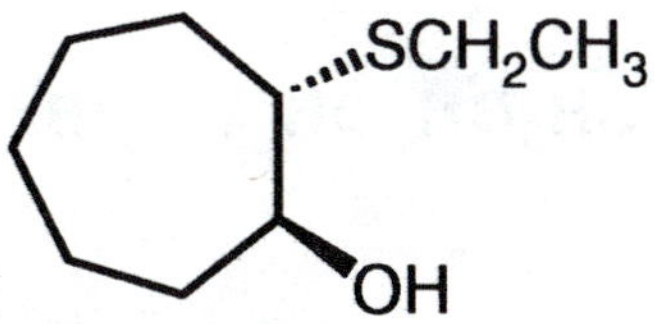 SCH_2CH_3, OH

Treatment of the epoxide from part (a) with either ethanethiol or with the sodium salt of ethanethiol.

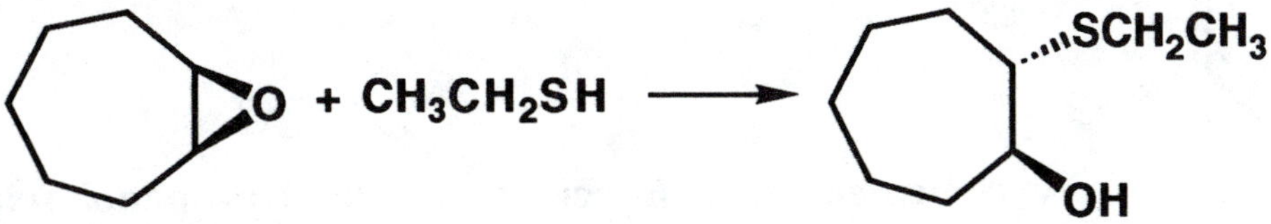

O + CH_3CH_2SH $\longrightarrow$ SCH_2CH_3, OH

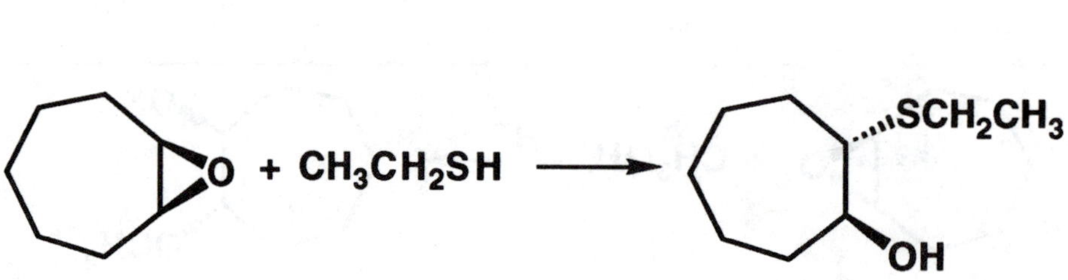

$$\text{cycloheptene oxide} + CH_3CH_2SH \longrightarrow \text{trans-2-(ethylthio)cycloheptanol } (SCH_2CH_3,\ OH)$$

(h) 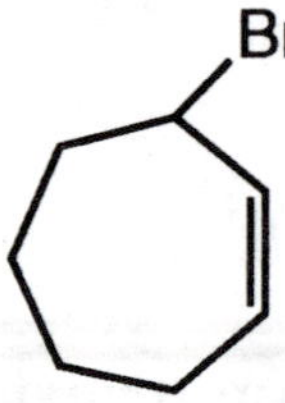 3-Bromocycloheptene (Br)

Allylic bromination of cycloheptene with N-bromosuccinimide (NBS) or with bromine at high temperature. In each case, reaction is by a radical chain mechanism.

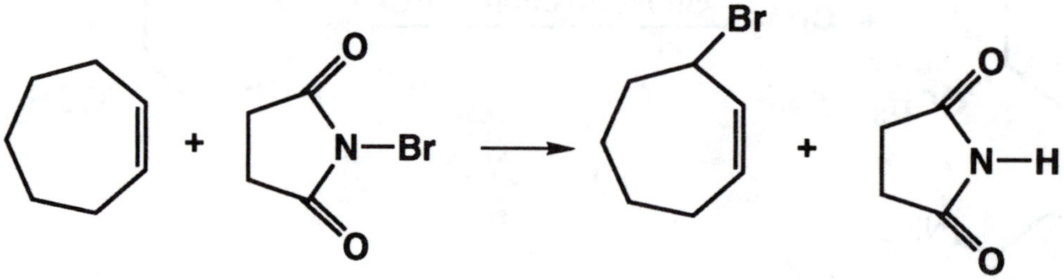

$$\text{cycloheptene} + \text{N-bromosuccinimide (N–Br)} \longrightarrow \text{3-bromocycloheptene} + \text{succinimide (N–H)}$$

(i) 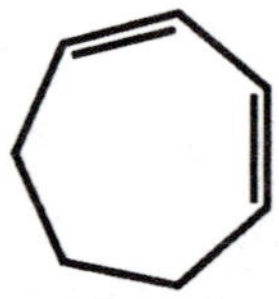 1,3-Cycloheptadiene

Dehydrohalogenation of product (h) with a strong base such as sodium ethoxide in ethanol.

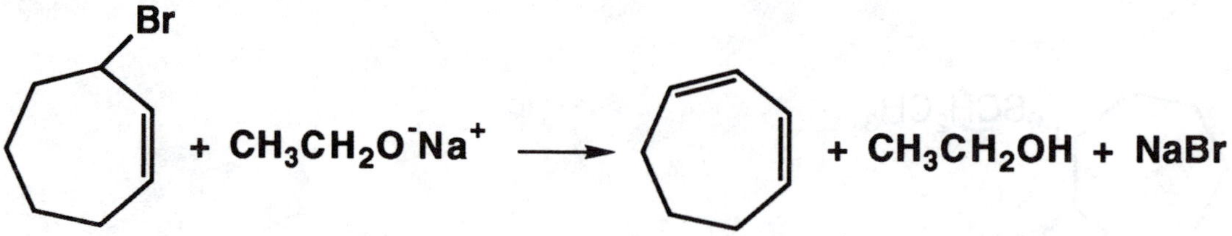

$$\text{3-bromocycloheptene} + CH_3CH_2O^-Na^+ \longrightarrow \text{1,3-cycloheptadiene} + CH_3CH_2OH + NaBr$$

(j) 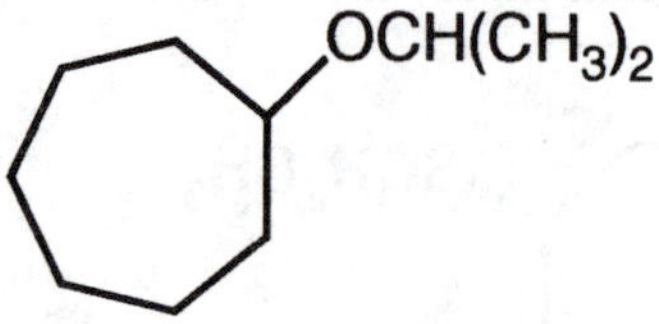 Cycloheptyl isopropyl ether ($OCH(CH_3)_2$)

Alkoxymercuration of cycloheptene with mercuric acetate, 2-propanol followed by reduction of the organomercury intermediate with sodium borohydride.

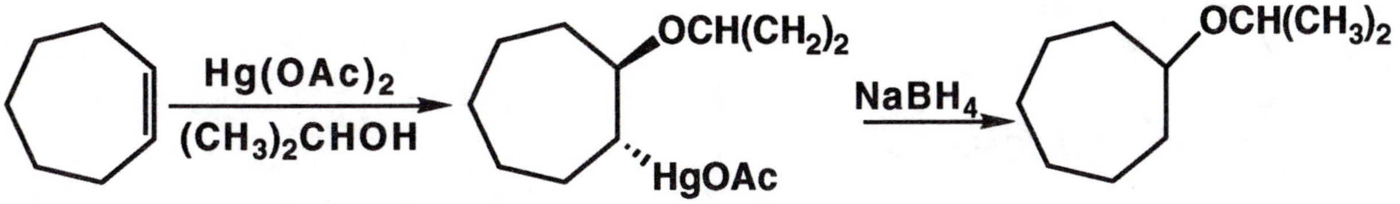

Hg(OAc)$_2$ / (CH$_3$)$_2$CHOH → OCH(CH$_2$)$_2$, HgOAc; NaBH$_4$ → OCH(CH$_3$)$_2$

An alternative is a Williamson ether synthesis. Using either combination of reagents involves an S$_N$2 reaction on a secondary halide, in which case E2 and S$_N$2 are competing reactions.

(k) SH

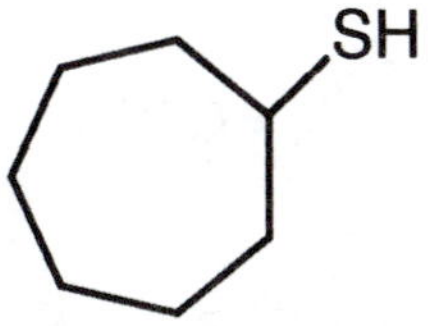

As a first step, hydration of cycloheptene to cycloheptanol. Hydration can be accomplished by (1) acid-catalyzed hydration, (2) oxymercuration followed by reduction with sodium borohydride, or (3) hydroboration followed by oxidation of the organoborane intermediate with alkaline hydrogen peroxide. Treatment of cycloheptanol with thionyl chloride and then sodium hydrosulfide gives the product.

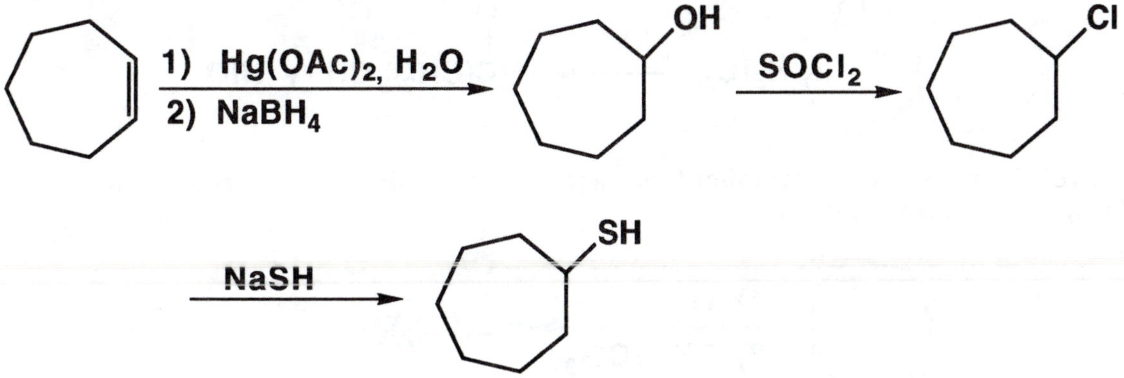

1) Hg(OAc)$_2$, H$_2$O / 2) NaBH$_4$ → OH; SOCl$_2$ → Cl

NaSH → SH

(l) OH, CN

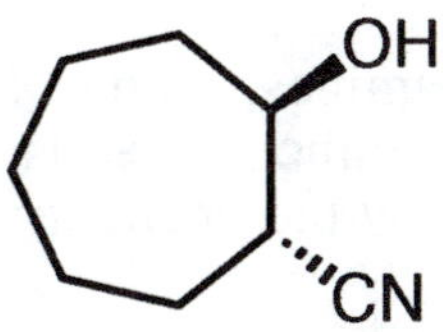

Treatment of the epoxide from part (a) with aqueous sodium cyanide.

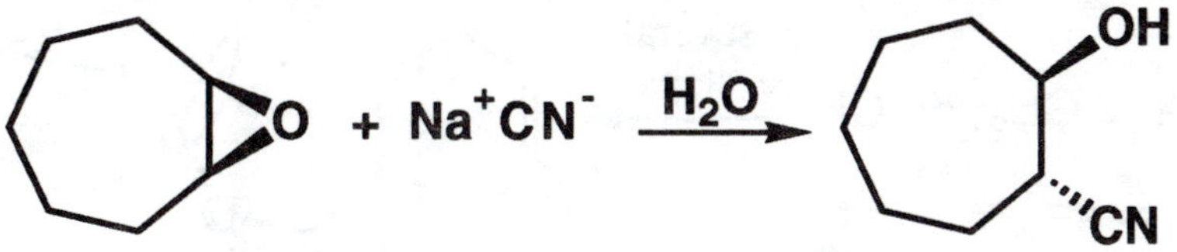

O + Na$^+$CN$^-$ H$_2$O → OH, CN

(m) [structure: cycloheptanol with OH and $C\equiv CCH_3$ substituents]

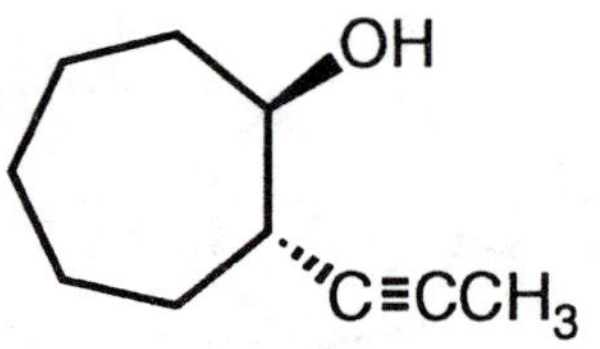

Treatment of the epoxide from part (a) with the sodium salt of propyne followed by treatment with H_2O.

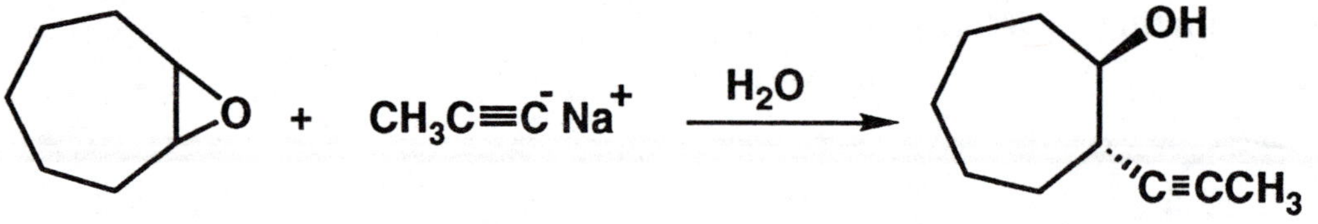

$$\text{cycloheptene oxide} + CH_3C\equiv\bar{C}\,Na^+ \xrightarrow{H_2O} \text{trans-2-(1-propynyl)cycloheptanol}$$

(n) $HC(=O)(CH_2)_5CH(=O)$

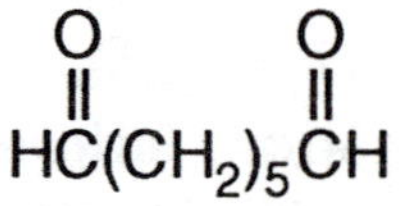

Oxidation of the glycol from part (b) or (c) with periodic acid or with lead tetraacetate.

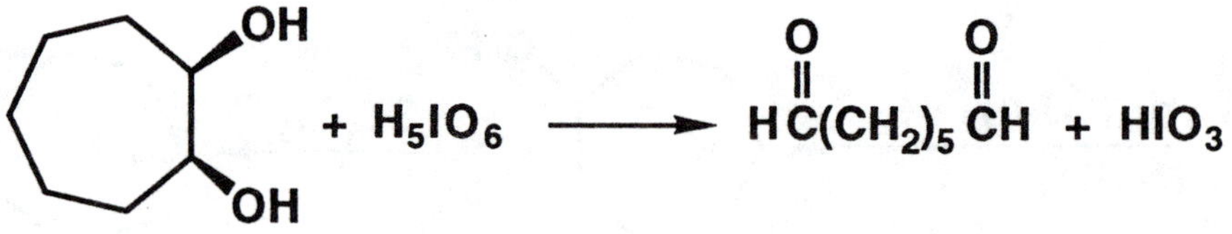

$$\text{cis-1,2-cycloheptanediol} + H_5IO_6 \longrightarrow HC(=O)(CH_2)_5CH(=O) + HIO_3$$

Alternatively, oxidation of cycloheptene with ozone followed by treatment of the ozonide with dimethyl sulfide.

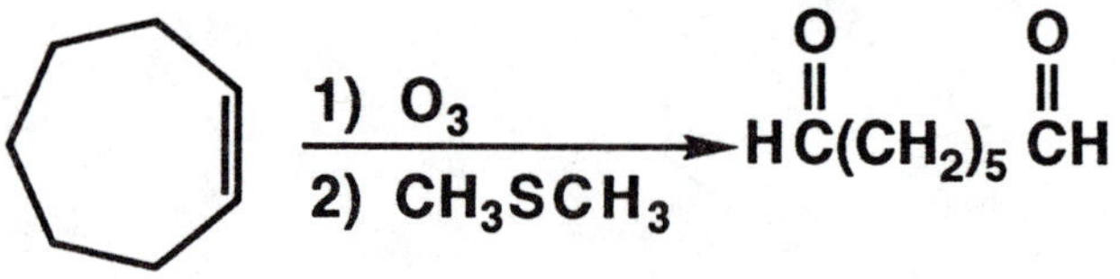

$$\text{cycloheptene} \xrightarrow[\text{2) } CH_3SCH_3]{\text{1) } O_3} HC(=O)(CH_2)_5CH(=O)$$

Problem 11.35 The following compound is one of a group of β-chloroamines, many of which have anti-tumor activity. Describe a synthesis of this compound from anthranilic acid and ethylene oxide. Hint: To see how the seven-member ring might be formed, review the chemistry of the nitrogen mustards (Section 10.7H).

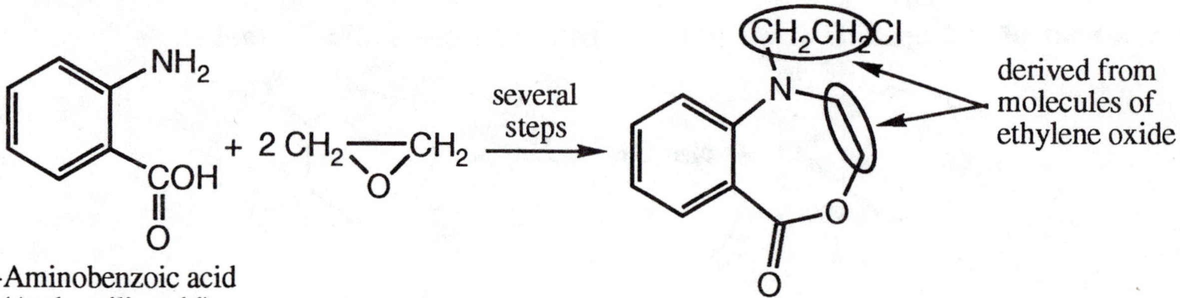

$$\text{2-Aminobenzoic acid} + 2\,CH_2\text{–}CH_2\text{(O)} \xrightarrow{\text{several steps}} \text{product with } N\text{–}CH_2CH_2Cl$$

NH$_2$, COH, O

2-Aminobenzoic acid
(Anthranilic acid)

CH_2CH_2Cl, N, O, O

derived from molecules of ethylene oxide

The nitrogen atom of the amine in anthranilic acid is a nucleophile and reacts in succession with two moles of ethylene oxide to give a molecule with two -CH_2CH_2OH groups on nitrogen. Treatment if this dialcohol with concentrated HCl converts each -OH to -Cl.

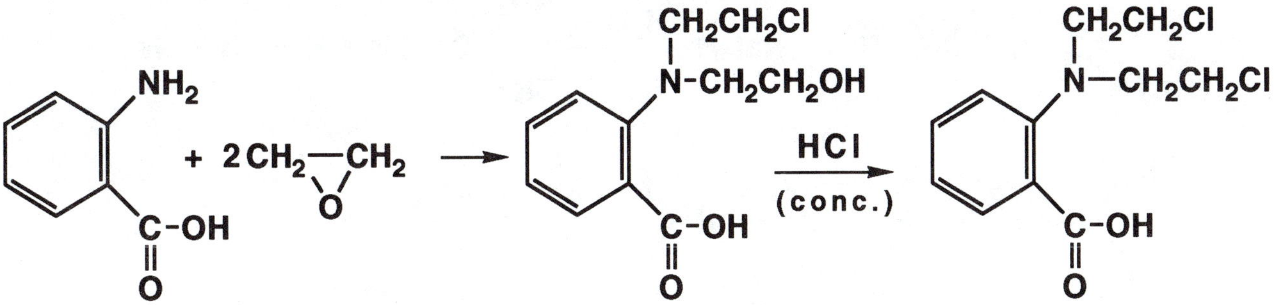

NH_2 C-OH O + 2 CH_2—CH_2 O → CH_2CH_2Cl N-CH_2CH_2OH C-OH O —HCl (conc.)→ CH_2CH_2Cl N—CH_2CH_2Cl C-OH O

Treatment of the carboxylic acid with a base such as sodium carbonate or sodium hydroxide converts it to a carboxylate anion. This anion is a nucleophile and displaces chloride by an intramolecular S_N2 reaction to give the product.

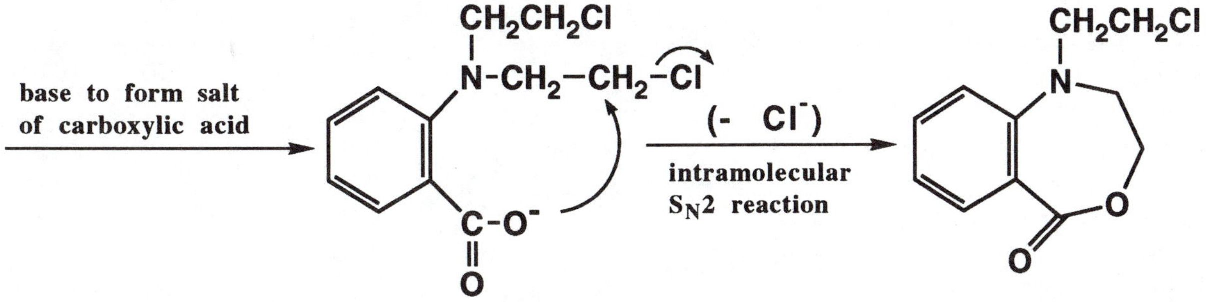

base to form salt of carboxylic acid → CH_2CH_2Cl N-CH_2-CH_2-Cl C-O⁻ O —(- Cl⁻) intramolecular S_N2 reaction→ CH_2CH_2Cl N O O

Problem 11.36 The following compound was wanted for the study of the Claisen rearrangement (Section 15.5E). Describe the synthesis of this compound from styrene and 2-methyl-2-butene.

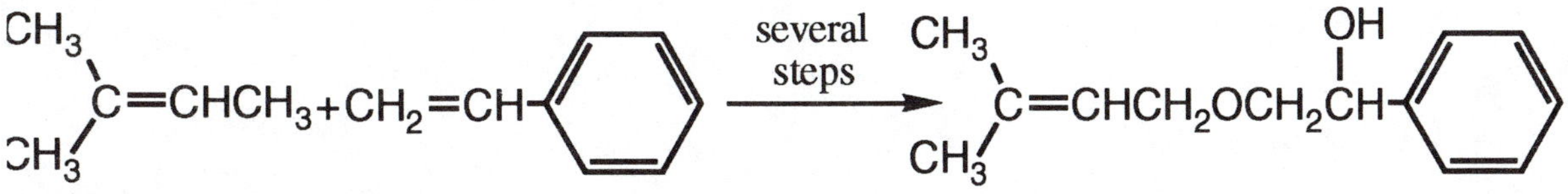

CH_3 CH_3 C=$CHCH_3$+CH_2=CH-C_6H_5 —several steps→ CH_3 CH_3 C=$CHCH_2OCH_2$CH(OH)-C_6H_5

The step by which the carbon skeletons of these two starting materials are joined involves treatment of styrene oxide with an alkoxide in an S_N2 type reaction. First, convert styrene to its epoxide by oxidation with a peroxycarboxylic acid.

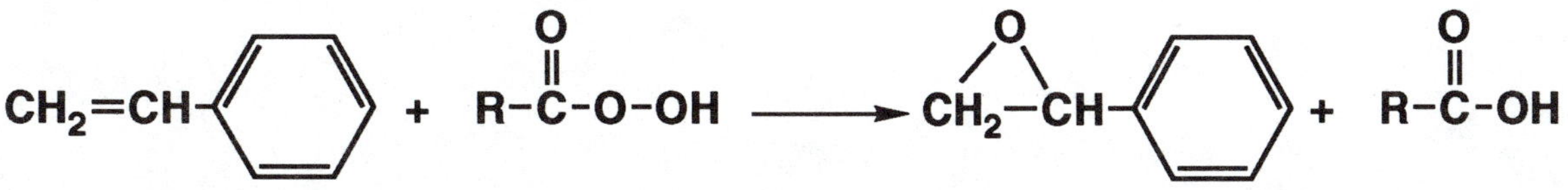

CH_2=CH-C_6H_5 + R-C(=O)-O-OH → CH_2—CH(O)-C_6H_5 + R-C(=O)-OH

Then convert 2-methyl-2-butene to 2-methyl-2-buten-1-ol and then to its sodium salt.

$$(CH_3)_2C{=}CH{-}CH_3 + Br_2 \xrightarrow{\text{heat}} (CH_3)_2C{=}CH{-}CH_2{-}Br + HBr$$

$$(CH_3)_2C{=}CH{-}CH_2{-}Br + NaOH \xrightarrow{H_2O} (CH_3)_2C{=}CH{-}CH_2{-}OH + NaBr$$

$$(CH_3)_2C{=}CHCH_2OH + \underset{\text{epoxide}}{CH_2{-}CH}{-}C_6H_5 \xrightarrow{\text{base}} (CH_3)_2C{=}CHCH_2OCH_2\overset{OH}{\overset{|}{C}H}{-}C_6H_5$$

CHAPTER 12: MASS SPECTROMETRY

12.0 OVERVIEW

- Today, instrumental methods are used to determine molecular structure. **Mass spectrometry** is an important instrumental method used routinely by organic chemists. ✳

12.1 MASS SPECTROMETRY

- In a **mass spectrometer**, electrons are removed from atoms or molecules to produce a stream of positive ions, which is accelerated in an electric field then passed through a magnetic field. The extent of curvature of the path of individual ions depends on the ratio of **mass-to-charge** (***m/e***) of the ion. Differences in curvature are measured. These measurements are then used to determine the mass of the ions, thereby providing important structural information. ✳

12.2 A MASS SPECTROMETER

- In a **mass spectrometer**:
  - The neutral atoms or molecules are converted to a beam of positively-charged ions in the **ionization chamber**. The atoms or molecules are bombarded with a stream of **high energy electrons**, and the resulting collisions result in loss of electrons from the sample atoms or molecules to produce the positive ions.
    The species formed by removal of a single electron from a molecule are called **molecular ions**. These species are actually **radical cations**, since they contain both an odd number of electrons and a positive charge.
  - Once molecular ions are formed, they are transformed into a rapidly traveling ion beam by a positively-charged **repeller plate** and negatively-charged **accelerator plates**.
  - The ion beam is focused by focusing slits and passed into the **mass analyzer**. The mass analyzer has a magnetic field, so the accelerated ions are deflected along a circular path. Ions with larger mass are deflected less than those with smaller mass. By varying the accelerating voltage or the strength of the magnetic field, ions can be focused on a detector where the ion current is detected and counted.
- A **mass spectrum** is a scan of relative ion abundance versus mass-to-charge ratio, and is normally plotted as a bar graph. The tallest peak in the mass spectrum is called the **base peak**.

12.3 FEATURES IN A MASS SPECTRUM

- **Resolution** is an important operating characteristic of a mass spectrometer that refers to how well it separates ions of differing mass. **Low resolution** mass spectrometers are capable of separating ions differing by **nominal mass** (ions that differ by one or more mass unit), and **high resolution** mass spectrometers are capable of separating ions differing in mass by as little as 0.001 **atomic mass units (amu).**
- Most common elements found in organic molecules (H, C, N, O, S, Cl, and Br) are found as a **mixture of isotopes**, which occur in characteristic ratios. Of particular interest are chlorine and bromine because they each have two predominant isotopes. Chlorine is found in nature as ^{35}Cl and ^{37}Cl in a 75.77 to 24.23 ratio, and bromine is found as ^{79}Br and ^{81}Br in a 50.69 to 49.31 ratio. Common elements that occur in nature only as single isotopes include F, P, and I.
- The electrons used in the ionization chamber are of such high energy that additional reactions can occur with the molecular ion that is initially formed. A common type of reaction is **fragmentation**, that is breaking up of the molecular ion into smaller fragments. Some of these fragments can break up further into even smaller pieces. ✳
  - Cation radicals usually break into a cation fragment and a radical fragment. Only the cation is observed in the mass spectrum, because a charge is required for the fragment to be

accelerated into the mass analyzer. An uncharged radical fragment will not be detected in the mass spectrum. *[This would be a good time to review the definitions of a cation, a radical, and a cation radical.]*

- When it comes to ion fragmentation, more stable cations are formed in preference to less stable cations. Thus, the probability of fragmentation to form a new carbocation increases in the order CH_3<1°<2°<3°=allylic.

12.4 INTERPRETING MASS SPECTRA

- When a chemist interprets a mass spectrum, identification of the compounds comes primarily from the mass of the molecular ion, and on the appearance of (M+1) and (M+2) peaks as evidence for the presence of heteroatoms such as Br and Cl. Important structural information can be obtained by analysis of the fragmentation patterns in mass spectra, even though they can be quite complicated (many peaks). Chemists rarely attempt a total analysis of a complicated mass spectrum, only certain key peaks are analyzed. This is in contrast to NMR spectra described in the next chapter, in which every peak is usually scrutinized. ✳
- Different types of molecules give characteristic fragmentation patterns, and these will be discussed in turn.
- **Alkanes**
  - Straight chain hydrocarbons fragment by breaking carbon-carbon bonds to form a homologous series of cations differing by 14 amu (a CH_2 group). Fragmentation tends to occur in the middle of unbranched chains and the most stable carbocation tends to be formed in preference to the more stable radical. Recall that the initially formed molecular ion fragments to give a cation and a radical, but only the cation is detected in the mass spectrum.
  - Branched hydrocarbons fragment to form secondary and tertiary cations.
  - Cycloalkanes fragment by losing side chains and also ethylene via fragmentation of the ring.
- **Alkenes**
  - Alkenes show a strong molecular ion peak due to removal of one electron from the pi bond.
  - Alkenes also fragment readily to form stable allylic cations.
  - Cyclohexenes fragment in a characteristic pattern that is a reverse Diels-Alder reaction.
- **Alkynes**
  - Alkynes show a strong molecular ion peak due to removal on one electron from a pi bond.
  - In general, alkynes fragment in ways that are analogous to alkenes.
  - The propargyl cation or substituted propargyl cations are usually prominent peaks in the mass spectra of alkynes.
- **Alcohols**
  - The molecular ion peak for primary and secondary alcohols is usually small, and usually not detectable for tertiary alcohols.
  - A common fragmentation pattern for alcohols is the loss of water to give a peak corresponding to the molecular ion minus 18.
  - Another common pattern is loss of an alkyl group from the carbon bearing the -OH group to from an oxonium ion and an alkyl radical. The oxonium ion is stable because of resonance delocalization of charge.
  - A **McLafferty rearrangement** also occurs with some alcohols. In this fragmentation reaction, the oxygen atom abstracts a hydrogen atom five atoms away via a six-member ring transition state. The McLafferty rearrangement results in formation of an alkene, water, and a new cation radical.
- **Ethers**
  - Ethers can fragment via cleavage of the carbon-carbon bond adjacent to the ether oxygen atom to give an oxonium ion.
  - The oxonium ion can fragment further by migration of a hydrogen atom beta to the oxygen, resulting in elimination of an alkene.

CHAPTER 12
Solutions to the Problems

Problem 12.1 Calculate the nominal mass of the following ions:
(a) $[CH_3\ ^{79}Br]^{+\cdot}$ (b) $[CH_3\ ^{81}Br]^{+\cdot}$ (c) $[^{13}CH_3\ ^{79}Br]^{+\cdot}$ (d) $[^{13}CH_3\ ^{81}Br]^{+\cdot}$

94 **96** **95** **97**

Problem 12.2 Propose a structural formula for the cation of *m/e* 41 observed in the mass spectrum of methylcyclopentane.

The cation of *m/e* 41 must have a molecular formula of C_3H_5. This corresponds to the remarkably stable allyl cation:

$$CH_2{=}CH{-}CH_2^+$$

The allyl cation could be formed by a process such as the one shown below that starts with the cation radical formed as described in the Example 12.2.

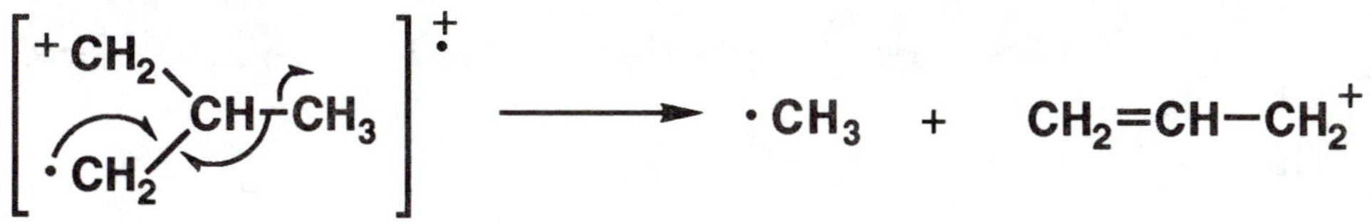

$$\left[\ {}^+CH_2{-}CH(CH_3){-}\dot{C}H_2\ \right]^{+\cdot} \longrightarrow \cdot CH_3 + CH_2{=}CH{-}CH_2^+$$

Problem 12.3 The low-resolution mass spectrum of 2-pentanol shows 15 peaks. Account for the formation of the peaks at *m/e* 73, 70, 55, 45, 43, 42, 41, and 28. (Hint: consider (1) the loss of water to form an alkene and the fragmentations the resulting alkene might then undergo, (2) the fragmentation of bonds to the carbon bearing the -OH group, and (3) McLafferty rearrangement along with subsequent fragmentation its products might undergo.

2-Pentanol has a molecular formula of $C_5H_{12}O$ and thus a nominal mass of 88.

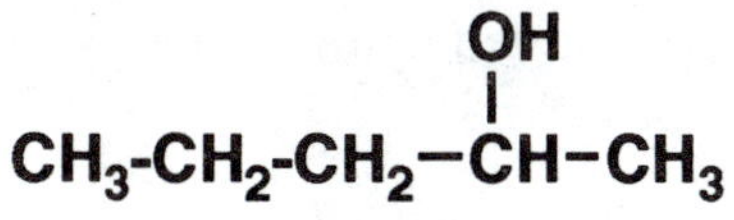

$$CH_3\text{-}CH_2\text{-}CH_2{-}CH(OH){-}CH_3$$

2-pentanol

Loss of a methyl radical would leave a fragment of mass 73:

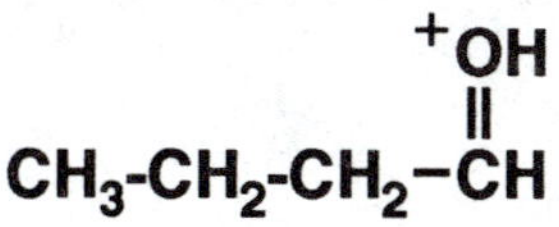

$$CH_3\text{-}CH_2\text{-}CH_2{-}CH{=}\overset{+}{O}H$$

Loss of water results in a fragment of mass 70:

$$[CH_3\text{-}CH_2\text{-}CH_2{-}CH{=}CH_2]^{+\cdot} \quad [CH_3\text{-}CH_2\text{-}CH{=}CH{-}CH_3]^{+\cdot}$$

The above alkene cation radicals could then lose a methyl radical to give the following fragments of mass 55:

$$[\,CH_2\text{-}CH_2\text{-}CH{=}CH_2\,]^+ \quad [\,CH_2\text{-}CH{=}CH\text{-}CH_3\,]^+$$

An alkyl radical could break off from the alcohol to give a fragment of mass 45:

$$\overset{+OH}{\overset{\|}{CH}}\text{-}CH_3$$

The original alcohol cation radical could have fragmented in such as way as to generate the following cation with mass 43:

$$CH_3\text{-}CH_2\text{-}CH_2^+$$

The following McLafferty rearrangement is possible with this alcohol, and the resulting alkene cation radical has a mass of 42:

$$CH_3\text{-}CH(\text{-}\ddot{O}^{\cdot}\text{-}H)\cdots H\text{-}CH_2\text{-}CH_2\text{-}CH_2 \longrightarrow CH_2{=}CH_2 + H_2O + [CH_3\text{-}CH{=}CH_2]^{+\cdot}$$

The stable allyl cation has a mass of 41:

$$CH_2{=}CH\text{-}CH_2^+$$

The ethenyl cation radical has a mass of 28:

$$[CH_2{=}CH_2\,]^{+\cdot}$$

Problem 12.4 Draw acceptable Lewis structures for a molecular ion (radical cation) formed by the following molecules when each is bombarded by high-energy electrons in a mass spectrometer.

(a) C_6H_6 (benzene) $\longrightarrow$ $C_6H_6^{+\cdot}$ (ring with C^+ and adjacent $C\cdot$)

(b) 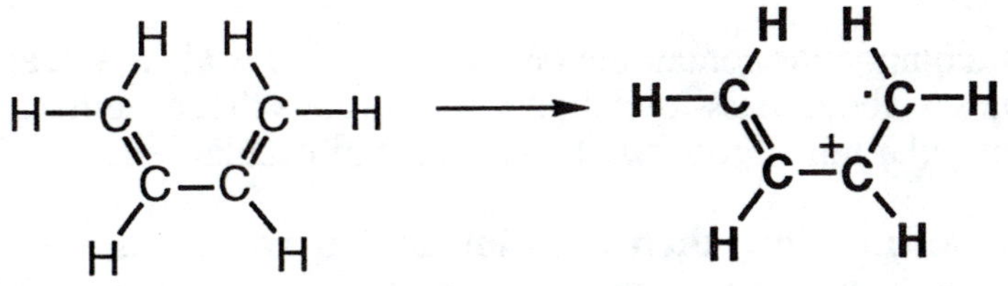 [reaction scheme]

(c)  [reaction scheme]

(d) H—C≡C—H ⟶ H—Ċ=Ċ⁺—H

Problem 12.5 Some organic molecules can add a single electron to form a normally unstable class of species called radical anions. A radical anion possesses both a negative charge and an unpaired electron. For example:

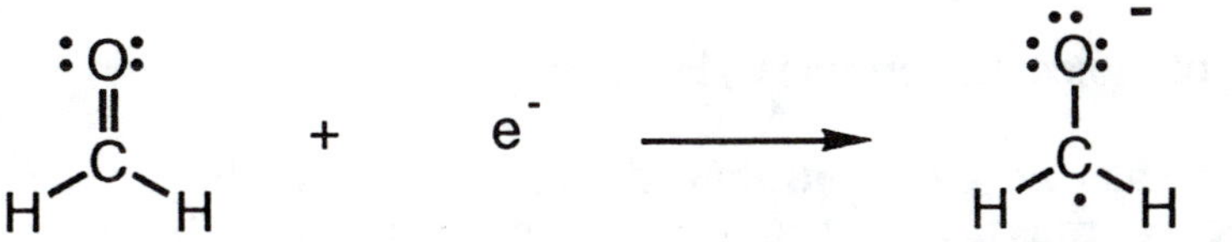

[reaction scheme: $H_2C=O$ + e^- ⟶ radical anion]

a radical anion

Draw an acceptable Lewis structure for a radical anion formed from the following molecules:

(a) 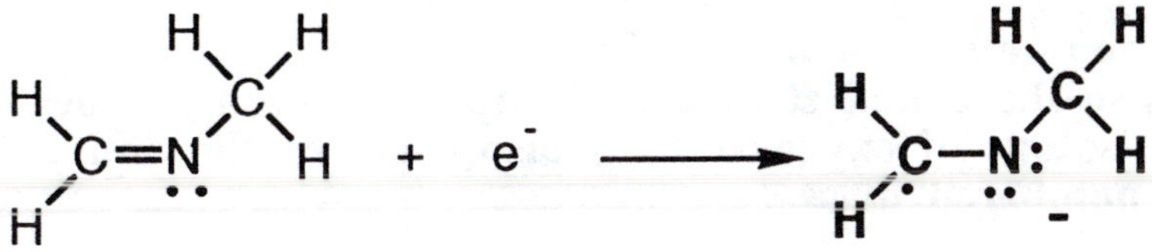 [reaction scheme] + e^- ⟶

(b) 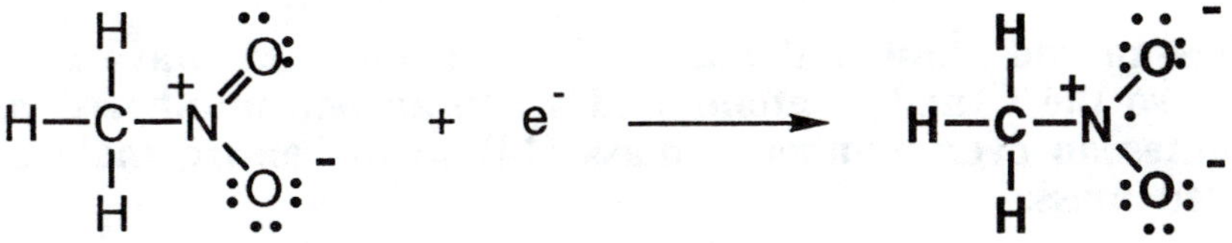 [reaction scheme] + e^- ⟶

(c) 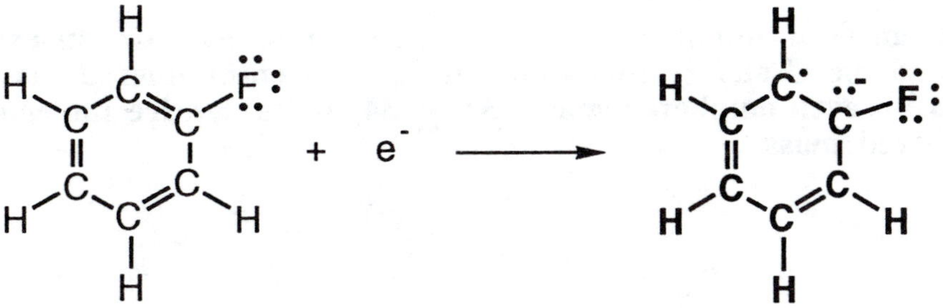 [reaction scheme] + e^- ⟶

Problem 12.6 The molecular ion for compounds containing only C, H, and O is always at an even mass-to-charge value. What can you say about mass-to-charge values of ions that arise from fragmentation of one bond in the molecular ion? From two bonds in the molecular ion?

If one bond is broken in the molecular ion, then a cation and a radical are formed. The cation is the only species observed in the mass spectrum, and this will have an odd mass-to-charge ratio. If the molecular ion fragments so that two bonds are broken, then a new radical cation is formed, and the new radical cation will have an even mass-to-charge ratio.

Problem 12.7 For which compounds containing a heteroatom (an atom other than carbon or hydrogen) does a molecular ion have an even-numbered mass and for which does it have an odd-numbered mass?

(a) A chloroalkane of molecular formula $C_nH_{2n+1}Cl$

There will be an overall even-numbered mass. The C and H atoms will add up to an odd-numbered mass, since there will be an odd number of hydrogen atoms. The chlorine provides an additional odd-numbered mass as either of the two most abundant isotopes (35 and 37, respectively).

(b) A bromoalkane of molecular formula $C_nH_{2n+1}Br$

The molecular ion has an even-numbered mass. Again, the C and H atoms add up to an odd-numbered mass, and bromine provides an additional odd-numbered mass as either of the two most abundant isotopes (79 and 81, respectively).

(c) An alcohol of molecular formula $C_nH_{2n+1}OH$

The molecular ion has an even-numbered mass. There is an even-numbered number of hydrogen atoms, so the C and H atoms add up to an even-numbered mass. The most abundant isotope of oxygen (16) is also even, so the entire molecular ion has an even-numbered mass.

(d) A primary amine of molecular formula $C_nH_{2n+1}NH_2$.

The molecular ion has an odd-numbered mass. These compounds have an odd number of hydrogens, so the C and H atoms add up to an odd-numbered mass. The nitrogen contributes an even-numbered mass (14) so the entire molecule will have an odd-numbered mass.

(e) A thiol of molecular formula $C_nH_{2n+1}SH$.

The molecular ion has an even-numbered mass. These compounds have an even number of hydrogens, so the C and H atoms add up to an even-numbered mass. The sulfur contributes an even-numbered mass (32 or 34) so the entire molecular ion has an even-numbered mass.

Problem 12.8 The so-called nitrogen rule states that if a molecular ion has an even mass value, then it contains either no nitrogen atom or an even number of nitrogen atoms. Explain the basis of this rule.

Nitrogen atoms have an even-numbered mass, but one lone pair of electrons in the neutral state. Thus, they make an odd number of bonds, namely three, to other atoms. As a result, compounds with nitrogen in the neutral state will have an odd-numbered mass.

Problem 12.9 The molecular ion of both $C_5H_{10}S$ and $C_6H_{14}O$ appear at *m/e* 102 in low-resolution mass spectrometry. Show how determination of the correct molecular formula can be made from appearance and relative intensity of the ($M^{+\cdot}$+2) peak of each compound.

As shown in Table 12.1, ^{16}O occurs in greater than 99.7 % abundance, so no ($M^{+\cdot}$+2) peak will be observable for the case of $C_6H_{14}O$. On the other hand, sulfur has one isotope, ^{34}S, that is 95 % abundant and another isotope 2 amu higher, namely ^{32}S, that has an abundance of 4.2 %. Thus, if the low-resolution mass spectrum has an ($M^{+\cdot}$+2) peak that is 4.2 % the height of the molecular ion peak, the compound must be $C_5H_{10}S$.

Problem 12.10 Water-^{18}O in which enrichment is 10 atom % ^{18}O is available commercially. Also available, at a considerably higher price, is water-^{18}O in which enrichment is 97 atom % ^{18}O. The oxygen-18 label can be transferred from water to acetone by establishing the following acid-catalyzed equilibrium. (We will discuss the chemistry of this equilibration in Chapter 17).

$$CH_3-\overset{\overset{O}{\|}}{C}-CH_3 + H_2{}^{18}O \underset{}{\overset{H^+}{\rightleftharpoons}} CH_3-\overset{\overset{^{18}O}{\|}}{C}-CH_3 + H_2O$$

Suppose 5.00 g of water-^{18}O, 97 atom % ^{18}O, is mixed with 11.6 g of acetone-^{16}O in the presence of an acid catalyst until equilibration is established. Assume for the purposes of this problem that the value of K_{eq}, the equilibrium constant for this reaction, is 1.00.
(a) Calculate the atom percent enrichment in acetone recovered after equilibration.

This problem is best answered in the units of moles.

$$\frac{5.00 \text{ gm}}{18 \text{ gm/mole}} = 0.28 \text{ mole total } H_2O \qquad \frac{11.6 \text{ gm}}{58 \text{ gm/mole}} = 0.20 \text{ mole total acetone}$$

After equilibrium has been established, the following amounts of material will be present, written in terms of the amount of ^{18}O-containing acetone (as "X").

$$\left[H_2{}^{18}O\right] = (0.28)(0.97) - X \qquad \left[H_2O\right] = (0.28)(0.03) + X$$

$$\left[CH_3-\overset{\overset{O}{\|}}{C}-CH_3\right] = 0.2 - X \qquad \left[CH_3-\overset{\overset{^{18}O}{\|}}{C}-CH_3\right] = X$$

$$K_{eq} = 1.00 = \frac{\left[CH_3-\overset{^{18}O}{\overset{\|}{C}}-CH_3\right]\left[H_2O\right]}{\left[CH_3-\overset{O}{\overset{\|}{C}}-CH_3\right]\left[H_2{}^{18}O\right]}$$

so at equilibrium $\left[CH_3-\overset{O}{\overset{\|}{C}}-CH_3\right]\left[H_2{}^{18}O\right] = \left[CH_3-\overset{^{18}O}{\overset{\|}{C}}-CH_3\right]\left[H_2O\right]$

Substituting the values into the equation gives:

$$(0.20 - X)(0.27 - X) = (X)(0.01 + X)$$
$$0.054 - 0.47\,X + X^2 = 0.01\,X + X^2$$

The X^2 terms cancel, so we are left with:

$$0.054 - 0.47\,X = 0.08\,X$$
$$0.054 = 0.48\,X$$
$$0.113 \text{ mole} = X$$

Thus, the atom percent (%) enrichment is (0.113 mole) / (0.20 mole)

= 57 atom %

(b) Show how the ratio of $(M^{+\cdot} + 2)/M^{+\cdot}$ can be used to verify the atom percent enrichment.

A 57 atom percent enrichment means that the ratio of $(M^{+\cdot} + 2)/M^{+\cdot}$ will be (0.57 / 0.43). This ratio measured in the mass spectrum would thus verify the atom percent enrichment.

Interpretation of Mass Spectra

Problem 12.11 Carboxylic acids often give strong fragment ions at *m/e* $(M^{+\cdot}-17)$. What is the likely structure of these ions and how might they be formed?

Loss of 17 results from losing the -OH group of the carboxylic acid. This produces an acylium ion:

$$R-\overset{+}{C}{=}\ddot{O}$$

These ions are probably derived from fragmentation of the molecular ion as follows:

$$R-C(=\ddot{O}:)-\ddot{O}\text{-}H + e^- \longrightarrow R-\overset{+}{C}(\cdot\ddot{O}:)-\ddot{O}\text{-}H \longrightarrow R-\overset{+}{C}{=}\ddot{O} + \cdot\ddot{O}H$$

molecular ion

Problem 12.12 The molecular ion in the mass spectrum of 2-methyl-1-pentene appears at *m/e* 84. Propose structural formulas for the prominent peaks at *m/e* 69, 55, 41, and 29.

$$CH_2{=}C(CH_3)\text{-}CH_2\text{-}CH_2\text{-}CH_3$$

2-methyl-1-pentene

The peak at 69 results from loss of a methyl radical:

$$CH_2{=}\overset{+}{C}\text{-}CH_2\text{-}CH_2\text{-}CH_3 \quad \text{or} \quad CH_2{=}C(CH_3)\text{-}CH_2\text{-}CH_2\text{-}CH_2^{+}$$

The peak at 55 results from loss of an ethyl group to generate an allylic cation:

$$CH_2{=}C(CH_3)\text{-}\overset{+}{C}H_2$$

The peak at 41 corresponds to the simple allyl cation:

$$CH_2{=}CH\text{-}\overset{+}{C}H_2$$

The peak at 29 corresponds to the ethyl cation:

$$CH_3\text{-}\overset{+}{C}H_2$$

Problem 12.13 Following is the mass spectrum of 1,1-dichloroethane. The molecular ion appears at *m/e* 98.

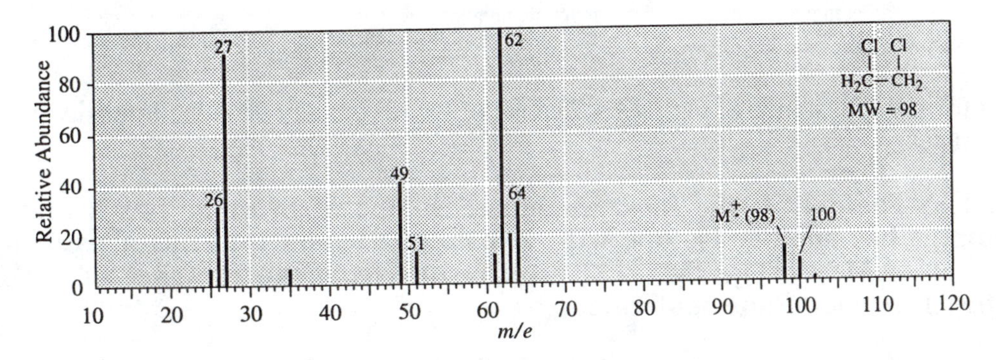

(a) Account for the appearance of an ($M^{+\cdot}$+ 2) peak with approximately two thirds the intensity of the molecular ion peak.

As detailed in table 12.1, the most abundant isotope of chlorine has a mass of 35 amu. Chlorine has another isotope with a mass of 37 amu that has a natural

abundance of 24.23%. The ($M^{+\cdot}$+ 2) peak results from the presence of a ^{37}Cl atom in the 1,1-dichloroethane molecule.

(b) Propose structural formulas for the cations of *m/e* 64, 62, 51, 49, 27, and 26.

$$CHCl_2\text{-}CH_3$$

1,1-dichloroethane

The peaks at *m/e* 64 and 62 correspond to the following structures:

$$[^{37}Cl\text{—}CH{=}CH_2]^{+\cdot} \qquad [^{35}Cl\text{—}CH{=}CH_2]^{+\cdot}$$

m/e = 64 *m/e* = 62

The peaks at *m/e* 51 and 49 correspond to the following structures:

$$^{37}Cl\text{—}CH_2^+ \qquad ^{35}Cl\text{—}CH_2^+$$

m/e = 51 *m/e* = 49

The peaks at *m/e* 27 and 26 correspond to the following structures:

$$^+CH{=}CH_2 \qquad [HC{\equiv}CH]^{+\cdot}$$

m/e = 27 *m/e* = 26

Problem 12.14 Following is the mass spectrum of 1-bromobutane. The molecular ion appears at *m/e* 136.

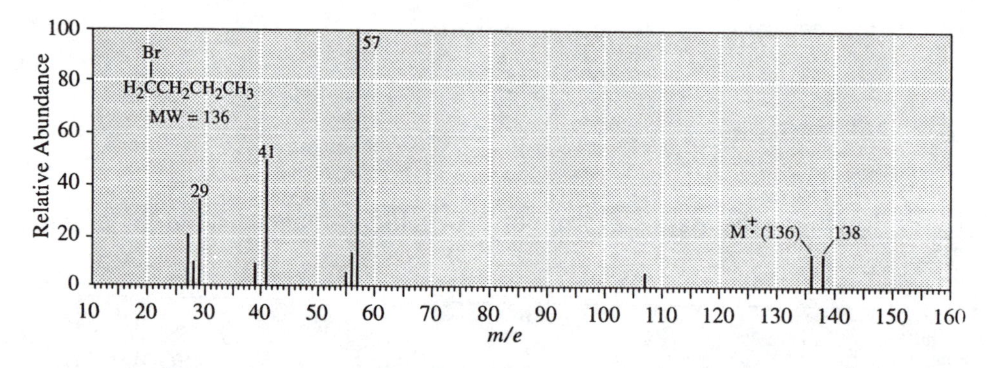

(a) Account for the appearance of the ($M^{+\cdot}$+ 2) peak of approximately 90% the intensity of the molecular ion peak.

As detailed in Table 12.1, the most abundant isotope of bromine has a mass of 79 amu. Bromine has another isotope with a mass of 81 amu that has a natural abundance of 49.31%. The ($M^{+\cdot}$+ 2) peak results from the presence of a ^{81}Br atom in the 1-bromobutane molecule.

(b) Propose structural formulas for the cations of *m/e* 57, 41, and 29.

The peak at 57 results from loss of a bromine atom:

$$\overset{+}{C}H_2\text{-}CH_2\text{-}CH_2\text{-}CH_3$$

The peak at 41 corresponds the stable allyl cation:

$$CH_2{=}CH\text{-}\overset{}{\underset{+}{CH_2}}$$

The peak at 29 corresponds to the ethyl cation:

$$CH_3\text{-}\underset{+}{CH_2}$$

Problem 12.15 Following is the mass spectrum of bromocyclopentane. The molecular ion *m/e* 148 is of such low intensity that it does not appear in this spectrum. Assign structural formulas for the cations of *m/e* 69 and 41.

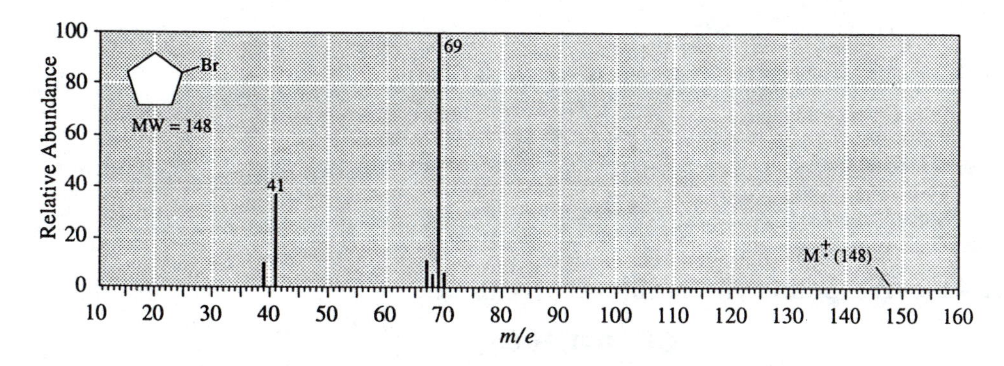

The peak at 69 corresponds to the loss of the bromine atom to give the cyclopentyl cation:

+

The peak at 41 corresponds the stable allyl cation:

$$CH_2{=}CH\text{-}\underset{+}{CH_2}$$

Problem 12.16 Following is the mass spectrum of 3-methyl-2-butanol. The molecular ion *m/e* 88 does not appear in this spectrum. Propose structural formulas for the cations of *m/e* 45, 43, and 41.

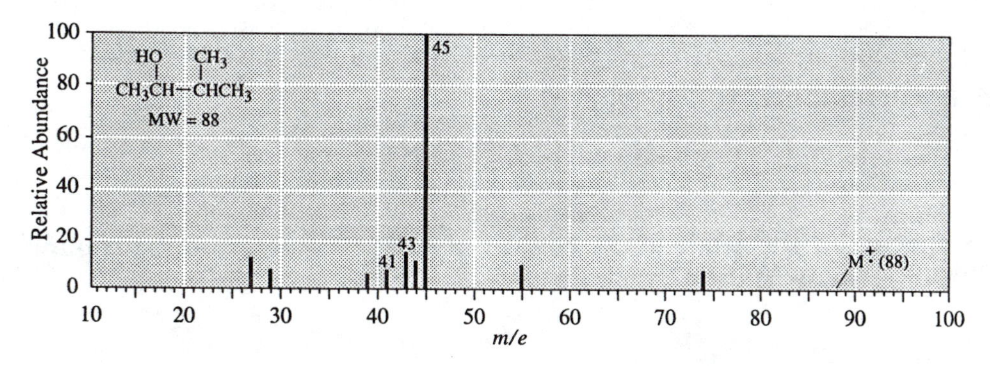

$$CH_3-CH(OH)-CH(CH_3)-CH_3$$

3-methyl-2-butanol

The peak at 45 corresponds to the following resonance stabilized cation:

$$CH_3-CH{=}\overset{+}{O}H$$

The peak at 43 corresponds to the isopropyl cation:

$$\overset{+}{C}H(CH_3)-CH_3$$

The peak at 41 corresponds the stable allyl cation:

$$CH_2{=}CH{-}\overset{+}{C}H_2$$

Problem 12.17 The following is the mass spectrum of a compound A, C_3H_8O. Compound A is infinitely soluble in water, undergoes reaction with sodium metal with the evolution of a gas, and undergoes reaction with thionyl chloride to give a water-insoluble chloroalkane. Propose a structural formula for compound A, and write equations for each of its reactions.

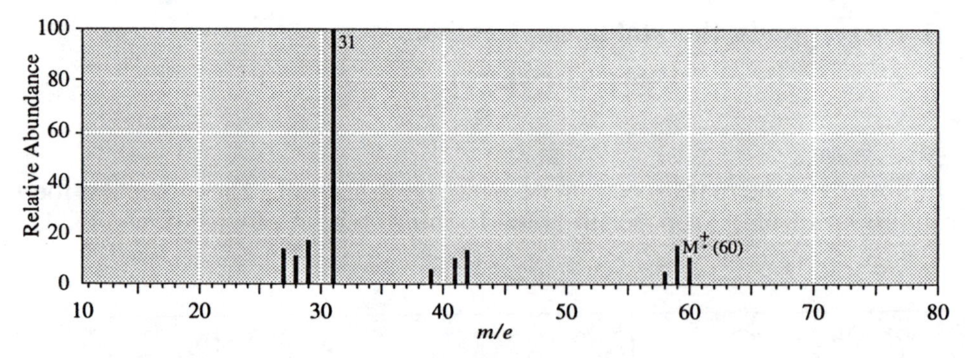

Compound A corresponds to 1-propanol.

$$CH_3\text{-}CH_2\text{-}CH_2OH$$

Compound A

In the mass spectrum, the peak at 31 could only have come from 1-propanol due to the following fragmentation.

$$CH_3\text{-}CH_2\text{—}\overset{\overset{\large OH}{|}}{C}H_2 \longrightarrow CH_3CH_2\cdot \quad \overset{+}{O}H\text{-}CH_2$$

m/e 31

The reaction with sodium liberates molecular hydrogen:

$$2\ CH_3\text{-}CH_2\text{-}CH_2OH \ + \ 2Na \longrightarrow 2\ CH_3\text{-}CH_2\text{-}O^-\ Na^+ \ + \ H_2$$

The reaction with thionyl chloride produces 1-chloropropane, SO_2, and H-Cl.

$$CH_3\text{-}CH_2\text{-}CH_2OH \ + \ SOCl_2 \longrightarrow CH_3\text{-}CH_2\text{-}CH_2Cl \ + \ SO_2 \ + \ H\text{-}Cl$$

Problem 12.18 Following are mass spectra for the constitutional isomers 2-pentanol and 2-methyl-2-butanol. Assign each isomer its correct spectrum.

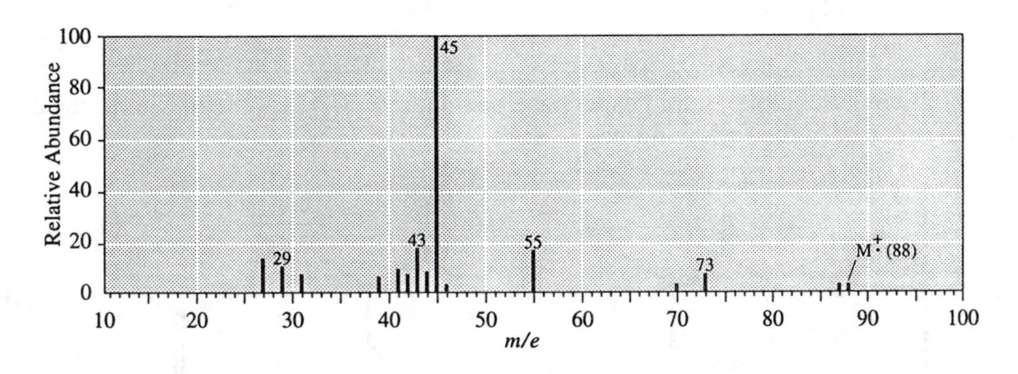

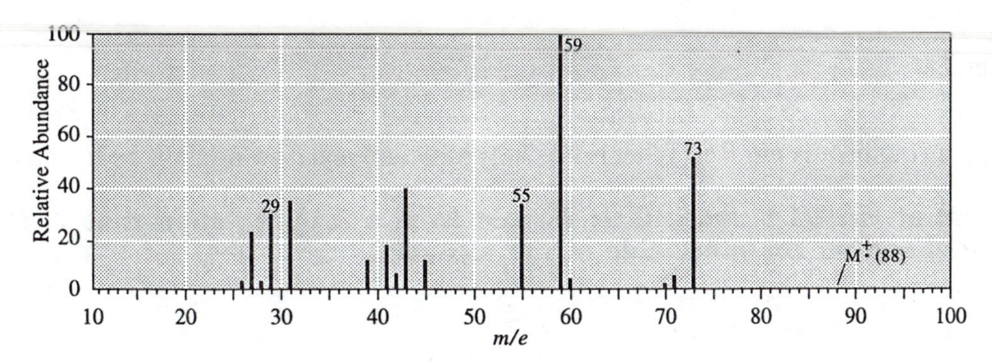

$$CH_3\text{—}\overset{\overset{\large OH}{|}}{C}H\text{—}CH_2\text{-}CH_2\text{-}CH_3$$

2-pentanol

$$CH_3\text{—}\underset{\underset{\large CH_3}{|}}{\overset{\overset{\large OH}{|}}{C}}\text{—}CH_2\text{-}CH_3$$

2-methyl-2-butanol

The main difference in the above spectra is the base peak at 45 in the first spectrum that is only about 10% of the base peak in the second spectrum. This peak corresponds to the following resonance stabilized cation:

$$CH_3-CH=\overset{+}{O}H$$

This species can be readily produced by loss of the propyl group from 2-pentanol. It is unlikely that 2-methyl-2-butanol could produce such a cation, thus the upper mass spectrum is identified as belonging to 2-pentanol and the lower mass spectrum must then belong to 2-methyl-2-butanol.

Also, the base peak in the second spectrum has a *m/e* of 59 that corresponds to the following fragmentation.

$$CH_3-C(OH)(CH_3)-CH_2-CH_3 \longrightarrow [CH_3-C(OH)(CH_3)]^+ + \cdot CH_2CH_3$$

Problem 12.19. 2-Methylcyclohexanol has its base peak at *m/e* 57. Fragmentation to produce this ion is thought to proceed in the following way:

$$[\text{2-methylcyclohexanol}]^{+\cdot} \longrightarrow \cdots \longrightarrow CH_2=CH-CH=\overset{+}{O}H + C_4H_9\cdot$$

2-methyl-cyclohexanol; *m/e* 57

A second major fragment is *m/e* 71. What is its likely structure and how might it be formed?

The fragment at *m/e* 71 is most likely formed from a fragmentation that starts by breaking the bond on the other side of the alcohol.

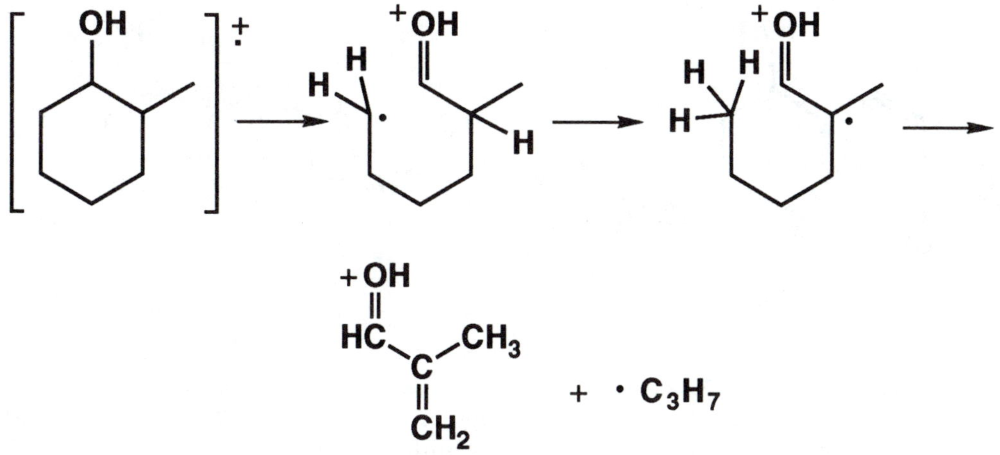

$$\left[\text{cyclohexanol derivative}\right]^{+\cdot} \longrightarrow \longrightarrow \longrightarrow$$

$$HC(=\overset{+}{O}H)-C(CH_3)=CH_2 \quad + \quad \cdot C_3H_7$$

m/e 71

Problem 12.20 Examination of many mass spectra usually show rather large peaks at *m/e* 28, 32, and 40. What is the likely source of these peaks?

This is a very tricky question. The peaks are due to air being present. The peak at 28 is from N_2, the peak at 32 is from O_2, and the peak at 40 is from argon.

Problem 12.21 Because of the sensitivity of mass spectrometry, it is often used to detect the presence of drugs in blood, urine, or other biological fluids. Tetrahydrocannabinol, a component of marijuana, exhibits two strong fragment ions at *m/e* 246 and 231 (the base peak). What is the likely structure of each ion?

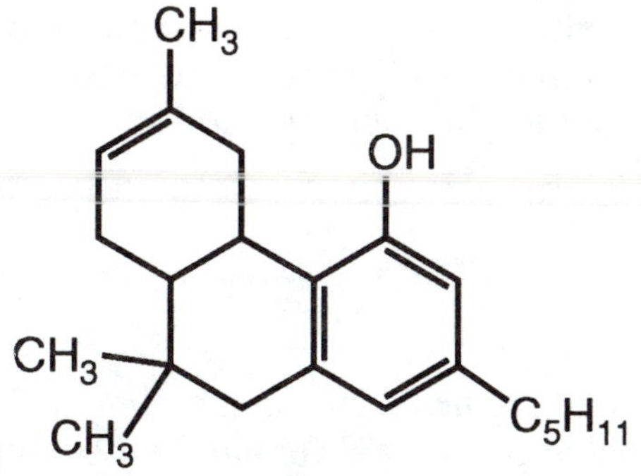

Tetrahydrocannibinol
($C_{21}H_{30}O$)

The peak at 246 is probably the result of a reverse Diels-Alder reaction as described in section 12.4B. In this case, the cyclohexene ring in the top left-hand corner of the structure is split apart from the rest of the molecule to leave the radical cation shown below. The peak at 231 results from loss of a methyl group from the *m/e* 246 radical cation to give a tertiary cation as shown.

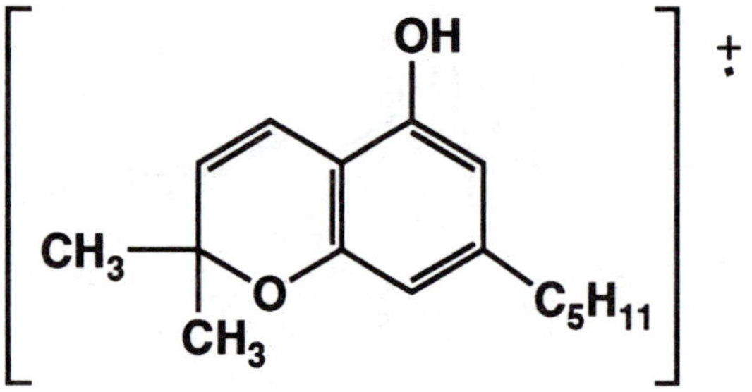

m/e **246** 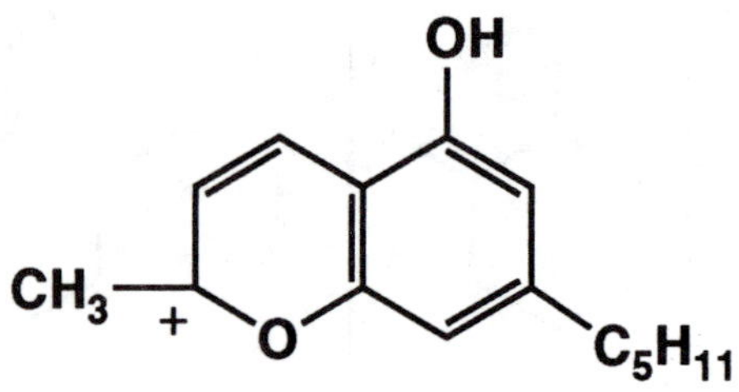 m/e **231**

Problem 12.22 Ion-spray mass spectrometry is a recently developed technique for looking at large molecules with a mass spectrometer. In this technique, molecular ions, each normally associated with one or more H^+ ions, are prepared under mild conditions in the mass spectrometer. As an example, a protein (P) with a molecular weight of 11,812 amu gives clusters of the type $(P+8H)^{8+}$, $(P+7H)^{7+}$, and $(P+6H)^{6+}$. At what mass-to-charge values do these three clusters appear in the mass spectrum?

The key to this question is to notice that these ions have multiple charges. Since ions are recorded in a mass spectrum according to their mass divided by their total charge, the mass-to-charge values for these ions are calculated by dividing their total mass (11,812 + the number of protons) by their total charge (8, 7, and 6, respectively). The final answers are thus (11,812 + 8)/8 = 1477.5, (11,812 + 7)/7 = 1688.4, and (11,812 + 6)/6 = 1969.7

Problem 12.23 Occasionally, weak, broad peaks are observed in a mass spectrum. These often have fractional mass-to-charge values, for example, 46.3 or 30.2. These are called metastable ion peaks and arise when an ion fragments *after* exiting the ionization chamber while it is passing into the analyzer region of the mass spectrometer. The observed mass, m*, of a metastable ion depends on the mass of the precursor ion (m_1) and the product ion (m_2) according to the following equation.

$$m^* = \frac{(m_2)^2}{m_1}$$

An ion of mass 59, for example, that fragments into an ion of mass 41 while passing into the analyzing chamber gives a metastable at $(41)^2/(59)$ or m/e 28.49. Metastable ions link two peaks together, as for example the peaks at m/e 59 and 41 and thus are useful in analyzing proposed fragmentation patterns. What metastable ion results from the fragmentation of $(CH_2OCH_3)^+$ to $(OCH_3)^+$?

In this case, the ion with mass 45 fragments to the ion with mass 29. As a result, the metastable ion shows up in the mass spectrum as $(29)^2/45$ = 16.3 *m/e*.

CHAPTER 13: NUCLEAR MAGNETIC RESONANCE SPECTROSCOPY

13.0 OVERVIEW

- **Nuclear magnetic resonance (NMR)** spectroscopy was developed in the early 1960's, and is now the most important technique for the determination of molecular structure. *[NMR is a very complicated type of spectroscopy, because there are so many difficult concepts involved. The best way to learn about NMR is to read through the concepts, then work through as many spectra as possible.]* ✱

13.1 ELECTROMAGNETIC RADIATION

- **Electromagnetic radiation** can be described like a wave, in terms of its wavelength and frequency. ✱
  - **Wavelength** is the distance between any two identical points on a wave and is given the symbol λ **(lambda)** and usually expressed in **meters (m)**.
  - **Frequency** is the number of full cycles of a wave that pass a given point in a fixed period of time. Frequency is given the symbol ν **(nu)** and is usually expressed in **hertz (Hz)** and given the units **sec^{-1}**. For example, 1 Hz corresponds to one cycle per second and 1 MHz corresponds to 10^6 Hz.
  - Wavelength and frequency are related to each other, and one can be calculated from the other using the expression $\nu = c/\lambda$. Here **c** is the **speed of light**, equal to **3.00×10^8 m/sec.**
- **Electromagnetic radiation** can also be described as a particle, and the particle is called a **photon**. ✱
  - The **energy (E)** measured in **kcal** of one mole of photons is related to wavelength and frequency according to the following relationships: **$E=h\nu=hc/\lambda$**. Here **h** is **Planck's constant** and is equal to **9.537×10^{-14} kcal-sec-mol^{-1}**.

13.2 MOLECULAR SPECTROSCOPY

- In general, an atom or molecule can be made to **undergo a transition** from energy state E_1 to a higher energy state E_2 by irradiating it with electromagnetic radiation corresponding to the energy difference between states E_1 and E_2. During this process, the atom or molecule **absorbs the energy** of the electromagnetic radiation. When the atom or molecule **returns to the ground state** E_1, an equivalent amount of **energy is emitted.** ✱
- **Molecular spectroscopy** involves measuring the frequencies of electromagnetic radiation that are absorbed or emitted by a molecule, then correlating the observed patterns with the details of molecular structure.
  - Regions of the electromagnetic spectrum are particularly interesting to the chemist because they represent the energies involved in transition between important types of molecular energy levels. In particular, **radiofrequency electromagnetic radiation** corresponds to transitions between **nuclear spin** energy levels, **infrared electromagnetic radiation** corresponds to transitions between **vibrational levels of chemical bonds** and **ultraviolet-visible electromagnetic radiation** corresponds to **electronic energy levels** of pi and nonbonding electrons. *[It may prove helpful to review how these different regions of electromagnetic radiation fit into the entire spectrum shown in Table 13.2.]*
  - The nuclear spin transitions caused by absorbed radiofrequency electromagnetic radiation form the basis for **nuclear magnetic resonance (NMR) spectroscopy**. ✱

13.3 NUCLEAR SPIN STATES

- Like electrons, certain nuclei have spin. That is, they behave as if they are spinning on an axis and thus have an associated **magnetic moment**. This is only true for nuclei that have **spin quantum numbers** that are not zero. *[It might help to review the concept of spin quantum numbers in a General Chemistry text.]*
  - Both **^{1}H** and **^{13}C** nuclei have spin quantum numbers of 1/2, so they have two allowed spin states with values of +1/2 and -1/2. These are the nuclei that are most often studied by NMR spectroscopy, but other nuclei such as ^{31}P and ^{15}N can also be used because they too have spin quantum numbers that are not zero. ✱
  - Some common nuclei such as ^{12}C and ^{16}O have a spin quantum number of zero, so they are not observed by NMR.

13.4 ORIENTATION OF NUCLEAR MAGNETIC MOMENTS IN AN APPLIED MAGNETIC FIELD

- Ordinarily, nuclear spins are oriented in a completely random fashion. However, in an **applied magnetic field (B_o), nuclei with non-zero spin interact with the applied field.** Recall that the nuclear spin produces it own magnetic moment, so this interaction is between the magnetic moments of the nuclei with the applied magnetic field. For ^{1}H and ^{13}C nuclei, there are two allowed orientations in the field. By convention, **nuclei** that have **spin +1/2** are **aligned with the applied field** and are in the **lower energy state**, and **nuclei** with **spin -1/2** are **aligned against the applied field** and are in the **higher energy state.** ✱
- The **difference** in **energy** between the **nuclear spin states increases** with **increasing applied field strength.** ✱ Nevertheless, these energy differences are small compared to other types of energy levels such as vibrational and electronic energy levels in molecules.

13.5 NUCLEAR MAGNETIC "RESONANCE"

- When the nuclei are placed in an applied magnetic field, a majority of spins are aligned with the field and are thus in the lower spin state. **Electromagnetic radiation** can **cause a transition from the lower spin state to the higher spin state.** An NMR spectrum is a plot of how much and of which energies are absorbed by a molecule as its atoms undergo these transitions from the lower to the higher nuclear spin state. ✱
  - The amount of energy required depends on the strength of the applied field and the type of nuclei being used, but this energy corresponds to electromagnetic radiation somewhere in the radiofrequency range. For example, in an applied magnetic field of strength 7.05 Tesla (T), the energy between the spin states of ^{1}H is around 0.0286 cal/mol, corresponding to electromagnetic energy of approximately 300 MHz (300,000,000 Hz).
  - The electromagnetic energy is absorbed because the nuclei **precess** in the applied magnetic field. That is, the nuclei that are lined up with or against the applied magnetic field actually precess just like a spinning top or gyroscope precesses and traces out a cone in the earth's gravitational field. The **rate of precession** can be expressed as a frequency in hertz. **When the precessing nuclei are irradiated with electromagnetic radiation of the same frequency as the rate of precession, then the two frequencies couple, energy is absorbed, and the nuclear spin flips from +1/2 to -1/2.** ✱
  - **Resonance** is the term used to describe the absorption of electromagnetic radiation when the precession frequency and the electromagnetic radiation frequencies couple. That is why the entire process is called nuclear magnetic resonance. The absorption of energy by a molecule is measured and plotted to give the NMR spectrum.
- The different types of atomic nuclei in a molecule, for example ^{13}C or ^{1}H, do not absorb energy at the same frequency. If they did, NMR would not be a useful probe of molecular structure. The fact is that there are usually different local chemical environments in a molecule that change the resonance frequencies of the different nuclei. By looking at the **different resonance**

frequencies measured in the NMR spectrum, information is obtained about the **different chemical environments in the molecule**, leading to an understanding **of the molecular structure.** ✱

- Atomic nuclei from different elements resonate at different frequencies in the same applied field. For example, in the presence of an applied field of 7.05 T, ^{1}H nuclei resonate at about 300 MHz, while ^{13}C nuclei resonate at about 75 MHz.
- The different nuclei in a molecule are in different chemical environments, if they are surrounded by electrons to varying degrees. The electrons themselves have spin and thereby create their own **local magnetic field**. These local magnetic fields **shield the nucleus** from the applied magnetic fields. **The greater the electron density around a nucleus**, the **greater the shielding**. This means that at constant magnetic field strength, **the greater the shielding of a nucleus, the lower the frequency of electromagnetic radiation required to bring about a spin flip**. Another way to look at it is from the point of view of constant electromagnetic radiation frequency. At constant frequency of electromagnetic radiation, **the greater the shielding of a nucleus, the higher the magnetic field strength required to bring the nucleus into resonance.** ✱ *[It is important to be able to think about shielding and spin flipping from the point of view of constant frequency of electromagnetic radiation as well as at constant magnetic field strength.]*
- The local magnetic fields are small compared to the applied magnetic fields. The differences in resonance frequencies for different nuclei caused by the local magnetic fields are usually on the order of 1×10^{-6} times as large as the original resonance frequencies. In other words, different ^{1}H nuclei in the same molecule have resonance frequencies that are different by an amount on the order of **parts per million (ppm)**. Thus, ppm is a convenient measurement unit for NMR spectroscopy. For example, a difference of 100 Hz is 1 ppm of 100 MHz, and a difference of 300 Hz is 1 ppm of 300 MHz.
- In order to increase the precision of resonance frequency measurements, a reference compound is used. By convention, the ^{1}H resonance in **tetramethylsilane (TMS)** is used as the reference against which the frequency of other ^{1}H resonances are measured. Similarly, the ^{13}C resonance in TMS is used as reference by which the frequency of other ^{13}C resonances are measured against.
- A unit called **chemical shift** is used to standardize reporting of NMR data. **Chemical shift** (δ) is the frequency shift from TMS divided by the operating radiofrequency of the spectrometer. **Chemical shift is reported in** the units **ppm.**

$$\delta = \frac{\text{shift in frequency from TMS (Hz)}}{\text{frequency of spectrometer (Hz)}} \times 10^{6}$$

13.6 AN NMR SPECTROMETER

- The essential features of an NMR spectrometer are a powerful magnet, a radiofrequency generator, a radiofrequency detector, and a sample tube.
  - Older **"field sweep" NMR spectrometers** operated by holding the radiofrequency constant and sweeping the magnetic field to determine where nuclei resonate.
  - Newer machines operate by a different principle and are called **Fourier transform NMR (FT-NMR) spectrometers**. In FT-NMR machines, the magnetic field strength is held constant, and the sample is irradiated with a short pulse of radiofrequency energy that flips all of the spins at once. The NMR spectrum is determined as the spins return to their equilibrium states. A mathematical process called Fourier transformation is used to convert the intensity versus time information produced in the experiment into the intensity versus frequency information that is actually plotted. Two advantages of FT-NMR are that it takes

much less time to run a scan, and multiple scans can be averaged to produce more intense signals from a dilute sample.

13.7 EQUIVALENT HYDROGENS

- All **equivalent hydrogens** have the same chemical environment within a molecule and **have identical chemical shifts**. For example, all of the hydrogens in dimethyl ether (CH_3-O-CH_3) are equivalent, so there is only one signal in the ^{1}H-NMR spectrum of this compound.
- Hydrogens that are not equivalent give rise to different signals with different chemical shifts. For example, there are two different signals in the ^{1}H-NMR spectrum of ethyl bromide (CH_3-CH_2-Br).

13.8 SIGNAL AREAS

- **The area under each signal** is **proportional to the number of hydrogens** giving rise to that signal in an ^{1}H-NMR spectrum. All modern NMR spectrometers can integrate the area under each signal. Please note that ^{13}C signals cannot be integrated accurately (Section 13.13) in ^{13}C-NMR spectra.

13.9 CHEMICAL SHIFT

- Each type of **equivalent hydrogen** within a molecule has only a **limited range of δ values**, and thus the value of the chemical shift for a signal in a ^{1}H-NMR spectrum gives valuable information about the type of hydrogen giving rise to that absorption. For example, the three hydrogens on methyl groups bonded to sp^3 hybridized carbons resonate near δ 1.0 ppm, while the three hydrogens on methyl groups bonded to an sp^2 hybridized carbonyl carbon atom resonate near δ 2.0 ppm. ✱
- A signal is considered **downfield** if it is shifted toward the left (weaker applied field) on the chart paper, and it is **upfield** if it is shifted to the right (stronger applied field).
- The chemical shift of a particular hydrogen depends primarily on the following three factors, but the influence of these factors on chemical shift falls off very quickly with distance.
  - **Electronegativity of adjacent atoms**. The greater the electronegativity of atoms near a particular hydrogen, the greater the chemical shift because the hydrogen is deshielded.
  - **Hybridization of adjacent atoms.** Hydrogens attached to an sp^3 hybridized carbon typically resonate at δ 0.8 to 1.7 ppm, hydrogens attached to an sp^2 hybridized carbon typically resonate at δ 4.6 to 5.7 ppm, and hydrogens attached to an sp hybridized carbon typically resonate at δ 2.0 to 3.0 ppm. These shifts are caused by a combination of the percent s character of the carbon orbital taking part in the C-H bond, and magnetic induction through the pi system.
  - **Magnetic induction in a pi system.** Pi electrons induce a magnetic field that can either increase or decrease the chemical shift of adjacent hydrogens in alkenes and alkynes, respectively. Carbonyl group pi electrons increase the chemical shift of aldehyde hydrogens all the way to near δ 10 ppm.

13.10 THE (n+1) RULE

- Signals can be split into several peaks and this phenomenon is called **spin-spin splitting**. In **spin-spin splitting**, the ^{1}H-NMR signal from one set of equivalent hydrogens is split by the influence of neighboring nonequivalent hydrogens. ✱
  - If a hydrogen has a set of **n** nonequivalent hydrogens on the same or adjacent atoms, its NMR signal will be split into **(n+1)** peaks. The nuclei of all adjacent hydrogens couple, but it is only between nonequivalent hydrogens that the coupling results in spin-spin splitting. For example, in ethyl bromide (CH_3-CH_2-Br) the CH_3- signal is split into three peaks by the two hydrogens on the adjacent -CH_2- group, and the -CH_2- signal is split into four peaks by

the three hydrogens on the adjacent CH_3- group. *[Understanding spin-spin splitting is absolutely essential for the interpretation of ^{1}H-NMR spectra, so it is essential that these ideas are understood before going on.]*

13.11 THE ORIGINS OF SPIN-SPIN SPLITTING

- The chemical shift of a given hydrogen is influenced by the magnetic field derived from the spin of an adjacent nonequivalent nucleus. When this happens, the nuclei are said to be **coupled.** For coupling to be important, the nuclei must usually be no more than three bonds away form each other. In other words, hydrogens on adjacent carbon atoms can couple.
  - The magnetic field derived from the nuclear spin of one nucleus is what influences the adjacent nucleus. For example, if there are two coupled hydrogens that we call H_a and H_b, then H_a has an equal probability of finding the nucleus of H_b in the +1/2 or the -1/2 spin state. Because the magnetic fields of the two different spin states are different, H_a could feel either of two total magnetic fields. (The total magnetic field felt by a given nucleus such as that in H_a is a sum of the applied field plus contributions from electrons and nearby nuclear spin magnetic fields such as that from H_b.) The bottom line is that H_a is split into two ^{1}H-NMR peaks because the spin state of H_b could be either +1/2 or -1/2. ✱
  - H_b is also coupled to H_a to the same extent, so it is also split into two peaks.
  - When a hydrogen nucleus is coupled to three equivalent hydrogens such as those in a methyl group, the nucleus can feel (n+1) or 4 different magnetic fields. This is because each of the three hydrogen nuclei in the methyl group can be in the +1/2 or -1/2 spin state, and these can add up in four different ways (+1/2,+1/2,+1/2; +1/2,+1/2,-1/2, etc.), in a 1:2:2:1 ratio.

13.12 COUPLING CONSTANTS

- A **coupling constant (J)** is the distance, in ppm, between adjacent peaks in a multiplet. In other words, **coupling constants** are a quantitative measure of the shielding/deshielding influence of induced magnetic fields from adjacent nuclei. ✱
  - Coupling constants are usually in the range of a few Hz. Because they are a property of the molecule, they are independent of applied field strength. For this reason, spectra recorded at higher field strength are easier to interpret because the signals (many of which are multiplets due to spin-spin splitting) are separated by more Hz and the individual peaks are less likely to overlap. *[This is a very important yet complicated idea, and it is worth understanding this before going on.]*
  - When two nuclei couple, the coupling constant is the same for both of their signals. Using the same example discussed in Section 13.11, both H_a and H_b will be doublets. Furthermore, the distance between the two peaks in each of these two doublets will be the same number of Hz, and this is the coupling constant for this interaction. ✱
  - When a single signal is coupled to two or more different sets of nuclei, then both sets of interactions lead to splitting according to the (n+1) rule. This can lead to some very complex multiplets. For example, in a molecule with a hydrogen that is coupled to two nonequivalent adjacent hydrogen atoms, the signal is split into 2 x 2 or 4 peaks. Such a signal is referred to as a doublet of doublets.

13.13 ^{13}C-NMR SPECTROSCOPY

- Carbon-12 (^{12}C) is the most abundant natural isotope of carbon (98.89%), but it is not seen in an NMR spectrum because its nucleus has only one allowed spin state. On the other hand, **carbon-13 (^{13}C)** (natural abundance of 1.11%) has two allowed nuclear spin states and it **can be detected by NMR.** ✱
- The NMR signals from ^{13}C are only about 10^{-4} times as strong as those from ^{1}H-NMR. This is because of the relatively low abundance of ^{13}C and the relatively small magnetic moment of the

^{13}C nuclei. For this reason, only the modern FT-NMR machines are able to routinely measure ^{13}C-NMR spectra.

- ^{1}H nuclei couple to the ^{13}C nuclei. Unfortunately, this can make the spectra very difficult to interpret. As a result, ^{13}C spectra are usually measured in the **hydrogen-decoupled mode** in which the sample is irradiated such that all hydrogens are in the same spin state and thus spin-spin splitting is prevented. The ^{13}C spectra can then be measured without interference from complex signals due to ^{1}H-^{13}C spin-spin splitting.
- As stated above, ^{13}C nuclei are very rare, so ordinarily only ^{12}C carbon atoms are adjacent to a given ^{13}C nucleus in a molecule. Thus, ^{13}C-^{13}C coupling and/or spin-spin splitting are usually not observed.
- Because of the way ^{13}C nuclei return to equilibrium states, ^{13}C signals cannot be integrated accurately in ^{13}C-NMR spectra.
- ^{13}C-NMR spectra provide important structural information, because each different carbon atom in a molecule gives rise to a different signal. Thus, by simply looking at a ^{13}C-NMR spectra, a chemist can tell how many different types of carbon atoms are in a molecule.
- Most important, the observed chemical shift of a ^{13}C-NMR signal provides important information. For example; sp^3 alkyl carbon atoms, sp^2 atoms in an alkene, or sp_2 carbonyl carbon atoms all have characteristic chemical shifts. Please see table 13.10 for a detailed list of ^{13}C chemical shifts. ✱

13.15 SOLVING NMR PROBLEMS

- Before analyzing the NMR spectrum of a given molecule, it is helpful to analyze the molecular formula. This could be determined by elemental analysis or mass spectrometry. An important piece of information contained in the molecular formula is the **index of hydrogen deficiency**. The **index of hydrogen deficiency** is the **number of rings and/or pi bonds in a molecule**. This is determined by comparing the number of hydrogens in the molecular formula of a compound of unknown structure with the number of hydrogens in a **reference compound**, a compound with the same number of carbon atoms and with no rings or pi bonds. In particular, the **index of hydrogen deficiency** is defined according to the following formula: ✱

$$\text{index of hydrogen deficiency} = \frac{\#\text{ of hydrogen atoms}_{(\text{reference})} - \#\text{ of hydrogens}_{(\text{molecule of interest})}}{2}$$

  - For reference compounds that contain only C and H atoms, the molecular formula is C_nH_{2n+2}.
  - For each atom of F, Cl, Br, or I subtract one hydrogen.
  - No correction is necessary for O, S, or Se.
  - For each atom of N or P add one hydrogen.
- **After the index of hydrogen deficiency has been determined**, the following steps should be followed when solving a spectral problem. *[Practice is the best and possibly only way to become good at this.]* ✱
  - **Count the number of signals** to determine how many different types of hydrogens are present.
  - **Examine the pattern of chemical shifts** and correlate them with the known characteristic chemical shifts for different types of hydrogen atoms.
  - **Analyze the integration of each signal**, to see how many hydrogen atoms of each type are present.
  - **Analyze the spin-spin splitting patterns**. This is usually the most difficult task by far, but also the most informative.
  - **Write the formula** that is consistent with all of the above information.

CHAPTER 13
Solutions to the Problems

Problem 13.1 Calculate the energy of red light (680 nm) in kilocalories per mole. Which form of radiation carries more energy, infrared radiation of wavelength 2.50 μm or red light of wavelength 680 nm?

Combining the two equations given in the text give:

$$E = h\nu = h\left(\frac{c}{\lambda}\right)$$

Plugging in the appropriate values gives the desired answer:

$$E = \frac{(9.537 \times 10^{-14}\ \text{kcal-sec-mol}^{-1})(3.00 \times 10^{8}\ \text{m-sec}^{-1})}{680 \times 10^{-9}\ \text{m}} = \boxed{42.1\ \text{kcal-mol}^{-1}}$$

Notice how the units canceled to give the final answer in kcal-mol^{-1}. As can be seen from the equations, the longer the wavelength, the lower the energy, thus red light carries more energy.

Problem 13.2 Each of the following compounds gives rise to a single absorption signal in its ^{1}H-NMR spectrum. On the basis of this information, propose a structural formula for each compound.

In order for these molecules to give a single absorption peak, each of the hydrogen nuclei must be in an identical environment. This will only occur in symmetrical molecules.

(a) C_3H_6O

CH_3–C(=O)–CH_3

(b) $C_2H_4Cl_2$

CH_2Cl-CH_2Cl

(c) C_5H_{12}

CH_3–C(CH_3)(CH_3)–CH_3

(d) $C_4H_6Cl_4$

CH_3-CCl_2-CCl_2-CH_3

Problem 13.3 The line of integration of the two signals in the ^{1}H-NMR spectrum of a ketone of molecular formula $C_7H_{14}O$ shows a vertical rise of 62 chart divisions and 10 chart divisions, respectively. Calculate the number of hydrogens giving rise to each signal, and propose a structural formula for this ketone.

The ratio of signals is approximately 1:6, which corresponds to a 2:12 ratio of hydrogens. Thus, the larger signal represents 12 hydrogens and the smaller

signal represents 2 hydrogens. A structure consistent with this assignment is 2,4-dimethyl-3-pentanone as shown below:

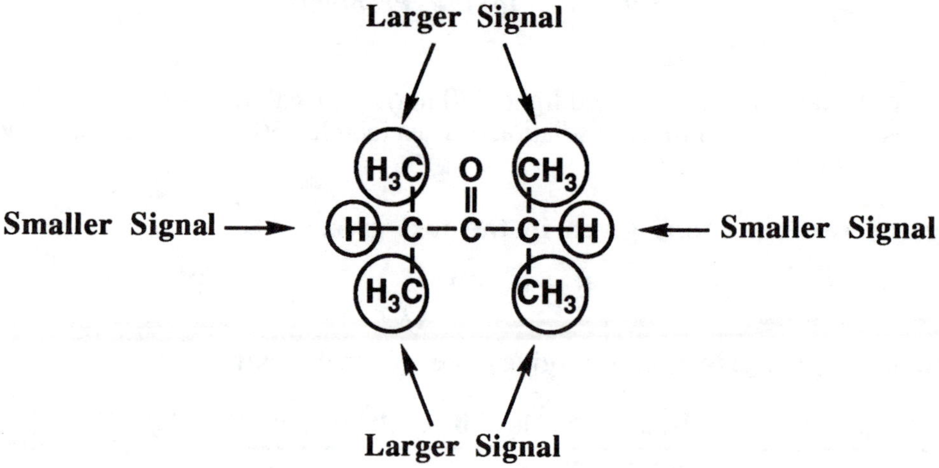

Larger Signal

$$\begin{array}{ccccccc} & H_3C & & O & & CH_3 & \\ & | & & \| & & | & \\ H & - C & - & C & - & C - & H \\ & | & & & & | & \\ & H_3C & & & & CH_3 & \end{array}$$

Smaller Signal → (H) ... (H) ← **Smaller Signal**

Larger Signal

Problem 13.4 Following are structural formulas for two constitutional isomers of molecular formula $C_5H_{10}O_2$:

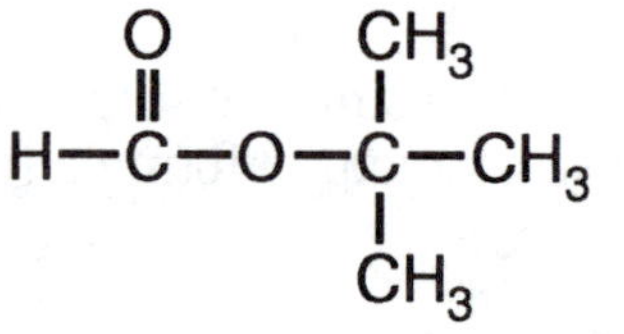

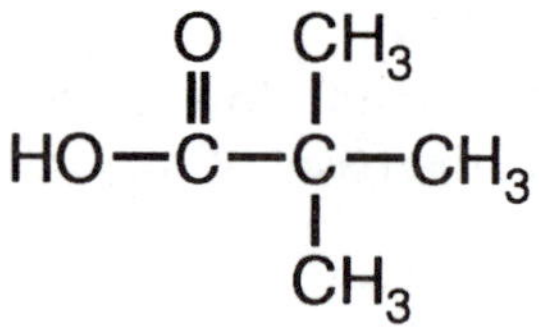

$$H-\overset{\overset{\displaystyle O}{\|}}{C}-O-\overset{\overset{\displaystyle CH_3}{|}}{\underset{\underset{\displaystyle CH_3}{|}}{C}}-CH_3 \qquad HO-\overset{\overset{\displaystyle O}{\|}}{C}-\overset{\overset{\displaystyle CH_3}{|}}{\underset{\underset{\displaystyle CH_3}{|}}{C}}-CH_3$$

(a) Predict the number of signals in the ^1H-NMR spectrum of each isomer.
(b) Predict the ratio of areas of the signals in each spectrum.

In both cases, there are only two different singlet signals; one that integrates to 9 hydrogens corresponding to the methyl groups, and another that integrates to 1 hydrogen corresponding to the aldehyde or carboxylic acid hydrogen, respectively.

(c) Show how you can distinguish between these isomers on the basis of chemical shift.

The easiest way to distinguish these compounds is on the basis of the signal corresponding to either the aldehyde or carboxylic acid hydrogen. If the molecule is the aldehyde drawn on the left, then there will be a sharp singlet integrating to 1 hydrogen that appears at δ 9.5-9.6. If the molecule is the carboxylic acid drawn on the right, then there will be a singlet integrating to 1 hydrogen that appears at δ 10-13.

Problem 13.5 Following are pairs of constitutional isomers. State the number of signals to be expected in the ^1H-NMR spectrum of each isomer and the splitting pattern of each signal.

(a) 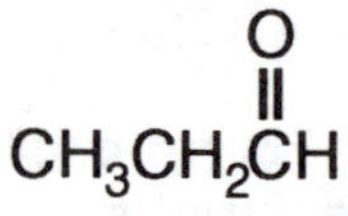 $CH_3CH_2\overset{\overset{\displaystyle O}{\|}}{C}H$ and $CH_3\overset{\overset{\displaystyle O}{\|}}{C}CH_3$

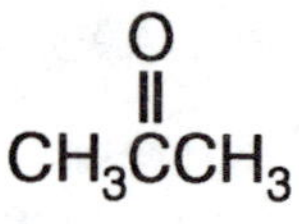

The aldehyde on the left will have three different signals and the ketone on the right will have one signal with splitting patterns as indicated.

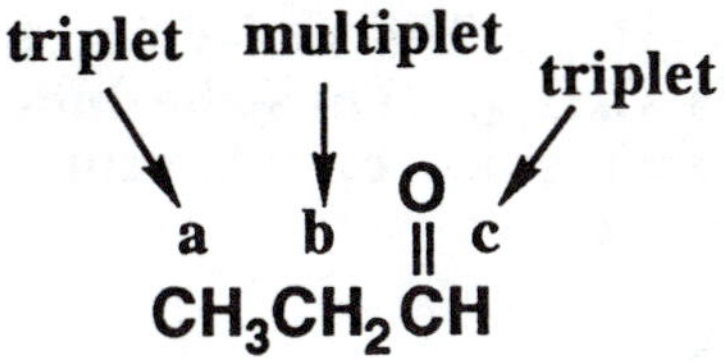

triplet multiplet triplet singlet

$\overset{a}{CH_3}\overset{b}{CH_2}\overset{c}{\underset{}{C}H}$ (with C=O) and $\overset{a}{CH_3}\overset{O}{\overset{\|}{C}}\overset{a}{CH_3}$

(b) 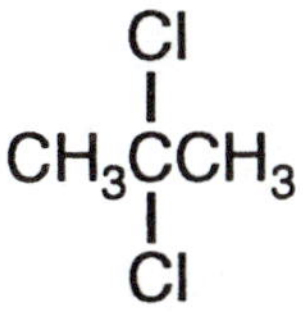 $CH_3CCl_2CH_3$ and $ClCH_2CH_2CH_2Cl$

The molecule on the left will have one signal and the molecule on the right will have two signals with splitting patterns as indicated.

singlet: $\overset{a}{CH_3}CCl_2\overset{a}{CH_3}$

triplet quintet triplet: $Cl\overset{a}{CH_2}\overset{b}{CH_2}\overset{a}{CH_2}Cl$

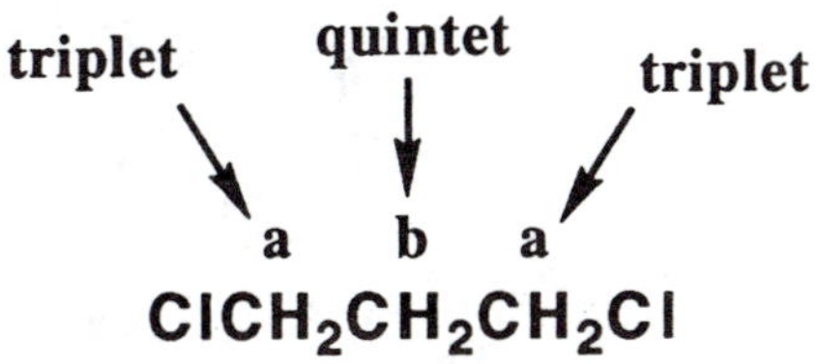

Problem 13.6 Explain how to distinguish between members of each pair of constitutional isomers based on the number of signals in the ^{13}C-NMR spectrum of each member.

(a) 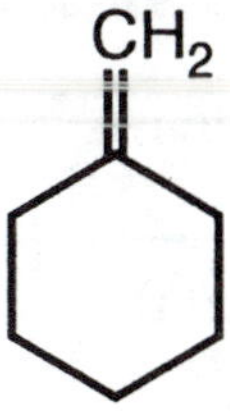 methylenecyclohexane (CH_2=) and 1-methylcyclohexene (CH_3)

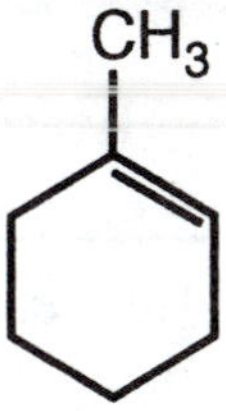

These molecules can be distinguished because they have different numbers of nonequivalent carbon nuclei and thus will have different numbers of ^{13}C-NMR signals. The molecule on the left has higher symmetry and will have 5 different signals, while the molecule on the right has less symmetry and will have 7 different signals.

(b) $CH_3CH{=}CHCH_2CH_2CH_3$ and $CH_3CH_2CH{=}CHCH_2CH_3$

The molecule on the left has lower symmetry and will have 6 different signals, while the molecule on the right has more symmetry and will only have 3 different signals.

Problem 13.7 Calculate the index of hydrogen deficiency of cyclohexene, and account for this deficiency by reference to its structural formula.

The molecular formula for cyclohexene is C_6H_{10}. The molecular formula for the reference compound with 6 carbon atoms is C_6H_{14}. Thus the index of hydrogen deficiency is (14-10)/2 or 2. This makes sense since cyclohexene has one ring and one pi bond.

cyclohexene

Problem 13.8 The index of hydrogen deficiency of niacin is 5. Account for this index of hydrogen deficiency by reference to the structural formula of niacin.

O
C
NH_2
N

niacin

The index of hydrogen deficiency of niacin is 5 because there are four pi bonds and one ring in the structure.

Index of Hydrogen Deficiency

Problem 13.9 Complete the following table.

Class of compound	Molecular formula	Index of hydrogen deficiency	Reason for hydrogen deficiency
alkane	C_nH_{2n+2}	0	(reference hydrocarbon)
alkene	C_nH_{2n}	1	one pi bond
alkyne	C_nH_{2n-2}	2	**two pi bonds**
alkadiene	C_nH_{2n-2}	2	**two pi bonds**
cycloalkane	C_nH_{2n}	1	**one ring**
cycloalkene	C_nH_{2n-2}	2	**one ring and one pi bond**
bicycloalkane	C_nH_{2n-2}	2	**two rings**

Problem 13.10 Calculate the index of hydrogen deficiency of the following compounds:

(a) Aspirin: $C_9H_8O_4$

(20-8)/2 = 6

(b) Ascorbic acid (vitamin C): $C_6H_8O_6$

(14-8)/2 = 3

(c) Pyridine: C_5H_5N

(13-5)/2 = 4 (nitrogen correction)

(d) Urea: CH_4N_2O

(6-4)/2 = 1 (nitrogen correction)

(e) Cholesterol: $C_{27}H_{46}O$

(56-46)/2 = 5

(f) Trichloroacetic acid: $C_2HCl_3O_2$

(3-1)/2 = 1 (halogen correction)

Interpretaion of ^{1}H-NMR and ^{13}C-NMR Spectra

Problem 13.11 A radiofrequency of 282 MHz is required to spin-flip a fluorine-19 nucleus in an applied field of 7.05 T. Calculate the energy (in calories per mole) associated with this transition.

$$E = h\nu = (9.537 \times 10^{-14} \text{ kcal-sec-mol}^{-1})(282 \times 10^{6} \text{ sec}^{-1}) = 2.69 \times 10^{-5} \text{ kcal-mol}^{-1} = 2.69 \times 10^{-2} \text{ cal-mol}^{-1}.$$

Problem 13.12 Complete the following table. Which nucleus requires the least energy to flip its spin at this applied field? Which nucleus requires the most energy.

Nucleus	Applied field (T)	Radiofrequency (MHz)	Energy (cal/mol)
^{1}H	7.05	300	**2.86×10^{-2}**
^{13}C	7.05	75.5	**7.20×10^{-3}**
^{19}F	7.05	282	**2.69×10^{-2}**

Based on the entries in the table, the ^{13}C requires the least energy to flip and the ^{1}H requires the most.

Problem 13.13 The natural abundance of ^{13}C is only 1.1%. Furthermore, its sensitivity in NMR spectroscopy (a measure of the energy difference between a spin aligned with or against an external magnetic field) is only 1.6% that of ^{1}H. What are the relative signal intensities expected for the ^{1}H-NMR and ^{13}C-NMR spectra of the same sample of $Si(CH_3)_4$?

A given ^{13}C signal is (0.011)(0.016) = 0.000176 as strong as a given ^{1}H signal. There are three times as many H atoms as C atoms in $Si(CH_3)_4$, so overall the ratio of H to C signals is 1 : (0.000176/3) = 1 : 0.000059. (Note that this is the same as 17,000 to 1)

Problem 13.14 The percent s character of carbon participating in a C-H bond can be established by measuring the ^{13}C-^{1}H coupling constant and using the relationship

$$\text{percent s character} = 0.2\, J(^{13}\text{C-}^{1}\text{H})$$

The ^{13}C-^{1}H coupling constant observed for methane, for example, is 125 Hz, which gives 25% s character, the value expected for an sp^3 hybridized carbon atom.

(a) Calculate the expected ^{13}C-^{1}H coupling constant in ethylene and acetylene.

For ethylene and acetylene the carbon atoms are sp^2 and sp hybridized and thus 33% and 50% s character, respectively. Using the above equation gives coupling constants of 165 Hz and 250 Hz, respectively.

(b) In cyclopropane, the ^{13}C-^{1}H coupling constant is 160 Hz. What is the hybridization of carbon in cyclopropane?

The carbon atoms in cyclopropane are (0.2)(160) = 32% s character. This corresponds roughly to an sp^2 hybridized carbon atom.

Problem 13.15 ^{13}C-NMR spectroscopy of "labeled bonds" can be used to follow connectivity changes in chemical reactions. Suppose you carried out a Diels-Alder reaction with $H_2C{=}^{13}CH$-$^{13}CH{=}CH_2$ and ethylene. Show how ^{13}C-NMR spectroscopy can be used to establish that the 2,3-bond of 1,3-butadiene remains intact in the cyclohexane product?

The product of the Diels-Alder reaction between butadiene and ethylene is cyclohexene. Since the 2,3-bond remains intact during the reaction, the ^{13}C atoms will be sp^2 hybridized as shown below. This will be easy to determine with ^{13}C-NMR spectroscopy because they will have a chemical shift (δ) in the 100-150 range. (If for some reason the 2,3-bond were broken during the reaction, then at least one of the labeled ^{13}C atoms would be sp^3 hybridized and thus exhibit a chemical shift in the 15-55 range.)

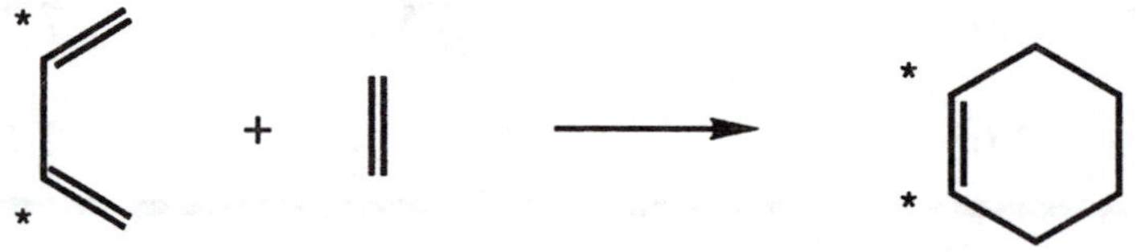

* * + ‖ ⟶ * *

Problem 13.16 Following are structural formulas for three constitutional isomers of molecular formula $C_7H_{16}O$ and three sets of ^{13}C-NMR spectral data. Assign each constitutional isomer its correct spectrum.

(a) $CH_3CH_2CH_2CH_2CH_2CH_2CH_2OH$

(b) $CH_3C(OH)(CH_3)CH_2CH_2CH_2CH_3$

(c) $CH_3CH_2C(OH)(CH_2CH_3)CH_2CH_3$

Spectrum 1:	Spectrum 2:	Spectrum 3:
74.66	70.97	62.93
30.54	43.74	32.79
7.73	29.21	31.86
	26.60	29.14
	23.27	25.75
	14.09	22.63
		14.08

These constitutional isomers are most readily distinguished by the number of sets of nonequivalent carbon atoms and thus different ^{13}C signals. Using the following analysis, it can be seen that compound (a) has 7 sets of nonequivalent carbon atoms corresponding to spectrum 3, compound (b) has 6 sets of nonequivalent carbon atoms corresponding to spectrum 2, and compound (c) has 3 sets of nonequivalent carbon atoms corresponding to spectrum 1.

g f e d c b a
$CH_3CH_2CH_2CH_2CH_2CH_2CH_2OH$

OH
e a | b c d f
$CH_3CCH_2CH_2CH_2CH_3$
|
CH_3
e

OH
c b a | b c
$CH_3CH_2CCH_2CH_3$
|
CH_2CH_3
b c

Problem 13.17 Following are structural formulas for the *cis* isomers of 1,2-, 1,3-, and 1,4-dimethylcyclohexanes and three sets of ^{13}C-NMR spectral data. Assign each constitutional isomer its correct spectrum.

(a) (b) (c)

Spectrum 1:	Spectrum 2:	Spectrum 3:
31.35	34.20	44.60
30.67	31.30	35.14
20.85	23.56	32.88
	15.97	26.54
		23.01

Again, these constitutional isomers are most readily distinguished by the number of sets of nonequivalent carbon atoms and thus different ^{13}C signals. Using the analysis shown below, it can be seen that compound (a) has 4 sets of nonequivalent carbon atoms corresponding to spectrum 2, compound (b) has 5 sets of nonequivalent carbon atoms corresponding to spectrum 3, and compound (c) has 3 sets of nonequivalent carbon atoms corresponding to spectrum 1.

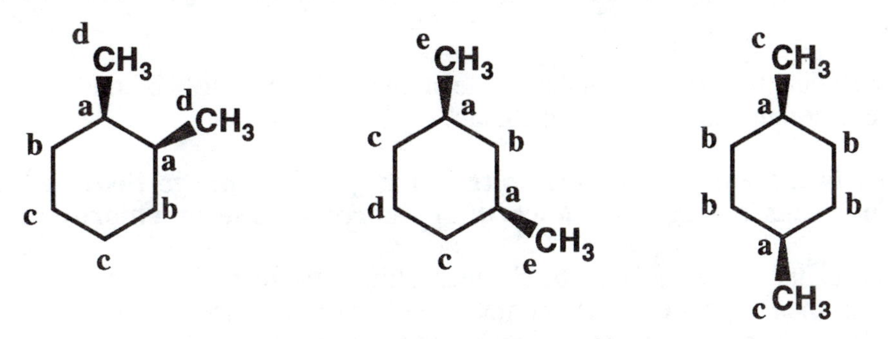

Problem 13.18 The ^{13}C-NMR spectrum of 3-methyl-1-butanol shows four signals, indicating that there are four sets of nonequivalent carbon atoms. The ^{13}C-NMR spectrum of 3-methyl-2-butanol shows five signals, indicating five nonequivalent carbon atoms. How do you account for the presence of four signals in the spectrum of 3-methyl-1-butanol but five signals in the spectrum of 3-methyl-2-butanol?

$$\underset{\text{3-methyl-1-butanol}}{\overset{CH_3}{\overset{|}{CH_3CHCH_2CH_2OH}}}$$

^{13}C-NMR (δ)	
61.14	24.73
41.71	22.62

3-methyl-1-butanol

$$CH_3\overset{CH_3}{\overset{|}{C}}H\overset{*}{\underset{OH}{\underset{|}{C}}}HCH_3$$

^{13}C-NMR (δ)	
72.73	18.16
35.07	17.90
19.99	

3-methyl-2-butanol

The key to this question is that 3-methyl-2-butanol contains a stereocenter at carbon number 2 while 3-methyl-1-butanol does not have any stereocenters. The stereocenter in 3-methyl-2-butanol makes the two methyl groups on carbon 3 nonequivalent so each methyl group gives rise to a different ^{13}C signal. Because 3-methyl-1-butanol does not have a stereocenter, the two methyl groups are equivalent and give rise to a single ^{13}C signal.

Problem 13.19 Following are structural formulas, dipole moments, and ^{1}H-NMR chemical shifts for acetonitrile, fluoromethane, and chloromethane.

$CH_3—C≡N$	$CH_3—F$	$CH_3—Cl$
Acetonitrile	Fluoromethane	Chloromethane
3.92 D	1.85 D	1.87 D
δ 1.97	δ 4.26	δ 3.05

(a) How do you account for the fact that the dipole moments of fluoromethane and chloromethane are almost identical even though fluorine is considerably more electronegative than chlorine?

Recall that dipole moment is proportional to the partial charge separation times the distance of that charge separation. Fluorine is a much smaller atom than chlorine, so it makes shorter bonds leading to relatively short charge separation distances. Thus, the differences in bond lengths exactly offsets the differences in electronegativities between fluorine and chlorine and the dipole moments come out the same.

(b) How do you account for the fact that the dipole moment of acetonitrile is considerably greater than that of either fluoromethane or chloromethane?

Again, the key is distance. The acetonitrile has partial charge distributed over more atoms and thus a larger distance than fluoromethane or chloromethane.

(c) How do you account for the fact that the chemical shift of the methyl hydrogens in acetonitrile is considerably less than that for either fluoromethane of chloromethane? Hint: Consider the magnetic induction in the pi system of acetonitrile and the orientation of this molecule in an applied magnetic field.

As described in Figure 13.10, a magnetic field is induced in the pi system of the nitrile that is against the applied field, thus decreasing the chemical shift.

Problem 13.20 Following are three compounds of molecular formula C_4H_8O, and three ^{1}H-NMR spectra:

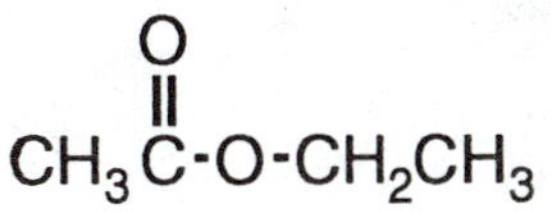

$$\underset{(1)}{CH_3\overset{O}{\overset{\|}{C}}\text{-}O\text{-}CH_2CH_3} \qquad \underset{(2)}{H\overset{O}{\overset{\|}{C}}\text{-}O\text{-}CH_2CH_2CH_3} \qquad \underset{(3)}{CH_3\text{-}O\text{-}\overset{O}{\overset{\|}{C}}CH_2CH_3}$$

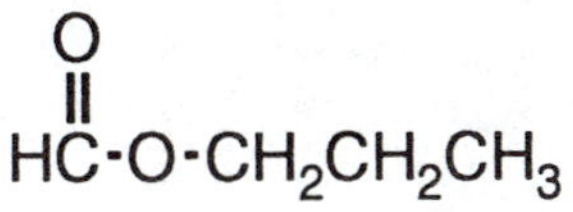

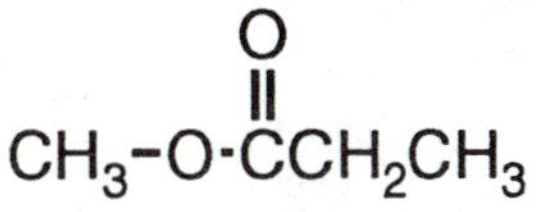

(a) Assign each compound its correct spectrum and assign all signals to their corresponding hydrogens.

For the spectral interpretations in the rest of this chapter the chemical shift (δ) is given followed by the relative integration, the multiplicity of the peak (singlet, doublet, triplet, etc.) and finally the identity of the hydrogens giving rise to the signal are shown in bold.

Spectrum A corresponds to compound 2: ^{1}H-NMR δ 8.1 (1H, singlet, H-C(O)-), 4.1 (2H, triplet, -O-CH_2-), 1.7 (2H, multiplet; a doublet of triplets, -CH_2-), 1.0 (3H, triplet, -CH_3).

Spectrum B corresponds to compound 3: ^{1}H-NMR δ 3.7 (3H, singlet, CH_3-C(O)-), 2.3 (2H, quartet, -C(O)-CH_2-), 1.2 (3H, triplet, -CH_3).

Spectrum C corresponds to compound 1: ^{1}H-NMR δ 4.1 (3H, singlet, CH_3-O-), 2.0 (2H, quartet, -C(O)-CH_2-), 1.25 (3H, triplet, -CH_3).

(b) From your study of these three compounds and spectra, begin to develop your own table of chemical shifts. For example, assign δ values to CH_3-O-, to -CH_2-O-, and so forth.

Preparing such a table is a good way to learn what chemical shifts can be expected for different types of hydrogen atoms.

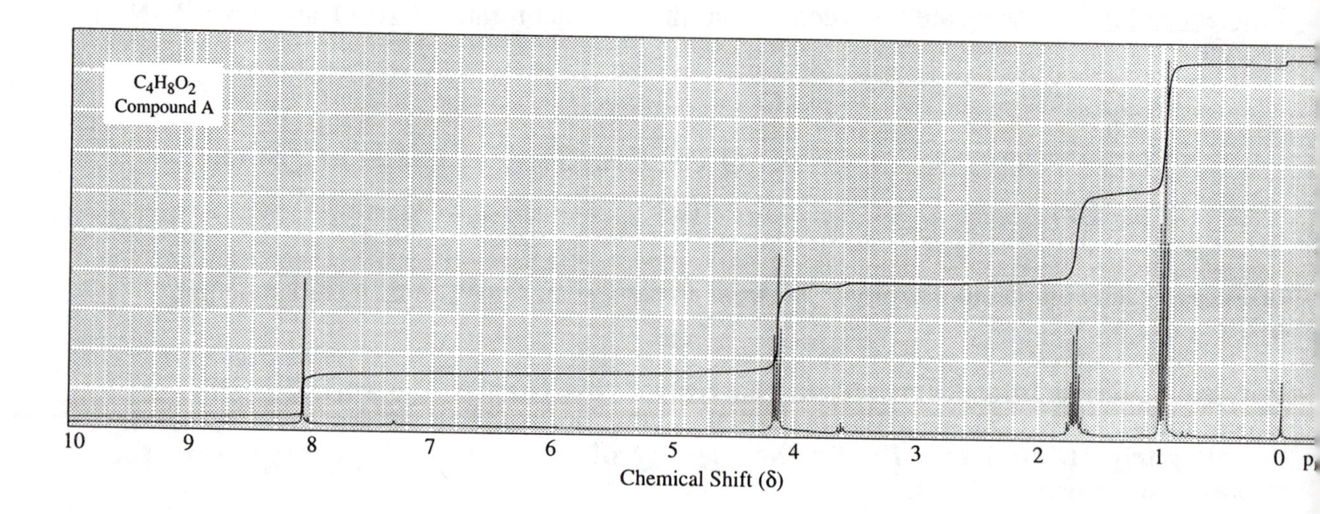

$C_4H_8O_2$
Compound A
10
9
8
7
6
5
4
3
2
1
0
Chemical Shift (δ)

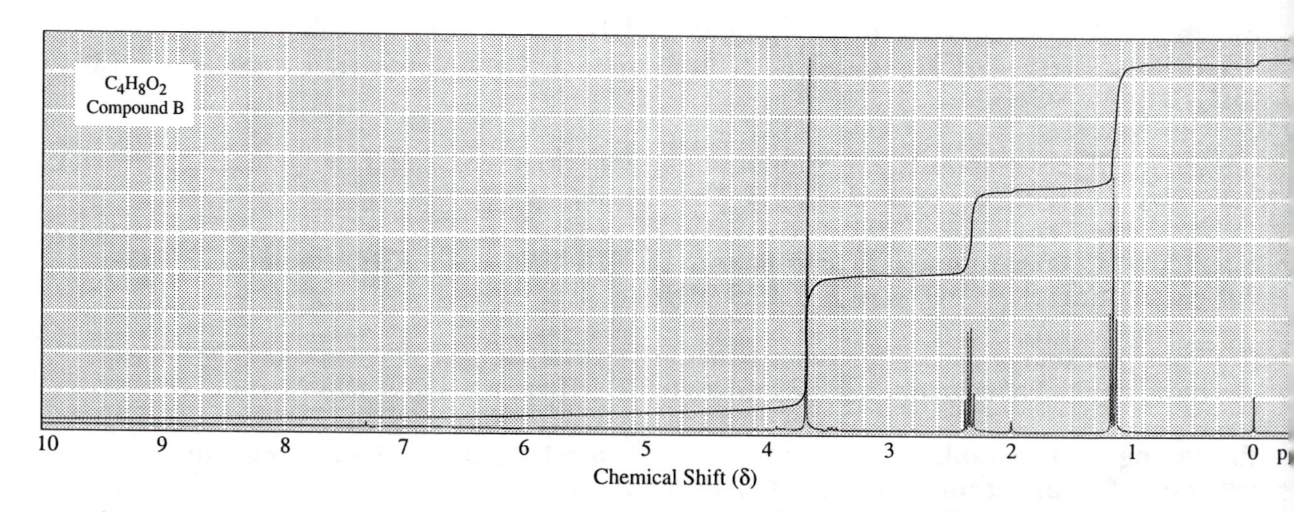

$C_4H_8O_2$
Compound B
10
9
8
7
6
5
4
3
2
1
0
Chemical Shift (δ)

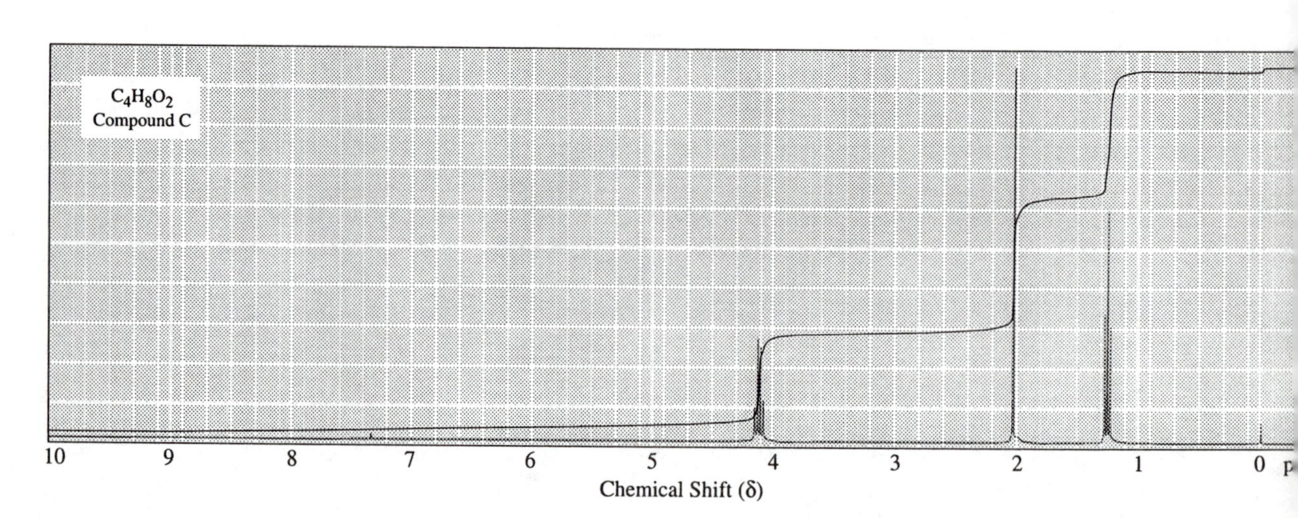

$C_4H_8O_2$
Compound C
10
9
8
7
6
5
4
3
2
1
0
Chemical Shift (δ)

<u>Problem 20.21</u> Following are ^{1}H-NMR spectra for compounds A,B, and C, each of molecular formula C_6H_{12}. Each readily decolorizes a solution of Br_2 in CCl_4 and also decolorizes aqueous $KMnO_4$ with formation of a brown precipitate of MnO_2. Deduce the structural formulas of compounds A,B, and C and account for the observed patterns of spin-spin splitting.

Each of the compounds has an index of hydrogen deficiency of 1, in the form of a double bond as evidenced by the reaction with Br_2 and $KMnO_4$. The rest of the detailed structures can be deduced from the spectra.

Compound A:

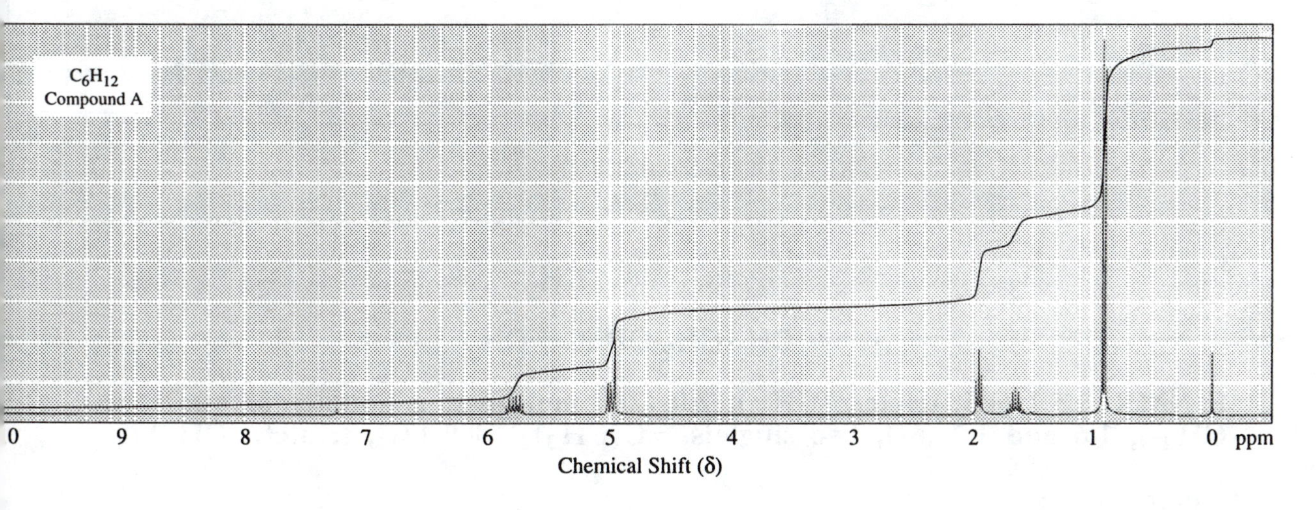

$$CH_3-CH(CH_3)-CH_2-CH{=}CH_2$$

^{1}H-NMR δ 5.8 (1H, multiplet; this is more complex than expected because the adjacent vinylic hydrogens are not equivalent, -CH=), 5.0 (2H, multiplet; this is asymmetric because these two vinylic hydrogens are not equivalent, $=CH_2$), 1.9 (2H, multiplet; doublet of doublets, $-CH_2-$), 1.6 (1H, multiplet; a triplet of septets, -CH-), 0.9 (6H, one doublet, $-CH_3$).

Compound B:

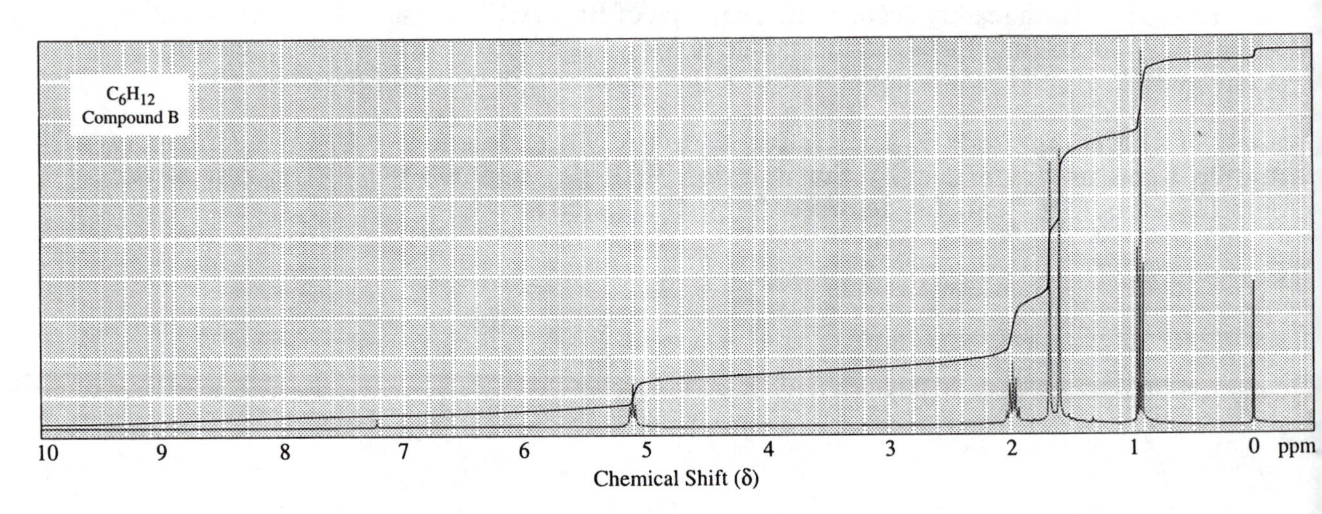

$$\begin{array}{c} \phantom{CH_3\text{-}CH_2\text{-}CH=}CH_3 \\ \phantom{CH_3\text{-}CH_2\text{-}CH=}| \\ CH_3\text{-}CH_2\text{-}CH=C-CH_3 \end{array}$$

^{1}H-NMR δ 5.1 (1H, triplet, -CH=), 2.0 (2H, multiplet; a doublet of quartets, -CH_2-), 1.6 and 1.7 (6H, two singlets, =$C(CH_3)_2$), 0.9 (3H, triplet, -CH_3)

Compound C:

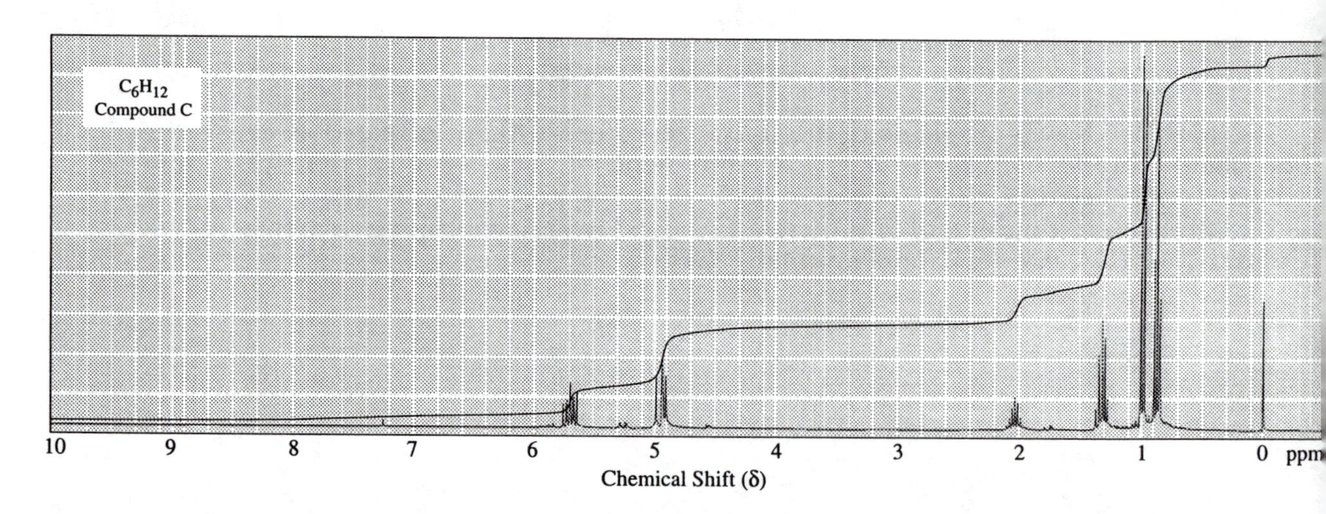

$$\begin{array}{c} CH_3 \\ | \\ CH_3\text{-}CH_2\text{-}CH\text{-}CH=CH_2 \end{array}$$

^{1}H-NMR δ 5.7 (1H, multiplet; this is more complex than expected because the adjacent vinylic hydrogens are not equivalent, -CH=), 4.9 (2H, multiplet; this is asymmetric because these two vinylic hydrogens are not equivalent, =CH_2), 2.0

(1H, multiplet; a doublet of a triplet of a quartet, -CH-), 1.3 (2H, multiplet; a doublet of a quartet, -CH_2-), 1.0 (3H, doublet, -CH-CH_3), 0.8 (3H, triplet, -CH_2-CH_3)

Problem 13.22 Following are ^{1}H-NMR spectra for compounds D, E, and F, each of molecular formula $C_5H_{12}O$. Each is a liquid at room temperature, slightly soluble in water, and reacts with sodium metal with the evolution of a gas. Deduce the structural formulas of compounds D, E, and F.

The index of hydrogen deficiency is 0 for these molecules, so there are no rings or double bonds. The fact that the compounds are slightly soluble in water and react with sodium metal indicate that each molecule has an -OH group.

Compound D:

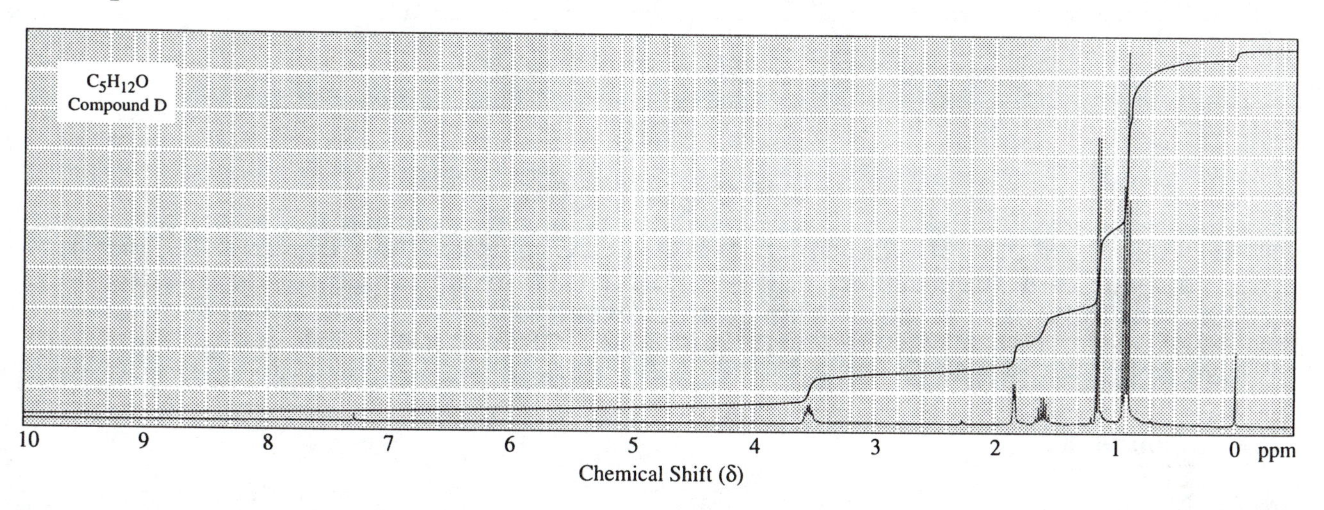

$$CH_3-CH(CH_3)-CH(OH)-CH_3$$

^{1}H-NMR δ 3.5 (1H, multiplet, -CH-OH-), 1.85 (1H, doublet, -OH), 1.6 (1H, multiplet, -CH-$(CH_3)_2$), 1.15 (3H, doublet, -C(OH)-CH_3), 0.9 (6H, triplet, -CH-$(CH_3)_2$).

Compound E:

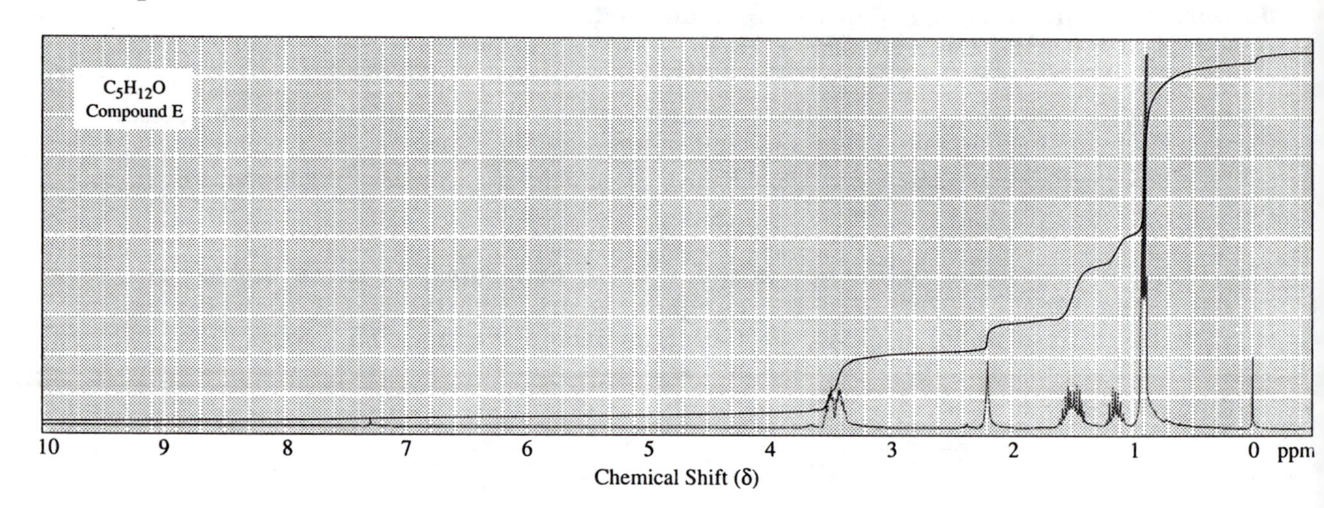

$$CH_3CH_2\overset{\overset{\displaystyle CH_3}{|}}{C}HCH_2OH$$

^{1}H-NMR δ 3.4-3.5 (2H, multiplet; this is more complex than expected because it is adjacent to a stereocenter, $-CH_2-OH$), 2.2 (1H, broad triplet, -OH), 1.4-1.6 (2H, multiplet; this is more complex than expected because it is adjacent to a stereocenter, CH_3-CH_2-), 1.1 (1H, multiplet, -CH-), 0.8-0.9 (6H, broad multiplet, both $-CH_3$ groups).

Compound F:

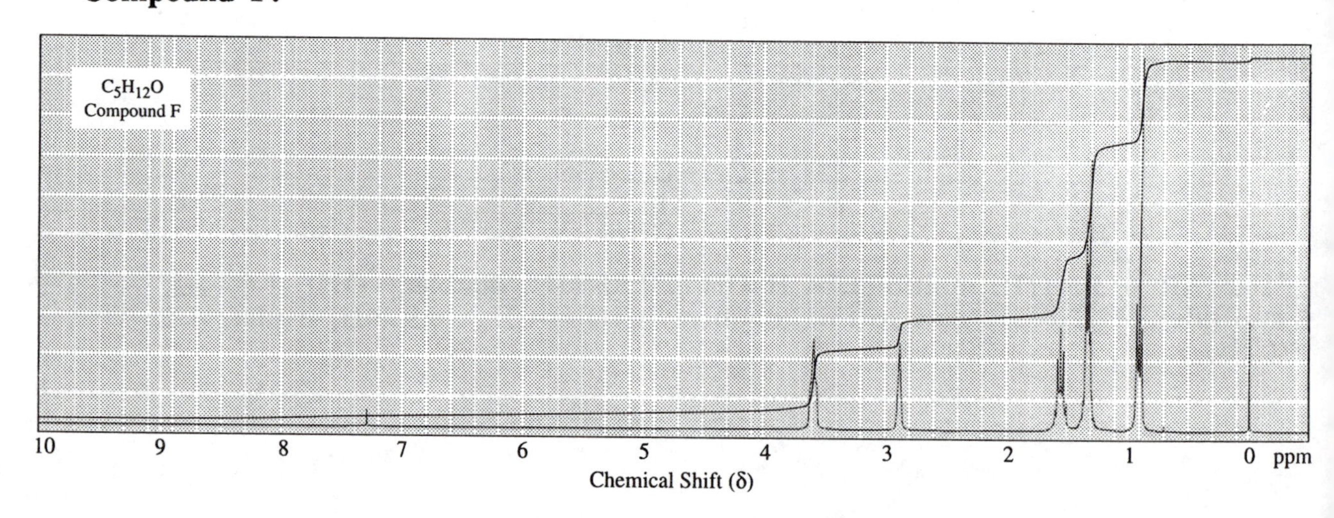

$$CH_3CH_2CH_2CH_2CH_2OH$$

^{1}H-NMR δ 3.6 (2H, broad multiplet, $-CH_2-OH$), 2.9 (1H, broad peak, -OH), 1.55 (2H, multiplet; a triplet of triplets, $-CH_2-CH_2-OH$), 1.4 (4H, multiplet, $CH_3-CH_2-CH_2-$ and $CH_3-CH_2-CH_2-$), 0.9 (3H, triplet, $-CH_3$)

Problem 13.23 Propose a structural formula for compound G, molecular formula C_3H_6O, consistent with the following ^{1}H-NMR spectrum.

Compound G:

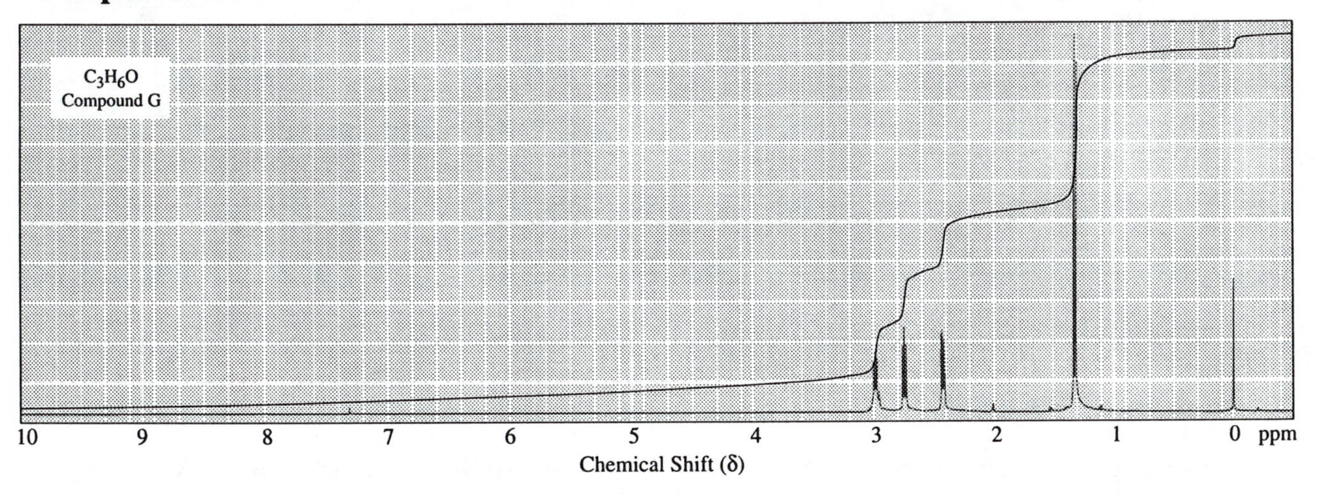

From the molecular formula, there is an index of hydrogen deficiency of 1 indicating that there is one ring or pi bond.

^{1}H-NMR δ 3.0 (1H, multiplet, H_c), 2.4 and 2.75 (2H, multiplets; these hydrogens are not equivalent because they cannot rotate freely, H_a and H_b), 1.3 (3H, doublet, $-CH_3$)

Problem 13.24 Compound H, molecular formula $C_6H_{14}O$, readily undergoes acid-catalyzed dehydration when warmed with phosphoric acid to give compound I, molecular formula C_6H_{12}, as the major organic product. Following is the ^{1}H-NMR spectrum of compound H. The ^{13}C-NMR spectrum of compound H shows peaks at 72.98, 33.72, 25.85, and 8.16. Deduce the structural formulas of compounds H and I.

Compound H:

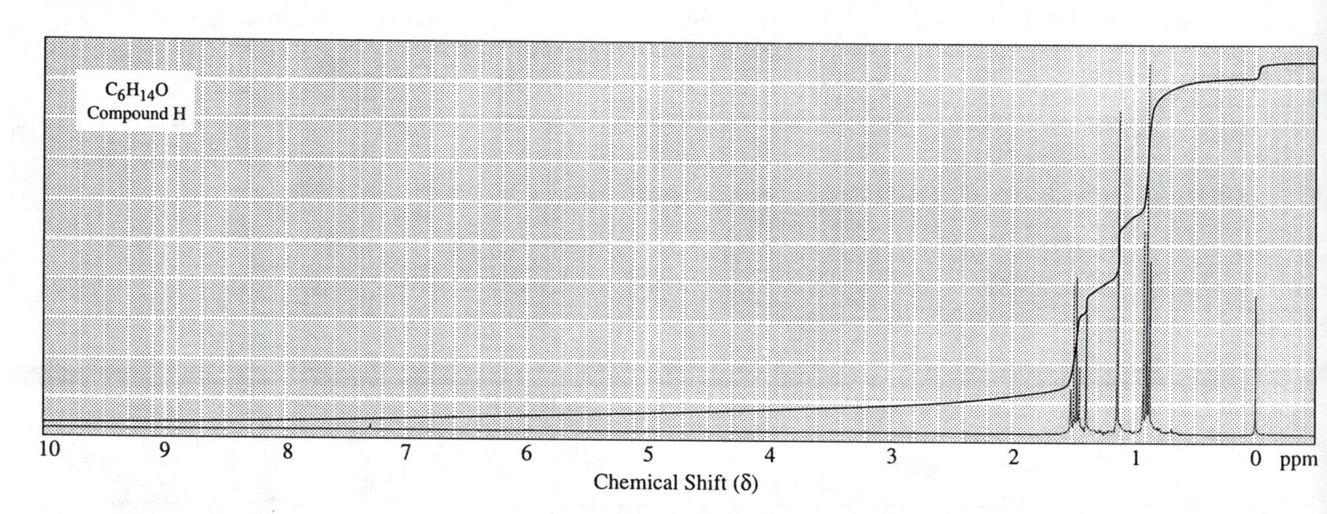

From the molecular formula, there is a hydrogen deficiency index of 0, so there are no rings or pi bonds in compound H. From the ^{13}C-NMR peak at 72.98 there is a carbon bonded to an -OH group. The rest of the structure can be deduced from the ^{1}H-NMR spectrum.

```
          OH
          |
CH3CH2CCH2CH3
          |
          CH3
```

^{1}H-NMR δ 1.6 (4H, quartet, $-CH_2$), 1.4 (1H, singlet, -OH), 1.1 (3H, singlet, $-C(OH)-CH_3$), 0.9 (6H, triplet, both $-CH_2-CH_3$ groups)

Dehydration of compound H gives the following alkene as compound I:

```
CH3       CH3
   \     /
    C=C
   /     \
  H       CH2CH3
```

Problem 13.25 Compound J, molecular formula $C_5H_{10}O$, readily decolorizes Br_2 in CCl_4, reacts with sodium metal with the evolution of a gas, and is converted by H_2/Ni into compound K, molecular $C_5H_{12}O$. Following is the ^{1}H-NMR spectrum of compound J. The ^{13}C-NMR spectrum of compound J shows peaks at 146.12, 110.75, 75.05, and 29.38. Deduce the structural formulas of compounds J and K.

Compound J:

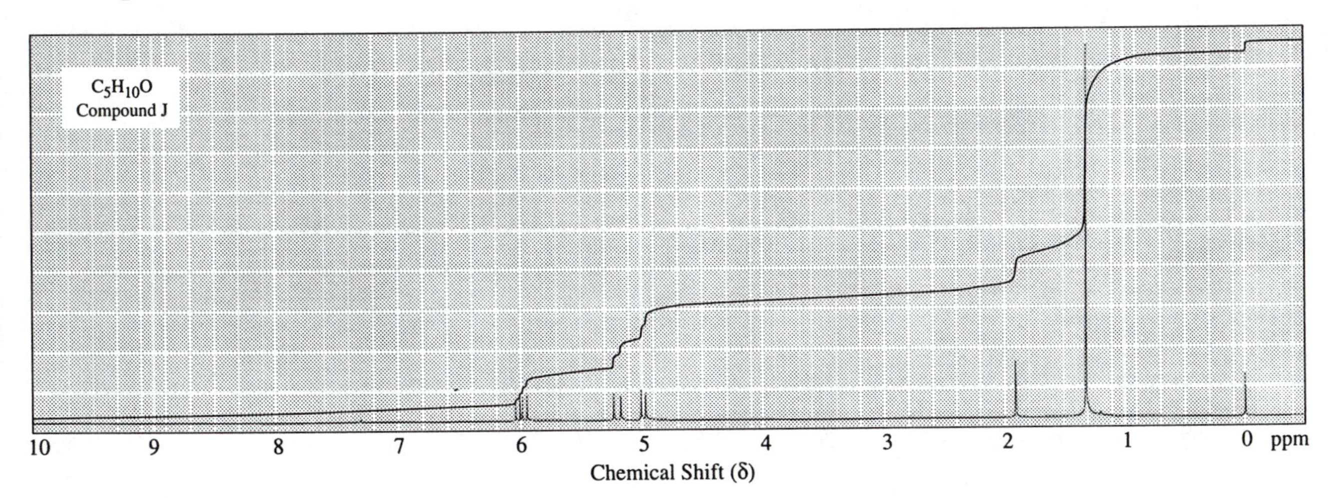

From the reaction with Br_2 and H_2/Ni it is clear the compound J has a double bond, and because it reacts with sodium metal, compound J must also have an -OH group. These conclusions are supported by the ^{13}C-NMR peaks corresponding to the sp^2 carbons at 146.12 and 110.75 as well as the carbon bonded to the -OH group at 71.05. The rest of the structure is deduced from the ^{1}H-NMR spectrum.

$CH_2{=}CH{-}C(OH)(CH_3){-}CH_3$ ≡ $(H_a)(H_b)C{=}C(H_c){-}C(OH)(CH_3)_2$

^{1}H-NMR δ 6.0 (1H, doublet of doublets, H_a), 4.9 and 5.2, (2H, two doublets; the vinylic hydrogens are not equivalent, H_b and H_c), 1.9 (1H, singlet, -OH), 1.3 (6H, singlet, C-$(CH_3)_2$)

Upon hydrogenation, compound J is reduced to the alcohol shown below as compound K:

$CH_3{-}CH_2{-}C(OH)(CH_3){-}CH_3$

Problem 13.26 Following is the ^{1}H-NMR spectrum of compound L, molecular formula C_7H_{12}. The ^{13}C-NMR spectrum of this compound shows peaks at 150.12, 106.32, 35.44, 28.36, and 26.36. Deduce the structural formula of compound L.

Compound L:

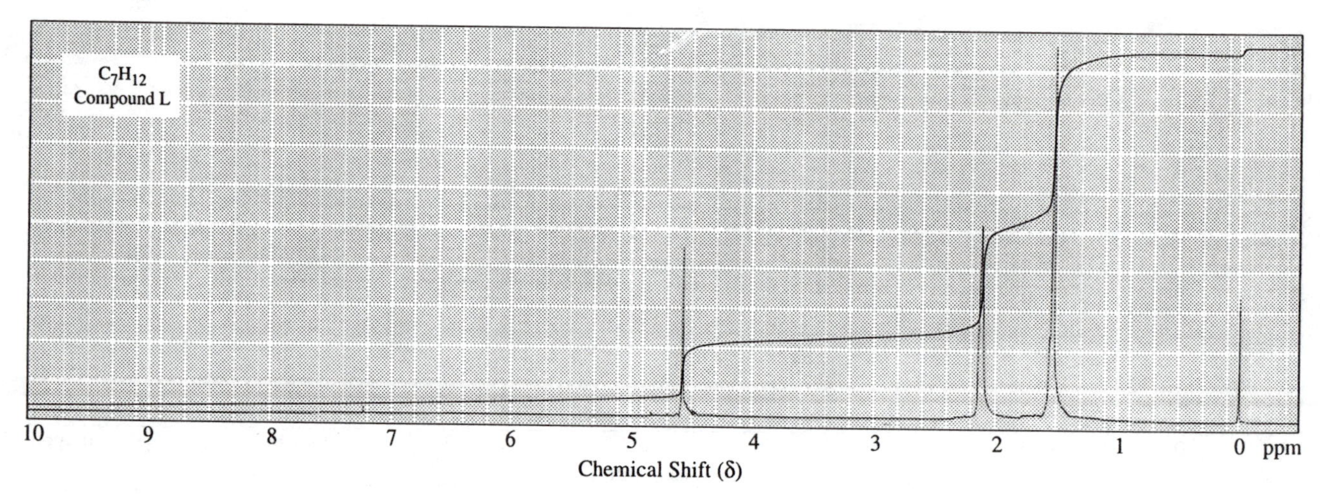

The molecular formula indicates that there is an index of hydrogen deficiency of 2, so there are two rings and/or pi bonds. The ^{13}C-NMR indicates there is only one double bond because there are only two resonances corresponding to sp^2 carbon atoms (150.12 and 106.43). Therefore, compound L must have one ring and one pi bond.

b a
b $=CH_2$
b a

^1H-NMR δ 4.6 (2H, singlet, $=CH_2$), 2.1 (4H, broad peak, the $-CH_2-$ groups adjacent to the sp^2 carbon atom on the ring labeled as "a" on the structure above), 1.6 (6H, broad peak, the three $-CH_2-$ groups labeled as "b" on the structure above)

Problem 13.27 Treatment of compound M with BH_3 followed by H_2O_2 and NaOH gives compound N. Following are ^1H-NMR spectra for compounds M and N along with the ^{13}C-NMR spectral data. From this information, deduce structural formulas for compounds M and N.

$$\underset{(M)}{C_7H_{12}} \xrightarrow[\text{2) } H_2O_2\text{, NaOH}]{\text{1) } BH_3} \underset{(N)}{C_7H_{14}O}$$

^{13}C-NMR	
(M)	(N)
132.38	72.71
32.12	37.59
29.14	28.13
27.45	22.68

Compound M:

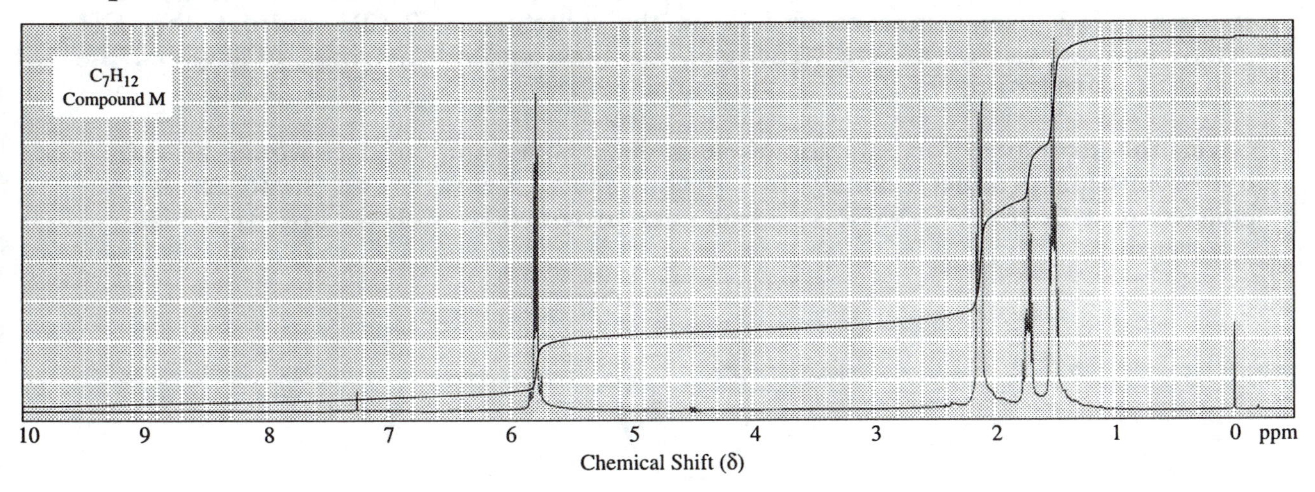

Compound N:

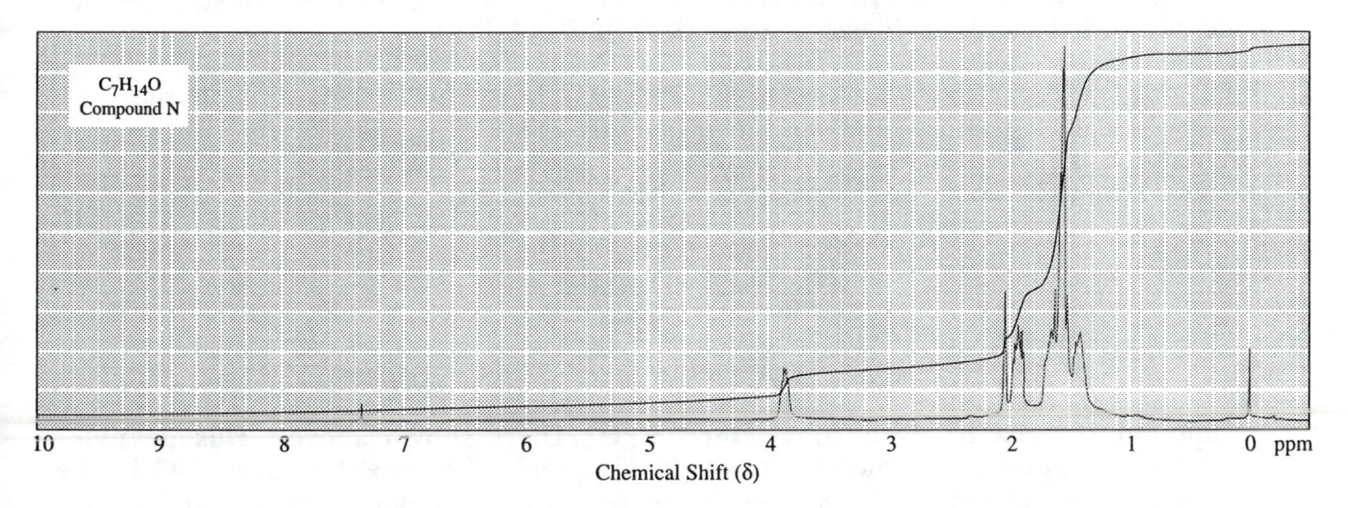

The molecular formula for M indicates that it has an index of hydrogen deficiency of 2 so that is has two rings and/or pi bonds. The ^{13}C-NMR spectral data shows that there is an sp^2 carbon atom (132.38). Since there must be two sp^2 carbon atoms to make a pi bond, then the molecule must be symmetric so that both sp^2 carbon atoms are equivalent. This also explains why there are so few other ^{13}C-NMR signals. Since there is presumably only one pi bond, then there must be one ring in the molecule. The rest of the structure can be deduced from the ^{1}H-NMR spectrum.

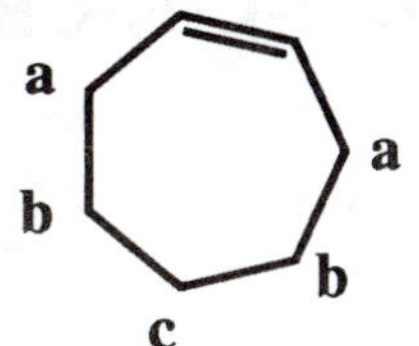

^{1}H-NMR δ 5.8 (2H, triplet, both =CH-), 2.1 (4H, multiplet; doublet of triplets, the two -CH_2- groups marked as "a" on the structure), 1.7 (2H, quintet, the -CH_2- group marked as "c" on the structure), 1.5 (4H, multiplet; triplet of triplets, the two -CH_2- groups marked as "b" on the structure)

Given the structural formula for M, it is clear that compound N would be the hydroboration product, namely the alcohol shown below. This structure is consistent with the ^{13}C-NMR spectral data provided as well as the ^{1}H-NMR spectrum.

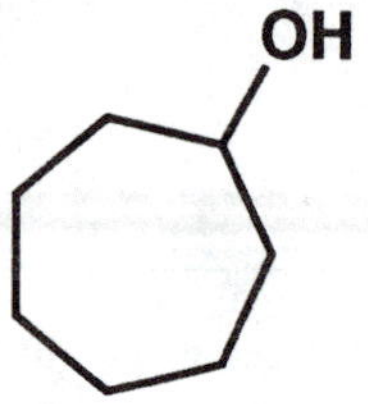

OH

^{1}H-NMR δ 3.8 (1H, broad peak, -C(OH)H-), 2.0 (1H, sharp singlet, -OH), 1.4-1.9 (12H, broad multiplets, all the remaining hydrogens on the ring. The peaks are so broad and the patterns so complex because this ring does not have a double bond to hold it rigid, so it has a great deal of flexibility)

Problem 13.28 Compound O is known to contain only C, H, and O. Its mass spectrum shows a weak molecular ion peak at M(102) and prominent peaks at *m/e* 87, 45, and 43. Its ^{1}H-NMR spectrum consists of two signals: δ 1.1 (doublet, 12H) and δ 3.6 (septet, 2H). Propose a structural formula for compound O consistent with this information.

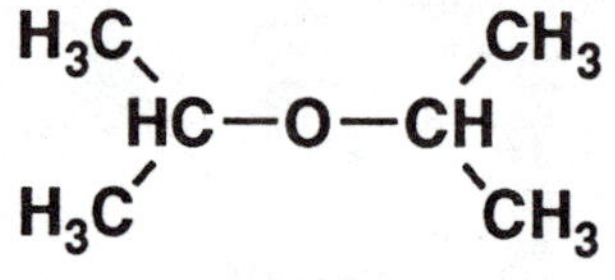

$$(H_3C)_2HC{-}O{-}CH(CH_3)_2$$

Compound O

Compound O must be diisopropyl ether, $C_6H_{14}O$, as shown above. Diisopropyl ether has the appropriate molecular weight to explain the parent peak of 102 in the mass spectrum as well as the proper symmetry to explain the relatively simple ^{1}H NMR spectrum.

Problem 13.29 Write structural formulas for the following compounds:
(a) $C_2H_4Br_2$: δ 2.5 (d, 3H) and 5.9 (q, 1H)

$$CH_3{-}CHBr_2$$

(b) $C_4H_8Cl_2$: δ 1.60 (d, 3H), 2.15 (m, 2H), 3.72 (t, 2H), and 4.27 (m, 1H)

$$CH_3{-}CHCl{-}CH_2{-}CH_2Cl$$

(c) $C_5H_8Br_4$: δ 3.6 (s, 8H)

$$\begin{array}{c} CH_2Br \\ | \\ CH_2Br-C-CH_2Br \\ | \\ CH_2Br \end{array}$$

(d) C_4H_8O: δ 1.0 (t, 3H), 2.1 (s, 3H), and 2.4 (q, 2H)

$$CH_3CH_2-\overset{\overset{O}{\|}}{C}-CH_3$$

(e) $C_4H_8O_2$: δ 1.2 (t, 3H), 2.1 (s, 3H), and 4.1 (q, 2H); contains an ester group.

$$CH_3-\overset{\overset{O}{\|}}{C}-OCH_2CH_3$$

(f) $C_4H_8O_2$: δ 1.2 (t, 3H), 2.3 (q, 2H), and 3.6 (s, 3H); contains an ester group.

$$CH_3CH_2-\overset{\overset{O}{\|}}{C}-OCH_3$$

(g) C_4H_9Br: δ 1.1 (d, 6H), 1.9 (m, 1H), and 3.4 (d, 2H)

$$CH_3-\overset{\overset{CH_3}{|}}{CH}-CH_2Br$$

(h) $C_6H_{12}O_2$: δ 1.5 (s, 9H) and 2.0 (s, 3H)

$$CH_3-\overset{\overset{O}{\|}}{C}-O-\underset{\underset{CH_3}{|}}{\overset{\overset{CH_3}{|}}{C}}-CH_3$$

(i) $C_7H_{14}O$: δ 0.9 (t, 6H), 1.6 (m, 4H), and 2.4 (t, 4H)

$$CH_3\text{-}CH_2\text{-}CH_2-\overset{\overset{O}{\|}}{C}-CH_2\text{-}CH_2\text{-}CH_3$$

(j) $C_5H_{10}O_2$: δ 1.2 (d, 6H), 2.0 (s, 3H), and 5.0 (septet, 1H)

$$CH_3-\overset{\overset{O}{\|}}{C}-O-\overset{\overset{CH_3}{|}}{CH}-CH_3$$

(k) $C_5H_{11}Br$: δ 1.1 (s, 9H) and 3.2 (s, 2H)

$$CH_3-C(CH_3)_2-CH_2Br$$

(l) $C_7H_{15}Cl$: δ 1.1 (s, 9H) and 1.6 (s, 6H)

$$CH_3-C(CH_3)_2-C(CH_3)_2-Cl$$

CHAPTER 14: INFRARED AND ULTRAVIOLET-VISIBLE SPECTROSCOPY

14.0 OVERVIEW

- Infrared and ultraviolet-visible spectroscopy give information about functional groups. **Infrared spectroscopy provides information about molecular vibrations,** while **ultraviolet-visible spectroscopy provides information about electronic transitions of pi and non-bonding electrons.** ✳
  - The main use of infrared spectroscopy is to determine the presence or absence of certain functional groups in a molecule.
  - The main use of ultraviolet-visible spectroscopy is to study molecules with double bonds, especially conjugated double bonds. Ultraviolet-visible spectroscopy is also used to analyze aromatic molecules and molecules with non-bonding electrons.

14.1 INFRARED SPECTROSCOPY

- The **vibrational infrared** region of electromagnetic radiation has wavelengths that extend from 2.5 μm to 25 μm. For the purposes of **infrared spectroscopy (IR spectroscopy)** it is useful to describe infrared radiation in terms of a unit called the **wavenumber** that is defined by the following:

$$\text{wavenumber} = \frac{10{,}000}{\lambda}$$

where λ is wavelength measured in micrometers

Wavenumbers are reported in the units of $\mathbf{cm^{-1}}$.

- **Atoms** joined by covalent bonds can **vibrate in a quantized fashion**, that is only specific vibrational energy levels are allowed. The **energy of these vibrations** corresponds to that of the **vibrational infrared region**. Therefore, absorption of infrared radiation of the appropriate wavelength results in a vibrationally excited state. ✳
- **Infrared radiation is absorbed** if the **frequency** of radiation **matches** that of an **allowed vibrational transition**. Furthermore, the **bond(s)** undergoing the vibrational transition **must have a dipole moment**, and **absorption** of radiation must result in a **change of** that **dipole moment**. The **greater** this **change in dipole moment**, the **more intense** the **absorption. Infrared active** vibrations meet the above criteria. Note that symmetrical bonds such as those in homonuclear diatomics (Br_2, O_2, etc.) do not absorb infrared radiation, because they do not have a dipole moment. ✳
- **Nonlinear molecules** (molecules with branches, etc.) that have **n** atoms will have **3n-6 allowed fundamental vibrations**. The simplest vibrations are **stretching** and **bending**. Stretching vibrations can be symmetric or asymmetric. Bending vibrations can be relatively complicated motions such as scissoring, rocking, wagging, or twisting. ✳
- Useful information may be obtained from a simplified calculation that is based upon analyzing one bond at a time in a molecule, and thus ignoring the other bonds. The two atoms are assumed to be two masses on a spring, so a form of **Hooke's law** is used:

$$\text{frequency of vibration (in cm}^{-1}) = \frac{1}{2\pi c}\sqrt{\frac{NK}{\mu}} \qquad \mu = \frac{m_1 \times m_2}{m_1 + m_2} = \text{reduced mass}$$

Here **c** is the **speed of light** (2.998×10^{10} cm/sec), **N** is **Avogadro's number** (6.022×10^{23} atoms/mol), **K** is the **force constant of the bond** in dynes/cm, and μ is the **reduced**

mass calculated as shown above using the mass of each atom calculated as grams per atom. The mass in grams per atom is calculated by dividing the mass of the element in grams per mole by Avogadro's number.

- An **infrared spectrophotometer** is the instrument that measures which frequencies of infrared radiation are absorbed by a given sample.
- The output is recorded in the form of a chart called an **infrared spectrum**. The **horizontal axis** of the chart is in wavelength plotted as **wavenumbers (cm^{-1})**. The **vertical axis** of the chart is **percent transmittance**, with **100% transmittance at the top** and **0% transmittance at the bottom**. Remember that **100% transmittance corresponds to no absorption**. Thus, absorption peaks are actually recorded as inverted signals that start at the top, come down to a point corresponding to maximum absorption, then go back to the top. *[Understanding this now will save a lot of confusion when it comes to interpretation of IR spectra.]* ✳
  - IR spectra are recorded from samples that are usually neat if they are liquids, or compacted wafers mixed with KBr if they are solids.
  - Most **IR spectra are very complex**. This is because besides the fundamental vibration absorptions discussed above, there are **overtones** and **coupling peaks**. **Overtones** are **higher frequency harmonic vibrations** of fundamental vibrations that occur at integral multiples. For example, an absorption at 600 cm^{-1} can have weaker overtone peaks at 1200 cm^{-1}, 1800 cm^{-1}, and 2400 cm^{-1}. **Coupling peaks** result from the **coupling of two vibrations** by addition or subtraction in certain allowed combinations.
  - **Characteristic absorptions** for different functional groups are recorded in **correlation tables**. The **intensity** of a particular absorption is referred to as being **strong (s)**, **medium (m)**, or **weak (w)**.
  - In general, most attention is paid to the region from 4000 cm^{-1} to 1600 cm^{-1} of an infrared spectrum, because most functional groups have characteristic absorptions in this area. Absorptions in the 1600 cm^{-1} to 400 cm^{-1} are much more complex and difficult to analyze, so this area is referred to as the **fingerprint region**.

14.2 INTERPRETING INFRARED SPECTRA

- **Alkanes** have **C-H stretching** vibrations near **2900 cm^{-1}** and **methylene bending** at **1450 cm^{-1}**.
- **Alkenes** have a vinylic **C-H stretch** near **3000 cm^{-1}**. The **C=C stretching** near **1600-1660 cm^{-1}** is often **weak** and may not be visible.
- **Alkynes** with a **hydrogen on the triple bonded carbon** have a **C-H stretching** vibration near **3300 cm^{-1}** that is usually sharp and strong. The **C≡C stretching** occurs at **2150 cm^{-1}**.
- **Alcohols** have an **OH stretch** that **depends on the amount of hydrogen bonding** present in the sample. With **no hydrogen bonding** such as when a dilute solution of the alcohol is used, the **OH stretch** is a **sharp peak** of weak intensity at **3625 cm^{-1}**. Normally, alcohol samples show **extensive hydrogen bonding** and a **broad absorption** near **3350 cm^{-1}**. The **C-O stretch** occurs at **1050 cm^{-1}, 1100 cm^{-1}, and 1150 cm^{-1}** for **primary, secondary, and tertiary alcohols**, respectively.
- **Ethers** have a **strong C-O stretch** near **1000 to 1300 cm^{-1}**.
- **Carbonyl**-containing species such as aldehydes, ketones, carboxylic acids, and carboxylic acid derivatives also have **very strong** and **characteristic IR absorptions** in **1630 cm^{-1}** to **1800 cm^{-1}** region. As a result, infrared spectroscopy is especially useful for characterizing these species.

14.3 ULTRAVIOLET-VISIBLE SPECTROSCOPY

- **Absorption of ultraviolet-visible region electromagnetic radiation** corresponds to **transitions** in molecules **between electronic energy levels of pi and nonbonding electrons.** As a result, ultraviolet-visible spectroscopy is most useful for only those molecules with pi or nonbonding electrons. ✱
- The vacuum ultraviolet region refers to wavelengths below 200 nm and this region is not generally used. The near ultraviolet extends from 200 to 400 nm, and the visible region extends from 400 nm (violet light) to 800 nm (red light).
- **Ultraviolet and visible spectra** are **recorded** as a chart of **wavelength** along the **horizontal axis** and **absorbance (A)** on the **vertical axis.** Absorbance is a surprisingly complicated measure, and is defined as:

$$\textbf{Absorbance (A)} = \log \frac{I_0}{I}$$

Here I_0 is the **intensity of radiation incident** on the sample, and **I** is the **intensity of radiation leaving** the sample. ✱

- The **relationship between absorbance, concentration, and length of sample** is known as the **Beer-Lambert law**:

$$\textbf{Absorbance (A)} = \varepsilon c l$$

Here ε is a proportionality constant called the **extinction coefficient, c** is the **concentration of the solute** in mol/liter and **l** is the **length of the cell** in cm.
  - This relationship means that absorbance is linearly dependent on concentration.
  - The value of ε ranges from 0 to 10^6 for different molecules, and absorptions with ε greater than 10^4 are considered high-intensity absorptions.
- **Absorption** of radiation in the **near ultraviolet-visible** spectrum results in a **transition of electrons** from a **lower energy occupied molecular orbital** to a **higher energy unoccupied molecular orbital.** The energy of this radiation can cause $\pi \rightarrow \pi^*$ **transitions** and $\mathbf{n} \rightarrow \pi^*$ **transitions,** involving interactions with pi and nonbonding electrons, respectively. Important examples of these include the excitation of pi electrons in molecules with conjugated pi systems, and excitation of the nonbonding electrons on the oxygen atom of carbonyl groups. On the other hand, **ultraviolet-visible radiation is not of sufficient energy to affect electrons in sigma bonding molecular orbitals.**
- In general, **ultraviolet-visible absorptions are broad,** extending over a range of wavelengths. This is because at room temperature, **many modes of vibration and rotation** are also excited, and **transitions between the various vibrations and rotations are superposed on the electronic absorptions.** These are so closely spaced, that they are not resolved but appear as a broad band. This phenomenon emphasizes that ultraviolet-visible radiation is of higher energy than infrared radiation, and the corresponding absorption of ultraviolet-visible radiation involves higher energy transitions as well.
- The **more double bonds** that are **conjugated** to each other, the **smaller the energy spacing** between the **highest occupied** π **orbital,** and the **lowest unoccupied** π^* **orbital.** Thus, the **greater the number of the conjugated double bonds,** the **lower the wavelength of the observed absorption.**

CHAPTER 14
Solutions to the Problems

Problem 14.1 Which is higher in energy?
(a) Infrared radiation of wavelength 1705 cm^{-1} or of 2800 cm^{-1}?

The higher the wavenumber, the higher the energy. As a result, 2800 cm^{-1} is higher energy than 1705 cm^{-1}.

(b) Microwave radiation of frequency 300 MHz or of 60 Hz?

Energy is directly proportional to frequency, so 300 MHz is higher energy than 60 Hz.

Problem 14.2 Without doing the calculation, which member of each pair do you expect to occur at the higher wavenumber?
(a) C=O or C=C stretching?

We must assume that C=O and C=C have similar force constants, since no actual values were provided. Using this assumption, the C=C bond vibrational absorption is predicted to have a higher wavenumber. The atomic weight of C is lower than O, and the frequency of vibration is predicted to be higher for bonds with atoms of lower atomic weight.

(b) C=O or C-O stretching?

Double bonds have higher force constants than single bonds, so the C=O bond will have a stretching frequency that occurs at higher wavenumber than C-O.

(c) C≡C and C=C stretching?

Triple bonds have higher force constants than double bonds, so the C≡C bond will have a stretching frequency that occurs at higher wavenumber than C=C.

(d) C-H or C-Cl stretching?

Assuming that C-H and C-Cl have similar force constants, then the C-H will have a higher wavenumber because the atomic weight of H is much smaller than Cl.

Problem 14.3 Wavelengths in ultraviolet-visible spectroscopy are commonly expressed in nanometers; wavelengths in infrared spectroscopy are commonly expressed in micrometers. Carry our the following conversions:
(a) 2.5 μm to nanometers. **2.5 μm is equal to 2500 nm.**
(b) 200 nm to micrometers. **200 nm is equal to 0.2 μm.**

Problem 14.4 The visible spectrum of the tetraterpene β-carotene dissolved in hexane shows intense absorption maxima at 463 nm and 494 nm, both in the blue-green region. Because light of these wavelengths is absorbed by β-carotene, we perceive the color of this compound as that of the complement to blue-green, namely, red-orange.

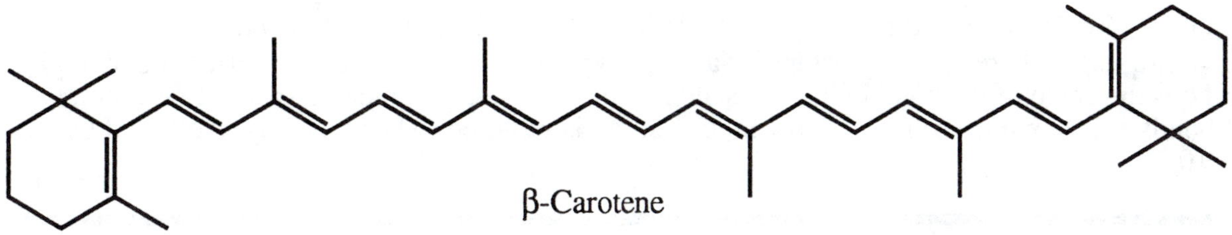

β-Carotene

λ_{max} 463 (log ε 5.10): 494 (log ε 4.77)

(a) β-Carotene contains 40 carbons. The molecular formula of the reference hydrocarbon of 40 carbons is $C_{40}H_{82}$. Given the fact that β-carotene contains 11 pi bonds and two rings, calculate the index of hydrogen deficiency (Section 13.15A) of this molecule, and write its molecular formula.

The index of hydrogen deficiency is 11 + 2 = 13. 13 = (82 - x)/2 so x = 56. Thus, the correct molecular formula is $C_{40}H_{56}$.

(b) Calculate the concentration in milligrams per milliliter of β-carotene that gives an absorbance of 1.8 at λ_{max} 463.

$$c = 1.8/(1.00 \text{ cm})(10^{5.10}) = 1.43 \times 10^{-5} \text{ mol/liter}$$

The molecular weight of β-carotene, $C_{40}H_{56}$, is (12 x 40) + (1 x 56) = 536 g/mol. Concentration in units of milligrams per milliliter is equal to the value of concentration of grams per liter, so:

$$c = (1.43 \times 10^{-5} \text{ mole/liter})(536 \text{ g/mole}) = 7.7 \times 10^{-3} \text{ g/liter} = 7.7 \times 10^{-3} \text{ milligram / milliliter.}$$

Problem 14.5 In molecular mechanics calculations, the energy required to stretch or compress a bond is given by $E_b = k_b(r-r_o)^2$, where k_b is a constant for a given type of bond, r_o is the equilibrium bond length, and r is the length of the bond in the stretched or compressed state. Values of these parameters for some common types of bonds are shown in the table. How much energy is required to stretch each type of bond by 5% of its length, by 10% of its length.

Bond type	k_b (kcal/mol·nm^2)	r_o (nm)	$E_{5\%\ stretch}$ kcal/mol	$E_{10\%\ stretch}$ kcal/mol
C=O	57.9 x 10^3	0.123	**2.19**	**8.78**
C(sp^3)-C(sp^3)	20.0 x 10^3	0.153	**1.17**	**4.68**
C(sp^3)-H	30.0 x 10^3	0.108	**0.875**	**3.50**
O-H	45.1 x 10^3	0.096	**1.04**	**4.15**

The answers in the table were calculated using the equation given in the question. Notice that the values of k_b are correct in this version of the table.

Problem 14.6 In molecular mechanics calculations, the energy required to bend a bond is given by $E_q = k_q(q-q_o)^2$, where k_q is a constant for a given type of bond, q_o is the equilibrium bond angle, and q is the angle of the bond in its bent state. Values of these parameters for some common types of bonds are shown in the table. How much energy is required to bend each type of bond by 5%, by 10%?

Bond type	k_q (kcal/mol·radians^2)	q_o (degrees)	**(radians)**	**$E_{5\%\ bend}$ kcal/mol**	**$E_{10\%\ bend}$ kcal/mol**
C(sp^3)-C=O	85.9	121.6	**2.12**	**0.966**	**3.86**
C(sp^3)-C(sp^3)-C(sp^3)	69.9	109.5	**1.91**	**0.638**	**2.55**
H-C(sp^3)-H	40.0	109.5	**1.91**	**0.365**	**1.46**
C(sp^3)-C(sp^3)-O	49.9	109.5	**1.91**	**0.456**	**1.82**

The answers are included in the table. The angle must be converted from degrees to radians (radians = degrees x 0.01745), since the bending constant is given in radians.

Problem 14.7 Given your answers to the two previous problems, is it easier to bend bonds or to stretch bonds?

As can be seen in the tables, it is easier (takes less energy) to bend bonds than to stretch them.

Infrared Spectra

Problem 14.8 Following are infrared spectra of methylenecyclopentane and 2,3-dimethyl-2-butene. Assign each compound its correct spectrum.

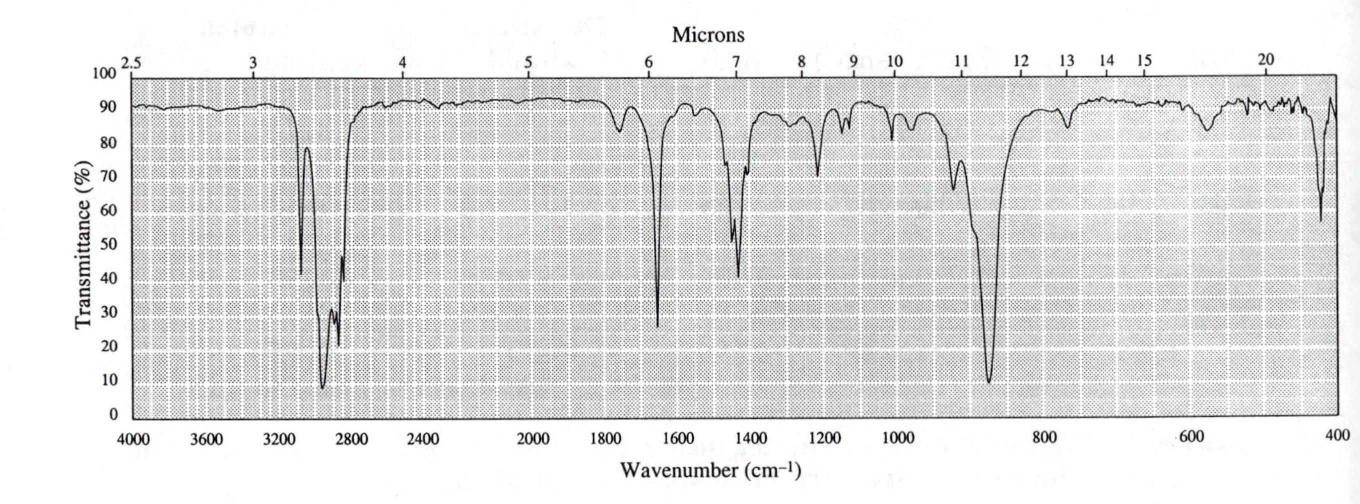

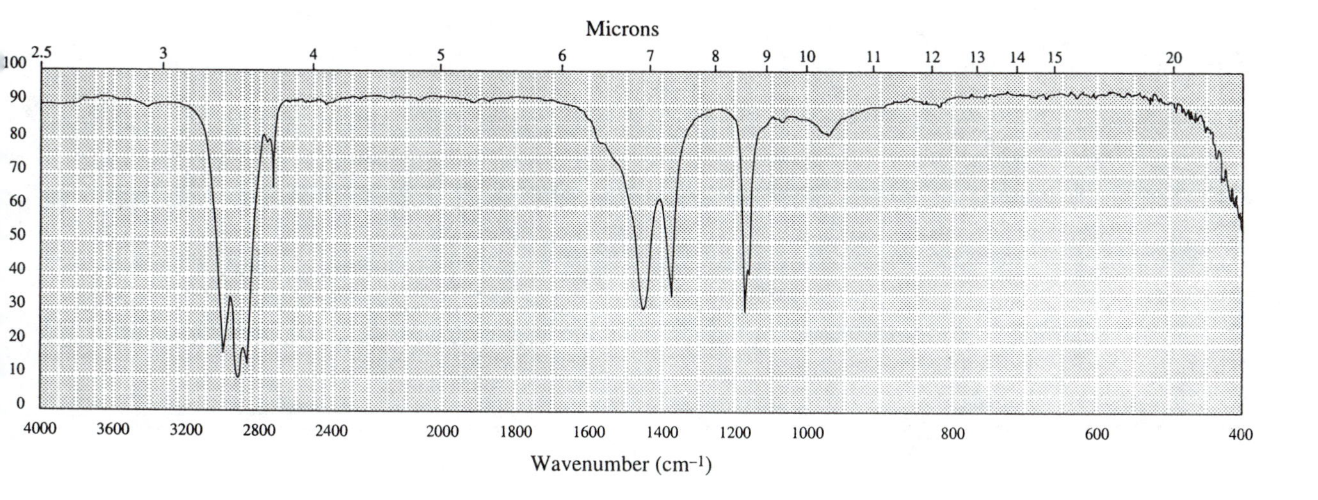

Both molecules have several C-H bonds and thus both spectra have C-H stretches and C-H bending vibrations at 2900 cm^{-1} and 1450 cm^{-1}, respectively. The alkene in methylenecyclopentane is unsymmetrical and therefore has a permanent dipole, so the C=C stretching will have a prominent band at 1654 cm^{-1} as seen in the first spectrum. On the other hand, 2,3-dimethyl-2-butene has a symmetrically substituted carbon-carbon double bond with no permanent dipole, so no C=C stretching should be prominent. In addition, the four methyl groups of 2,3-dimethyl-2-butene should give a prominent CH_3 bending band at 1375 cm^{-1} as is seen in the second spectrum. Thus, the first spectrum corresponds to methylenecyclopentane and the second spectrum corresponds to 2,3-dimethyl-2-butene.

Problem 14.9 Following are infrared spectra of nonane and 1-hexanol. Assign each compound its correct spectrum.

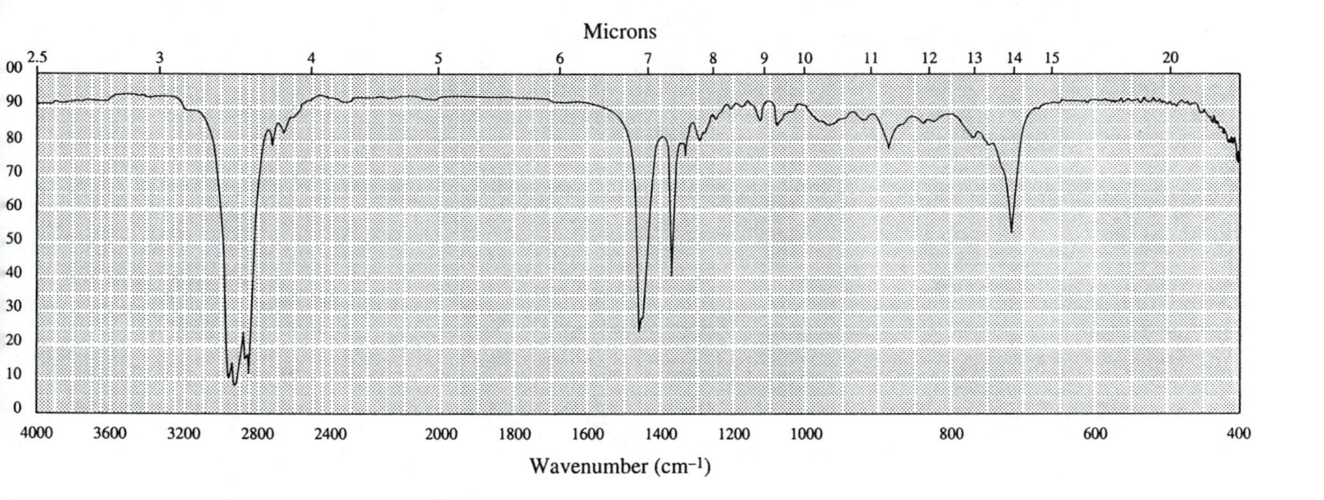

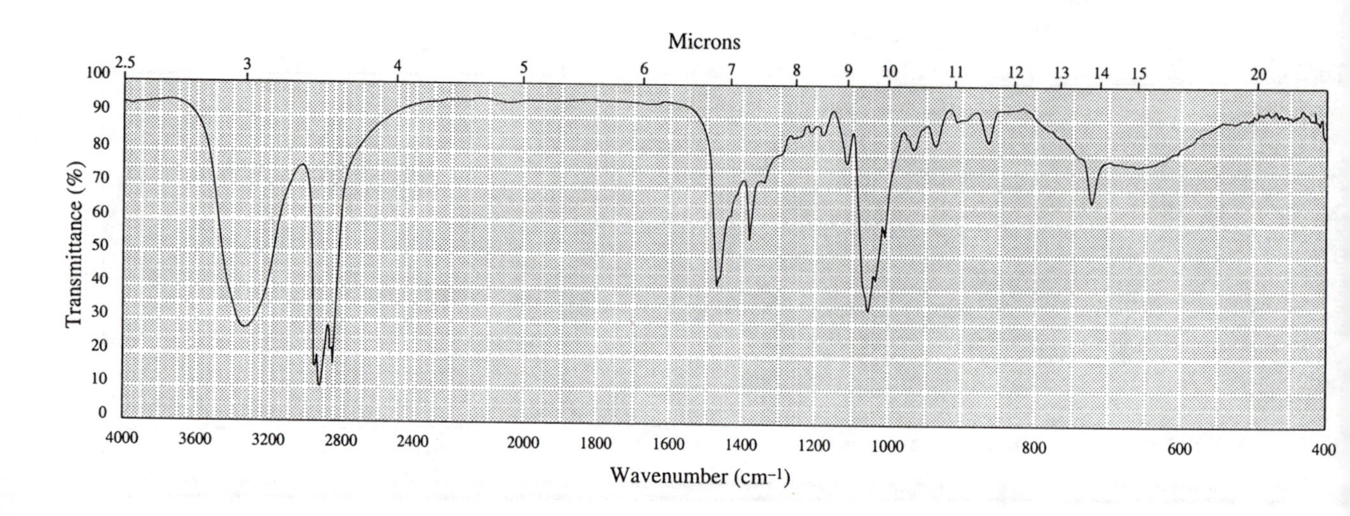

Both compounds have C-H bonds, so both spectra have C-H stretches and bends at 2900 cm^{-1} and 1450 cm^{-1}, respectively. Furthermore, both compounds have methyl groups so both spectra have methyl bending vibrations at 1375 cm^{-1}. On the other hand, the 1-hexanol has an OH group, that will give rise to an O-H and C-O stretching vibrations at 3340 cm^{-1} and 1050 cm^{-1}, respectively. These two features are in the second spectra, so the second spectra must correspond to the 1-hexanol and the first spectra must correspond to nonane.

Problem 14.10 Following are infrared spectra of 2-methyl-1-butanol and *tert*-butyl ethyl ether. Assign each compound its correct spectra.

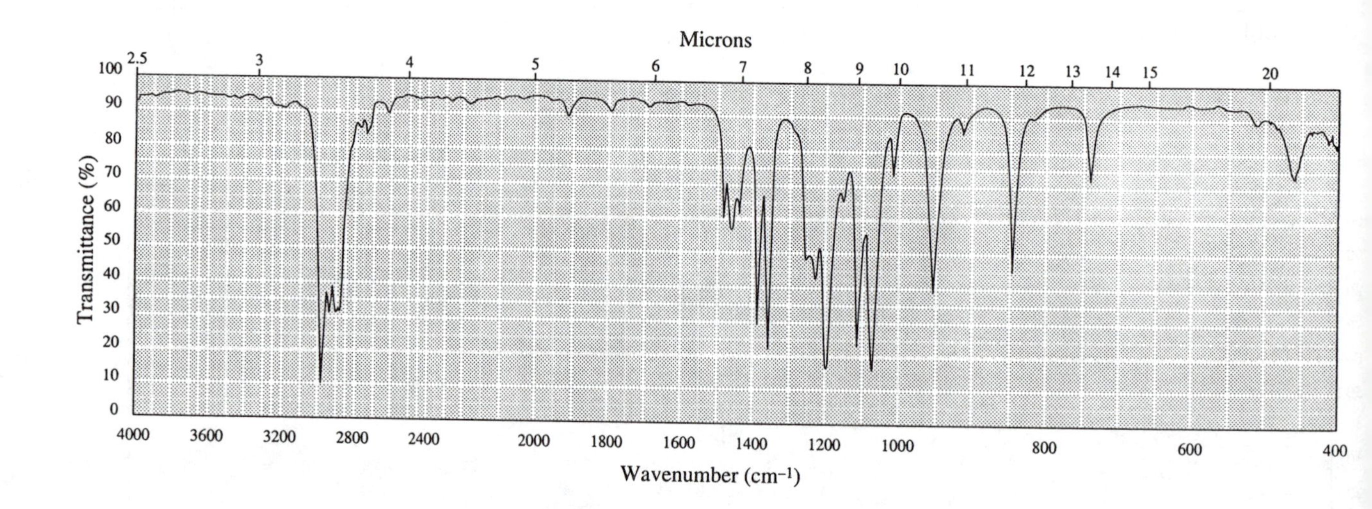

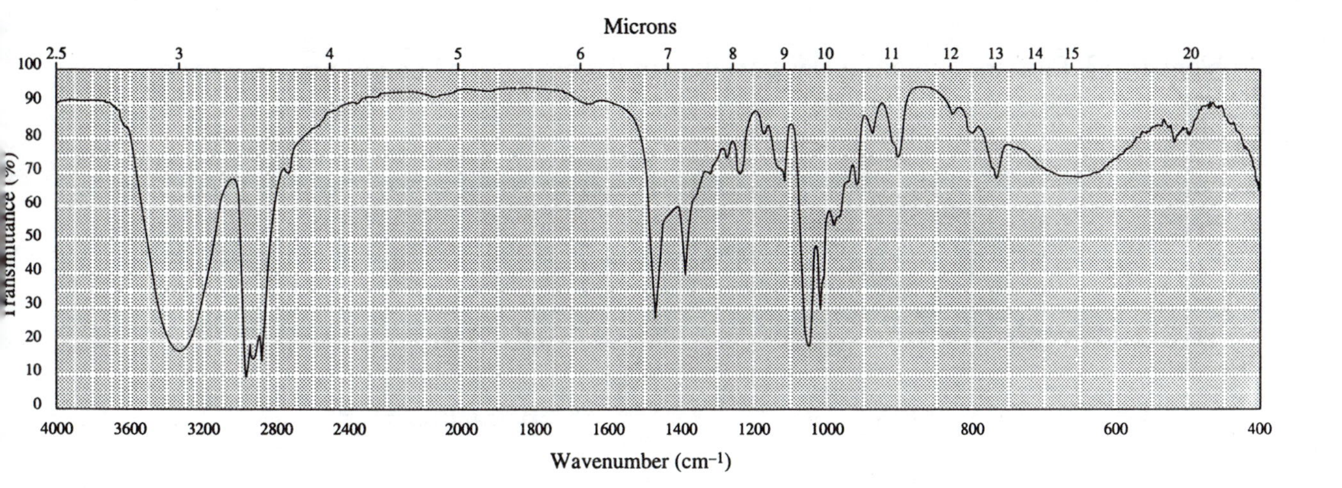

The molecules are extremely similar except for the -OH group present in the 2-methyl-1-butanol. Since the characteristic O-H stretch is present at 3625 cm^{-1} in the second spectrum, this verifies that the second spectrum corresponds to 2-methyl-1-butanol. Therefore, the first spectrum corresponds to *tert*-butyl ethyl ether.

Problem 14.11 The IR intensity of the C≡C stretch in internal alkynes is frequently very weak. Why is this?

For a bond vibration to have a corresponding absorption of high intensity in the IR, that bond must have a dipole moment, and absorption of energy must result in a large change in that dipole moment. For symmetrical alkynes, there is no significant dipole moment associated with the C≡C bond, and no large change in dipole moment associated with bond stretching. Thus, the IR band is very weak.

Ultraviolet-Visible Spectra

Problem 14.12 Show how to distinguish between 1,3-cyclohexadiene and 1,4-cyclohexadiene by ultraviolet spectroscopy.

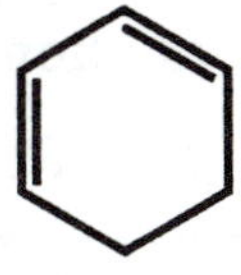

1,3-Cyclohexadiene

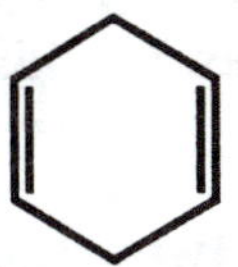

1,4-Cyclohexadiene

In 1,3-cyclohexadiene, the pi bonds are conjugated so the absorption will occur near 217 nm compared with the 165 nm or so absorption that will be observed for the 1,4-cyclohexadiene that has only isolated (unconjugated) pi bonds.

Problem 14.13 Pyridine exhibits a UV transition of the type n → π^* at 270 nm. In this transition, the unshared electron pair is promoted from a nonbonding MO to a pi antibonding MO. What is the effect on this UV peak if pyridine is protonated?

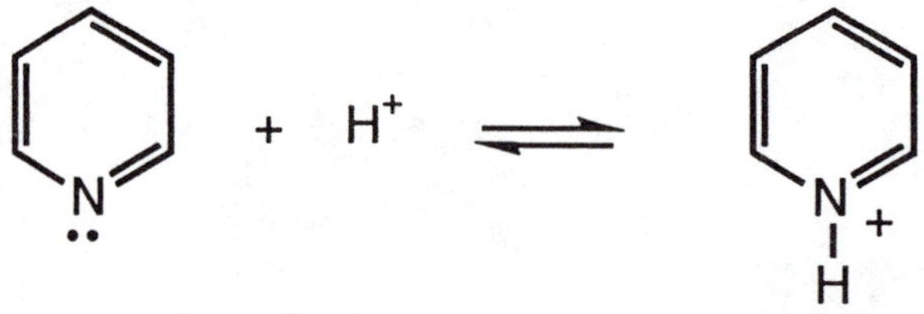

Pyridine + H^+ ⇌ Pyridinium ion

When the pyridium ion is protonated, the lone pair of electrons on nitrogen is tied up in a bond to hydrogen. As a result, they are lower in energy compared with the unprotonated form, so it takes more energy to promote one of these electrons into the π^* orbital. Therefore, protonation shifts the absorbance peak to lower wavelength (higher energy).

Problem 14.14 Many organic reactions occur on absorption of UV radiation. An important photochemical reaction of ketones is the Norrish type II cleavage of ketones:

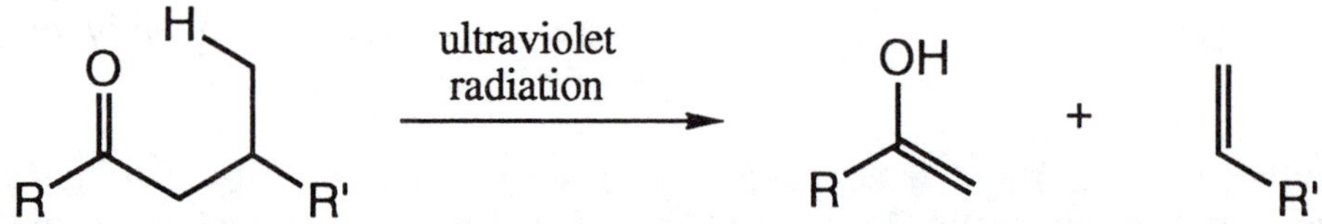

ultraviolet radiation

Compare this photochemical reaction to the McLafferty rearrangement seen in mass spectrometry. Why does absorption of UV light give reactions similar to fragmentations seen in mass spectrometry?

As described in section 12.4D, the McLafferty rearrangement involves a radical on a heteroatom like oxygen abstracting a hydrogen five atoms away. Like the Norrish Type II reaction shown above, the McLafferty rearrangement occurs via a six-member ring transition state. On the other hand, the McLafferty rearrangement involves a radical, and results in the production of three pieces instead of just two.

The similarity of the two reactions is not surprising since in both cases the electrons on the oxygen atom are made more reactive. In the case of the Norrish type II reaction, absorption of light promotes an electron into an antibonding molecular orbital thus making it reactive. In the case of the McLafferty rearrangement, one of the electrons is removed altogether, leading to the resulting radical reaction.

Problem 14.15 The weight of proteins or nucleic acids in solution is commonly determined by UV spectroscopy using the Beer-Lambert law. For example, the ε of double-stranded DNA at 260 nm is 6670. The formula weight of the repeating unit in DNA (650) can be used as the molecular weight. What is the weight of DNA in 2.0 mL of aqueous buffer if the absorbance, measured in a 1-cm cuvette, is 0.75?

According to the Beer-Lambert law:

$$0.75 = (6670)(1)(x)$$

where x equals the unknown concentration of DNA in moles per liter. Rearranging gives:

$$x = 0.75/(6670)(1) = 1.12 \times 10^{-4} \text{ mole/liter}$$

The molecular weight used is that of a single base pair of DNA, namely 650 grams/mole. Furthermore, there is a total of 2.0 mL of solution, so the total weight of double stranded DNA in the sample is:

$$(1.12 \times 10^{-4} \text{ moles/liter})(650 \text{ grams/mole})(2.0 \times 10^{-3} \text{ liter}) =$$

$$\boxed{1.46 \times 10^{-4} \text{ gram}}$$

Problem 14.16 A sample of adenosine triphosphate (ATP)(MW 507, ε = 14,700 at 257 nm) is dissolved in 5.0 mL of buffer. A 250 μL aliquot is removed and placed in a 1-cm cuvette with sufficient buffer to give a total volume of 2.0 mL. The absorbance of the sample at 257 nm is 1.15. Calculate the weight of ATP in the original 5.0 mL sample?

The concentration of sample in the cuvette can be determined by using the Beer-Lambert law as follows:

$$1.15 = (14{,}700)(1.0 \text{ cm})(x \text{ moles/liter})$$

$$x = 1.15/14{,}700 = 7.82 \times 10^{-5} \text{ mole/liter}$$

The sample measured in the cuvette is actually diluted 0.250 in 2.0 or 1 in 8 compared to the original unknown sample, so the concentration of ATP in the original unknown sample is:

$$(8)(7.82 \times 10^{-5}) = 6.26 \times 10^{-4} \text{ mole/liter}$$

Since there is a total of 5.0 mL or 5 x 10^{-3} liter and ATP has a MW = 507 grams/mole then the weight of ATP in the original sample can be calculated as:

$$6.26 \times 10^{-4} \text{ mole/liter} = (x \text{ grams})/(506 \text{ grams/mole})(5 \times 10^{-3} \text{ liters})$$

$$x = (6.26 \times 10^{-4})(5 \times 10^{-3} \text{ liters})(506) =$$

$$\boxed{1.59 \times 10^{-3} \text{ grams} = 1.59 \text{ mg}}$$

Problem 14.17 Biochemical molecules are frequently sold by optical density (OD) units, where one OD unit is the amount of compound that gives an absorbance of 1.0 at its UV maximum in 1.0 mL of solvent in a 1 cm cuvette. If the cost of 10.0 OD units of a DNA polymer, $\varepsilon = 6600$ at 262 nm, is $51, what is the cost per gram of this biochemical?

According to the information given, 10.0 OD units is equal to the following amount of compound in 1.0 mL (1×10^{-3} liters). Recall from problem 14.15 that the molecular weight used for DNA is 650 g/mol:

$$10.0 = (6600)(1\ \text{cm})(1/650\ \text{g/mol})(x\ \text{grams}/1 \times 10^{-3}\ \text{liters})$$
$$x = (650\ \text{g/mol})(10.0)(1 \times 10^{-3}\ \text{liters})/(6600) = 9.8 \times 10^{-4}\ \text{grams}$$

Converting to dollars per gram gives:

$$(x\ \text{dollars})/\text{gram} = (\$51)/(9.8 \times 10^{-4}\ \text{gram})$$
$$\boxed{= \$52{,}000\ !}$$

Problem 14.18 The Beer-Lambert law applies to IR spectroscopy as well as UV. Whereas ultraviolet spectra are a plot of absorbance (A) versus wavelength, IR spectra are a plot of percentage transmittance (%T) versus frequency. Absorbance and percentage transmittance are related in the following way:

$$\text{Absorbance (A)} = \log \frac{T_o}{T}$$

Where T_o is the baseline (100%) transmittance and T = transmittance of a peak.
(a) What is the absorbance of a peak in an IR spectrum with 10% transmittance?

$$\textbf{Absorbance} = \log (100\%/10\%) = \log (10) = 1.0$$

(b) If the concentration of this sample is halved, how does the absorbance and percentage transmittance change?

If the concentration of the sample is halved, the absorbance is halved as well. On the other hand, the change in transmittance is harder to calculate and is given by:

$$0.5 = \log (T_o/T) \text{ so } 10.0^{0.5} = T_o/T = 3.16$$
$$\text{thus } T = T_o/3.16 \text{ so it is decreased by } 31\%$$

Combined Spectral Problems

Problem 14.19 Compound A, a hydrocarbon, bp 81° C, decolorizes a solution of bromine in carbon tetrachloride. Following are its mass spectrum, ^{1}H-NMR spectrum, and infrared spectrum. Compound A is transparent to ultraviolet-visible radiation.

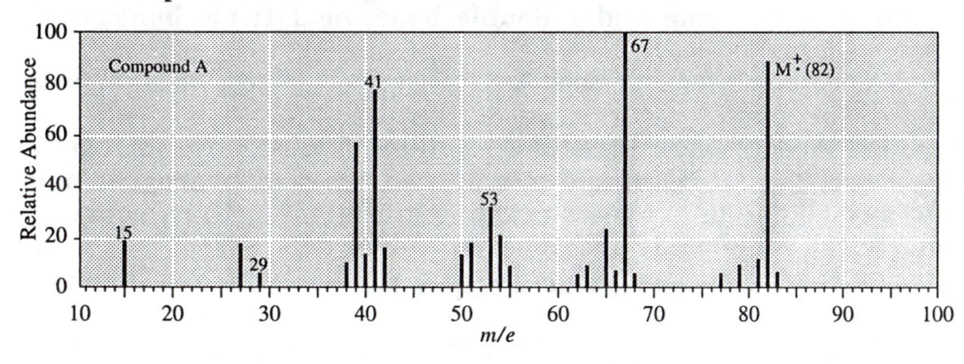

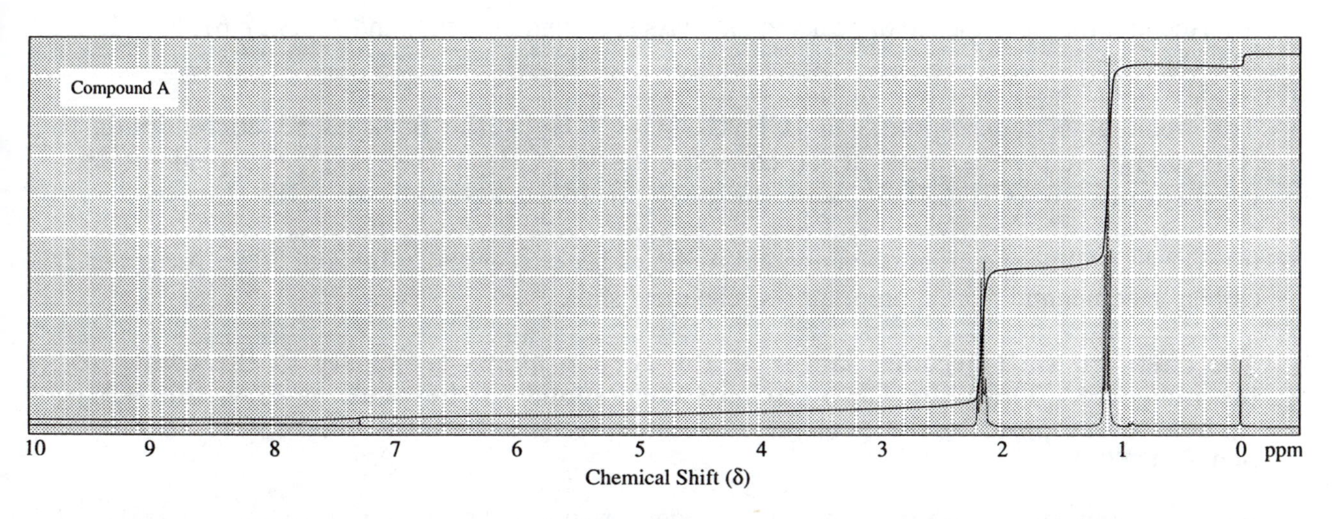

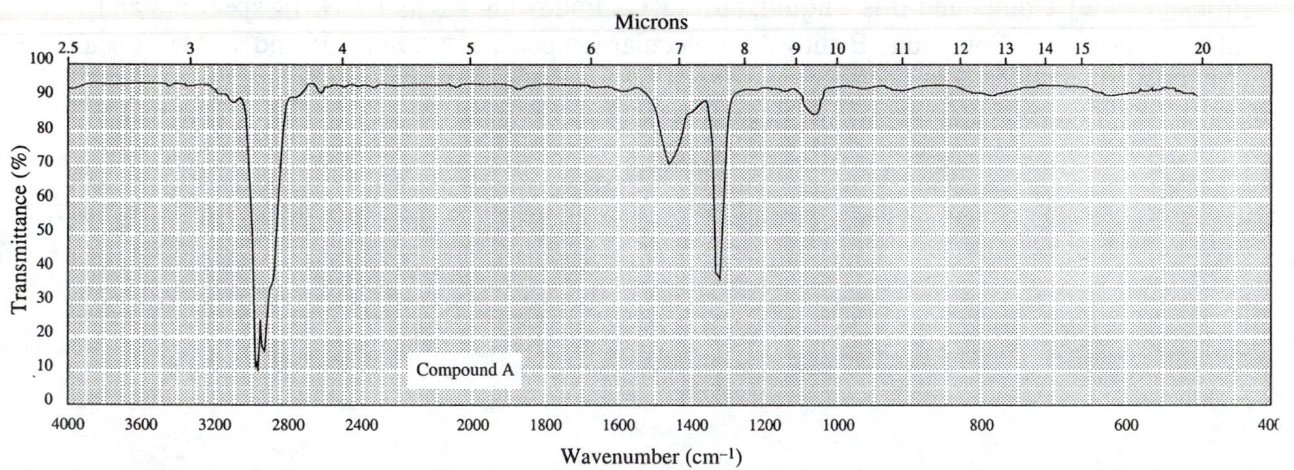

(a) What is the molecular formula of compound A?

Based on the mass spectrum, compound A has a MW of 82. This is only possible for a hydrocarbon with a molecular formula of C_6H_{10}.

(b) From its molecular formula, calculate the index of hydrogen deficiency of compound A. How many rings are possible for this compound? How many double bonds? How many triple bonds?

The index of hydrogen deficiency for compound A is 2. Compound A could have 2 rings, 2 double bonds, 1 ring and 1 double bond, or 1 triple bond.

(c) Propose a structural formula for compound A consistent with the spectral information.

Because compound A decolorizes bromine solution, it must have at least 1 pi bond. Furthermore, the ^{1}H-NMR shows only two types of hydrogens and the IR does not show any alkene or alkyne stretching vibrations. Thus, compound B is a highly symmetrical compound, and must be 3-hexyne.

$$CH_3CH_2C{\equiv}CCH_2CH_3$$

3-hexyne

(d) Account for the presence of peaks in the mass spectrum of compound A at *m/e* 67, 53, 41, 29, and 15.

The peak at 67 corresponds to $[CH_3CH_2C{\equiv}CCH_2]^+$, the peak at 53 corresponds to $[CH_3CH_2C{\equiv}C]^+$, the peak at 41 corresponds to $[CH_2{=}CH{-}CH_2]^+$, the peak at 29 corresponds to $[CH_3CH_2]^+$, and the peak at 15 corresponds to $[CH_3]^+$.

(e) The ^{13}C-NMR spectrum of compound A shows peaks at δ 12.4, 14.4, and 80.9. Assign each carbon in compound A its appropriate ^{13}C chemical shift.

The chemical shift becomes larger as the carbon atoms get closer to the center of the molecule. Thus, the CH_3- carbons correspond to the peak at 12.4, the $-CH_2-$ carbons correspond to the peak at 14.4, and the -C≡ carbons correspond to the peak at 80.9.

Problem 14.20 Compound B is a liquid, bp 122°C. Following are its ^{1}H-NMR spectrum and infrared spectrum. Compound B shows a molecular ion peak, M^+ at *m/e* 136 and an $M^+{+}2$ peak of almost equal intensity at *m/e* 138.

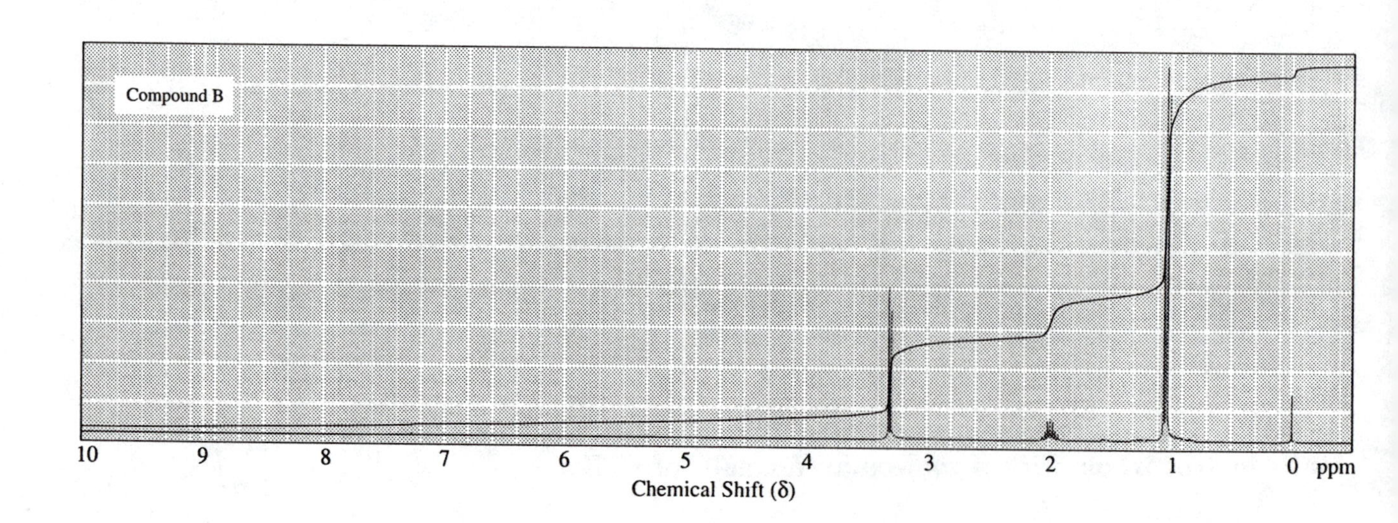

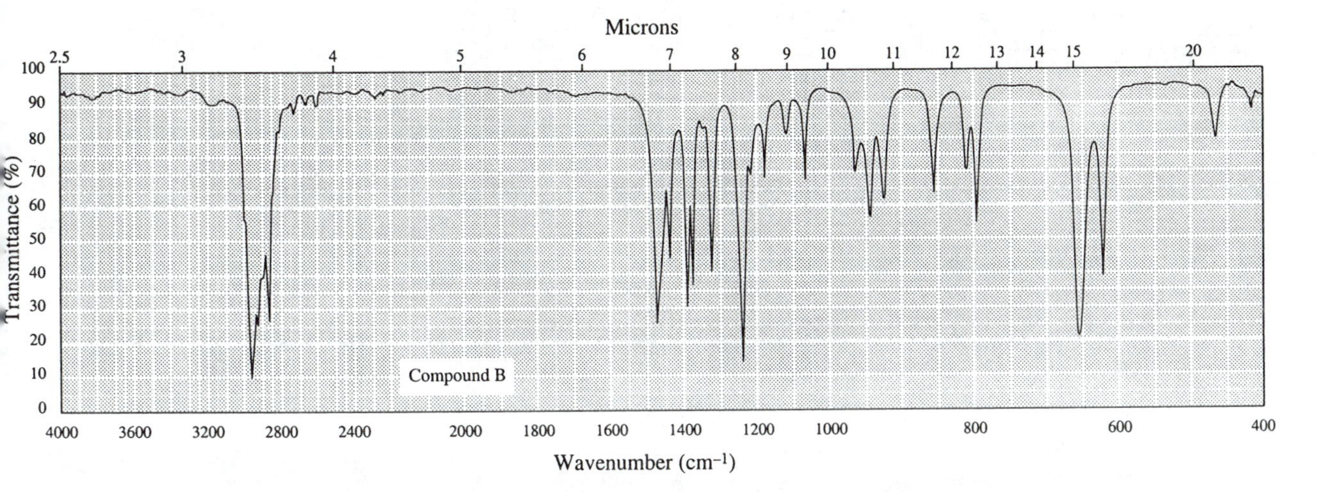

(a) From this information, deduce the structural formula of compound B.

From the mass spectrum, the MW is 136. The IR shows no alkenes or alcohols. The ^{1}H-NMR shows that there are three types of hydrogens in a 2:1:6 ratio. These ratios along with the observed splitting pattern correspond to a $(CH_3)_2CH$-CH_2- group. This piece has a molecular weight of 57 leaving 79 in the formula. Since bromine corresponds to 79, this means the spectra corresponds to isobutyl bromide.

CH_3
|
CH
CH_3 CH_2Br

isobutyl bromide

(b) Account for the presence of peaks in its mass spectrum at *m/e* 138, 136, 123, 121, 43, and 41.

The peak at 138 corresponds to $C_4H_9{}^{81}Br$, the peak at 136 corresponds to $C_4H_9{}^{79}Br$, the peak at 123 corresponds to loss of a methyl group $[CH_3\text{-}CH\text{-}CH_2\text{-}Br]^+$, the peak at 121 corresponds to $[CH_2{=}C\text{-}CH_2\text{-}Br]^+$, the peak at 43 corresponds to $[CH_3\text{-}CH\text{-}CH_3)]^+$, and the peak at 41 corresponds to $[CH_2{=}CH\text{-}CH_2]^+$.

(c) The ^{13}C-NMR spectrum of compound B shows peaks at δ 21.0, 30.47, and 42.45. Assign each carbon in compound B its appropriate ^{13}C chemical shift.

The chemical shift of the carbon atoms increase as they get closer to the bromine atom. Thus the methyl groups correspond to the peak at 21.0, the CH carbon atom corresponds to 30.47, and the -CH_2-Br carbon atom corresponds to the peak at 42.45.

Problem 14.21 Compound C, C_4H_8O is a liquid, bp 97°C. Following are its ^{1}H-NMR spectrum and infrared spectrum.

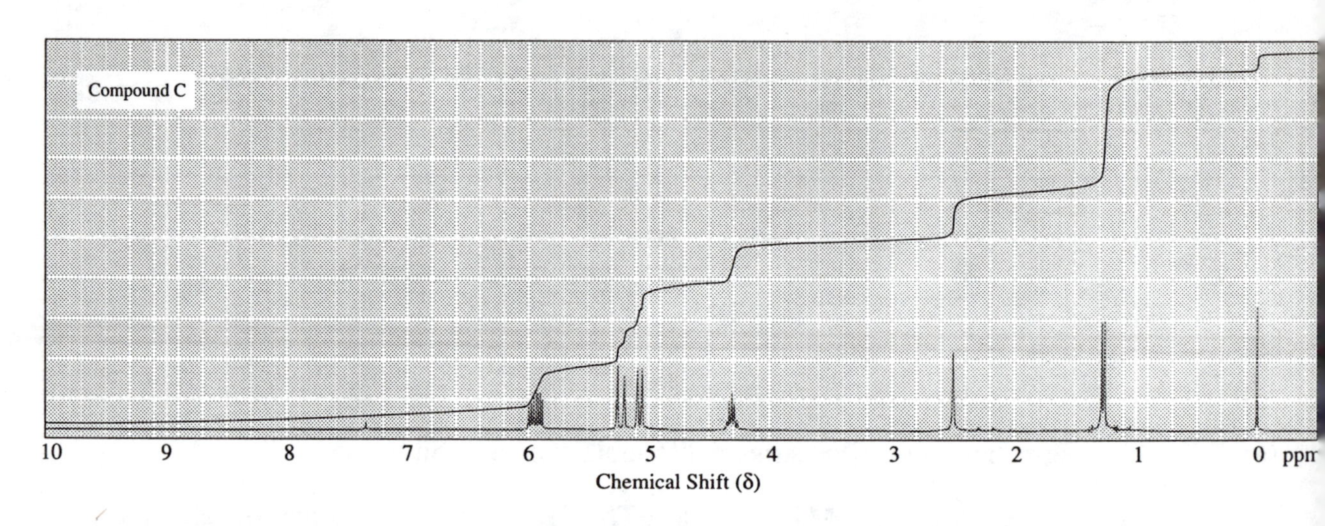

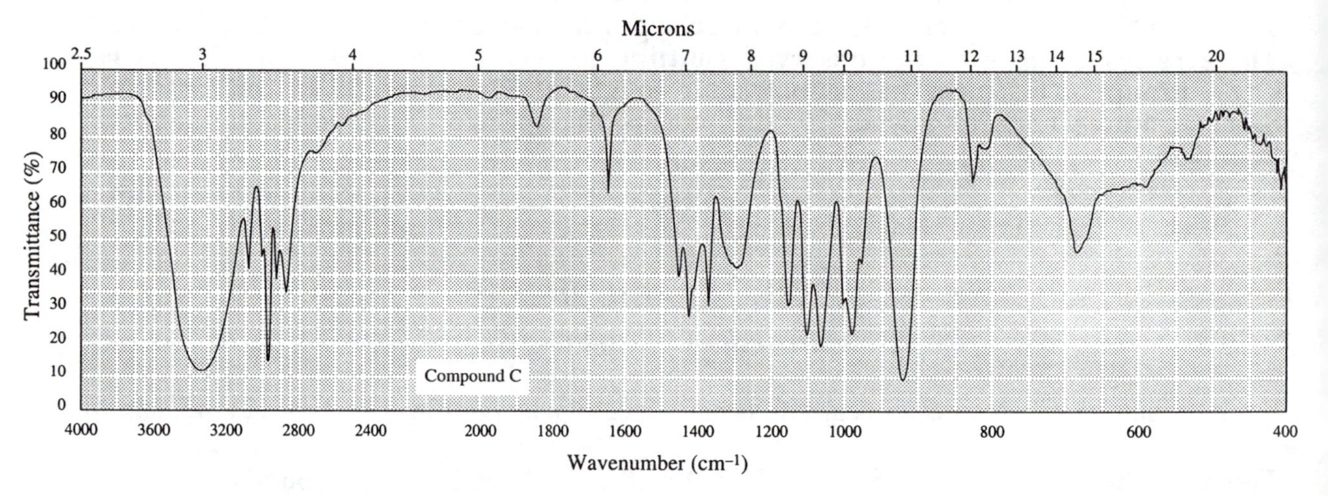

(a) What is the index of hydrogen deficiency of compound C? How many rings or double bonds can it be contain?

The index of hydrogen deficiency is 1, so compound C must contain 1 ring or double bond.

(b) What information can you learn from the infrared spectrum about the oxygen-containing functional group?

The broad peak at 3400 cm^{-1} indicates the oxygen atom is contained in an -OH group.

(c) What information can you learn from the ^{1}H-NMR spectrum about the presence or absence of vinylic hydrogens (hydrogens on a carbon-carbon double bond)?

The multiplet at 5.9 and two doublets near 5.1 are classic vinylic hydrogen peaks, indicating the presence of a double bond in the molecule. Since there are a total

of three of them, the double bond must be on the end of the molecule as opposed to being in the middle of a chain.

(d) Propose a structural formula for compound C consistent with the spectral and chemical information.

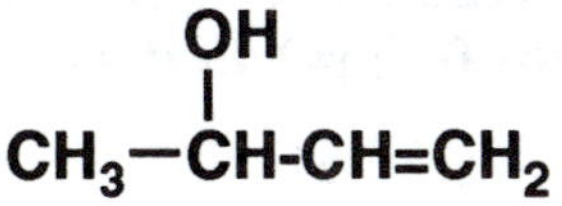

$$CH_3-\underset{\displaystyle |}{\overset{\displaystyle OH}{C}}H-CH=CH_2$$

3-butenol

This structure is confirmed by the ^{1}H-NMR since there is a methyl peak integrating to 3 hydrogens at 1.3. Since this peak is a doublet, it must be adjacent to a single hydrogen, and these considerations limit the possibilities to the structure shown above. The singlet integrating to 1 hydrogen at 2.5 is the alcohol hydrogen.

(e) Account for the presence of peaks in the mass spectrum of compound C at m/e 71, 57, 45, 27, and 15.

Consistent with other alcohols, the peak at 71 is assigned as:

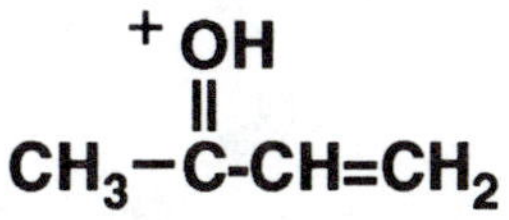

$$CH_3-\overset{\overset{+}{O}H}{\overset{\|}{C}}-CH=CH_2$$

The peak at 57 results from loss of a methyl group:

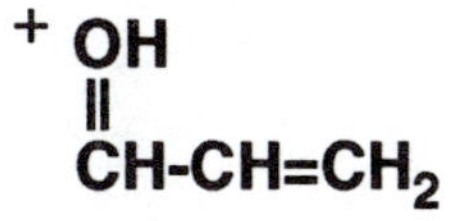

$$\overset{\overset{+}{O}H}{\overset{\|}{C}}H-CH=CH_2$$

The peak at 45 results from cleavage on the other side of the alcohol:

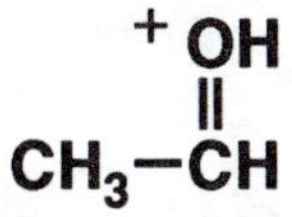

$$CH_3-\overset{\overset{+}{O}H}{\overset{\|}{C}}H$$

The peak at 27 is the other part of that fragmentation, namely $[CH_2=CH]^+$ and the peak at 15 is the methyl group $[CH_3]^+$.

(f) Account for the splitting pattern of single hydrogens as δ 5.1, 5.3 and 5.8. (*Hint:* Review the ^{1}H-NMR of vinyl acetate (Figure 13.17).

The two terminal hydrogens on the alkene are in different environments, so they both couple to the =CH- hydrogen. This same =CH- hydrogen is also split by the adjacent hydrogen on the alcoholic carbon atom. Thus, the peak at 5.8, the =CH- hydrogen, is actually a doublet of doublet of doublets. The coupling between the two terminal hydrogens is very small and not observed in this spectrum. Thus, each of the terminal hydrogens is a doublet, split by the =CH- hydrogen.

(g) The ^{13}C-NMR spectrum of compound C shows peaks at δ 23.1, 68.92, 113.5, and 143.3. Assign these peaks to the appropriate carbon on compound C.

The peak at 23.1 corresponds to the methyl carbon atom, the peak as 68.92 corresponds to the carbon with the -OH group attached, the peak at 113.5 corresponds to the terminal sp^2 carbon ($=CH_2$), and the peak at 143.3 corresponds to the sp^2 carbon atom (-CH=) that is adjacent to the carbon with the -OH group attached.

CHAPTER 15: AROMATICS 1: BENZENE AND ITS DERIVATIVES

SUMMARY OF REACTIONS

Starting Material ↓ / Product →	2-Allylphenols	Aryl Alkyl Ethers	Benzoic Acids	Cyclohexanes	1-Haloalkylbenzenes	Hydroquinones	Phenols, Alkyl Halides	Phenoxides	Phthalic Acids	Quinones
Allyl Phenyl Ethers	**15A** 15.5E*									
Alkylbenzenes			**15B** 15.6A		**15C** 15.6B					
Aryl Alkyl Ethers							**15D** 15.5D			
Benzenes				**15E** 15.1C						
Naphthalene									**15F** 15.6A	
Phenols								**15G** 15.5C		
Phenols,Catechols Hydroquinones										**15H** 15.5F
Phenoxides Alkyl Halides		**15I** 15.5D								
Quinones						**15J** 15.5F				

*Section in book that describes reaction.

REACTION 15A: CLAISEN REARRANGEMENT OF ALLYL PHENYL ETHERS (Section 15.5E)

O—C—C'=C" → (Heat) → OH, C"—C'=C

- The **Claisen rearrangement** transforms allyl phenyl ethers to 2-allylphenols, and involves an **intramolecular rearrangement**. In this case, the bond making and bond breaking are simultaneous to give a substituted cyclohexadienone intermediate that undergoes keto-enol tautomerism to give the final product. ✱

REACTION 15B: OXIDATION AT A BENZYLIC POSITION (Section 15.6A)

$$\text{PhC(H)} \xrightarrow[\text{or } Na_2Cr_2O_7/H_2SO_4]{\text{1) } KMnO_4/HO^- \text{ 2) } H^+} \text{PhC(=O)OH}$$

- Compounds with at least **one benzylic hydrogen** react with $KMnO_4$ in aqueous base or $Na_2Cr_2O_7$ in aqueous sulfuric acid to produce **benzoic acid**. Notice that other groups attached to the benzylic carbon atom are removed in the process. ✱

REACTION 15C: HALOGENATION AT A BENZYLIC POSITION (Section 15.6B)

$$\text{PhC(H)} \xrightarrow[\text{heat or light}]{X_2} \text{PhC(X)}$$

- **Benzylic hydrogens** can be **replaced by bromine or chlorine** in the presence of light or heat. ✱
- In compounds such as toluene, more that one of the benzylic hydrogens can be replaced when excess halogen in used.
- The reaction occurs via a radical chain mechanism initiated when the X_2 is converted to two X· radicals. An X· radical then abstracts the benzylic hydrogen to create a relatively stable benzylic radical that reacts with another molecule of X_2 to give the halogenated product and a new X· radical that continues the chain reaction.
- In molecules with alkyl groups attached to the benzylic carbon, the reaction is selective for replacement of the benzylic hydrogen because the benzylic radical is more stable than the other possible radicals.
- Bromine is more selective than chlorine in these reactions, because the transition state for bromination is reached later. The later transition state means more radical character, so the relative stability of the benzylic radical becomes more important.

REACTION 15D: CLEAVAGE OF ALKYL ARYL ETHERS WITH H-X (Section 15.5D)

$$\text{Ph–O–C} \xrightarrow{\text{H-X}} \text{Ph–OH} + \text{X–C}$$

- **Alkyl aryl ethers** are **cleaved by H-X** to form an alkyl halide and phenol. ✱

REACTION 15E: HYDROGENATION OF BENZENE WITH Ni AND HIGH PRESSURES OF H_2 (Section 15.1C)

Ni

H_2/200-300 atm

- **Benzene** can be converted into **cyclohexane** by **hydrogenation** over a **Ni catalyst.** Because benzene is aromatic, it reacts very slowly under normal conditions. Hydrogen pressures of several hundred atmospheres are usually used. ✱

REACTION 15F: OXIDATION OF NAPHTHALENE TO PHTHALIC ACID (Section 15.6A)

+ excess O_2 —V_2O_5 / heat→ CO_2H CO_2H + CO_2

- **Naphthalene** is oxidized to **phthalic acid** by molecular oxygen in the presence of a **vanadium(V) oxide catalyst.** Polynuclear aromatic hydrocarbons like napthalene are much easier to oxidize than simple benzene. ✱

REACTION 15G: ACID-BASE REACTIONS OF PHENOLS (Section 15.5C)

OH —NaOH→ O^- Na^+ + H_2O

- **Phenols** are **weak acids** that react with strong bases such as NaOH to form water-soluble salts. ✱

REACTION 15H: OXIDATION OF PHENOLS, CATECHOLS, AND HYDROQUINONES TO QUINONES (Section 15.5F)

OH —$K_2Cr_2O_7$→ O O

- **Phenols** are **susceptible to oxidation** because of the electron-donating -OH group on the ring. As a result, they react with strong oxidizing agents like potassium dichromate to give 1,4-benzoquinones. ✱
- **Catechols** (1,2-benzenediols) and **hydroquinones** (1,4-benzenediols) are also oxidized to quinones under these conditions.

REACTION 15I: FORMATION OF ALKYL ARYL ETHERS FROM PHENOXIDES AND ALKYL HALIDES (Section 15.5D)

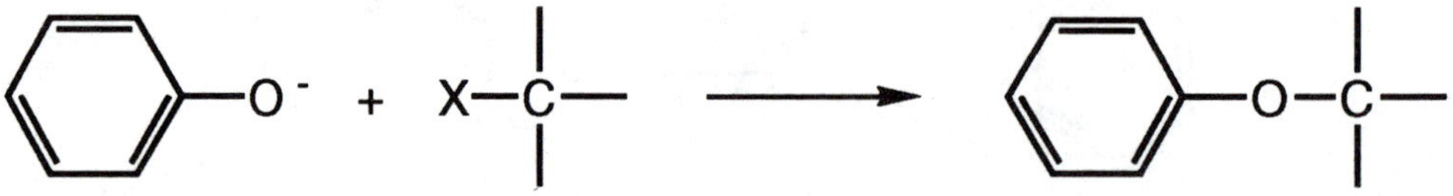

- The **Williamson ether synthesis** can be used to **prepare** certain **alkyl aryl ethers**. In these cases the weakly nucleophilic phenol must be converted to the much more nucleophilic phenoxide ion that then reacts with an alkyl halide. Ether synthesis is often accomplished using phase transfer catalysis. ✱
- Note that the reverse reagents cannot be used, since aryl halides are not reactive enough to nucleophilic attack.

REACTION 15J: REDUCTION OF QUINONES TO HYDROQUINONES (Section 15.5F)

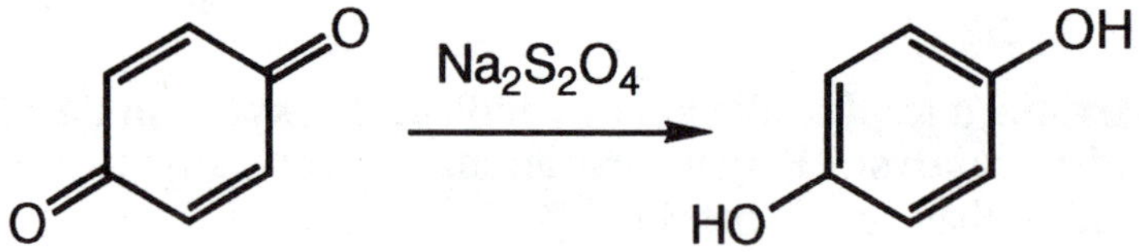

$Na_2S_2O_4$

- **Quinones** can be **reduced to hydroquinones** by reducing agents such as sodium dithionite in neutral or alkaline solutions. ✱

SUMMARY OF IMPORTANT CONCEPTS

15.0 OVERVIEW

- Benzene and its derivatives have marked distinctions from other types of molecules, and they are broadly classified as **aromatic**. They have certain similar physical properties as well as a remarkable stability that makes them unreactive toward reagents that attack other species such as alkenes and alkynes. **Aromaticity** is the term used to describe this special stability of benzene and its derivatives, and the term **arene** refers to aromatic hydrocarbons. ✱

15.1 THE STRUCTURE OF BENZENE

- An important development in the identification of benzene's structure was Kekulé's proposal that benzene is composed of six carbon atoms in a ring, with one hydrogen atom attached to each carbon. ✱
- The six carbon atoms of the ring are equivalent, and the carbon-carbon bond lengths are all intermediate between a single and double bond. Thus, it is not accurate to think of benzene as simply having alternating single and double bonds that are static, because this would predict alternating longer and shorter carbon-carbon bonds.
- **Valence bond description of benzene**. With the advent of more sophisticated models of chemical bonding, especially the concepts of **hybridization of atomic orbitals** and **resonance**, a more accurate picture of the bonding in benzene was proposed. ✱
  - Each carbon atom of the ring is sp^2 hybridized.
  - Each carbon atom of the ring makes sigma bonds by sp^2-sp^2 overlap with the two adjacent carbon atoms and sp^2-1s overlap with a hydrogen atom.
  - Each carbon atom also has a single unhybridized 2p orbital containing one electron. These six 2p orbitals overlap to form a continuous pi system that extends over all six carbon atoms.

The electron density of this pi system thus lies in two bagel-shaped regions, one above and one below the plane of the ring.

- Benzene can be represented as a resonance hybrid composed of two resonance forms in which the locations of the double bonds are reversed. Alternatively, benzene is denoted as a hexagon with a circle drawn on the inside.

- **Resonance energy** is the difference in energy between a resonance hybrid and the most stable hypothetical contributing structure in which electron density is localized on particular atoms and on particular bonds. The resonance energy for benzene is large, namely 36.0 kcal/mol. What this means is that the pi system of benzene is extremely stable, and dramatically less reactive than would be expected for a normal alkene under conditions like catalytic hydrogenation (reaction 15E, Section 15.1C09,).

• **Molecular orbital description of benzene.** According to the molecular orbital approach, the six 2p orbitals give a set of six molecular orbitals. These molecular orbitals are arranged in a 1:2:2:1 pattern with respect to energy. The six pi electrons fill the three pi bonding molecular orbitals, all of which are at lower energy than the six isolated 2p orbitals. Thus, the molecular orbital approach also explains the extra stability of benzene and its derivatives. ✱

• To be **aromatic** like benzene a molecule must be cyclic, planar, have one 2p orbital on each atom of the ring, and contain **(4n + 2)** pi electrons, where **n** is a positive integer (0, 1, 2, 3, 4, 5, . . .). That is, a total of 2, 6, 10, 14, . . . pi electrons. These criteria are referred to as **Hückel's criteria for aromaticity**, named after the chemist who first described them. Note that all of the atoms in an aromatic ring must be sp^2 hybridized, so there cannot be any $-CH_2-$ groups on the ring. ✱

• An **antiaromatic hydrocarbon** is the same as an aromatic hydrocarbon described above in that they are cyclic, planar, and have one 2p orbital on each atom of the ring. On the other hand, antiaromatic hydrocarbons are different because they have **4n** pi electrons. ✱

- Unlike aromatic hydrocarbons that have extremely stable pi systems, antiaromatic hydrocarbons are less stable than an acyclic analog with the same number of pi electrons.
- This instability can be explained for antiaromatic hydrocarbons such as cyclobutadiene by using molecular orbital theory. The four 2p orbitals of the pi system form four molecular orbitals in a 1:2:1 pattern. The four pi electrons fill these orbitals to give one filled bonding pi molecular orbital and two half-filled degenerate nonbonding molecular orbitals. It is the presence of the two unpaired electrons that makes cyclobutadiene so reactive and unstable relative to aromatic hydrocarbons. See Figure 15.6 for the molecular orbital energy diagram of cyclobutadiene.
- Larger antiaromatic structures like cyclooctatetraene are only stable in a non-planar geometry. In this case, there are alternating double and single bonds, and as a result there are two different carbon-carbon bonds lengths observed by experiment corresponding to single and double bonds, respectively.

• In order to predict the pattern of molecular orbitals to be found on a molecular orbital energy diagram, it is helpful to use the **inscribed polygon method**. Here the shape of the polygon being analyzed (for example a hexagon for benzene) is drawn in a ring with one point down, and the relative energies of the molecular orbitals are indicated by the points of the polygon that touch the circle. A horizontal line is drawn through the center of the figure. Bonding molecular orbitals are below the line, nonbonding molecular orbitals (if any) are on the line, and antibonding molecular orbitals are above the line.

- An **annulene** is a planar, cyclic hydrocarbon with a continuous overlapping pi system. Thus, cyclobutadiene and benzene are annulenes, namely [4]annulene and [6]annulene, respectively. Annulenes can be much larger, such as [14]annulene and [18]annulene. Annulenes that have (4n+2) pi electrons are aromatic as long as the ring can accommodate a planar structure. For example, [14] annulene and [18]annulene are aromatic. [10]annulene would be aromatic if the relatively small ring could adopt a planar geometry. ✱
- A **heterocyclic compound** is one that contains more than one kind of atom in a ring. Certain heterocycles can be aromatic if the Hückel criteria are met. Nature is filled with aromatic heterocycles such as indoles, purines, and pyrimidines.
- An important parameter to keep track of in aromatic heterocycles is whether lone pairs of electrons are part of the aromatic pi system or not. ✱
  - For example, in pyridine (C_5H_5N) the lone pair of electrons on nitrogen is perpendicular to the six 2p orbitals of the aromatic 6 pi electron system. Thus, the lone pair of electrons on the nitrogen is not part of the aromatic pi system, and is free to take part in interactions with other species.
  - On the other hand, in compounds such as pyrrole (C_4H_5N), the lone pair electrons on nitrogen is part of the pi system to allow for a total of 6 pi electrons and aromaticity. Thus, this lone pair of electrons is not as available to take part in interactions with other species. *[This is an important but subtle point, so drawing the appropriate structures may be helpful.]*
- Charged ring systems can also be aromatic. For example, the **cyclopropenyl cation** has 2 pi electrons, is planar, cyclic, and all of the carbon atoms are sp^2 hybridized. Thus this species satisfies the Hückel criteria for aromaticity. Other aromatic ions include the **cyclopenta-dienyl anion** and **cycloheptatrienyl cation**. These ionic species are of course not anywhere near as stable as benzene or other neutral aromatic compounds, but they are highly stabilized compared to other nonaromatic cations or anions. ✱

15.3 NOMENCLATURE OF AROMATIC COMPOUNDS

- The **IUPAC system retains certain common names** for several of the simpler benzene derivatives including **toluene**, **cumene**, **styrene**, **xylene**, **phenol**, **aniline**, **benzoic acid**, and **anisole**.
- In more complicated molecules, the benzene ring is named as a substituent on a parent chain.
  - The C_6H_5 group is given the name **phenyl**. For example, the IUPAC name for $C_6H_5CH_2CH_2OH$ is 2-phenylethanol.
  - The $C_6H_5CH_2$- group is given the name **benzyl**. These compounds are derivatives of toluene $C_6H_5CH_3$. *[The terms phenyl and benzyl are often confused by students. Make sure you know when each should be used.]*
- For benzene rings with **two substituents**, the three possible constitutional isomers are named as **ortho** (1,2 substitution), **meta** (1,3 substitution), and **para** (1,4 substitution). These are abbreviated as ***o***, ***m***, and ***p***, respectively. It is also acceptable to name these species with numbers as locators. When one of the substituents has a special name (if NH_2 is present the molecule is an aniline, etc.) then the molecule is named after that parent molecule. For example, 3-chloroaniline and *m*-chloroaniline are both acceptable names for the same molecule. If neither group imparts a special name, then the substituents are listed in alphabetical order. For example, 1-chloro-4-ethylbenzene and *p*-chloroethylbenzene are acceptable names for the same molecule. ✱
- **Polynuclear aromatic hydrocarbons (PAH)** contain more than one benzene ring, each pair of which shares two carbon atoms. For example, napthalene is two benzene rings fused together and anthracene is three benzene rings fused together in a linear fashion. Other common PAH's include phenanthrene, pyrene, coronene, and benzo[*a*]pyrene. Benzo[*a*]pyrene has been especially well-studied because it is a potent carcinogen. ✱

15.4 SPECTROSCOPIC PROPERTIES

- The **mass spectra** of aromatic hydrocarbons generally have a **strong molecular ion peak.** Alkyl-substituted aromatic hydrocarbons also contain fragments derived from **cleavage at the benzylic carbon.** Interestingly, there is some evidence that the benzyl cation (*m/e* 91) is not the compound observed in the mass spectrum, but rather a **rearrangement occurs** to generate the **more stable tropylium cation** (*m/e* 91).
- **Hydrogens** attached to benzene rings come into resonance at about **δ 6.5 to δ 8.5** in the **^{1}H-NMR spectrum.** These signals are this far downfield because of the **ring current** present in aromatic pi systems.
  - The **ring current** in the pi system of an aromatic compound is **induced by the applied magnetic field** when the plane of the aromatic ring is perpendicular to that of the applied magnetic field. The pi electrons are induced to circulate around the aromatic ring, which in turn sets up a local magnetic field that reinforces the applied field on the outside of the ring. As a result, the aromatic hydrogen atoms come into resonance at a lower applied field (larger chemical shift). Note that hydrogens on the inside of a large aromatic annulene such as [18]annulene exhibit the opposite effect and these peaks are actually at negative values relative to TMS.
  - With multiple substituents, the aromatic hydrogen peaks may be difficult to interpret due to complex splitting patterns. One splitting pattern that is easy to recognize is the so-called **para pattern** that is the **pair of doublets** that result from **1,4 disubstitution.**
- In **^{13}C-NMR**, carbon atoms of aromatic rings give peaks in the range of **δ 110 to δ 160.** The substitution patterns of aromatic rings can usually be discerned by simply counting the aromatic peaks in ^{13}C-NMR. Note that alkene carbons come into resonance in the same region as the aromatic carbons, so care must be taken when interpreting spectra for molecules that contain both types of carbon atoms.
- In **infrared spectra**, the **C-H hydrogen stretching vibration** shows up as a moderate peak near **3030 cm^{-1}.** There are **C=C stretching** peaks at **1600 cm^{-1}** and **475 cm^{-1}**, and a strong **out-of-plane C-H bend** in the region **690 cm^{-1} to 900 cm^{-1}.**
- Molecules with aromatic rings show **strong absorptions** in **ultraviolet-visible absorption spectra** as a result of **π to π* transitions.** There are **usually two peaks**, the first **high intensity peak** is **near 205 nm** and a second, **less intense peak** near **270 nm.**

15.5 PHENOLS

- The characteristic feature of **phenols** is a **hydroxyl group attached to a benzene ring.**
- **Phenols are more acidic** than simple alcohols, because the **phenoxide anion is more stable** than an alkoxide anion. This **increased stability** is due to **resonance stabilization of the phenoxide anion.** In other words, upon deprotonation of a phenol, the resulting phenoxide is more stable, because the phenoxide can be considered a resonance hybrid of three contributing resonance structures that delocalize the negative charge onto three different carbon atoms of the ring. This charge delocalization means that no one atom must absorb the entire negative charge, and the delocalized anion is thus more stable. There is no similar resonance stabilization possible for alkoxide anions. ✱
  - **Substituents on the ring** can have a **dramatic influence** on the **acidity** of phenols. The fundamental concept is that **anything** on the ring that **leads to a further stabilization of the phenoxide anion increases acidity** of a phenol, while **anything that destabilizes the phenoxide decreases acidity of a phenol.** Two major effects are important for the stabilization or destabilization of phenoxide anions by ring substituents.

    The **inductive effect** is due to **electron polarization** caused by **differences in relative electronegativities of bonded atoms.** Atoms or groups that are **more electronegative than the sp^2 carbon atoms of the ring** are said to be **electron-**

withdrawing, while atoms or groups that are **less electronegative than the sp^2 carbon atoms of the ring** are said to be **electron-releasing.** Electron-withdrawing groups stabilize a phenoxide anion by absorbing some of the negative charge, while electron-releasing groups destabilize a phenoxide anion by dumping even more electron density into the ring. The bottom line is that **phenols with electron-withdrawing groups like fluorine atoms are more acidic** than phenol, while **phenols with electron-releasing alkyl groups are less acidic** than simple phenol. ✻
Another important effect is the **resonance effect**. Certain groups, especially at the ortho and para positions, also stabilize phenoxide anions because the **negative charge can be further delocalized** into the group. For example, an ortho or para nitro (-NO_2) group increases phenol acidity in part due to a resonance effect, since the phenoxide negative charge can be distributed partially onto the nitro oxygen atoms. ✻

CHAPTER 15
Solutions to the Problems

Problem 15.1 Write names for these molecules.

(a) 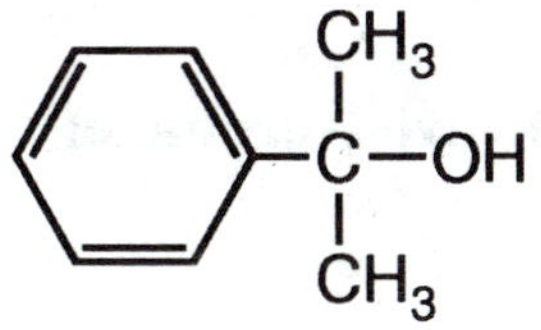
CH3
C—OH
CH3

(b) 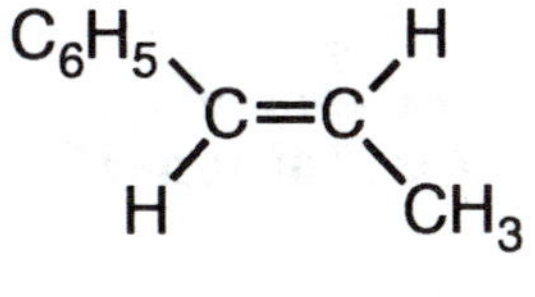
C6H5
H
C=C
H
CH3

(c) 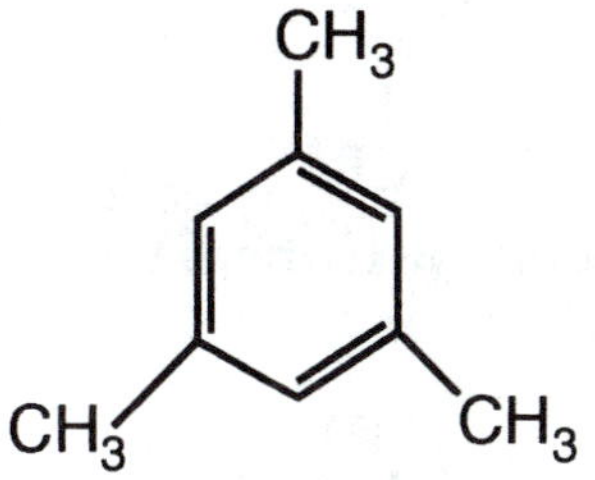
CH3
CH3
CH3

2-phenyl-2-propanol **(E)-1-phenylpropene** **1,3,5-trimethylbenzene (mesitylene)**

Problem 15.2 Which compound gives a signal in the ^{1}H-NMR spectrum at lower applied field (with a larger chemical shift), furan or cyclopentadiene? Explain.

Furan. Furan is an aromatic compound. Hydrogens on this aromatic ring are deshielded by the induced ring current and appear farther downfield compared to those of cyclopentadiene.

Problem 15.3 Predict the products resulting from vigorous oxidation of the following compounds by $K_2Cr_2O_7$ in aqueous H_2SO_4.

(a) 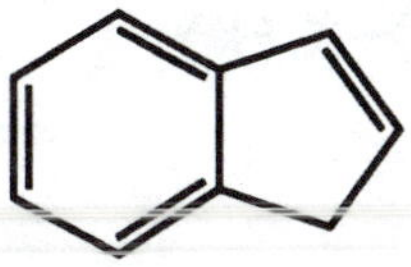

(b) 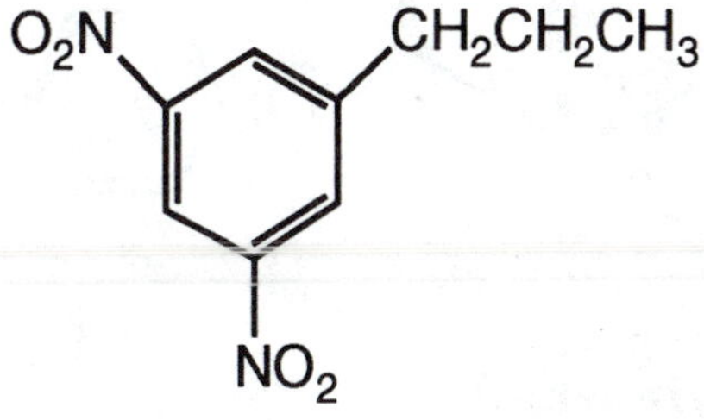
O2N
CH2CH2CH3
NO2

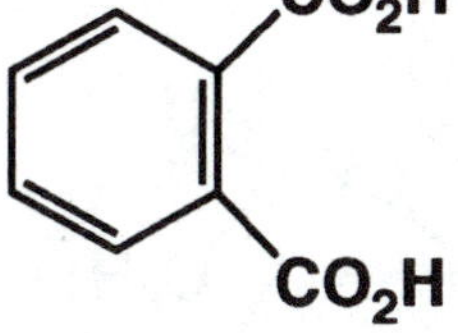
CO2H
CO2H

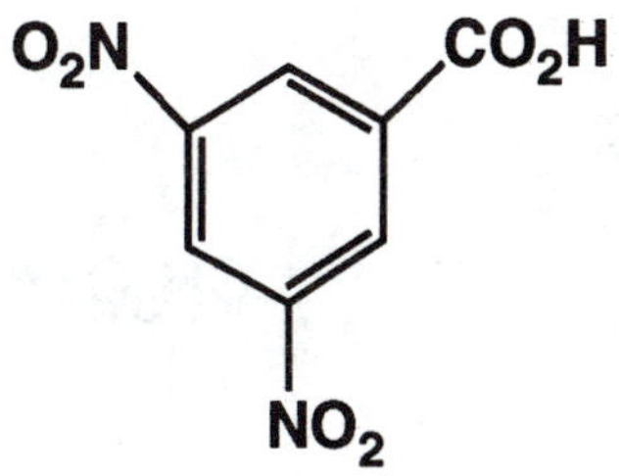
O2N
CO2H
NO2

Problem 15.4 Name the following molecules and ions.

(a) NO_2, Cl
4-chloronitrobenzene

(b) CH_3, Br
2-bromotoluene
(*o*-bromotoluene)

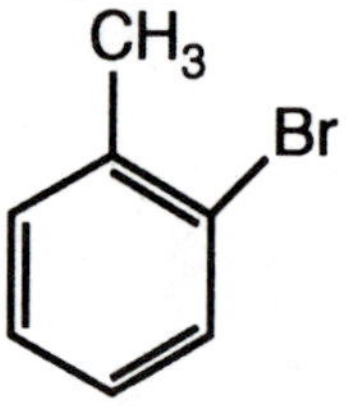

(c) $CH_2CH_2CH_2OH$
3-phenyl-1-propanol

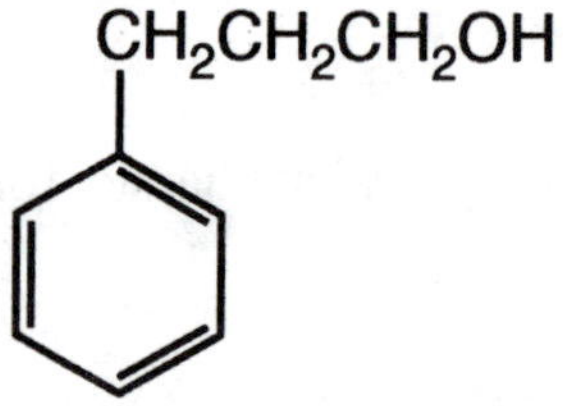

(d) NO_2, NO_2
1,5-dinitronapthalene

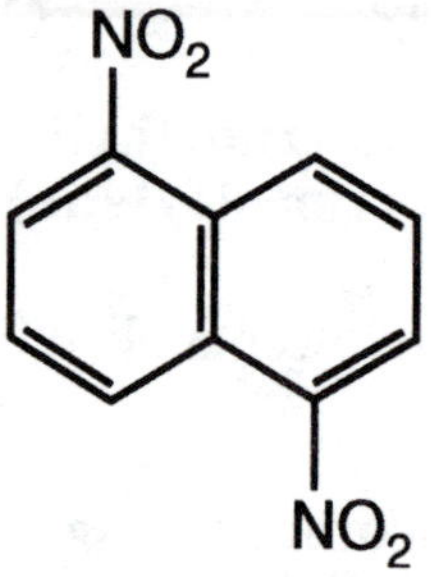

(e) OH, $CH_3CCH{=}CH_2$
2-phenyl-3-buten-2-ol

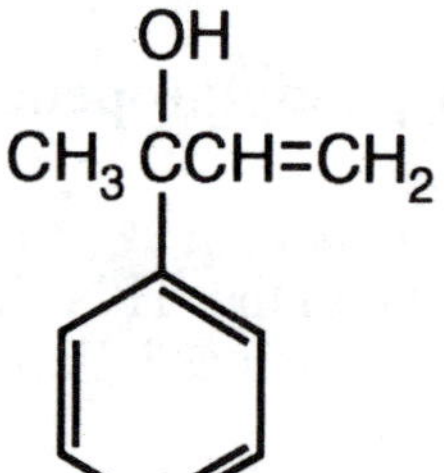

(f) O_2N, $-C{\equiv}CH$
3-nitrophenylethyne
(*m*-nitrophenylacetylene)

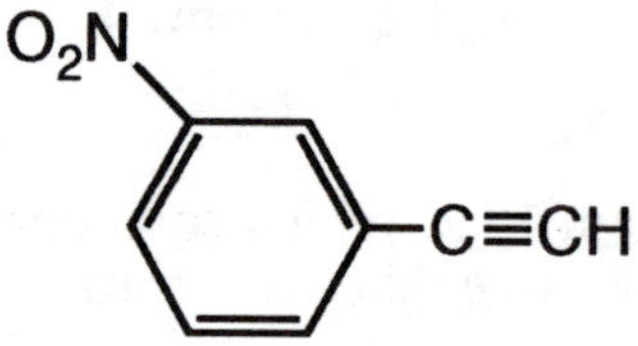

(g) OH
2-phenylphenol
(*o*-phenylphenol)

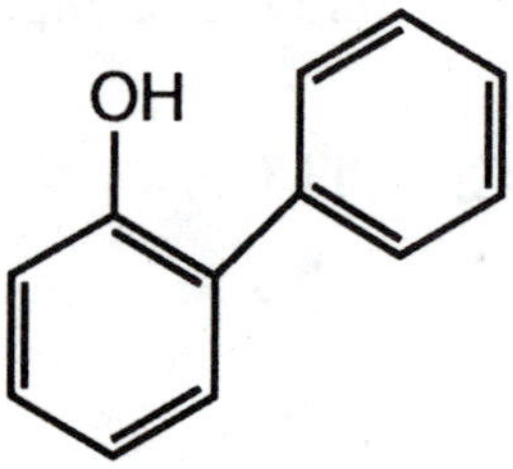

(h) $-CH_2-$
benzylcyclopentane

(i) $-(CH_2)_4-$
1,4-diphenylbutane

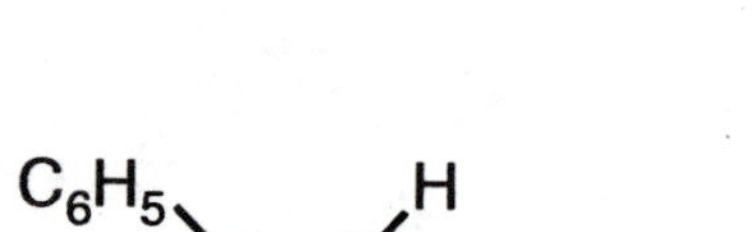

(j) CH_3, Cl, Cl
2,4-dichlorotoluene

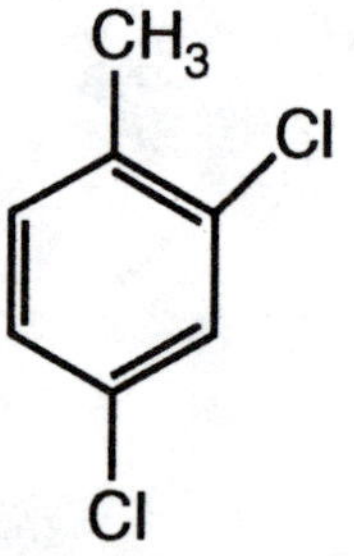

(k) CH_3O-, $-\overset{+}{C}H_2$
4-methoxybenzyl cation

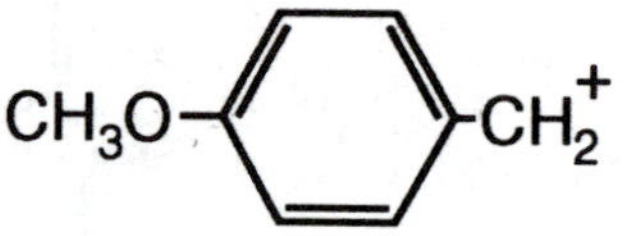

(l) C_6H_5, H, C=C, H, C_6H_5
(E)-1,2-diphenylethene
(*trans*-1,2-diphenylethylene)

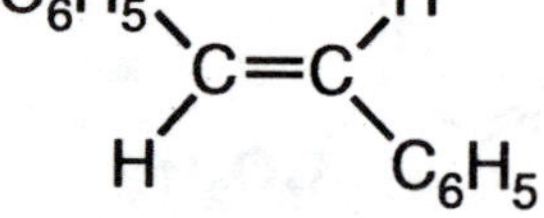

(m) 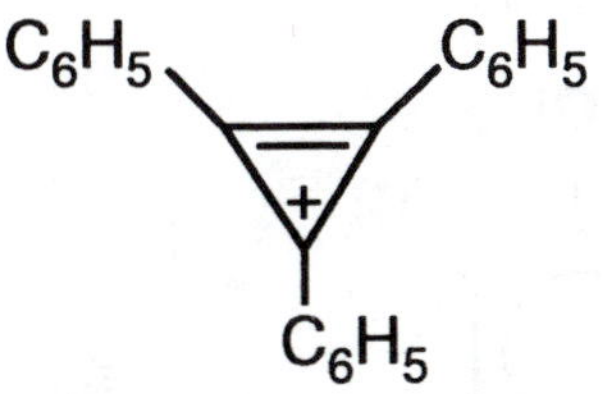 C_6H_5, C_6H_5, C_6H_5 (+)

triphenylcyclopropenium ion
(triphenylcyclopropenyl cation)

Problem 15.5 Draw structural formulas for the following molecules.

(a) 1-Bromo-2-chloro-4-ethylbenzene

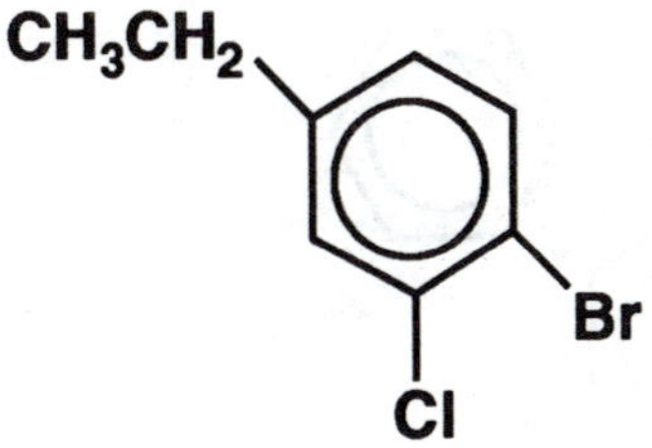

CH_3CH_2, Br, Cl

(b) *m*-Nitrocumene

CH_3CHCH_3, NO_2

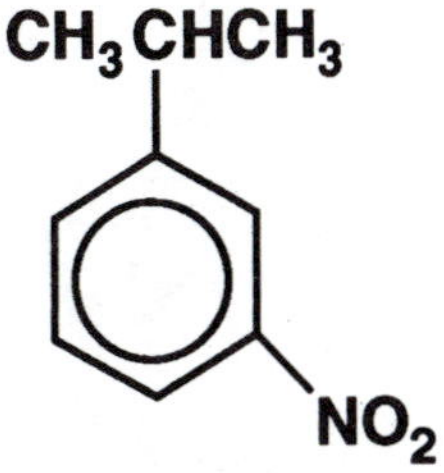

(c) 1,2-Dimethyl-4-iodobenzene

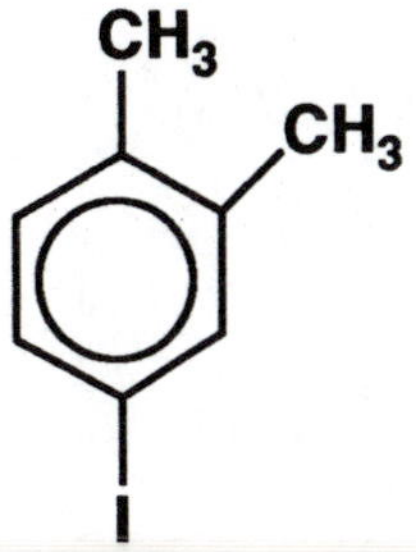

CH_3, CH_3, I

(d) 3,5-Dinitrotoluene

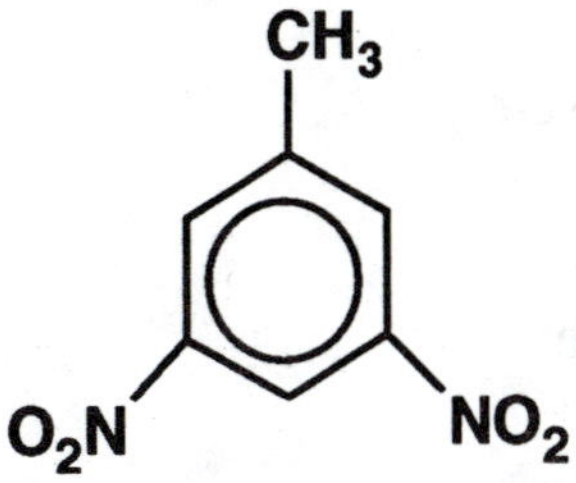

CH_3, O_2N, NO_2

(e) 2,4,6-Trinitrotoluene

CH_3, O_2N, NO_2, NO_2

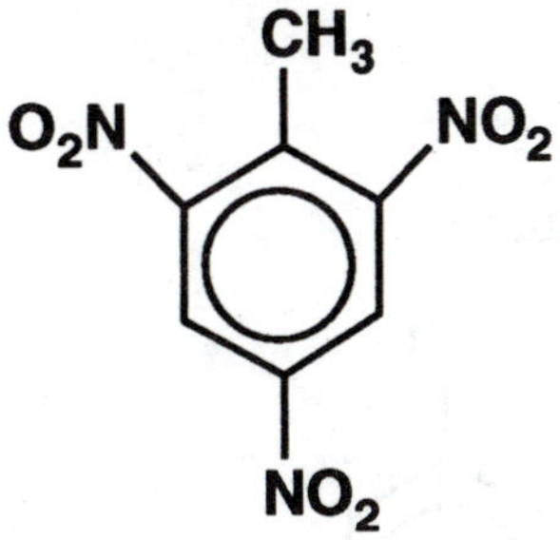

(f) 4-Phenyl-2-pentanol

OH, CH_3CHCH_2CH–, CH_3

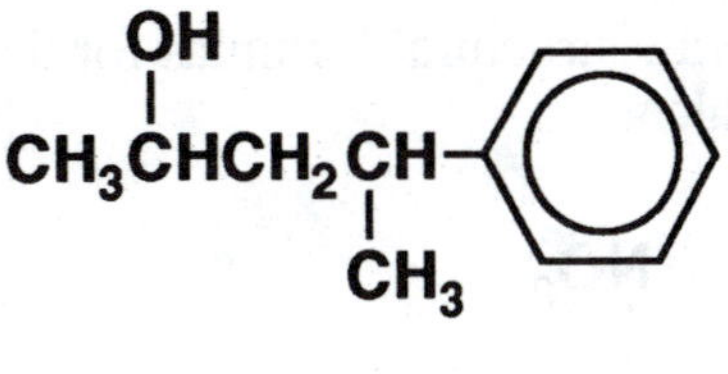

(g) *p*-Cresol

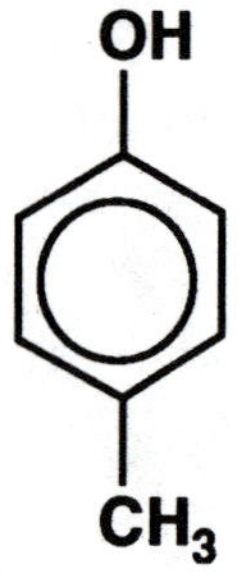

OH
CH$_3$

(h) Pentachlorophenol

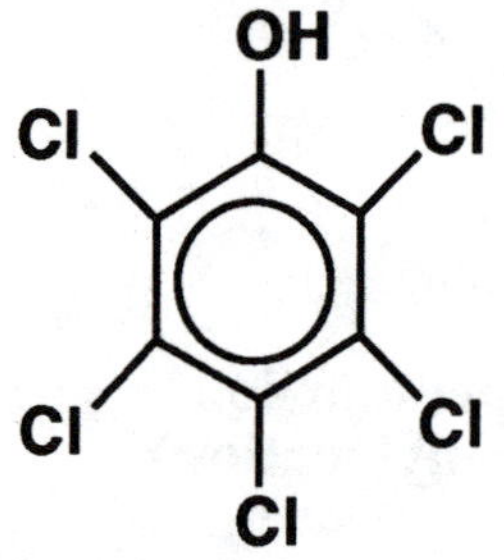

OH
Cl Cl
Cl Cl
Cl

(i) 1-phenylcyclopropanol

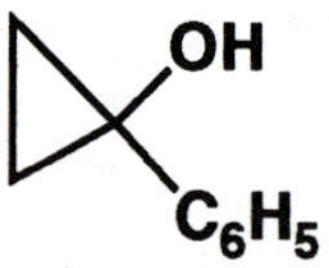

OH
C_6H_5

(j) triphenylmethane

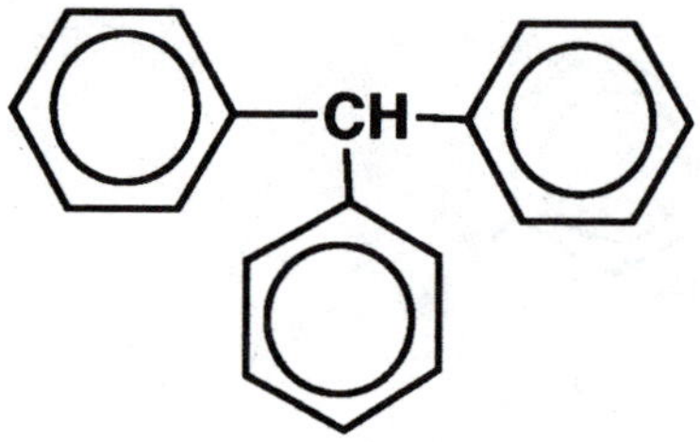

CH

(k) phenylethylene

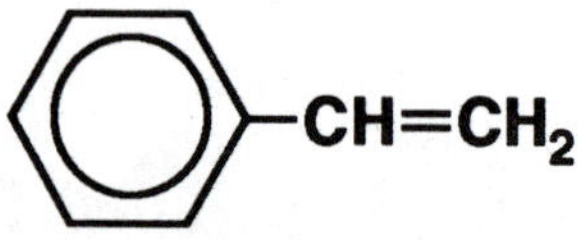

$CH=CH_2$

(l) benzyl bromide

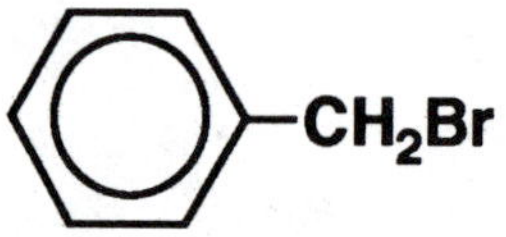

CH_2Br

(m) 1-phenyl-1-butyne

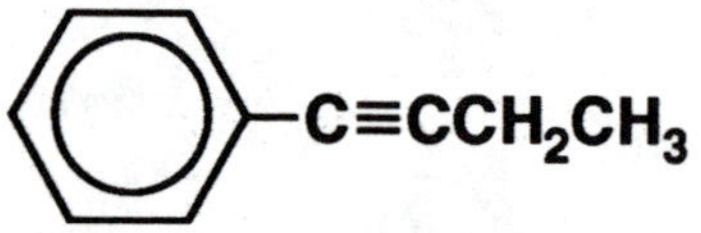

$C{\equiv}CCH_2CH_3$

(n) 3-phenyl-2-propen-1-ol

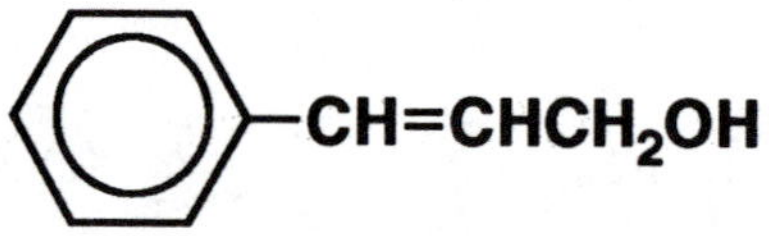

$CH=CHCH_2OH$

Problem 15.6 Draw structural formulas for the following molecules.

(a) 1-nitronapthalene

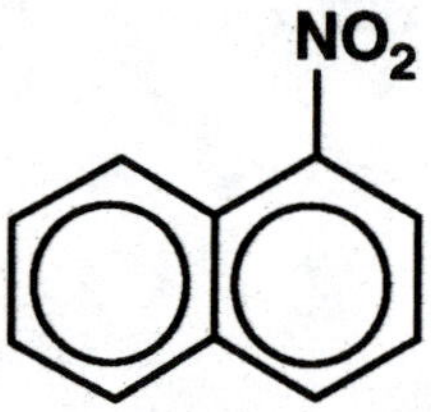

NO_2

(b) 1,6-dichloronapthalene

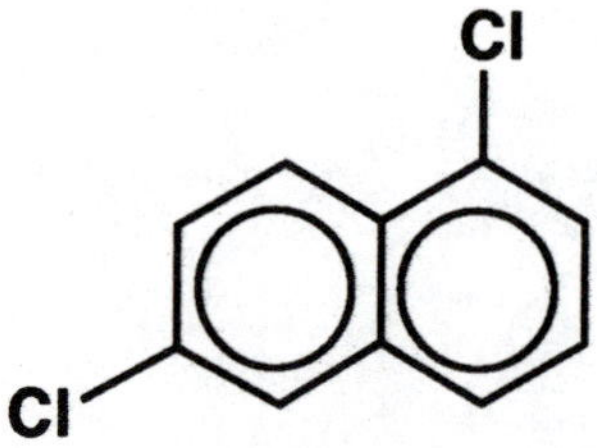

Cl
Cl

(c) 9-bromoanthracene

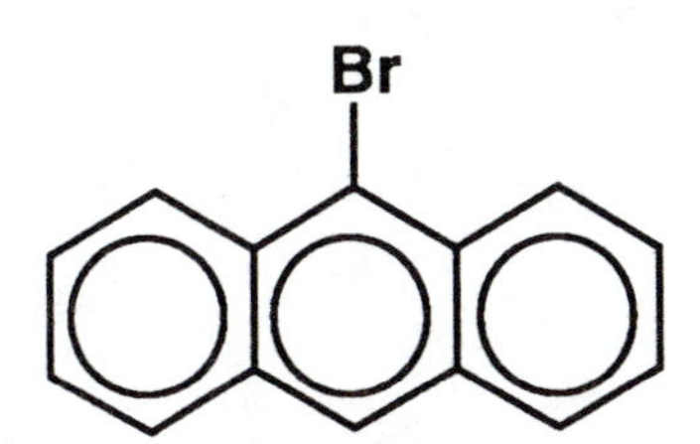

(d) 2-methylphenanthrene

Problem 15.7 Molecules of 6,6'-dinitrophenyl-2,2'dicarboxylic acid have no tetrahedral stereocenter, and yet this compound can be resolved to a pair of enantiomers. Account for this chirality. *Hint:* it will help to build a model and study the ease of rotation about the single bond joining the two benzene rings.

6,6'-dinitrobiphenyl-2,2'-dicarboxylic acid

The key here is that the central bond between the benzene rings cannot rotate freely at room temperature due to the steric hindrance provided by the nitro and carboxyl groups. In other words, these groups run into each other as the molecule attempts to rotate around the central bond, so it is prevented from rotating. As a result, the molecule is chiral because, like a propeller, there are two different orientations possible as shown below. The two orientations represent non-superimposable mirror images so they are a pair of enantiomers.

Resonance in Aromatic Compounds

Problem 15.8 Following each name is the number of Kekulé structures that can be drawn for it. Draw these Kekulé structures, and show using curved arrows, how the first contributing structure is converted to the second and so forth.

(a) napthalene (3)

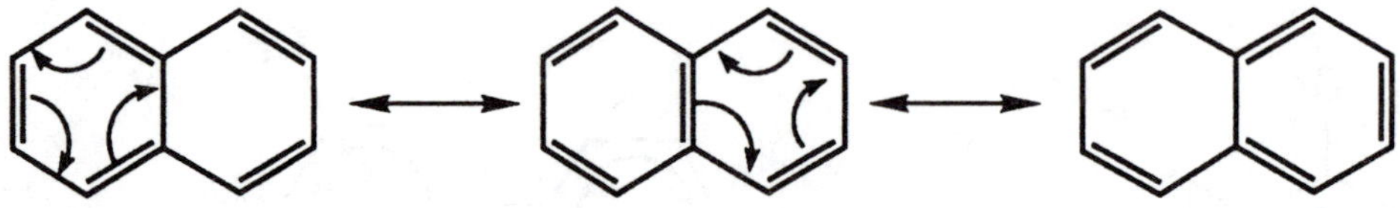

(b) phenanthrene (4)

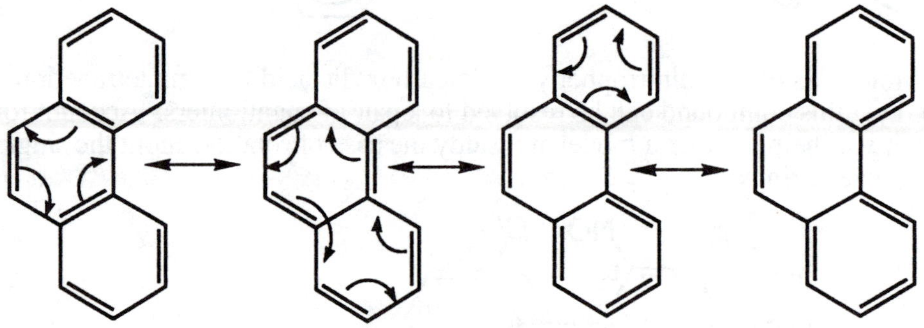

Problem 15.9 Each molecule below can be drawn as a hybrid of five contributing structures; two Kekulé structures and three that involve creation and separation of unlike charges. For (a) chlorobenzene and (b) phenol, the creation and separation of unlike charge place a formal positive charge on the substituent and a formal negative charge on the ring. For nitrobenzene (c) a positive charge is placed on the ring and an additional negative charge is placed on the $-NO_2$ group. Draw these five contributing structures for each molecule.

(a)

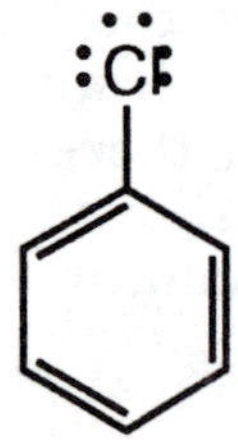

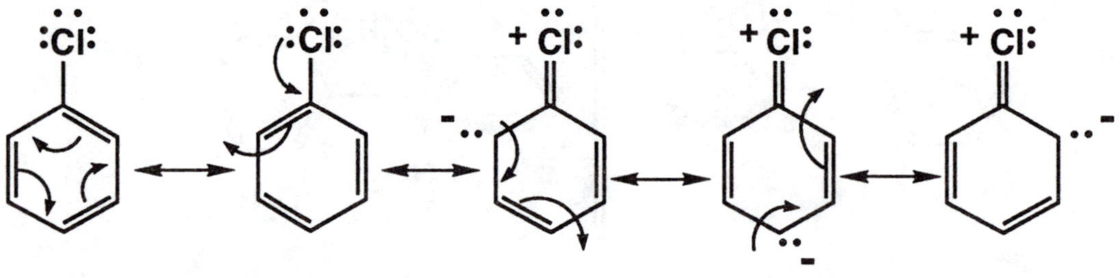

(b)

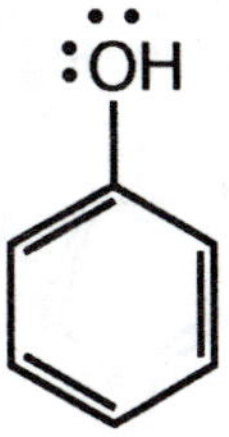

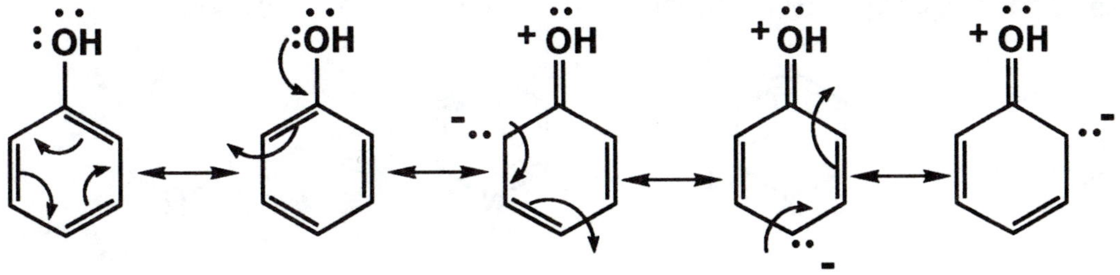

(c)

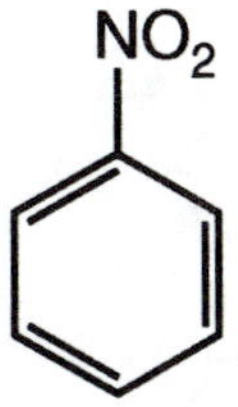

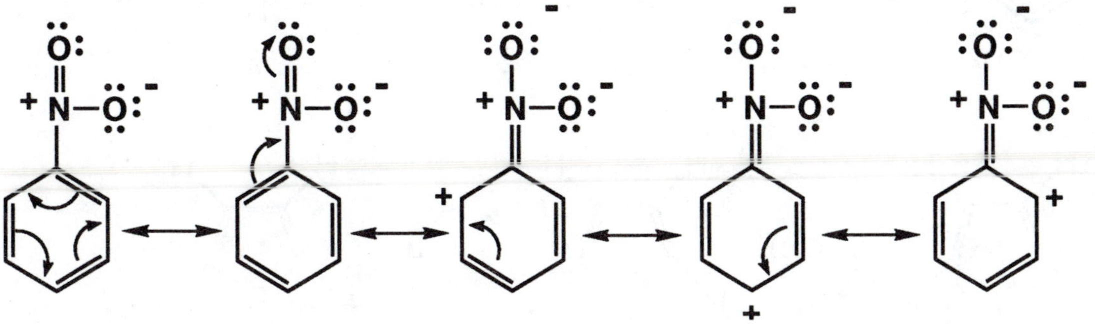

Problem 15.10 Following are structural formulas for furan and pyridine.

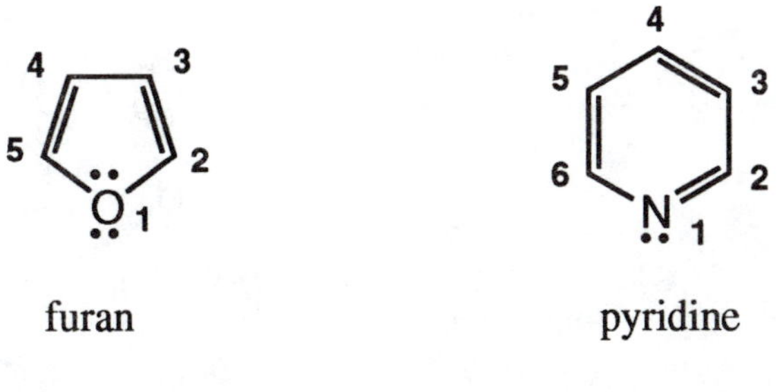

furan

pyridine

(a) Write four contributing structures for the furan hybrid which place a positive charge on oxygen and a negative charge first on carbon 3 of the ring and then on each other carbon of the ring.

(b) Write three contributing structures for the pyridine hybrid which place a negative charge on nitrogen and a positive charge first on carbon 2 and then on carbon 4 and carbon 6.

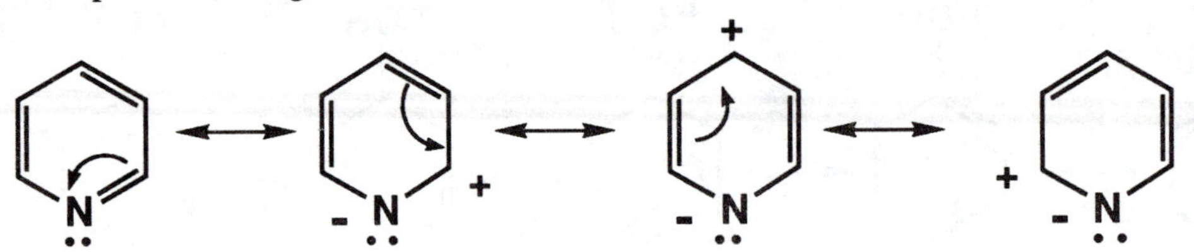

The Concept of Aromaticity

Problem 15.11 State the number of p orbital electrons in each of the following.

(a) 10

(b) 12

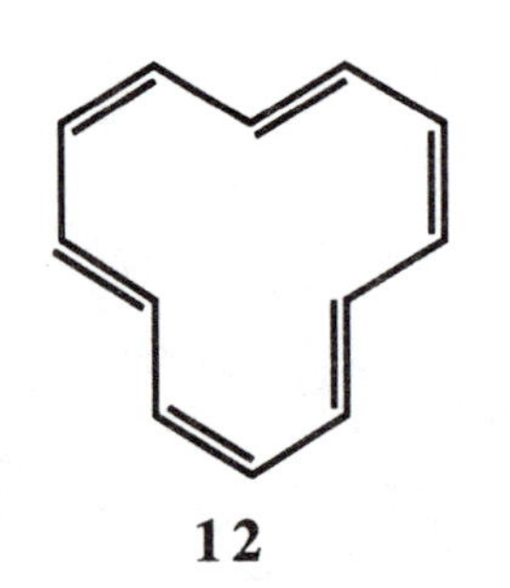

(c) 4

(d) 5

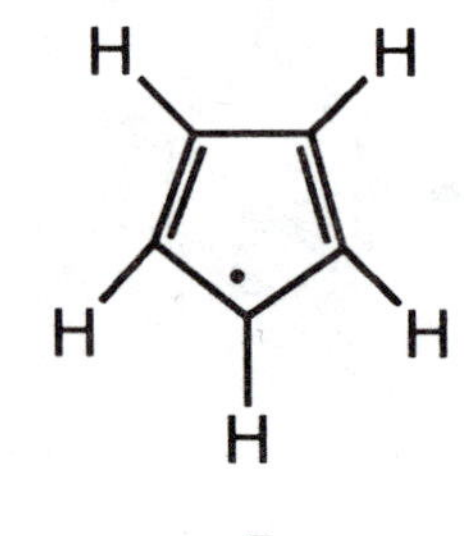

(e) 6

(f) 4

(g) 6

(h) 6

(i) 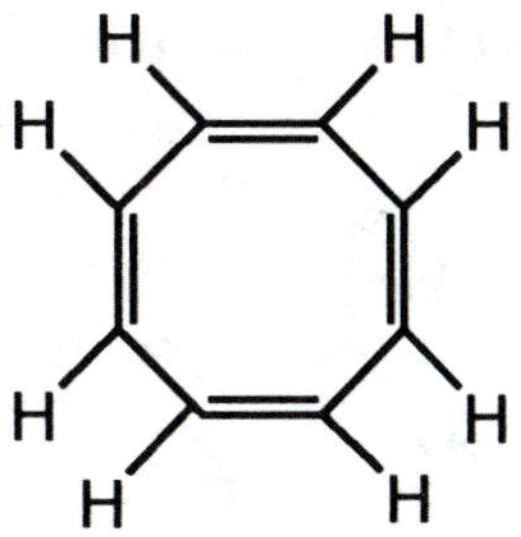 H H H H H H H H

8

(j) 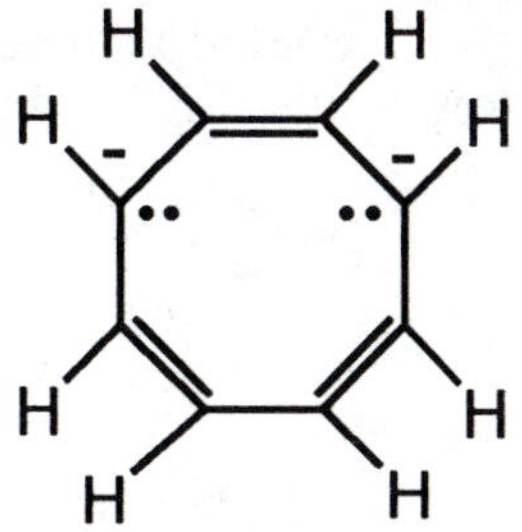 H H H H H H H H

10

Problem 15.12 Which of the molecules and ions given in the previous problem are aromatic according to the Huckel criteria? Which, if planar, would be antiaromatic?

The following molecules are aromatic because they have 4n + 2 π electrons: a, e, g, h, and j.

The following molecules would be antiaromatic if planar because they have 4n π electrons: b, c, f, and i.

Problem 15.13 All attempts to synthesize cyclopentadienone yield only a Diels-Alder adduct. Cycloheptatrienone, however, has been prepared by several methods and is stable.

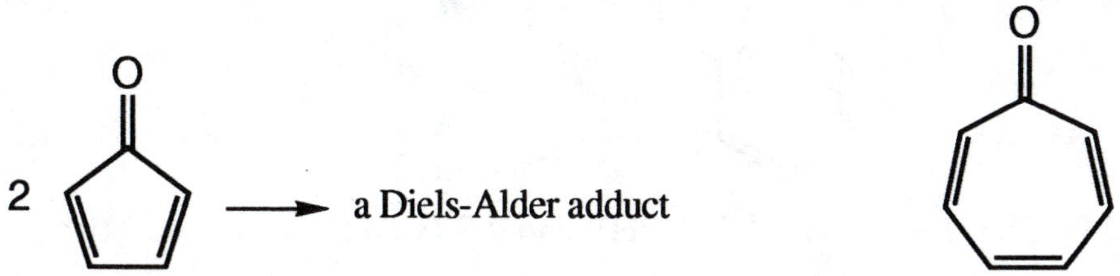

2 O → a Diels-Alder adduct O

(a) Draw a structural formula for the Diels-Alder adduct formed by cyclopentadienone.

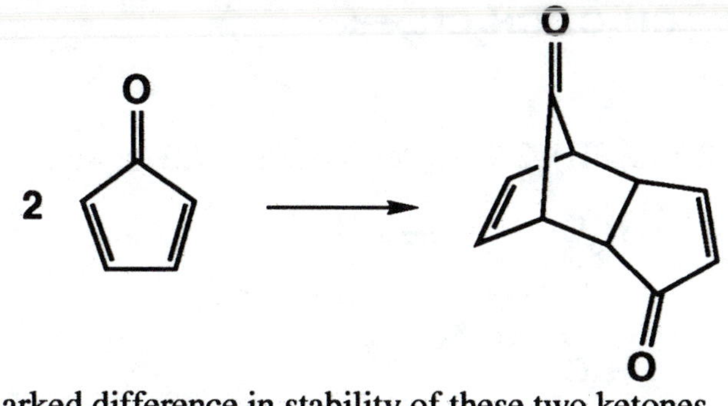

2 O → O O

(b) Account for the marked difference in stability of these two ketones.

The ring of the cyclopentadienone has 4 π electrons, an antiaromatic number. On the other hand, the more stable cycloheptatrienone ring has 6 π electrons, an aromatic number. Also, the following structure has aromatic character.

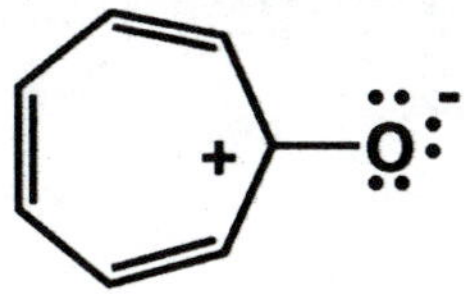

+ O -

Spectroscopy

Problem 15.14 1-Phenylpentane shows a strong fragment ion at *m/e* 92, which has the following structure. How might this ion arise?

H H H_2C +

1-phenylpentane *m/e* 92

H H (-e⁻) H H + H H + +

H H H H H H H H

Problem 15.15 Propose a structural formula consistent with each ^{1}H-NMR spectrum.

(a) $C_9H_{10}O$; δ 1.2 (t, 3H), 3.0 (quartet, 2H), 7.4-8.0 (m, 5H)

O

(b) $C_{10}H_{12}O_2$; δ 2.0 (s, 3H), 2.9 (t, 2H), 4.3 (t, 2H), and 7.3 (s, 5H)

$CH_3\overset{O}{\overset{\|}{C}}OCH_2CH_2-$

(c) $C_{10}H_{14}$; δ 1.2 (d, 6H), 2.3 (s, 3H), 2.9 (septet, 1H), and 7.0 (s, 4H)

H_3C HC CH_3 H_3C

(d) C_8H_9Br; δ 1.8 (d, 3H), 5.0 (quartet, 1H), 7.3 (s, 5H)

Br C CH_3 H

Problem 15.16 Compound A, molecular formula C_9H_{12} shows prominent peaks in its mass spectrum at *m/e* (120) and 105. Compound B, also of molecular formula C_9H_{12}, shows prominent peaks at *m/e* (120) 92, and 91. On vigorous oxidation with alkaline potassium permanganate followed by acidification, both compounds give benzoic acid. From this information, deduce the structural formulas of Compounds A and B.

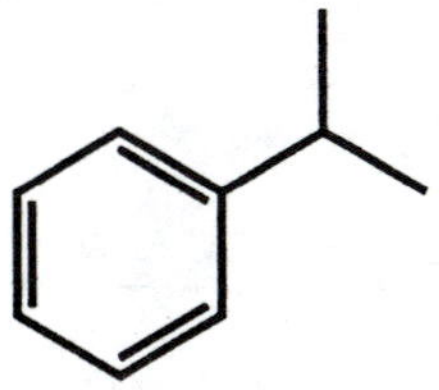

A 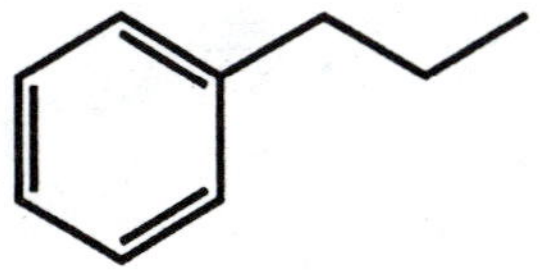 B

The compounds can be identified by the major fragments in the mass spectrum, both of which are benzylic cations. For compound A, the peak at *m/e* 105 corresponds to the cation shown below:

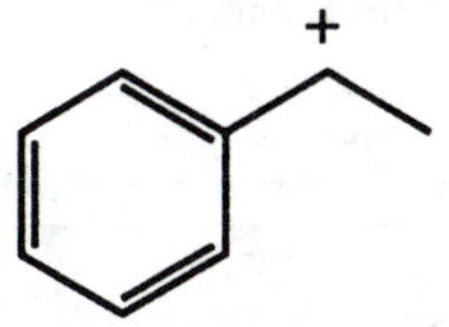

Peak at *m/e* 105 for compound A

For compound B, the peak at *m/e* 91 corresponds to the tropylium ion that was produced by rearrangement of the benzyl cation (Section 15.4)

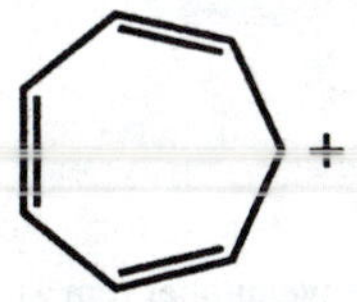

Peak at *m/e* 91 for compound B

Both of the compounds will produce benzoic acid upon oxidation with $KMnO_4$.

Problem 15.17 Compound C shows strong peaks in its mass spectrum at *m/e* ·(148), 105 and, 77. Following are its infrared and ^{1}H-NMR spectra.

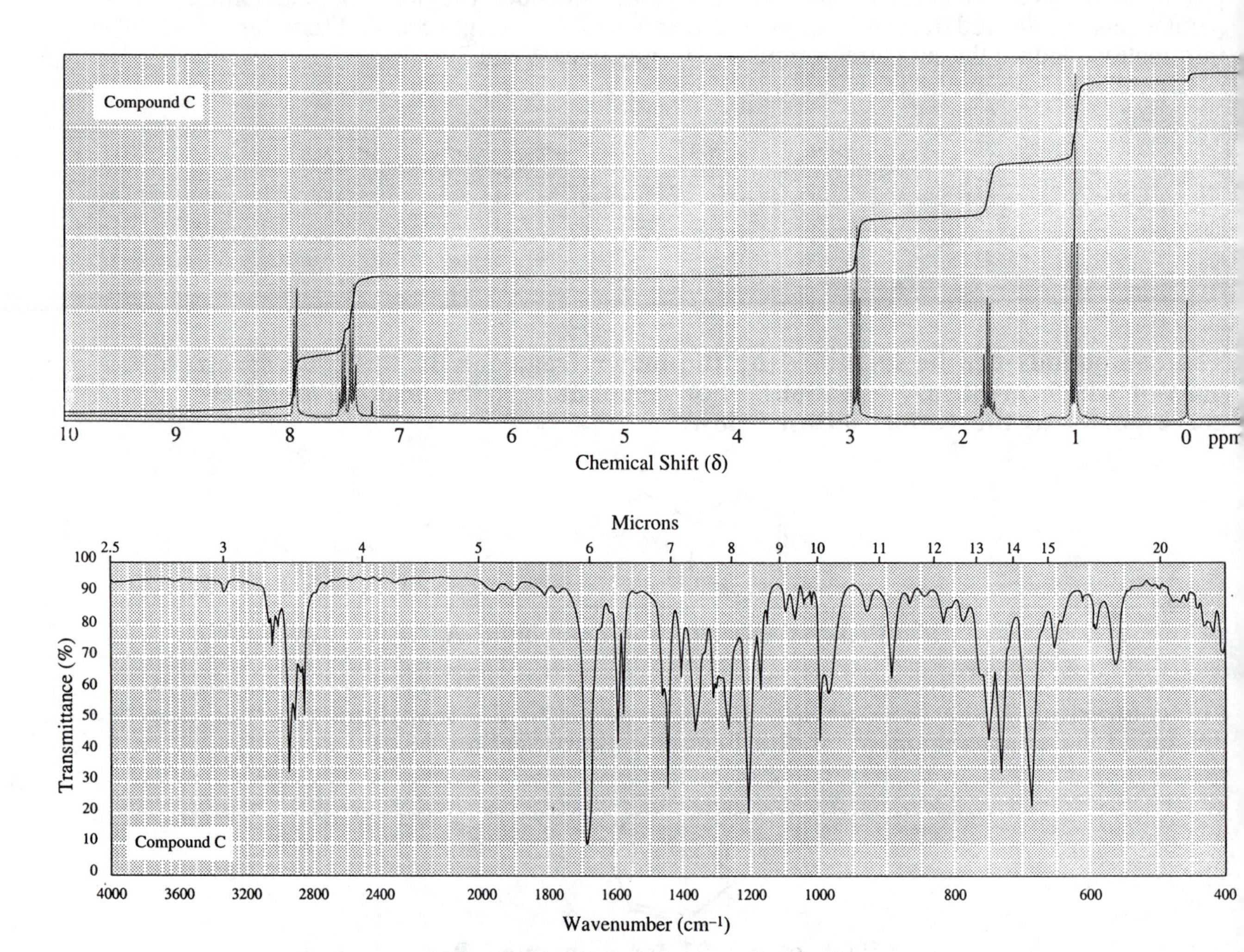

(a) Deduce the structural formula of Compound C.

$\overset{\text{O}}{\overset{\|}{\text{C}}}\text{CH}_2\text{CH}_2\text{CH}_3$

This compound, $C_{10}H_{12}O$, has the correct molecular formula weight of 148. This compound also has a carbonyl corresponding to the peak at 1680 cm^{-1} in the IR spectrum and the correct pattern of hydrogens to explain the ^{1}H-NMR spectrum.

(b) Account for the appearence of peaks in its mass spectrum at m/e 105 and 77.

The peaks at 105 and 77 correspond to the following fragments produced by α-cleavage (Section 17.4) on either side of the ketone.

m/e 105 m/e 77

Problem 15.18 Following are IR and ^{1}H-NMR spectra of Compound D. The mass spectrum of Compound D shows a molecular ion peak at *m/e* 136, a base peak at *m/e* 107, and other prominant peaks at *m/e* 118 and 77.

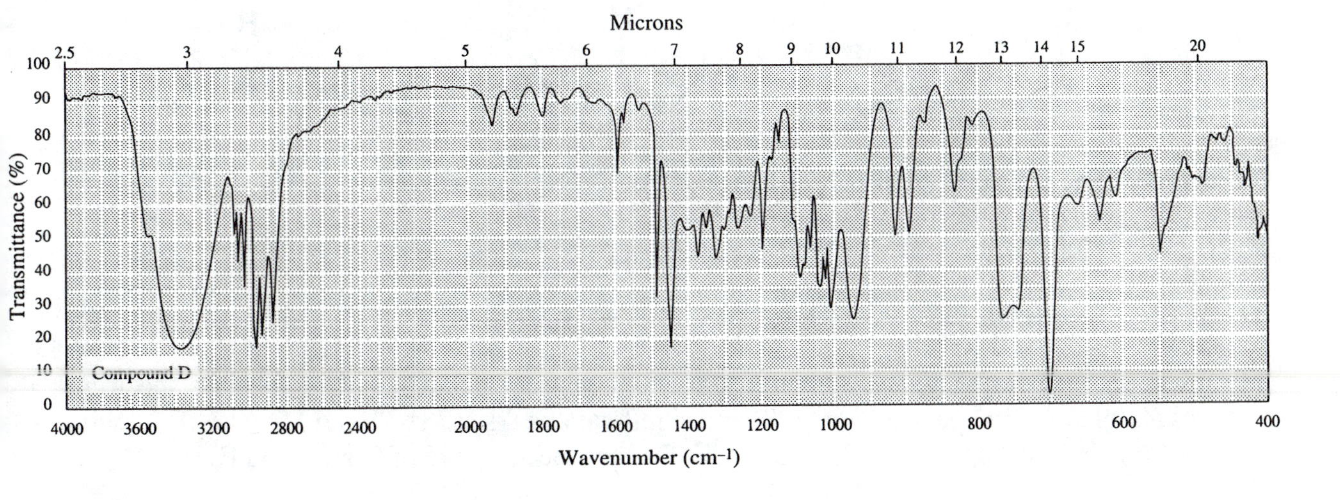

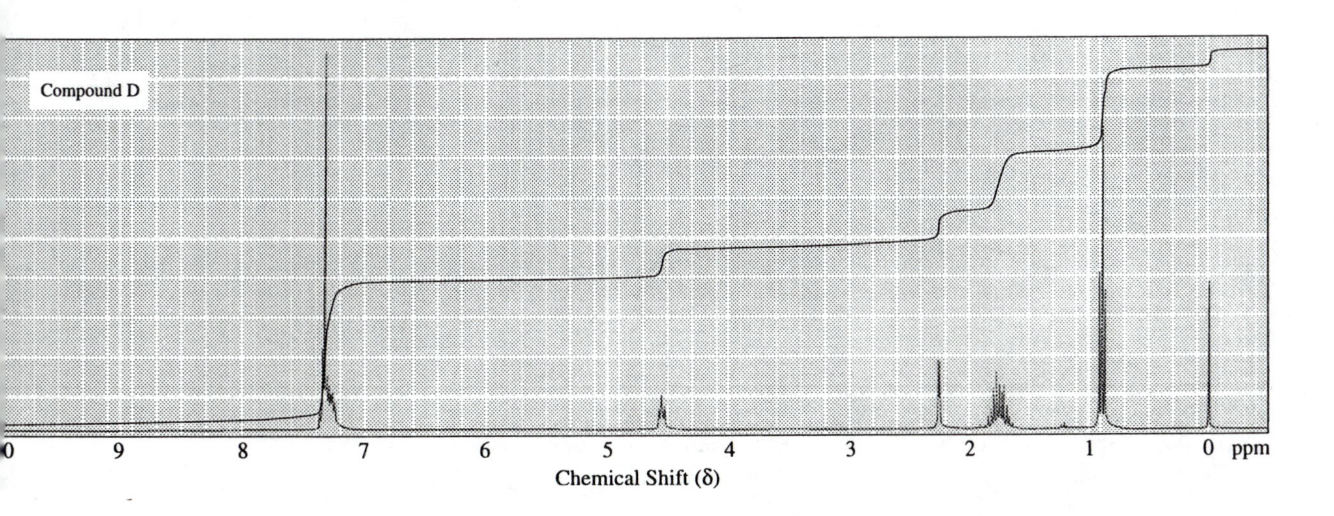

(a) Propose a structural formula for Compound D based on this information.

This compound, $C_9H_{12}O$, has the correct molecular weight of 136. An alcohol function corresponds to the broad peak at 3300 cm^{-1} in the IR spectrum and the pattern of hydrogens present explains the ^{1}H-NMR spectrum.

(b) Propose structural formulas for peaks in the mass spectrum at *m/e* 118, 107 and 77.

The peaks at 118, 107 and 77 correspond to the following fragments:

m/e **118** *m/e* **107** *m/e* **77**

Problem 15.19 Compound E is a neutral solid, molecular formula $C_8H_{10}O_2$. Its mass spectrum shows a molecular ion peak at *m/e* (138) and significant peaks at $M^{+\cdot}$-1 and $M^{+\cdot}$-17. Following are IR and ^{1}H-NMR spectra of Compound E. Deduce the structure of Compound E.

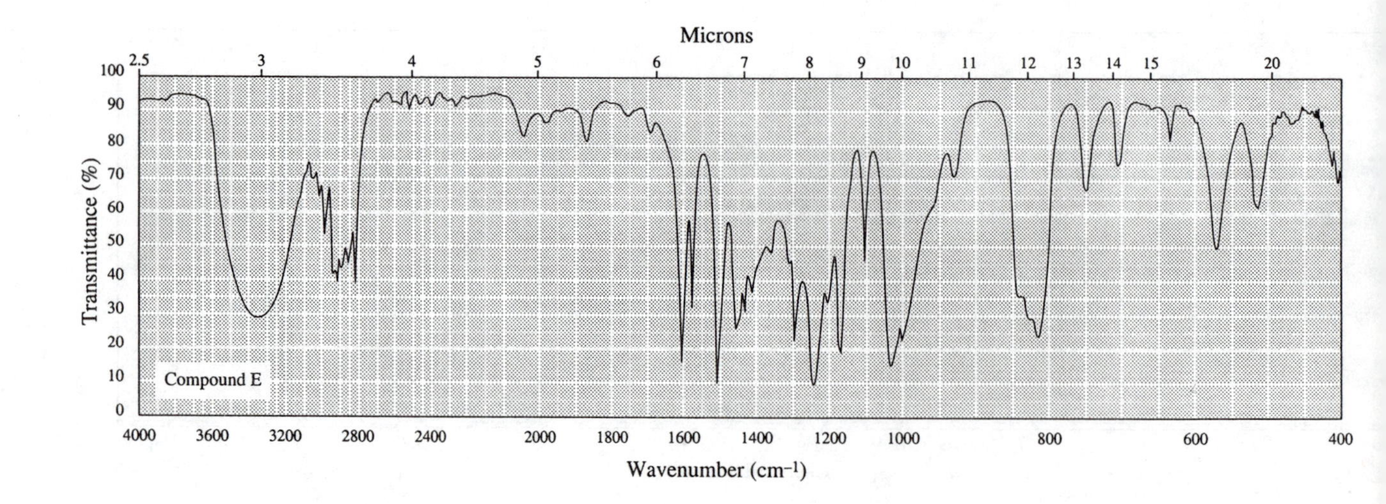

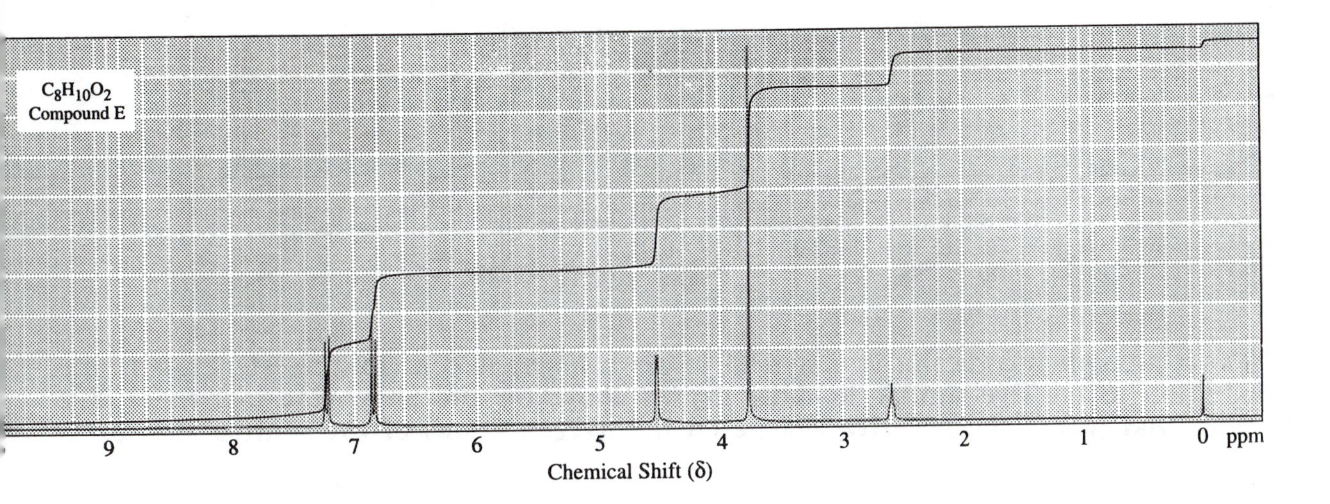

This compound has the correct molecular formula, $C_8H_{10}O_2$, an alcohol function corresponding to the broad peak at 3350 cm^{-1} in the IR spectrum, and the correct pattern of hydrogens to explain the ^{1}H-NMR spectrum. Notice especially the distinctive *para* pattern near δ 7.0 and the methyl group peak at 3.78.

Acidity of Phenols

Problem 15.20 Use the resonance theory to account for the fact that p-nitrophenol is a stronger acid than phenol.

$K_a = 1.0 \times 10^{-10}$ $K_a = 7.0 \times 10^{-8}$

As seen from the acid ionization constants, *p*-nitrophenol is the stronger acid. To account for this fact, draw resonance contributing structures for each anion and compare the degree of delocalization of negative charge (i.e., the resonance stabilization of each anion). Phenoxide ion is a resonance hybrid of five important contributing structures, two of which place the negative charge on the phenoxide oxygen, and three of which place the negative charge on the atoms of the ring.

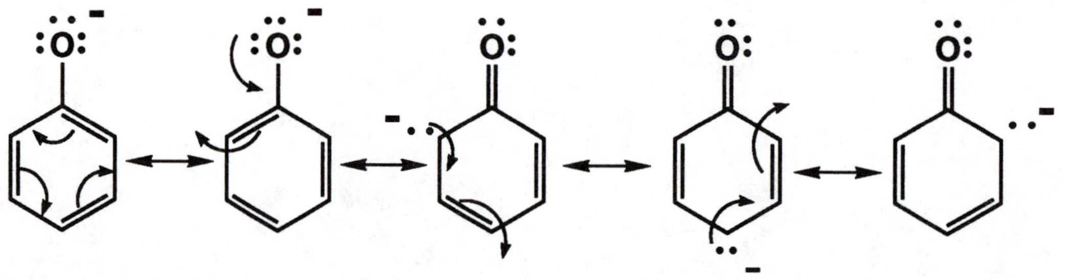

The *p*-nitrophenoxide ion is a hybrid of six important contributing structures. In addition to the five similar to those drawn above for the phenoxide ion, there is a sixth that places the negative charge on the oxygen atoms of the *p*-nitro group.

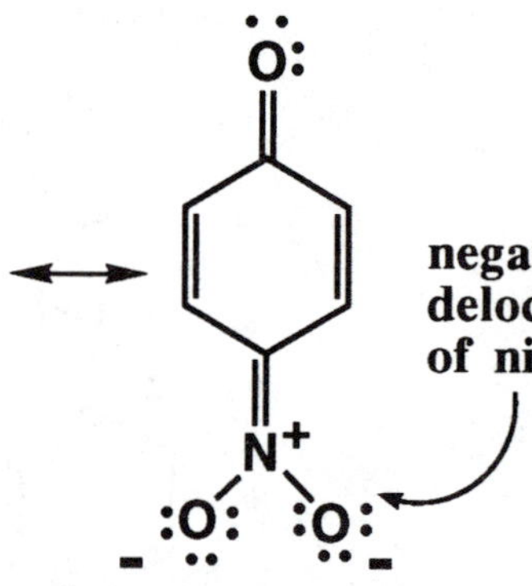

five contributing structures similar to those drawn for the phenoxide ion

negative charge delocalized to oxygens of nitro group

Thus, because of the greater delocalization of the negative charge onto the more electronegative oxygen atoms of the nitro group, *p*-nitrophenol is a stronger acid than phenol.

Problem 15.21 Explain the trends in the acidity of phenol and the monofluoro isomers of phenol:

OH OH F OH F OH F

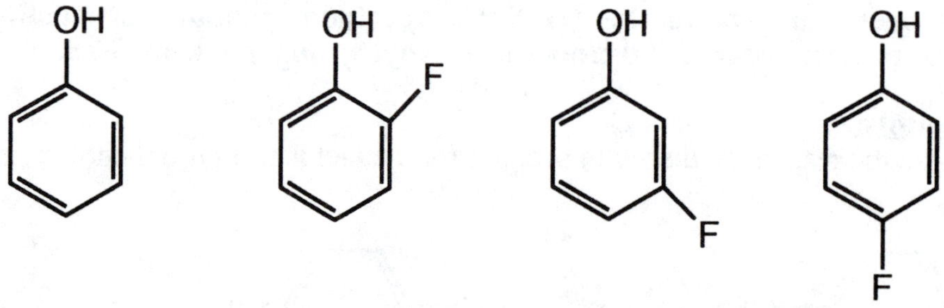

$pK_a = 10.0$ $pK_a = 8.81$ $pK_a = 9.28$ $pK_a = 9.81$

The electronegative fluoro substituent increases the acidity of the phenol through an inductive effect, so all of the monofluoro isomers of phenol are more acidic than phenol itself. Because this is an inductive effect, the closer the fluorine atom to the phenolic OH group, the stronger the effect and the greater the acidity. Thus, the *ortho* fluoro derivative is most acidic, followed by the *meta* fluoro derivative, then finally the *para* flouro derivative.

Problem 15.22 You wish to determine the inductive effects of a series of functional groups, for example Cl, Br, CN, CO_2H, and C_6H_5. Is it best to use a series of *ortho*-substituted, *meta*-substituted, or *para*-substituted phenols? Explain your answer.

The question to be addressed involves inductive effects only. It would be best to use the derivatives with the substituents in the *meta* position, because this would minimize any contributions from resonance effects. The resonance of effects are maximal when substituents are in the *ortho* and *para* positions.

Problem 15.23 Arrange the molecules and ions in each set in order of increasing acidity (from least acidic to most acidic).

(a) C_6H_5-OH C_6H_{11}-OH CH_3COOH

To arrange these in order of increasing acidity, refer to Table 4.2. For those compounds not listed in Table 3.2, estimate pKa using values for compounds that are given in the table.

C_6H_{11}-OH	C_6H_5-OH	CH_3COOH
pKa ~ 18	**pKa 9.95**	**pKa 4.76**

(b) C_6H_5-OH HCO_3^- H_2O

H_2O	HCO_3^-	C_6H_5-OH
pKa 15.7	**pKa 10.33**	**pKa 9.95**

(c) C_6H_5-C≡CH C_6H_5-OH C_6H_5-CH_2OH

C_6H_5-C≡CH	C_6H_5-CH_2OH	C_6H_5-OH
pKa ~ 25	**pKa ~ 18**	**pKa 9.95**

Problem 15.24 From each pair select the stronger base.

To estimate which is the stronger base, first determine which conjugate acid is the weaker acid. The weaker the acid, the stronger its conjugate base.

(a) $C_6H_5O^-$ or OH^- — OH^-

stronger base (anion of weaker acid)

(b) $C_6H_5O^-$ or $C_6H_{11}O^-$ (cyclohexoxide) — $C_6H_{11}O^-$

stronger base (anion of weaker acid)

(c) $C_6H_5O^-$ or HCO_3^- — $C_6H_5O^-$

stronger base (anion of weaker acid)

(d) $C_6H_5O^-$ or CH_3COO^- — $C_6H_5O^-$

stronger base (anion of weaker acid)

Problem 15.25 Propranolol is a β-adrenergic receptor antagonist. Members of this class have received enormous clinical attention because of their effectiveness in treatment of hypertension (high blood pressure), migraine headaches, glaucoma, ischemic heart disease, and certain cardiac arrhythmias. Starting materials for the synthesis of propranolol are propene, 1-naphthol, and isopropylamine. Show how to convert propene to epichlorohydrin in stage 1, and then complete the synthesis of propranolol in stage 2.

Stage 1: synthesis of epichlorohydrin

$CH_3CH{=}CH_2 \longrightarrow ClCH_2CH{=}CH_2 \longrightarrow ClCH_2CH(O)CH_2$

propene → 3-chloropropene (allyl chloride) → 3-chloro-1,2-epoxypropane (epichlorohydrin)

The reaction sequence involves allylic halogenation to give the allyl chloride followed by epoxidation to give the epichlorohydrin.

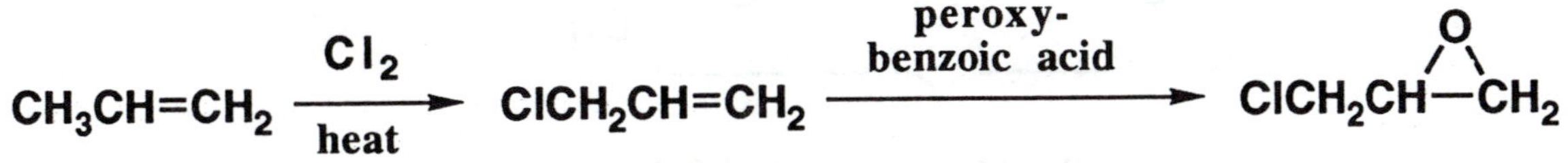

$$CH_3CH{=}CH_2 \xrightarrow[\text{heat}]{Cl_2} ClCH_2CH{=}CH_2 \xrightarrow{\text{peroxy-benzoic acid}} ClCH_2\overset{O}{CH{-}CH_2}$$

Stage 2: synthesis of propranolol

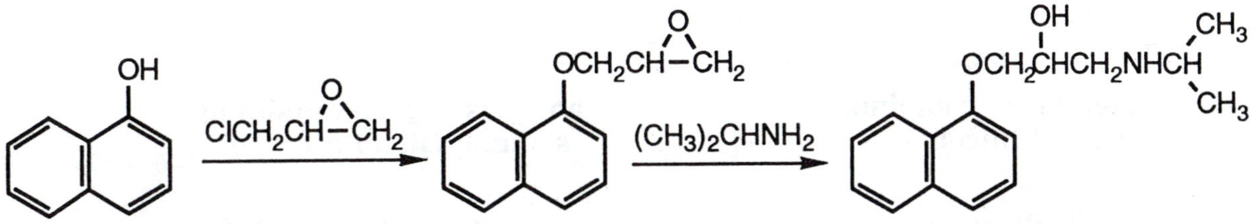

OH

$ClCH_2CH{-}CH_2$ (epoxide)

$OCH_2CH{-}CH_2$ (epoxide)

$(CH_3)_2CHNH_2$

OH

$OCH_2CHCH_2NHCH(CH_3)_2$

1-naphthol propranolol

1-Naphthol, a poor nucleophile, is converted to its anion (a good nucleophile) which then displaces Cl from epichlorhydrin by an S_N2 mechanism. Finally, S_N2 opening of the epoxide by the nucleophilic nitrogen atom of isopropylamine completes the synthesis.

Problem 15.26 Describe a procedure to separate a mixture of benzyl alcohol and *o*-cresol and recover each in pure form.

CH_2OH CH_3

OH

benzyl alcohol (bp 205°C) *o*-cresol (bp 191°C)

Following is a flow chart for an experimental method for separating these two compounds. Separation is based on the facts that each is insoluble in water, soluble in diethyl ether, and that *o*-cresol reacts with 10% NaOH to form a water-soluble phenoxide salt while benzyl alcohol does not.

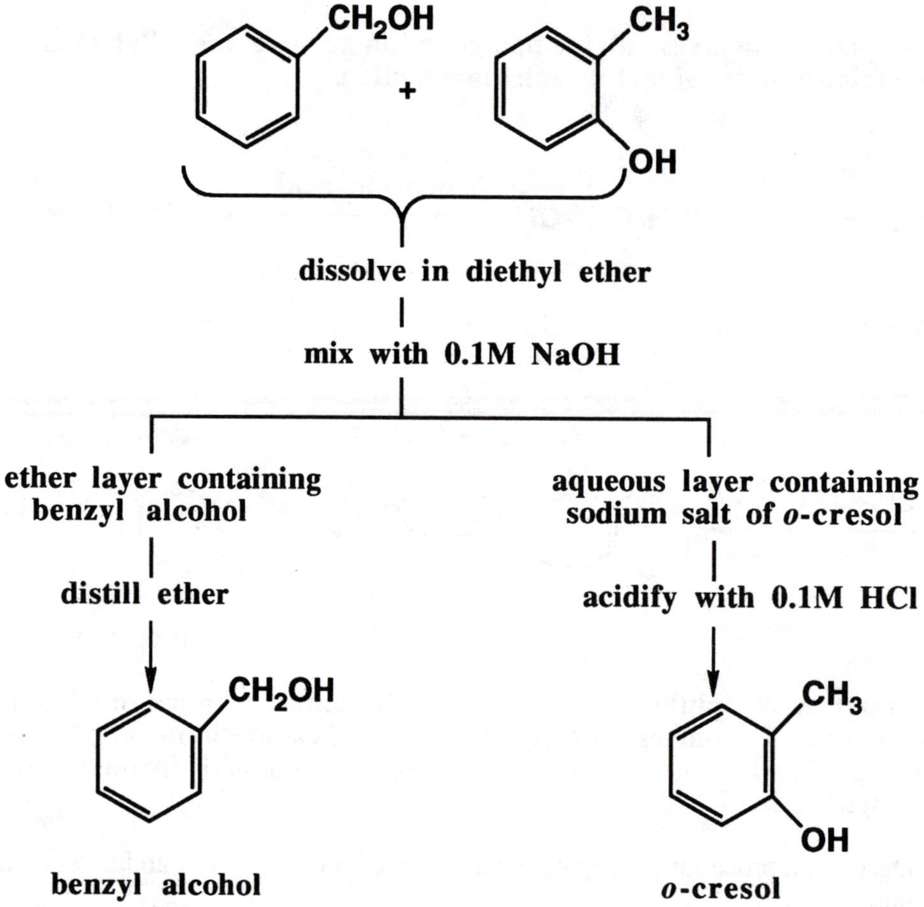

CH_2OH + CH_3 / OH

dissolve in diethyl ether

mix with 0.1M NaOH

ether layer containing benzyl alcohol — **distill ether** → CH_2OH **benzyl alcohol**

aqueous layer containing sodium salt of *o*-cresol — **acidify with 0.1M HCl** → CH_3 / OH ***o*-cresol**

Problem 15.27 The molecule 2-hydroxypyridine, a derivative of pyridine, is in equilibrium with 2- pyridone. 2-Hydroxypyridine is aromatic. Does 2-pyridone have comparable aromatic character? Explain.

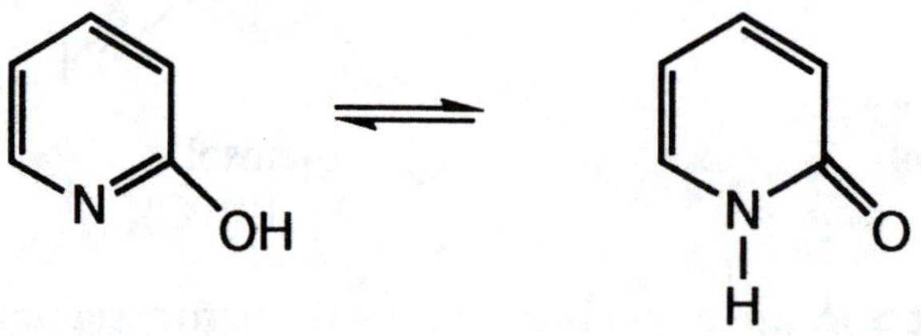

N OH ⇌ N O / H

2-hydroxypyridine 2-pyridone

2-Pyridones have aromatic character because of the contribution from the resonance form shown on the right.

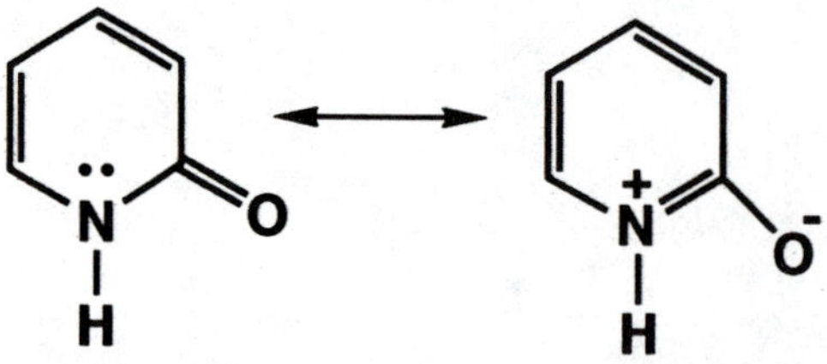

N O / H ⟷ N+ O- / H

Problem 15.28 Draw the structural formula of the product you would expect from Claisen rearrangement of the following allyl phenyl ether.

$OCH_2CH{=}CHCH_3$

Reactions at the Benzylic Position

Problem 15.29 Write a balanced equation for the oxidation of *p*-xylene to 1,4-benzenedicarboxylic acid (terephthalic acid) using potassium permanganate in aqueous sodium hydroxide. How many milligrams of $KMnO_4$ are required to oxidize 250 mg of *p*-xylene to terephthalic acid?

Following are balanced equations for the oxidation half-reaction and the reduction half-reaction. Because oxidation takes place in aqueous NaOH, each half-reaction is balanced using OH^- and H_2O to balance oxygens and hydrogens.

$$CH_3C_6H_4CH_3 + 14OH^- \longrightarrow {}^-O_2CC_6H_4CO_2^- + 10H_2O + 12e^-$$ **oxidation half-reaction**

$$4 \times [\ MnO_4^- + 2H_2O + 3e^- \longrightarrow MnO_2 + 4OH^-\ ]$$ **reduction half-reaction**

$$CH_3C_6H_4CH_3 + 4MnO_4^- \longrightarrow {}^-O_2CC_6H_4CO_2^- + 4MnO_2 + 2H_2O + 2OH^-$$ **balanced redox equation**

$$\mathbf{0.250\ g\ Xyl \times \frac{mol\ Xyl}{106g\ Xyl} \times \frac{4\ mol\ KMnO_4}{1\ mol\ Xyl} \times \frac{158\ g\ KMnO_4}{mol\ KMnO_4}}$$

$$\mathbf{= 1.49\ g\ KMnO_4 = 1490\ mg\ KMnO_4}$$

Problem 15.30 Each of the following reactions occurs by a radical chain mechanism.

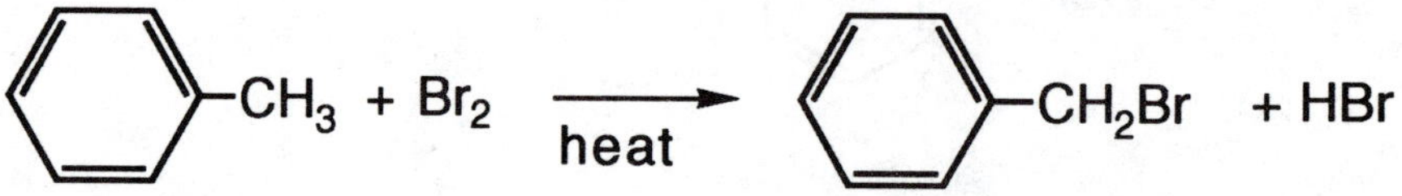

C_6H_5–CH_3 + Br_2 $\xrightarrow{heat}$ C_6H_5–CH_2Br + HBr

toluene → benzyl bromide

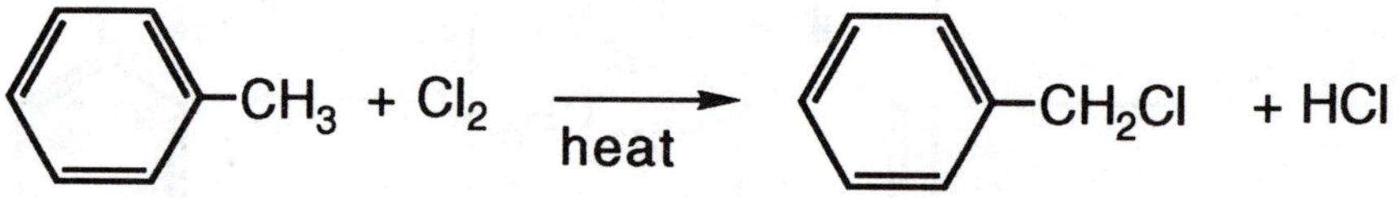

C_6H_5–CH_3 + Cl_2 $\xrightarrow{heat}$ C_6H_5–CH_2Cl + HCl

toluene → benzyl chloride

(a) Calculate the enthalpy of reaction, ΔH, in kcal/mol for each reaction.

Following are the calculations of the enthalpies of reaction.

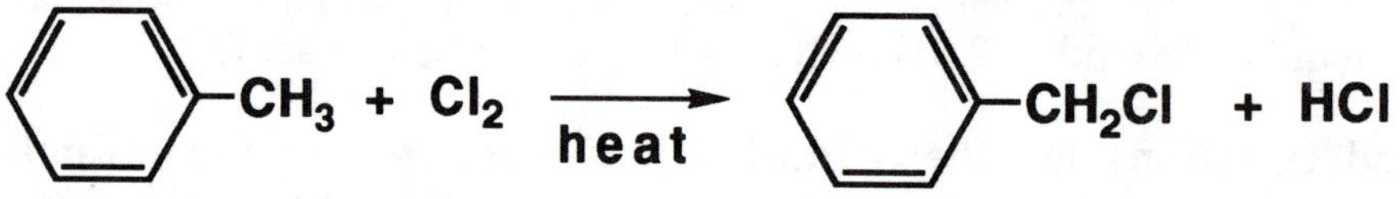

C_6H_5–CH_3	+ Cl_2	$\xrightarrow{heat}$	C_6H_5–CH_2Cl	+ HCl	ΔH (kcal/mol)
+88	**+58**		**-72**	**-103**	**-29**

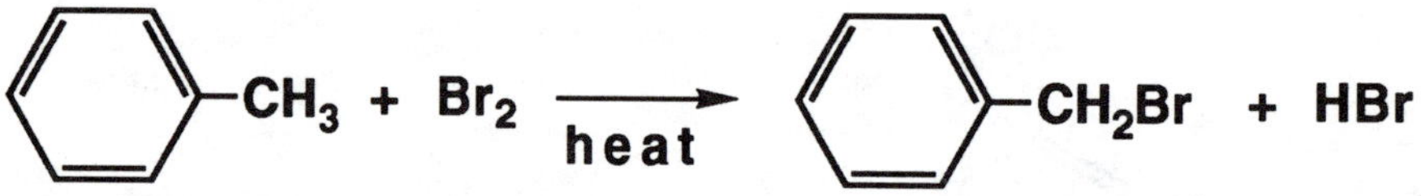

C_6H_5–CH_3	+ Br_2	$\xrightarrow{heat}$	C_6H_5–CH_2Br	+ HBr	ΔH (kcal/mol)
+88	**+46**		**-55**	**-88**	**-9**

(b) Write a pair of chain propagation steps for each mechanism and show that the net result of the chain propagation steps is the observed reaction.

(c) Calculate ΔH for each chain propagation step and show that the sum for each pair of chain propagation steps is identical with the value of ΔH calculated in part (a).

Shown below are pairs of chain propagation steps for each reaction. Each pair adds up to the observed reaction and the observed enthalpy of reaction. This answers both (b) and (c).

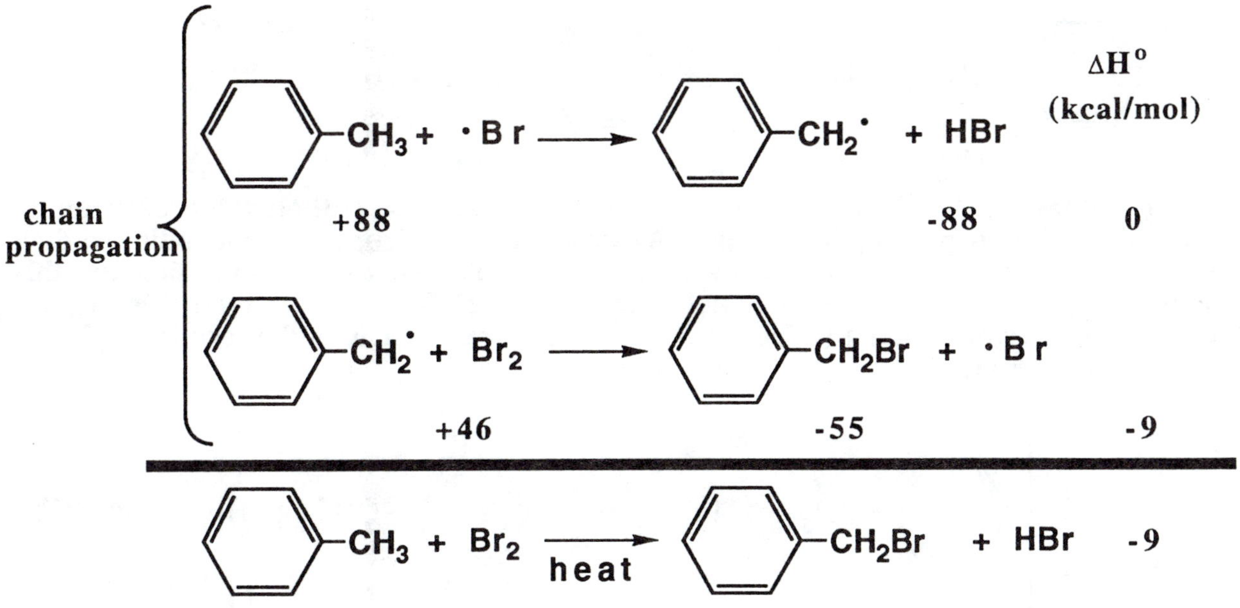

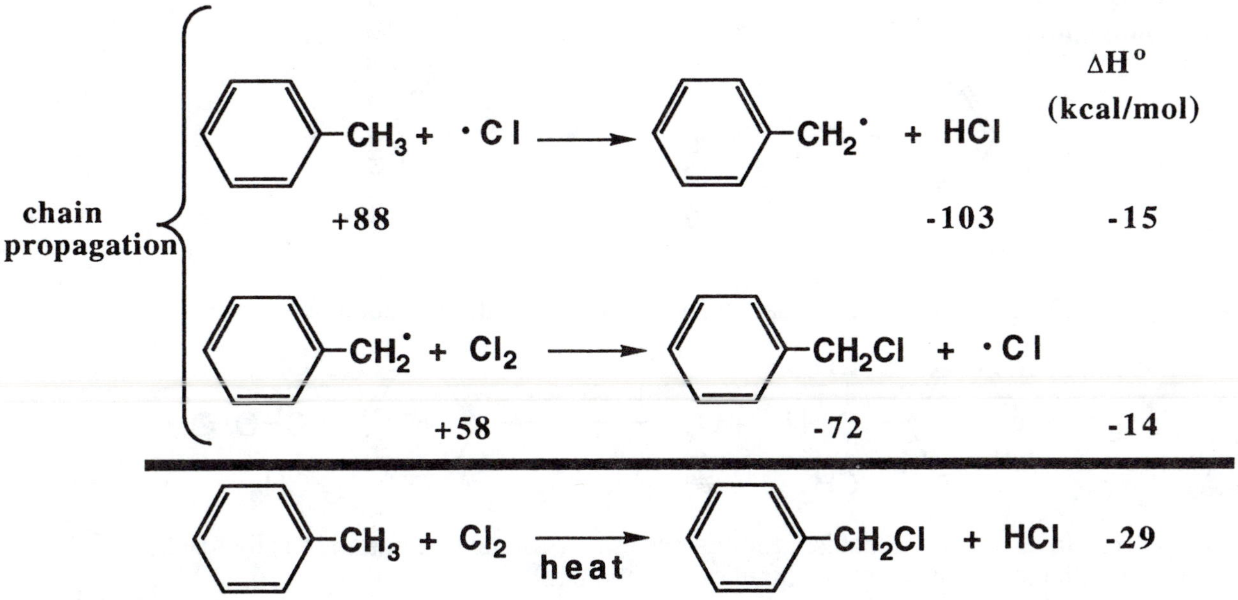

chain propagation

Reaction			ΔH° (kcal/mol)
C_6H_5–CH_3 + ·Br ⟶ C_6H_5–CH_2· + HBr	+88	-88	0
C_6H_5–CH_2· + Br_2 ⟶ C_6H_5–CH_2Br + ·Br	+46	-55	-9
C_6H_5–CH_3 + Br_2 —heat→ C_6H_5–CH_2Br + HBr			-9

chain propagation

Reaction			ΔH° (kcal/mol)
C_6H_5–CH_3 + ·Cl ⟶ C_6H_5–CH_2· + HCl	+88	-103	-15
C_6H_5–CH_2· + Cl_2 ⟶ C_6H_5–CH_2Cl + ·Cl	+58	-72	-14
C_6H_5–CH_3 + Cl_2 —heat→ C_6H_5–CH_2Cl + HCl			-29

Problem 15.31 Following is an equation for iodination of toluene.

C_6H_5–CH_3 + I_2 ⟶ C_6H_5–CH_2I + HI

toluene benzyl iodide

This reaction does not take place. All that happens under experimental conditions for the formation of radicals is initiation to form iodine radicals, I•, followed by termination to reform I_2. How do you account for these observations?

Reaction of toluene and iodine to form benzyl iodide and HI is endothermic.

$$C_6H_5\text{-}CH_3 + I_2 \longrightarrow C_6H_5\text{-}CH_2I + HI$$

	$C_6H_5CH_3$	I_2	$C_6H_5CH_2I$	HI	ΔH (kcal/mol)
	+88	+36	-45	-71	+8

Using values for bond dissociation energies, calculate the enthalpy for each of the most likely chain propagation steps. Abstraction of hydrogen by an iodine radical (an iodine atom) is endothermic by 17 kcal/mol. The energy of activation for this step is approximately a few kcal/mol greater than 17 kcal/mol. Given this large energy of activation and the fact that the overall reaction is endothermic, it does not occur as written.

chain propagation

$$C_6H_5\text{-}CH_3 + I\cdot \longrightarrow C_6H_5\text{-}\dot{C}H_2 + HI \qquad \Delta H\ (kcal/mol)$$

+88 -71 +17

$$C_6H_5\text{-}\dot{C}H_2 + I_2 \longrightarrow C_6H_5\text{-}CH_2I + I\cdot$$

+36 -45 -9

Problem 15.32 Following is an equation for hydroperoxidation of cumene.

$$C_6H_5\text{-}C(CH_3)_2\text{-}H + O_2 \xrightarrow{\text{light}} C_6H_5\text{-}C(CH_3)_2\text{-}O\text{-}OH$$

Propose a radical chain mechanism for this reaction. Assume that initiation is by reaction of cumene with molecular oxygen, a diradical (Section 1.9).

The stability of the benzyl radical, especially with the added methyl groups, facilitates the reaction with oxygen according to the following mechanism.

Step 1: Initiation

$$C_6H_5\text{-}C(CH_3)_2\text{-}H + \cdot\ddot{O}\text{-}\ddot{O}\cdot \longrightarrow C_6H_5\text{-}\dot{C}(CH_3)_2 + H\text{-}\ddot{O}\text{-}\ddot{O}\cdot$$

Step 2: Propagation

Step 3: 2nd Propagation Step

Problem 15.33 Para-substituted benzyl halides undergo reaction with methanol by an S_N1 mechanism to give a benzyl ether. Account for the following order of reactivity under these conditions.

$$R-C_6H_4-CH_2Br + CH_3OH \xrightarrow{\text{methanol}} R-C_6H_4-CH_2OCH_3 + HCl$$

Rate of S_N1 reaction: $R = CH_3O- > CH_3- > H- > NO_2-$

The rate-limiting step in this process is formation of the benzylic cation, so anything that stabilizes the benzylic cation will speed up the S_N1 reaction. Electron-donating groups such as the methoxy group are stabilizing to a benzylic cation. Electron-withdrawing groups such as the nitro group are destabilizing to a benzylic cation. Thus, the groups are listed in order from most electron-donating to most electron-withdrawing.

Problem 15.34 When warmed in dilute sulfuric acid, 1-phenyl-1,2-propanediol undergoes dehydration and rearrangement to give 2-phenylpropanal.

$$C_6H_5-CH(OH)-CH(OH)-CH_3 \xrightarrow{H_2SO_4} C_6H_5-CH(CH_3)-C(=O)-H + H_2O$$

1-phenyl-1,2-propanediol　　　2-phenylpropanal

(a) Propose a mechanism for this example of a pinacol rearrangement (Section 9.5F).

Step 1: proton transfer, loss of water and formation of resonance-stabilized benzylic carbocation.

$$C_6H_5\text{-}CH(\ddot{O}H)\text{-}CH(OH)\text{-}CH_3 + H\text{-}\ddot{O}^+(H)\text{-}H \longrightarrow C_6H_5\text{-}CH(\ddot{O}^+H_2)\text{-}CH(OH)\text{-}CH_3 + :\ddot{O}(H)\text{-}H$$

$$C_6H_5\text{-}CH(\ddot{O}^+H_2)\text{-}CH(OH)\text{-}CH_3 \longrightarrow C_6H_5\text{-}\overset{+}{C}H\text{-}CH(OH)\text{-}CH_3 + H_2\ddot{O}:$$

Step 2: migration of a pair of electrons and methyl to form a new, resonance-stabilized cation.

$$C_6H_5\text{-}\overset{+}{C}H\text{-}C(CH_3)(:\ddot{O}H)\text{-}H \longrightarrow C_6H_5\text{-}CH(CH_3)\text{-}\overset{+}{C}(:\ddot{O}\text{-}H)\text{-}H \longleftrightarrow C_6H_5\text{-}CH(CH_3)\text{-}C(=\overset{+}{\ddot{O}}\text{-}H)\text{-}H$$

Step 3: proton transfer to give the observed aldehyde.

$$C_6H_5\text{-}CH(CH_3)\text{-}C(=\overset{+}{\ddot{O}}\text{-}H)\text{-}H + :\ddot{O}(H)\text{-}H \longrightarrow C_6H_5\text{-}CH(CH_3)\text{-}C(=\ddot{O}:)\text{-}H + H\text{-}\ddot{O}^+(H)\text{-}H$$

(b) Account for the fact that 2-methylpropanal is formed rather than either of the following ketones.

$$C_6H_5\text{-}\overset{O}{\overset{\|}{C}}\text{-}CH_2\text{-}CH_3 \qquad C_6H_5\text{-}CH_2\text{-}\overset{O}{\overset{\|}{C}}\text{-}CH_3$$

In the observed reaction that leads to the aldehyde product, protonation of the benzylic hydroxyl followed by loss of H_2O gives a benzylic carbocation. The ketone on the right would result from protonation and loss of the other -OH group, but that would give a less stable secondary carbocation so it is not observed.

Problem 15.35 In the chemical synthesis of DNA and RNA, hydroxyl groups are normally converted to triphenylmethyl (trityl) ethers to protect the hydroxyl group from reaction with other reagents.

$$RCH_2OH + Ph_3CCl \xrightarrow{\text{tertiary amine}} RCH_2OCPh_3 + HCl$$

HCl: neutralized by the tertiary amine

Ph_3CCl: triphenylmethyl chloride (trityl chloride)

RCH_2OCPh_3: a triphenylmethyl ether (a trityl ether)

Triphenylmethyl ethers are stable to aqueous base, but are rapidly cleaved in aqueous acid.

$$RCH_2OCPh_3 + H_2O \xrightarrow{H^+} RCH_2OH + Ph_3COH$$

(a) Why are triphenylmethyl ethers so readily hydrolyzed by aqueous acid?

The triphenylmethyl ethers are hydrolyzed according to the following mechanism.

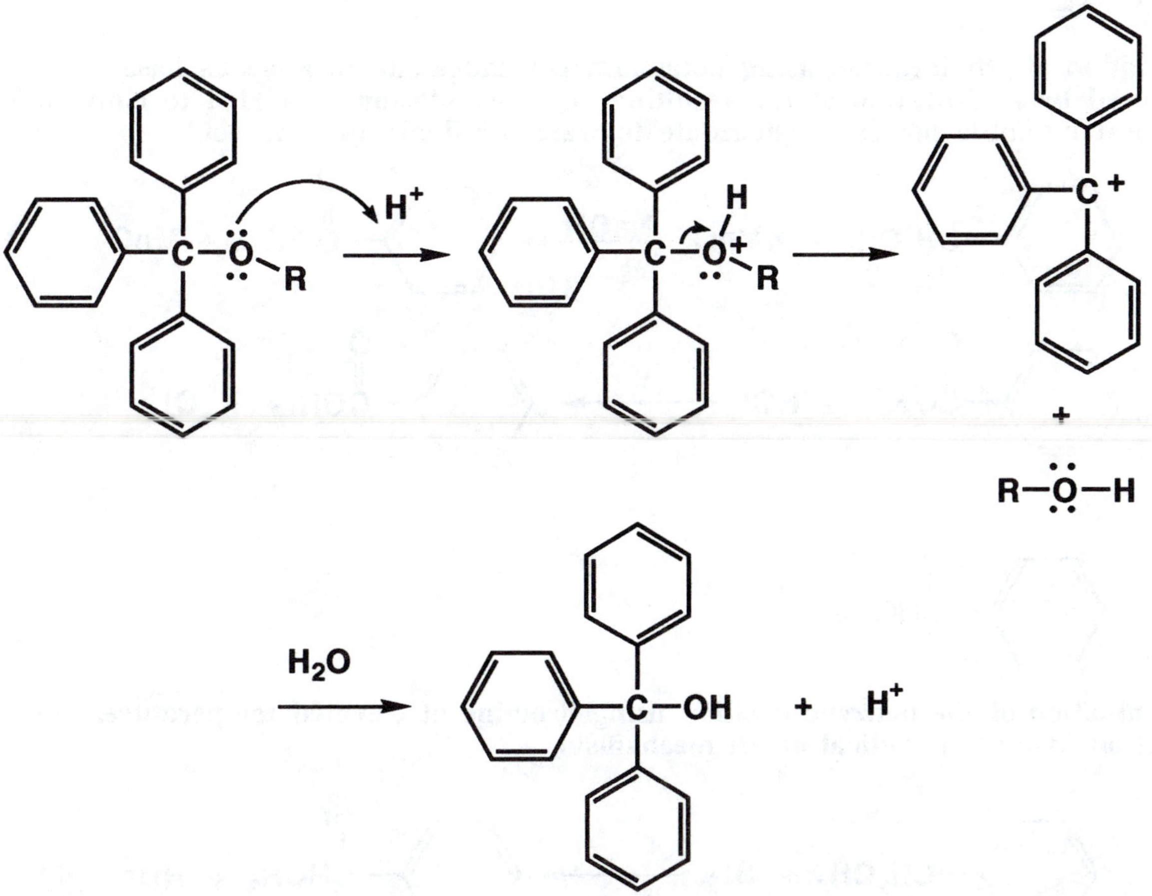

H⁺ → C–O(H⁺)–R → C+ + R–O–H

H₂O → C–OH + H⁺

The key step in this mechanism is the heterolytic bond cleavage that generates the triphenylmethyl cation and the alcohol product. The reaction is facilitated because this cation is stabilized by all three phenyl rings and thus more stable than a simple benzyl cation.

(b) How might the structure of the triphenylmethyl group be modified in order to increase or decrease its acid sensitivity?

Electron-releasing substituents like methoxy groups stabilize the triphenylmethyl cation and thereby increase the sensitivity of these triphenyl methyl ethers to acid. On the other hand, electron-withdrawing groups like nitro groups destabilize the triphenyl methyl cation and thereby decrease the sensitivity of these triphenylmethyl ethers to acid.

Synthesis

Problem 15.36 Using ethylbenzene as the only aromatic starting material, show how to synthesize the following compounds. In addition to the indicated starting material, use any other necessary organic or inorganic chemicals. Note that any compound already synthesized in one part of this problem may then be used to make any other compound in the problem.

(a) C_6H_5COOH

Oxidation of ethylbenzene using potassium permanganate in aqueous base followed by acidification of the resulting aqueous solution with HCl to convert the water-soluble potassium benzoate to water-insoluble benzoic acid.

$$C_6H_5CH_2CH_3 + KMnO_4 \xrightarrow{NaOH} C_6H_5CO_2^-K^+ + MnO_2$$

$$C_6H_5CO_2^-K^+ + HCl \longrightarrow C_6H_5COOH + K^+Cl^-$$

(b) $C_6H_5CHBrCH_3$

Bromination of the benzylic position using bromine at elevated temperature. The reaction involves a radical chain mechanism.

$$C_6H_5CH_2CH_3 + Br_2 \xrightarrow[heat]{} C_6H_5CHBrCH_3 + HBr$$

(c) 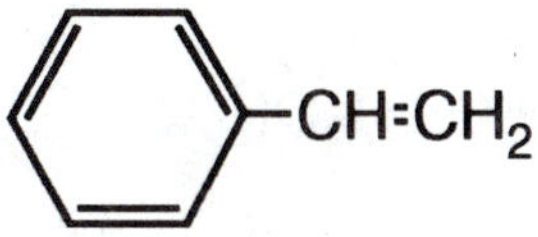 C_6H_5-CH=CH$_2$

Dehydrohalogenation of the alkyl bromide from (b), brought about by a strong base such as KOH. This reaction is an example of a beta-elimination reaction.

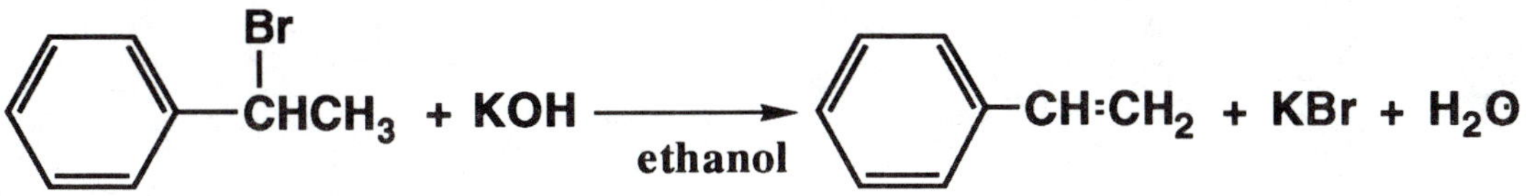

$$C_6H_5\text{-}CH(Br)CH_3 + KOH \xrightarrow[\text{ethanol}]{} C_6H_5\text{-}CH{=}CH_2 + KBr + H_2O$$

(d) 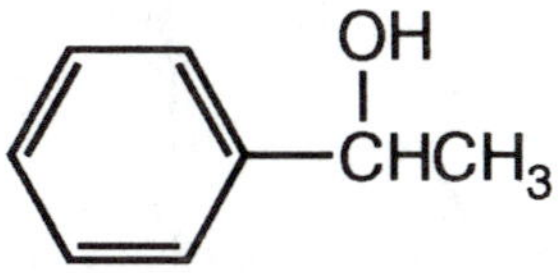 C_6H_5-CH(OH)CH$_3$

Acid-catalyzed hydration of a carbon-carbon double bond of (c). The same reaction may also be brought about by oxymercuration followed by reduction with $NaBH_4$.

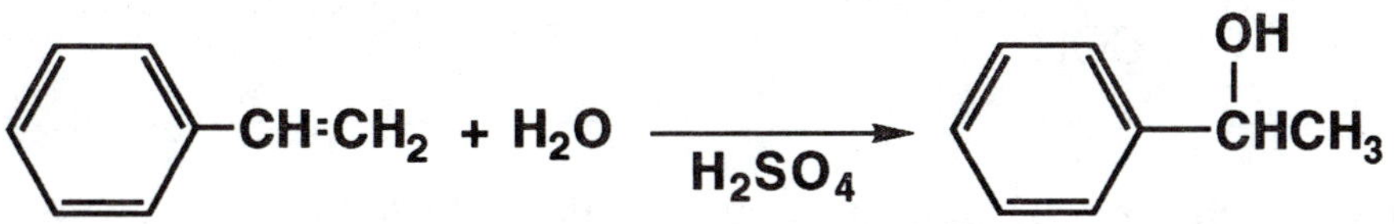

$$C_6H_5\text{-}CH{=}CH_2 + H_2O \xrightarrow[H_2SO_4]{} C_6H_5\text{-}CH(OH)CH_3$$

(e) 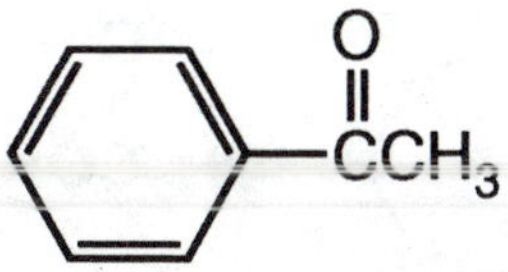 C_6H_5-C(=O)CH$_3$

Oxidation of the secondary alcohol of (d) using chromic acid in aqueous sulfuric acid-acetone. The same oxidation may be brought about using the more selective oxidizing agents chromium trioxide in pyridine or pyridinium chlorochromate.

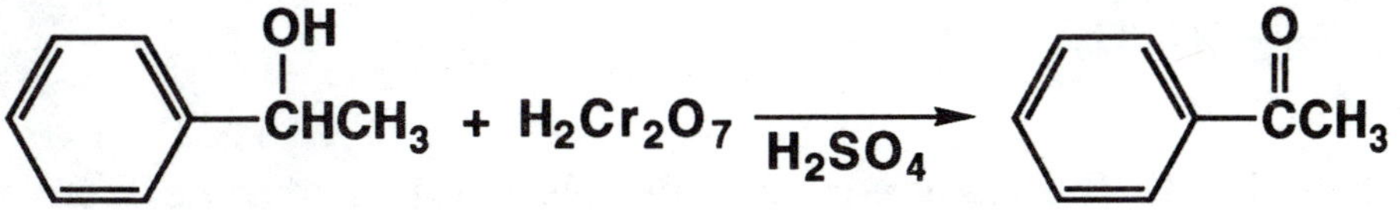

$$C_6H_5\text{-}CH(OH)CH_3 + H_2Cr_2O_7 \xrightarrow[H_2SO_4]{} C_6H_5\text{-}C(=O)CH_3$$

(f) 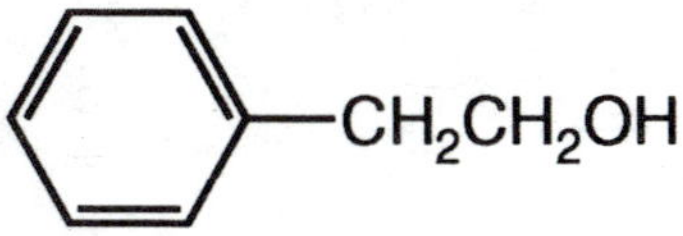 C_6H_5-CH$_2$CH$_2$OH

Hydroboration/oxidation of styrene from part (c).

$$C_6H_5-CH{=}CH_2 \xrightarrow[\text{2) } H_2O_2,\ NaOH]{\text{1) } B_2H_6} C_6H_5-CH_2CH_2OH$$

(g) $C_6H_5-CH_2\overset{O}{\overset{\|}{C}}H$

Oxidation of the primary alcohol of (f) using pyridinium chlorochromate.

$$C_6H_5-CH_2CH_2OH + PCC \xrightarrow[CH_2Cl_2]{} C_6H_5-CH_2\overset{O}{\overset{\|}{C}}H$$

(h) $C_6H_5-CH_2\overset{O}{\overset{\|}{C}}OH$

Oxidation of the primary alcohol of (f) using chromic acid. The same product may be formed by similar oxidation of the aldehyde (g).

$$C_6H_5-CH_2CH_2OH + H_2Cr_2O_7 \xrightarrow[\text{acetone}]{H_2SO_4} C_6H_5-CH_2\overset{O}{\overset{\|}{C}}OH$$

(i) $C_6H_5-\overset{Br}{\overset{|}{C}}H\overset{Br}{\overset{|}{C}}H_2$

Addition of bromine to the carbon-carbon double bond of styrene from part (c).

$$C_6H_5-CH{=}CH_2 + Br_2 \xrightarrow[CCl_4]{} C_6H_5-\overset{Br}{\overset{|}{C}}H\overset{Br}{\overset{|}{C}}H_2$$

(j) C_6H_5–C≡CH

A double dehydrohalogenation of product (i) using sodium amide as the base.

$$C_6H_5\text{-}\underset{Br}{CH}\underset{Br}{CH_2} + 2NaNH_2 \longrightarrow C_6H_5\text{-}C{\equiv}CH + 2NaBr + 2NH_3$$

(k) C_6H_5–$C{\equiv}CCH_2CH{=}CH_2$

The terminal alkyne from (j) is deprotonated with sodium amide to produce the anionic species that reacts with allyl chloride to produce the desired product.

$$C_6H_5\text{-}C{\equiv}CH \xrightarrow{NaNH_2} C_6H_5\text{-}C{\equiv}C{:}^- \xrightarrow{ClCH_2CH=CH_2} C_6H_5\text{-}C{\equiv}C\text{-}CH_2\text{-}CH{=}CH_2$$

(l) C_6H_5–$C{\equiv}C(CH_2)_5CH_3$

The deprotonated terminal alkyne from (k) reacts with hexyl chloride to produce the desired product.

$$C_6H_5\text{-}C{\equiv}C{:}^- \xrightarrow{ClCH_2\text{-}(CH_2)_4\text{-}CH_3} C_6H_5\text{-}C{\equiv}C\text{-}CH_2\text{-}(CH_2)_4\text{-}CH_3$$

(m) $C_6H_5(H)C{=}C(H)(CH_2)_5CH_3$ (trans)

The alkyne produced in part (l) is reduced with sodium metal in liquid ammonia to produce the desired *trans* alkene.

$$C_6H_5\text{-}C{\equiv}C\text{-}CH_2\text{-}(CH_2)_4\text{-}CH_3 \xrightarrow[NH_3(l)]{2Na} \text{trans-}C_6H_5CH{=}CH(CH_2)_5CH_3$$

(n) $C_6H_5CH{=}CH(CH_2)_5CH_3$ (cis; C_6H_5 and $(CH_2)_5CH_3$ on the same side, H and H on the same side)

The alkyne produced in part (l) is reduced with hydrogen and the Lindlar catalyst to produce the desired *cis* alkene.

$$C_6H_5\text{-}C{\equiv}C\text{-}CH_2\text{-}(CH_2)_4\text{-}CH_3 \xrightarrow[\text{Lindlar Catalyst}]{H_2} \text{cis-}C_6H_5CH{=}CH(CH_2)_5CH_3$$

Problem 15.37 Show how to convert 1-phenylpropane into each of the following. In addition to this starting material, use any necessary inorganic reagents. Any compound synthesized in one part of this problem may then be used to make any other compound in the problem.

(a) $C_6H_5\text{-}CHBrCH_2CH_3$

Radical chain bromination of 1-phenylpropane. Bromination is highly regioselective for the benzylic position.

$$C_6H_5\text{-}CH_2CH_2CH_3 + Br_2 \xrightarrow{\text{heat}} C_6H_5\text{-}CHBrCH_2CH_3 + HBr$$

(b) $C_6H_5\text{-}CH{=}CHCH_3$

Dehydrohalogenation of product (a) using KOH or other strong base.

$$C_6H_5\text{-}CHBrCH_2CH_3 + KOH \xrightarrow{\text{ethanol}} C_6H_5\text{-}CH{=}CHCH_3 + KBr + H_2O$$

(c) $C_6H_5\text{-}CHCl\text{-}CHClCH_3$

Addition of chlorine to the double bond of product (b) by electrophilic addition.

$$C_6H_5\text{-}CH{=}CHCH_3 + Cl_2 \longrightarrow C_6H_5\text{-}\overset{\overset{Cl}{|}}{C}H\text{-}\overset{\overset{Cl}{|}}{C}HCH_3$$

(d) $C_6H_5\text{-}C{\equiv}CCH_3$

Double dehydrohalogenation of product (c) using sodium amide as the base.

$$C_6H_5\text{-}\overset{\overset{Cl}{|}}{C}H\text{-}\overset{\overset{Cl}{|}}{C}HCH_3 + 2NaNH_2 \longrightarrow C_6H_5\text{-}C{\equiv}CCH_3 + 2NaBr + 2NH_3$$

(e) $C_6H_5(H)C{=}C(H)CH_3$ (C_6H_5 and CH_3 on opposite sides)

Chemical reduction of the alkyne from part (d) to a *cis*-alkene using sodium metal in liquid ammonia.

$$C_6H_5\text{-}C{\equiv}CCH_3 + Na \xrightarrow{NH_3(l)} C_6H_5(H)C{=}C(H)CH_3$$

(f) $C_6H_5(H)C{=}C(H)CH_3$ (C_6H_5 and CH_3 on the same side)

Catalytic reduction of the alkyne using the specially prepared Lindlar catalyst to reduce the alkyne to the alkene, but not further reduce the alkene.

$$C_6H_5\text{-}C{\equiv}CCH_3 + H_2 \xrightarrow{\text{Lindlar catalyst}} C_6H_5(H)C{=}C(H)CH_3$$

(g) $C_6H_5\text{-}\overset{\overset{OH}{|}}{C}H\text{-}\overset{\overset{OH}{|}}{C}HCH_3$

Oxidation of the alkene from part (b) to a glycol using potassium permanganate in cold, alkaline solution at pH 11.8. Alternatively oxidize the alkene with osmium tetroxide in the presence of hydrogen peroxide.

$$C_6H_5\text{-}CH{=}CHCH_3 + KMnO_4 \xrightarrow[\text{pH } 11.8]{} C_6H_5\text{-}\overset{\overset{OH}{|}}{C}H\text{-}\overset{\overset{OH}{|}}{C}HCH_3$$

(h) $C_6H_5-CH(OH)CH_2CH_3$

Acid-catalyzed hydration of the alkene from part (b). The reaction is highly regioselective because of the stability of the benzylic carbocation formed by protonation of the alkene. Alternatively, oxymercuration followed by reduction with NaBH4 forms the same secondary alcohol.

$$C_6H_5-CH{=}CHCH_3 + H_2O \xrightarrow[H_2SO_4]{} C_6H_5-CH(OH)CH_2CH_3$$

(i) $C_6H_5-C(=O)CH_2CH_3$

Oxidation of the secondary alcohol of (h) using chromic acid in aqueous sulfuric acid and the alcohol dissolved in acetone. Alternatively use chromium trioxide in pyridine or pyridinium chlorochromate as the oxidizing agent.

$$C_6H_5-CH(OH)CH_2CH_3 + H_2Cr_2O_7 \xrightarrow[H_2SO_4]{} C_6H_5-C(=O)CH_2CH_3$$

Problem 15.38 Benzylic bromination followed by loss of HBr by heating at high temperature can be used to generate reactive intermediates such as (1). How do you take advantage of this observation to synthesize hexaradialene. (See L.G. Harruff, M. Brown, and V. Boekelheide, *J. Am. Chem. Soc.*, **100**, 2893 (1978)).

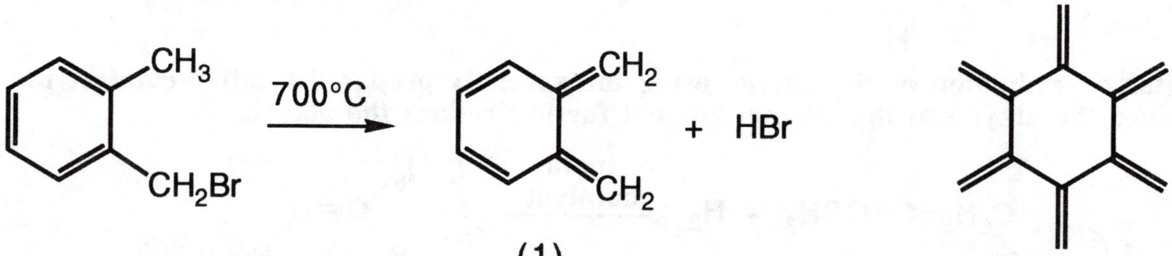

CH_3, CH_2Br — 700°C → CH_2, CH_2 + HBr

(1)

hexaradialene

Using the above observation as a starting point, hexaradialene could be synthesized from the starting material shown below.

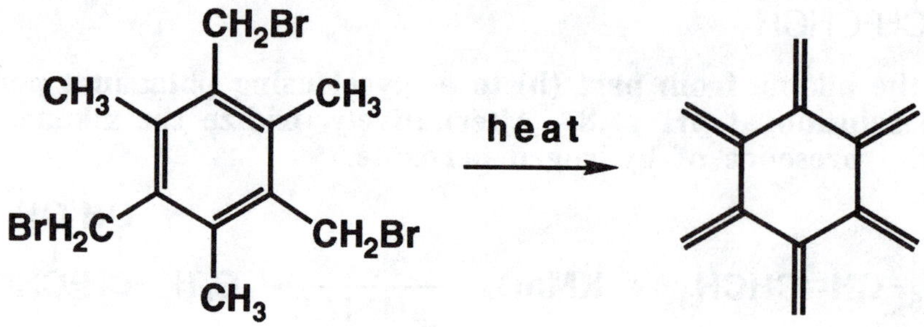

CH_2Br, CH_3, CH_3, BrH_2C, CH_2Br, CH_3 — heat →

CHAPTER 16: AROMATICS II: REACTIONS OF BENZENE AND ITS DERIVATIVES

SUMMARY OF REACTIONS

Starting Material ↓ / Product →	Alkyl Benzenes	Anilines	Aryl Halides	Aryl Hydrazines	Aryl Ketones	Aryl Sulfonic Acids	Phenols		Nitro Aromatics
Aromatic Rings			**16A** 16.1A*			**16B** 16.1B			**16C** 16.1B
Aromatic Rings Acid Chlorides					**16D** 16.1C				
Aromatic Rings Alcohols	**16E** 16.1D								
Aromatic Rings Alkenes	**16F** 16.1D								
Aromatic Rings Alkyl Halides	**16G** 16.1C								
Aryl Halides		**16H** 16.3A		**16I** 16.3B			**16J** 16.3A	**16K** 16.3B	

*Section in book that describes reaction.

REACTION 16A: BROMINATION AND CHLORINATION (Section 16.1A)

$$\text{benzene} \xrightarrow[X\,=\,Br,\ Cl]{X_2,\ FeX_3} \text{C}_6\text{H}_5\text{–X}$$

- **Aromatic rings** react with **Br_2** in the presence of the **Lewis acid catalyst $FeBr_3$** to give an **aryl bromide**. This is an example of **electrophilic aromatic substitution.** ∗
- The mechanism involves an initial reaction between Br_2 and $FeBr_3$ to generate a molecular complex that can rearrange to give a Br^+ $FeBr_4^-$ ion pair. This reacts as a very strong electrophile with the weakly nucleophilic aromatic pi cloud to form a resonance-stabilized cation that loses a proton to give the final product.
- An analogous reaction can be carried out with **Cl_2 and $FeCl_3$** to give an **aryl chloride**.

REACTION 16B: SULFONATION (Section 16.1B)

$$\text{benzene} \xrightarrow[H_2SO_4]{SO_3} \text{C}_6\text{H}_5\text{–}SO_3H$$

- **Aromatic rings** react with **SO_3** in the presence of **sulfuric acid** to yield **aryl sulfonic acids** via **electrophilic aromatic substitution.** ✱
- The mechanism involves reaction of SO_3 as a very strong electrophile with the weakly nucleophilic aromatic pi cloud to form a resonance-stabilized cation that loses a proton to give the final product.

REACTION 16C: NITRATION (Section 16.1B)

HNO_3, H_2SO_4; NO_2

- **Aromatic rings** react with **nitric acid** in the presence of **sulfuric acid** to yield **nitro aromatic compounds** via **electrophilic aromatic substitution.** ✱
- The mechanism involves an initial reaction between nitric acid and sulfuric acid to yield the nitronium ion NO_2^+. This reacts as a very strong electrophile with the weakly nucleophilic aromatic pi cloud to form a resonance-stabilized cation that loses a proton to give the final product.

REACTION 16D: FRIEDEL-CRAFTS ACYLATION (Section 16.1C)

+ —C—C'(=O)—Cl; $AlCl_3$; O, C'—C—

- **Aromatic rings** react with **acyl chlorides** in the presence of a Lewis acid catalyst like **$AlCl_3$** to produce an aryl ketone via **electrophilic aromatic substitution.** ✱
- The mechanism involves an initial reaction between the acyl chloride and $AlCl_3$ to yield the acylium ion R-C+=O. This reacts as a very strong electrophile with the weakly nucleophilic aromatic pi cloud to form a resonance stabilized cation that loses a proton to give the final product.
- Rearrangement is not a problem with acylium ions like it is with carbocations.

REACTION 16E: REACTIONS OF ALCOHOLS WITH AROMATIC RINGS IN THE PRESENCE OF STRONG ACID (Section 16.1D)

+ —C—OH; H^+; C

- **Aromatic rings** react with **alcohols** in the presence of a strong **acid catalyst** like **H_3PO_4, H_2SO_4, and HF** to produce an alkyl benzene via **electrophilic aromatic substitution.** ✱

- The mechanism involves an initial reaction between the alcohol and strong acid to yield a carbocation. This reacts as a very strong electrophile with the weakly nucleophilic aromatic pi cloud to form a resonance stabilized cation that loses a proton to give the final product.
- Because carbocations are involved in the mechanism, rearrangements can be a problem, especially with primary or secondary alcohols.

REACTION 16F: REACTIONS OF ALKENES WITH AROMATIC RINGS IN THE PRESENCE OF STRONG ACID OR LEWIS ACID (Section 16.1D)

$$\text{C}_6\text{H}_6 + \text{C=C'} \xrightarrow{H^+ \text{ or } AlCl_3} \text{C}_6\text{H}_5\text{–C–C'H}$$

- **Aromatic rings** react with **alkenes** in the presence of a **strong acid catalyst** like **H_3PO_4, H_2SO_4,** and **HF** or **Lewis acid** like **$AlCl_3$** to produce an **alkyl benzene** via **electrophilic aromatic substitution.** ✱
- The mechanism involves an initial reaction between the alkene and strong acid or Lewis acid to yield a positively charged carbocation. This reacts as a very strong electrophile with the weakly nucleophilic aromatic pi cloud to form a resonance stabilized cation that loses a proton to give the final product.

Because carbocations are involved in the mechanism, rearrangements can be a problem.

REACTION 16G: FRIEDEL-CRAFTS ALKYLATION (Section 16.1C)

$$\text{C}_6\text{H}_6 + \text{–C–X} \xrightarrow{AlCl_3} \text{C}_6\text{H}_5\text{–C} + HX$$

- **Aromatic rings** react with **alkyl halides** in the presence of a **Lewis acid** like **$AlCl_3$** to produce an **alkyl benzene** via **electrophilic aromatic substitution.** ✱
- The mechanism involves an initial reaction between the alkyl halide and Lewis acid to yield an intermediate that can be thought of as a carbocation. This reacts as a very strong electrophile with the weakly nucleophilic aromatic pi cloud to form a resonance stabilized cation that loses a proton to give the final product.
- Because carbocations are involved in the mechanism, rearrangements can be a problem, especially with primary or secondary alkyl halides.

REACTION 16H: REACTION OF AN ARYL HALIDE WITH SODIUM AMIDE (Section 16.3A)

$$\text{C}_6\text{H}_5\text{X} + NaNH_2 \xrightarrow{NH_3\ (l)} \text{C}_6\text{H}_5NH_2$$

- **Aryl halides** react with **strongly basic nucleophiles** such as **sodium amide** to yield **anilines**. The -NH_2 group ends up on the ring carbon atom that was originally bonded to the halogen, as well as positions adjacent (ortho) to it. ✱
- The mechanism involves an initial reaction between the aryl halide and strong base to give a benzyne intermediate. This undergoes addition at either sp carbon atom to give the aniline products.

REACTION 16I: REACTION OF AN ARYL HALIDE WITH HYDRAZINE (Section 16.3B)

X + NH_2NH_2 ⟶ $NHNH_2$

- **Activated aryl halides** react with **strong nucleophiles** such as **hydrazine** to give **aryl hydrazines** in a regioselective manner. The -$NHNH_2$ group ends up on the ring carbon atom that was originally bonded to the halogen. This reaction does not occur unless there are electron withdrawing groups ortho and/or para to the halogen. The electron withdrawing groups activate the ring toward nucleophilic attack. ✱
- Unlike reaction 16H that involves a benzyne intermediate, this reaction involves a nucleophilic attack of the ring carbon containing the halogen to give a negatively charged Meisenheimer complex. Loss of halogen results in formation of the product.

REACTION 16J: REACTION OF AN ARYL HALIDE WITH SODIUM HYDROXIDE (Section 16.3A)

X ⟶ OH

1) NaOH Heat, Pressure
2) HCl, H_2O

- **Aryl halides** react with **strongly basic nucleophiles** such as **sodium hydroxide** to yield **phenols**. The -OH group ends up on the ring carbon atom that was originally bonded to the halogen, as well as positions adjacent (ortho) to it. ✱
- The mechanism involves an initial reaction between the aryl halide and strong base to give a benzyne intermediate. This undergoes addition at either sp hybridized carbon atom to give the aniline products.

REACTION 16K: REACTION OF AN ACTIVATED ARYL HALIDE WITH AQUEOUS BASE (Section 16.3B)

X ⟶ OH

1) Na_2CO_3, H_2O Heat, Pressure
2) HCl, H_2O

- **Activated aryl halides** react with **aqueous base** to give **phenols**. The -OH group ends up on the ring carbon atom that was originally bonded to the halogen. This reaction does not occur unless there are electron withdrawing groups ortho and/or para to the halogen. The electron withdrawing groups activate the ring toward nucleophilic attack. ✱

- This reaction involves a nucleophilic attack of the ring carbon containing the halogen to give a negatively charged Meisenheimer complex. Loss of halogen results in formation of the product.

SUMMARY OF IMPORTANT CONCEPTS

16.0 OVERVIEW

- **Aromatic rings,** such as benzene, **react with very strong electrophiles** in a reaction that results in substitution of a ring hydrogen. **Nucleophilic attack** on aromatic rings is **much less common.** ✻

16.1 ELECTROPHILIC AROMATIC SUBSTITUTION

- A variety of electrophiles react with aromatic rings via **electrophilic aromatic substitution.** The electrophiles are usually positively charged and examples include reactions 16A-16G. The general **mechanism involves attack on the electrophile** by the weakly nucleophilic aromatic pi cloud to form **a resonance-stabilized cation intermediate** that loses a proton to give the final product. ✻

H + E^+ → [+ H E] → E + H^+

16.2 DISUBSTITUTION AND POLYSUBSTITUTION

- **Substituents** other than hydrogen on an aromatic ring can have a **profound influence** on the **reaction rate** and **substitution pattern** of **electrophilic aromatic substitution.** In particular, groups can either speed up or slow down the reaction, and can direct new groups meta or ortho-para. ✻
- Substituents can be divided into three broad classes:
  - **Alkyl groups** and all **groups in which the atom bonded to the ring** has an **unshared pair of electrons** are **ortho-para directing**. These groups are **electron-donating** and are thus **activating** toward electrophilic aromatic substitution.
  - Groups in which the **atom bonded to the aromatic ring** bears a **partial or full positive charge** are **meta directing**. These groups often have **multiple bonds such as =O on the atom bonded to the aromatic ring**. These groups are **electron-withdrawing** and are thus **deactivating** toward electrophilic aromatic substitution.
  - **Halogens** are exceptions in that they are **ortho-para directing**, but **electron-withdrawing** and thus **weakly deactivating** toward electrophilic aromatic substitution.
- These effects have a large practical significance, since in synthesizing polysubstituted aromatics, the **order of addition of the substituents must be taken into account**. For example, when making *m*-bromonitrobenzene from benzene, the nitro group (meta directing) must be added before the bromine atom (ortho-para directing). Adding the bromine atom first followed by the nitro group would result in the majority of product being the unwanted ortho and para isomers.
- These effects are the result of two types of interactions:
  - 1) **An inductive effect** in which atoms that are **more electronegative** than the sp^2 ring carbons **pull electron density out** of the aromatic ring, or conversely atoms that are **less electronegative** than the sp^2 ring carbons **add electron density** into the aromatic ring. Since the aromatic ring is acting as a weak nucleophile in the electrophilic substitution reactions, the **more electronegative atoms or groups (electron-withdrawing) reduce** electron density and thus the **nucleophilicity of the ring**, so they are **deactivating**. The **less electronegative atoms or groups (electron-**

releasing) increase the electron density and thus the **nucleophilicity of the ring**, so they are **activating.** ✳
- **2) A resonance effect** in which the **positively-charged cation intermediate** in an electrophilic aromatic substitution is **stabilized**, or **destabilized**. Groups such as $-NH_2$ or -OH in which **different resonance forms reveal that electron density is added to the ring** help to **stabilize the cation intermediate**, thereby **lowering the activation barrier** of the reaction and **activating the ring**. Other groups like $-NO_2$ or -C≡N in which different resonance forms reveal that **electron density is removed from the ring destabilize the cation intermediate**, thereby **raising the activation barrier** of the reaction and **deactivating the ring.** ✳

• These inductive and resonance effects have **different levels of influence depending on their position** relative to the incoming electrophile. ✳
- The effects of substituents that ***activate*** **the ring** are **most *activating* ortho and para.** For this reason, the **incoming electrophile reacts predominantly ortho and para** to the existing activating substituent. Such activating substituents are referred to as being **ortho-para directing.**
- The effects of substituents that ***deactivate*** **the ring** are **most *deactivating* ortho and para.** For this reason, the **incoming electrophile reacts predominantly meta** to the existing activating substituent, that is the less deactivated positions. Such deactivating substituents are referred to as being **meta directing**..

• The three basic types of substituents can be understood in the context of the above ideas. ✳
- Electron-releasing groups are activating, and thus always ortho-para directing.
- Electron-withdrawing groups are deactivating and thus meta directing.
- The halogens show deactivation, but ortho-para substitution preference. This is not in violation of the above rules, because they show overall **inductive deactivation** of the **aromatic ring**, but **resonance stabilization** of the **cation intermediate** and thus ortho-para direction.

CHAPTER 16
Solutions to the Problems

Problem 16.1 Write the stepwise mechanism for sulfonation of benzene. In this reaction, the electrophile is SO_3 formed as shown in the following equation.

$$H_2SO_4 \rightleftharpoons SO_3 + H_2O$$

Hint: In thinking about a mechanism for this reaction, consider formal charges on sulfur and oxygen in the Lewis structure of sulfur trioxide.

In sulfonation of benzene, the electrophile is sulfur trioxide. In step 1, reaction of benzene with the electrophile yields a resonance-stabilized cation. In step 2, this intermediate loses a proton to complete the reaction: water is shown as the base accepting the proton. A second-proton transfer reaction gives benzene sulfonic acid.

Step 1:

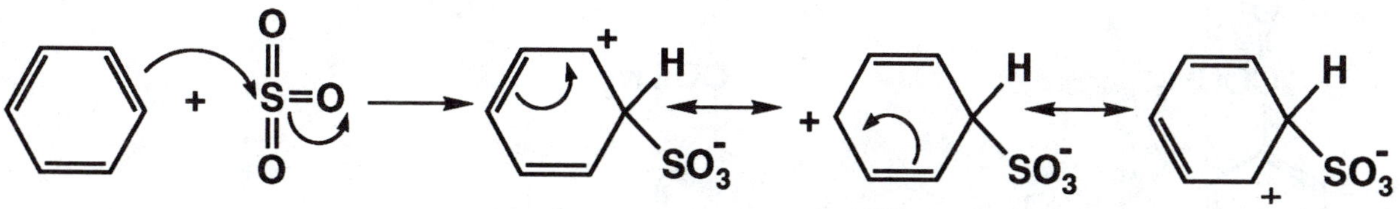

(a resonance-stabilized cation intermediate)

Step 2:

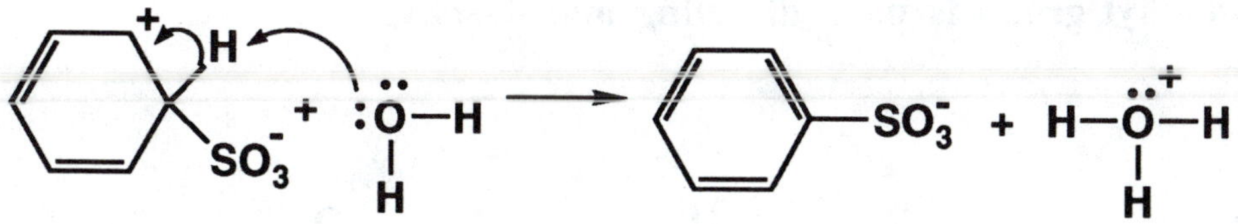

Step 3:

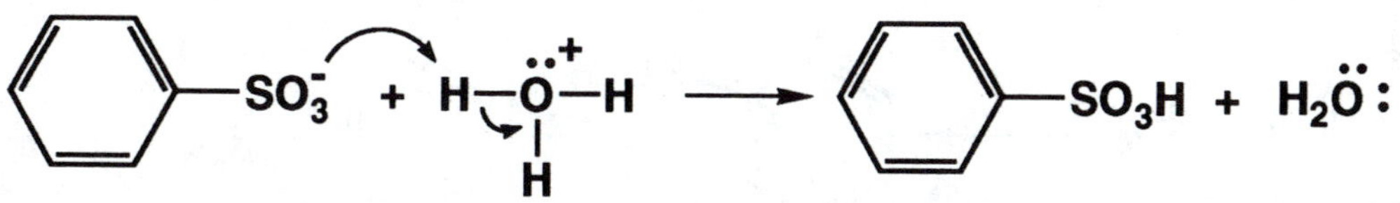

benzenesulfonic acid

Problem 16.2 Write structural formulas of the products you would expect from Friedel-Crafts alkylation or acylation of benzene with:

(a) 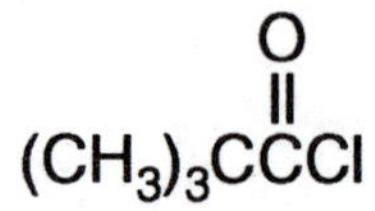 $(CH_3)_3CCCl$ (with C=O)   (b) $CH_3CH_2CH_2Cl$ (c) 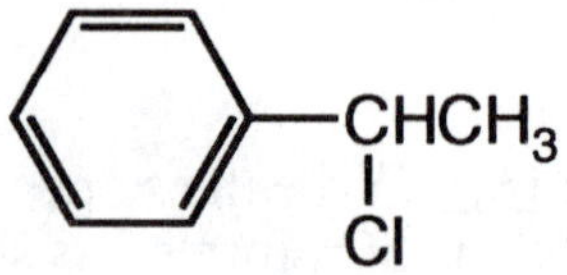 C_6H_5–CHCH$_3$ (with Cl)

C_6H_5–C(=O)–C$(CH_3)_3$ C_6H_5–CH(CH_3)CH$_3$ C_6H_5–CH(CH_3)–C_6H_5

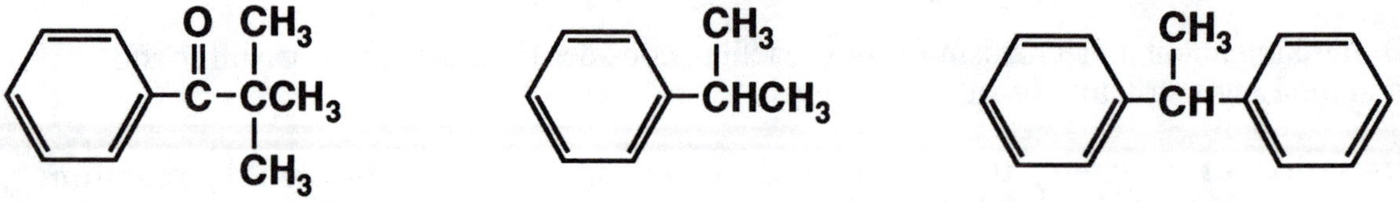

Problem 16.3 Complete the following electrophilic aromatic substitution reactions. Where you predict meta substitution, show only the meta product. Where you predict ortho/para substitution, show both products.

(a) 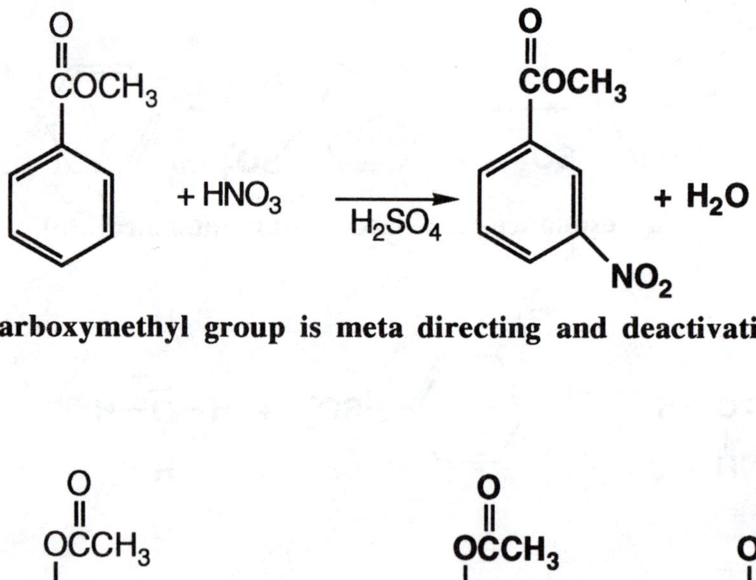 $C_6H_5COCH_3$ (with C=O) + HNO_3 $\xrightarrow{H_2SO_4}$ m-$NO_2C_6H_4COCH_3$ + H_2O

The carboxymethyl group is meta directing and deactivating.

(b) 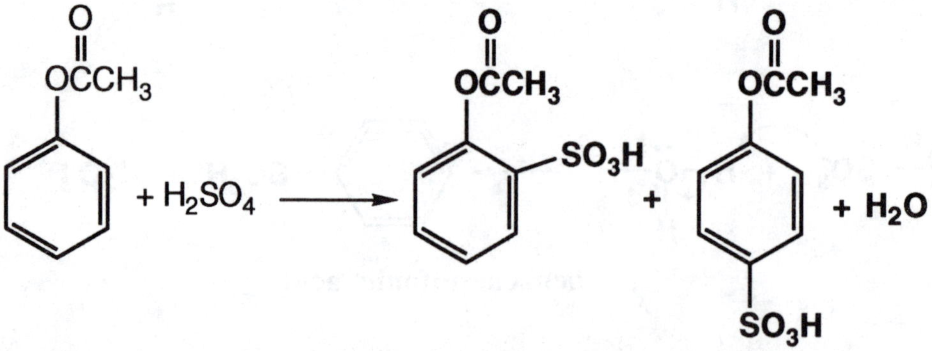 $C_6H_5OCCH_3$ (with C=O) + H_2SO_4 $\longrightarrow$ o-$HO_3SC_6H_4OCCH_3$ + p-$HO_3SC_6H_4OCCH_3$ + H_2O

The acetoxy group is ortho/para directing and activating.

(c) 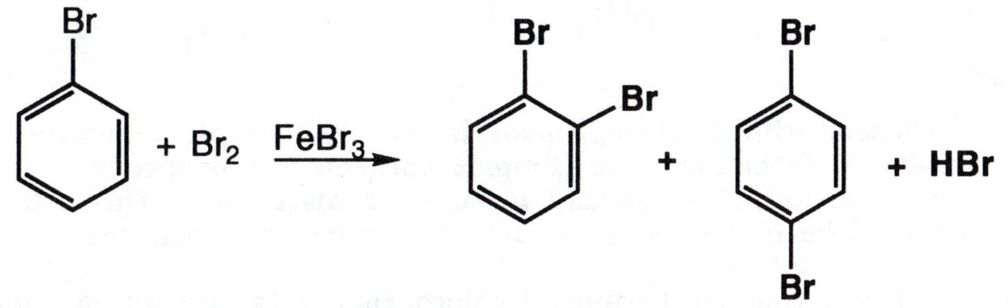 Bromobenzene + Br_2 $\xrightarrow{FeBr_3}$ o-dibromobenzene + p-dibromobenzene + HBr

Bromine is ortho/para directing and weakly deactivating.

Electrophilic Aromatic Substitution: Monosubstitution

Problem 16.4 Write a stepwise mechanism for each of the following reactions. Use curved arrows to show the flow of electrons in each step.

(a) 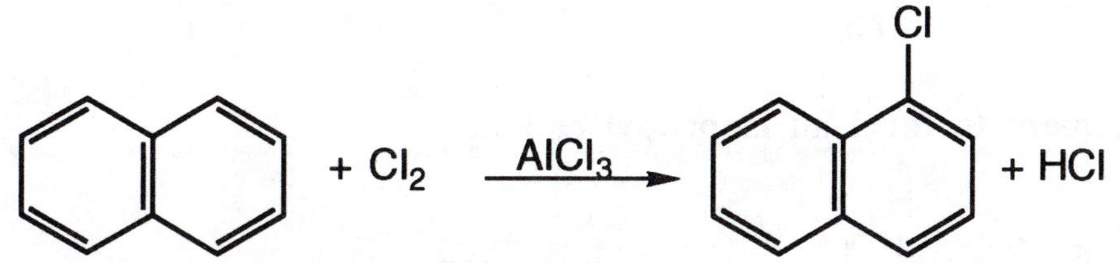 Naphthalene + Cl_2 $\xrightarrow{AlCl_3}$ 1-chloronaphthalene + HCl

Chlorination of naphthalene by chlorine in the presence of aluminum chloride is an example of electrophilic aromatic substitution. It is shown in three steps.

Step 1: Activation of chlorine to form an electrophile.

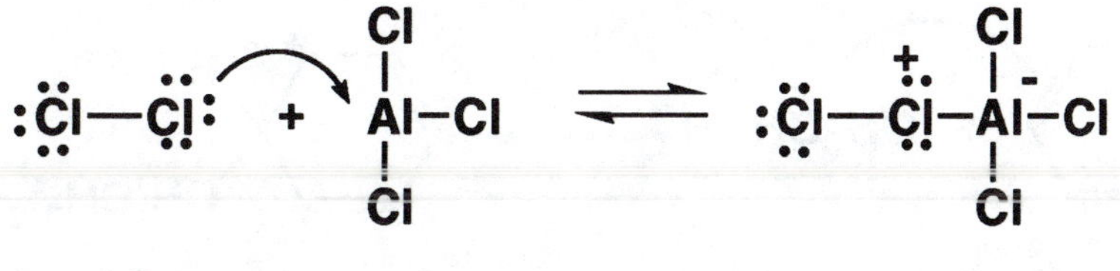

$$:\ddot{Cl}-\ddot{Cl}: + AlCl_3 \rightleftharpoons :\ddot{Cl}-\overset{+}{\ddot{Cl}}-\overset{-}{Al}Cl_3$$

Step 2: Reaction of the electrophile with the aromatic ring, a nucleophile.

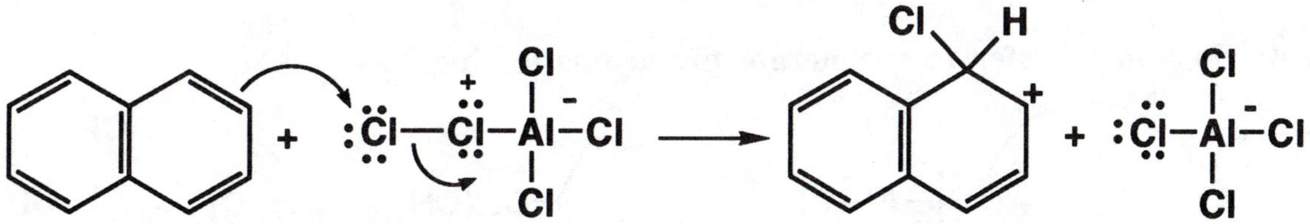

$$\text{Naphthalene} + :\ddot{Cl}-\overset{+}{\ddot{Cl}}-\overset{-}{Al}Cl_3 \longrightarrow \text{cation intermediate} + :\ddot{Cl}-\overset{-}{Al}Cl_3$$

a resonance stabilized cation intermediate

Step 3: Proton transfer and reformation of the aromatic ring.

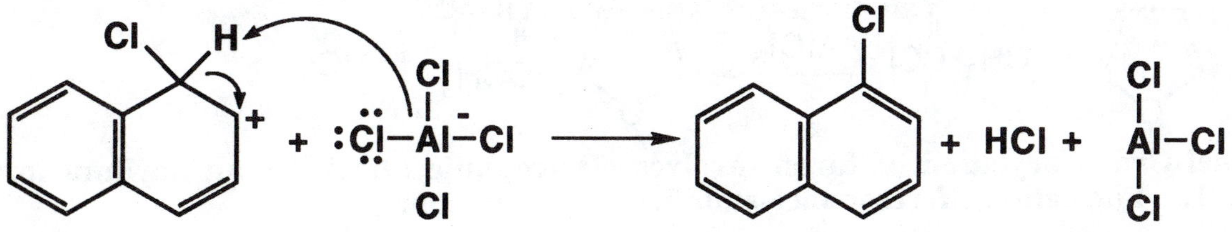

$$\text{cation intermediate} + :\ddot{Cl}-\overset{-}{Al}Cl_3 \longrightarrow \text{1-chloronaphthalene} + HCl + AlCl_3$$

(b) C_6H_6 + $ClCH_2CH_2CH_3$ $\xrightarrow{AlCl_3}$ C_6H_5–$CH(CH_3)_2$

The reaction of benzene with 1-chloropropane in the presence of aluminum chloride involves initial formation of a complex between 1-chloropropane and aluminum chloride, and its rearrangement to an isopropyl cation. This cationic species is the electrophile that undergoes further reaction with benzene.

Step 1: Formation of a complex between 1-chloropropane (a Lewis base) and aluminum chloride (a Lewis acid).

$$CH_3CH_2CH_2-\ddot{C}l: + AlCl_3 \rightleftharpoons CH_3CH_2\overset{+}{C}H_2\ :\ddot{C}l-\overset{-}{Al}Cl_3$$

Step 2: Rearrangement to form an isopropyl cation.

$$CH_3CH(H)\overset{+}{C}H_2 \longrightarrow CH_3\overset{+}{C}HCH_3$$

Step 3: Electrophilic attack on the aromatic ring.

$$C_6H_6 + CH_3\overset{+}{C}HCH_3 \longrightarrow \text{(cyclohexadienyl cation with H and } CH(CH_3)_2)$$

resonance-stabilized cation intermediate

Step 4: Proton transfer to regenerate the aromatic ring.

$$\text{(cation with H and } CH(CH_3)_2) + :\ddot{C}l-\overset{-}{Al}Cl_3 \longrightarrow C_6H_5-CH(CH_3)_2 + H-\ddot{C}l: + AlCl_3$$

(c) furan + $CH_3\overset{O}{\overset{\|}{C}}Cl$ $\xrightarrow{AlCl_3}$ 2-furyl–$\overset{O}{\overset{\|}{C}}CH_3$ + HCl

Friedel-Crafts acylation of furan involves electrophilic attack by an acylium ion.
Step 1: Formation of resonance-stabilized acylium ion.

Step 1: Formation of resonance-stabilized acylium ion.

$$CH_3-\overset{O}{\overset{\|}{C}}Cl: + AlCl_3 \longrightarrow CH_3-\overset{+}{C}=O: \longleftrightarrow CH_3-C\equiv\overset{+}{O}: + :Cl-\overset{-}{Al}Cl_3$$

a resonance-stabilized acylium ion

Step 2: Reaction of acylium ion (an electrophile) and furan (a nucleophile).

$$\text{furan} + CH_3-\overset{+}{C}=O: \longrightarrow \text{(cation intermediate: furan ring with H and } -\overset{O}{\overset{\|}{C}}-CH_3\text{)}$$

Step 3: Proton transfer and regeneration of aromatic ring.

$$\text{intermediate} + :Cl-\overset{-}{Al}Cl_3 \longrightarrow \text{2-furyl}-\overset{O}{\overset{\|}{C}}CH_3 + HCl: + AlCl_3$$

(d) $$2\,C_6H_6 + CH_2Cl_2 \xrightarrow{AlCl_3} C_6H_5-CH_2-C_6H_5 + 2HCl$$

Formation of diphenylmethane involves two successive Friedel-Crafts alkylations.

$$ClCH_2-Cl: + AlCl_3 \rightleftharpoons ClCH_2-\overset{+}{Cl}-\overset{-}{Al}Cl_3$$

$$C_6H_6 + ClCH_2-\overset{+}{Cl}-\overset{-}{Al}Cl_3 \longrightarrow \text{(cation: ring with H and } CH_2Cl\text{)} + :Cl-\overset{-}{Al}Cl_3$$

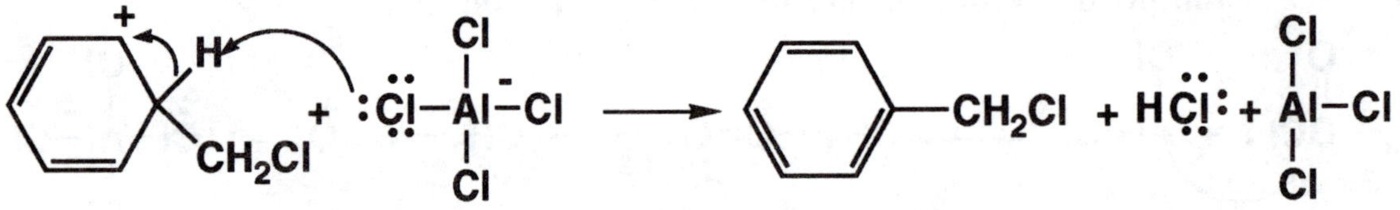

Formation of benzyl chloride completes the first Friedel-Crafts alkylation. This molecule then is a reactant for the second Friedel-Crafts alkylation.

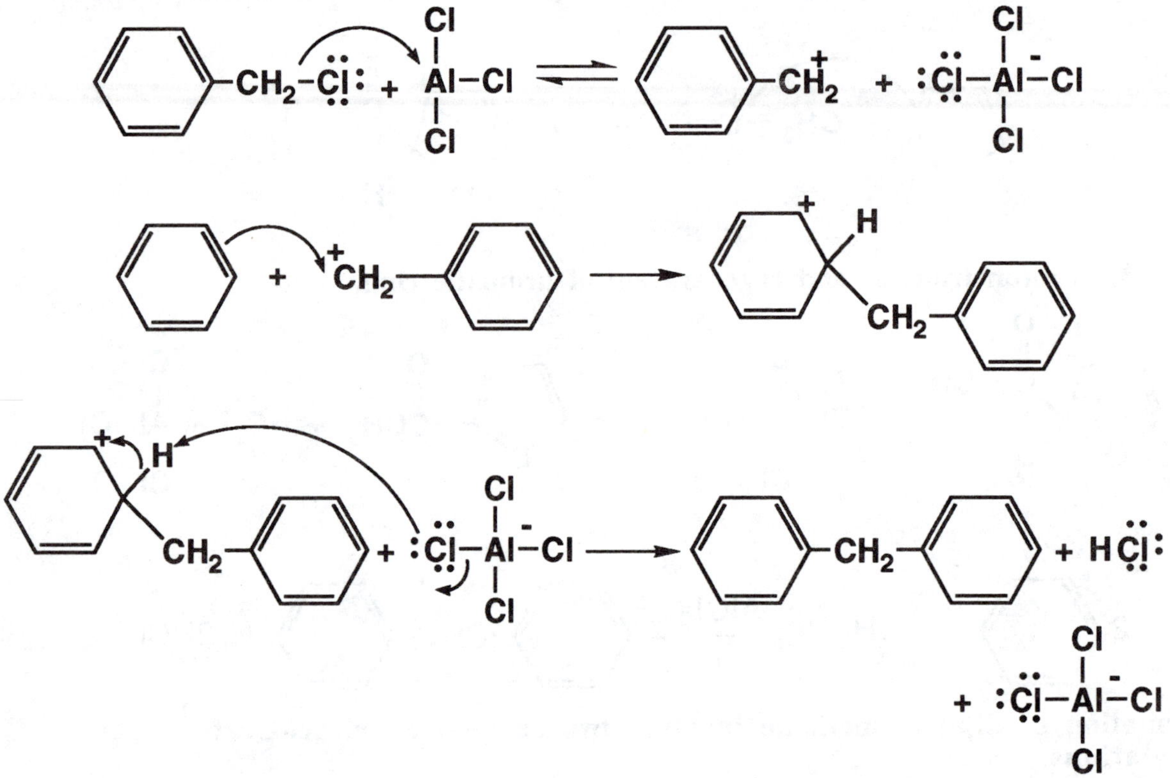

Problem 16.5 Pyridine undergoes electrophilic aromatic substitution preferentially at the 3-position as illustrated by the synthesis of 3-nitropyridine. Pyrrole undergoes electrophilic aromatic substitution preferentially at the 2-position as illustrated by the synthesis of 2-nitropyrrole.

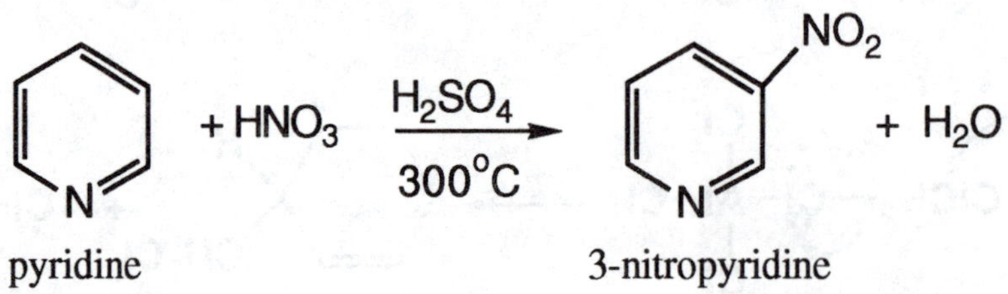

$$\text{pyridine} + HNO_3 \xrightarrow[300^{\circ}C]{H_2SO_4} \text{3-nitropyridine} + H_2O$$

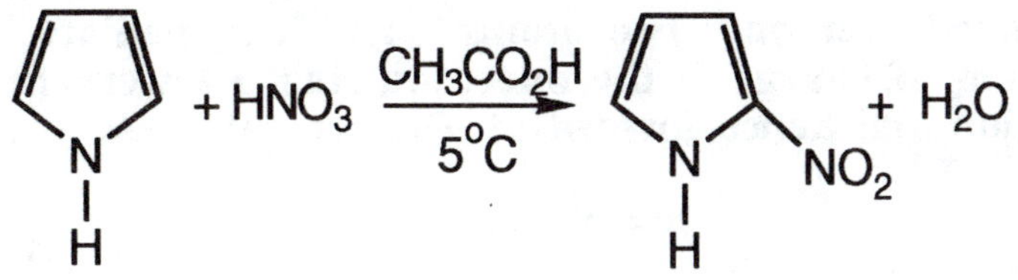

pyrrole + HNO_3 $\xrightarrow[5^{o}C]{CH_3CO_2H}$ 2-nitropyrrole + H_2O

(a) Write resonance contributing structures for the intermediate formed by the attack of NO_2^+ at the 2-, 3-, and 4-positions of pyridine. From examination of these intermediates, offer an explanation for preferential nitration at the 3-position.

Pyridine is a base, and in the presence of a nitric acid-sulfuric acid mixture, it is protonated. It is the protonated form which must be attacked by the electrophile NO_2^+. For nitration at the 3-position, the additional positive charge in the cation intermediate may be delocalized on three carbon atoms of the pyridine ring. None of the contributing structures, however, places both positive charges on the same atom.

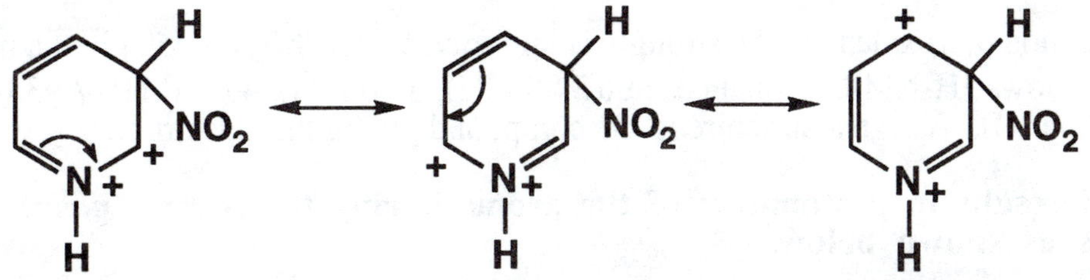

For nitration at the 4-position or the 2-position, the additional positive charge in the cation intermediate is also delocalized on three atoms of the pyridine ring, but one of these contributing structures has a charge of +2 on nitrogen. This situation is thus less stable than that which occurs for nitration at the 3-position.

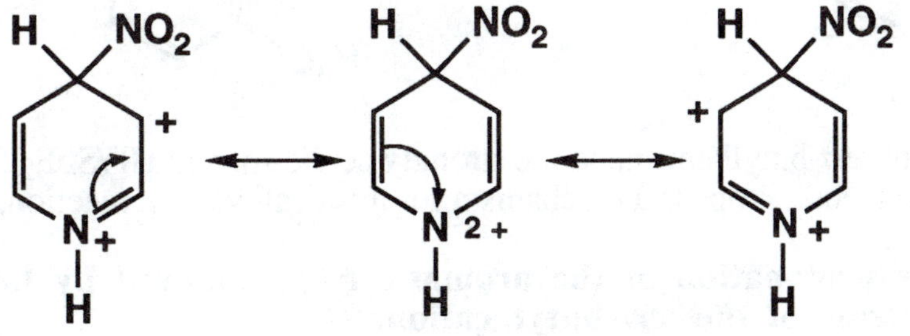

a very poor contributing structure because it places a charge of +2 on nitrogen

(b) Write resonance contributing structures for the intermediate formed by attack of NO_2^+ at the 2- and 3-positions of pyrrole. From examination of these intermediates, offer an explanation for preferential nitration at the 2-position.

Pyrrole is nitrated under considerably milder conditions than pyridine. For nitration at the 2-position, the positive charge on the cation intermediate is

delocalized over three atoms of the pyrrole ring whereas for nitration at the 3-position, it is delocalized over only two atoms. The intermediate with the greater degree of delocalization of charge is the one requiring a lower energy of activation for formation and hence formed at a faster rate.

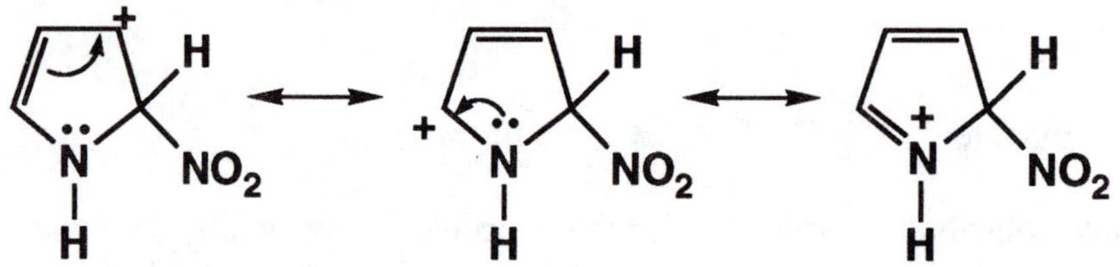

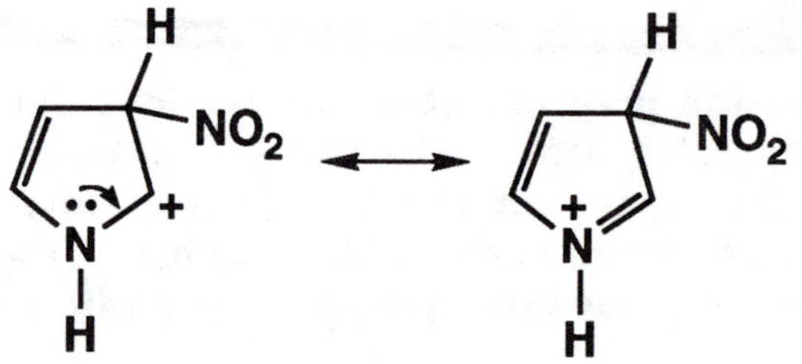

H NO2 H NO2 H NO2

H NO2 H NO2

Problem 16.6 Addition of *m*-xylene to the strongly acidic solvent HF/SbF_5 at -45° gives a new compound which shows ^{1}H-NMR resonances at δ 2.88 (3H), 3.00 (3H), 4.67 (2H), 7.93 (1H), 7.83 (1H), and 8.68 (1H). Assign a structure to the compound giving this spectrum.

The strong acid results in protonation of the aromatic ring to create a positively charged species as shown below.

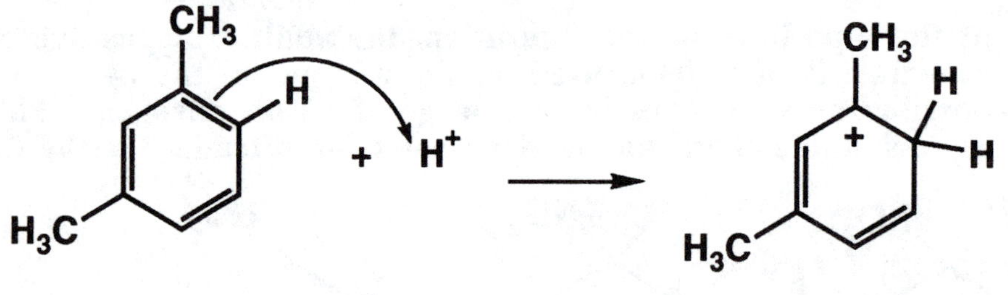

CH_3 H + H^+ ⟶ CH_3 H H H_3C H_3C

Problem 16.7 Addition of *tert*-butylbenzene to the strongly acidic solvent HF/SbF_5 followed by aqueous workup gives benzene. Propose a mechanism for this dealkylation reaction.

The reaction involves protonation of the aromatic ring, followed by heterolytic bond cleavage and release of the *tert*-butyl cation.

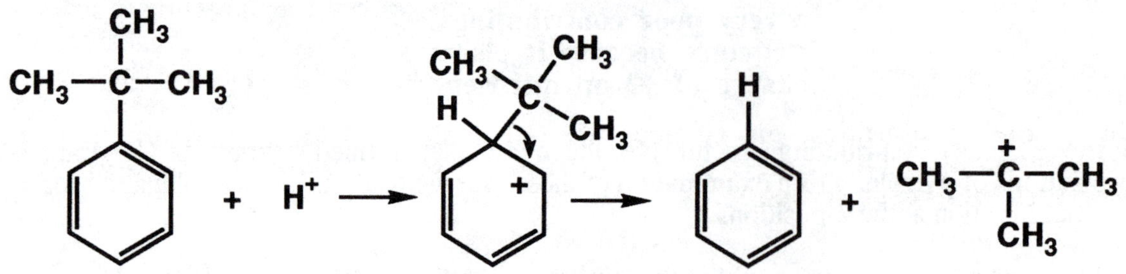

CH_3–C(CH_3)$_2$–Ph + H^+ ⟶ [arenium ion] ⟶ H–Ph + CH_3–$\overset{+}{C}$(CH_3)–CH_3

Problem 16.8 What product do you predict from the reaction of SCl_2 with benzene in the presence of $AlCl_3$? What product results if diphenyl ether is treated with SCl_2 and $AlCl_3$?

The Lewis acid, $AlCl_3$, facilitates departure of one of the chlorine atoms from SCl_2, and the resulting electrophile takes part in electrophilic aromatic substitution reaction to create the -SCl derivative that then reacts again to create diphenyl sulfide. The diphenyl sulfide is activated compared to benzene, so this will react further to generate a polymeric product as shown.

$$C_6H_6 + SCl_2 \xrightarrow{AlCl_3} [C_6H_5SCl] \xrightarrow{AlCl_3} [C_6H_5SC_6H_5] \xrightarrow{AlCl_3} -S-C_6H_4-[C_6H_4-S]_n-$$

If diphenyl ether were treated in a similar manner, the resulting polymeric species will have alternating ether and thioether functions.

$$C_6H_5OC_6H_5 + SCl_2 \xrightarrow{AlCl_3} -S-[C_6H_4-O-C_6H_4-S]_n-$$

Problem 16.9 Other groups besides H^+ can act as leaving groups in electrophilic aromatic substitution reactions. One of the best leaving groups is the trimethylsilyl group (Me_3Si-). For example, treatment of $Me_3SiC_6H_5$ with CF_3CO_2D rapidly forms DC_6H_5. What are the properties of a silicon-carbon bond that allows you to predict this kind of reactivity?

Based on simple electronegativities, the C-Si bond is polarized such that a partial positive charge is on the Si atom. Furthermore, the heterolytic bond cleavage is facilitated because the $(CH)_3Si^+$ cation is so stable.

Disubstitution and Polysubstitution

Problem 16.10 The following groups are ortho-para directors. Draw a contributing structure for the resonance-stabilized aryl cation formed during electrophilic aromatic substitution that shows the role of each group in stabilizing the intermediate and further delocalizing its positive charge.

In the following solutions, the electrophile (E) is shown adding para to the substituent group. The same type of contributing structure can be drawn for ortho attack of the electrophile.

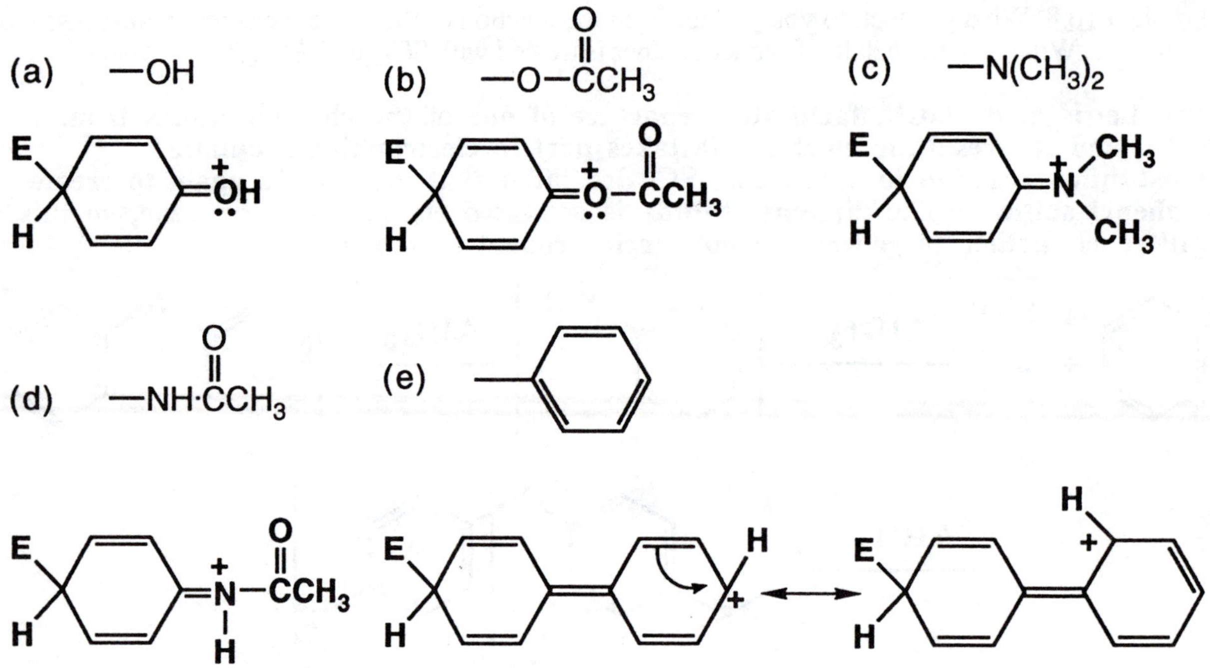

(a) —OH (b) $-O-\overset{O}{\overset{\|}{C}}CH_3$ (c) $-N(CH_3)_2$

(d) $-NH\cdot\overset{O}{\overset{\|}{C}}CH_3$ (e) —C₆H₅

Problem 16.11 Predict the major product from treatment of each molecule with HNO_3, H_2SO_4.

(a)

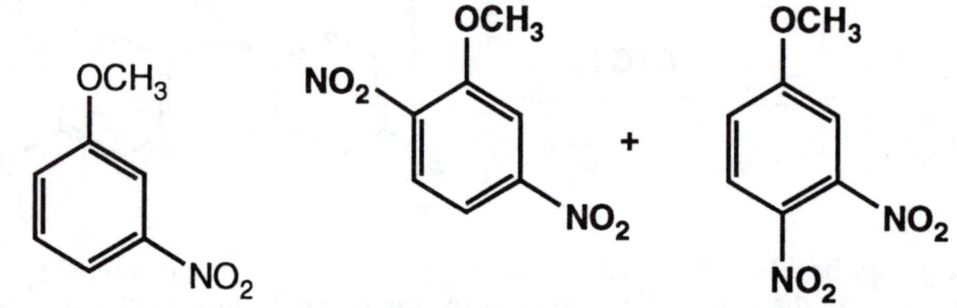

When there is more than one substituent on a ring, the predominant product is derived from the orientation preference of the most activating substituent.

(b)

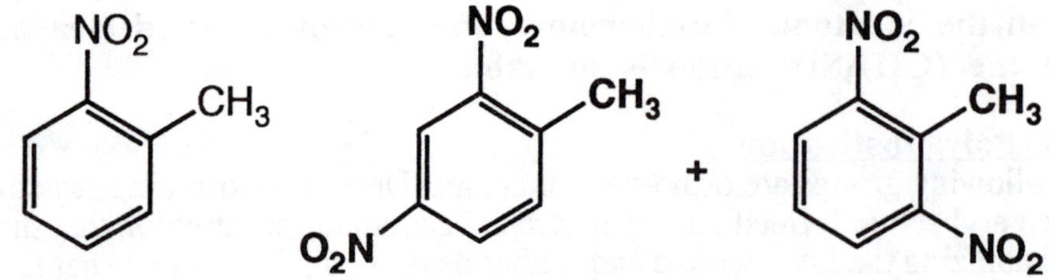

(c)

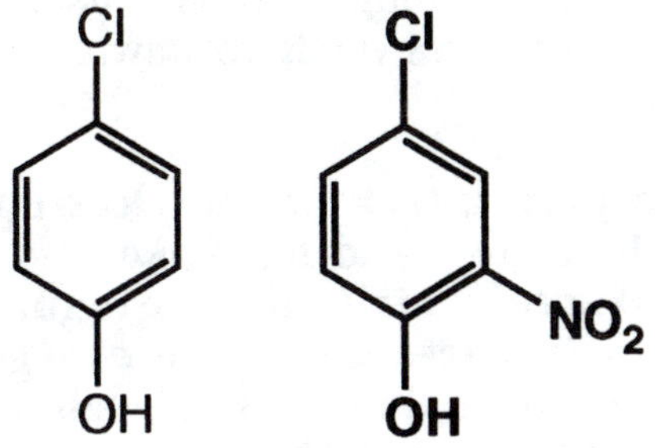

(d) +

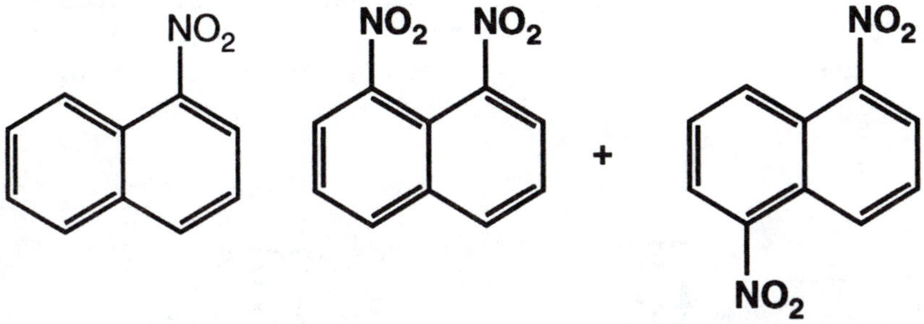

Nitration occurs as shown above because the ring without the nitro group is less deactivated and reacts at the α-positions. The α-positions are more reactive, because in this case more of the resonance forms leave the other ring intact and thus aromatic.

Problem 16.12 How do you account for the fact that N-phenylacetamide (acetanilide) is less reactive toward electrophilic aromatic substitution than aniline?

N-phenylacetamide (acetanilide) aniline

The unshared pair of electrons on the nitrogen atom of acetanilide is involved in a resonance interaction with the carbonyl group of the amide, and, therefore, less available for stabilization of an aryl cation intermediate.

Problem 16.13 The trifluoromethyl group is almost exclusively meta directing as shown in the following example:

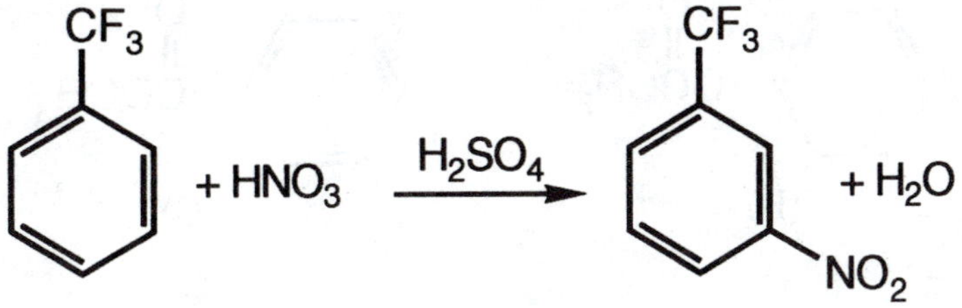

$$+ HNO_3 \xrightarrow{H_2SO_4} + H_2O$$

Draw contributing structures for the aryl cation intermediate formed during nitration. First assume para attack and then meta attack. By reference to contributing structures you have drawn, account for the fact that nitration is essentially 100% in the meta position.

Following are contributing structures for meta and para attack of the electrophile. For meta attack, three contributing structures can be drawn and all make approximately equal contributions to the hybrid. Three contributing structures can also be drawn for ortho/para attack, one of which places a positive charge on carbon bearing the trifluoromethyl group; this structure makes only a negligible contribution to the hybrid. Thus, for meta attack, the positive charge on the aryl cation intermediate can be delocalized almost equally over three atoms of the ring giving this cation's formation a lower energy of activation. For ortho/para attack, the positive charge on the aryl cation intermediate is delocalized over only two carbons of the ring, giving this cation's formation a higher energy of activation.

meta attack:

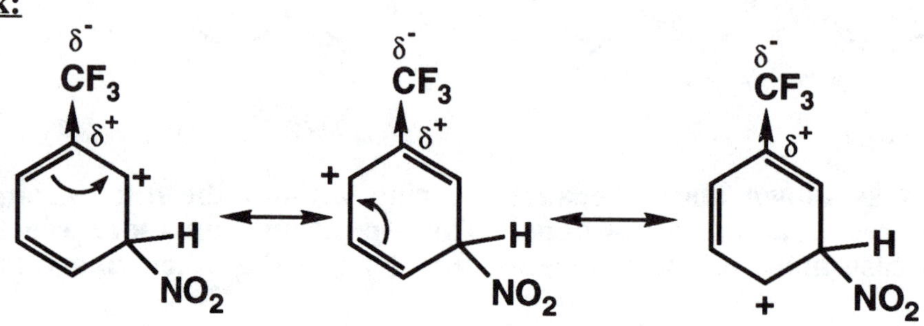

δ^- CF_3 δ^+ ... H, NO_2 ⟷ ... ⟷ ...

ortho/para attack:

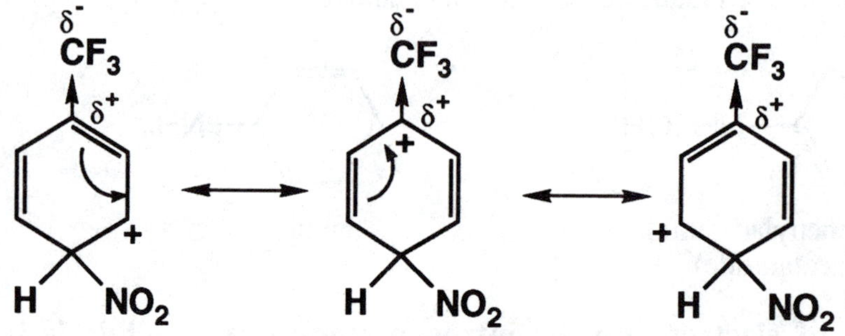

δ^- CF_3 δ^+ ... H, NO_2 ⟷ ... ⟷ ...

adjacent positive charges

Problem 16.14 Arrange the following in order of decreasing reactivity (fastest to slowest) toward electrophilic aromatic substitution.

(a) (A) benzene (B) $C_6H_5OCCH_3$ (C=O) (C) $C_6H_5COCH_3$ (C=O)

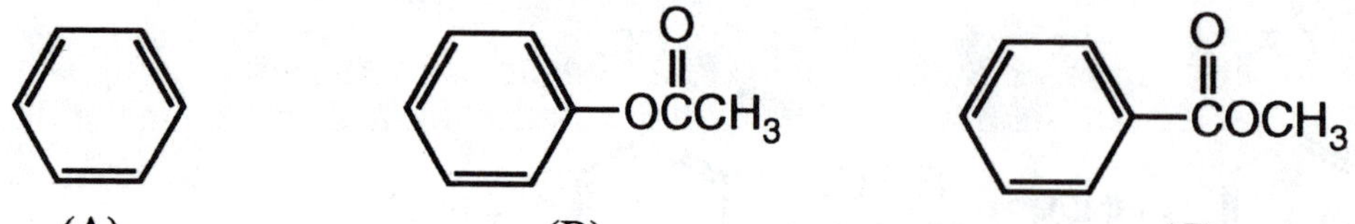

B>A>C

(b) $C_6H_5{-}NO_2$ (A) $C_6H_5{-}CO_2H$ (B) C_6H_6 (C)

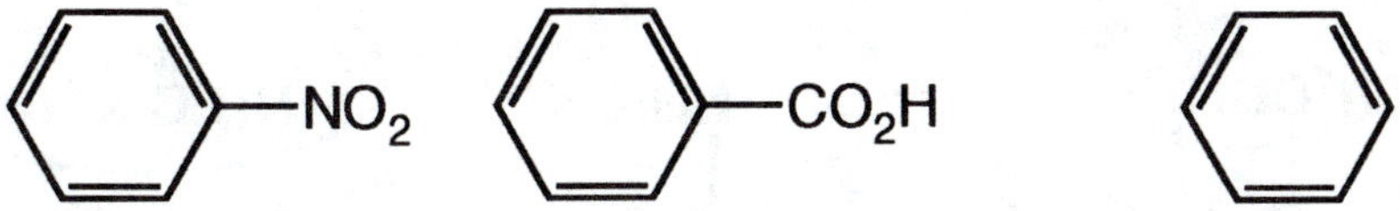

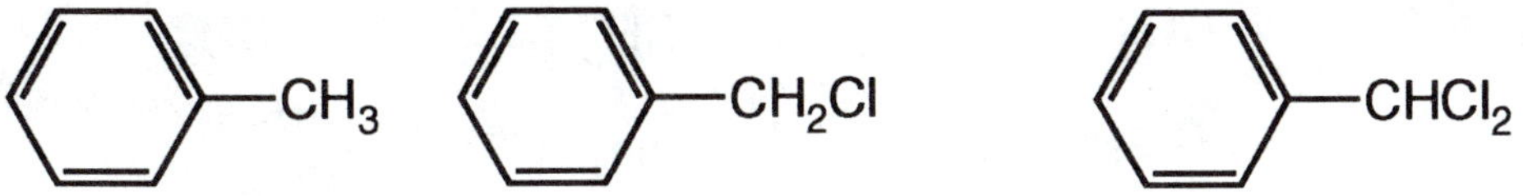

C>B>A

(c) $C_6H_5{-}CH_3$ (A) $C_6H_5{-}CH_2Cl$ (B) $C_6H_5{-}CHCl_2$ (C)

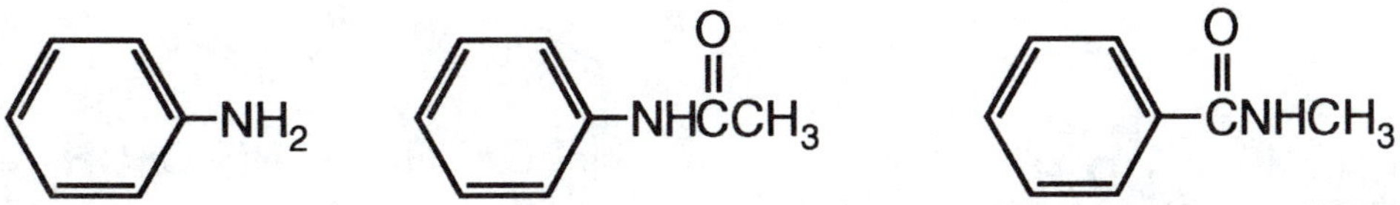

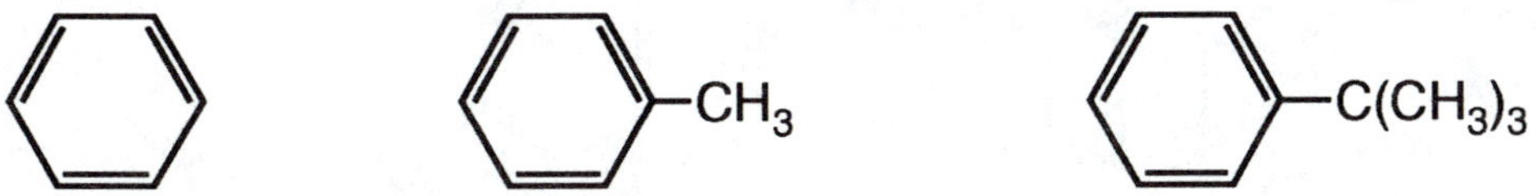

A>B>C

(d) $C_6H_5{-}SO_3H$ (A) $C_6H_5{-}C{\equiv}N$ (B) $C_6H_5{-}OCH_2CH_3$ (C)

C>B>A

(e) $C_6H_5{-}NH_2$ (A) $C_6H_5{-}NHC(=O)CH_3$ (B) $C_6H_5{-}C(=O)NHCH_3$ (C)

A>B>C

(f) C_6H_6 (A) $C_6H_5{-}CH_3$ (B) $C_6H_5{-}C(CH_3)_3$ (C)

C>B>A

Problem 16.15 For each compound, indicate which group on the ring is the more strongly activating and then draw the structural formula of the major product formed by nitration of that ring.

In the following structures, the more strongly activating group is circled and arrows show the position(s) of nitration. Where both ortho and para nitration are possible, two arrows are shown. A broken arrow shows a product formed in only negligible amounts.

(a) OCH_3 / CH_3 → **OCH_3** / **CH_3**

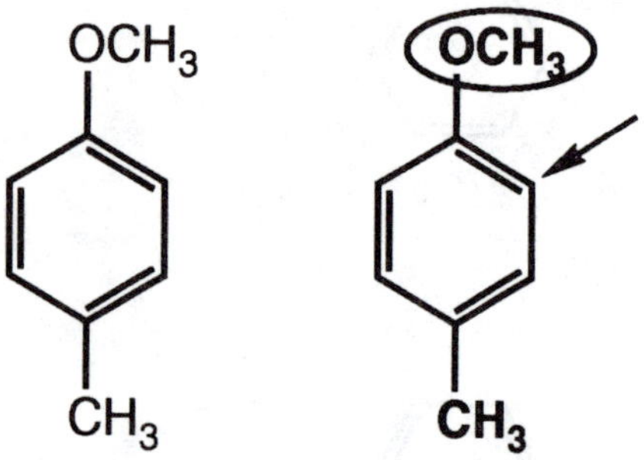

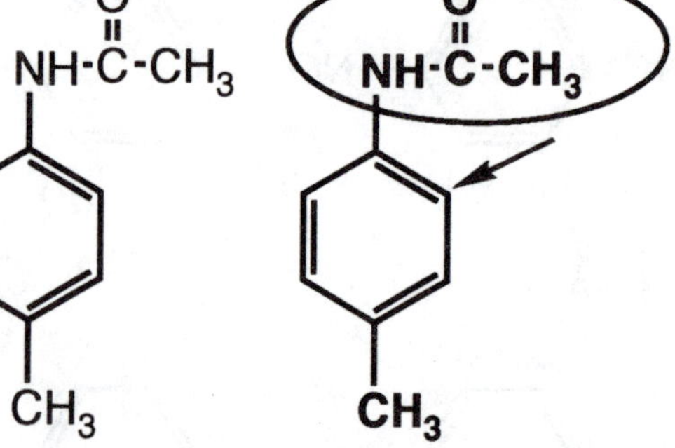

(b) $NH\text{-}C(=O)\text{-}CH_3$ / CH_3 → **$NH\text{-}C(=O)\text{-}CH_3$** / **$CH_3$**

(c) CO_2H / Cl → **CO_2H** / **Cl**

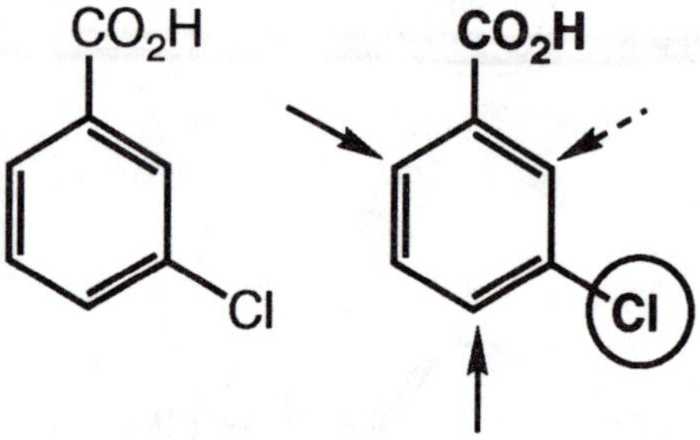

(d) CH_3 / SO_3H → **CH_3** / **SO_3H**

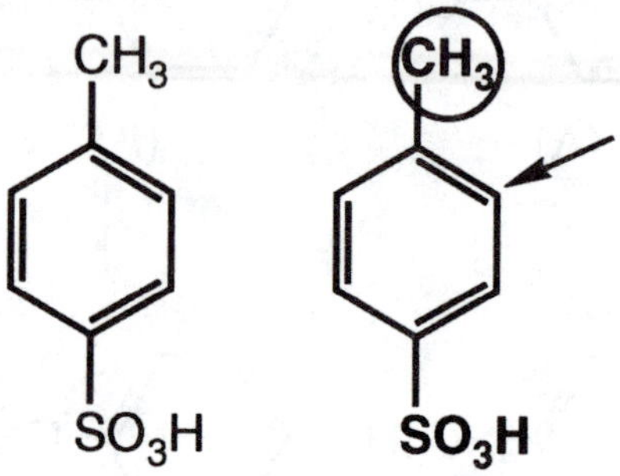

(e) CH_3 / CO_2H → **CH_3** / **CO_2H**

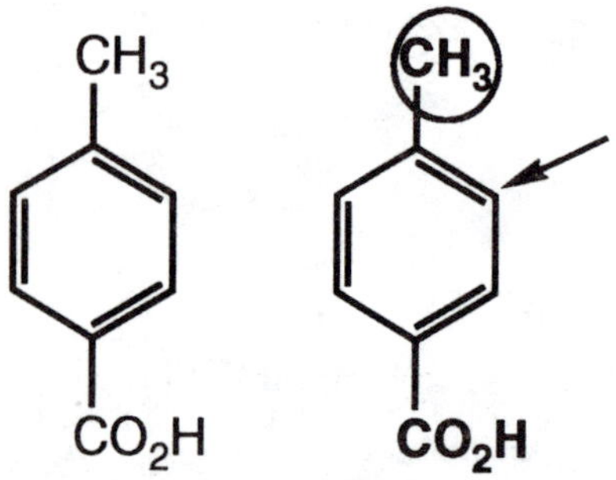

(f) $C(=O)\text{-}CH_3$ / $NH\text{-}C(=O)\text{-}CH_3$ → **$C(=O)\text{-}CH_3$** / **$NH\text{-}C(=O)\text{-}CH_3$**

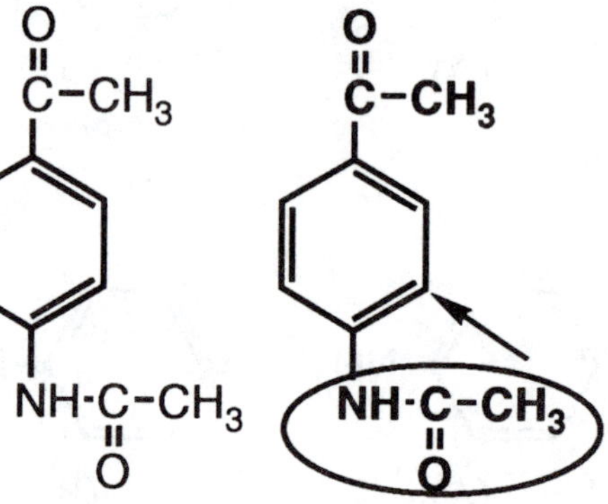

(g) OCH_3 / CO_2H → **OCH_3** / **CO_2H**

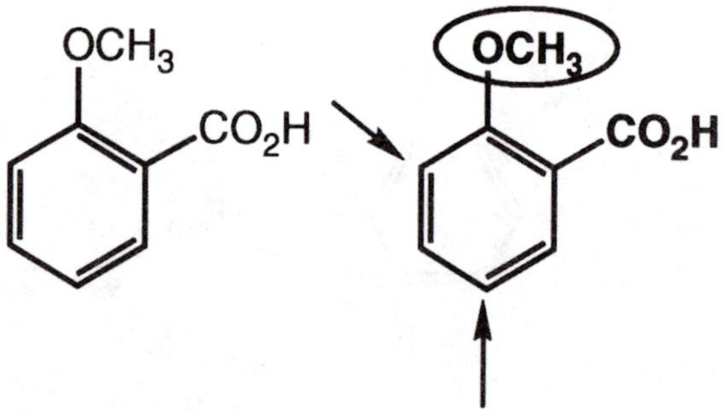

(h) NO_2 / Cl → **NO_2** / **Cl**

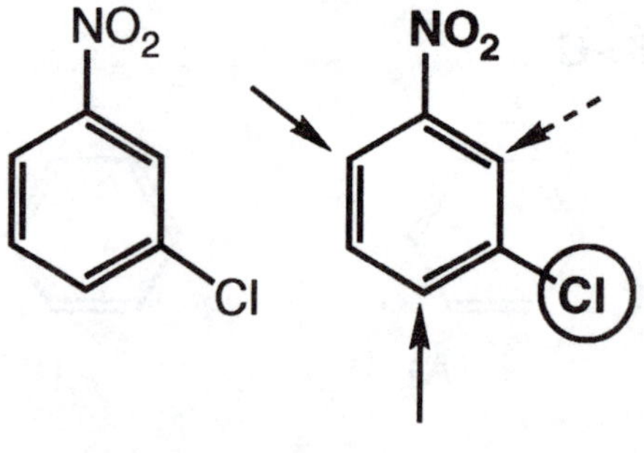

<u>Problem 16.16</u> Show how you might convert toluene to each of the following.

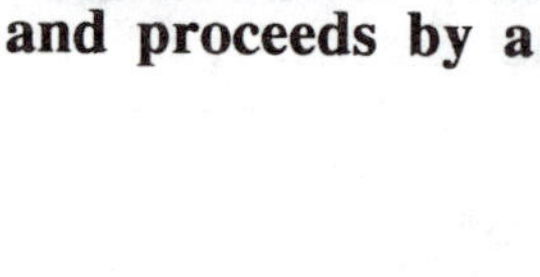

C_6H_5–CH_2Br

Bromination at the benzylic position uses bromine in the presence of heat or light, and proceeds by a radical chain mechanism.

$$C_6H_5-CH_3 + Br_2 \xrightarrow{\text{heat}} C_6H_5-CH_2Br + HBr$$

$$Br-C_6H_4-CH_3$$

Bromination of the aromatic ring by electrophilic aromatic substitution uses bromine in the presence of ferric bromide or other Lewis acid catalyst.

$$C_6H_5-CH_3 + Br_2 \xrightarrow{FeBr_3} Br-C_6H_4-CH_3$$

Problem 16.17 Show how to convert toluene to the following carboxylic acids.

(a) 2,4-dinitrobenzoic acid (C(=O)–OH, NO_2, NO_2)

Methyl is ortho/para directing. Therefore, toluene can be nitrated twice then oxidized with potassium permanganate in aqueous NaOH to convert the methyl group into the carboxyl group.

$$C_6H_5CH_3 \xrightarrow[H_2SO_4]{HNO_3} 2,4\text{-}(NO_2)_2C_6H_3CH_3 \xrightarrow[2)\ H_3O^+]{1)\ KMnO_4, NaOH} 2,4\text{-}(NO_2)_2C_6H_3COOH$$

(b)

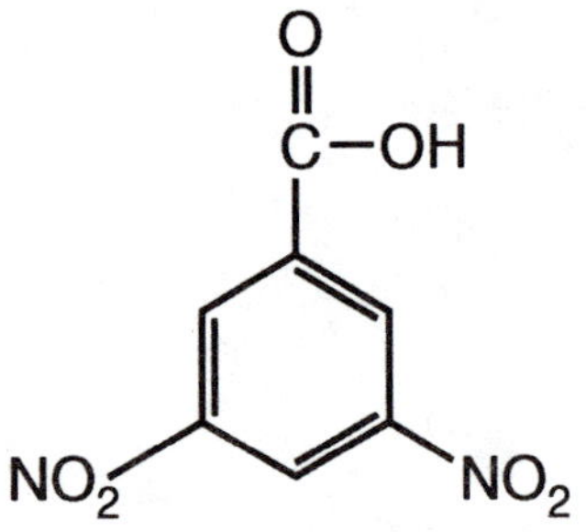

The reaction sequence is very similar to the last one, except now the order of the reactions is reversed because the carboxylic acid group is a meta director.

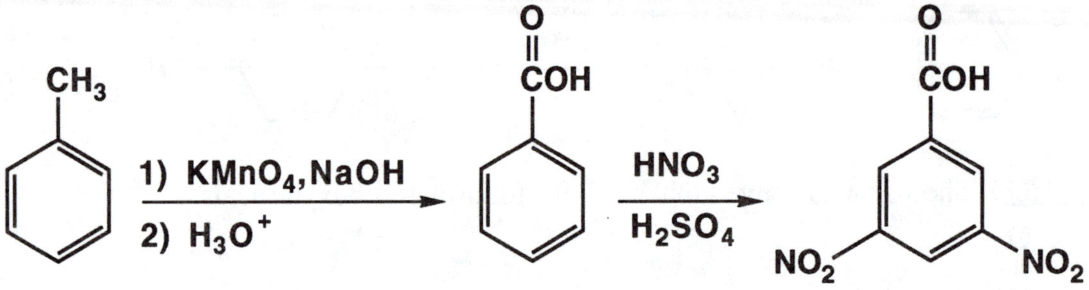

1) $KMnO_4$, NaOH
2) H_3O^+

HNO_3
H_2SO_4

Problem 16.18 The following molecules each contain two rings. Which of the two rings undergoes electrophilic aromatic substitution more readily. Draw the major product formed on nitration.

(a)

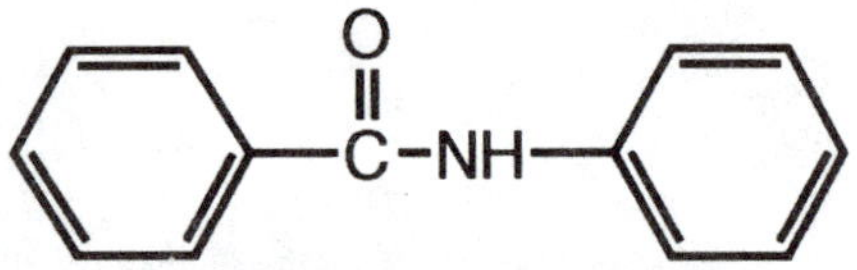

The nitrogen side of the amide group is more activating, so nitration produces the ortho/para products on this ring.

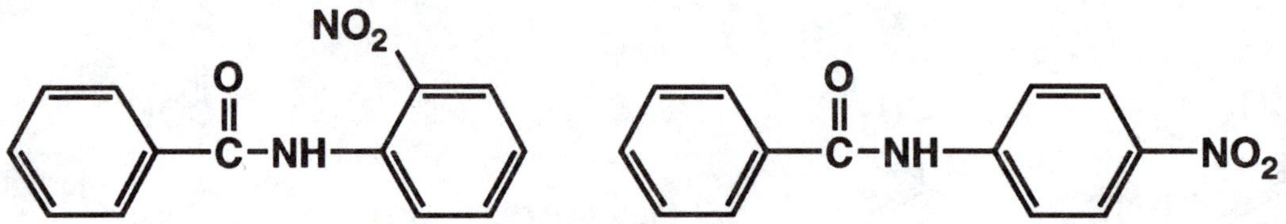

(b)

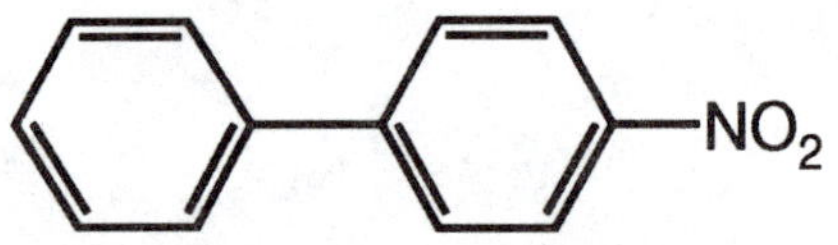

The *p*-nitrophenyl group is deactivating and meta directing. Note that nitration takes place on the ring that does not already possess the nitro group.

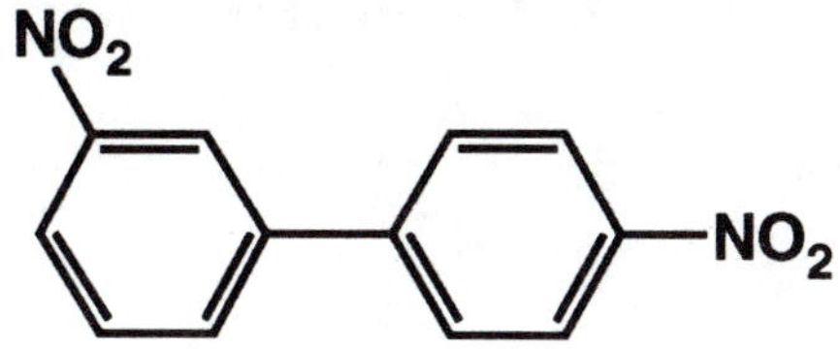

NO_2 NO_2

(c) 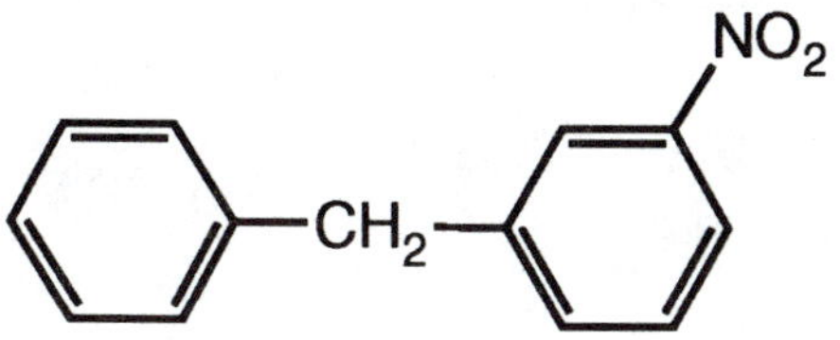 CH_2 NO_2

As above, the *m*-nitrobenzyl group is deactivating and meta directing. Again, nitration takes place on the ring that does not already possess the nitro group.

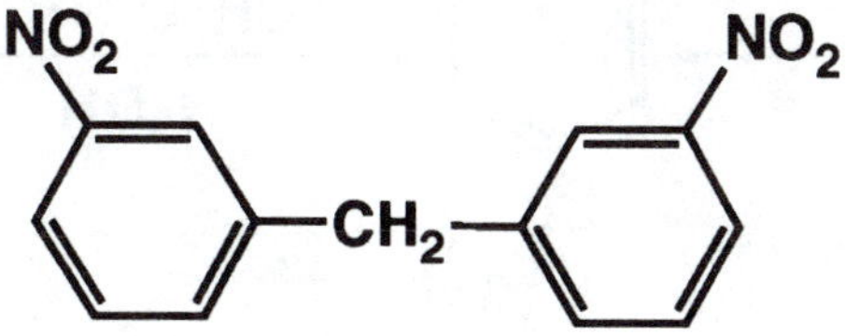

NO_2 CH_2 NO_2

(d) 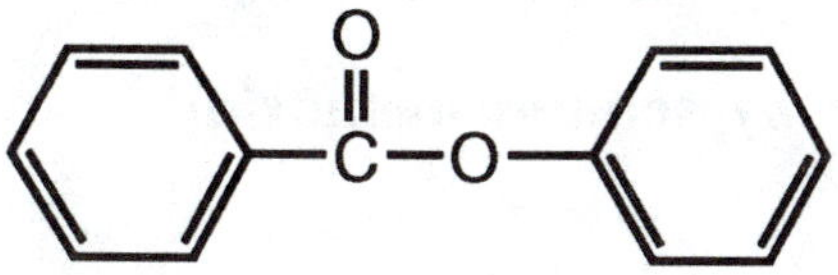 O ‖ C—O

The oxygen side of the ester is more activating, so ortho/para nitration takes place on this ring.

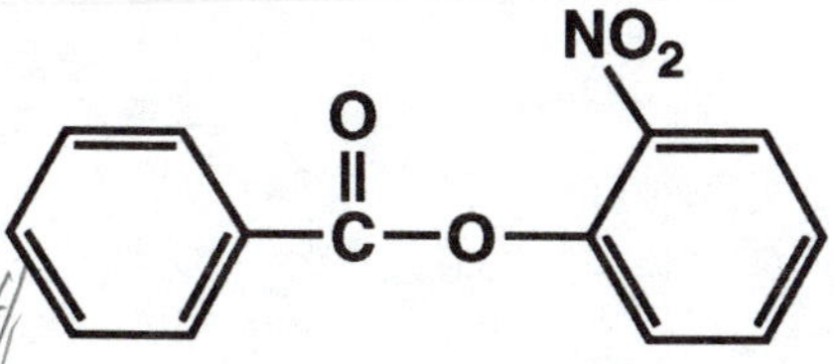

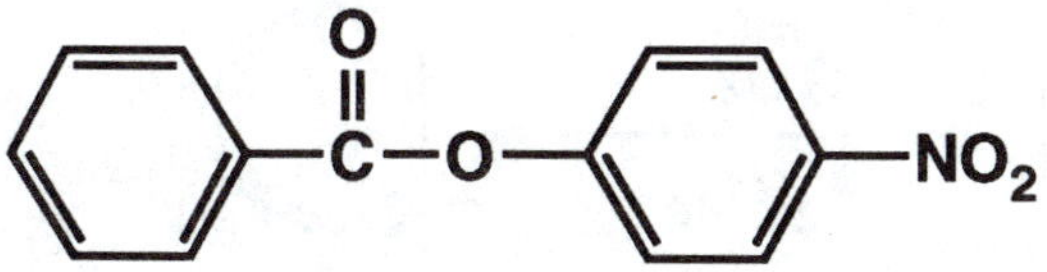

NO_2 O ‖ C—O   O ‖ C—O NO_2

Problem 16.19 Show reagents and conditions to bring about the following conversions. Each reaction arrow indicates one step; the numbers over the arrows are to number successive steps.

(a) 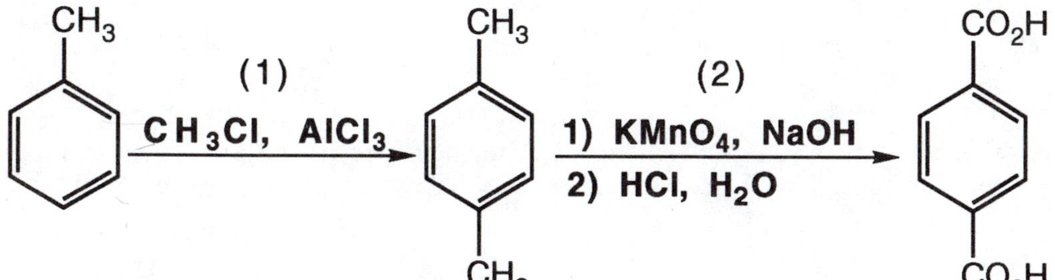 CH_3 —(1) CH_3Cl, $AlCl_3$→ CH_3, CH_3 —(2) 1) $KMnO_4$, NaOH 2) HCl, H_2O→ CO_2H, CO_2H

1. Friedel-Crafts alkylation using chloromethane (Section 16.1C).

2. Alkaline permanganate oxidation of each benzylic carbon to a carboxyl group (Section 15.6).

(b) phenol (OH) —(1) NaOH, then CH_3CH_2Br→ ethoxybenzene (OCH_2CH_3)

(c) chlorobenzene (Cl) —(1) HNO_3, H_2SO_4→ 1-chloro-2,4-dinitrobenzene (Cl, NO_2, NO_2) —(2) $CH_3O^-Na^+$, CH_3OH→ 2,4-dinitroanisole (OCH_3, NO_2, NO_2)

1. Nitration of chlorobenzene (Section 16.1B). Chlorine is ortho/para directing and determines both the location of the first nitration as well as the position of the second nitration.
2. Nucleophilic aromatic substitution of chlorine by sodium methoxide in methanol (Section 16.3A).

(d) acetanilide ($NHCCH_3$, C=O) —(1) SO_3, H_2SO_4→ p-acetamidobenzenesulfonic acid ($NHCCH_3$, SO_3H)

Sulfonation of acetanilide (Section 16.1 B).

(e) Anisole $\xrightarrow[\text{(1)}]{CH_3CCl(=O),\ AlCl_3}$ 4-methoxyacetophenone (OCH_3, CCH_3 with $=O$) $\xrightarrow{HNO_3,\ H_2SO_4}$ O_2N-substituted product (OCH_3, O_2N, CCH_3 with $=O$)

Friedel-Crafts acylation of anisole using acetyl chloride in the presence of aluminum chloride as a Lewis acid catalyst (Section 16.1C), followed by nitration.

(f) CH_2CH_3 (ethylbenzene) $\xrightarrow[\text{(1)}]{Cl_2,\ AlCl_3}$ $CHCH_3$ / Cl $\xrightarrow[\text{(2)}]{Br_2,\ heat}$ Br-$CHCH_3$ / Cl $\xrightarrow[\text{(3)}]{KOH}$ $CH{=}CH_2$ / Cl

1. Chlorination of the aromatic ring using chlorine in the presence of aluminum chloride as a Lewis acid catalyst (Section 16.1A).
2. Bromination at the benzylic position using bromine at high temperature by a radical chain reaction mechanism (Section 16.6B).
3. Dehydrohalogenation with potassium hydroxide, sodium ethoxide, or other strong base to form the alkene (Section 4.5).

Problem 16.20 Propose a synthesis of triphenylmethane from benzene as the only source of aromatic rings, and any other necessary inorganic reagents.

Reaction of three moles of benzene with one mole of trichloromethane (chloroform) will give triphenylmethane.

$$3\ C_6H_6 + CHCl_3 \xrightarrow{AlCl_3} (C_6H_5)_3CH + 3HCl$$

Problem 16.21 Reaction of phenol with acetone in the presence of an acid catalyst gives a compound known as bisphenol A. Bisphenol A is used in the production of epoxy resins and polycarbonate resins. Propose a mechanism for the formation of bisphenol A.

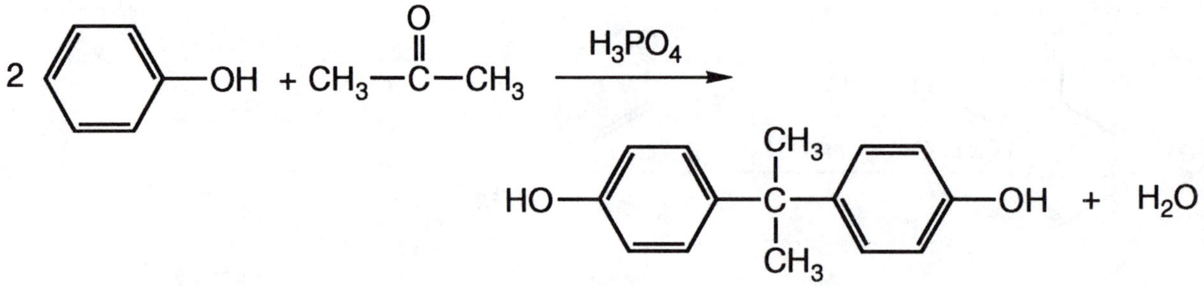

$$2\ C_6H_5OH + CH_3COCH_3 \xrightarrow{H_3PO_4} HO\text{-}C_6H_4\text{-}C(CH_3)_2\text{-}C_6H_4\text{-}OH + H_2O$$

Bisphenol A

The reaction begins with protonation of acetone to form its conjugate acid which may be written as a hybrid of two contributing structures.

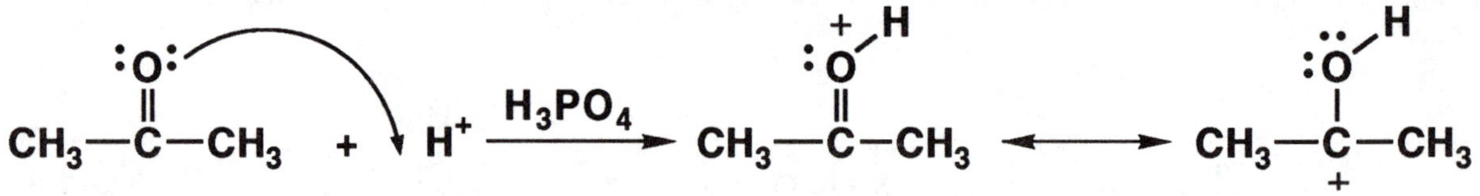

$$CH_3-\overset{\overset{\ddot{O}}{\|}}{C}-CH_3 + H^+ \xrightarrow{H_3PO_4} CH_3-\overset{\overset{+O-H}{\|}}{C}-CH_3 \longleftrightarrow CH_3-\overset{\overset{\ddot{O}-H}{|}}{\underset{+}{C}}-CH_3$$

The conjugate acid of acetone is an electrophile and reacts with phenol at the para position by electrophilic aromatic substitution to give 2-(4-hydroxyphenyl)-2-propanol. Protonation of the tertiary alcohol in this molecule and departure of water gives a resonance-stabilized cation that reacts with a second molecule of phenol to give bisphenol A.

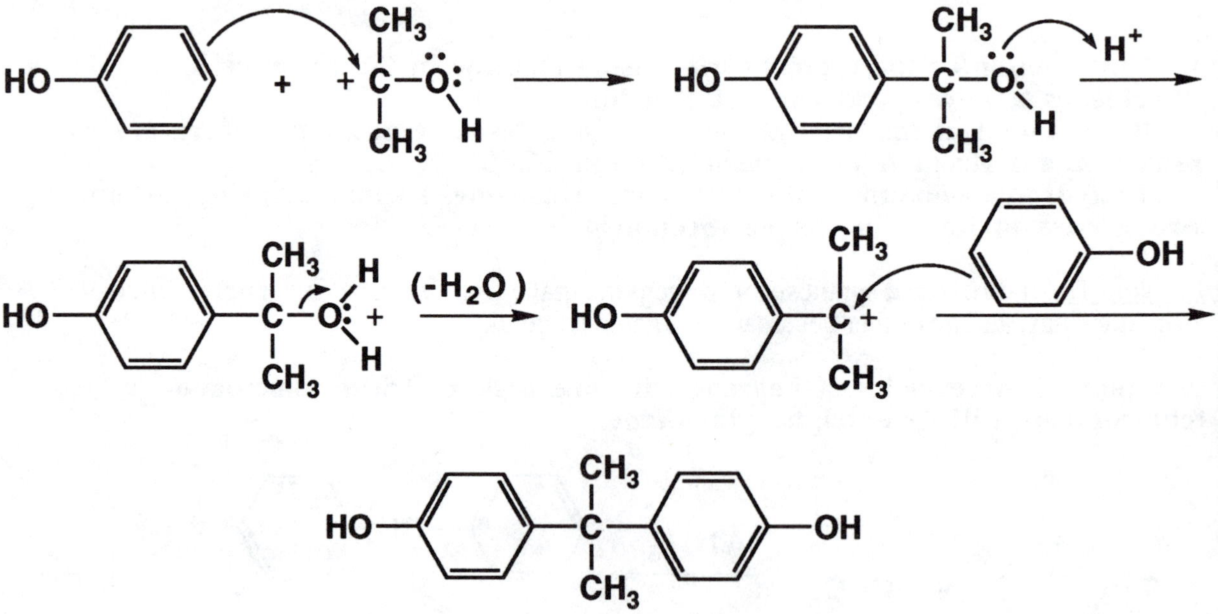

$$HO\text{-}C_6H_5 + {}^+C(CH_3)_2\text{-}\ddot{O}H \longrightarrow HO\text{-}C_6H_4\text{-}C(CH_3)_2\text{-}\ddot{O}H \xrightarrow{H^+}$$

$$HO\text{-}C_6H_4\text{-}C(CH_3)_2\text{-}\overset{+}{O}H_2 \xrightarrow{(-H_2O)} HO\text{-}C_6H_4\text{-}\overset{+}{C}(CH_3)_2 \xrightarrow{C_6H_5OH}$$

$$HO\text{-}C_6H_4\text{-}C(CH_3)_2\text{-}C_6H_4\text{-}OH$$

Bisphenol A

Problem 16.22 2,6-Di-*tert*-butyl-4-methylphenol, alternatively known as butylated hydroxytoluene (BHT), is used as an antioxidant in foods to "retard spoilage". BHT is synthesized industrially from 4-methylphenol (*p*-cresol) by reaction with 2-methylpropene in the presence of phosphoric acid. Propose a mechanism for this reaction.

$$\text{4-methylphenol} + 2CH_3-C(CH_3)=CH_2 \xrightarrow{H_3PO_4} \text{2,6-di-}tert\text{-butyl-4-methylphenol}$$

4-methylphenol
(*p*-cresol)

2,6-di-*tert*-butyl-4-methylphenol
"butylated hydroxytoluene"
(BHT)

The reaction involves an initial proton transfer from phosphoric acid to 2-methylpropene to give an electrophilic *tert*-butyl cation that then reacts with the aromatic ring ortho to the strongly activating -OH group to form 2-*tert*-butyl-4-methylphenol. A second electrophilic aromatic substitution gives the final product.

$$CH_3-C(CH_3)=CH_2 + H^+ \xrightarrow{H_3PO_4} CH_3-C^+(CH_3)-CH_3$$

$$\text{4-methylphenol} + CH_3-C^+(CH_3)-CH_3 \longrightarrow \text{2-}C(CH_3)_3\text{-4-methylphenol} + H^+$$

$$\text{2-}C(CH_3)_3\text{-4-methylphenol} + CH_3-C^+(CH_3)-CH_3 \longrightarrow \text{2,6-di-}C(CH_3)_3\text{-4-methylphenol} + H^+$$

Problem 16.23 The first widely used herbicides for control of weeds were 2,4-dichlorophenoxyacetic acid (2,4-D) and 2,4,6-trichlorophenoxyacetic acid (2,4,6-T). Show how each might be synthesized from phenol and 2-chloroacetic acid by way of the given chlorinated phenol intermediate.

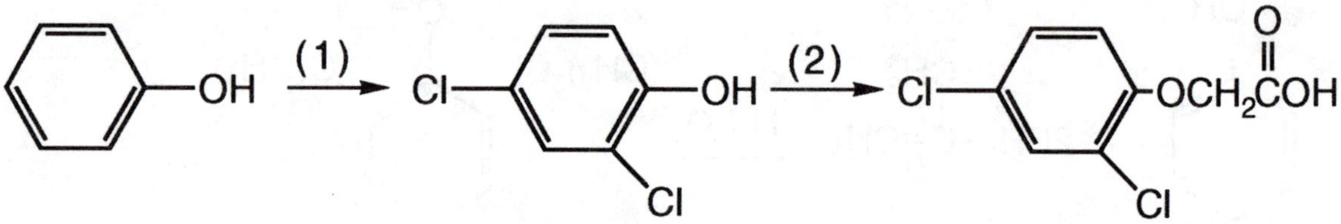

phenol $\xrightarrow{(1)}$ 2,4-dichlorophenol $\xrightarrow{(2)}$ 2,4-dichlorophenoxyacetic acid (2,4-D)

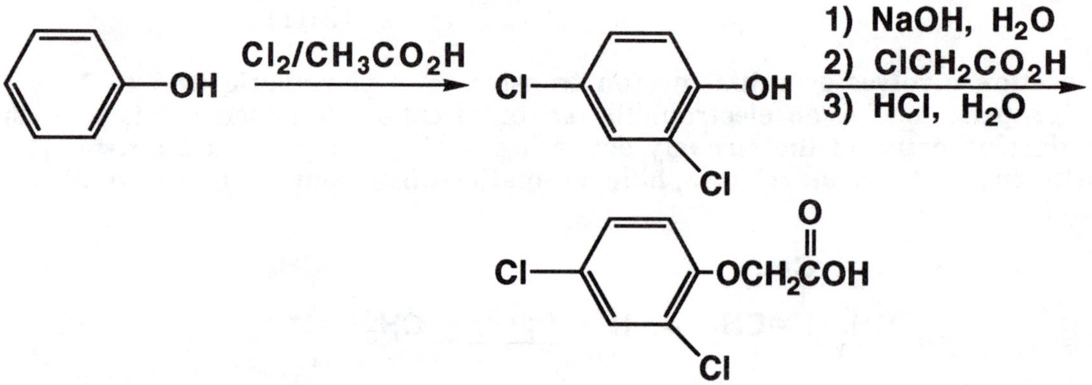

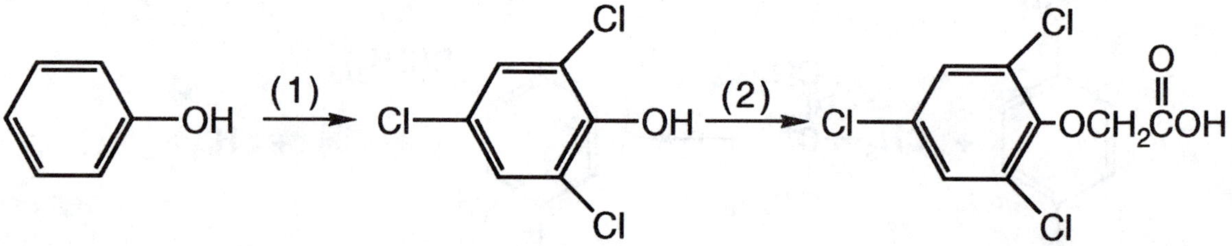

-OH, Cl, OH, Cl, Cl, OCH_2COH, O

Phenol $\xrightarrow{Cl_2/CH_3CO_2H}$ 2,4-dichlorophenol $\xrightarrow{\text{1) NaOH, } H_2O \text{; 2) } ClCH_2CO_2H \text{; 3) HCl, } H_2O}$ 2,4-dichlorophenoxyacetic acid (OCH_2COH)

phenol $\xrightarrow{(1)}$ 2,4,6-trichlorophenol $\xrightarrow{(2)}$ 2,4,6-trichlorophenoxyacetic acid (2,4,6-T)

-OH, Cl, Cl, Cl, OH, Cl, Cl, Cl, OCH_2COH, O

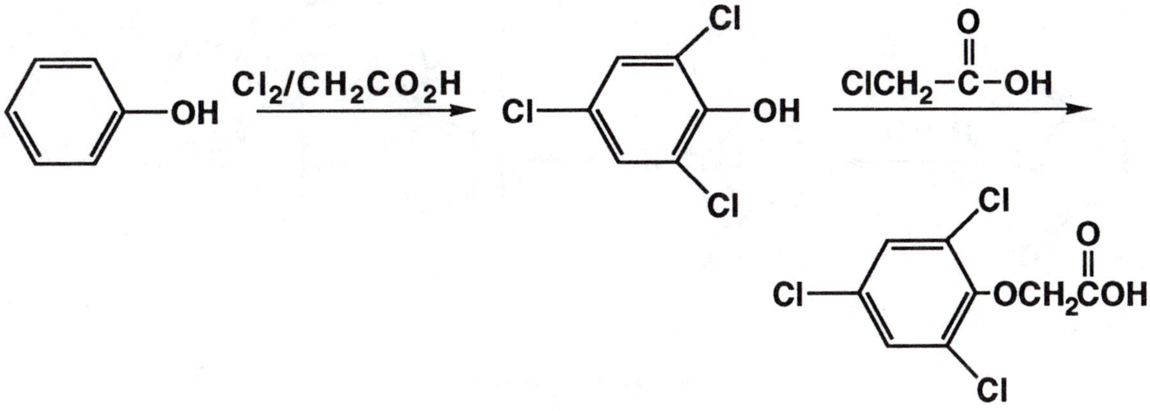

$\xrightarrow{Cl_2/CH_2CO_2H}$ Cl, Cl, Cl, OH $\xrightarrow{ClCH_2-C(=O)-OH}$ Cl, Cl, Cl, $OCH_2C(=O)OH$

Problem 16.24 Phenol is the starting material for the synthesis of 2,3,4,5,6-pentachlorophenol, known alternatively as pentachlorophenol or more simply as "penta". At one time, penta was widely used as a wood preservative for decks, sidings, and outdoor wood furniture. Draw the structural formula for pentachlorophenol and describe its synthesis from phenol.

"Penta" is synthesized industrially from phenol. Because the -OH group is strongly activating, reaction of phenol with chlorine in a polar solvent such as acetic acid at room temperature gives 2,4,6-trichlorophenol. Further chlorination of this intermediate gives pentachlorophenol.

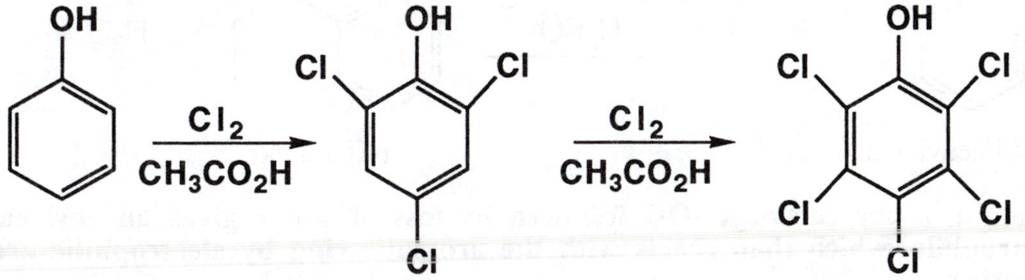

$\xrightarrow[CH_3CO_2H]{Cl_2}$ **2,4,6-trichlorophenol** $\xrightarrow[CH_3CO_2H]{Cl_2}$ **2,3,4,5,6-pentachlorophenol**

Problem 16.25 Treatment of benzene with succinic anhydride in the presence of an acid catalyst, most often phosphoric acid or sulfuric acid, gives the following γ-ketoacid. Propose a mechanism for this reaction.

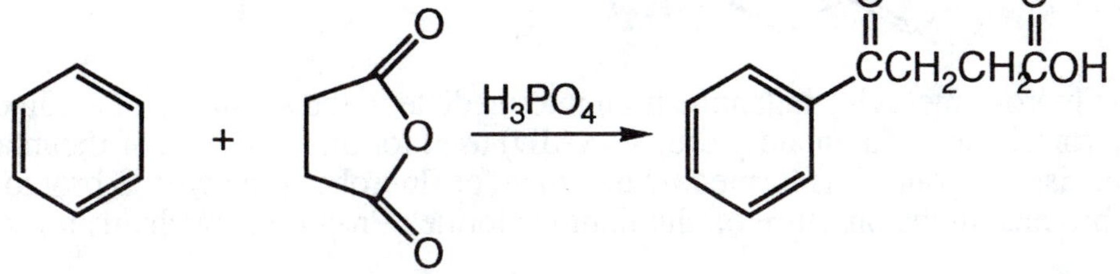

Benzene + Succinic anhydride $\xrightarrow{H_3PO_4}$ 4-Oxo-4-phenylbutanoic acid ($C_6H_5CCH_2CH_2COH$, with C=O groups)

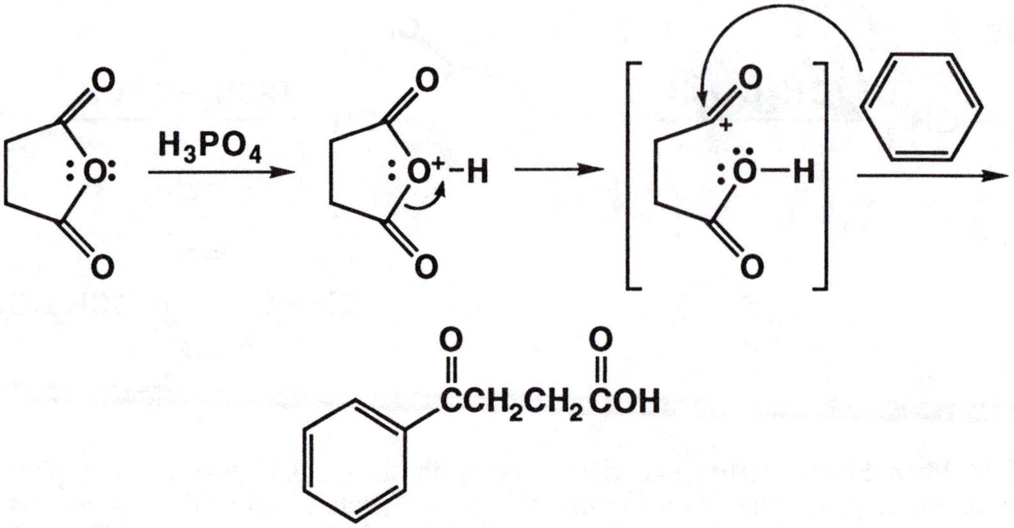

H_3PO_4 :O: → :O⁺-H → [:Ö-H] → CCH_2CH_2COH

Problem 16.26 Reaction of 4-phenylbutanoic acid in the presence of concentrated sulfuric acid gives a compound known most commonly as α-tetralone. Propose a mechanism for this cyclodehydration.

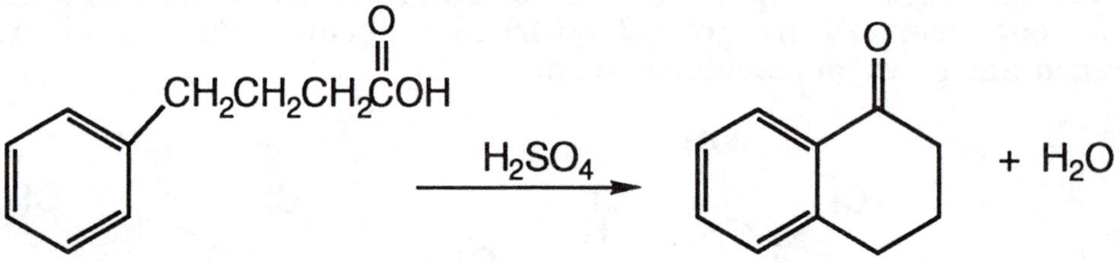

$CH_2CH_2CH_2COH$ $\xrightarrow{H_2SO_4}$ + H_2O

4-Phenylbutanoic acid α-Tetralone

Protonation of the carboxyl -OH followed by loss of water gives an acyl cation, an electrophile, which then reacts with the aromatic ring by electrophilic aromatic substitution.

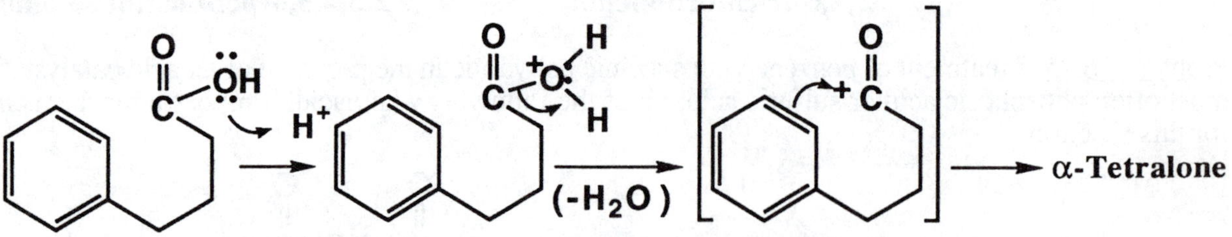

H^+ → $\xrightarrow{(-H_2O)}$ → **α-Tetralone**

Problem 16.27 5-Fluoro-2-methyl-1-indanone is an intermediate in the synthesis of suldindac (Clinoril), a nonsteroidal anti-inflammatory drug (NSAID) used for the treatment of rheumatoid arthritis, osteoarthritis, and gout. It is formed by treatment of fluorobenzene with 2-bromo-2-methylpropanoyl bromide in the presence of aluminum chloride. Propose a mechanism for this transformation.

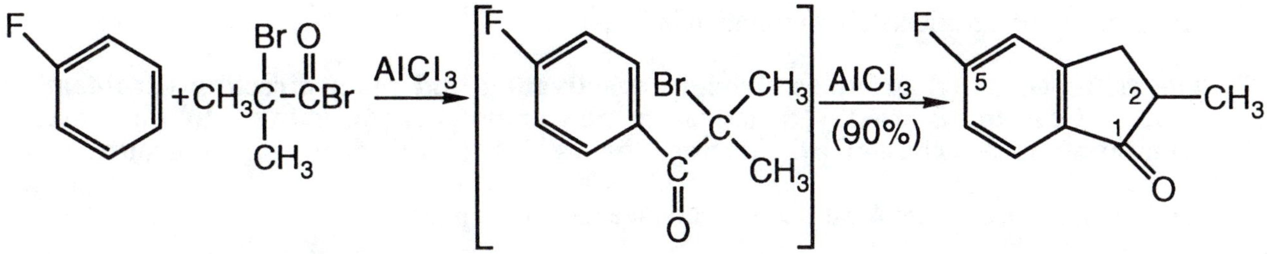

F, Br O, +CH_3C-CBr, CH_3, $AlCl_3$, Br, CH_3, C, C, O, CH_3, $AlCl_3$ (90%), F, 5, 2, 1, CH_3, O

fluoro-benzene 2-bromo-2-methyl-propanoyl bromide (an intermediate; not isolated) 5-fluoro-2-methyl-1-indanone

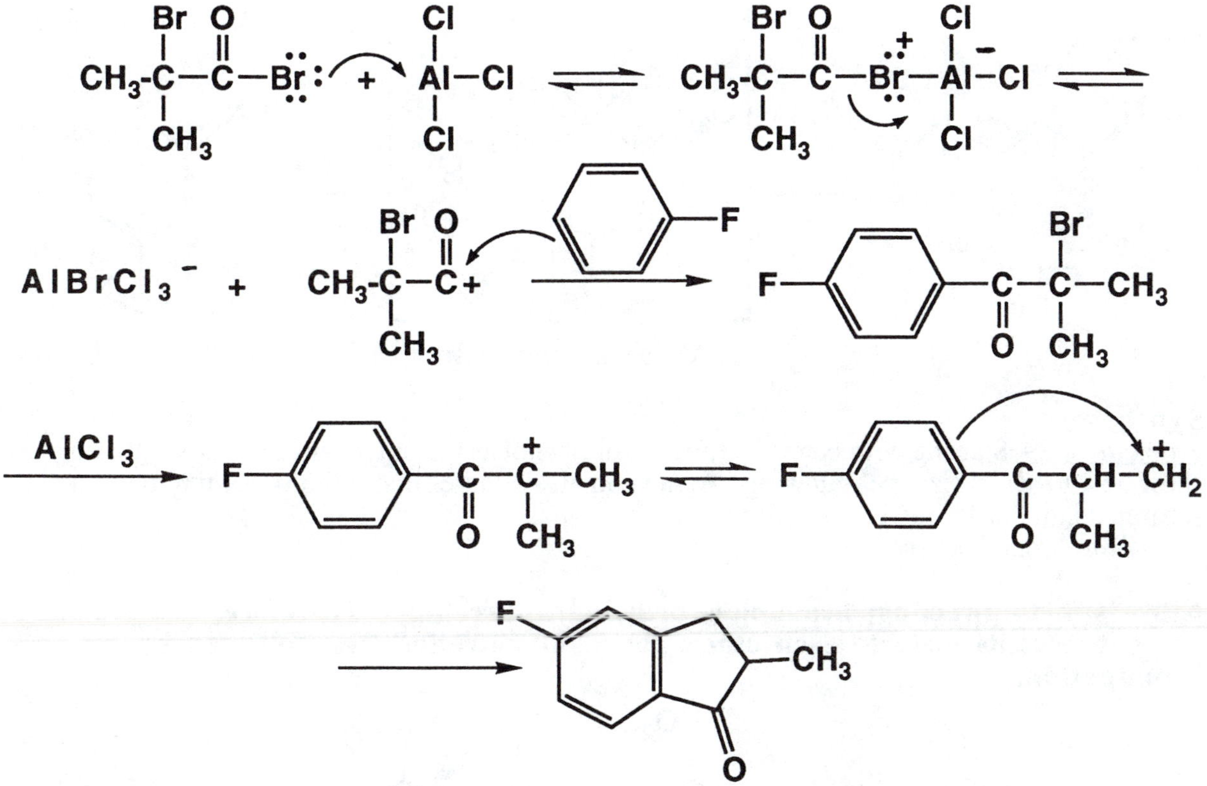

$AlBrCl_3^-$ + CH_3-C-C+ ... $AlCl_3$

Nucleophilic Aromatic Substitution

Problem 16.28 Following are the final steps in the synthesis of trifluralin B, a pre-emergent herbicide.

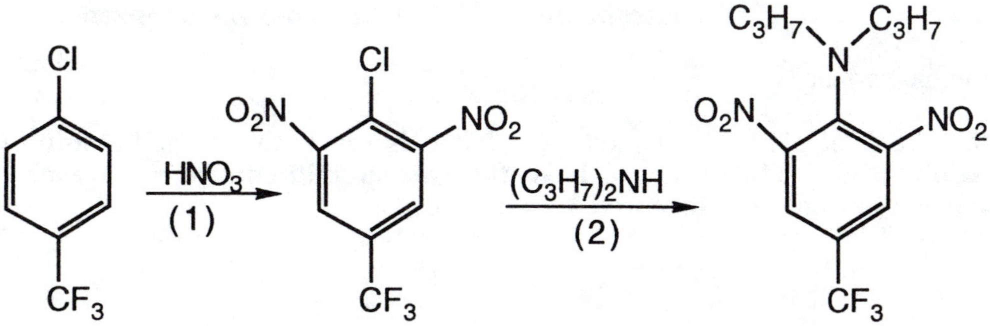

Cl, CF_3, HNO_3 (1), O_2N, Cl, NO_2, CF_3, $(C_3H_7)_2NH$ (2), C_3H_7, C_3H_7, N, O_2N, NO_2, CF_3

(a) Account for the orientation of nitration in Step (1).

The trifluoromethyl group is strongly deactivating and meta directing (Problem 16.13). Chlorine is weakly deactivating and ortho/para directing. In this case, orientation is determined by chlorine, the weaker of the deactivating groups.

(b) Propose a mechanism for the substitution reaction in Step (2).

The second step of this transformation is an example of nucleophilic aromatic substitution. In the following structural formulas, the propyl group of dipropylamine is abbreviated R.

R_2NH Cl O_2N N O O CF$_3$ → [Cl NR_2 O_2N N O O CF$_3$] $\xrightarrow{(-Cl^-)}$ C_3H_7 N C_3H_7 O_2N NO_2 CF$_3$

(a Meisenheimer complex) **Trifluralin B**

Syntheses

Problem 16.29 Starting with benzene, toluene, or phenol as the only sources of aromatic rings, show how to synthesize the following. Assume in all syntheses that ortho-para mixtures can be separated into the desired isomer.

(a) *m*-Nitrobromobenzene

Nitro is meta directing; bromine is ortho/para directing. Therefore, to have the two substituents meta to each other, carry out nitration first followed by bromination.

$$\text{benzene} \xrightarrow[H_2SO_4]{HNO_3} \text{nitrobenzene} \xrightarrow[FeBr_3]{Br_2} \text{1-bromo-3-nitrobenzene}$$

nitrobenzene **1-bromo-3-nitrobenzene**

(b) *p*-Nitrobromobenzene

Reverse the order of steps from part (a). Nitro is meta directing; bromine is ortho-para directing. Therefore, to have the two substituents para to each other, carry out bromination first followed by nitration.

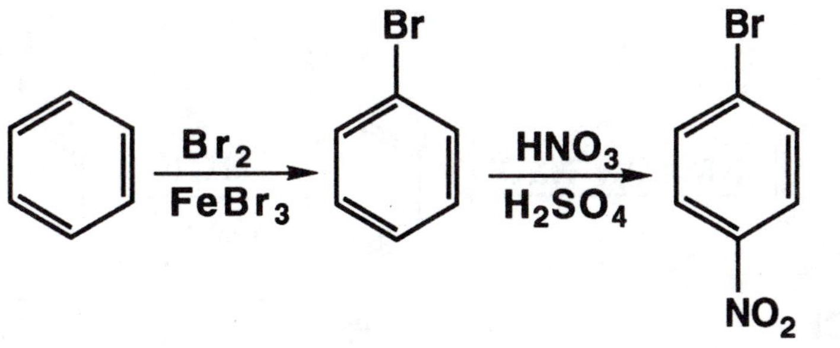

Br_2, $FeBr_3$ → Br-benzene; HNO_3, H_2SO_4 → 1-bromo-4-nitrobenzene (NO_2)

Bromobenzene 1-Bromo-4-nitrobenzene

(c) 2,4,6-Trinitrotoluene (TNT)

The methyl group is ortho/para directing. Therefore, nitrate toluene three successive times.

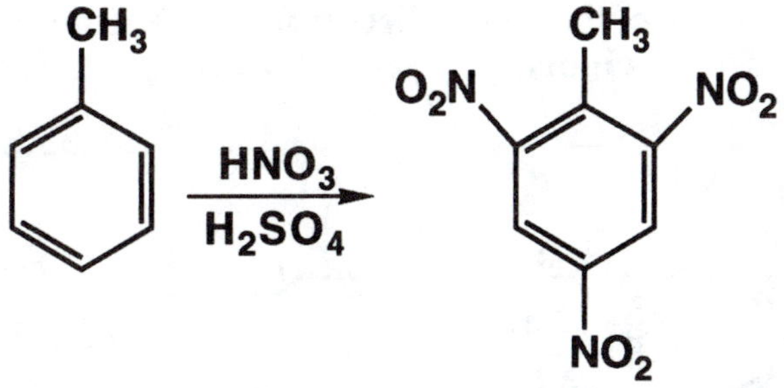

CH_3 — HNO_3, H_2SO_4 → CH_3, O_2N, NO_2, NO_2

2,4,6-trinitrotoluene

(d) *m*-Chlorobenzoic acid

The carboxyl group and chlorine atom are meta to each other, an orientation best accomplished by chlorination of benzoic acid (the carboxyl group is meta-directing). Oxidation of toluene with alkaline potassium permanganate followed by acidification gives benzoic acid. Treatment of benzoic acid with chlorine in the presence of ferric chloride gives the desired product.

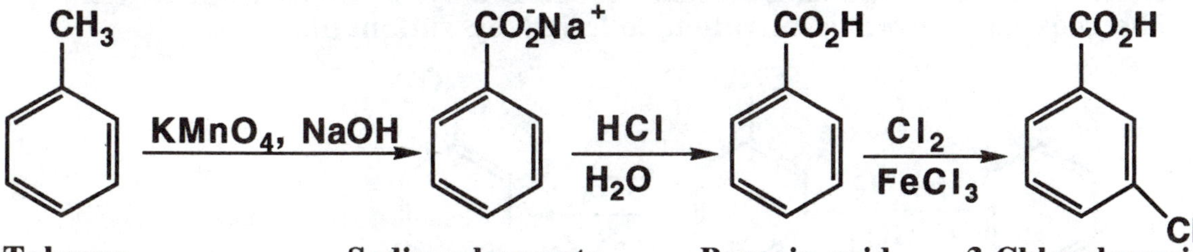

CH_3 — $KMnO_4$, NaOH → $CO_2^-Na^+$ — HCl, H_2O → CO_2H — Cl_2, $FeCl_3$ → CO_2H, Cl

Toluene Sodium benzoate Benzoic acid 3-Chlorobenzoic acid

(e) *p*-Chlorobenzoic acid

Start with toluene. Methyl is weakly activating and directs chlorination to the ortho/para positions. Separate the desired para isomer and then oxidize the methyl group to a carboxyl group using potassium permanganate in aqueous sodium hydroxide. Acidification of the reaction mixture gives 4-chlorobenzoic acid.

CH_3 (toluene) $\xrightarrow[AlCl_3]{Cl_2}$ 4-chlorotoluene $\xrightarrow{KMnO_4,\ NaOH}$ sodium 4-chlorobenzoate ($CO_2^-Na^+$, Cl) $\xrightarrow[H_2O]{HCl}$ 4-chlorobenzoic acid (CO_2H, Cl)

toluene 4-chlorotoluene sodium 4-chloro- benzoate 4-chlorobenzoic acid

(f) *p*-Dichlorobenzene

Treatment of benzene with chlorine in the presence of aluminum chloride gives chlorobenzene. The chlorine atom is ortho/para directing. Treatment with chlorine in the presence of aluminum chloride a second time gives 1,4-dichlorobenzene.

benzene $\xrightarrow[AlCl_3]{Cl_2}$ chlorobenzene (Cl) $\xrightarrow[AlCl_3]{Cl_2}$ *p*-dichlorobenzene (Cl, Cl)

benzene chlorobenzene *p*-dichlorobenzene

(g) *m*-Nitrobenzenesulfonic acid

Both the sulfonic acid group and the nitro group are meta directors. Therefore, the two electrophilic aromatic substitution reactions may be carried out in either order. The sequence shown is nitration followed by sulfonation.

benzene $\xrightarrow[H_2SO_4]{HNO_3}$ nitrobenzene (NO_2) $\xrightarrow[H_2SO_4]{SO_3}$ *m*-nitrobenzene sulfonic acid (NO_2, SO_3H)

benzene nitrobenzene *m*-nitrobenzene sulfonic acid

Problem 16.30 Following is the structure of the orris odor ketone. Describe its synthesis from benzene.

$$(CH_3)_2CH-C_6H_4-\overset{O}{\overset{\|}{C}}CH_3$$

4-isopropylacetophenone

The isopropyl group is weakly activating and ortho-para directing; the carbonyl group of the acetyl group is deactivating and meta directing. Therefore, start with benzene, convert it to isopropylbenzene (cumene) and then carry out a Friedel-Crafts acylation using acetyl chloride in the presence of aluminum chloride. Friedel-Crafts alkylation of benzene can be accomplished using a 2-halopropane, 2-propanol, or propene, each in the presence of an appropriate catalyst.

$C_6H_6 + CH_3CH(Cl)CH_3 \xrightarrow{AlCl_3}$

$C_6H_6 + CH_3CH(OH)CH_3 \xrightarrow{H_2SO_4}$

$C_6H_6 + CH_3CH{=}CH_2 \xrightarrow{H_2SO_4}$

$\longrightarrow C_6H_5CH(CH_3)_2$ **isopropylbenzene (cumene)** $\xrightarrow[AlCl_3]{CH_3\overset{O}{\overset{\|}{C}}Cl}$ $(CH_3)_2CH-C_6H_4-\overset{O}{\overset{\|}{C}}CH_3$ **4-isopropyl-acetophenone**

Problem 16.31 Following is the structural formula of musk ambrette, a synthetic musk, essential in perfumes to enhance and retain odor. Describe its synthesis from *m*-cresol (3-methylphenol).

(Structure: benzene ring bearing CH_3, two NO_2 groups, OCH_3, and $CH(CH_3)_2$)

Both methyl and methoxyl groups are ortho/para directing. The methoxyl group is a moderately strong o,p-directing group while the methyl group is only weakly o,p-directing. Introduction of the isopropyl group by Friedel-Crafts alkylation gives a mixture of 4-isopropyl-3-methoxytoluene and 2-isopropyl-5-methoxytoluene. Following separation of the desired isomer, nitration both ortho and para to the methoxyl group gives the desired product.

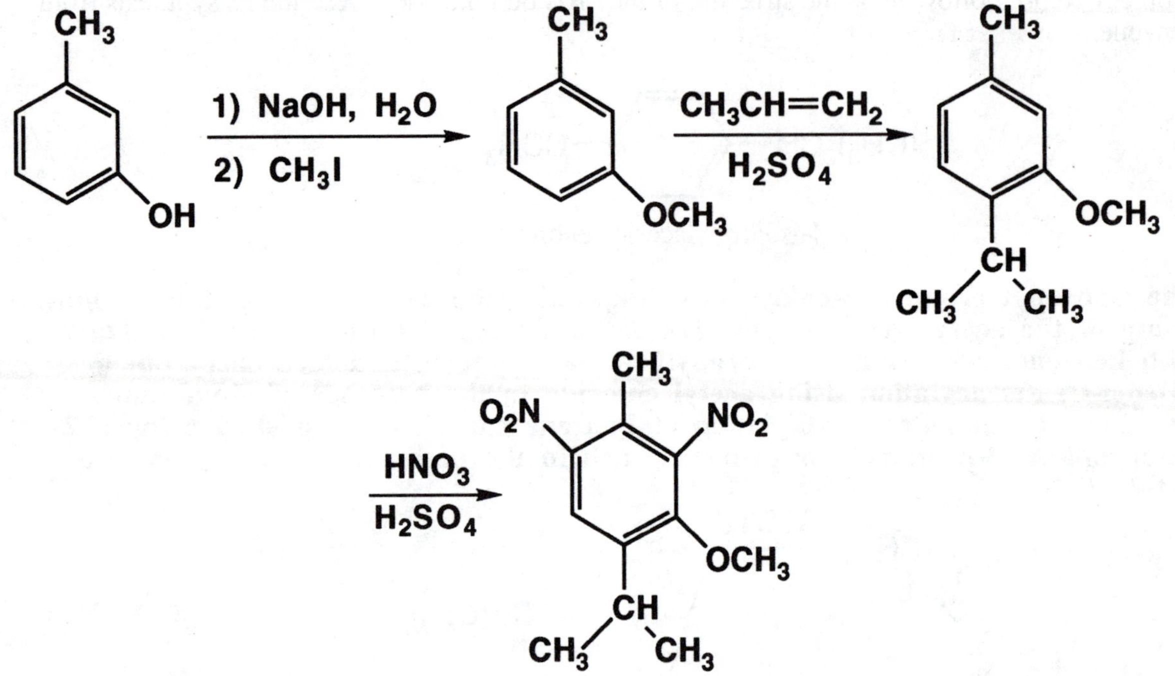

CH_3, OH $\xrightarrow[\text{2) } CH_3I]{\text{1) } NaOH,\ H_2O}$ CH_3, OCH_3 $\xrightarrow[H_2SO_4]{CH_3CH{=}CH_2}$ CH_3, OCH_3, CH, CH_3, CH_3

$\xrightarrow[H_2SO_4]{HNO_3}$ CH_3, O_2N, NO_2, OCH_3, CH, CH_3, CH_3